CLOUDS AND STORMS

CLOUDS AND STORMS

The Behavior and Effect
of Water in the Atmosphere

F. H. Ludlam

THE PENNSYLVANIA STATE UNIVERSITY PRESS

University Park and London

The Pennsylvania State University Press wishes to acknowledge the assistance and encouragement of the American Meteorological Society, Boston, Massachusetts, The National Center for Atmospheric Research, Boulder, Colorado, and the Department of Meteorology of the College of Earth and Mineral Sciences at Penn State during the preparation and publication of this book.

Library of Congress Cataloging in Publication Data

Ludlam, F H
   Clouds and storms: The behavior and effect of water in the atmosphere.

  Includes bibliography and index.
  1. Cloud physics. I. Title.
QC921.5.L83   551.5'76   77-22281
ISBN 0-271-00515-7

Copyright © 1980 The Pennsylvania State University
All rights reserved

Index by Chester and Harriet Newton

Designed by Glenn Ruby

Composition by Asco Trade Typesetting Limited, Hong Kong
Printed in the United States of America

Contents

Dedication—Frank Henry Ludlam, David Atlas and John Mason — xiii
Professor Frank Ludlam—Science of the Clouds, R.S. Scorer — xiv

Preface — xv

1 The influence of water on the general behavior of the atmosphere — 1

    1.1. Solar radiation and terrestrial temperatures — 1
        Lunar surface temperatures · The range of terrestrial surface temperatures

    1.2. The influence of water vapor on terrestrial long-wave radiation and temperature — 2
        The concept of radiative equilibrium · The impossibility of radiative equilibrium in the lower atmosphere · The troposphere, tropopause, and stratosphere · The mean rate of radiative energy loss in the troposphere · The latitudinal distribution of the terrestrial radiation into space

    1.3. General characteristics of atmospheric convection — 6
        Ordinary and slope convection · Ordinary, including cellular, convection in the laboratory · Atmospheric convection

    Appendix 1.1. The attenuation of electromagnetic radiation in the atmosphere: long-wave absorption by clouds — 9
        Atmospheric attenuation · Absorption of long-wave radiation by clouds

    References — 11

2 The physical properties of the moist atmosphere and its particles — 12

    2.1. Equations of state — 12

    2.2. Saturation — 12

    2.3. The specification of humidity — 13

    2.4. The measurement of humidity — 14

    2.5. The theory of the psychrometer — 14

Contents

2.6.	Properties of atmospheric particles	16
	Settling and Brownian motion of microscopic and submicroscopic particles · The aggregation of submicroscopic particles by Brownian motion · The deposition of submicroscopic particles on cloud droplets · Settling speeds of larger particles · Fall speeds of hailstones · Settling speeds of ice crystals and snowflakes · Aggregation of particles by differential settling; efficiency of catch · Cloud virtual temperature	
2.7.	Atmospheric refraction	25
2.8.	Scattering of radiation in the atmosphere	26
	The theory of scattering · Scattering functions · Scattering cross sections and efficiency factors · Atmospheric extinction · Backscattering; the use of radar · Common optical phenomena associated with scattering from clouds and precipitation	
Appendix 2.1.	Physical data	41
Appendix 2.2.	Relations between air saturated with respect to liquid water and with respect to ice	42
Appendix 2.3.	The performance of hygrometers in soundings	43
Appendix 2.4.	The drag coefficient of spheres and water drops	45
Appendix 2.5.	The paths of rays through stratified atmospheres	45
Appendix 2.6.	The scattering of starlight by turbulent layers	47
References		48

3 Thermodynamic reference processes — 51

3.1.	Isobaric mixing of moist air masses	51
3.2.	Production of clouds by the isobaric mixing of unsaturated air masses	52
3.3.	Dissipation of cloud by isobaric mixing with clear air	54
3.4.	Thermodynamic wet-bulb temperature	56
3.5.	Change of state of unsaturated air during expansion and compression: the dry adiabatic reference process, and potential temperature θ	56
3.6.	The lapse rate of the dew point; the saturation level	57
3.7.	Change of state of saturated air during expansion and compression: saturated reference processes, and saturation potential temperature θ_s	58
3.8.	Aerological diagrams	59
3.9.	Wet-bulb potential temperature θ_w	60
3.10.	Comparison of theoretical and practical wet-bulb temperatures	61
Appendix 3.1.	The tephigram	61
Appendix 3.2.	The rate of condensation in ascending saturated air	63
References		63

4 Atmospheric aerosol — 64

- 4.1. Sources of atmospheric trace substances — 64
  Sea spray · Dust · Smokes · Gases and vapors · The average mass concentration of aerosol particles in the lower troposphere
- 4.2. The size distribution of tropospheric aerosol particles — 66
- 4.3. The nomenclature of tropospheric aerosol particles — 68
- 4.4. The composition of tropospheric aerosol particles — 68
- 4.5. Extraterrestrial sources of aerosol particles — 69
  Meteorites, micrometeorites, and interplanetary dust
- 4.6. Stratospheric aerosol particles — 70
- 4.7. The equilibrium of solution droplets — 71
  Conditions for the equilibrium of solution droplets
- 4.8. The scattering of light by atmospheric aerosol; visual range — 72
  Visibility and visual range · Visual range near the ground · Visual range in strongly polluted air · Visual range in "cleaned" air · Slant visual range

Appendix 4.1. Twilight phenomena — 75
  Civil twilight; the purple light · Astronomical twilight; the zodiacal light · Abnormal twilight phenomena

References — 79

5 The formation, growth, and evaporation of cloud particles — 82

- 5.1. The growth of solution droplets — 82
- 5.2. The deviation of solution droplets from equilibrium — 83
- 5.3. The formation of a cloud from a population of solution droplets — 85
- 5.4. The supersaturation well above the saturation level — 88
- 5.5. Comparison of the contributions of condensation and coalescence to the rate of growth of an individual droplet — 90
- 5.6. The formation of clouds by isobaric mixing — 91
- 5.7. The evaporation of drops; precipitation — 91
- 5.8. The ventilation coefficients in the transfer equations — 92
- 5.9. The temperature of evaporating water drops — 94
- 5.10. The distance of fall of drops before complete evaporation; the distinction among cloud droplets, drizzle drops, and raindrops — 95
- 5.11. The formation of ice particles — 95
  Homogeneous nucleation of ice particles · Heterogeneous nucleation of ice particles · Observations of ice crystal concentration in artificial clouds produced in samples of atmospheric aerosol · The meteorological significance of the observed concentrations of ice nuclei

Contents

	5.12. The growth of ice crystals by condensation	99
	The form of ice crystals · The fall speeds of ice crystals · The growth of ice crystals by condensation · The growth of ice crystals on sublimation nuclei · The growth of ice crystals at saturation with respect to liquid water	
	5.13. The growth of ice particles by accretion	105
	The temperature of rime ice · The density and structure of rime ice · The formation of soft hail · The evaporation of ice particles	
	5.14. The multiplication of ice particles by fragmentation	109
	References	111
6	**Atmospheric motion systems**	**113**
	6.1. The identification of motion systems	113
	The physical description of motion systems	
	6.2. The scales of motion systems	114
	6.3. The development of the theory of motion systems	115
	6.4. The classification of motion systems	116
	The time scale of motion systems · The kinetic energy of motion systems	
	6.5. The classification and nomenclature of clouds	123
	Morphological classifications; the International Classification · Physical classifications	
	References	127
7	**Cumulus convection**	**129**
	7.1. General aspects of transfers of heat and water vapor between the atmosphere and the underlying medium	130
	7.2. The simple parcel theory of convection	132
	7.3. Fluxes and gradients near the surface	134
	7.4. The estimation of convective transfers; the condensation level	135
	The Bowen ratio H/LE · Transfers inferred from synoptic data · The height of cumulus bases	
	7.5. The stratification during cumulus convection	141
	The superadiabatic layer · The adiabatic layer · The transition layer · The cumulus layer and the cumulus tower layer	
	7.6. The arrangement of convective circulations	147
	Convective circulations shown by radar; mantle echoes · Cumulus rows · Large cellular patterns of cumulus · Intermediate-scale circulations associated with topographic features	
	7.7. Experimental forms of convection	154
	The buoyant plume · The buoyant thermal · The starting plume · Buoyant ascent in stably stratified surroundings; erosion · Thermals with increasing total buoyancy · The mushroom clouds of nuclear bombs: atmospheric thermals	

7.8. Observations of the properties of cumulus clouds 158
Visual observations; properties of cumulus populations · The shape of individual cumulus; cumulus towers · The rate of rise of cumulus towers · Flight observations of cumulus properties · Cloud temperature, liquid water concentration, and buoyancy · The vertical velocity of air in cumulus

7.9. The interpretation of observed cumulus properties 166

7.10. Theories of cumulus convection 169
Perturbation theories · Steady-state theories; mixing between cloudy and clear air; shelf clouds · Modifications of the simple parcel theory · The numerical modeling of cumulus convection · The requirements of a theory of cumulus convection

Appendix 7.1. Cloud mensuration 179

References 179

8 Cumulonimbus convection 182

8.1. The formation of showers 182
The size distribution of cumulus droplets · The size distribution and radar reflectivity factor Z of raindrops · Visual aspects of shower formation · Shower formation as observed by radar · The variation on individual days of the minimum thickness of shower-producing cumulus · The variation of the minimum thickness of shower clouds with the large-scale meteorological situation · The interpretation of the observations of shower formation · Observed and calculated evolutions of the size distribution of cloud droplets · The appearance of the ice phase in cumulus

8.2. The vertical development of cumulonimbus 204
The growth surge following glaciation · The stratification associated with cumulonimbus convection

8.3. The properties of ordinary cumulonimbus 207
Visual observations · Observations from aircraft · The weather and the downdraft at the ground · Experimental studies of dense outflows · Cold outflows in the atmosphere; haboobs and andhis · Numerical studies of cold outflows · The formation of cumulonimbus cells over the squall front · The speed of ascent of air over the squall front

8.4. Intense cumulonimbus; severe local storms 216
An intense traveling storm over England

8.5. General characteristics of intense traveling storms; cumulonimbus dynamics 220
Early studies of intense traveling storms · The intensification of cumulonimbus convection in the presence of wind shear; cumulonimbus dynamics

8.6. The distribution over the world of (intense) organized cumulonimbus 226
The squall storms of low latitudes · Intense cumulonimbus in the southern hemisphere · Intense cumulonimbus over Asia · Intense cumulonimbus over Europe · Intense cumulonimbus over North America · The summer monsoon over northern Mexico and the arid southwest of the United States

8.7. Conditions for the formation of intense storms 239
The formation of intense storms near the dry line over the United States · The formation of intense storms over western Europe

8.8. Characteristics of severe local storms 246
The visual appearance of intense cumulonimbus · The velocity of severe local storms

Contents

- 8.9. Tornadoes and waterspouts — 250
  The funnel and other clouds associated with the tornado · The generation of the rotation in the tornado and other atmospheric vortices · Waterspouts

- 8.10. The evaporation of rain in cumulonimbus downdrafts — 252
  The departure from the saturated state in updrafts and downdrafts · The estimation of the vertical temperature distribution in downdrafts · The dependence of the downdraft magnitude on the rain characteristics and the atmospheric stratification

- 8.11. Hail — 256
  Small hail · Large and giant hailstones · The internal structure of hailstones · The growth of large and giant hailstones · The consolidation of hail with an outer layer of spongy rime · The melting of hailstones · The rarity of hail in low latitudes

- 8.12. The precipitation intensity and radar reflectivity in cumulonimbus — 270
  The amount, duration, and intensity of precipitation from cumulonimbus · Rain "gushes" · The radar reflectivity of cumulonimbus · The depletion of small cloud particles by the growth of hail

- Appendix 8.1. Features of cumulus and cumulonimbus convection observed in special studies over Europe — 281

- Appendix 8.2. Observations of shower formation in and near the United States — 282
  Studies with radar and double-theodolite ranging or cloud stereophotogrammetry · Studies using ground-based radar alone · Studies from aircraft, using airborne radar · Observations using a tracking radar and time-lapse photography

- Appendix 8.3. The digital representation of the distribution of radar echoes and their intensity — 284

- Appendix 8.4. Descriptions of arched squalls — 286

- Appendix 8.5. The seasonal variation of hail damage to crops in the United States — 287

- Appendix 8.6. The boundary layer wind jet — 287

- Appendix 8.7. Nocturnal thunderstorms — 288

- Appendix 8.8. Castellanus clouds — 289

- Appendix 8.9. Cumulonimbus details: mamma — 290

- Appendix 8.10. Some visual observations of tornadoes — 290

- Appendix 8.11. Evidence of the breaking of large hailstones in the air — 291

- Appendix 8.12. Reports of giant hailstones of extreme size — 291

- Appendix 8.13. The recycling of objects in cumulonimbus updrafts — 293

- Appendix 8.14. Stationary local storms — 295
  The Selden hailstorm · Prolonged hailstorms reported from Austria and southern England · Other prolonged hailstorms and rainstorms · Exceptional rainstorms in southern England · Exceptional rainstorms in northwest France · Exceptional rainstorms in the United States · World record rainfalls

- Appendix 8.15. Visual range in intense rains — 300

- Appendix 8.16. Radar observations of cumulonimbus — 301
  The height of cumulonimbus tops · Details in the distribution of radar reflectivity; the echo-free vault

References — 303

		Contents
9	**Large-scale slope convection**	**310**
9.1.	General aspects of the large-scale flow in the troposphere outside the tropics	310
9.2.	Scale analysis and simple states of large-scale atmospheric motion Geostrophic and thermal winds · Vorticity; long waves	312
9.3.	The stability of large-scale sheared flows Studies by the perturbation method · The structure of the dominant wave · The evolution of baroclinic disturbances; frontal zones · The role of water in slope convection · The flow relative to a baroclinic disturbance	315
9.4.	The observed properties of large-scale slope convection Long waves and cyclone families · The Norwegian cyclone model · Observed properties of cyclones · The general circulation	319
9.5.	The large-scale cloud and precipitation systems of slope convection The relative flow and bands of frontal cloud in long waves · Ascent to the upper troposphere in frontal zones · Deep ordinary convection in frontal zones · Cloud and precipitation systems in cyclones	331
9.6.	Small-scale clouds and cloud systems in large-scale flows Small-scale features in the precipitation systems of slope convection · Cirrus clouds · Wave clouds · Billows	359
Appendix 9.1.	The mean annual latitudinal distribution of components of the energy balance of the atmosphere	383
Appendix 9.2.	The relative proportions and typical rates of middle-latitude precipitation from frontal zones and from showers	384
Appendix 9.3.	The frequency distribution of the heights of shower-cloud tops over the North Atlantic	385
Appendix 9.4.	The location of thunderstorms by observations of atmospherics	386
Appendix 9.5.	Experience of hazardous mountain wave phenomena	386
Appendix 9.6.	Climatological data	388
References		389
CONCLUSION		395
INDEX		397
PLATES		407

Dedication—Frank Henry Ludlam

Frank Ludlam died at his home near Ascot in England on June 3, 1977, at the age of 57, after an extended illness. His death is a great personal loss to his family and friends and to all of us who were privileged to work with him. Indeed, his students and colleagues invariably became his friends because his warmth and openness quickly overcame any formal barriers in professional relationships. But his passing will also be felt by the meteorological community at large because of the formidable impact which he had on the development of the science.

We are deeply saddened that Frank should not have seen this volume, undoubtedly his most important contribution, in final form. However, he was especially comforted in the knowledge that it was well on its way toward publication and delighted to have seen early galley proofs just a few weeks before his death.

This masterful volume, to which he dedicated several years of unremitting effort, represents the culmination of Frank's life work, the study of clouds and storms and their roles in the overall mechanics of the atmosphere. It was completed during a period when he was suffering great pain and was indeed immobilized by his illness. Under these conditions, lesser men might have abandoned such an ambitious and all-consuming project.

We doubt that he was driven by the desire for some measure of immortality, although this book will clearly remain a living memorial to him for generations to come. Rather, we believe that Frank was motivated by his characteristic need to synthesize and draw together the apparently unrelated bits and pieces of knowledge into some coherent and unified picture. He had an uncanny ability to perceive such links which so often eluded the rest of us. Indeed, it was this unique perception and ability to synthesize which characterized most of Frank's work and which is broadly recognized as the essence of science, which, wrote Sir Lawrence Bragg, "lies not in discovering of facts, but in discovering new ways of thinking about them." Similarly applicable to Frank's work are Arthur Koestler's words: "Creativity is the bisociation of two apparently independent matrices of knowledge."

This volume amply demonstrates Frank Ludlam's preeminence as a synthesizer. It ties together, in a way that no other author has succeeded in doing, the many interrelated aspects of dynamical and synoptic meteorology and cloud physics. Virtually all of the relevant physical processes ranging from radiation to cloud microphysics, along with methods of observation and their interpretation, are treated in a meaningful, lucid, and compact manner. Observational, analytical, and physical aspects are interwoven in his inimitable style. The overall theme is the interconnection of meteorological phenomena on all scales, from "bubble" convection through cumulonimbus, cyclones, on up to the general circulation. This kind of treatment is especially significant at the present stage of development of meteorology, since we have only recently begun to appreciate the importance of the two-way interactions between phenomena of varying scales, a recurrent theme of this book.

Frank's penetrating insight into the workings of the atmosphere and his ability to piece together a physical picture of a complex physical mechanism from a few observational facts were matched by his great qualities as a lecturer and writer. In both his talks and his papers, he was able to translate intricate physical processes into elegantly simple descriptions which call up images in the mind of the reader. He was also able to convey his own appreciation of the beauty of clouds and weather through equally beautiful word pictures of unique clarity. And where words alone were inadequate, he resorted to illustrations of his own making which combined his skills as scientist and artist. In fact, the bulk of the drawings in this volume were meticulously done by Frank himself.

Our words are hardly a fitting memorial to Frank Ludlam, beloved friend and colleague, and admired scientist. Let this volume stand as both his memorial and heritage to future generations.

David Atlas
Laboratory for Atmospheric Sciences
NASA Goddard Space Flight Center
Greenbelt, Maryland

John Mason
Director-General
Meteorological Office
United Kingdom
Bracknell, England

Professor Frank Ludlam—Science of the Clouds

The death of Frank Ludlam on June 3 at the age of 57 robs the international meteorological community of one of its most valued and individual thinkers. He joined the Meteorological Office shortly before the war rather than seek a place at a university because none offered an undergraduate course in the study of clouds. He was impatient to get on with his life's work.

Throughout the 1940s he was formulating ideas about ice clouds which earned him a Leverhulme research fellowship at Imperial College in 1949. Without a bachelor's degree he was appointed lecturer in 1951, later reader, and in 1965 professor of meteorology. He was awarded the D.Sc. in 1960. Because of his great influence on research into the mechanics of rain visitors came from many parts of the world to test out ideas on him. For two years he edited the magazine *Weather* and was Honorary Secretary of the Royal Meteorological Society for three.

His co-authorship of *Cloud Study* in 1957 displayed his freshly simple descriptions of clouds. Not for him obfuscation or pretension: let the drama and beauty of clouds speak for themselves! One of his dream-jobs was to man a lighthouse for the sheer pleasure of watching storms from within. He was a great friend of Bergeron, and Rossby invited him to spend a year in Sweden early in his career. He made several visits to Italy where he was always affectionately received, for he understood the farmers' need to "do something" when hail was destroying their crops, even though he knew that we must remain mere observers of storms.

He disliked the growth of large research teams and was anxious about the influence of computers which befuddled the mind with excessive detail and obscured essentials. His best work was done with slide rule and graph paper aided by an uncanny judgment of the actual magnitude of the forces at work. He enjoyed the very choosing of words, which he sometimes obtusely mispronounced.

He was freely iconoclastic; but the icons in question were modern, some even of his own creation. He possessed enormous respect for the early mariners and country gentlemen who knew more about storms than they are usually given credit for, and he would tell simple facts about the history of meteorology which were a pleasure to hear and which one felt ashamed not to have known already. His own name will be specially associated with hail and self-propagating showers.

His illness progressed over several years. He always displayed the same calm courage as he had displayed once during the war, in Iceland, when he was ordered, in an emergency, to leave the aircraft in which he had obtained a ride to observe the weather. His parachute opened inside the aircraft, but he firmly gathered it in his arms and climbed out through the upper gun turret which bristled with hooks and knobs, and let it go after jumping.

His students and colleagues owe a great debt to his wife on whose devotion and efforts he depended to continue working until the end. It is to be hoped that the publication of [this] definitive treatise on clouds will not be long delayed for it will consolidate the truly seminal character of his work.

R. S. Scorer
Imperial College of Science and Technology
London
—*The Times* (London)

Preface

The subject of this book is the behavior of water in the atmosphere, and its influence on the motion of the atmosphere. It aims to be a reference and guide for teaching and research in this and other aspects of meteorology, as well as an account readily intelligible to those who have an interest in the variety and spectacle of atmospheric phenomena but no special training as meteorologists.

In the last two decades related studies have developed at a great pace, stimulated in the beginning by the discovery of techniques of freezing supercooled cloud droplets, which at that time opened dramatic prospects for the artificial control of weather. Ensuing work was concentrated on processes affecting cloud *particles*, and although it became increasingly apparent that in the physics of atmospheric clouds the air motion on a great range of scales is an inseparable and often dominant part, through lack of knowledge of motions on other than the largest scales the subject developed as a kind of microphysics, applied to simple kinematical models of natural clouds. Microphysical processes are considered also in this book, but only so far as necessary for general discussion; for a comprehensive account of experiment, instrumental technique, and theory in this field reference can be made to a specialized text such as Mason's *The Physics of Clouds*.

Recently substantial improvements in the dynamical modeling of clouds have sprung from more powerful observational techniques, deployed during special investigations, and from a growing awareness among dynamicists concerned mainly with large-scale and long-period weather prediction that success depends partly on appropriate treatment of processes of smaller scale. The book should promote this trend by presenting the general background of the subject and its treatment of some principal phenomena (to limit the length there is no discussion of, for example, the organized systems of clouds and rains peculiar to low latitudes). It should illustrate relations between microphysical, thermodynamic, and dynamical aspects of the subject, which eventually must be integrated into a coherent cloud physics and an essential part of meteorology.

Measures have been taken to simplify and shorten the text. First, there is usually no discussion of historical development. Second, references are by number, and mainly to those of recent origin which themselves contain references to earlier publications. Third, very few names are mentioned, mainly because it is difficult to attribute responsibility for development and to identify the sources of ideas; I prefer to regard the construction of the subject as a collective enterprise in which all individuals, including many never quoted, play important parts. On the other hand, I have included many original diagrams and pictures, and some asides (appendices) intended to provide data not always easy to locate and to direct attention to related or neglected topics which deserve further study. The pictures can be regarded as composing a comprehensive atlas, accompanied for the first time by physical interpretations.

In the discussion of various processes the assessment of their importance rests generally upon magnitudes and not upon accurate values, so that terms such as "about," "a few," "several," and "of order" appear frequently. (By "a few" I mean probably between one and three, and by "several" probably between four and ten; by "of the order of" a quoted power of ten I mean that the value of a variable is a few or several units of that power of ten.)

I am grateful to colleagues who have instructed me, but am aware that in some places I have adopted a view still controversial, and that in others I may have overlooked work already published, while elsewhere fresh information and theory will eventually put matters in a new light.

F.H. Ludlam

CLOUDS AND STORMS

1 The influence of water on the general behavior of the atmosphere

1.1 Solar radiation and terrestrial temperatures

The radiation emitted by the sun has a maximum intensity at a wavelength of about 0.5 μ in the blue-green of the visible spectrum; about 0.99 of the energy is radiated at wavelengths less than about 4 μ (of which about 0.4 is in the near infrared, between about 0.8 and 4 μ). An approximate mean value of terrestrial temperatures can be inferred from eye observations of the color of the sun and its apparent angular diameter of about $\frac{1}{2}°$. With more specific information the inference can be made as follows.

The spectral variation and the total intensity of the energy radiated by the sun are approximately those of a black body with a temperature of 6000°K. Variations in the intensity of individual lines in the spectrum can be detected, and there are very large fluctuations in the ultraviolet. However, the ultraviolet comprises only about 10^{-5} of the total energy flux, which, when determined from the ground (where measurements are hampered by the difficulty of estimating attenuation by the atmosphere), varies by less than about 2% (1). At the earth's mean distance from the sun the mean energy flux in the solar radiation, the *solar constant f*, is very nearly 2 cal min^{-1} across each square centimeter normal to the beam (1.40 × 10^6 erg cm^{-2} sec^{-1}; 140 mw cm^{-2}). Of the radiation which reaches the earth a fraction a (the *albedo*, whose mean value is usually estimated as about 0.4) is scattered back into space without absorption. The average annual rate of absorption of solar energy over the globe is therefore

$$S = f\pi r^2(1 - a) \qquad (1.1)$$

where r is the radius of the earth.

Since other sources of energy are negligible and changes of mean terrestrial temperature over such a period are inappreciable, the earth radiates at the same mean rate, with an effective (black-body) emission temperature T_e given by

$$4\pi r^2 \delta T_e^4 = f\pi r^2(1 - a) \qquad (1.2)$$

where δ is the Stefan-Boltzmann constant (5.7 × 10^{-5} erg cm^{-2} °K^{-4} sec^{-1}; 8.1 × 10^{-11} cal cm^{-2} °K^{-4} min^{-1}). Thus $T_e \approx 250$°K, considerably less than the mean temperature at the surface of the earth (where the abundance of life depends on temperatures generally somewhat above the freezing point of water), and implying that a substantial part of the terrestrial radiation into space is emitted by the cooler atmosphere. At this temperature the radiation has its maximum intensity at a wavelength of about 15 μ, and so little is emitted at wavelengths less than 4 μ that it is usual to distinguish the solar as *short-wave radiation* and the terrestrial as *long-wave radiation*.

Lunar surface temperatures

On the moon there is virtually no atmosphere and the thermal conductivity of the surface material is very small. Where the sun is in the zenith the temperature of level surfaces attains a maximum value which can be estimated as that at which black-body emission equals the rate of absorption of solar energy. The albedo of the moon's surface is only about 0.07, close to the lowest terrestrial values (associated with oceans and forests), but less than that of dry terrestrial surface materials, perhaps on account of a fine porous structure also responsible for the poor thermal conductivity. Accepting this value the calculated maximum surface temperature is 388°K, somewhat lower than the value of about

1   The influence of water on the general behavior of the atmosphere

405°K inferred from measurements on earth of the radiation received from the moon in a part of the infrared spectrum which suffers little atmospheric attenuation (2). During the long lunar night the rapid fall of surface temperature is checked only by the upward conduction of heat which penetrates below the surface during the day. The minimum temperature reached is about 105°K, so that the surface temperature over most of the moon varies by considerably more than 100°K on either side of the freezing point of water.

The range of terrestrial surface temperatures

Excessive temperature variation does not occur on the earth's surface principally because of the abundance of water and the presence of a moist atmosphere. Where the sun is at a high elevation a large fraction of the solar radiation passes through the atmosphere and is absorbed at the surface, but because of the comparatively great heat capacity and thermal conductivity of the surface materials, and also because generally a large fraction of the radiation is used to evaporate water (Chapter 7), it produces a smaller rise of temperature than on the moon. Moreover, water vapor in the atmosphere, and especially low-level clouds, strongly absorb and emit long-wave terrestrial radiation, and thereby markedly reduce the fall of surface temperature during the night. Accordingly, the diurnal variation in the surface temperature is generally only about 10°C over land, and a small fraction of 1°C over the ocean. The greatest diurnal temperature ranges occur during cloudless weather in low-latitude deserts (1), where there is little water either in the atmosphere or in the ground. The surface temperature of the ground may reach 350°K and that of black cloth or wood exposed to the sun has been measured as 360–375°K, not much less than the maximum lunar surface temperature. The minimum terrestrial surface temperatures, in the range of about 180–200°K, occur in the interiors of the continents (in Asia and Antarctica) in high latitudes during the long polar night, and are considerably greater than those on the moon, because air continually arrives at low levels from warmer and moister regions.

1.2   The influence of water vapor on terrestrial long-wave radiation and temperature

The concept of radiative equilibrium

Absorption bands in the infrared (Appendix 1.1) make water vapor the atmospheric constituent with the most important influence on the radiative energy fluxes in the lower atmosphere. At temperatures near the mean radiative terrestrial temperature its density is sufficiently great for it to attenuate appreciably the part of the solar radiation in the near infrared. A further fraction of the solar radiation is scattered back into space by the atmosphere and the particles suspended in it, but nevertheless most of the energy in this short-wave radiation passes through a cloudless atmosphere and is absorbed at the earth's surface. The water vapor absorbs long-wave radiation much more strongly, thus the return of this energy through the atmosphere and into space by radiation demands that the temperature in the atmosphere decreases rapidly with height.

The vertical temperature distribution representing a state of radiative equilibrium, in which at all levels between the earth's surface and the outer limit of the atmosphere the net upward flux of long-wave radiation is equal to the net downward flux of short-wave radiation, can be calculated for some specified vertical distribution of atmospheric constituents and their radiative properties. The result shown by the curve R in Fig. 1.1 is based on the following specifications:

1. The solar radiation is the annual mean in latitude 35°N, and the albedo of the earth's surface is 0.1.
2. The atmosphere is cloudless, with the mean vertical distribution of water vapor observed at 35°N in April, and with a constant proportion by volume (3×10^{-4}) of carbon dioxide, another constituent radiatively of some, but much lesser, importance.

The calculated decrease of temperature with height (the *lapse rate* of temperature) is most marked near the earth's surface, where the density of water vapor is greatest, and decreases with height to become small above about 25 km, where the total quantity of water vapor remaining above is radiatively insignificant. The calculated temperatures are not realistic; the temperature of the surface is higher than the mean radiative terrestrial temperature, mainly, as might be anticipated, because the surface receives not only the solar radiation but also long-wave radiation from an adjacent atmospheric layer whose mean radiative temperature is nearly as high. The temperature, however, is some 40°K above the mean value observed in this latitude. Further, the temperatures are not compatible with the assumed distribution of water vapor, since they imply that away from the surface the vapor density much exceeds the

The influence of water vapor on terrestrial long-wave radiation and temperature

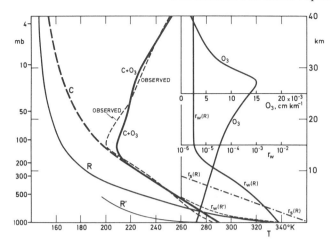

Fig. 1.1 Observed and calculated mean vertical temperature distributions for latitude 35°N in April, based on work described in references 3 and 4.

Radiative equilibrium states. The mean distribution of water vapor, expressed as a proportion r_w by weight of water vapor to dry air (the *mixing ratio*, defined in Chapter 2), as observed up to about 10 km and inferred above, is shown by the curve $r_w(R)$ on the right of the diagram. The curve R on the left is a calculated temperature profile under which the mean solar radiation in this latitude and season, partly scattered from or absorbed in a clear atmosphere, but mainly absorbed at the earth's surface (albedo 0.1), is steadily transferred upward by long-wave radiation through the atmosphere and into space. The corresponding profile of saturation mixing ratio $r_s(R)$ shows it to be less than the assumed mixing ratio $r_w(R)$ above about 3 km, and therefore inconsistent with the assumption of a cloudless atmosphere. This inconsistency can be avoided only by assuming a much more rapid decrease upward of the mixing ratio, such as shown by the curve $r_w(R')$, which results in a calculated temperature profile R'.

Radiation-convection equilibrium states. If the calculations are repeated, again assuming the observed mixing ratios, but preventing the lapse rate from exceeding 6.5°C km^{-1} by in effect introducing an upward (convective) heat transport other than by radiation, then the temperature profile C is obtained: lapse rates less than 6.5°C km^{-1} and heat transport by radiation alone occur only above 11 km (dashed curve). If, further, the presence of ozone O_3 is taken into account (whose concentration, here expressed as cm depth at NTP km^{-1}, is shown on the right), then the profile C changes to that marked ($C + O_3$), close to that observed (dashed line).

away from the surface has to be reduced, with the result that near the surface the calculated lapse rate becomes even greater. This result is illustrated by the curve R' in Fig. 1.1, obtained by assuming the distribution $r_w(R')$ of water vapor shown on the right of the diagram. In both this and the previous calculation the temperature decreases by about 90°K in the lowermost 5 km of the atmosphere, whereas the observed decrease is less than 30°K. It may be inferred that the presence of water vapor in the atmosphere makes a state of radiative equilibrium impossible, except perhaps at low temperatures (and great heights), at which its concentration is very small. The observed moderate decrease of temperature with height in the lowest several kilometers of the atmosphere implies that a process other than radiation—convection—contributes to the upward transport of energy there. Since it will later be shown that ordinary convective overturning prevents the establishment in a deep layer of any lapse rate exceeding about 10°C km^{-1}, and considerably less if condensation occurs, the atmospheric water vapor can be regarded as responsible for the continual presence of convection in the lowermost few kilometers.

The calculation of temperature distribution can be repeated with the following two conditions:

1. The lapse rate must not exceed the observed mean value of about 6.5°C km^{-1}; in the lower atmospheric layers there may then be a net loss of energy by radiation.

2. At the earth's surface the excess of absorbed radiation over the net upward long-wave radiation must equal the total net loss of energy in the atmosphere.

By these means the calculation in effect includes an upward heat transfer by convection without specifying its mechanics, but recognizing that its presence results in a mean lapse rate of only several degrees per kilometer. The result is shown by the curve C in Fig. 1.1, which up to a height of about 12 km is a more realistic representation of the observed temperature distribution and is therefore not inconsistent with the observed, and assumed, distribution of vapor density.

saturated vapor density, so that the presence of a thick cloud of condensed water must be anticipated, which would drastically increase the terrestrial albedo and diminish the solar radiation entering the lower atmosphere and reaching the earth's surface.

The impossibility of radiative equilibrium in the lower atmosphere

If this inconsistency is to be removed, the concentration of water vapor assumed to be present in the atmosphere

The troposphere, tropopause, and stratosphere

If the observed distribution of ozone is introduced into the calculation, an abrupt reversal of the lapse rate arises near the top of the layer in which the convection is assumed to maintain a constant lapse rate of 6.5°C km^{-1}, as shown by the curve ($C + O_3$) in Fig. 1.1. The trace gas ozone is produced photochemically at heights somewhat above about 25 km, in concentrations which although small very strongly absorb solar radiation in the ultraviolet (it thereby shields the earth's surface from radiation at wavelengths less than about 0.3 μ

which damages organisms). Ozone represents an important local heat source, which causes the temperature in the uppermost layers included in Fig. 1.1 to increase with height and reach a value of nearly 260°K at 40 km, in close agreement with the observed mean value. Thereby the considered atmosphere can be divided into a lowermost layer in which convection is present, beneath a layer of small and reversed lapse rate, in a state of radiative equilibrium. Observations show that this division into a *troposphere* of pronounced lapse rate beneath a *stratosphere* of small or reversed lapse rate is a characteristic feature of the atmosphere; soundings frequently show that the transition between these regions is remarkably sharply defined, occurring in a layer less than 100 m deep and therefore so shallow that in most analyses it can be regarded as a discontinuity, the *tropopause*.

[The atmosphere is conventionally divided into several characteristic layers, of which those considered in this book are defined by the vertical profile of temperature (16). A level of temperature maximum—about 270°K—near 50 km, the *stratopause*, separates the stratosphere from the *mesosphere*. Another level of temperature minimum—about 180°K—the *mesopause*, is found at about 80 km, and separates the mesosphere from the *thermosphere*.]

The mean rate of radiative energy loss in the troposphere

Figure 1.2 shows the vertical distribution of the principal mean components of the heat balance according to the calculations whose result is shown by the curve $(C + O_3)$ in Fig. 1.1. In the stratosphere the state of radiative equilibrium is maintained as a balance between the heat sources associated with the absorption of short-wave radiation by ozone (and to a minor extent by water vapor and CO_2) and the heat sinks associated with a net loss of energy by long-wave radiation from water vapor and CO_2. In the troposphere the absorption of short-wave radiation by water vapor is a considerable heat source which somewhat reduces the pronounced heat sink, equivalent to a rate of cooling of between 1 and 2°C day^{-1}, which represents the net loss of energy by long-wave radiation, and which is balanced by convection of heat from the earth's surface. When the presence of an average amount of cloud is taken into account the equivalent rate of cooling becomes approximately 1°C day^{-1}, which other calculations show to be a characteristic mean value in the troposphere generally (Fig. 1.3). The magnitude of this mean value is set simply by the rate of absorption of solar radiation in the troposphere and at the earth's surface. Of that absorbed at the surface only a small percentage is returned directly into space as long-wave radiation not strongly absorbed by water vapor. Assuming that the

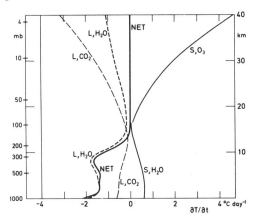

Fig. 1.2 Vertical distributions of the rate of energy loss or gain by radiation which is associated with various atmospheric constituents, in the equilibrium state represented by the curve $(C + O_3)$ of Fig. 1.1 (from reference 3). The rate of energy gain is expressed as a rate of change of temperature $\partial T/\partial t$ in the absence of other sources or sinks. Only the principal components of the energy balance are indicated: the absorption of solar radiation S by ozone O_3 and water vapor H_2O (solid lines), and net loss of energy by long-wave radiation L associated with carbon dioxide CO_2 and with H_2O (dashed lines). Minor absorption of solar radiation by CO_2, and by H_2O in the stratosphere, is omitted.

The net rate of energy loss (heavy line) is zero in the stratosphere, indicating a state of radiative equilibrium, and equivalent to a rate of cooling of 1–2°C day^{-1} in the troposphere, which is compensated by a convective heat transport from the earth's surface.

mass of air in the troposphere is 800 g over each unit area of 1 cm^2 (corresponding to an average tropopause height at about the 200-mb level), the mean rate of radiative cooling must be about $f(1 - a)\pi r^2 \times 12 \times 60/4\pi r^2 \times 800 c_p$ °C day^{-1}, where $c_p = 0.24$ cal g^{-1} °C^{-1} is the specific heat of air at constant pressure, or 1°C day^{-1}.

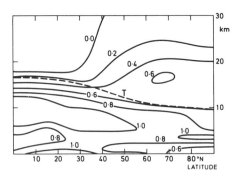

Fig. 1.3 An example of a calculated annual mean distribution of rate of energy loss by radiation, expressed as an equivalent rate of cooling (isopleths labeled in °C day^{-1}), from reference 5. The dashed line T marks the approximate mean position of the tropopause. Near and above the tropopause the result of the calculation is sensitive to the assumed distribution of water vapor, which has not been observed carefully.

The latitudinal distribution of the terrestrial radiation into space

In the calculations of the heat balance for latitude 35°N in April, leading to the curve $(C + O_3)$ in Fig. 1.1, the net downward flux of solar radiation at the top of the (cloudless) atmosphere, and therefore also the net outward flux of terrestrial radiation there, was about 0.43 cal cm^{-2} min^{-1}. Of the latter flux, about 0.03 cal cm^{-2} min^{-1} was radiated directly from the ground into space in parts of the infrared where the atmospheric absorption is small, and the large remainder from the atmosphere itself. Now the atmospheric emission into space can be regarded as determined effectively by an average temperature in the highest atmospheric layer containing sufficient water vapor to be considered as practically opaque in the infrared. Since the vapor density in the troposphere decreases rapidly with height this layer is well-defined, and, moreover, since the vapor density, by its limitation to the saturated vapor density, is strongly related to the temperature (Fig. 1.4), the temperature of this layer might be anticipated to be about the same everywhere, so that the mean annual long-wave terrestrial emission into space would be about the same in all latitudes.

This near uniformity appears in calculations based on the mean observed distributions with height of vapor density and temperature, as seen in Fig. 1.5. In this diagram the curve L indicates that the mean annual flux of long-wave radiation into space diminishes with increase of latitude, but only from about 0.35 cal cm^{-2} min^{-1} near the equator to 0.25 cal cm^{-2} min^{-1} near the pole. Curve L' in Fig. 1.5 confirms the small latitudinal variation of the flux, even in the midwinter season at the time of the greatest latitudinal variations of solar radiation, and of tropospheric temperature and vapor density.) This is a considerable variation, but notably smaller than that in the incoming solar radiation, due to the obliquity of the solar beam to the earth's surface in high latitudes: the mean net flux of the solar radiation, shown by the curve S, varies from about 0.4 cal cm^{-2} min^{-1} in low latitudes to only about 0.1 cal cm^{-2} min^{-1} near the pole. Thus only in some middle latitude can there be a mean local heat balance: in lower latitudes the supply of solar radiation exceeds the terrestrial emission, whereas in high latitudes it is the reverse. To preserve the mean temperature distribution in the troposphere there must be a substantial transport of heat from low into high latitudes; this is accomplished mainly by a large-scale convection manifest as great systems of wind and weather, and partly by ocean currents which warm the lower troposphere in the higher latitudes by ordinary convection.

If, contrary to this reasoning, a balance could be maintained in every latitude between the influx of solar radiation and the terrestrial long-wave emission, assisted in the troposphere by only an *upward* convective heat transport, then the variation of tropospheric temperature with latitude about the mean effective radiative temperature T_e of 250°K could be anticipated to be about bT_e. The fraction b is the fourth root of the ratio between the solar radiation received in a given latitude and the hemispheric mean value of about 0.33 cal cm^{-2} min^{-1}. According to Fig. 1.5 this ratio is about 1.2 in low latitudes and 0.3 in high latitudes, implying mean temperatures of about 260 and 185°K, respectively. The corresponding mean latitudinal temperature difference *in the horizontal* is more than 75°K, and thus more than

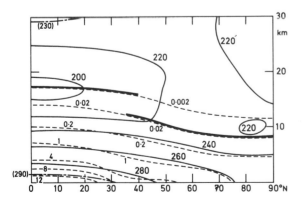

Fig. 1.4 Approximate mean annual distribution of temperature (isopleths labeled in °K) and water vapor mixing ratio (dashed isopleths labeled in g kg^{-1}). The mean positions of the tropical and extratropical tropopauses are shown by heavy lines.

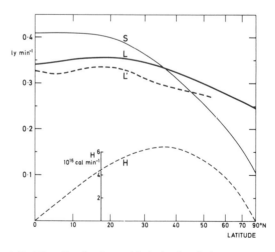

Fig. 1.5 The distribution with latitude of the mean annual downward flux of solar radiation S and the upward flux of terrestrial radiation L (estimated from observed mean annual atmospheric state). The dashed line L' shows the distribution of terrestrial radiation according to observations from the meteorological satellite TIROS II, made between late November 1960 and early January 1961 (7). The implied poleward heat transport H is shown by the lowermost curve, against the inset scale.

1   The influence of water on the general behavior of the atmosphere

twice the difference observed at all levels within the troposphere (Fig. 1.4). This discrepancy again reveals the presence of radiative imbalance and the large-scale convection of heat from low to high latitudes; in low latitudes the troposphere is a little cooler, and in high latitudes considerably warmer, than would be expected in the absence of convection.

Figure 1.3 shows that in latitude 35°N the stratosphere is approximately in a state of radiative equilibrium, as was inferred in the calculations whose results are illustrated in Fig. 1.1. On the other hand the stratosphere in higher latitudes is not in radiative equilibrium, and in its lower part must contain heat transports of large magnitude. It is not immediately evident whether these are associated with the distribution of water vapor or other radiatively significant components, or whether they are due to motions whose energy is provided by the tropospheric convection.

It is clear, however, that the abundant supply of water vapor at the general terrestrial temperatures necessitates the continual presence of convection transferring heat from the earth's surface through a troposphere not only upward but also on a hemispheric scale from low into high latitudes.

1.3   General characteristics of atmospheric convection

Ordinary and slope convection

There are two principal types of convection in the troposphere. The first, which will be called *ordinary* because of its familiar form, consists of individual circulations whose vertical and horizontal dimensions are about the same. It occurs overland by day, when the ground is warmed by sunshine, and by day and night in air which flows over the oceans toward lower latitudes and higher sea surface temperatures. When water vapor condenses in the upper parts of the circulations the clouds which are formed are usually about as broad as they are tall, and appear in scattered populations covering very large areas. Occasionally they extend upward through the whole troposphere, but especially in middle latitudes the ordinary convection is generally confined to the lowest 2–3 km. It evidently transfers energy in the form of latent and sensible heat from the earth's surface into this lowermost layer, which then supplies the warm currents of the second kind of convection, called *slope convection*, because although its flows extend through the whole depth of the troposphere they reach over much greater distances, so that they are only slightly inclined to the horizontal. At least outside the tropics this kind of convection is responsible not only for the upward transport of heat through the upper part of the troposphere but also for the poleward transport.

Each kind of convection has its own characteristic properties, which will later be described in detail. At present, considering only the magnitudes of the heat transports which they effect, it is possible to make plausible estimates of the typical magnitude of the associated vertical (and horizontal) air speeds. These are required in the following discussion of the microphysical processes attending the formation of the corresponding two principal kinds of clouds: the ordinary convection (*cumulus* or *heap*) clouds, and the clouds of the slope convection, which because of their great extent and hardly appreciable inclination (and because of the unfamiliar form of the slope convection) are generally described as *layer* clouds and not specifically associated with another kind of convection.

The layer of air whose properties have recently been modified by the ordinary convection is considered the *planetary boundary layer*, in contrast to the *free atmosphere* above.

Ordinary, including cellular, convection in the laboratory

Convection is produced in fluids by a warming from below. In liquids, which are virtually incompressible, the density throughout even a deep layer is constant when the temperature is uniform, but generally decreases when the temperature rises. Consequently, the warming generally produces a decrease in the density of the lower layers of the fluid and a rise in the center of gravity of the whole mass. In this state the fluid can be said to be mechanically unstable: if as a result of some disturbance the less dense liquid rises without significant temperature change to occupy the top of the layer, the center of gravity is lowered and the decrease in potential energy is available to provide the energy of the motion which effects the overturning.

If, however, the warming of the base of a shallow layer of liquid is only slight, the liquid may remain undisturbed; if it is prolonged, a state may be attained in which the temperature decreases uniformly with height and the energy supplied below is transferred steadily upward by the agency of molecular conduction alone. A small vertical displacement within a liquid in this condition causes a local temperature and therefore density change, and a buoyancy force which tends to increase the displacement and produce an acceleration, but this tendency is resisted and may be overwhelmed by internal friction (molecular viscosity) and by the enhanced internal diffusion of heat (molecular conductivity). If the heat flux and the temperature gradient in the liquid are increased, these restraints are eventually overcome and the convection begins at a critical value of a parameter known as the Rayleigh number after the original contribution of Lord Rayleigh to a comprehensive theory of the motion (8). The Rayleigh number Ra is

$$\text{Ra} = \frac{g\alpha(\partial T/\partial z)d^4}{kv} \quad (1.3)$$

where α = coefficient of volume expansion (°C^{-1})
k = thermometric conductivity (cm^2 sec^{-1})
v = kinematic viscosity (cm^2 sec^{-1})
d = depth of the layer in which the vertical temperature gradient = $(\partial T/\partial z)$

Expressed another way,

$$\text{Ra} = \frac{g\alpha\Delta T d^3}{kv} \quad (1.4)$$

where ΔT = temperature difference between the top and bottom of the layer

The critical value of the Rayleigh number depends on the nature of the bounding surfaces at the top and bottom of the layer of fluid. Theoretically it is 1101 for one rigid and one free boundary, and 1708 for a layer confined between two rigid surfaces. The first experimental determinations for the latter condition gave a value of 1770 ± 140, and subsequent more refined experiments provided an even better agreement with the theoretical value, for example, 1700 ± 50 in each of several different liquids.

There is also a satisfactory theory for the separation of the upcurrents and downcurrents when the convection proceeds under a Rayleigh number at or just beyond the critical value. It is observed that the fluid often becomes divided into cells whose horizontal sections are hexagons nearly three times as wide as the depth of the layer, within each of which there is a steady circulation, usually with ascent in the center and descent near the walls. The cells are often called *Bénard cells* after the French scientist who was the first to make quantitative experiments with this kind of convection. When the motion is established the original uniform decrease of temperature with height is altered; throughout most of the fluid the temperature gradient is decreased, but approaching the boundaries the motion becomes more nearly horizontal and the temperature gradient increases to allow the heat transport to occur eventually by molecular conductivity only.

The very regular patterns of hexagonal convection cells are most readily produced in shallow layers of liquid with a free upper surface, in which it has been found that the variation of surface tension associated with the variation of temperature in the surface has a dominating influence on the form of the convection. If this influence is removed, the form of the convection observed in laboratory experiments appears to be determined by the conditions at the lateral walls of the fluid container (9); the cells may appear as long horizontal cells as a result of an instability not yet satisfactorily explained. Also in a layer of fluid moving over a solid boundary cells and their circulations are replaced by opposing spiral circulations in long parallel rolls lying usually in the direction of the mean *shear* throughout the layer (the vector difference between the mean velocities at its top and bottom). If the magnitude of the shear is sufficiently small, the rolls may lie transverse to the shear.

The patterns of motion are made visible by introducing into the fluid various tracers, such as smokes in gases, and small suspended flakes in liquids (the flakes become aligned along the direction of the local shear, which is practically the direction of motion, so that viewed from above they are easily visible where the flow is horizontal and almost invisible near the centers and sides of the cells). The patterns thus made apparent have a strong superficial resemblance to the dappled and banded structures often produced by convection in atmospheric clouds, but steady cellular convection of the kind observed in the laboratory, in which the heat transfer is very small and the motion is only just maintained against the molecular diffusion of heat and momentum, cannot be expected in the atmosphere, except perhaps very close to the ground. Steady cellular convection can be produced readily only in very shallow layers; it proceeds under a temperature difference of 1°C between the top and bottom of a layer which in water is only about 2 mm and in air only about 2 cm deep. In much deeper layers the corresponding temperature differences, which are inversely proportional to the cube of the depth, become too small to contrive and maintain; alternatively regarded, the steady convection demands a heat flux of a particular very small magnitude which is difficult to provide and control. In laboratory experiments it is seen that if the heat flux is increased and the Rayleigh number is only slightly greater than the critical value, the cellular pattern persists; if it is several times greater, the circulations become intermittent and turbulent, and their arrangement becomes irregular. In the atmosphere the upward convective heat flux is typically of order 10^{-1} cal cm^{-2} min^{-1} (a large fraction of the mean total upward flux; see Fig. 1.5), and the corresponding depth of a layer in which the steady cellular convection could occur is only about 1 cm. The convective flux in a layer of cloud of sufficient thickness to behave as a black body in the infrared (about 100 m in the middle and low troposphere; see Appendix 1.1) can be anticipated to have about the same magnitude, and therefore also implies an unsteady, turbulent convection. [However, a steady kind of convection in an atmospheric layer up to about 1 km deep can be envisaged if it is regarded as reasonable to replace the molecular transfer coefficients by turbulent eddy transfer coefficients larger by several orders of magnitude. Some arrangements of ordinary convection clouds observed from meteorological satellites are suggestive of a regular pattern of cells whose width is about 30 times as great as the depth (Chapter 7). Such cells have been attributed to a steady convection in which at least one of the eddy transfer coefficients is greater in the horizontal than in the vertical (10, 11). There is no theory of their actual or relative magnitude.]

Atmospheric convection

The poleward transfer of heat in the troposphere accompanies systems of cool currents which descend from high levels in high latitudes to low levels in low latitudes, and warm currents which ascend as they flow toward the poles. It can reasonably be anticipated that in middle latitudes a typical temperature difference ΔT between these warm and cool currents is a small fraction of the difference of 30–40°C which exists between low and high latitudes. Overlooking the complications that a large part of the heat transfer is due to a net poleward flux of latent heat (associated with the condensation and precipitation of water vapor evaporated in a lower latitude), the middle-latitude horizontal heat flux can be written as

$$H_h = mc_p v \Delta T \qquad (1.5)$$

where m = the mass of a column of unit area of that part of the atmosphere involved
c_p = the specific heat at constant pressure
v = the mean meridional component of the speed of the air currents

From Fig. 1.5, the mean value of H_h in middle latitudes is about $10^{15}/l$ cal sec^{-1}, where l is the length of a mid-latitude parallel, say 2.8×10^9 cm (corresponding to latitude 45°N). If the troposphere up to about the 300-mb level is involved, then $m \approx 700$ g cm^{-2}, and with $c_p = 0.24$ cal g^{-1} °K^{-1}, it follows that

$$H_h = 700 \times 0.24 v \Delta T = 10^6/2.8 \text{ cal cm}^{-1} \text{ sec}^{-1} \qquad (1.6)$$

Thus

$$v \Delta T \approx 2 \times 10^3 \text{ cm sec}^{-1} \text{ °K} \qquad (1.7)$$

If ΔT is taken as 2°C, the mean *meridional* wind speed v becomes 10 m sec^{-1}, reasonably consistent with the observed mean kinetic energy of the troposphere, which corresponds to a mean wind speed of about 17 m sec^{-1}. Accordingly, the radiative imbalance of the troposphere can be regarded as determining the magnitude of the mean wind speed in the slope convection; near the earth's surface the mean speed is much less because of friction, and on the other hand the mechanics of the convection leads to the production locally of much greater wind speeds, especially in the high troposphere, where the maximum values attained are about 100 m sec^{-1}.

The magnitude of the vertical components of the mid-tropospheric winds follows from the mean inclination of the currents in the slope convection, which can be anticipated to be about the depth of the troposphere over the distance between a low and a high latitude, say 10 km over about 5000 km between latitudes 25 and 65°, or 1/500. Thus the magnitude of the mean vertical speeds is only about 2 cm sec^{-1}, which can be taken as about the minimum ever associated with the formation and development of layer clouds, since they appear in parts of the motion systems where both the horizontal and vertical air speeds tend locally to be intensified.

The magnitudes of the vertical heat transport H_v effected by the slope convection and, in the lower troposphere, by the ordinary convection are about the same (less than 0.1 cal cm^{-2} min^{-1}, as might be anticipated from Fig. 1.5), and therefore correspond to a temperature difference of about 2°C between ascending and descending currents of speeds about 2 cm sec^{-1}. In the ordinary convection of much smaller scale, however, the temperature difference is typically much smaller than this, the ascending currents are narrow, and the upward vertical speeds are correspondingly much larger. An approximate relation between the temperature difference and the upward speeds is provided by the concept of buoyancy, the density difference $\Delta \rho$ at a particular level from the mean value $\bar{\rho}$ being regarded as producing the vertical acceleration $g \Delta \rho/\bar{\rho}$ or $g \Delta T/\bar{T}$ which would be experienced by a solid body moving through a fluid without resistance. Again overlooking complications associated with the possible release of latent heat and inefficiencies in the motion systems, on this assumption a value of ΔT of only 1/20°C maintained over an ascent of 1 km would provide a speed of about 2 m sec^{-1} in upcurrents occupying one-tenth of a horizontal area, and a heat flux at the upper level of the required intensity. These values for the temperature difference, upward velocity, and heat flux are about those observed several hundred meters above the ground during ordinary convection (11). Thus the upward air speed during the formation of ordinary convection clouds has the magnitude of 1 m sec^{-1}. In the ordinary development of the clouds the proportion of a horizontal area occupied by those which reach into the middle or upper troposphere is still further reduced, and the typical upward air speeds inside the clouds attain the magnitude of 10 m sec^{-1}. Accordingly, the typical ranges of values of the upward air speeds encountered in the layer clouds and the cumulus are, respectively, 1–10 cm sec^{-1} and 1–10 m sec^{-1}.

Appendix 1.1 The attenuation of electromagnetic radiation in the atmosphere; long-wave absorption by clouds

Atmospheric attenuation

Figure 1.6, compiled from a variety of sources, shows the relative importance of various atmospheric constituents in attenuating a parallel beam of electromagnetic radiation, according to its wavelength λ. The attenuation is expressed in decibels, as $10 \log_{10} (I_\lambda)_0/(I_\lambda)_1$, where $(I_\lambda)_0$ and $(I_\lambda)_1$ are the intensities of the radiation (the fluxes of energy in a unit interval of wavelength across unit areas normal to the direction of the beam, per unit solid angle) at the ends of a horizontal path of length 1 km along which the pressure is 800 mb and the temperature is 10°C (representative of the lower troposphere). The atmosphere at this level is supposed to contain various absorbing gases, water vapor, or scattering and absorbing water particles, in the following representative concentrations:

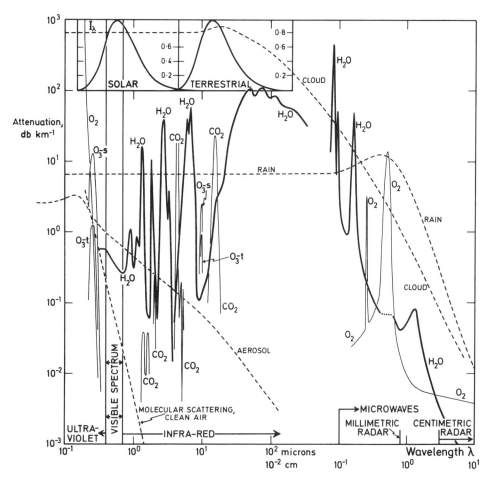

Fig. 1.6 The relative importance at various wavelengths of atmospheric constituents which attenuate electromagnetic radiation by absorption or scattering. The attenuation in a horizontal path of 1 km, in the presence of constituents whose assumed concentrations are specified in the text, is expressed in decibels, that is, as $10 \log_{10} \kappa$, where κ is the ratio of the intensities of the radiation at the beginning and end of the path. The curves in the inset in the upper left of the diagram represent the intensities of black-body radiation at temperatures of 6000 and 250°K, representative of short-wave solar radiation and long-wave terrestrial radiation (normalized so that the areas under the curves imply equal energy fluxes).

1   The influence of water on the general behavior of the atmosphere

1. Gases and water vapor (solid lines in Fig. 1.6):

O_2:   23% by mass   (23 g cm^{-2} km^{-1})
CO_2:   22.8 atmo-cm km^{-1} (4.4×10^{-2} g cm^{-2} km^{-1})
H_2O: 5 g kg^{-1} (4.9×10^{-1} g cm^{-2} km^{-1})
O_3-t
(tropo-
spheric): 10^{-3} atmo-cm km^{-1} (2.1×10^{-6} g cm^{-2} km^{-1})

In the stratosphere a representative concentration is an order of magnitude greater:

O_3-s
(strato-
spheric): 10^{-2} atmo-cm km^{-1} (2.1×10^{-5} g cm^{-2} km^{-1})

2. Water droplets (dashed lines in Fig. 1.6):

Small droplets (representing aerosol particles), radius = 0.1 μ: 4×10^{-5} g m^{-3}
Cloud droplets, radius = 10 μ: 1 g m^{-3}
Rain drops, radius = 1 mm: 1 g m^{-3} (corresponding to a radar reflectivity factor Z of 1.5×10^4 mm^6 m^{-3} and a rainfall rate of about 30 mm hr^{-1})

Comparison of the curves drawn in the inset in the upper left of Fig. 1.6 with the attenuation curves shows the comparative opacity of the lower troposphere to the terrestrial long-wave radiation, principally because of the absorption by water vapor, except in a narrow range of wavelengths between about 8 and 12 μ, where there is said to be a "window" [it is now thought that this window is less transparent than indicated, perhaps because of absorption by $(H_2O)_2$ (17), which at the high vapor pressures found at low levels in the tropics may lead to a doubling of the attenuation at a wavelength of 10 μ]. Also apparent is the greater transparency of atmospheric aerosol at the long wavelengths, which is exploited in photography in the (near) infrared. The significant attenuation by oxygen and water vapor in the microwave region constrains the choice of wavelengths for meteorological radar to about 8 mm (used at short ranges, often with beams fixed in the vertical), and to more than about 3 cm (for use at ranges of up to 100 km or even more).

Absorption of long-wave radiation by clouds

Liquid water absorbs long-wave radiation very strongly (to a degree not particularly sensitive to temperature in the meteorological range); a layer of liquid only about 5 μ thick absorbs half of the normally incident long-wave radiation (13).

In considering absorption in clouds it is customary to neglect the complications associated with the diffuseness of the incident radiation, multiple scattering from the cloud particles, and large variations in their scattering cross section which arise because their size is comparable with the wavelengths of the radiation. It appears that the absorptivity of a cloud is not overestimated if it is taken to be that of a uniform layer of liquid produced by its precipitation, and is about 0.9 for a cloud only 200 m thick if the mean concentration of liquid water is about 0.1 g m^{-3}. Since this mean concentration is about that produced in the lower troposphere by an adiabatic ascent 200 m beyond the saturation level, it is reasonable to regard extensive droplet clouds of greater thickness as black bodies; a similar thickness is sufficient to obscure the sun (requiring an attenuation of the direct sunlight exceeding about 20 db).

The absorptivity of ice particles integrated over the spectrum of long-wave radiation probably does not differ materially from that of water droplets, but clouds formed at the low temperatures of the high troposphere do not usually have the density or thickness to behave as black bodies. With a saturation level at 300 mb, $-40°C$, for example, an adiabatic ascent in a cloud of thickness about 1 km is needed to produce the required amount of precipitable water, and substantially more is needed at higher levels and lower temperatures. High clouds of the required thickness and density are produced only in narrow belts or small regions by the ascent of air from the middle or low troposphere (in frontal zones and cumulonimbus), and those which are more widespread and frequently seen from the ground in fair weather are either the tenuous slowly evaporating residues of such dense clouds or scattered shallow clouds with comparatively small concentrations of condensed water. When they pass in front of the sun the attenuation of both the short- and long-wave radiation in direct sunlight is between about 5 and 10 db (14), much too little to obscure the position of the sun's disc. Nevertheless, the presence of high clouds too tenuous and uniform to be recognizable from the ground is thought to be responsible for balloon-borne radiometer observations which sometimes imply that the rate of radiative loss of energy from a layer a few kilometers deep near the tropopause is significantly less than would be calculated assuming the absence of cloud (15). Dense high cloud in an otherwise mainly clear atmosphere can be expected to experience a radiative warming, which could materially assist its eventual evaporation.

References

1. *Handbook of Geophysics*, rev. ed. 1960. Macmillan, New York.
2. Sinton, W.M. 1962. Temperature on the lunar surface, in *Physics and Astronomy of the Moon*. Z. Kopal, ed. Academic Press, New York, ch. 11.
3. Manabe, S., and Strickler, R.F. 1964. Thermal equilibrium of the atmosphere with a convective adjustment, *J. Atmos. Sci.*, **21**, 361–85.
4. Manabe, S., and Wetherald, R.T. 1967. Thermal equilibrium of an atmosphere with a given distribution of relative humidity, *J. Atmos. Sci.*, **24**, 241–59.
5. Manabe, S., and Möller, F. 1961. On the radiative equilibrium and heat balance of the atmosphere, *Mon. Weather Rev.*, **89**, 503–32.
6. Smagorinsky, J., Manabe, S., and Holloway, J.L. 1965. Numerical results from a nine-level general circulation model of the atmosphere, *Mon. Weather Rev.*, **93**, 727–68; Manabe, S., Smagorinsky, J., and Strickler, R.F. 1965. Simulated climatology of a general circulation model with a hydrological cycle, *Mon. Weather Rev.*, **93**, 769–98.
7. Winston, J.S., and Krishna Rao, P. 1963. Temporal and spatial variations in the planetary-scale outgoing long-wave radiation as derived from Tiros II measurements, *Mon. Weather Rev.*, **91**, 641–57.
8. Chandrasekhar, S. 1961. *Hydrodynamic and Hydromagnetic Stability*. Clarendon Press, Oxford.
9. Koschmeider, E.L. 1967. On convection under an air surface, *J. Fluid Mech.*, **30**, 9–15.
10. Priestley, C.H.B. 1962. The width-height ratio of large convection cells, *Tellus*, **14**, 123–24.
11. Ray, D. 1965. Cellular convection with non-isotropic eddies, *Tellus*, **17**, 434–39.
12. Telford, J.W., and Warner, J. 1964. Fluxes of heat and vapor in the lower atmosphere derived from aircraft observations, *J. Atmos. Sci.*, **21**, 539–48.
13. McDonald, J.E. 1960. Absorption of atmospheric radiation by water films and water clouds, *J. Meteorol.* **17**, 232–38.
14. Gates, D.M., and Shaw, C.C. 1960. Infrared transmission of clouds, *J. Opt. Soc. Am.*, **50**, 876–82.
15. Cox, S.K. 1969. Observational evidence of anomalous infrared cooling in a clear tropical atmosphere, *J. Atmos. Sci.*, **26**, 1347–49.
16. Craig, R.A. 1965. *The Upper Atmosphere*. Academic Press, New York.
17. Bignell, K.J. 1970. The water-vapour infra-red continuum, *Q. J. R. Meteorol Soc.*, **96**, 390–403.

2 The physical properties of the moist atmosphere and its particles

This chapter briefly summarizes some of the physical properties of moist air and of aerosol and water particles, which are considered in the discussion of condensation, evaporation, and precipitation processes. Numerical values of selected properties are listed in Appendix 2.1; more comprehensive and accurate specifications are found in standard meteorological tables (1, 2).

2.1 Equations of state

For all but the most accurate calculations it can be supposed that both dry air and unsaturated water vapor behave as ideal gases [over the range of meteorological conditions the values of the gas "constants" R and R_v in the following equations of state vary by only about 0.1 and 0.3%, respectively (2)], and that in moist air Dalton's law of partial pressures is obeyed. Then the principal variables of state (pressure, density, and temperature) are related by the expressions

$$p - e = R\rho_a T \qquad (2.1)$$

and

$$e = R_v \rho_v T \qquad (2.2)$$

where p = total pressure
e = partial pressure of water vapor
ρ_a, ρ_v = densities of dry air and water vapor, respectively
T = temperature

Using metric units of dynes per square centimeter ($= 10^{-3}$ mb) for pressure and grams per cubic centimeter for density, the gas constants are

$$R = 2.87 \times 10^6 \text{ erg g}^{-1} \text{ K}^{-1}$$
$$(6.86 \times 10^{-2} \text{ cal g}^{-1} \,°\text{K}^{-1})$$

and

$$R_v = 4.61 \times 10^6 \text{ erg g}^{-1} \text{ K}^{-1}$$
$$(1.10 \times 10^{-1} \text{ cal g}^{-1} \,°\text{K}^{-1})$$

implying that the ratio ε of the densities of water vapor and of air at the same temperature and pressure is

$$\varepsilon = \frac{R}{R_v} = 0.622$$

(If, as in this text, the millibar is used as the unit of pressure, the numerical values of the gas constants given above are reduced by a factor of 10^3.)

2.2 Saturation

Pure water vapor in equilibrium with a plane surface of the condensed phase at the same temperature is said to be *saturated*. The equilibrium is influenced slightly by the presence of air, but not by more than about 0.5% in any meteorological conditions (1). Although the effect is appreciable it is generally ignored; the expressions *saturated vapor* and *saturated air* are used indiscriminately, and the equilibrium vapor pressure e_s and density ρ_s are regarded as functions of temperature only. The dependence can be found from simple thermodynamic reasoning to obey approximately the (so-called Clausius-Clapeyron) relations

$$\frac{de_s}{dT} = \frac{e_s L_v}{R_v T^2} \tag{2.3}$$

and

$$\frac{d\rho_s}{dT} = \frac{\rho_s(L_v - R_v T)}{R_v T^2} \tag{2.4}$$

where $L_v(T)$ is the latent heat of evaporation.

More complicated formulas, considering deviation from the behavior of an ideal gas, have been used to calculate e_s at low meteorological temperatures at which it has not been measured, and are the basis for the standard tables and the extracts listed in Appendix 2.1. [There are other formulas which are more convenient to use in computations and which are sufficiently exact for practically all meteorological purposes; for example (3),

$$\ln e_s = \ln 6.108 + \frac{a(T - 273.16)}{(T - b)} \tag{2.5}$$

where $a = 17.27, b = 35.86$ and $a = 21.87, b = 7.66$ for saturation over liquid water and over ice, respectively. The percentage error in the value of e_s calculated from this formula is less than 0.1% at temperatures above $-5°C$, less than 1% above $-25°C$, and less than 3% above $-40°C$.]

In some problems concerning condensation and evaporation the effect of a small change of temperature upon e_s or ρ_s is conveniently taken into account by using Eq. 2.3 or 2.4 in the finite difference forms:

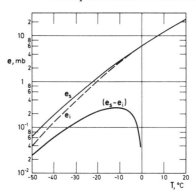

Fig. 2.1 The saturation vapor pressure e_s over liquid water, the saturation vapor pressure e_i over ice, and their difference $(e_s - e_i)$, all drawn on a logarithmic scale as a function of temperature T.

$$\Delta e_s = \frac{e_s L_v \Delta T}{R_v T^2} \tag{2.6}$$

$$\Delta \rho_s = \frac{\rho_s(L_v - R_v T)\Delta T}{R_v T^2} \tag{2.7}$$

The percentage error in the value of $\Delta \rho_s$ given by Eq. 2.7 is less than 3% for $\Delta T = 1°C$, but is nearly 6% for $\Delta T = 2°C$.

At temperatures below $0°C$ liquid water may occur in a metastable (supercooled) state, and it is necessary to distinguish between saturation over the liquid and over ice, and between the *dew point* and the *frost point*, the temperature at which moist air cooled isobarically becomes saturated. At the same temperature the saturation vapor pressure is higher over the liquid than over ice (Fig. 2.1), and accordingly the frost point is higher than the dew point: the difference increases as the temperature falls, amounting to about $1°C$ when the frost point is $-10°C$ and about $4°C$ when the frost point is $-40°C$ (Table 2.13, Appendix 2.1).

2.3 The specification of humidity

The physical state of atmospheric water vapor is determined by its density or partial pressure, and by the air temperature, which is the only property easily measured.

It is convenient to define the following properties:

Mixing ratio r, the ratio between the mass of water vapor and the mass of dry air with which it is associated.

Specific humidity q, the ratio between the mass of vapor and the mass of moist air with which it is associated (both ratios frequently are conveniently expressed in grams per kilogram and are then written as R and Q).

Partial pressure e of the water vapor in air of total pressure p, defined as $e = rp/(R/R_v + r)$; then

$$r = \frac{0.622e}{p - e}; \quad q = \frac{0.622e}{p - 0.378e}$$

The partial vapor pressure e as a fraction of p is at most about 2% near the ground, and only a small part of 1% in the upper troposphere, so that for many purposes to a good approximation

$$r \approx q \approx \frac{0.622e}{p}$$

2   The physical properties of the moist atmosphere and its particles

The *virtual temperature* T_v of moist air is defined by the relation

$$p = R\rho T_v \qquad (2.8)$$

where p and ρ are respectively the pressure and density of moist air. Accordingly,

$$\frac{T}{T_v} = 1 - \frac{(1-\varepsilon)e}{p} \qquad (2.9)$$

$$\frac{T_v}{T} = \frac{1 + r/\varepsilon}{1 + r} \qquad (2.10)$$

Then $\Delta T_v = (T_v - T)$, the *virtual temperature increment*, at temperatures not far from 0°C, is given to a good approximation by

$$\Delta T_v \approx \frac{R}{6} \approx \frac{Q}{6} \qquad (2.11)$$

where R and Q are expressed in grams per kilogram.

2.4   The measurement of humidity

Practical means of determining the state of atmospheric water vapor (4) do not measure directly its pressure or density. In a method simple in principle a suitable flat surface is cooled and its temperature measured when a thin deposit of dew or frost can be maintained without noticeable growth or evaporation. This temperature is the dew or frost point, which gives the vapor pressure by reference to tables of saturated vapor pressure. The method is useful at low temperatures, at which the accuracy of most other techniques is poor. Used on aircraft, it has provided the most reliable measurements made at heights near the level of the tropopause.

It has proved difficult to use a dew point hygrometer on balloon soundings, even in special studies, and routine balloon soundings, which have provided the basic data for the analysis of atmospheric motion systems, are often made with sensors of hair or skin. These expand slightly with increase in the relative humidity u of the surrounding air, the ratio e/e_s between the partial pressure of the water vapor and the saturation vapor pressure at the air temperature (usually expressed as a percentage U); in saturated air, for example, human hair is about 2% longer than in perfectly dry air. Remarkably, the extension is dependent upon the relative humidity with respect to liquid water, even at temperatures far below 0°C. This property is employed in making simple recording hygrometers for use at the ground, where they can easily be kept in good condition. It is also used for making light and cheap hygrometers for radiosondes, which require careful treatment and calibration, but still have a number of defects. In particular they suffer from hysteresis after exposure to low relative humidities or after becoming wet with liquid water in rain or dense cloud (and perhaps later becoming frozen), and their lag increases with falling temperature, to become excessive in the upper troposphere. Analysis of the air flow in the upper troposphere and low stratosphere could be helped greatly by more accurate observations of humidity; probably the disappointing inaccuracy of the present reports could be improved by making corrections based on more detailed study of the lag and hysteresis (Appendix 2.3), but such work seems to have been abandoned in anticipation of the more widespread adoption of superior sensors. The most promising are those which measure the electrical properties of very thin films of hygroscopic substances exposed to the atmosphere, as in the aluminum oxide hygrometer (5).

The hygrometer in general use at the ground is the *psychrometer*, consisting of two thermometers, one of which has a bulb covered with a thin cloth kept wet with distilled water; this indicates a wet-bulb temperature T_w somewhat below the air temperature T.

2.5   The theory of the psychrometer

An elementary theory (the convection theory of August) assumes that after the wet bulb has attained a steady temperature all the heat required to evaporate water from it is supplied by passing air, and that all the air affected is chilled to the wet-bulb temperature and saturated. This is expressed by the relation

$$(c_p + rc_{pv})(T - T_w) = L(r_s - r) \qquad (2.12)$$

where c_p, c_{pv} = specific heat at constant pressure of dry air and water vapor, respectively
L, r_s = the latent heat of evaporation and the saturation mixing ratio *at the wet-bulb temperature*, respectively
r = the mixing ratio of the surrounding air

By regarding rc_{pv} as negligible compared with c_p,

to a good approximation this can be written as

$$r = \frac{r_s - c_p(T - T_w)}{L} \quad (2.13)$$

showing that r, which with T defines the state of atmospheric water vapor, is a function only of the air pressure, T and T_w, and can be entered in tables provided for use with a psychrometer.

It is found empirically that a psychrometric constant A, which for a particular rate of ventilation of a given psychrometer varies slightly with temperature, can be defined by the relation

$$e = e_s - Ap(T - T_w) \quad (2.14)$$

where e_s is the saturated vapor pressure at the wet-bulb temperature.

With a further approximation $(p - e) \approx (p - e_s)$, Eq. 2.13 can be put into this form to give for A_C, the psychrometric constant according to the convection theory,

$$A_C = \frac{(1 - e_s/p)c_p}{\varepsilon L} \quad (2.15)$$

The values thus predicted are surprisingly close to the empirically found values, considering that in the vicinity of the wet bulb the flows of heat and vapor proceed under gradients of temperature and vapor density, and that away from its surface the air is neither chilled to the wet-bulb temperature nor saturated.

An alternative theory considers these flows to occur by steady-state molecular diffusion in air at rest. Then the flux F of water vapor across a surface of radius R from the center of a spherical wet bulb of radius r is

$$F = -4\pi R^2 D \, d\rho/dR$$

where D = coefficient of diffusion of water vapor in air (cm² sec⁻¹)
ρ = vapor density (g cm⁻³)

Thus

$$F \, dR/R^2 = -4\pi D \, d\rho \quad (2.16)$$

In the steady state $F = \text{const}$ and upon integration from $R = r$ to $R = \infty$,

$$F = 4\pi r D(\rho_r - \rho_v) \quad (2.17)$$

where ρ_r and ρ_v are respectively the vapor densities at and far from the surface of the wet bulb. Similarly, the steady-state heat flux H is

$$H = 4\pi r k \rho_a c_p (T - T_w) \quad (2.18)$$

where k (cm² sec⁻¹) and ρ_a are, respectively, the thermometric conductivity and density of the air. Since $H = FL_{T(w)}$,

$$\rho_r - \rho_v = \frac{k\rho_a c_p(T - T_w)}{DL_{T(w)}}$$

or

$$r_s - r = \frac{kc_p(T - T_w)}{DL_{T(w)}}$$

and

$$r = r_s - A_D(T - T_w)$$

where

$$A_D = \frac{kc_p}{DL_{T(w)}} \quad (2.19)$$

is the psychrometric constant according to the diffusion theory. From Eq. 2.13 its value A_C according to the convection theory is

$$A_C = \frac{c_p}{L_{T(w)}}$$

Now according to the kinetic theory of gases (6) both k and D are proportional to η/ρ_a, where η (g cm⁻¹ sec⁻¹) is the dynamic viscosity of air, with constants of proportionality about 1.4 and 1.5, respectively. Thus there is little difference between A_C and A_D. Evidently the degree of success of the convection theory is related to the approximate equality of these transfer coefficients, not only in the turbulent regime away from the surface of a ventilated wet bulb, but also in the boundary layer in which the fluxes are governed by molecular diffusion.

It is found that when a wet bulb is artificially ventilated its temperature falls as the ventilation is increased, suggesting that under weak or natural ventilation (due to the chilling of air near the bulb) it receives a significant flux of heat by radiation from the warmer surroundings. Upon examination, it is found that if the ventilation is small, this flux appreciably raises the wet-bulb temperature, and indeed if it is taken into account in the diffusion theory, the calculated values become closely comparable to those which standard tables indicate are read from the unshielded thermometers exposed inside a Stevenson screen, as in British practice (Table 3.2). With bulbs of ordinary dimensions the radiative flux does not become insignificant until the ventilation speed reaches a few meters per second; for accurate work a special psychrometer should be used, whose thermometer bulbs have silvered shields and are subjected to a draft of not less than 4 m sec⁻¹. The temperature of the wet bulb can then be regarded as a definite *psychrometric wet-bulb temperature* T_w.

Wet-bulb temperature, familiar as a routine observation, has a more general significance in meteorological thermodynamics, in which it is defined as a variable of state useful to consider in processes involving changes of phase of water. Appropriate definitions will be given later and corresponding values are compared with wet-bulb temperatures indicated by thermometers in Table 3.2, from which the differences that arise between wet-bulb temperatures indicated by screen and psychrometer thermometers, and as calculated according to the diffusion theory (with and without adjustment for radiative heat flux) and the convection theory, can be seen to be generally less than 1°C.

2.6 Properties of atmospheric particles

Particles are present in the atmosphere in a great range of sizes and concentrations. Most conspicuous are the clouds of droplets which arise by the chilling of water vapor; they have a mass concentration which is typically about 1 g m^{-3} or less (limited to a fraction of the saturated vapor density at which condensation begins), and a particle number concentration which will be shown later to be generally of order 10^2 cm^{-3}, so that the particle size soon after cloud formation is in microns. During the evolution of clouds, however, aggregation often produces much larger particles which eventually fall as precipitation, mostly raindrops and snowflakes of millimetric size, and more rarely hailstones of centimetric size. Many (trace) substances other than water enter the atmosphere, partly as dispersed solids or liquids but mainly as gases and vapors. Near very limited prolific sources (e.g., fires) these may have mass concentrations in grams per cubic meter, but after travel and dilution in the atmosphere they produce comparatively tenuous clouds of particles. The particles which arise by the condensation of vapors or the chemical reaction of trace gases are generally at first only of molecular size, but during accumulation from a variety of sources, into concentrations of around 1 mg m^{-3} or less, aggregation leads to the development of a large range of particle sizes extending up to microns or even more. The particles have hygroscopic components, and especially at high relative humidities their presence is often seen as haze.

The aggregation of the smallest particles with others of similar or larger size is usually very rapid, and consequently the lower limit of the size of atmospheric particles which need be considered is between about 10^{-3} and 10^{-2} μ. The upper limit is about 10 μ in hazes, and about 10 cm in clouds (the diameter of giant hailstones).

An important property of these particles is their motion relative to the air, which depends partly on particle density and shape, but more essentially, in most circumstances, on particle size. It will become apparent that particles may be divided according to size into three principal classes:

1. Size considerably less than about 10^{-4} cm (1 μ): submicroscopic *aerosol* particles of trace substance including some water, whose motion is mainly Brownian agitation.

2. Size between about several and 100 μ: microscopic *cloud* particles of nearly pure water, which settle under gravity at speeds of order centimeters per second.

3. Size greater than about 100 μ: *precipitation* particles, whose settling speeds are of order meters per second.

Settling and Brownian motion of microscopic and submicroscopic particles

Except for ice crystals, which have one dimension very different from the others, for example, long needlelike or thin platelike crystals, the motion of particles approximates sufficiently well for most discussions to that of rigid spheres of the same mass and density.

The important nondimensional numbers are then the *Reynolds number* Re, and the *drag-coefficient* C_D, defined by

$$\mathrm{Re} = \frac{2rV\rho_a}{\eta}$$

and

$$D = \frac{C_D A \rho_a V^2}{2} \qquad (2.20)$$

where r and V are the particle radius and fall speed; D, the *drag*, is the resistance of the fluid to the motion; and A is the area presented by the particle across the direction of motion.

When Re $\ll$ 1 the inertial forces on the air moving past the particle are negligible compared with the viscous forces, and D can be obtained theoretically as $D = 6\pi\eta rV$. If this is equated to the gravitational force on the particle, neglecting the air density in comparison with the particle density δ, Stokes's law is obtained:

$$V = \frac{2gr^2\delta}{9\eta} \qquad (2.21)$$

Thus the settling speed is proportional to the square of the particle radius and varies only with the temperature of the air. It is confirmed experimentally to be very accurate under appropriate conditions, and can be taken as correct to within 10% over the range of r from about 1 to 35 μ (7). To about the same degree of accuracy the settling speed of water droplets of these sizes over the meteorological range of temperature between 20 and $-40°C$ is

$$V = 1.3 \times 10^6 r^2 \qquad (2.22)$$

Thus a cloud droplet of typical radius about 10 μ has a settling speed of about 1 cm sec^{-1}.

When the particle size is so small as to be comparable with the mean free path of the air molecules (about 10^{-5} cm near the ground and 10^{-4} cm in the lower stratosphere) Cunningham's correction is applied to Stokes's law:

$$V = \frac{2gr^2\delta}{9\eta}\left(\frac{1 + b\lambda}{r}\right) \qquad (2.23)$$

where λ is the mean free path and the factor b has a value close to unity (7). With this correction the error in V does not reach 10% until r has decreased to about 5×10^{-6} cm, for which V is only about 1 μ sec^{-1}.

The motion of particles as small as this is predominantly Brownian agitation, and they can be regarded as diffusing like a gas with a coefficient of diffusion D_p given by the relations

$$\overline{x^2} = 2D_p t \quad \text{and} \quad D_p = kTB \qquad (2.24)$$

where $\overline{x^2}$ = the mean square displacement of a particle along any axis during a time t
k = the Boltzmann constant
B = the mobility of the particles (the ratio of the velocity of a particle to the force F causing steady motion: $B = 1/6\pi\eta r$ in the Stokes's law regime)

Values of D_p and of $|\bar{x}|$ over an interval of 1 sec are given in Table 2.1 as a function of the radius of particles of unit density, under ordinary conditions near the ground. The latter values are compared with the distance v fallen under gravity during 1 sec, to show that over such a period the motion of a particle is dominated by Brownian motion when $R < 10^{-5}$ cm, and by gravity when $r > 10^{-4}$ cm.

It will also appear that the mutual aggregation of particles of sizes smaller than this (and over short periods their deposition on surfaces, at any inclination to the vertical) results from diffusion and collision during Brownian motion, whereas the aggregation of larger particles is due to approach and collision while settling at different speeds. The removal of the smaller particles from the atmosphere depends on their deposition by Brownian motion on the surfaces of larger particles or the earth's surface, after they have been brought very close by eddy diffusion, whereas considerably larger particles fall out, mostly after incorporation into rain.

In the presence of a gradient of temperature or of vapor density the particles migrate in the direction of lower temperature or vapor density, exhibiting the phenomena known as thermophoresis and diffusiophoresis. Similarly, if the particles are electrically charged, they migrate under an electrical potential gradient (electrophoresis). Also, in the vicinity of a droplet which is evaporating or growing by condensation, particles are carried by a flow of moist air produced by the sink or source of water vapor (the *Stefan flow*). These phenomena are employed to clean air in the thermal and electrostatic precipitators, and are manifest in the dark space within a small fraction of 1 mm of the surfaces of warm bodies or evaporating drops; they may prevent the deposition of submicroscopic particles on such surfaces. However, in the gradients of temperature, vapor density, and electrical potential which occur naturally near droplets and other bodies in the atmosphere they are thought to have less effect on the deposition of submicroscopic particles than the ordinary Brownian motion (8).

The aggregation of submicroscopic particles by Brownian motion

According to simple theory the average rate of collisions between small particles of uniform size in Brownian motion is proportional to $rD_p n^2$, where n is their concentration (and D_p is approximately inversely proportional to r). Assuming that every collision results in aggregation (which is reasonable since the potential energy of molecular attraction between small touching particles generally exceeds their kinetic energy), the concentration then changes with time according to the relation

$$dn/dt = -K_0 n^2$$

where

$$K_0 = 8\pi r D_p$$

and varies little when $r > 10^{-5}$ cm. However, the diffu-

Table 2.1 The diffusion coefficient D_p, the mean displacement $|\bar{x}|$ along a given axis produced in 1 sec by Brownian motion, and the distance v settled under gravity in 1 sec, as functions of the radius r of a particle (of unit density, in still air at normal surface pressure and a temperature of 23°C) (7)

r(cm)	D_p(cm^2 sec^{-1})	$|\bar{x}|$(cm)	v(cm)	$v/|\bar{x}|$
5×10^{-7}	5.2×10^{-4}	2.6×10^{-2}	6.6×10^{-6}	3×10^{-4}
10^{-6}	1.4×10^{-4}	1.3×10^{-2}	1.4×10^{-5}	1×10^{-3}
5×10^{-6}	6.8×10^{-6}	3.0×10^{-3}	8.6×10^{-5}	3×10^{-2}
10^{-5}	2.2×10^{-6}	1.7×10^{-3}	2.2×10^{-4}	1×10^{-1}
5×10^{-5}	2.7×10^{-7}	5.9×10^{-4}	3.5×10^{-3}	6
10^{-4}	1.3×10^{-7}	4.0×10^{-4}	1.2×10^{-2}	3×10
5×10^{-4}	2.4×10^{-8}	1.7×10^{-4}	3.0×10^{-1}	2×10^3

2 The physical properties of the moist atmosphere and its particles

Table 2.2 Time T required to reduce an initial concentration n_0 cm^{-3} of small particles of uniform size to $n_0/10$ under ordinary conditions near the ground

n_0(cm^{-3})	T
10^{11}	0.2 sec
10^6	5 hr
10^5	2 days
10^4	3 wk
10^3	6 mo

sion equations used need modification when r is about or less than the apparent mean free path of the particles (which under ordinary conditions is about 6×10^{-6} cm over the whole range of particle radius considered and down to molecular size). Then K_0 appears as a limiting value, at large radii, of a coefficient K which varies by less than a factor of about 2 from an average value of 5×10^{-10} cm^3 sec^{-1} as r varies between 10^{-7} and 10^{-3} cm (7). Thus

$$dn/dt = -Kn^2$$

or

$$1/n = 1/n_0 + Kt \qquad (2.25)$$

where n_0 and n are the particle concentrations initially and after a time t, and

$$K \approx 5 \times 10^{-10} \text{ cm}^3 \text{ sec}^{-1}$$

under ordinary conditions near the ground. Accordingly, the time required to reduce an initial concentration by an order of magnitude varies as shown in Table 2.2, while Table 2.3 shows how an initial concentration is changed after one week, illustrating how after such a period the concentration tends to about 10^3 cm^{-3} however much larger was the initial value (as Kt becomes much greater than $1/n_0$).

If a population of particles is initially composed of equal concentrations of two sizes, the aggregation proceeds more rapidly than among the same total concentration of the smaller alone. However, in the important range of radii between about 10^{-5} and 10^{-4} cm the appropriate value of the coagulation

Table 2.3 Change of concentration of small particles of uniform size during one week, under ordinary conditions near the ground

n_0(cm^{-3})	n(cm^{-3})
10^{12}	2.9×10^3
10^8	2.9×10^3
10^4	2.2×10^3
2.2×10^3	1.3×10^3

constant K hardly changes, since the mobility of a particle decreases as its size increases. If the large particles are comparatively few, the rate of change of the total concentration is affected very little.

The deposition of submicroscopic particles on cloud droplets

The concentration n of submicroscopic particles of radius r in a cloud of droplets of concentration n_c and uniform radius r_c decreases with time, as the Brownian motion causes their deposition on the droplet surfaces, according to the relation (8)

$$dn/n = -4\pi D_p n_c r_c \, dt \qquad (2.26)$$

Thus the time T during which the concentration n is reduced to the fraction n/e is

$$T = 1/4\pi D_p n_c r_c \qquad (2.27)$$

Table 2.4 shows T as a function of r for typical values of n_c and r_c, assuming the values of D_p previously given in Table 2.1. From the table it appears that the smaller submicroscopic particles are rapidly incorporated into the cloud droplets; we shall see later that in the atmosphere the larger, which are not very readily aggregated among themselves on account of their small mobility and concentration, appear in cloud droplets by serving as nuclei for the condensation which produces clouds.

Table 2.4 Time T for deposition by Brownian motion on a cloud of droplets of concentration $n_c = 2.5 \times 10^2$ cm^{-3} and uniform radius 10^{-3} cm (corresponding to a condensed water concentration of 1 g m^{-3}) to reduce a concentration n of aerosol particles of radius r (of unit density, under ordinary conditions near the ground) to n/e

r(cm)	T
5×10^{-7}	10 min
10^{-6}	40 min
5×10^{-6}	0.5 day
10^{-5}	2 days

Settling speeds of larger particles

If the Reynolds number of a settling particle is not very small, there is no analytical expression for the drag coefficient and recourse has to be made to an empirical relation with the Reynolds number. Again assuming the particle to be spherical, with density δ, and neglecting buoyancy, we have

$$\pi r^2 \cdot \frac{1}{2} \rho_a V^2 \cdot C_D = \frac{4}{3} \pi r^3 \delta g$$

or

Properties of atmospheric particles 2.6

Table 2.5 Settling speeds of smooth rigid spheres (drag coefficient C_D) under typical tropospheric conditions, as a function of radius r and density δ, and the settling speeds of large raindrops of equivalent spherical radius r' derived from measured values of drag coefficient C'_D (10)

		$T = 10°C$ ($\theta_s = 23°C$) $p = 700$ mb (3 km)			$T = -20°C$ ($\theta_s = 20°C$) $p = 400$ mb (7 km)					
		$\delta = 1$ g cm^{-3}			$\delta = 0.3$ g cm^{-3}			$\delta = 0.9$ g cm^{-3}		
Re	C_D	r (μ)	V (cm sec^{-1})	$V/r\mu$	r (μ)	V (cm sec^{-1})	$V/r\mu$	r (μ)	V (cm sec^{-1})	$V/r\mu$
1	28.2	46	22	0.48	75	20	0.26	52	28	0.54
4	8.7	78	52	0.67	127	46	0.36	89	66	0.74
10	4.2	113	91	0.80	184	79	0.43	130	115	0.89
20	2.8	157	130	0.83	255	115	0.45	180	165	0.93
60	1.5	265	232	0.87	430	200	0.47	300	290	0.97
200	0.80	480	427	0.89	780	375	0.48	540	540	1.00
400	0.59	690	594	0.86	1,120	520	0.47	780	750	0.96
10^3	0.46	1,170	875	0.78	1,900	770	0.40	1,320	1,110	0.84
Re	C_D	r' (mm)	V (m sec^{-1})		C'_D	r (mm)	V (m sec^{-1})		r (mm)	V (m sec^{-1})
10^4	0.41	—	—			8.5	17.1		5.9	24.6
5×10^4	0.46	—	—			25.7	28.4		17.9	40.7
Raindrops (aspherical)										
10^3	0.46	1.21	8.5		0.51	1.9	7.7		1.3	11.1
2×10^3	0.41	1.95	10.5		0.53	2.9	10.1		2.0	14.5
3×10^3	0.41	2.75	11.2		0.66	3.8	11.5		2.6	16.6
3.5×10^3	0.40	3.18	11.3		0.75	4.2	12.2		2.9	17.6

$$r^3 = \frac{9\eta^2}{4\delta\rho_a g} \cdot \frac{C_D Re^2}{24} \quad (2.28)$$

For specified values of δ, ρ_a, and temperature, the right-hand side of Eq. 2.28 can be determined for selected values of Re, to provide values of r, and then of V from the relation $V = Re\eta/2r\rho_a$. Some values of C_D as a function of Re (for smooth rigid spheres) are listed in Table 2.5 and in Appendix 2.4. In the Stokes's law regime, at very small Reynolds numbers, $C_D = 24/Re$ and $V \propto r^2$, while at Reynolds numbers exceeding about 10^3, appropriate to large hailstones, C_D becomes approximately constant and $V \propto r^{1/2}$. Over the range of Re from about 1 to 10^3, appropriate to large cloud drops, raindrops, and small hailstones, V is very approximately directly proportional to r, as seen in Table 2.5.

For radii of up to about 1 mm and Reynolds numbers up to about 10^3 drops are nearly spherical, but a larger drop becomes flattened (9) and its drag coefficient substantially exceeds that of a rigid sphere of the same volume (10), as indicated in Table 2.5. Accordingly, as the equivalent radius (that of a spherical drop of the same volume) increases beyond 1 mm the fall speed increases only slowly toward a maximum value of about 10 m sec^{-1}. Eventually the distortion becomes so great that at some value of the equivalent radius near 4 mm, depending on the degree of small-scale turbulence in the airstream, a drop disrupts into several millimetric drops and a spray of more numerous but very much smaller droplets (11).

Fall speeds of hailstones

The drag coefficient of hailstones depends considerably on their mean density, shape, and surface texture as well as their size. Small hailstones of diameter less than about 0.5 cm may have mean densities as small as 0.1 g cm^{-3}, consisting of loosely packed droplets which froze rapidly and individually upon accretion, and are known by the German name *graupel*, or as *soft hail*, from their friable structure. They often have the form of a cone with a rounded base, and fall with the base lowermost.

As will be discussed in Chapter 5, the density of rime ice produced by the accretion of supercooled droplets depends in a complicated way on the surface temperature of the collector, the size of the droplets, and the speed with which they strike the surface of the collector (12). When a hailstone commences its growth as an ice particle or frozen droplet whose diameter is only a small fraction of a millimeter, rather than as a frozen raindrop, the density of the accreted ice is likely to be as little as 0.1 g cm^{-3} until the diameter of the hailstone reaches

2 The physical properties of the moist atmosphere and its particles

about 1 mm (13; Section 8.11); thereafter the density increases, so that the mean density of hailstones several millimeters in diameter lies in the range 0.3–0.8 g cm^{-3} if they are grown in clouds with cold bases (base temperatures about 0°C; most of the accretion then occurs at low values of both air and hailstone temperature), and in the range 0.7–0.9 g cm^{-3} in clouds with warm bases (base temperatures between about 10 and 20°C). Hailstones whose diameters approach or exceed 1 cm have a mean density nearly that of pure ice and a glazed translucent appearance.

In considering the fall speeds of hail at Reynolds numbers up to about 10^3 there is more uncertainty about the mean density than the drag coefficient, which is probably a little greater than that of smooth spheres and becomes almost constant at about 0.8 in the range of Re from about 10^3 to 10^4 (14). At Reynolds numbers greater than 10^4 (radii about 1 cm or more) the density can be assumed to be close to that of pure ice, but the drag coefficient of a hailstone is appreciably affected not only by its shape and attitude but also by its surface texture; the surface may consist of a rough coat of rime ice, or of smooth ice, perhaps with a film of liquid water even at air temperatures well below 0°C. Large hailstones tend to acquire the general shape of oblate spheroids (15) covered with knobbly protuberances (Pl. 8.21). Information on the drag coefficients of similar bodies and artificially grown hailstones, obtained from observations made when they were held in wind tunnels or allowed to fall freely in the atmosphere, are summarized in Fig. 2.2. These show that at Reynolds numbers between about 10^4 and 10^5 it may be justifiable to regard the drag coefficient of hailstones as a constant, considering the uncertainties concerning the shape and surface properties. Equation 2.28 assumes the simpler form

$$V^2 = \frac{8r\,\delta g}{3\rho_a C_D} \qquad (2.29)$$

For the typical conditions under which large hailstones are grown in the upper troposphere ($\theta_s \approx 20°C$, $T \approx -20°C$) the density of air ρ_a is about 5.5×10^{-4} g cm^{-3} while δ and C_D are, respectively, about 0.9 g cm^{-3} and 0.6, so that for hailstones of radius more than about 1 cm the fall speed is given approximately by

$$V \approx 30 r^{1/2} \text{ m sec}^{-1} \qquad (r \text{ in cm})$$

Usually when the Reynolds number reaches some value near 10^5 the flow in the boundary layer becomes turbulent and the drag coefficient suddenly decreases. The threshold value of Re at which the transition occurs is influenced by the surface roughness of the body and the intensity of the small-scale turbulence in the airstream, and for a smooth sphere in a smooth flow is about 2×10^5. In the atmosphere the small-scale

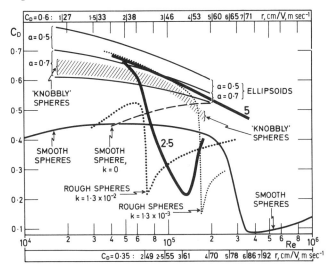

Fig. 2.2 A selection of experimental data on the drag coefficient C_D of various materials as a function of the Reynolds number Re:

Smooth spheres (from H. Schlichting, 1960, *Boundary Layer Theory*, McGraw-Hill, New York).

Smooth ellipsoids, of ice (in free fall and in a wind tunnel) and plasticine (in a wind tunnel), from reference 17; ratio of axes 1:1:a; The individual values mostly fell within the indicated bounds.

Knobbly spheres, of plasticine (in a wind tunnel; 17). The knobs were blunt pointed projections of base diameter about 1 cm, randomly placed 3–4 cm apart. The values of C_D mostly fell within the hatched zone.

Rough spheres, of roughness parameter (ratio of diameter of close-packed surface grains to sphere diameter) $k = 1.3 \times 10^{-2}$, $k = 1.3 \times 10^{-3}$, and $k = 0$ (smooth sphere), from wind tunnel measurements (16). A value of $k \approx 10^{-2}$ was deduced from radar observations of a frost-coated ice sphere in free fall in the upper troposphere.

Artificially grown hailstones, from wind tunnel measurements (18). The heavy lines labeled 2.5 and 5 refer to two series of measurements with, respectively, a hailstone of moderate surface roughness and equivalent spherical radius of 2.5 cm, and a knobbly large hailstone of equivalent spherical radius 5 cm. They have been selected from a number of such measurements specified in the reference.

The scales of radius r cm, and fall speed V m sec^{-1} at the top and bottom of the diagram are appropriate to spheres in the upper troposphere (height 7 km, pressure 400 mb, temperature $-20°C$), with drag coefficients C_D of 0.6 and 0.35, respectively.

motions produced by neighboring falling particles of a range of lesser sizes or otherwise are likely to lower the threshold value, while experiments in smooth flow with single particles having a surface roughness similar to that of some natural hailstones show that it may be reduced to as little as 6×10^4 (16); in the upper troposphere this corresponds to a hailstone radius of about 2 cm and a fall speed of about 50 m sec^{-1}. A series of measurements made in a wind tunnel, in which artificially grown hailstones were held in a fixed position under a range of Reynolds numbers, suggest that for natural hailstones the transition to a low value of the drag coefficient is most likely to occur when their surface texture is moderately rough and their equivalent spherical radii reach values between about 2 and 3 cm (18; one of the most striking experimental examples is shown by the curve marked 2.5 in Fig. 2.2). These measurements showed, however, that no two of the various artificial hailstones provided closely similar relations between Re and C_D, and, in particular, the large and knobbly hailstones (diameter 7–10 cm) showed no sudden fall of C_D as Re was increased from about 3×10^4 to 3×10^5, but only a gradual change to a value of about 0.4 at the high values of Re. An example is shown in Fig. 2.2 by the curve marked 5 (representing an equivalent spherical radius of 5 cm and a mass of 0.5 kg; the fall speed of such a hailstone in the upper troposphere would be about 75 m sec^{-1}). It is not known how the tumbling motion of hailstones evidently characteristic of their free fall (Chapter 8) might influence their fall speeds, but the uncertainty is probably no greater than that due to variations in overall shape and surface configuration. The maximum size of hailstones reliably reported, corresponding to an equivalent spherical radius of about 7 cm, implies fall speeds in the upper troposphere of between about 70 and 90 m sec^{-1}.

Settling speeds of ice crystals and snowflakes

Ice crystals occur with markedly aspherical shapes, principally as long hexagonal columns or needles, and as thin hexagonal plates or stars, sometimes more or less heavily coated with frozen droplets (rimed), and sometimes aggregated into snowflakes.

When the crystals are very small they settle stably in any orientation, but at Reynolds numbers approaching and greater than 1 they tend to become aligned with the greatest area or longest edges horizontal, so that the resistance of the fluid is a maximum. The drag coefficients of crystals in the form of hexagonal plates and columns can be assumed to be equal to those measured for thin discs and long cylinders, which are summarized in Fig. 2.3. Platelike crystals settle more slowly than spheres of the same diameter, and those whose diameter varies from a small fraction of a millimeter up to several millimeters are observed to have fall speeds of about 0.5 m sec^{-1} almost independently of their diameter, implying that their thickness is almost constant.

Snowflakes, which consist of loosely clustered aggregates of as many as 100 individual crystals, have vertical dimensions which generally are little less than the horizontal dimensions, and have low mean densities of about 0.01–0.1 g cm^{-3}. Their drag coefficient is about unity, and their fall speeds are usually in the range 1–2 m sec^{-1}, varying more with mean density than size in the range of diameters from 2 mm to 1 cm or more

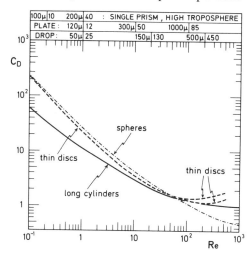

Fig. 2.3 The drag coefficients of spheres and of thin discs and long cylinders falling with their greatest dimensions horizontal (from references 19 and 20). At values of the Reynolds number Re greater than 10^2 the discs fall unstably and the drag coefficient C_D becomes variable between the approximate limits indicated. In the panels at the top approximate values of half-diameter or half-length (μ) and fall speed (cm sec^{-1}) are written beside the appropriate Reynolds number for water drops and thin platelike crystals in the middle troposphere, and for single long-prism crystals in the high troposphere (see Chapter 5).

(21). The fall speeds increase considerably when the snowflakes begin to melt and become wet, and the ratio V/V_s, between the fall speeds V of a raindrop and V_s of a dry snowflake, which produces it by complete melting, varies from about 2 to about 6 as the drop radius increases from 0.5 to 2 mm.

Aggregation of particles by differential settling; efficiency of catch

It can be anticipated from Table 2.1 that when particle sizes exceed about 1 μ the differential settling of particles of different size is the dominant cause of their collision and aggregation. To estimate its effectiveness we make the simplest reasonable assumptions: particles which are droplets of equal radius r and fall speed v are considered to be uniformly dispersed in a cloud of average concentration m of condensed water, and a single spherical particle of fall speed V and radius R is assumed to collect condensed water from the volume swept by its cross section. Then

$$4\pi R^2 \delta\, dR = \pi R^2 (V - v) m\, dt$$

or

$$\frac{dR}{dt} = \frac{(V - v)m}{4\delta} \quad (2.30)$$

where δ is the density of the accreted water. For the typical meteorological concentration of condensed water

Table 2.6 Time T for a particle of initial radius R_0 to increase to $10R_0$ by aggregation due to differential settling in a cloud of particles of radius $R_0/2$ and concentration 1 g m^{-3}

R_0 (cm)	T
10^{-5}	4 days
10^{-4}	1 days
10^{-3}	1 hr
10^{-2}	$\frac{1}{3}$ hr

($m = 1$ g m^{-3}) the time T for an initial radius $R_0 = 2r$ (representing a plausible variation of particle size) to increase by an order of magnitude can be obtained as a function of R_0, and for a particle of unit density under ordinary conditions in the lower troposphere is seen in Table 2.6.

Evidently the process can lead to precipitation from clouds in which at least some particles have radii exceeding several microns and whose persistence is of the order of an hour. On closer examination it appears that conditions are more stringent: not all collisions result in aggregation, and if the small particles have little inertia they may so closely follow the streamlines of the air around the collecting particles that there is a negligible frequency of collisions (for this reason hovering midges are not dashed to the ground by a shower of raindrops). Equation 2.30, for a spherical collector, therefore must be replaced by

$$\frac{dR}{dt} = \frac{EE'(V-v)m}{4\delta} \quad (2.31)$$

where E is a *coefficient of collision*, defined as that fraction of the particles lying within the projected cross section of the collecting particle with which collisions occur, and E' is another coefficient representing the fraction of particles captured after collision. It appears that in meteorological conditions (even in ice clouds at temperatures well below 0°C) collisions between particles virtually always result in aggregation (27), so that the efficiency of collision is effectively an efficiency of catch.

There is no satisfactory theory for the efficiency of catch over the range of circumstances of meteorological interest (including impaction on collectors of other than spherical shape; e.g., ice crystals and aircraft). However, the most important of such circumstances in cloud physics are those in which water droplets collide and coalesce or hailstones grow by accretion in supercooled clouds; the collected particles are droplets of radius predominantly between a few and about 20 μ, and the collectors are larger droplets of radius extending from similar values up to a few millimeters, or approximately spherical hailstones of radius up to several centimeters. In these conditions the motion of the small particles through the air can be considered to be subject to the drag appropriate to the Stokes's law regime, and theoretical solutions can be found for E_V, the efficiency of catch for $Re(R, V) \ll 1$ [$Re(R, V)$ is hereafter written simply as Re], corresponding to viscous flow around the collector, and E_A, the efficiency of catch for $Re \gg 1$, when the air flow at least in front of the collector and outside a boundary layer corresponds to potential (aerodynamic) flow. In viscous flow the relative streamlines in front of the collector diverge more gradually (Fig. 2.4), so that small particles of given size and inertia are less readily caught.

The factors which determine E are R, V, r, v, δ (the density of the small particles), ρ_a, and η; the relative velocity of the particles when they are far apart is $(V - v)$ and is often regarded simply as V when R is large compared with r. These factors can be combined into three important nondimensional numbers: (r/R), Re, and $s/R = K$, the *Stokes number* or *particle parameter*, where s is the *stopping distance* of the small particles, that distance traveled before they are brought to rest after projection with an initial velocity $(V - v)$. Under the condition specified above concerning the small particles

$$s = \frac{2\delta r^2(V-v)}{9\eta}$$

and

$$K = \frac{2\delta r^2(V-V)}{9\eta R} \quad (2.32)$$

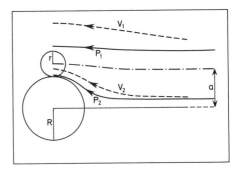

Fig. 2.4 The form of the relative streamlines near a sphere of radius R moving through a fluid in the viscous (dashed lines V_1, V_2) and potential (solid lines P_1, P_2) flow regimes. The dot-dash line represents the trajectory of a spherical particle of radius r which, because of the inertia of the particle, crosses the streamlines and makes a grazing contact with the large sphere. Far from the sphere this trajectory is at a distance a from the axis drawn through the center of the sphere in the direction of motion; the efficiency of collision E between the sphere and particles of the same size and density is defined as $E = (a/R)^2$. In the potential flow, which in front of the sphere resembles real flows for which $Re \gg 1$, the streamlines approaching the sphere diverge more rapidly than in the viscous flow, and accordingly for particles of the same size and inertia E is greater.

In the simplest case for which $r \ll R$ the small particles can be regarded as points describing trajectories some of which reach the surface of the collector. A grazing trajectory is found which far from the collector lies at a distance a_0 from the axis through its center (Fig. 2.4); then

$$E = a_0^2/R^2$$

For this case $E < 1$, and, as could be anticipated from inspection of the streamlines in Fig. 2.4, E is a large fraction only if s is comparable with R, and the Stokes number K is of order 1. Langmuir, who made the first accurate computations, found that the following empirical formulas express fairly well the calculated values of E for spherical collectors (7):

Viscous flow, Re $\ll 1$:

$$E_V = \left(1 + \frac{0.75 \ln 2K}{K - 1.214}\right)^{-2} \quad (2.33)$$

$r \ll R$ ($E_V = 0$ for $K < 1.214$)

Potential flow, Re $\gg 1$:

$$E_A = \frac{K^2}{(K + 1/2)^2} \quad \text{for} \quad K \geq 0.2 \quad (2.34)$$

$r \ll R$ ($E_A = 0$ for $K < 1/12$)

For intermediate values of Re he proposed the following interpolation formula, apparently intuitively:

$$E = \frac{E_V + E_A \text{Re}/60}{1 + \text{Re}/60} \quad (2.35)$$

The expressions 2.33–2.35, representing the only simple theoretical formulas available, are summarized in Fig. 2.5. Experimental tests of their accuracy are difficult, and may be complicated by several factors not considered in the theory and of doubtful relevance to natural conditions. These include the presence of electric charges on droplets produced by atomization, which are known to have appreciable effects on the collision and coalescence efficiency of small particles (of radius up to several microns; however, in natural clouds it is believed that the effects of particle charge and electric fields are unimportant until the field strengths have the magnitude of those occurring in thunderstorms). Other complications arise from intermittent locally intense small-scale shears (of magnitude 10 cm sec^{-1} over distances of about 1 cm), which are probably characteristic of the turbulent flow in clouds of ordinary convection, and which may appreciably increase the average efficiency of collisions between small droplets in a cloud (22), from wake eddies behind a collector, which may lead to collisions on its rear surface, and from the aspherical shape of large drops. On the whole,

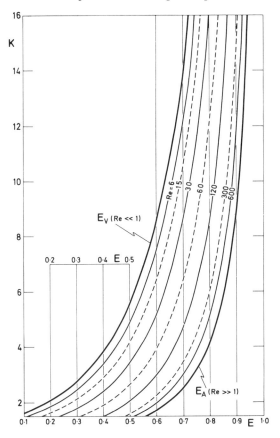

Fig. 2.5 Efficiency of collision as a function of the Stokes number K, and the Reynolds number Re, as originally obtained by Langmuir by neglecting the size of the small particles and by interpolating between the solutions for viscous flow (E_V, Re $\ll 1$) and potential flow (E_A, Re $\gg 1$) (7). For a drop of radius R greater than about 100 μ settling among droplets of radius r about 10 μ $K(\alpha\, r^2) \approx 10$. For drops of radius R considerably less than 100 μ, for which the ratio r/R may not be very small, E may be given more reliably by Fig. 2.6.

however, experiments show that over a large range of conditions the formulas are approximately correct [giving values of E not substantially different from those indicated by the formulas (23, 24), although rather large discrepancies have been found in some apparently well-executed experiments outdoors with artificial rains (25)]. Considering the uncertainties about variability of droplet size and air flow on large scales the formulas are the best available for the study of the growth of rain and small hail by the capture of cloud droplets, although in some particular conditions they certainly need modification. First, in the earliest stages of aggregation when its rate is slow and crucially influences the total time required for the production of large particles, r is comparable with R and the particle paths may be altered significantly in unsteady flow and also when the smaller particle approaches the collector. The latter case is meteorologically important only within the Stokes's law regime, and so it has been possible

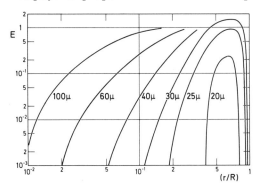

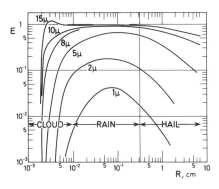

Fig. 2.6 An approximate summary of the collision efficiency E as a function of (r/R), for small particles of unit density, from theoretical and experimental data given in reference 25.

Fig. 2.7 The approximate value of the efficiency of catch E of cloud droplets (radii 1–15 μ) by larger particles of radius R (cloud or rain drops for $R <$ about 3 mm—conditions appropriate to lower troposphere: pressure 700 mb and temperature 10°C; spherical hailstones of density 0.9 g cm^{-3} and drag coefficient 0.6 for $R > 3$ mm—conditions appropriate to the upper troposphere: pressure 400 mb, temperature $-20°$C).

to examine it theoretically (26) as well as experimentally (27), with results which agree fairly well (Fig. 2.6).

Moreover, when V and R become large the relative trajectories of small droplets depart significantly from those calculated assuming only a viscous drag, and the Reynolds number Re(r, V) becomes relevant, appearing in another dimensionless number ϕ defined as

$$\phi = \frac{[\text{Re}(r, V)]^2}{K}$$
$$= \left(\frac{18\rho_a^2}{\eta\delta}\right) RV \qquad (2.36)$$

According to Langmuir's calculations, E_A is considerably reduced as ϕ becomes large. For millimetric raindrops $\phi < 30$ and does not appreciably affect E_A, but for a hailstone of radius 3 cm ϕ is approximately 2500, and if $K = 5$, $E_A \approx 0.60$ compared with $E_A \approx 0.83$ for $\phi = 0$.

Experiments have been made in a wind tunnel (28) on the rate of growth of collectors (of diameter between 1 and 9 cm) by the accretion of supercooled droplets (with median-volume radii of about 15 μ and maximum radii of about 30 μ). The speed V of the airstream was varied between 4 and 40 m sec^{-1}, giving values of Re(R, V) in the range between 2×10^4 and 2×10^5. It was found that the inferred collection efficiency E agreed closely with the theoretical value E_A for $R = 0.6$ cm and $V \approx 30$ m sec^{-1} (Re $\approx 2 \times 10^4$). However, with V constant and R increasing to about 4.5 cm, E_A steadily decreased to about 0.38, whereas E decreased to about 0.26. In another set of experiments with $R = 0.6$ cm and V diminishing from 40 to 4 m sec^{-1}, E_A decreased to about 0.5 as E decreased to about 0.25. Although the reasons for the discrepancies are not understood, they suggest that the collection efficiencies of large and giant hailstones may be appreciably less than indicated by E_A and by Fig. 2.7. (In these experiments the collection efficiency of artificially grown hailstones with irregular surfaces and distinctly oblate shapes was not significantly different from that of smooth spheres of ice if equivalent radii were determined for the former from the areas of their cross sections normal to the airstream.)

Figure 2.6 has the important implication that E is extremely small unless both the small particles and the collector particle exceed certain sizes. Thus for droplets of water

$$E \approx 0 \quad \text{if} \quad R < 18\,\mu$$
$$\text{if} \quad R = 20\,\mu \quad \text{and} \quad r < 8 \text{ or } > 19\,\mu$$
$$\text{and if} \quad R = 30\,\mu \quad \text{and} \quad r < 5 \text{ or } > 28\,\mu$$

while appreciably away from these limiting values E becomes a large fraction, and as defined above may even slightly exceed unity.

There is thus a notable gap between the range of radii over which aggregation proceeds rapidly by the Brownian motion, extending up to about 0.1 μ (Tables 2.1, 2.4), and that over which differential settling is effective among atmospheric cloud particles, beginning when there is a spread of particle sizes such that the radii of many exceed about 5 μ and the radii of some exceed about 20 μ. (In the atmosphere this gap is bridged intermittently but effectively by the condensation of water vapor, which thereby plays a vital part in the production of precipitation and an important part in removing atmospheric trace substances and determining their typical concentrations.)

These limiting values of the radii are in the first process determined by the intensity of the Brownian agitation, and therefore by the density of the particles and more especially by the atmospheric temperature, and in the second kind of process by the mobility and inertia of particles settling under gravity, and therefore

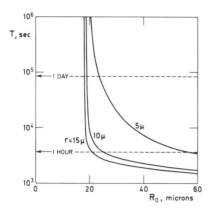

Fig. 2.8 Time T required by drops of original radius R_0 to reach a radius of 1 mm, by aggregation with droplets of uniform radius 5, 10, or 15 μ in a concentration of 1 g m^{-3}. Values of efficiency of catch taken from Fig. 2.7.

again primarily by their density and the temperature, and also by the value of the gravitational constant.

Figure 2.7 is a tentative summary of the efficiency of catch over a range of conditions of meteorological interest. The implications of this diagram concerning the formation of rain by the aggregation of droplets are illustrated in Fig. 2.8, which shows the time taken by a drop of original radius R_0 to reach a radius of 1 mm by settling through a population of uniform droplets composing a cloud with a condensed water concentration of 1 g m^{-3}. It is apparent that the small droplets must have a radius of at least several microns, and the large drop an initial radius of about 20 μ or more, if its growth is to be completed in a reasonable time. However, if these lower limits are exceeded even slightly, the growth can be completed in less than an hour.

Cloud virtual temperature

A cloud of particles settling at their terminal velocities exerts a drag on the air containing it which effectively increases the density of the air by the particles' mixing ratio. Under this force a cloud of microscopic droplets with a diameter of more than a few meters may settle through its surroundings at a speed greater than that of the individual droplets in still air. The effect on the air density of condensed water present in a mixing ratio of r_c may be taken into account by defining a *cloud virtual temperature* T_{vc} which is less than the virtual temperature T_v by the amount $T_v r_c$.

The effect of very large concentrations of particles in causing rapid motion of the medium in which they are suspended is occasionally manifest in the sea near the edges of continental shelves by *turbidity currents*, which may break submarine cables, and in the atmosphere by avalanches and the even more hazardous volcanic *nuées ardentes* (fiery clouds), which rush down slopes in spite of their high temperature. In the eruption of Mt. Pelee in Martinique in 1902 such a cloud led to the incineration of the port of St. Pierre.

2.7 Atmospheric refraction

The refraction and scattering of radiation in the atmosphere have many interesting effects, some of which very much enhance the splendor of open-air scenes.

The refractive index n of air generally depends on its composition, pressure, and temperature, and on the wavelength of the radiation. It has been studied at optical wavelengths between about 0.2 and 20 μ and at microwave or radio wavelengths of more about 1 cm. In the optical range the refractive index is strongly dependent on the wavelength but practically independent of the presence of water vapor, the major variable atmospheric constituent, whereas in the longer range the reverse is true. Since n is very nearly 1, it is convenient to refer to the *refractive modulus* N defined by

$$N = 10^6(n - 1) \qquad (2.37)$$

Then the formulas recommended (32) for N as sufficiently accurate for most purposes are

$$N = 77.6\frac{p}{T} + 0.584\frac{p}{T\lambda^2}; \quad \lambda < 20\,\mu \qquad (2.38)$$

and

$$N = 77.6\frac{p}{T} + 3.7 \times 10^5\frac{e}{T^2}; \quad \lambda > 1 \text{ cm} \qquad (2.39)$$

where p = total atmospheric pressure (mb)
T = temperature (°K)
e = water vapor pressure (mb)
λ = wavelength (μ; Eq. 2.38)

In the atmosphere N usually decreases with height (from about 300 near the earth's surface to about 100 at 10 km and 20 at 20 km). Consequently, rays of radiation are curved; usually the visible horizon is beyond the geometrical horizon, and objects seen at low elevation angles are above their geometrical elevations. The curvature has to be taken into account in relating the height, range, and apparent elevation of distant clouds and radar targets (Appendix 2.5). A large decrease of vapor pressure with height, which is not uncommon near the ground or between the planetary boundary layer and the free atmosphere when the boundary layer is only 1–2 km deep, sometimes causes an abnormal curvature of radar beams which may "trap" the radiation from a radar on the ground in a "duct," and allow the detection of low-level targets at ranges of several hundred kilometers, far beyond the ordinary radar horizon. Similarly, a rapid increase of temperature with height in a shallow zone at the top of the boundary layer sometimes causes the disc of the sun when it is seen near the horizon to be intersected by deep horizontal notches, or even to be split into two separate parts; these notches remain at the same elevation as the sun rises or sets. However, the very large vertical gradients of temperature which are required to produce abnormal optical refraction are most readily produced close to the earth's surface; they are responsible for mirages and other spectacular phenomena discussed in texts on atmospheric optics (33).

The importance of the refractive index in radio propagation has led to the construction of airborne instruments which measure its mean value and its variation over small time intervals, and distances as small as 0.1–1 m (34). The observations have interesting implications concerning atmospheric motions and the distribution of water vapor on small scales, and will be mentioned in Section 7.8.

2.8 Scattering of radiation in the atmosphere

The theory of scattering

Generally light does not reach the eye directly by a single smooth path from its source, but arrives as *scattered radiation*. In addition to the scattering from small-scale irregularities of refractive index of the air itself, discussed in Appendix 2.6, scattering from extensive surfaces on the ground and from molecules and particles in the atmosphere produces many phenomena, but only those that provide useful information about the constitution of clouds and precipitation will be mentioned here.

Incident and scattered radiation are specified by intensity I, the flux of energy per unit solid angle in the direction of propagation which is contained in a small interval of frequency (watts/steradian). In most meteorological studies the distance between the scattering elements and both the source and sensor are so large that the radiation is contained within a very small solid angle around some direction. It therefore becomes convenient, at the risk of some confusion, to regard the radiation as a parallel beam and as monochromatic, that is, within some small range of frequency about a particular value. The irradiance may then be specified in terms of the flux density E, the flux of energy across a unit area normal to the direction of propagation (watts/m^2) in a particular direction and at a particular frequency or wavelength.

A general (Mie) theory exists for the properties of the radiation scattered from a single particle of spherical or other simple shape (37). The important parameters are the size and shape of the particle, the refractive index of its material m and of the surrounding medium, and the wavelength λ of the incident radiation (unchanged by the scattering). In most meteorological applications the particle can be regarded as a sphere of radius r in a medium of refractive index 1, and the two important nondimensional numbers are m, or $(m - 1)$, and $x = kr = 2\pi r/\lambda$. Of principal concern is the flux density of the incident radiation which is absorbed, or scattered at an angle θ to the direction of propagation of the incident radiation (0° in that direction).

A beam of radiation is attenuated as it passes through a medium, suffering *extinction*, which is often noticed, for example, as a dimming when looking directly at a familiar source of light. Absorption is usually defined as that part of the extinction not due to scattering. In the atmosphere it may occasionally make a large contribution, as it does locally in many smokes and clouds of ash or soot, and more generally in some hazes whose particles consist mainly of dust, or which, especially at low humidities, have considerable solid components. In ordinary hazes and clouds, however, absorption is thought to be insignificant, simplifying many studies in atmospheric optics.

Often another simplification can be made in considering scattering, because the particles are sufficiently well separated for the radiation incident on each particle, and the scattering from it, to be considered independent of the presence of other particles. Thus when the extinction is small the intensity of the (*single*) scattering from a population of particles can be found simply by adding the contributions from individual particles. When the extinction is large, however, the irradiation of particles may be less from the direct beam than by scattering from other particles: there is said to be *multiple* scattering, in which the interpretation of observations is very difficult if not impossible.

Scattering functions

If the flux density of the incident radiation is E_0, the intensity I and flux density E of the radiation scattered by a single particle are given by

$$\frac{I}{R^2} = E = \frac{E_0 \, f(m, x, \theta, \phi)}{k^2 R^2} \quad (2.40)$$

where the direction of scattering is defined by the angle θ which it makes with the direction of propagation of the incident radiation and an azimuth ϕ, R is distance from the particle, and f is a *scattering function*. For natural (unpolarized) incident light, the scattering function evaluated from the general theory is written as

$$f = \frac{\{i_1(m, x, \theta) + i_2(m, x, \theta)\}}{2} \quad (2.41)$$

where i_1 and i_2 refer to the components of the scattered (polarized) light whose electric vector vibrates perpendicular and parallel, respectively, to the plane through the directions of propagation of the incident and scattered beams. For water m varies in visible light from 1.330 at $\lambda = 0.7\,\mu$ to 1.342 at $\lambda = 0.4\,\mu$, and most calculations of the scattering functions of water droplets have been made for a mean value of 1.33 or 4/3, and tabulated for various values of θ and x.

Table 2.7 gives the scattering function f for a range of droplet radii up to about the wavelength of visible light. It shows that the energy in the radiation scattered from a single particle decreases rapidly with its radius; for the smallest particle size included in the table the energy is distributed almost uniformly with θ and is very strongly polarized at right angles to the incident beam. The same feature appears in the Rayleigh theory of scattering by small particles, valid when $x \ll 1$. According to this theory the scattering function for spherical particles in unpolarized radiation is

$$f = \frac{(1 + \cos^2 \theta)\, x^6 K^2}{2} \quad (2.42)$$

where

$$K = \left|\frac{(m^2 - 1)}{(m^2 + 2)}\right|$$

(a relation which is also valid for $\theta = 0$ or $180°$ in incident radiation of arbitrary polarization). Then the flux density of the scattered radiation is proportional to the sixth power of the particle radius, and (from Eq. 2.40) to k^4, or λ^{-4}.

For $x \gg 1$, on the other hand, the polarization of the scattered radiation is small; this condition is fulfilled in almost all atmospheric clouds, the only exception being in the faint mesospheric noctilucent clouds. These clouds are sometimes seen when the troposphere is in the earth's shadow. The light which they scatter is polarized in the same sense as, though rather less strongly than, the light from the neighboring sky scattered by the molecules of air above the troposphere. Thus the contrast between these clouds and the sky cannot be increased by using a polarizing filter, a device often exploited in the photography of ordinary clouds in parts of the sky well away from the sun or

Table 2.7 Values of the scattering functions i_1 and i_2, $2f = (i_1 + i_2)$, and the degree of polarization $P = (i_1 - i_2)/(i_1 + i_2)$, for unpolarized light incident upon a spherical water droplet of radius r and refractive index $m = 1.33$, for various values of the angle of scattering θ and of $x = 2\pi r/\lambda$ (38)

x		θ (deg.)							r for $\lambda = 0.55\mu$ (μ)
		0	30	60	90	120	150	180	
0.5	$10^5 i_1$	71	70	67	63	60	58	57	0.04
	$10^5 i_2$	71	52	17	0	15	43	57	
	$10^5(i_1 + i_2)$	141	112	84	63	75	101	114	
	P	0	0.16	0.59	1.00	0.60	0.24	0	
1.0	$10^5 i_1$	5,260	4,969	4,242	3,395	2,693	2,259	2,115	0.09
	$10^5 i_2$	5,260	3,797	1,171	4	597	1,654	2,115	
	$10^5(i_1 + i_2)$	10,520	8,766	5,413	3,399	3,290	3,913	4,230	
	P	0	0.13	0.57	1.00	0.64	0.15	0	
2.0	$10^3 i_1$	3,937	3,056	1,408	350	30	22	44	0.17
	$10^3 i_2$	3,937	2,543	668	77	10	24	44	
	$10^3(i_1 + i_2)$	7,874	5,599	2,076	427	41	45	87	
	P	0	0.09	0.36	0.64	0.48	0.04	0	
4.0	$10 i_1$	1,977	583	15	14	12	7	9	0.35
	$10 i_2$	1,977	571	26	10	5	5	9	
	$10(i_1 + i_2)$	3,954	1,154	41	25	17	13	17	
	P	0	−0.01	−0.28	0.17	0.40	0.15	0	
6.0	i_1	1,253	43	7	3	4	6	4	0.52
	i_2	1,253	70	12	4	1	3	4	
	$(i_1 + i_2)$	2,506	113	19	7	5	9	7	
	P	0	0.24	0.25	0.03	−0.50	−0.29	0	

2 The physical properties of the moist atmosphere and its particles

the antisolar point. The pronounced polarization indicates an upper limit to the predominant size of the particles in noctilucent clouds, and provides a means of distinguishing them from tenuous clouds in the high troposphere which may be observed at the same time. Moreover, since the intensity of the radiation scattered in the Rayleigh regime is inversely proportional to λ^4, noctilucent clouds seen well above the horizon have a silvery color, often with a distinct blue tinge.

Molecular scattering in the atmosphere is responsible for the blue color of the sky and a yellow tinge in the color of the sun, which changes gradually into orange and eventually into red as the sun approaches the horizon and the path of the sunlight through the atmosphere rapidly lengthens. (The optically effective parameter is the mass of air in a cylinder of unit cross section along the path. If this is unity when the sun is directly overhead, it is 2 units when the solar elevation is $30°$, increasing to 10 units at an elevation of $5°$ and about 27 units at an elevation of $1°$.) Similarly, molecular scattering produces the characteristic yellowish tinge of distant high clouds observed near the horizon (noctilucent clouds, and cirrus and cumulonimbus tops seen during the day). Usually it is scattering from aerosol particles between the observer and distant dark objects which is predominantly responsible for the veiling of their colors and contrasts, until at some distance (the *visual range*, discussed in Chapter 4) they become indistinguishable from the background of the sky near the horizon. However, when the air is particularly clean and the visual range is great, the molecular scattering has comparable importance, and it is noticeable that hills farther and farther away have an ever paler *blue* color.

In the Rayleigh regime the scattering has equally intense shallow maxima in both the forward and backward directions, but as x increases so the forward scattering, in directions close to that of the propagation of the incident radiation ($0°$), becomes increasingly pronounced. This is already apparent in Table 2.7, and is shown more clearly in Table 2.8, which includes values of x extending up to 120. When x is very large it is customary to regard scattering as a combination of phenomena included under *geometrical optics*, in which the paths of rays are subject to *refraction* and *reflection* (dependent on m but not on x), but near and within particles are rectilinear, and under *diffraction*, representing departures from rectilinear propagation which are important at very small scattering angles (and dependent on x but not m). The values of the scattering function specified in Table 2.8 for $x \gg 1$ (large) are therefore the sums of the entries in the last two rows.

A consequence of the strong forward scattering as x becomes large is that near the sun shallow or isolated droplet clouds are strikingly brilliant, since extinction

Table 2.8 Values of the scattering function $f/x^2 = \{i_1(\theta) + i_2(\theta)\}/2x^2$, for unpolarized light incident on a spherical water droplet of radius r and refractive index $m = 1.33$, for various values of the angle of scattering θ and of $x = 2\pi r/\lambda$ (37, 39). In the last rows D represents the contribution according to the theory of diffraction, valid as θ approaches zero, and G the contribution according to the theory of geometrical optics

	θ (deg.)					r for $\lambda = 0.55\,\mu$ (μ)
x	0	10	30	60	90–180 (average)	
1	0.053	0.052	0.044	0.027	0.018	0.09
2	0.99	0.95	0.70	0.26	0.015	0.17
3	4.62	4.30	2.33	0.13	0.03	0.26
4	12.3	10.8	3.6	0.13	0.05	0.35
5	23.4	19.1	3.1	0.41	0.07	0.44
6	34.8	25.7	1.57	0.26	0.08	0.52
8	46.0	24.4	2.4	0.40	0.10	0.70
10	36	8.6	2.2	0.37	0.01	0.87
12	23	4.2	1.1	0.14	—	1.05
15	104	16.9	0.7	0.24	0.02	1.31
20	120	3.0	0.5	0.15	0.32	1.75
25	244	0.2	0.9	0.13	0.06	2.2
30	231	14.9	1.2	0.12	0.06	2.6
35	445	2.3	1.5	0.18	0.09	3.1
40	401	1.2	2.0	0.20	0.07	3.5
80	1,680	6.3	1.8	0.17	0.04	7.0
120	4,230	1.2	2.0	0.12	0.05	11.5
large D	$x^2/4$	0	0	0	0	
G	4.09	3.37	1.15	0.11	0.05	

in them is small and their droplets have radii typically of several microns, for which $x \gtrsim 50$. The particles of ice clouds are commonly even larger, but have prismatic shapes which are more or less randomly oriented (Section 5.12), so that the scattering from ice clouds is much more nearly isotropic. They can therefore often be distinguished from droplet clouds by their comparative lack of brilliance when seen near the sun, and their greater brightness at large angles from the sun. (Examples of the former distinction can be seen in the upper parts of Pl. 8.10, 8.6, and 8.25.) On the other hand, droplet clouds which are not optically dense appear dark when seen at a large angle from the sun against a background of ice cloud or of the sky near the horizon, whose brightness is due to scattering from aerosol particles.

Scattering cross sections and efficiency factors

The total energy scattered from a particle in all directions, that absorbed within it, and that removed from the incident beam of radiation can be expressed as the energy of the beam which is intercepted on normal areas C_s, C_{ab}, and C_{ext}, the *cross sections* of the particle for scattering, absorption, and extinction, respectively. By definition

$$C_{ext} = C_{ab} + C_s$$

and for particles of nonabsorbent material

$$C_{ext} = C_s$$

The cross sections can be related to the geometrical cross sections normal to the beam by efficiency factors, or *normalized cross sections*, σ. Thus for nonabsorbent spheres

$$C_{ext} = C_s = \sigma_s \pi r^2$$

Of particular interest in cloud physics are the extinction and scattering of visible light by clouds, and the *backscatter* (at $\theta = 180°$) of energy to a radar emitting microwave radiation (or a lidar emitting short-wave radiation) from a volume of cloud or precipitation particles.

Atmospheric extinction

If a cloud is optically thin, so that every particle is illuminated by the same intensity I_0 of incident light, then it is possible to obtain the scattering by the cloud as a whole by summing the contributions of the individual particles. For example, in a cloud of spherical nonabsorbent particles of the same composition, in which the concentration of particles of radius between r and $(r + dr)$ is $n(r)dr$, and in which the attenuation of a beam from a distant source follows the law

$$E = E_0 e^{-\tau s} \qquad (2.43)$$

where s = distance
τ = an attenuation or extinction coefficient
then by earlier definition

$$\tau = \int_0^\infty C_s n(r) dr \qquad (2.44)$$

$$= \int_0^\infty \pi r^2 \sigma_s(r) n(r) dr \qquad (2.45)$$

Now for a spherical particle, from Eq. 2.40, the energy scattered in all directions is

$$C_s = \frac{1}{k^2} \int f(m, x, \theta, \phi) d\omega$$

where

$$d\omega = \sin\theta \, d\theta \, d\phi$$

is the element of solid angle, or

$$C_s = \frac{1}{k^2} \int_0^\pi \int_0^{2\pi} f(m, x, \theta, \phi) \sin\theta \, d\theta \, d\phi \qquad (2.46)$$

For the Rayleigh regime ($x \ll 1$)

$$f = \frac{(1 + \cos^2\theta) x^6 K^2}{2} \qquad (2.42)$$

Thus

$$C_s = \frac{8\pi k^4 r^6 K^2}{3} \qquad (2.47)$$

and

$$\sigma_s = \frac{8 x^4 K^2}{3} \qquad (2.48)$$

These expressions can be used to estimate the extinction coefficient in the atmosphere due to scattering not only from small aerosol particles but also from the air molecules. Neither their validity for molecules nor the appropriate value for K is obvious. However, it can be shown (37) that in a medium consisting of randomly distributed scattering molecules the expressions are valid, and if the apparent refractive index n of the medium as a whole is close to 1, then K and n are related by

$$(n - 1)^2 = 4N^2 \pi^2 r^6 K^2$$

where

$$N = \int n(r) dr$$

is the number concentration of the molecules, and the extinction coefficient τ_0 due to the molecular scattering is

$$\tau_0 = \frac{32\pi^3}{3N\lambda^4 (n - 1)^2} \qquad (2.49)$$

and is approximately proportional to the density of the gas. Over the range of wavelengths of visible light λ varies by a factor of about 1.5, and $\tau_0(\lambda)$ therefore varies by a factor of about 5.

In the lower atmosphere attenuation in clear air normally is due mainly to scattering by aerosol particles rather than to molecular scattering, and τ has an associated component τ_p which has to be found by integrating Eq. 2.45, given the size distribution, shape, and composition of the particles. At high relative humidities the particles are watery and can be regarded as spheres with a refractive index $m = 1.33$; however, at relative humidities near and below 70% the presence of undissolved salts or insoluble components causes the refractive index to be appreciably greater, about 1.50 (40). For particles with either value of m Fig. 2.9 shows the efficiency factor σ_s as a function of x; at small values of x, in the Rayleigh regime, it increases very rapidly with x but reaches a maximum value of about 4 when x is between 4 and 6, and subsequently oscillates with decreasing amplitude about a value of 2 which is a limiting value for $x \gg 1$. The maxima are farther from the value 2 than the minima, and the curves have many

2 The physical properties of the moist atmosphere and its particles

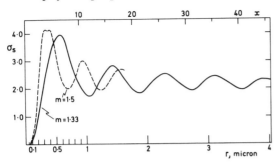

Fig. 2.9 The total scattering (or extinction) efficiency factor σ_s of water droplets ($m = 1.33$; solid line) and of aerosol particles at relative humidities below about 70% ($m = 1.50$; dashed line), shown as a function of x. Some corresponding values of droplet radius for $\lambda = 0.55\,\mu$ are indicated along the base of the diagram.

minor oscillations which have been removed from those drawn in the diagram. The general form of the curves, and in particular the approach of σ_s to 2 as x becomes large, is typical of particles with m not very close to 1 ($m - 1 \ll 1$). For values of $x < 1$, σ_s for spherical particles with $m = 1.50$ is greater than for those with $m = 1.33$ by a factor of about 3, but this factor falls to about 2 at $x = 4$, and to an average value of only about 1.1 for $30 > x > 4$, and still nearer 1.0 for $x > 30$. The dominant contribution to attenuation in the atmosphere by scattering will be shown in Chapter 4 to be from aerosol particles for which x lies in the range between about 1 and 10, and in estimating the attenuation at ordinary relative humidities uncertainty about the appropriate value of their refractive index is usually less important than that in the specification of their size distribution. Considering both uncertainties it will be sufficiently accurate to write $\sigma_s \approx 3$ for typical size distributions of natural aerosol particles, and

$$\tau_p \approx 3\pi \int_0^\infty r^2 n(r)\,dr \qquad (2.50)$$

since the contribution to τ_p by particles for which x lies outside the range between 1 and 10 is usually insignificant.

Backscattering; the use of radar

Radar technique

The technique of radar is explained in specialized texts (41, 42). The essentials of a radar are a *transmitter* which generates electromagnetic energy in pulses of about 1 μsec in duration, an *antenna* which radiates the energy (usually plane-polarized) and intercepts that scattered back during intervals of about 1 msec between successive pulses, a *receiver* which detects and amplifies the received signal (*echo*), and an *indicator* on which it is displayed in visual form. A radar is characterized by the frequency and power of the radiated energy, the duration of the pulses (which determines the resolution in the measured range of the *targets* that return the radiation), the pulse repetition frequency (which determines the maximum range that can be measured unambiguously), the kind of antenna (which determines the shape of the beam and the resolution in the measurement of the angular position of the targets), and the maximum sensitivity of the receiver. Some compromises are always made in the choice of these several characteristics in order to obtain the best or most economical performance for a particular set of purposes. For example, a narrower beam improves angular resolution and increases the minimum detectable echo intensity, but it requires a larger antenna, which is likely to be more expensive and also more unwieldy, increasing to an unacceptable degree the interval required to survey in detail a large part of the sky.

Conventional meteorological radars operate at wavelengths of about 1, 3, 5, or 10 cm, have a peak transmitted power P_t of order 100 kw, and a minimum detectable average received power $\bar{P}_r$ of about 10^{-13} watt (often quoted in dbm, decibels below 1 mw). The pulse of energy from the transmitter is led to a small source at the focal point of an antenna which is paraboloidal, producing a beam in which the energy is concentrated near the axis of a main lobe, but which has a number of side lobes directed at various angles away from the main axis, in which the radiated power is about 25 db or more below that of the main lobe. The *beam width* is specified as twice the angle at which the power in the main lobe is half that on its axis; usually a width of between 1 and 2° provides acceptable resolution, but it may be reduced to less than 1° in either elevation (ϕ) or azimuth (θ) in order to improve the resolution in those directions.

The indicator generally is the face of a cathode-ray tube, on which the following kinds of display can be presented:

1. Amplitude of echo in some fixed direction as a function of range (*A*-scope).
2. Position of echo as a function of range and azimuth at a fixed elevation (generally small, to give a *plan-position indicator*, or PPI display; the antenna makes complete revolutions about a vertical axis with a period of a few seconds).
3. Position of echo as a function of range and elevation at a fixed azimuth (to give a *range-height indicator*, or RHI display; the antenna nods between elevations of 0 and about 20° with a period of a few seconds).
4. Other kinds of display, for example, of echo position on a surface of fixed height (*constant altitude plan-position indicator*, or CAPPI display), or echo

position as a function of azimuth and elevation or height at a fixed range (*C*-scope).

All of these types of display have been used to assist in the representation of the distribution of the targets in space and their movement and evolution.

Further valuable information is given by the power of the echo received from targets, which has usually been measured by reducing the receiver gain until the echo just disappears (as observed by eye or camera) from the display. Great care has always to be taken to operate under standard and continually calibrated conditions.

The theory of radar echoes from particle populations

Meteorological targets are irregularities of refractive index in turbulent air, or randomly distributed populations of particles in relative motion due to their differing sizes and turbulent air motions, and their echoes are said to be *incoherent*, exhibiting rapid fluctuations of intensity similar to those that characterize receiver noise. The intensity of such echoes must be specified by some statistical property, such as the average power $\bar{P}_r$ received from a number of consecutive pulses, which has been shown, when measured with appropriate precautions, to be with acceptable accuracy (within about 2 db) equal to the sum of the powers received from individual particles regarded as independent scatterers (42).

One method of calibration uses a measurement of the reduction of receiver gain which is required just to remove from a display the signal from a distant metal sphere of known size and backscatter cross section. Such a target gives a *coherent echo* of well-defined power. If a similar measurement is made on an incoherent echo, such as that received from precipitation, by using a manual control to reduce the gain until the operator sees on the display only an occasional flash from the peak amplitudes of the fluctuating signal, then referring to the calibration the operator overestimates the average received power by as much as 5–6 db (42). Some measurements mentioned in Chapter 8 which were made in this way therefore have been adjusted by this amount. Evidently measurements of echo intensity, which can be made in several ways, should not be discussed without a statement of the technique by which they were obtained.

Considering a population of scattering particles (of homogeneous composition) whose dimensions are provisionally specified by values of a radius r, the backscatter cross section C_b is defined as

$$C_b(m, r) = \frac{4\pi R^2 E(R, 180°)}{E_0}$$

$$= \frac{2\pi\{i_1(180°) + i_2(180°)\}}{k^2} \quad (2.51)$$

so that if a particle scatters isotropically without absorption its backscatter efficiency factor is $\sigma_b = 1$.

The *reflectivity* η of the scattering inhomogeneities of refractive index in a unit volume is then the total backscatter cross section

$$\eta = \int_0^\infty C_b(m, r) n(r) dr \quad (2.52)$$

where $n(r) dr$ is the concentration of particles of radii in the interval between r and $(r + dr)$. If the particles occupy the whole volume occupied by the transmitted pulse, with a length h in the direction of the beam and at a range R from the radar, the relation between the peak transmitted power P_t (which is practically uniform throughout the pulse duration) and the average power $\bar{P}_r$ scattered back to the radar (42) is

$$\bar{P}_r = \frac{[P_t h G^2 \lambda^2 \theta \phi] \eta \kappa}{32\pi^2 R^2 \log_e 2} \quad (2.53)$$

where κ is an attenuation coefficient defined by

$$10 \log \kappa = 2 \int_0^R (g + c + p) dR$$

and g, c, and p represent the attenuation (one-way) in decibels per kilometer due to atmospheric gases, cloud particles, and precipitation, respectively. The term enclosed in square brackets is a characteristic of the particular radar: θ and ϕ are the conventionally defined horizontal and vertical beam widths, and G is the antenna gain, that is, the ratio of the radiated energy at a given distance on the beam axis to that which would be provided by an isotropic radiator of the same source power. It is related to the beam widths by $G = \pi^2 k_1^2/\theta\phi$, where k_1^2 is an efficiency factor, which is 1 for most antennas of circular cross section.

In clouds and rains examined with centimetric radar the particles frequently are composed wholly of water or of ice, and can be regarded as spherical and as scattering in the Rayleigh regime for radii up to some value r_1 for which $x \lesssim 0.2$, with backscatter cross sections given to sufficient accuracy by Eq. 2.41 and 2.42 as

$$C_b = \frac{4\pi}{k^2} f(x, 180°)$$

$$= \frac{4\pi K^2 x^6}{k^2}$$

$$= \frac{64 K^2 \pi^5 r^6}{\lambda^4} \quad (2.54)$$

where K^2 is practically independent of wavelength (0.93 for water and $0.20/\delta^2$ for ice particles such as soft hail and snowflakes which contain air enclosures reducing their mean density to δ).

Then

2 The physical properties of the moist atmosphere and its particles

$$\eta = \frac{64K^2\pi^5}{\lambda^4} \int_0^{r_1} r^6 n(r) dr \qquad (2.55)$$

$$= \frac{K^2\pi^5}{\lambda^4} \int_0^{D_1} D^6 n(D) dD \qquad (2.56)$$

where D is drop diameter.

In some circumstances the attenuation coefficient κ is significant, and in others the particles occupy only a fraction of the pulse volume. Assuming that some estimate can be made of these factors, the property of a population of particles which can be inferred (using Eq. 2.53) from a measurement of the power which they scatter back to a centimetric radar is η, or, for rains ($K^2 \approx 0.93$), in which $x \lesssim 0.2$, that is, $r < 1$ mm for λ as small as 3 cm,

$$Z = \int_0^{D_1} D^6 n(D) dD \qquad (2.57)$$

Z is known as the *radar reflectivity factor* and is usually expressed in units of mm^6 m^{-3}. Empirical relations have been sought between Z, M, and R, where M is the mass concentration and R the rate of rainfall associated with the scattering drops (Section 8.1).

When rains include hail η cannot be simply represented: the hailstones have a different refractive index, have a variable content of liquid water, and may have aspherical shapes as well as sizes for which x is no longer small. Then η has to be calculated from the more general expression (2.52), but an *equivalent radar reflectivity factor* Z_e can be defined as that value of Z for which η would have the same value. Figure 2.10 compares theoretical values of the backscatter efficiency factor δ_b of spheres of pure ice and of liquid water, and shows that whereas in the Rayleigh regime drops backscatter more efficiently than ice particles, the reverse is true when x exceeds about 2. Droplets do not exist for such values of x, but ice spheres coated with unfrozen water may behave effectively as liquid spheres. Figure 2.10 is the basis of inferences made in Chapter 8 from quantitative observations of the equivalent radar reflectivity factor in hailstorms.

With most radars there are manual controls for adjusting θ, ϕ, and the receiver gain, and for selecting the kind of display. For a meteorological observer it is a fascinating pursuit to use these controls in examining various features of precipitation systems within a range of 100–200 km, and this kind of exploration is necessary to gain a full appreciation of the potential value of radar. However, it is now recognized that to collect the greatest amount of useful information, to ensure that unexpected developments do not pass unobserved, and to maintain observation over the several hours or more during which systems evolve, it is necessary to adopt automatic scanning and recording procedures (and even to use more than one radar). In particular, techniques have been developed to display echoes bounded by contours of Z_e, as illustrated in the RHI and PPI displays of Pl. 2.1. For most purposes other than the study of the detailed structure of precipitation systems such displays contain irrelevant detail, and it is preferable to transform them to digital displays of the average or maximum values of Z_e over small areas, as shown in Pl. 2.2 and Fig. 8.36 (Appendix 8.3). Such displays are not only individually more comprehensible, but they represent only one of a variety of statistics, useful in research and in forecasting practice, which can be provided rapidly by an electronic computer (e.g., the distribution of accumulated rainfall).

Doppler radar

The Doppler radar can measure not only the power received from a target, but also a small difference between the frequency of the transmitted and received radiation, which is a measure of the radial velocity of the target. In meteorological applications the signal received by the radar is from a large number of targets. Their individual radial velocities vary because they have differing fall speeds and because they move with a wind which, because of the angular spread of the beam, generally has a radial velocity with an apparent variation even if it is steady and uniform, and with a real variation if it is steady but has significant shear, or if it fluctuates because of the presence of small-scale motion systems or turbulence.

However, with suitable technique the Doppler radar can analyze the received power as a function of frequency shift, to provide information about the mean radial motion of the targets and the intensity of the fluctuations (42). In particular, in the presence of widely distributed artificial or natural targets (precipitation,

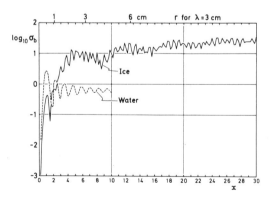

Fig. 2.10 Calculated values, as a function of x, of σ_b, the backscatter efficiency factor of spherical particles of water ($m = 7.14 - 2.89i$) and of ice ($m = 1.78 - 0.00241i$), valid for microwave radiation of wavelengths between 1 and 10 cm. Experimental values obtained with individual spheres of ice in the laboratory, and outdoors (using radar under operational conditions) are consistent with the calculated values (42), which have been used in the construction of Fig. 8.57.

or other passive wind-borne targets such as small insects) it can measure mean winds (with an error of less than 0.5 m sec^{-1} and a few degrees) over the perimeter of circles of radii about 10 km centered over the radar, at heights up to several kilometers with a height resolution of about 100 m, and with a time resolution of a few minutes. Such wind soundings are valuable in the study of motion systems accompanying prolonged precipitation.

Further, a Doppler radar with the beam directed vertically through precipitation can provide useful data on the concentration in selected height intervals of the particles whose vertical velocities lie between a succession of chosen values, usually at intervals of about 1 m sec^{-1}. In widespread rain the vertical component of the air motion is generally small compared with the fall speeds of the smallest detectable raindrops, and they can be regarded as scattering in the Rayleigh regime, so that their size distribution as a function of height may be obtained directly; the interpretation of echoes from snow is less certain, but the Doppler analysis is still informative. In showers the vertical air speeds (typically of several meters per second or more) are comparable with the fall speeds of the largest precipitation particles, but in those showers which can be presumed to contain detectable concentrations of small drops, ice crystals, or snowflakes with fall speeds of only about 1 m sec^{-1}, the subtraction of this value from the observed range of approach velocities in a selected height interval allows the mean vertical air speed and the fall speeds relative to the air of the larger particles there to be inferred to a useful degree of accuracy.

Common optical phenomena associated with scattering from clouds and precipitation

The corona

As x increases beyond 1 the intensity of the light scattered by particles not only becomes greater in the forward direction, with an increasing maximum at $\theta = 0°$, but its variation with θ develops increasingly pronounced subsidiary maxima and minima, as seen in Fig. 2.11. The number of the maxima is approximately x, and they are therefore mostly separated by angles of about $180°/x$. The consequent complication of the dependence of intensity upon angle is responsible for the erratic values of the scattering function at fixed angles which are entered in Table 2.8 for large x. This complication adds greatly to the labor of calculating scattering functions for large values of x, because they must be made at very small intervals of θ if they are to provide a satisfactorily complete representation, and to allow, for example, an accurate average value of practical interest to be obtained over a moderately large interval of θ.

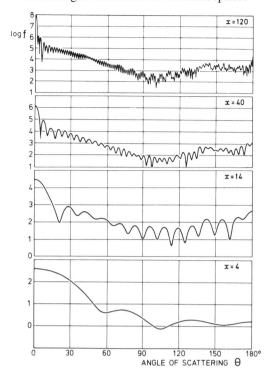

Fig. 2.11 The scattering function $(i_1 + i_2)$ as a function of the scattering angle θ, when unpolarized light is incident on spherical water droplets ($m = 1.33$) with the following values of x and the corresponding droplet radius r for $\lambda = 0.55$ μ: (a) $x = 4$ ($r = 0.35$ μ); (b) $x = 14$ ($r = 1.2$ μ); (c) $x = 40$ ($r = 3.5$ μ); (d) $x = 120$ ($r = 11.5$ μ) (39).

The angular positions of the first few minima in the intensity of the light scattered forward are not well defined at all values of x when x is not very large; according to diffraction theory ($x \gg 1$) they lie at $x \sin \theta = a_n$, where $a_1 = 3.8$, $a_2 = 7.0$, and $a_3 = 10.2$ for the first three minima. However, where the first minimum is well defined the complete Mie theory gives values of a_1 which vary between about 3.0 and 4.2 for $x \lesssim 25$.

In a cloud in which the extinction is small, and which consists of droplets of uniform size, the oscillation in amplitude of the scattering function and the inverse relation between x and λ can lead to color separation. Thus there arises the very common corona around the sun or moon covered with thin cloud. Usually this consists only of the *aureole*, an adjacent zone which is blue-white close to the edge of the luminary, then a yellowish white, surrounded by a brown-red border at a radius which varies between about 1 and 5°. Occasionally, however, the aureole is surrounded by more strikingly colored rings of blue, and especially of green and red (in that order), sometimes repeated once or twice; very rarely there may even be a fourth red ring. Coronas seen around the moon can be observed comfortably; if they are sought around the sun with aids such as a black mirror and polarizing filters, they are found to be brighter, more extensive, and more common,

2 The physical properties of the moist atmosphere and its particles

at least in fragmentary form, rarely being absent in thin droplet clouds which come near the sun.

No calculations appear to have been made of the characteristics of the size distribution spectra of clouds of droplets under which the corona can be expected to be well developed. It seems likely that the important condition is that there be little variation about a mean droplet size in a path through the cloud along the line of sight. However, it has been shown that the color of light scattered by aerosols with particles of radius about 1 μ (x about 10), which is successfully exploited to measure particle size, may not be markedly affected as a very narrow size distribution is broadened a little, and that the color may be dominated by the particles of nearly the largest size present, which scatter light most efficiently (43). Nevertheless, directly observed droplet spectra in natural clouds are broad, especially in clouds which have aged (such as cumulus clouds, discussed in Chapter 8), and it still appears necessary to discuss circumstances in which the spectra are comparatively narrow and bright coronas may be expected.

The discussion in Chapter 5 indicates that droplets of predominantly about the same size may be grown in air which ascends smoothly (adiabatically) above its saturation level, as happens almost only in wave motions produced by the flow of stably stratified airstreams over hills. The mean droplet size depends on the speed of ascent and the size distribution of the aerosol particles at the saturation level (which determine the droplet number concentration), and on the displacement above the saturation level and the temperature there (which determine the droplet mass concentration). Of all the properties whose variations are most likely to have a significant effect on the *local* mean droplet size $\bar{r}$ the most important is the displacement z' above the saturation level; in clouds formed in the troposphere the size increases rapidly with height in the first few hundred meters of ascent beyond this level to become several microns, and thereafter increases more slowly ($\bar{r}^3$ is approximately proportional to z'). Accordingly, the variation of mean droplet size throughout the thickness of one part of a wave cloud depends on the depression ΔT_D of the dew point below the air temperature on the undisturbed streamlines (which determines the saturation level; Section 3.6) and the vertical displacement of the streamlines in the wave cloud. In most ordinary wave clouds the vertical displacement is a few hundred meters and probably does not vary much with height; the condition favoring nearly uniform droplet size is then that there should be little variation of ΔT_D with height in the layer of air in which condensation occurs. This condition is not typical of deep layers and would not be anticipated from any familiar meteorological process, but it may be more nearly satisfied or relaxed when a wave cloud forms only in a shallow moist layer (with comparatively dry air above and below), in which all the air is lifted considerably above its saturation level. Such a cloud seen from the side, across the wind direction, would have an arched base or top, or the form of an inverted saucer if limited in lateral extent by a decrease in the amplitude of the wave motion (as may be expected if the responsible obstacle is an isolated hill rather than a long ridge). Arched wave clouds have been described in observations from aircraft flying at about the same height (44), but their form could hardly ever be recognizable from the ground, even when seen at low elevations. However, even in such clouds the mean droplet size must be expected to increase significantly from the edges toward the thicker middle parts of the cloud. Since the heights and horizontal dimensions of these clouds are usually a few kilometers, they cannot be expected to produce complete coronas several degrees in diameter unless the sun or moon is at a high elevation (45° or more), so that the coronas occupy parts of the clouds not more than about 1 km in horizontal extent. Even so, across such a distance some considerable variation of a locally nearly uniform mean droplet size should not be uncommon, so that the rings of a corona are probably not always exactly circular; pronounced asymmetry has been noted by a persistent observer on as many as one seventh of all occasions (45). It is interesting that small coronas are distinctly flattened on the side of a moon which is not full.

A corona, in addition to showing that the particles in a cloud are of nearly uniform size, allows this size to be estimated from a measurement of the angular radius of the first or second red or red-brown ring, which is usually assumed to be where there is a minimum in the intensity of the green light to which the eye is most sensitive ($\lambda \approx 0.55\,\mu$). If the radius of the first ring is θ_1, then $x \sin \theta_1 \approx 4$, or

$$\bar{r} \approx 20/\theta_1 \qquad (2.58)$$

where $\bar{r}$ is the mean radius of the droplets in microns, and θ_1 is in degrees (a correction for the finite width of the sun or moon seems hardly justifiable; this width, however, presumably at least contributes to the poor color separation in the aureole). The normal range of angular radii observed therefore implies mean droplet radii or between about 4 and 20 μ.

There has been some debate whether cloud particles of other than spherical shape, for example, the prismatic forms of ice crystals, can produce coronas. The variation with θ in the intensity of the light scattered forward at small angles by a cloud of randomly oriented long cylinders, all with a radius sufficiently large to ensure that $x \gg 1$, is not much different from that in a cloud of spherical particles with the same radius (37). However, individual crystals of the size and shape considered tend to fall with their long axes horizontal; only those particles which consist of a cluster of prisms radiating from a center (mentioned in Section 5.12 and illustrated in Pl. 5.1) are likely to have the required variety of orientations. The prisms of such clusters grow rapidly to attain within a few minutes of their formation lengths of 200–500 μ, and half-widths of 30–80 μ. The radii of the rings in coronas which they may then produce is therefore small: even the second red ring is likely to be within about $\frac{1}{2}°$ of the rim of the sun or moon. Moreover, the clouds in which this kind of ice particle grows [the fallstreaks (46) and, perhaps, the "shred" (47) and "puff" (48) forms of cirrus in the upper troposphere] typically have small angular dimensions—not more than about 1°—and high speeds of travel (in contrast with the almost stationary wave clouds of droplets in the middle and lower troposphere); consequently any coronas which they produce are likely to be incomplete and rarely and fleetingly visible; indeed, they are unlikely to be noticed at all without careful and diligent search. The widespread opinion that ice particles cannot produce *diffraction* color phenomena is based on three kinds of evidence. First, there is a striking absence of such phenomena in the trails of ice particles that occasionally fall from shallow clouds of supercooled droplets, which usually produce complete or partial coronas. Second, they are generally not noticed in those thin high clouds which by a characteristic texture are thought to consist of ice particles, or which are shown to contain ice crystals by the occurrence of *haloes* (refraction phenomena, mentioned below). Third, in observations made in the Antarctic, where both clouds of droplets (supercooled to temperatures as low as $-29°C$) and ice clouds occurred at the ground, as well as in the free atmosphere, it was found that coronas and *haloes* were never observed simultaneously in the same cloud (49). Nevertheless, according to one assiduous observer (44), a very small lunar aureole, with a narrow faint red ring of radius less than 1°, is commonly present in extensive thin high clouds which produce haloes. Moreover, I have on a few occasions seen in extensive clouds with the texture typical of aged ice clouds (and with no apparent intervening droplet clouds) a persistent complete double corona, with a first dull red ring at a radius of less than 2°, surrounded by rings of blue-white, faint green, and purple-red (at about 3°). It therefore seems that ice clouds may produce coronas very close to the luminary, which are not ordinarily noticed because near the moon they are too faint, and near the sun they are lost to the unaided eye in the dazzling glare. Their narrowness indicates a predominant crystal thickness of about 40 μ, which is characteristically much larger than the predominant particle diameter of several microns in the droplet clouds which cause ordinary coronas.

The glory

The glory consists of a series of colored rings, much like those of the corona, seen around the shadow of an observer on the top of a fog or layer of droplet cloud in bright sunshine. The outer rings are more pronounced and therefore often more numerous (up to five orders may be present) than in ordinary coronas [their intensity decreases as $(180 - \theta)^{-1}$, rather than approximately as θ^{-3}], and there are other differences: their light is strongly polarized, and the separation of successive rings is a little different; however, the diameter of the rings is inversely proportional to the effective droplet size, in about the same ratio as for the rings of the corona (37). Nevertheless, the glory evidently cannot be regarded as a diffraction phenomenon, or considered in geometrical optics without recourse to the hypothesis that some rays which reach a drop at grazing incidence propagate around its surface before escaping (50); the detailed properties of the glory cannot be explained satisfactorily other than by calculations of the backscatter based on the comprehensive Mie theory (for values of x of order 100, droplet radius about 10 μ or more).

Iridescence

Diffraction colors which are seen on clouds whose angular dimensions are too small for the display of complete coronas are known as iridescence; however, iridescence cannot be regarded as simply a diffraction phenomenon, because the colors often can be seen at large angles from the luminary and even in the opposite part of the sky. In casual observation they are therefore noticed much more frequently in daylight than moonlight, in which they are very faint. Like many other meteorological phenomena, when carefully sought iridescence is frequent and sometimes very beautiful. On a small cloud there may be one color only, whereas on more extensive clouds bands of colors tend to be arranged parallel to the cloud edges. The colors often are distinct at angular distances as great as 20° from the sun on fragmentary low clouds, and on other clouds, probably wave clouds, even as far as 50° (45).

The most spectacular iridescence is that associated with the *nacreous*, or *mother-of-pearl*, *clouds*, which occur very infrequently in the lower stratosphere during the winter season, near mountain ranges in high latitudes. The most detailed observations have been made in Norway, where it appears that the clouds may be seen with an average frequency of one or two days each winter, and where reliable photogrammetric measurements have given heights varying from 17 to 31 km (and by 3 to 8 km on individual occasions), with a mean value of about 24 km (51). No simultaneous measurements of temperature have been made at the cloud levels, but it is thought that it is generally about

2 The physical properties of the moist atmosphere and its particles

−80°C, and that the concentration of water vapor present is such that condensation is not likely to occur unless locally the air is lifted about 1 km and thus chilled by about 10°C. At the saturation level the concentration of water vapor is then only about 10^{-4} g m^{-3} (Appendix 2.1), and consequently this is the magnitude of the mass concentration of condensed water that may be produced by the further ascent of several hundred meters which occurs in wave clouds. In a rapid ascent the number concentration of particles is likely to be that of all the condensation nuclei present (Chapter 5), which according to Fig. 4.2 is about 3×10^5 m^{-3}. The corresponding mean radius of the largest cloud particles can therefore be anticipated to be not more than about 2 μ ($x \approx 20$), considerably less than is characteristic of tropospheric clouds. This estimate appears to be supported by a single observation of a distinct corona around the moon in a cloud of this kind; the inner region of the corona was lilac-yellow, and part of a surrounding red ring extended between 15 and about 18° from the moon, which according to Eq. 2.58 implies a particle radius of more than 1 μ.

The range of particle radii in nacreous clouds thus is likely to vary across the clouds from considerably less than 1 μ near their edges to about 2 μ in their central parts, representing a range of values of x between about 5 and 20. At cloud temperatures of about −90°C the particles must be expected to consist of ice, whose stable crystal form is the hexagonal prism to considerably lower temperatures. It is not clear that particles which are so small will develop well-defined crystal faces and become markedly aspherical. Particles almost equally small, with radii in the range between about 1 and several microns, are characteristic of ice fogs which form at low temperatures (generally between about −30 and −40°C; Section 3.2); most of these particles are roughly spherical, with poorly developed crystal faces (53), and they produce coronas rather than halo phenomena. Their mode of formation is essentially different, however, from that of the nacreous cloud particles, for they are due to the mixture of cold air with warm, vapor-laden industrial exhausts, and they may reach practically their final size as droplets before freezing. It has also to be considered that a proportion of the particles form by the growth of large condensation nuclei as supercooled solution-droplets containing sulfates or sulfuric acid in concentrations which may prevent freezing until the particle radii approach 1 μ (Section 4.6, Fig. 4.2; 54). The brilliance of the iridescent colors in nacreous clouds certainly suggests that the particles are not only of locally uniform size, but that if they are aspherical the degree of asymmetry is not important in the range of sizes concerned, comparable with the wavelength of light. Moreover, although they are frozen, the refractive index of ice differs little from that of liquid water (the indices are, respectively, 1.307 and 1.330 for red light, and 1.317 and 1.342 for violet light). Consequently, it can be expected that for values of x between about 5 and 20 the scattering function $f(\theta)$, which has been tabulated for spheres with $m = 1.33$ (Tables 2.7 and 2.8), and illustrated in Fig. 2.11, is applicable also to the scattering from the particles of nacreous clouds. The scattering function $f(\theta)$ has few maxima and minima, allowing a marked color separation over large angles well away from the luminary, but a low intensity. Observed in the middle of the day, when the troposphere also is illuminated by sunlight, the clouds consistently are not bright and the iridescent colors are faint. For up to about an hour before sunrise and after sunset, however, the clouds remain sunlit in a twilight sky and then often show remarkably bright and apparently pure colors, sometimes in individual clouds arranged in several bands parallel to sharply defined edges. In the Norwegian observations the colors are said to have been observed at angles of up to about 44° from the sun, and again at angles between 144 and 166° (52). On one occasion in Scotland bright iridescence was observed at angles of up to between 70 and 80° from the setting sun (56).

Those fortunate enough to see bright and extensive colored clouds have expressed wonder and admiration. Frequently particularly bright clouds have been called dazzling, and the following extracts from some descriptions show how impressive they have been:

At 1010 MET on 20 January 1932, at Lofoten, 68°N, 14°E, when the sun would have been just below the south-southeast horizon:

> I was working in my office and had the light on because it was so dark. On going out I noticed the sky became lighter ... and saw that the illumination came from a cloud in the SE about 30 to 35° above the horizon. This cloud was oval, not very large, and a little irregular. The light increased rapidly and changed from golden to light green, deep green, light red, deep red, light blue and deep blue and the upper border shone in a sharp golden color. The colors were arranged in horizontal bands in the order given above. Some minutes later the colors faded.... The cloud was, for a time, so brilliant that one could not look at it without being dazzled. (55)

On the same occasion an observer farther east reported that at 0830h the light from nacreous clouds whose upper border was in the east at an elevation of 10°

> was very strong; the landscape was illuminated by the clouds with a strange golden-red tinge. The sight was like a fairy tale.
> Gradually, with the failing of daylight ... following the sunset, the red colors predominated altogether and the blue and yellow fields disappeared. About 17 o'clock the reflections of these clouds on the roof of an out-house, thatch, with slates, gleamed quite like fire. (55)

> Two shell-like clouds ... had the most pure color—rose,

yellow-green and bright violet.... More to the west was a flat shining stripe, bright yellow and yellow-green with porcelain-blue color along the upper border; the cloud was quite dazzling like silver.

It is impossible to describe their splendor. They showed all the colors of the spectrum and were at some places shining like a sun. (56)

During an afternoon display at Anchorage, in Alaska, "in the city many people actually stopped and gaped at the spectacle, little groups forming along the sidewalks to discuss the colors" (57).

Half-tone reproductions of photographs of nacreous clouds, such as Pl. 2.3, can show some of the characteristics of their form and give an impression of their brightness but of course can only hint at their splendor. Some illustrations which have been published in color, listed at the end of the references, are a little more successful.

The edges of the bright clouds are usually sharply defined, but frequently tenuous clouds with no iridescent colors extend in long trails from their downwind sides. A similar form is typical of many wave clouds formed in the upper troposphere at temperatures near $-40°C$ (below which all cloud droplets and dilute solution-nuclei freeze; Section 5.11). The rate of ascent of air on the upwind side of waves is typically several meters per second, and no visible cloud forms until the air temperature falls to the *dew point* and there is a rapid growth of supercooled droplets (Section 5.12). If some of these droplets freeze, the crystals which develop do not evaporate with the remaining droplets at the downwind edge of the wave cloud but persist until the air has descended a sufficient further distance for the air temperature to rise above the frost point, then appearing as a more or less faint trail depending on the crystal concentration, which never shows iridescence and sometimes extends far from its parent cloud (58).

The explanation of the trails from tropospheric wave clouds thus depends essentially on a change of phase from supercooled liquid water to ice, which is experienced by a proportion of the cloud particles and which is a familiar process at temperatures near and above $-40°C$. It cannot be relevant to the trails from nacreous clouds, whose temperature is very much lower. However, at their level it is possible that in large condensation nuclei which do not grow to radii exceeding about $1\,\mu$ there may be a crystallization of various solid hydrates of sulfuric acid, producing some particles with an equilibrium vapor pressure different from that of ice and giving them a distinctive behavior (54). A close inspection of pictures of nacreous clouds shows that the tenuous noniridescent parts sometimes appear beyond their *upwind* edges, although they are usually much more extensive on their downwind sides. Moreover, they are reported generally to have some uniform tinge, usually grey-blue, suggesting that the reason why they are not iridescent is that their particles are too small: if the radii are only a fraction of a micron, corresponding to a value of x about 1, the scattering function $f(\theta)$ may have a magnitude great enough to make them just visible under favorable illumination (when the atmosphere along the line of sight beneath the clouds lies in the earth's shadow), but with insufficient variation or oscillation with scattering angle to allow a distinct color separation. It is not infrequently remarked when nacreous clouds occur that very thin veils of grey-blue or grey-green cloud give large parts of the twilight sky a characteristic unusual color; the same effect is suggested by the irregular form and intensity of the twilight arch in Pl. 2.3. It therefore appears that the tenuous ("veil") clouds, which are not iridescent, are much more widespread than bright nacreous clouds, which are generally small in extent. The particles of the veil clouds are probably formed independently of the nacreous clouds, but may be modified and made more easily visible near and downwind of them.

The rainbow

For raindrops x is of order 10^3, and the scattering diagrams of i_1 or i_2 against θ have successive maxima and successive minima at intervals of about $180°/x$. These would correspond to a conspicuous pattern in a cloud of uniform particles illuminated by monochromatic light from a small source, but in natural conditions they are smoothed beyond perception, except near especially pronounced fluctuations which appear as bows centered on the luminary or the position in the celestial sphere opposite to it. The positions of these bows and some of their properties are given reasonably accurately by geometrical optics. The principal rainbows can frequently be seen by moonlight but are then faint, with hardly perceptible colors, and accordingly they are subsequently described as seen under favorable conditions in sunlight.

Brightly colored rainbows are among the most familiar and beautiful of natural optical phenomena. Their main features were satisfactorily explained as long ago as 1611. However, the detailed properties of the various bows and their variability have continued into recent years to exercise the ingenuity of physicists, who have used geometrical optics with successive refinements. Even so, a comprehensive description requires recourse to elaborate computations based on the general Mie scattering theory (37). More extensive and detailed observations than yet available might prove a stimulus, but neither kind of work is likely to be undertaken in view of the apparently very limited potential outcome of practically useful information.

The scattering pattern of a large water drop ($x \gg 1$) is very insensitive to its size, and according to geometrical

2 The physical properties of the moist atmosphere and its particles

optics can be regarded as a combination of four principal components, associated with rays which are (a) reflected from the outer surface and (b) refracted twice in passing through the drop, and those which experience (c) one internal and (d) two internal reflections. Much of the incident light emerges from the drop near angles of minimum deviation associated with the latter two kinds of path, especially (c), which leads to the bright *primary rainbow* in an illuminated shaft of rain. Because of the variation of refractive index with wavelength the angles of minimum deviation also vary a little with wavelength, and there is a striking color separation in which the red border of the primary bow is about $1\frac{1}{2}°$ farther from the antisolar point than the violet border.

Other rays experience more than two internal reflections, and each has an angle of minimum or of maximum deviation from the direction of the incident light, for which the scattered light has a maximum intensity and at which there is therefore a bow at a definite radius from the sun. None of these other bows, however, has a significant intensity and is likely to be perceptible, except possibly that associated with rays having three internal reflections (the *tertiary rainbow*), seen toward the sun at a radius θ of about $42°$, where the general intensity of the light scattered forward is about two orders of magnitude greater than in the bow. Table 2.9 gives the scattering pattern of the large drop with $m = 4/3$, which shows a peak of intensity at $140°$, near the position of the primary rainbow (at $138.0°$; containing rays with one internal reflection) and the *secondary rainbow* (at $128.7°$; containing rays with two internal reflections). A detailed graph has recently been given for four different wavelengths and a drop diameter of 1 mm (59).

The primary rainbow is the most brilliant; it has a spectrum of colors in which observers usually distinguish in succession orange, yellow, green, blue, and violet. The breadths and relative intensities of the colors vary from occasion to occasion (and in eye observations from one observer to another). The angular positions of the maximum intensities for the red and the violet, and the difference between them, increase significantly as the drop radius becomes smaller than about 100 μ; the difference eventually becomes several degrees for a radius as small as 20 μ. Accordingly, in a cloud with a distribution of drop sizes the colors overlap and the bow becomes the *fogbow*, much wider than the ordinary rainbow, mainly white and tinged only with orange on the outside and blue on the inside. Such bows are common in fogs at air temperatures below 0°C, showing that the droplets are sometimes strongly supercooled (e.g., to temperatures as low as $-28°$C in one series of observations in Antarctica; 60). The fogbow and the primary rainbow are brighter on their inner (violet) sides, consistent with the entries in Table 2.9.

Table 2.9 The intensity of red light scattered from a large drop of water ($m = 4/3$, $x \gg 1$), as a function of the angle θ between the direction of propagation of the scattered and incident light, and expressed as the ratios G_1 and G_2 between the intensities i_1 and i_2 and those from an isotropic scatterer (37)

θ (deg)	Contributions due to reflection from outer surface		Contributions due to refraction		Contributions due to refraction with one and two internal reflections	
	G_1	G_2	G_1	G_2	G_1	G_2
0	1.000	1.000[a]	15.35	15.35		
10	0.674	0.493	13.27	13.49		
20	0.457	0.239	8.09	8.61		
30	0.314	0.111	4.15	4.77		
40	0.220	0.047	1.951	2.501		
50	0.157	0.017	0.838	1.257	Very weak	
60	0.115	0.004	0.270	0.482		
70	0.086	0.000	0.051	0.112		
80	0.067	0.000	0.000	0.001	0.001	0.000
90	0.053	0.003			0.001	0.000
100	0.043	0.006			0.002	0.000
110	0.036	0.009			0.005	0.000[b]
120	0.031	0.012			0.026	0.016[b]
130	0.027	0.015	Empty		Almost empty[b]	
140	0.025	0.017			1.000	0.090[b]
150	0.023	0.018			0.264	0.183[b]
160	0.021	0.020			0.111	0.093
170	0.021	0.020			0.082	0.075
180	0.020	0.020			0.078	0.078

[a] Associated with diffraction.
[b] Near the primary and secondary rainbows, these intensities are significantly affected by drop size.

The sky is comparatively dark between the primary and the less strongly polarized secondary rainbow, whose colors are broader, weaker, and in the reverse order. Just inside the primary bow, and often also outside the secondary bow, are narrow bands of color, usually of red and green, called *supernumerary bows*, marking the positions of adjacent intensity maxima and minima in the green light to which the eye is most sensitive. Their presence cannot be explained simply by geometrical optics: they are usually attributed to interference between coincident rays with angles of incidence on the drops slightly on either side of that which provides the ray of minimum deviation. Up to several such bows may be visible, whose clarity or number is an indication of the narrowness of the spectrum of droplet sizes.

The supernumerary bows are characteristic of the more brilliant rainbows produced when there is a predominant contribution from large raindrops. Usually

Table 2.10 The appearance of rainbows according to the predominant size of the scattering drops (after Minnaert; 33)

Predominant droplet radius (mm)	Appearance of rainbow
1–2	Numerous supernumerary bows merging into a primary bow with very bright violet and green, and a pure red but little blue
0.5	Fewer supernumerary bows, and the red of the primary bow weaker
0.2–0.3	Red of primary bow absent; the supernumerary bows are more yellow; a gap appears between the supernumerary bows and between the primary and first supernumerary bow as the drop diameter decreases below 0.2 mm
0.08–0.10	The primary bow is broader and paler, with only the violet vivid; there is a wide gap between the primary and the first supernumerary bow, which clearly shows white tints
0.06	The primary bow has a distinct white stripe
< 0.05	The primary bow becomes the mainly white fogbow

they fade toward the feet of the bow, an effect which has been attributed (61) to the aspherical shape of drops with radii larger than about 1 mm (noted at the end of Section 2.6). (Moreover, a shuddering of rainbows, especially of the secondary bow whose intensity is more sensitive to the presence of large drops, which has occasionally been observed to accompany thunder, has been tentatively associated with oscillations in the shape of the large drops which may be induced by intense sound waves; 61.) The properties of rainbows are related to the dominant effective drop size, as seen in Table 2.10.

The most spectacular rainbows are seen in receding intense rain showers when the sun has a low elevation. In middle latitudes the showers usually move from the west, and the rainbows are therefore best seen in the eastern sky in the late afternoon. The rain often falls from a cloud whose lower surface is steeply inclined toward the rear (Fig. 8.46), so that at the back of the shower there is a tall curtain of precipitation on which the bows can be displayed in their full vertical extent about or soon after the time when the rain ceases at the ground. The bows are best seen when this precipitation contains only rain, and little or no hail, and therefore on the flanks of intense showers or in weak showers during warm weather. It is interesting that in such weak showers the bows are sometimes clearly produced in precipitation which can be inferred from its position with respect to the cloud base level to be somewhat above the 0°C level, and therefore to contain supercooled rain. More often a bow in a shower ends abruptly beneath a brighter region, in a narrow zone which evidently marks where snow melts into rain (coincident with the radar bright band, discussed near the end of Section 9.5).

Haloes

Halo phenomena are produced by the refraction of light during its passage through ice crystals, or by reflection from their plane surfaces. They consist of arcs or patches of light, called halo *components*, in characteristic positions relative to the luminary. In sunlight some can be seen to be colored, but in moonlight the colors are faint and practically imperceptible.

Usually only one or two components are visible at the same time; several together are said to form a *composite* or complex halo. On rare occasions composite haloes have been observed to contain nearly all the components described in texts on atmospheric optics. Various crystal shapes have been postulated to account for some of the rarely seen components (33, 62), but at least the common can be explained with reference to the two principal forms of atmospheric ice crystals: the hexagonal prisms whose principal axes tend to be considerably longer or considerably shorter than the secondary axes, resulting in forms described as columns or as plates (Section 5.12). One or the other of these has been found to be the predominant form in simultaneous observations of haloes and of the particles in the responsible clouds, both at the ground in polar regions (60) and during flights in the upper troposphere over central Europe (63).

The dimensions of the crystals in ice clouds reach values which are typically much larger than those of the majority of cloud droplets; even their shortest dimension is generally in the range between 20 and 40 μ, so that $x > 100$ and the theory of haloes can be based on geometrical optics. The corresponding Reynolds number is less than 1 for the smaller crystals (Fig. 2.3), so that they may settle stably in any orientation, but the largest crystals tend to settle in a particular attitude (although probably with some tumbling or fluttering), for which the air resistance is a maximum, that is, long columns with their principal axis horizontal. However, the long columns often occur in radiating clusters (Pl. 5.1), so that in a cloud of such particles the principal axes have random orientations.

The most common halo components are the ring of the *ordinary 22° halo*, often seen complete in the extensive high clouds associated with frontal zones, and the *parhelia*, or "sun dogs," bright spots at 22° or a little more from the sun and at the same elevation. Both the ordinary halo and the parhelia are sharply bounded and colored red or red-brown on their inner borders. A parhelion may be so bright as to be mistaken for showing the position of the sun when its true position is obscured by denser cloud, and frequently is strikingly colored, the inner red part being succeeded by zones of yellow and green, and finally a barely perceptible tinge of blue.

In the production of haloes the most important paths of rays are those which enter one of the rectangular sides of a prism and leave by an alternate side, experiencing a refraction as though by a prism of angle 60°, or those which pass through a side and the flat hexagonal top or base of a prism, experiencing a refraction as though by a prism of angle 90°. Rays of the former kind which are normal to the principal axis are deviated from their original direction by an angle δ, which depends on the refractive index of ice and the angle α between this direction and the normal to the prism face upon which they are incident. The relation is shown in Table 2.11, which includes values for the deviation δ_{85} when the incident ray is at an angle of 85° to the direction of the principal axis.

From the table it is evident that if the sun is seen horizontally through a uniform cloud of ice prisms whose principal axes are vertical, as can be expected if they consist of large plates, refracted light can reach the observer only at angles of between 21.8 and 43.5° from the sun. Moreover, if the prisms have a random orientation about their principal axes, the refracted light will produce ordinary parhelia, whose brightness has maxima at angles of about 22° from the sun and decreases rapidly at angles beyond about 25°. From the last column of the table it is evident that small variations of the inclination of the principal axes to the vertical increase the deviations of the rays and diffuse the parhelia; a more quantitative analysis of the distribution of intensity in the parhelia would require a specification of the frequency distributions of the various crystal orientations and of the effectiveness of each in intercepting and refracting the sunlight.

Because of the variation of the refractive index of ice with the wavelength of light the angle of minimum deviation is 21°34' for red light and 22°22' for violet, a difference sufficiently large to allow distinct color separation, but because of the diffusion produced by slight variations in the inclination of the principal axes of the crystals only the red of the innermost border of the parhelion is likely to be consistently visible. Moreover, as the elevation of the sun increases, the angles of minimum and maximum deviation also increase, but the latter less rapidly, so that the total extent of the parhelia in azimuth decreases, from about 22° when the elevation of the sun is zero to about 15° when the elevation is 40° and to zero when the elevation is about 61° (above which there is no refraction to the observer through prisms with vertical principal axes). The most striking parhelia are seen when the sun's elevation is low, but not so much so that it has lost its brilliance. The plate crystals grow in a range of temperatures between about -10 and -25°C (Fig. 5.6), which occur at the ground in high latitudes, but which elsewhere are characteristic of the middle troposphere. Usually the crystals form in isolated droplet clouds, from which they fall in trails or are left behind in patches of limited extent. The parhelia and other half components produced by plate crystals do not last long, because the clouds soon move out of the required position, but often they are brilliant and colorful, particularly when the observer and most of the line of sight lie in the shadow of dense droplet clouds.

When the sun is seen through an extensive cloud of prisms whose principal axes are randomly inclined, the ordinary 22° ring halo is produced. It seems to be a characteristic of clouds containing small columnar crystals, or long columns in clusters, and at least a part of it is virtually always visible, generally without any other halo components, in the ice clouds of the high troposphere. It is sometimes brightest directly above and below the sun, presumably when at least a proportion of the crystals are sufficiently large to tend to settle with their principal axes horizontal. The most brilliant of these haloes, however, are seen in extensive clouds which are optically thin, show virtually no detail, and consist mainly of small, short prisms. These clouds typically spread gradually across the sky and thicken very slowly; it is often suggested that they are formed by widespread unusually slow ascent of air in the upper troposphere, but it is equally probable that they are diffuse and slowly evaporating residues of ordinary ice clouds, on the downwind fringes of slow-moving large-scale cloud systems.

A ring halo with a radius of about 46°, and faint

Table 2.11 The deviation δ, δ_{85} of rays of yellow light which are normal or at 85° to the direction of the principal axis of a prism of ice of angle 60°, and incident upon one face at the angle α (64); it is assumed that the refractive index of ice is $m = 1.31$

α (deg.)	δ (deg.)	δ_{85} (deg.)
90	43.5 (max)	44.7 (max)
80	34.8	36.3
70	28.7	31.0
60	24.8	26.8
50	22.5	24.9
41.3		24.3 (min)
40.9	21.8 (min)	
40	21.8	24.3
30	22.0	25.3
20	27.6	30.1
14.5		44.7 (max)
13.5	43.5 (max)	
< 13.5	total reflection	

parhelia at radii between about 46 and 90°, depending on the sun's elevation, are produced by the refraction of rays which pass through a side face and a face normal to the principal axis of a prism, whose angle of minimum deviation for yellow light is 45.8°. These and other halo components due to refraction are rare, and we shall here mention only two other components, the colorless under-sun and sun-pillar, both due to reflection, which are more often useful in revealing the presence of small concentrations of crystals otherwise difficult to detect.

The *under-sun* is a bright patch of light seen from an aircraft or a high place, centered about as much below the horizon as the sun is above it, produced by the reflection of light from horizontal crystal faces. It is nearly always colorless and usually elongated in the vertical because of small oscillations in the orientation of the crystals (Pl. 2.4).

The *sun-pillar* is a vertical column of light through the sun most often seen when it is at a low elevation or even below the horizon; it extends sometimes below the sun, but more often and more extensively—occasionally up to about 15°—above it. It has the same color as the sun, and evidently is also produced by reflection, probably from the side faces of columns with horizontal principal axes, but perhaps sometimes from plates; however, the typical distributions of brightness in both the under-sun and sun-pillar have not been explained satisfactorily (60). There may be some refraction associated with reflection from internal crystal faces.

The under-sun sometimes appears in very tenuous clouds of crystals which grow in the air near the ground during slow prolonged cooling, without the previous formation of a droplet fog (Section 5.12). The sun-pillar is occasionally the only evidence in observation from the ground that small concentrations of crystals have fallen from apparently stable shallow droplet clouds in the middle troposphere.

Appendix 2.1 Physical data

Table 2.12 Properties of air

T	\multicolumn{5}{c}{Air density at pressure of}	η	κ	$k(p_0)$	$v(p_0)$
	1,000	700	500	300	100				
40	1.11 0					1.91 −4	6.46 −5	2.42 −1	1.72 −1
30	1.15					1.86	6.28	2.28	1.62
20	1.19	8.32 −1	5.94 −1			1.81	6.09	2.15	1.53
10	1.23	8.61	6.15			1.77	5.92	2.02	1.44
0	1.28	8.93	6.38			1.72	5.74	1.89	1.35
−10	1.32	9.27	6.62			1.67	5.54	1.77	1.26
−20	1.38	9.63	6.88	4.13 −1		1.62	5.33	1.65	1.17
−30	1.43	1.00 0	7.16	4.30	1.43 −1	1.56	5.14		
−40	1.49	1.05	7.47	4.48	1.49	1.51	4.94		
−50			7.81	4.68	1.56	1.46	4.75		
−60				4.90	1.63	1.40	4.54		
70				5.14	1.71	1.35	4.35		
−80					1.80	1.29	4.18		

Specific heat capacity of dry air (at constant pressure) $c_p = 0.240 \text{ cal g}^{-1} \text{ °K}^{-1} = 1.005 \times 10^3 \text{ J kg}^{-1} \text{ °K}^{-1}$. More detailed and accurate specifications are given in standard tables (1, 2).

Key:

ρ_a = density of pure air at temperature T and pressure p (see Eq. 2.1)
η = dynamic viscosity
κ = thermal conductivity
$k(p_0) = \kappa/\rho_a c_p$ = thermometric conductivity at a pressure p_0 of 1000 mb
$v(p_0) = \eta/\rho_a$ = kinematic viscosity at a pressure p_0 of 1000 mb
Temperature T is in °C
Pressure p is in mb
Density ρ_a is in kg m^{-3}
Kinematic viscosity v and thermometric conductivity k are in cm^2 sec^{-1} ($= 10^{-4}$ m^2 sec^{-1})
Dynamic viscosity η is in g cm^{-1} sec^{-1} ($= 0.1$ N sec m^{-2})
Thermal conductivity κ is in cal cm^{-1} sec^{-1} °K^{-1} ($= 4.19 \times 10^2$ J m^{-1} sec^{-1} °K^{-1}).

The numbers to the right of a value give the power of 10 by which that and succeeding values have to be multiplied.

2   The physical properties of the moist atmosphere and its particles

Table 2.13 Properties of water substance (see also Appendix 2.2)

T (°C)	e_s (mb)	ρ_s (g m^{-3})	L_v (cal g^{-1})	L_s (cal g^{-1})	L_f (cal g^{-1})	$D(p_0)$ (cm^2 sec^{-1})
40	7.38	5.11 1	575			2.84 −1
30	4.24	3.04	580			2.69
20	2.34	1.73	586			2.54
10	1.15	9.40 0	592			2.39
0	6.11 0	4.85	597	677	80	2.25
−10	2.86	2.36	603	677	74	2.11
−20	1.25	1.07	609	678	69	1.98
−30	5.09 −1	4.53 −1	615	678	63	1.84
−40	1.89	1.76	622	678	56	1.71
−50	6.35 −2	6.18 −2	629	678	49	1.59
−60	1.90	1.93		677		1.47
−70	4.93 −3	5.26 −3		677		1.36
−80	1.07	1.20		676		1.24

Specific heat capacities:
of water vapor (at constant pressure):

$$c_{pv} = 0.441 \text{ cal g}^{-1}\,{}^\circ\text{K}^{-1} = 1.85 \times 10^3 \text{ J kg}^{-1}\,{}^\circ\text{K}^{-1}$$

of liquid water:

$$c = 1.000 \text{ cal g}^{-1}\,{}^\circ\text{K}^{-1} = 4.19 \times 10^3 \text{ J kg}^{-1}\,{}^\circ\text{K}^{-1}$$

of ice:

$$c_i = 0.5 \quad \text{cal g}^{-1}\,{}^\circ\text{K}^{-1} = 2.09 \times 10^3 \text{ J kg}^{-1}\,{}^\circ\text{K}^{-1}$$

More carefully defined specifications and values to more significant figures than in Table 2.13 are given in standard tables (1, 2); however, values at temperatures more than a few degrees below 0°C have not been experimentally confirmed.

Key:

- e_s = saturated vapor pressure over the liquid
- ρ_s = saturated vapor density over the liquid
- L_v = latent heat of evaporation
- L_s = latent heat of sublimation
- L_f = latent heat of fusion
- $D(p_0)$ = coefficient of diffusion of water vapor in air at pressure $p_0 = 1000$ mb; at pressure p the coefficient is $D = p_0 D(p_0)/p$

Pressure is in mb ($= 10^3$ dyn cm^{-2} $= 10^2$ N m^{-2})
Density is in g m^{-3} ($= 10^{-3}$ kg m^{-3})
Latent heats are in cal g^{-1} ($= 4.19 \times 10^7$ erg g^{-1} $= 4.19 \times 10^3$ J kg^{-1})
D is in cm^2 sec^{-1} ($= 10^{-4}$ m^2 sec^{-1})

The numbers to the right of a value give the power of 10 by which that and succeeding values have to be multiplied.

Appendix 2.2   Relations between air saturated with respect to liquid water and saturated with respect to ice

Table 2.14 Pressure versus air temperature

T (°C)	e_s (mb)	e_i (mb)	ρ_s (g m^{-3})	ρ_i (g m^{-3})	e_s/e_i (%)	e_i/e_s (%)	$\Delta\rho'$ (g m^{-3})	$(T_f - T_d)$ (°C)	$\Delta z'$ (m)
−2	5.27 0	5.17 0	4.21 0	4.13 0	102	98	0.81 −1	0.26	30
−4	4.54	4.37	3.66	3.52	104	96	1.39	0.52	65
−6	3.90	3.68	3.17	2.99	106	94	1.80	0.76	95
−8	3.34	3.09	2.74	2.53	108	93	2.05	1.00	120
−10	2.86	2.59	2.36	2.14	110	91	2.19	1.23	150
−11	2.64	2.37	2.19	1.96	111	90	2.22	1.34	165
−12	2.44	2.17	2.03	1.80	112	89	2.24	1.45	180
−13	2.25	1.98	1.88	1.65	114	88	2.23	1.55	190
−14	2.07	1.81	1.74	1.51	115	87	2.21	1.66	200
−15	1.91	1.65	1.60	1.39	116	86	2.18	1.76	215
−16	1.75	1.50	1.48	1.27	117	86	2.15	1.86	225
−18	1.48	1.24	1.26	1.06	119	84	2.03	2.05	250
−20	1.25	1.03	1.07	8.83 −1	122	82	1.90	2.24	270
−22	1.05	8.50 −1	9.09 −1	7.33	124	81	1.76	2.42	290
−25	8.06 −1	6.32	7.05	5.52	128	78	1.52	2.67	325
−30	5.09	3.80	4.53	3.38	134	75	1.15	3.05	370
−35	3.14	2.23	2.85	2.03	141	71	0.82	3.39	410
−40	1.89	1.28	1.76	1.19	147	68	0.57	3.67	440
−45	1.11	7.20 −2	1.06	6.83 −2	154	65	0.37	3.91	470
−50	6.35 −2	3.93	6.17 −2	3.82	162	62	0.24	4.10	490
−55	3.53	2.09	3.51	2.08	169	59	0.14	4.23	505
−60	1.90	1.08	1.93	1.10	176	57	0.83 −2	4.31	510
−65	9.85 −3	5.40 −3	1.03	5.63 −3	182	55	0.46	4.34	510
−70	4.92	2.61	5.25 −3	2.79	188	53	0.25	4.31	505
−75	2.35	1.22	2.57	1.33	193	52	0.12	4.20	490

Table 2.14 (*continued*)

T (°C)	e_s (mb)	e_i (mb)	ρ_s (g m^{-3})	ρ_i (g m^{-3})	e_s/e_i (%)	e_i/e_s (%)	$\Delta\rho'$ (g m^{-3})	$(T_f - T_d)$ (°C)	$\Delta z'$ (m)
−80	1.07	5.47 −4	1.20	6.14 −4	196	51	0.59 −3	4.03	470
−85	4.62 −4	2.35	5.32 −4	2.71	196	51	0.26	3.77	440
−90	1.87	9.67 −5	2.21	1.14	193	52	0.11	3.41	395
−95	7.02 −5	3.78	8.53 −5	4.60 −5	186	54	0.39 −4	2.95	340
−100	2.41	1.40	3.01	1.75	172	58	0.13	2.36	270

Key:

e_s = saturated vapor pressure over the liquid
e_i = saturated vapor pressure over ice
ρ_s = saturated vapor density over ice liquid
ρ_i = saturated vapor density over ice
$\Delta\rho' = \rho_s - \rho_i$, as a function of the temperature $T(= T_d$, the dew point) of air saturated with respect to the liquid
$(T_f - T_d)$ = difference between the frost point and dew point
$e_s/e_i, e_i/e_s = (T_f - T_d), T = T_f$, expressed as percentages
$\Delta z'$ = the distance in adiabatic ascent (without condensation) between the levels of saturation with respect to ice and with respect to liquid water, as a function of the frost point $T = T_f$ (from Eq. 3.23)

The single figure to the right of columns 1–4 and 6 is the power of 10 by which the adjacent and succeeding values have to be multiplied.

Appendix 2.3 The performance of hygrometers in soundings

The accurate observation of the moisture content of the upper troposphere and lower stratosphere, which would be a great help in the study of high clouds and more generally in the analysis of circulations in those regions, is made difficult by the very small vapor densities at the low temperatures prevalent there. For example, at a temperature of −40°C the saturated vapor density with respect to ice is 10^{-1} g m^{-3}, while at −80°C it is less than 10^{-3} g m^{-3}, corresponding to a mixing ratio at high levels of only about 10^{-6}. Consequently, under these conditions sensors, such as skin and hair, whose responses depend on the transfer to or from the atmosphere of sufficient water substance to change their physical state, are characterized by a slow response to change of humidity and are practically useless for soundings, even though their thickness may be only a few tens of microns. The only satisfactory observations at low temperatures have been made with well-ventilated frost point hygrometers, which sense extremely thin condensed films, used by aircraft during periods of level flight (29), or carried aloft by special large balloons with great precautions to minimize contamination by water vapor exhaled from the balloons or their trains (30; see also Fig. 9.30).

It is generally thought that the poor response of ordinary light hygrometers used on routine radiosondes makes them useless for identifying the detailed stratification of moisture, and so far in British practice, for example, no corrections are applied for lag, and relative humidity is not reported at all when the air temperature is below −40°C. Nevertheless, the performance of these sensors in the lower troposphere is remarkably good in most conditions, and with corrections to the observations based on improved knowledge of their properties they could probably be made satisfactory throughout the upper troposphere of middle latitudes (31).

The sensor of gold-beaters' skin which is used in the British radiosonde unfortunately exhibits hysteresis after indicating relative humidities below 30%, which often happens in the lower troposphere above the layer of ordinary convection and complicates the interpretation of subsequently observed values. Nevertheless, the skin indicates relative humidity with respect to liquid water, and although its response is poor at very high and very low relative humidities, it is comparatively good at relative humidities between about 50 and 80%, which in the high troposphere include the maximum values likely to occur (in layers saturated with respect to ice). The response to a change of relative humidity is an approximately exponential function of time, so that a lag coefficient c, approximately proportional to the saturated vapor density, can be defined as the time required for the difference between an indicated and a constant ambient relative humidity to decrease by the factor $1/e$. The value of c is less than 10^{-1} min at temperatures above 0°C, about 1 min at −30°C, and about 10 min at −50°C; considering that in a radiosonde sounding two or three observations are made each minute (corresponding to an ascent of about 600 m), corrections to the indicated values of relative humidity

2 The physical properties of the moist atmosphere and its particles

are evidently unjustified at temperatures above about $-20°C$ (i.e., in the lower troposphere) and below about -50 or $-60°C$ (depending on the accuracy of the recording). Similar values of the lag coefficient are probably applicable to the sensors used on all the routine soundings illustrated in this book, on none of which, as far as is known, were any corrections made for lag. In interpreting the humidities implied by the plotted radiosonde soundings the following must be considered:

1. Recorded values of relative humidity of less than about 20%, and plotted values of large rates of decrease of dew point, are probably underestimates.

2. The dew point hardly ever reaches the air temperature, but a moist layer often is indicated, with dew point depressions of 5°C or less, where extensive droplet clouds are known or may be inferred to have been present. The failure of the sensor to indicate saturation may be due to its poor response either at high relative humidities or (in the middle troposphere) at low temperatures. On the other hand, many clouds cover a small fraction of the sky, even though seen from the ground, especially at low elevations, they seem to be widespread.

Figure 2.12 presents two routine British soundings known to have entered ice clouds in the upper troposphere; the relative humidities are those indicated, without any corrections for hysteresis or lag, but read from the original records. The soundings show the value of even the simple skin sensor (see also Fig. 3.13).

Recently it was found that the humidity sensor in standard radiosondes in the United States (which measures the electrical conductivity of a hygroscopic film) underestimated relative humidity because of warming in sunshine and inadequate ventilation. The error, especially in the daytime and in the upper troposphere, probably exceeded 15%; it affects many of the soundings quoted in this text. An improved sensor housing introduced in 1972 has greatly reduced these errors (76).

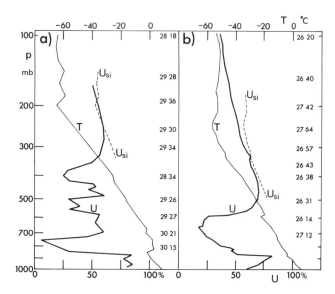

Fig. 2.12 Radiosonde soundings, from Crawley (southeastern England), at about 1200 GMT on (a) 12 November 1967 and (b) 12 March 1967. The profiles of relative humidity U and temperature T were taken from the original records, and the reported winds in tens of degrees and meters per second are entered to their right. The dashed line labeled U_{si} shows the relative humidity corresponding to saturation with respect to ice.

In (a) the maxima of relative humidity at about 650 and 470 mb probably indicate the levels of altocumulus clouds reported from most observing stations in southern England, and the layer between about 300 mb and the tropopause in which the relative humidity indicates saturation over ice is consistent with reported widespread dense high cloud. The slight supersaturation is probably an effect of the lag of the sensor and the decrease of U_{si} with height.

On the occasion of the sounding (b) the balloon entered an overcast of ice cloud in the middle troposphere (altostratus with mamma); there was no lower cloud.

Appendix 2.4 The drag coefficient of spheres and water drops

Table 2.15 The variation with the Reynolds number Re of the drag coefficient C_D of smooth rigid spheres, and of the drag coefficient C'_D of water drops (10)

Re	C_D	X	Re	C_D	X	Re	C_D	X	Re	C_D	X
0.1	244	4.64×10^{-1}	1×10^4	0.41	120	1×10^2	1.1	7.71			
0.2	125	5.92×10^{-1}	2	.44	194	2	0.78	10.9			
0.4	64.3	7.54×10^{-1}	4	.46	313	4	.59	15.8			
1	28.2	1.06	6	.46	410	6	.54	20.1			
2	15.4	1.37	1×10^5	.44	568						
4	8.7	1.80	2	.41	881				Re	C'_D	X'
6	6.4	2.13	3	.20	873						
8	5.0	2.37	4	.09	843	1×10^3	.46	26.8	1×10^3	0.51	27.7
10	4.2	2.60	6	.10	1140	2	.41	40.9	2	.53	44.5
20	2.8	3.59	1×10^6	.14	1700	3	.41	53.5	3	.66	62.8
40	1.8	4.93				3.5	.40	58.9	3.5	.75	72.6
60	1.5	6.08				4	.40	64.4			
80	1.2	6.84				6	.40	84.3			

The radius of the sphere of equal volume was used in the calculation of Re. For Re $\leq 6 \times 10^2$ there was no significant difference between C_D and C'_D. Also given are corresponding values of $X = (C_D \text{Re}^2/24)^{1/3}$.

Appendix 2.5 The paths of rays through stratified atmospheres

In Fig. 2.13 the path of a ray through successive thin atmospheric shells in which the refractive index has the values n_2, n_1, is indicated by the line ABC. At A and B the distances from the center of the earth, O, are r_2 and r_1, and the angular deviations of the ray from the vertical are ψ_2 and ψ_1. Then, referring to other angles indicated in the diagram,

$$n_2 \sin \theta = n_1 \sin \psi_1 \quad (2.59)$$

Now considering the triangle AOB

$$\sin \theta = \frac{r_2 \sin \psi_2}{r_1} \quad (2.60)$$

Thus

$$n_2 r_2 \sin \psi_2 = n_1 r_1 \sin \psi_1 \quad (2.61)$$

In general

$$nr \sin \psi = \text{const}$$

and in particular

$$nr \sin \psi = n_0 r_0 \cos \varepsilon_0 \quad (2.62)$$

where ε_0 is the elevation of the ray above the horizontal at the position of an observer or a radar on the ground. (Incidentally, if p is the normal from the center of the earth to the tangent drawn from the ray where the refractive index is n, $pn = \text{const.}$) Now

$$ds = dr/\cos \psi$$

where s is distance along the ray; thus

$$s = \int_{r_0}^{r_1} \left(\frac{1 - r_0^2 n_0^2 \cos^2 \varepsilon_0}{r^2 n^2} \right)^{-1/2} dr \quad (2.63)$$

If z is height above the level of an observer on the ground and R is the radius of the earth, $z \ll R$ in situa-

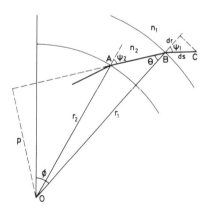

Fig. 2.13 The geometry of a ray refracted by the atmosphere, along a path AB and BC through layers of refractive index n_2 and n_1. Angular position from the center of the earth 0 is denoted by ϕ, the distance from the center of the earth by r, the local inclination from the vertical by ψ, distance along the ray by s, and the length of the line normal to the projected path and through the center of the earth by p. In a layer of infinitesimal thickness the path of the ray can be considered straight and the angle θ to be within a triangle ABO.

2 The physical properties of the moist atmosphere and its particles

tions of meteorological interest, and considering also that $n = 1 + 10^{-6}N$, where $10^{-6}N \ll 1$, Eq. 2.63 may be written to a sufficient accuracy as

$$s = \int_0^z \left\{ 1 - \cos^2 \varepsilon_0 \left[1 - \frac{2z}{R} + 2 \times 10^{-6}(N_0 - N) \right] \right\}^{-1/2} dz \qquad (2.64)$$

from which s (which is practically the distance $\int_0^s n\,ds$ measured in the detection of radar targets), z, and ε_0 may be related, given N as a function of z. Similarly,

$$\phi = \int_0^z \left[\frac{1 + 2z/R - 2 \times 10^{-6}(N_0 - N)}{\cos^2 \varepsilon_0 - 1} \right]^{-1/2} dz \qquad (2.65)$$

and the distance along a level surface through the observer to a place vertically below a point on the ray is $R\phi$.

In consequence of atmospheric refraction the visible horizon is farther than the geometrical horizon, and near the horizon the apparent elevation of the sun is greater than its geometrical elevation (normally by about 0.6° when the former is 0.0°, and by about 0.1° when it is 9°); similarly, the apparent elevation of a distant high cloud is a little greater than its geometrical elevation when it is seen near the horizon from a great distance.

Figure 2.14 shows the relation between the apparent angular elevation of a target observed from the ground, its height above the ground, and its slant range (along the path of the optical ray), for the average stratification of the atmosphere in central Sweden during several fine summer days, on which cloud heights were determined

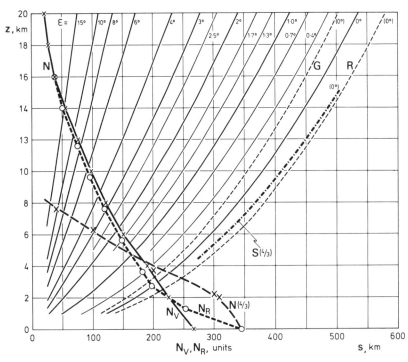

Fig. 2.14 The relation between the height z and the angular elevation ε of a cloud above an observer near sea level, and its range s along the optical path from the observer, calculated for the distribution with height of the refractive modulus shown by the line marked N_V. (This distribution is that for a wavelength $\lambda = 0.55\,\mu$ in the mean stratification during a period of measurements of cloud heights over central Sweden, specified in Fig. 7.36. Except at elevations below about 2°, the values of z and s in other ordinary stratifications are not likely to differ from those indicated by more than 1%.) The diagram can be used to infer the height of clouds whose elevation is observed visually and whose range is known approximately as that of some obviously related orographic feature.

The two dashed lines on the right represent the paths of rays with $\varepsilon = 0°$: that marked G when no atmosphere is present, and that marked R for a centimetric (radar) wavelength in a summer stratification observed over Oklahoma, in which the vertical gradient of the refractive modulus (shown by the line N_R) was in the first kilometer about three times as great as in a standard atmosphere (32).

The dot-dash line marked $S(4/3)$ shows the path of the radar ray for $\varepsilon = 0°$ when the profile of (centimetric) refractive modulus is given by the line marked $N(4/3)$; for this profile the normal assumption that radar rays can be regarded as straight over an earth of radius $4R/3$ is valid (Eq. 2.67). It can be seen that this assumption may lead to heights up to about 0.50 km too high; however, as ε is increased and the range of tropospheric targets decreases, this height error also decreases, and in most meteorological measurements it is negligible.

by triangulation with theodolites. The tolerable error with which the height of clouds is measured optically or by radar is about 1%, or 100 m at a height of 10 km. In all ordinary conditions the (slant) ranges indicated in Fig. 2.14 as functions of z and ε_0 are not likely to be in error by more than 1% until ε_0 decreases below 2° (but may sometimes reach 2% at $\varepsilon_0 = 1°$ and several percent at $\varepsilon_0 = 0°$). If d is the range on the plane tangent to the earth's surface at the position of an observer, determined in triangulation from the differing azimuths measured at the ends of a baseline, then in the absence of refraction

$$z = d\tan\varepsilon_0 + d^2/2R$$

in most ordinary conditions with optical and radar ray refraction, similar relations can be written with acceptable accuracy for ranges up to about 200 km and for values of $\varepsilon_0 \gtrsim 2°$, which effectively assume that the rays are linear and that the radius of the earth is increased by a factor f:

$$z = d\tan\varepsilon_0 + d^2/2fR, \qquad f \approx 1.15 \quad (2.66)$$

for optical wavelengths, and

$$z = s\sin\varepsilon_0 + s^2/2fR, \qquad f \approx 1.33 \quad (2.67)$$

for radar wavelengths. Equation 2.67 usually is employed to adjust the height and range scales on the RHI displays of radars.

Atmospheric inversions, in which the temperature increases with height, are characterized by rapid upward decreases of vapor pressure, and the combined effect is to produce abnormally large upward gradients of the refractive index for radio waves. The radius of curvature R' of a ray, which is given by

$$1/R' = -(dn/n\,dz)\cos\varepsilon \quad (2.68)$$

where ε is the local angle of elevation, then becomes abnormally small, and if $\varepsilon = 0$ is equal to the radius of curvature of the earth if $dN/dz = -157$ units/km. This is the minimum upward gradient under which the radio wave is said to be "trapped," and the inversion layer to form a radio *duct*; it is assumed that the conditions under which ray tracing is permissible are fulfilled, in particular that the depth of the inversion layer is much greater than the wavelength. Obviously such ducts are confined to the first few kilometers of the troposphere, and the most important are those close to the surface (usually between 100 and 300 m deep), where the rays from transmitters based at the surface may enter them with very small elevation angles. If the inversion base is at the height z, the minimum magnitude of dN/dz for duct formation is proportional to $\varepsilon_0^2/2z$. For an inversion layer 0.1 km deep the magnitude is 740 units/km for $\varepsilon_0 = 0°$ and about 1000 units/km for $\varepsilon_0 = 0.4°$ if its base is at 0.5 km, whereas these values are only 157 units/km and about 400 units/km, respectively, if its base is on the surface (65). In radiosonde data, even after correction for the lag of the humidity sensor (Appendix 2.3), a magnitude of 500 units km^{-1} is about the greatest observed.

Appendix 2.6   The scattering of starlight by turbulent layers

Familiar manifestations of the scattering of light by small-scale irregularities of refractive index are the shimmering of objects seen over strongly heated ground and the twinkling of stars, especially at low elevations. The irregularities are due to variations of temperature associated with small-scale turbulent motions, arising from the instability of sheared and perhaps some other kinds of flow. Such motions are virtually always present in the planetary boundary layer, where the shear is due to friction during flow over the earth's surface, and in the circulations of ordinary convection. They might be anticipated to be absent on many calm nights, but there seems always to be some twinkling, and examined with telescopes the scintillation is found to be little more intense by day than by night. However, the bumpy flight occasionally experienced by aircraft flying in clear air at levels well above the ground shows that turbulent motions arise also in shallow layers in the free atmosphere, especially in the sheared flows associated with strong winds in the high troposphere, and where shears are locally intensified during flow over topographical features. In contrast to the turbulent motions of the boundary layer, turbulence in the free atmosphere is thought to occur only *locally* and *intermittently*, although in the upper troposphere over land in middle latitudes aircraft experience appreciable bumpiness in clear air (fluctuations of vertical velocity exceeding about 0.5 m sec^{-1}) usually averaging a small percentage of their flying time.

The motion systems which arise in the sheared flows do not usually produce clouds (as discussed in Section 9.6), and so their forms are difficult to visualize and model. Probably they develop initially as amplifying waves with a length scale of the order of the depth of the unstable layer, typically up to about 1 km. However, as they grow they produce instability on smaller scales, so that whereas at first the motion has a definite form determined by the geometry of the large-scale flow, it soon becomes distorted and complicated. The kinetic energy is transferred rapidly (over periods of minutes) to successively smaller scales, until the motion is chaotic and the properties of the air can be given only a statistical description, and appear to be represented well by the theory of homogeneous and isotropic turbulence, in which the statistics of the properties in time (including a mean value) are independent of position, and along an axis are independent of its orientation. The properties are determined by the *rate of dissipation* of kinetic energy (into internal energy) by molecular viscosity, which occurs in the motions whose scale is less than a few centimeters. The spectrum of turbulent motions cannot persist for many minutes unless energy is continually provided by renewed development at the upper end of the spectrum of scales of motion.

Light from a star, after passing through irregularities of refractive index in the atmosphere, produces a variable pattern of illumination on the ground or the objective of a telescope. The elements of this pattern have a typical duration of only about 10^{-2} sec, but it can be demonstrated that there is normally a dominant spatial dimension present in the range between about 10 and 30 cm, and a well-defined horizontal velocity of translation which evidently corresponds to that of the wind at the height where the important turbulence occurs. It seems generally accepted that the pattern velocities, at least when the scintillation is pronounced, correlate well with strong winds at heights between about 6 and 16 km, often near the tropopause. For example, in a series of measurements of the velocities (and spatial power spectra) of the patterns made during two winter months at Philadelphia, reference to the winds observed at the two nearest routine sounding stations indicated that the height of the scattering layer varied between about 6 and 10 km (35). An analysis (36) of the observed power spectra, based on the theory of propagation of light through a layer containing small variations of refractive index associated with homogeneous isotropic turbulence, produced the striking result that the form of the spectra was consistent with a background scattering in a boundary layer extending up to about 1 km, together with a major contribution from a shallow layer (of depth about 100 m) at a height H, producing a spectral maximum (in the pattern from a star nearly overhead) at a wavelength $l = (2\pi\lambda H)^{1/2}$. The value of H corresponding to the average observed value of about 16 cm for l was therefore 8 km, in satisfactory agreement with the average height inferred from the pattern velocities. (In the observations as presented, however, there does not appear to be a close correlation between the heights inferred on individual occasions from l and from the velocity, perhaps because the observed winds were not always representative.) It is remarkable that both the dimensions and velocity of the patterns should thus confirm that the turbulence primarily responsible for the intensity of stellar scintillation is in the upper troposphere rather than in the boundary layer. In the series of observations mentioned the measurements were made during only a short period each night, but indicated one high turbulent layer on ten nights, two on seven nights, and three on one night, with velocities from the west with speeds between about 16 and 70 m sec^{-1}. It appeared that there was not always a marked shear of wind in the layer identified, and the frequency of the occurrence of turbulent layers shown by these observations, and by ordinary experience of the prevalence of stellar scintillation, is difficult to explain meteorologically. Other features of stellar scintillation, such as the occasional asymmetry of the illumination patterns, and marked variation of intensity at the same elevation but with change of azimuth, suggest that further studies with observations from a variety of places, especially if made with simultaneous soundings or thorough meteorological analyses, could provide information of great meteorological interest. (Attempts to relate the intensity of scintillation to such meteorological variables as surface pressure and relative humidity, of the kind ordinarily made, are evidently not likely to be rewarding.) That such studies have apparently not been undertaken may be due partly to the recent realization that turbulent layers can be detected and examined more directly (by both day and night) with powerful and sensitive radars. Since such radars are available in only a few places it would be desirable to have simultaneous radar and scintillation observations to improve the interpretation of scintillation alone, which could readily be made at many more places to establish a climatology of turbulence in the free atmosphere, perhaps with some application in the choice of the most favorable sites for astronomical telescopes.

References

1. List, R.J. 1958. *Smithsonian Meteorological Tables*, 6th ed., rev. Smithsonian Institution, Washington, D.C.
2. *International Meteorological Tables*. 1966. WMO 188, TP 94, World Meteorological Organization, Geneva.
3. Murray, F.W. 1967. On the computation of saturation vapour pressure, *J. Appl. Meteorol.*, **6**, 203–4.
4. The construction, properties, and uses of various hygrometers are discussed in the following references: Handbook

of Meteorological Instruments: pt. 1, *Instruments for Surface Observations*, 1965; pt. 2, *Instruments for Upper Air Observations*, 1961. H.M.S.O., London.

5. Chleck, D. 1966. Aluminium oxide hygrometer: Laboratory performance and flight results, *J. Appl. Meteorol.*, **5**, 878–86.
6. Jeans, J. 1946. *Kinetic Theory of Gases*. Cambridge University Press, Cambridge (see pp. 190, 216).
7. Fuchs, N.A. 1964. *The Mechanics of Aerosols*. Pergamon Press, London.
8. Goldsmith, P., Delafield, H.J., and Cox, L.C. 1963. Diffusiophoresis and scavenging, *Q. J. R. Meteorol. Soc.*, **89**, 43–61.
9. Magono, C. 1954. On the shape of water drops falling in stagnant air, *J. Meteorol.*, **11**, 77–79.
10. Gunn, R., and Kinzer, G.D. 1949. The terminal velocity of fall for water droplets in stagnant air, *J. Meteorol.*, **6**, 243–48; Beard, K.V., and Pruppacher, H.R. 1969. A determination of the terminal velocity and drag of small water drops by means of a wind tunnel, *J. Atmos. Sci.*, **26**, 1066–72.
11. Matthews, B.J., and Mason, B.J. 1964. Electrification produced by the rupture of large water drops in an electric field, *Q. J. R. Meteorol. Soc.*, **90**, 275–86.
12. Macklin, W.C. 1962. The density and structure of ice formed by accretion, *Q. J. R. Meteorol. Soc.*, **88**, 30–50.
13. Browning, K.A., Ludlam, F.H., and Macklin, W.C. 1963. The density and structure of hailstones, *Q. J. R. Meteorol. Soc.*, **89**, 75–84.
14. List, R. 1959. Zur Aerodynamik von Hagelkornern, *Z. angew. Math. Phys.*, **10**, 143–59.
15. Weickmann, H. 1953. Observational data on the formation of precipitation in cumulonimbus clouds, in *Thunderstorm Electricity*, H.R. Byers, ed. University of Chicago Press, Chicago, pp. 66–138.
16. Eng Young, R.G., and Browning, K.A. 1967. Wind tunnel tests of simulated spherical hailstones with variable roughness, *J. Atmos. Sci.*, **24**, 58–62.
17. Macklin, W.C., and Ludlam, F.H. 1961. The fall-speeds of hailstones, *Q. J. R. Meteorol. Soc.*, **87**, 72–81.
18. Bailey, I.H., and Macklin, W.C. 1968. The surface configuration and internal structure of artificial hailstones, *Q. J. R. Meteorol. Soc.*, **94**, 1–11.
19. Willmarth, W.W., Hawk, N.E., and Harvey, R.L. 1964. Steady and unsteady motions and wakes of freely falling discs, *Phys. Fluids*, **7**, 197–208.
20. Jayaweera, K.O.L.F., and Mason, B.J. 1965. The behaviour of freely falling cylinders and cones in a viscous fluid, *J. Fluid Mech.*, **22**, 709–20.
21. Magono, C., and Nakamura, T. 1965. Aerodynamic studies of falling snowflakes, *J. Meteorol. Soc. Japan*, ser. 2, **43**, 139–47.
22. Tennekes, H., and Woods, J.D. 1973. Coalescence in a weakly turbulent cloud, *Q. J. R. Meteorol. Soc.*, **99**, 758–63.
23. Gunn, K., and Hitschfeld, W. 1951. A laboratory investigation on the coalescence between large and small water-drops, *J. Meteorol.*, **8**, 7–16.
24. Walton, W., and Woolcock, A. 1960. The suppression of airborne dust by water spray, in *Aerodynamic Capture of Particles*. Pergamon Press, London; Starr, J.R., and Mason, B.J.; 1966. The capture of airborne particles by water drops and simulated snow crystals, *Q. J. R. Meteorol. Soc.*, **92**, 490–99.
25. Engelmann, R.A. 1965. Rain scavenging of zinc sulphide particles, *J. Atmos. Sci.*, **22**, 719–27.
26. Shafrir, U., and Neiburger, M. 1963. Collision efficiencies of two spheres falling in a viscous medium, *J. Geophys. Res.*, **68**, 4141–48; Hobbs, P.V. 1965. The aggregation of ice particles in clouds and fogs at low temperatures, *J. Atmos. Sci.*, **22**, 296–300; Smith-Johannsen, R.I. 1969. Ice crystal agglomeration: T formation, *J. Atmos. Sci.*, **26**, 532–34.
27. Woods, J.D., and Mason, B.J. 1964. Experimental determination of collection efficiencies for small water droplets in air, *Q. J. R. Meteorol. Soc.*, **90**, 373–81.
28. Macklin, W.C., and Bailey, I.H. 1968. The collection efficiency of hailstones, *Q. J. R. Meteorol. Soc.*, **94**, 393–96.
29. Helliwell, N.C., Mackenzie, J.K., and Kerley, J.J. 1957. Some further observations from aircraft of frost point and temperature up to 50,000 ft., *Q. J. R. Meteorol. Soc.*, **93**, 257–62.
30. Mastenbrook, H.J. 1964. Frost-point hygrometer measurements in the stratosphere and the problem of moisture contamination, in *Humidity and Moisture*, vol. 2. Reinhold, New York, pp. 480–85.
31. McIlveen, J.F.R., and Ludlam, F.H. 1969. The lag of the humidity sensor in the British radiosonde, *Meteorol. Mag.*, **98**, 233–46.
32. Electromagnetic wave propagation in the lower atmosphere. 1960. *Handbook of Geophysics*, rev. ed. Macmillan, New York, ch. 13.
33. Minnaert, M. 1954. *The Nature of Light and Colour in the Open Air*. Dover, New York; Pernter, J.M., and Exner, F.M. 1922. *Meteorologische Optik*. Wilhelm Braumüller, Leipzig; Humphreys, W.J. 1940. *Physics of the air*. McGraw-Hill, New York; Neuberger, H. 1951. General meteorological optics, in *Compendium of Meteorology*. American Meteorological Society, Boston, pp. 61–78.
34. Fowler, C.W., Champion, R.J.B., and Tyler, J.N. 1966. A three-cavity refractometer and associated telemetry equipment, *Radio Electronic Eng.*, **32**, 186–90.
35. Protheroe, W.M. 1964. The motion and structure of stellar shadow-band patterns, *Q. J. R. Meteorol. Soc.*, **90**, 27–42.
36. Townsend, A.A. 1965. The interpretation of stellar shadow-bands as a consequence of turbulent mixing, *Q. J. R. Meteorol. Soc.*, **91**, 1–9.
37. van de Hulst, H.C. 1957. *Light Scattering by Small Particles*. Wiley, New York.
38. Lowan, A.N. 1949. *Tables of Scattering Functions for Spherical Particles*. National Bureau of Standards, Applied Mathematics Series 4, Washington, D.C.
39. Howell, H.B. 1969. *Angular Scattering Functions for Spherical Water Droplets*. NRL Report 6955, Naval Research Laboratory, Washington, D.C. These comprehensive tables provide $i_1(\theta)$ and $i_2(\theta)$ for x at intervals of 1 between 1 and 20, of 2 between 22 and 50, and of 5 between 55 and 125, and for θ at intervals of 2° between 0 and 180°; diagrams are included based upon calculations with θ at intervals of 0.2°.
40. Bullrich, K. 1964. Scattered radiation in the atmosphere and the natural aerosol, *Adv. Geophys.*, **10**, 101–260.
41. Battan, L.J. 1959. *Radar Meteorology*. University of Chicago Press, Chicago.
42. Atlas, D. 1964. Advances in radar meteorology, *Adv. Geophys.* **10**, 318–478.
43. Kerker, M. 1969. *The Scattering of Light*. Academic Press, New York.
44. Austin, A.R.I. 1952. Wave clouds over southern England, *Weather*, **7**, 50–53; Ludlam, F.H., and Scorer, R.S. 1957.

Cloud Study. John Murray, London (see p. 68).
45. Brooks, C.F. 1925. Coronas and iridescent clouds, *Mon. Weather Rev.*, **53**, 49–58.
46. Ludlam, F.H. 1948. The forms of ice clouds, *Q. J. R. Meteorol. Soc.*, **74**, 39–56.
47. Ludlam, F.H. 1956. The forms of ice clouds: II, *Q. J. R. Meteorol. Soc.*, **82**, 257–65.
48. Ludlam, F.H. 1967. Characteristics of billow clouds and their relation to clear-air turbulence, *Q. J. R. Meteorol. Soc.*, **93**, 419–35.
49. Simpson, G.C. 1912. Coronae and iridescent clouds, *Q. J. R. Meteorol. Soc.*, **38**, 291–301.
50. Bryant, H.C., and Cox, A.J. 1966. Mie theory and the glory, *J. Opt. Soc. Am.*, **56**, 1529–32.
51. Hesstvedt, E. 1959. Mother of pearl clouds in Norway, *Geofys. Publ.* (Oslo), **20**, no. 10.
52. Störmer, C. 1948. Mother of pearl clouds, *Weather*, **3**, 13–18.
53. Benson, C.S. 1965. *Ice Fog: Low Temperature Air Pollution.* UAG R-173, University of Alaska, College; Thuman, W.C., and Robinson, E. 1954. Studies of Alaskan ice-fog particles, *J. Meteorol.*, **11**, 151–56.
54. Hallett, J., and Lewis, R.E.J. 1967. Mother-of-pearl clouds, *Weather*, **22**, 56–65.
55. Störmer, C. 1939. Observations and photographic measurements of mother-of-pearl clouds over Scandinavia 1930–1938, Part 1, *Geofys. Publ.* (Oslo), **12**, no. 11.
56. Störmer, C. 1940. Observations and photographic measurements of mother-of-pearl clouds over Scandinavia 1930–1938, Part 2, *Geofys. Publ.* (Oslo), **12**, no. 13.
57. Williams, G.C. 1950. Nacreous clouds observed in southeastern Alaska January 24, 1950, *Bull. Am. Meteorol. Soc.*, **31**, 322–23.
58. Ludlam, F.H. 1952. Orographic cirrus clouds, *Q. J. R. Meteorol. Soc.*, **78**, 554–62; Ludlam, F.H. 1952. Hill-wave cirrus, *Weather*, **7**, 300–306.
59. Querfeld, C.W. 1965. Mie atmospheric optics, *J. Opt. Soc. Am.*, **55**, 105–6.
60. Liljequist, G.H. 1956. Halo phenomena and ice crystals, Special studies, Part 2, Norwegian-British-Swedish Antarctic Expedition 1949–52, *Sci. Results*, **2** (Norsk Polar Institutet, Oslo).
61. Volz, F.E. 1960. Some aspects of the optics of the rainbow and the physics of rain, in *Physics of Precipitation.* Geophysics Monograph no. 5, American Geophysical Union, Washington, D.C., pp. 280–86.
62. Wegener, A. 1926. Optik der atmosphäre, in *Lehrbuch der Physik*, 2d ed., vol. 5 (1st half). Fr. Vieweg u. Sohn, Braunschweig.
63. Weickmann, H.K. 1945. Formen und Bildung der atmospärischer Eiskristalle, *Beitr. Phys. freien Atmos.*, **28**, 12–52; Weickmann, H.K. 1949. *Die Eisphase in der Atmosphäre.* Berlin Wetterdienst, no. 6.
64. Dietze, G. 1957. *Einführung in die Optik der Atmosphäre*, Akad. Verlags., Geest u. Portig K.-G., Leipzig.
65. Bean, B.R., and Dutton, E.J. 1968. *Radio Meteorology.* Dover, New York.
66. Ludlam, F.H., and Scorer, R.S. 1957. *Cloud Study.* John Murray, London.
67. Scorer, R.S., and Wexler, H. 1963. *A Colour Guide to Clouds.* Pergamon Press, London.
68. Scorer, R.S., and Wexler, H. 1967. *Cloud Studies in Colour.* Pergamon Press, London.
69. Bull, G.A. 1961. Reflection rainbow, *Weather*, **16**, 267.
70. Tyldesley, J.B. 1967. A thunderstorm in Morocco, *Weather*, **22**, 317–18.
71. *Weather.* 1964. **19**, plate facing p. 48.
72. Livingston, W.C. 1969. Aircraft and cloud formation, *Weather*, **24**, 56–59.
73. Sun pillar and arc of contact. 1960. *Weather*, **15**, 406–7.
74. Queney, P., et al. 1960. *The Airflow over Mountains.* Tech. Note 34, World Meteorological Organisation, Geneva.
75. Scorer, R. 1972. *Clouds of the World.* David and Charles, Newton Abbot (Great Britain).
76. Friedman, M. 1972. A new radiosonde case: the problem and the solution, *Bull. Am. Meteorol. Soc.*, **53**, 884–87.

Illustrations of optical phenomena. The periodical *Weather* is well known for its abundant series of illustrations, some in color, of clouds and optical and other meteorological phenomena. Illustrations of some of the optical phenomena mentioned in this chapter have been published (with variable success) in recent years in reference 75 and the following references: The corona: 54, 64; The glory: 66, 67, 68; Iridescence: 54, 68; The rainbow: 69, 70; Nacreous clouds: 52, 54, 66, 68, 74; Haloes: The 22°-halo: 64, 67, 68, 71; The parhelion: 68; The 46°-halo: 72; The sun-pillar: 64, 67, 68, 73; The circumzenithal arc: 67, 68.

3 Thermodynamic reference processes

Changes in the physical state of air moving through atmospheric motion systems are associated with radiative exchange, expansion or compression (predominantly during vertical displacement), phase changes of water, and diffusion (involving motions of scale too small to be defined by the available observations, with results ascribed to "mixing"). The radiative energy exchanges are difficult to evaluate accurately, being subject to complicated laws (1) and sensitive to the distribution of cloud which is usually inadequately observed, but they will not enter most of our discussions, because air passes through the cloudy parts of motion systems in periods of about a day or less, during which they have only subsidiary effects. The other processes are studied with reference to the ideal processes of adiabatic (strictly, isentropic) expansion or compression, and isobaric mixing.

Changes of state in real processes are greatly complicated by the presence of variable amounts of water. Generally, however, the air can be regarded as dry unless the water changes phase. This is because even at the higher atmospheric temperatures the proportion r of water present is barely 1%; during the passage of air through motion systems the energy changes associated with terms such as $rc_{pv}\Delta T$ or $r_c c \Delta T$ (where r_c and c are the mixing ratio and specific heat capacity of condensed water, respectively) are very small compared with $c_p \Delta T$ and $L_v \Delta r_c$. It is desirable to have the simplest possible reference processes, in which the changes of state are independent of the variables r or r_c; they are therefore defined for perfectly dry air or for saturated air in which r_c is considered to remain zero, so that the only independent variables of state are p and T. For most diagnosis and prediction these reference processes are sufficiently realistic, but of course it is often necessary to consider more complicated processes, regarded as compounded of successive small ideal adiabatic expansions and isobaric mixings.

3.1 Isobaric mixing of moist air masses

The mixing at constant pressure of two air masses of temperatures (T_1, T_2) and vapor pressures (e_1, e_2), together regarded as a closed system, can be considered with reference to Fig. 3.1, in which the abscissa is temperature and the ordinate vapor pressure. Here the conditions in the two masses (m_1, m_2) are represented by the points A and B. If the masses of vapor and the specific humidities are v_1, v_2 and q_1, q_2, respectively, and if the conditions in the mixture are written without subscripts,

$$q_1 = \frac{v_1}{m_1} \approx \frac{\varepsilon e_1}{p} \quad \text{and} \quad q_2 = \frac{v_2}{m_2} \approx \frac{\varepsilon e_2}{p}$$

then

$$q = \frac{v_1 + v_2}{m_1 + m_2} \approx \frac{\varepsilon(m_1 e_1 + m_2 e_2)}{p(m_1 + m_2)} \approx \frac{\varepsilon e}{p}$$

where p is the constant total pressure, used instead of terms like $[p - (1 - \varepsilon)e]$ to obtain the approximations. These may produce errors of up to 2% in calculated

3 Thermodynamic reference processes

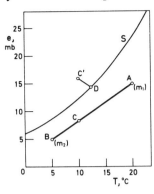

Fig. 3.1 Effect of isobaric mixing of masses m_1, m_2 of moist air, represented on a diagram of vapor pressure against temperature by the points A, B. The condition of the mixture is represented by the point C, where $AC/CB = m_2/m_1$ (in this example $m_2/m_1 = 2/1$). If the point C should lie above the saturated vapor pressure curve S, as at C', it is presumed that condensation and warming occur, so that the final state is represented by the point D on the curve S. In this example the atmospheric pressure is taken as 1000 mb, for which the slope of the line $C'D$ is 1.53 mb °C^{-1}.

changes of vapor pressure in conditions encountered near the surface, and much smaller and quite insignificant errors at low temperatures. Thus sufficiently accurately

$$e = \frac{m_1 e_1 + m_2 e_2}{m_1 + m_2} \quad (3.1)$$

Since the transformation of the whole system is adiabatic and isobaric, enthalpy is conserved,

$$c_p m_1 (1 - q_1)(T_1 - T) + c'_p m_1 q_1 (T_1 - T)$$
$$= c_p m_2 (1 - q_2)(T - T_2) + c'_p m_2 q_2 (T - T_2)$$

where c'_p is the specific heat at constant pressure of water vapor. To about the same degree of approximation as that used above this reduces simply to

$$T = \frac{m_1 T_1 + m_2 T_2}{m_1 + m_2} \quad (3.2)$$

It follows from Eq. 3.1 and 3.2 that the temperature and vapor pressure in the mixture are represented in Fig. 3.1 by a point C on the straight line AB for which $AC/CB = m_2/m_1$. A similar construction can be made to a similar degree of accuracy on a diagram of q or r against T.

Clearly for some initial conditions the point C may lie in a position C' above the curve showing the saturated vapor pressure; in the presence of suitable surfaces condensation can be presumed to reduce the vapor pressure and warm the mixture until equilibrium is attained at a point D on the curve. Since the heat capacity of the water substance is small by comparison with that of the air, and if Δe and ΔT are the adjustments to the vapor pressure and the temperature,

$$c_p \Delta T \approx L_v \Delta e \varepsilon / p \quad (3.3)$$

giving the slope of the line $C'D$ and allowing the position of the point D to be found graphically.

3.2 Production of clouds by the isobaric mixing of unsaturated air masses

The isobaric mixing of moist air masses is thus a process which can cause cloud formation. However, given the moderate temperature differences of up to a few degrees Celsius which occur naturally over small distances it can be effective only if the warmer or both air masses have already been brought close to saturation by some other process, so that generally it can play only a very minor part in the formation of natural clouds. Nevertheless, it can be regarded as responsible for the shallow *steam fogs* which sometimes form over wet ground, lakes and rivers, or the sea (*sea smokes*) (Pl. 3.1, 3.2; 2). In these fogs the temperature of the air about 1 m above the surface is found to be at least several and occasionally more than 10°C lower than the surface temperature. Such large differences arise in winter when air chilled over cold ground reaches unfrozen inland and coastal waters, and less frequently over the open ocean in strong winds toward warmer water.

More familiar examples of condensation by mixing are artificial clouds formed on cold mornings in the breath and in the plumes rising from cooling towers, and those formed at high levels in the exhaust of aircraft (condensation trails or contrails; Pl. 3.3). The formation of contrails has been studied in some detail (3) because of their importance in betraying the presence and position of aircraft in military operations.

Nearly all of the water condensed in the contrail is produced by the combustion of the aircraft fuel, so that the condition for its formation depends primarily upon the kind of fuel and the atmospheric temperature, and hardly at all on the atmospheric humidity. This can be illustrated with reference to Fig. 3.2, which contains a curve showing the saturated vapor pressure over liquid water and a point A representing a typical condition in the jet engine exhaust as it leaves the tailpipe, that is, a temperature of about 600°K and a vapor pressure of a few millibars. (The vapor pressure, e_j, is approximately pr_j/ε and can be inferred from the information that the

Production of clouds by the isobaric mixing of unsaturated air masses 3.2

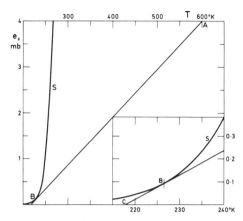

Fig. 3.2 Formation of condensation trails by jet aircraft. The curve S shows the saturation vapor pressure over liquid water and the point A represents the conditions of temperature and vapor pressure typical at the exhaust pipe of the jet engine (the exhaust is assumed to be at the atmospheric pressure of 200 mb). Condensation can occur in the diluted exhaust if the atmosphere is saturated with a temperature less than that at the point B where the straight line AB is a tangent to the curve S, and if the atmosphere is perfectly dry with a temperature less than that at the point C where the line AB meets the axis of the diagram. The inset diagram shows the line and curve near the point B on expanded scales.

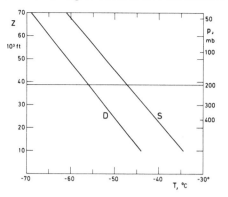

Fig. 3.3 Maximum atmospheric temperature T for condensation trail formation behind jet aircraft using JP-4 fuel, as a function of altimeter height Z (thousands of ft) or pressure p (mb), in an atmosphere which may be perfectly dry (curve D), or saturated with respect to liquid water (curve S) (3).

efficient combustion of 1 g of engine fuel produces about 1.35 g of water vapor and 10,000 cal of heat. Accordingly, the increase ΔT in the temperature of the air passing through the engine is related to the increase r_j in the mixing ratio of the vapor present by the relation $r_j = 0.24 \times 1.35 \Delta T/10^4$; for the temperature rise from about 220 to 600°K, r_j is about 0.123 g kg^{-1}, corresponding to values for e_j of about 4 mb at 200 mb and 20 mb at the ground.) Evidently a condition for a cloud to form by mixing between the exhaust and the atmosphere is that the air temperature should be below that at point B on the saturated vapor pressure curve where the straight line AB is a tangent. (It can be shown experimentally in the laboratory that contrail formation occurs when the saturation vapor pressure over *liquid* water is exceeded, even at very low temperatures. It has also been observed that atmospheric contrails contain liquid water at distances up to about 100 m behind the aircraft, for towed test bodies become rimed; 12.) Figure 3.2 shows that for an aircraft flying at the 200-mb level this limiting temperature is about $-47°C$, corresponding to a saturated atmosphere; clearly the limiting temperature in a perfectly dry atmosphere, given by the intersection of the line AB with the horizontal axis, is several degrees lower. Practically the same result is given by more elaborate analyses (2), as seen in Fig. 3.3.

As the atmospheric pressure and therefore e_j is increased, the position of the point A on Fig. 3.2 rises and the limiting temperature for contrail formation increases slightly. Although normally conditions favorable for contrail formation are found only at high levels, they may arise at the surface in cold winter climates. Thus in Fig. 3.2 the limiting temperature at a pressure of 1000 mb becomes about $-34°C$, in agreement with experience that contrails form at temperatures below about $-30°C$ on airfields in Alaska (5). Moreover, since most domestic and industrial fuels produce water vapor and heat in about the proportions assumed in Fig. 3.2, at such low temperatures unwelcome fogs form over and downwind of towns (6; see Fig. 3.4).

The persistence of contrails depends mainly on the atmospheric humidity and the rate at which the exhaust trail is diffused. Although it seems that contrails form by condensation into the liquid phase, the particles are crystalline except perhaps very close to the aircraft; the contrails become very long when the general humidity is close to that representing saturation over ice, and occasionally when there is supersaturation even in the absence of natural clouds contrails may thicken and expand by the condensation of atmospheric vapor.

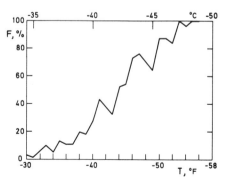

Fig. 3.4 Percentage frequency F of reports of fog (visibility less than 0.5 mi) as a function of temperature T, from observations made at Fairbanks, Alaska, during a period of five years (6).

3 Thermodynamic reference processes

3.3 Dissipation of cloud by isobaric mixing with clear air

Frequently clouds are evaporated by small-scale motions which stir them into surrounding clear air, and thus by a practically isobaric mixing process. If the cloudy and clear air have initially the same temperature, the mixture, following the evaporation of some condensed water, is always colder. Some clouds, however, are at first warmer than their surroundings, and it is interesting to consider the temperature at their edges, where the mixing has produced just-saturated air containing no condensed water.

The process can again by examined by the same approximate constructions used on Fig. 3.1, as in Fig. 3.5, in which specific humidity q has replaced vapor pressure e as ordinate.

In this diagram the conditions in the cloudy and clear air are represented by the points C and A, respectively. Mixing of cloudy air with increasing proportions of clear air determines points B which lie progressively nearer A, and points M on the curve S showing saturation, representing cloudy mixtures with diminishing amounts of condensed water. Eventually, for some proportion of clear air, at some point M_3 the mixture is saturated but no condensed water remains; thereafter, as the proportion is further increased, the point M leaves the saturation curve and moves directly toward the point A.

Clearly for some given initial conditions the temperature of the mixture at the stage when all the condensed water has evaporated is just equal to the initial clear air temperature. By making the constructions shown in Fig. 3.6, and using the properties of the similar triangles formed, it may readily be shown that for this final state the initial temperature-excess ΔT (°C) and specific humidity q'_0 of the condensed water obey the relation

$$\Delta T = q'_0(L_v/c_p) \qquad (3.4)$$
$$\approx 2.5 Q'_0$$

where Q'_0 is the condensed water concentration (g kg^{-1}), irrespective of the clear air temperature or relative humidity (i.e., independently of the height of the point A on any isotherm). The admixture of further clear air subsequently leaves the temperature unchanged. The air at and for some distance outside the edge of a cloud at which mixing occurs with unsaturated air can therefore be anticipated to be colder or warmer than the undisturbed clear air according as ΔT is less or greater than $\Delta T_0 = q'_0(L_v/c_p)$.

An expression for the result of mixing clear into cloudy air in a proportion insufficient to evaporate all the condensed water can be obtained if it is assumed that the temperatures in the cloudy air and the mixture differ by only a small amount, say 2°C or less, from that in the clear air, so that to a reasonable approximation the differences in the corresponding saturation specific humidities are given by the Clausius-Clapeyron equation in its finite difference form (Eq. 2.6).

The conditions in the air masses before and after

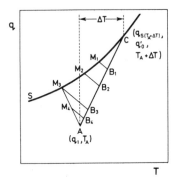

Fig. 3.5 Result of mixing clear air of temperature T_A and specific humidity q_1 (point A) with cloudy air of temperature $(T_A + \Delta T)$, containing q_0 g of condensed water in unit mass (point C), represented on a diagram of specific humidity q against temperature T. As the proportion of clear air in the mixture is increased more and more condensed water evaporates, but at first the mixture remains saturated and represented by points $M_1, M_2, \ldots$, on the curve S showing the saturation specific humidity, determined by the intersection of lines B_1M_1, B_2M_2 drawn through points $B_1, B_2, \ldots$, with the slope $(-c_p/L_v)$. Eventually the condensed water is completely evaporated and the condition of the mixture is represented by the point M_3; subsequently the mixture becomes unsaturated with a condition represented by a point such as M_4 on the straight line M_3A.

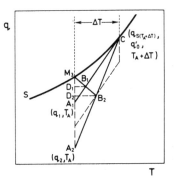

Fig. 3.6 The relation between the temperature-excess ΔT and content q'_0 of condensed water (g/g), when mixture with clear air of arbitrary temperature T_A and specific humidity q_A produces a saturated mixture at the same temperature, with no condensed water, represented by the point M_3. Two initial specific humidities $q_a = q_1, q_2$ are arbitrarily chosen for the clear air, corresponding to the two points A_1 and A_2, so that the proportions of clear air in the mixture are, respectively, CB_1/A_1B_1 and CB_2/A_2B_2 (the straight line $M_3B_1B_2$ has the slope $-c_p/L_v$). By drawing the dashed lines, including one through B_2 parallel to CA_1, it may be shown from the similarity of triangles present that $q'_0 = (CA_1/B_1A_1)M_3D_1 = (CA_2/B_2A_2)M_3D_2$, and therefore that $\Delta T = (L/c_p)q'_0$ and is independent of T_A, q_A.

mixing are written as follows:

	cloudy air	clear air	mixture
mass	1	m	$(1+m)$
temperature	$T + \Delta T$	T	$T + \Delta T'$
specific humidity:			
vapor	$q_{S(T+\Delta T)}$	uq_{ST}	$q_{S(T+\Delta T')}$
condensed water	$q_{ST} \cdot w_0$	0	$q_{ST} \cdot w_1/(1+m)$

Then
$$q_{S(T+\Delta T)} = q_{ST}(1 + a\Delta T)$$
$$q_{S(T+\Delta T')} = q_{ST}(1 + a\Delta T')$$
where
$$a = L_V/R_V T^2$$

The conservation of water substance then gives
$$w_0 q_{ST} + q_{ST}(1 + a\Delta T) + m u q_{ST}$$
$$= (1+m)q_{ST}(1 + a\Delta T') + w_1 q_{ST}$$
or
$$(w_0 - w_1) + a\Delta T = a(1+m)\Delta T' + m(1-u) \quad (3.5)$$

Regarding the specific heat of the various air masses as given sufficiently accurately by that of dry air, c_P, the conservation of enthalpy gives
$$c_p \Delta T + c w_0 q_{ST} \Delta T = c_p(1+m)\Delta T' + c w_1 q_{ST} \Delta T'$$
$$+ L_v q_{ST}(w_0 - w_1)$$
or, since $L_v \gg c \Delta T$, $L_v \gg c \Delta T'$,
$$\Delta T' = \frac{\Delta T - b(w_0 - w_1)}{1 + m} \quad (3.6)$$

where $b = q_{ST}(L_v/c_p)$. Substitution into Eq. 3.5 gives
$$w_1 = w_0 - \frac{m(1-u)}{1 + ab} \quad (3.7)$$
whence
$$\Delta T' = \frac{\Delta T - bm(1-u)/(1+ab)}{1+m} \quad (3.8)$$

Note that from Eq. 3.6, if $\Delta T' = w_1 = 0$, then
$$\Delta T = bw_0 = q'_0(L_v/c_p)$$
which is the relation previously obtained graphically (Eq. 3.4), and that from Eq. 3.7, if $w_1 = 0$,
$$m = \frac{w_0(1 + ab)}{1 - u}$$
$$= \frac{q'_0(1 + ab)}{q_{ST}(1 - u)} \quad (3.9)$$

That the complete evaporation of the condensed water requires the admixture of clear air in a proportion which is independent of the temperature excess of the cloudy air is a consequence of the assumed linearity of the relation between the saturation specific humidity and the temperature. This is readily confirmed graphically in Fig. 3.7, where it is apparent as a consequence of the parallelism of the saturation curve and the line through the points $B_1, B_2, \ldots$, which divide in constant proportion lines joining the points representing the states of the clear air A and of cloudy air with various temperature-excesses $C_1, C_2, \ldots$.

From Eq. 3.9 it follows that the proportion of clear air which must be mixed into cloudy air in order to evaporate all the condensed water increases as the temperature falls. If the relative humidity u of the clear air is 0.5 (50%) and if the concentration of condensed water is 1 g kg^{-1} of cloudy air, then m takes the following approximate values in conditions typical of various tropospheric levels:

low troposphere (800 mb, 10°C): $m \approx 0.5$
middle troposphere (600 mb, −5°C): $m \approx 1$
high troposphere (300 mb, −40°C): $m \approx 6$

The tops of tall clouds which reach into the high troposphere frequently contain as high a concentration of condensed water in the form of frozen particles too small to have fallen out. Particularly because the surrounding air there often has a high relative humidity, approximating to saturation with respect to ice, such clouds do not readily evaporate, and typically spread into anvil-like extensions tens or even hundreds of kilometers across.

Equation 3.4 provides a criterion for determining

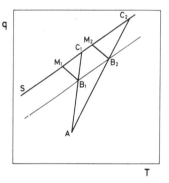

Fig. 3.7 If a segment of the saturation curve S is represented by a straight line (as when the finite difference form of the Clausius-Clapeyron equation is used), then the complete evaporation of the condensed water in cloudy air requires the admixture of a proportion of clear air which is independent of the temperature excess of the cloud air (the mixture is supposed saturated).

This is apparent from the diagram, for since the line through the points B_1, B_2—corresponding to the mixture in constant proportion, without any evaporation, of the clear air A and cloudy air masses of differing temperature excess C_1, C_2—is parallel to the line S, the paths B_1M_1, B_2M_2 described during subsequent evaporation have equal length, representing the evaporation of the same amount of condensed water.

3 Thermodynamic reference processes

whether a cloud mixed with clear air to the point of complete evaporation has a temperature above or below that of the clear air; it is interesting now to compare the densities, or virtual temperatures, rather than the actual temperatures.

From the approximate relation (2.11), the virtual temperatures of the unaffected clear air and of the saturated mixture in which no condensed water remains are, respectively, $(T + uQ_{ST}/6)$ and $[T + \Delta T' + Q_{ST}(1 + a\Delta T')/6]$. If these are equal,

$$uQ_{ST} = 6\Delta T' + Q_{ST}(1 + a\Delta T') \qquad (3.10)$$

Now if $w_1 = 0$, from Eq. 3.6 and 3.7,

$$\Delta T' = \frac{\Delta T - bw_0}{1 + m}$$

and

$$m = \frac{w_0(1 + ab)}{1 - u}$$

Thus

$$\Delta T' = \frac{(1 - u)(\Delta T - bw_0)}{1 - u + w_0(1 + ab)}$$

Substitution into Eq. 3.10 gives

$$\Delta T = \left(\frac{10^{-3}L_v}{c_p}\right) Q'_0$$
$$- \frac{1 - u + (1 + ab)Q'_0/Q_{ST}}{a + 6/Q_{ST}} \qquad (3.11)$$

Table 3.1 Relation between the temperature-excess ΔT and condensed water concentration Q'_0 of cloudy air, if complete evaporation of condensed water during mixture with clear air of relative humidity u produces a saturated mixture with the same density as that of the clear air

Q'_0 (g kg^{-1})	u	ΔT at 900 mb, 17°C (°C)	ΔT at 600 mb, −10°C (°C)
1	0.2	0.4	2.0
1	0.5	1.0	2.1
1	0.9	1.8	2.3
2	0.2	2.4	4.3
2	0.5	3.0	4.5
2	0.9	3.8	4.6

At low temperatures, when the virtual temperature increment is negligible, this expression becomes equivalent to (3.4), but at the higher temperatures found in the lower troposphere it appreciably reduces the value of ΔT corresponding to a given cloud water concentration, as shown in Table 3.1.

If for a given concentration of condensed water the temperature excess of cloudy air is less than that calculated above, then sufficient mixing with clear air to evaporate all the condensed water produces a mixture denser than the clear air. It will appear later that this is the usual circumstance at the edges of natural clouds; the induced sinking motions hasten evaporation and accentuate the sharpness of the edges.

3.4 Thermodynamic wet-bulb temperature

In another isobaric reference process air is cooled and eventually saturated by the introduction and evaporation of liquid water. The lowest temperature to which air can be cooled in this way (at which it becomes saturated) is most conveniently defined if the liquid is introduced with this temperature, and is known as the *thermodynamic wet-bulb temperature* T_{wt}; evidently it does not change during the course of the evaporation (as later illustrated, it is not quite the same as the pyschrometric wet-bulb temperature discussed in Section 2.5).

The recommended definition (7, p. 349) expresses the conservation of enthalpy during the process, and to a good approximation for most purposes can be written (8)

$$(c_p + rc_{pv})(T - T_{wt}) = L_{T_{wt}}(r_s - r) \qquad (3.12)$$

where T, r = temperature and mixing ratio of the unsaturated air, respectively
T_{wt}, r_s = temperature and mixing ratio of the saturated air, respectively

Some values of T_{wt} calculated from this expression are included in Table 3.2 for comparison with some other kinds of wet-bulb temperature.

3.5 Change of state of unsaturated air during expansion and compression: the dry adiabatic reference process, and potential temperature θ

The isobaric reference processes are useful in considering changes of state which accompany motions of small scale at a virtually constant pressure or height. Other kinds of reference process are needed in the discussion of changes of state accompanying motions over a considerable range of pressure or height, of which the most convenient are based on adiabatic (strictly, isentropic) compression or expansion.

During such compression of dry air, the conservation of energy requires that the work $-p\,dv$ done on the air is equal to the increase $c_v\,dT$ in its internal energy:

$$c_v\,dT + p\,dv = 0 \qquad (3.13)$$

where v is the specific volume. Combined with the equation of state, this gives

$$c_p\,dT - RT\,dp/p = 0 \qquad (3.14)$$

and the relations

$$pv^\gamma = \text{const} \qquad (3.15)$$

where γ is the ratio $c_p/c_v = 1.40$ of the specific heats at constant pressure and constant volume, and

$$T_1/T = (p/p_1)^\kappa \qquad (3.16)$$

where T_1, p_1 define some initial state and $\kappa = R/c_p = 2/7$.

The last relation is used to define the *potential temperature* θ which results from adiabatic compression of dry air of temperature T and pressure p mb to the standard pressure of 1000 mb:

$$\theta = T(1000/p)^\kappa \qquad (3.17)$$

In the atmosphere the principal changes of pressure experienced by air arise from vertical displacement. If it is assumed in the reference process that the change of pressure in air displaced vertically through an undisturbed atmosphere is the hydrostatic variation in the surroundings, we have

$$dp = -g\rho'\,dz = -gp\,dz/RT' \qquad (3.18)$$

where ρ' and T' are, respectively, the density and temperature in the surroundings.

Then from Eq. 3.14,

$$dT/dz = -(g/c_p)(T/T') \qquad (3.19)$$

and if $T = T'$,

$$dT/dz = -g/c_p \qquad (3.20)$$

a decrease of temperature at the rate of 9.8°C km^{-1}, known as the *dry adiabatic lapse rate*. This can be regarded as the lapse rate in a dry atmosphere of uniform potential temperature, or one in which air moved up and down adiabatically always has the temperature of its surroundings.

A similar reference process could be defined in which the air and its surroundings contain water vapor, requiring the replacement of c_p and R by $(c_p + rc_{pv})$ and $R(1 + r/\varepsilon)/(1 + r)$. The temperature changes accompanying expansion or ascent then obtained differ from those in the dry adiabatic process by at most about 2%.

3.6 The lapse rate of the dew point; the saturation level

Equation 2.3 written as

$$de/e = L_v\,dT_d/R_v T_d^2 \qquad (3.21)$$

can be regarded as a relation between the vapor pressure e and the dew point T_d, since T_d is the temperature at which e is the saturated vapor pressure.

In a vertical displacement of an element of moist air

$$de/e = dp/p$$

and if its pressure is given by that in the undisturbed surroundings

$$dp/p \approx -g\,dz/RT$$

whence to a good approximation

$$L_v\,dT_d/R_v T_d^2 = -g\,dz/RT$$

or

$$dT_d/dz = -gR_v T_d^2/L_v RT$$
$$= -gT_d^2/L_v \varepsilon T \qquad (3.22)$$

or

$$d(T - T_d)/dz = -(g/c_p - gT_d^2/L_v \varepsilon T) \qquad (3.23)$$

The rate of change of the depression of the dew point in moist air ascending adiabatically is therefore a slowly varying function practically of the temperature only. For $T = T_d$, $dz/d(T - T_d)$ is 0.127, 0.124, and 0.121 km °C^{-1}, respectively, at 30, 0, and -30°C. Thus to a good approximation an adiabatic *ascent* of h km will bring to its *saturation level* air in which the depression of the dew point is ΔT_d°C, where

$$h = \Delta T_d/8 \qquad (3.24)$$

That saturation should be approached during adiabatic *ascent* rather than descent evidently depends on the inequality $L_v > Tc_p/\varepsilon$. During the ascent the vapor pressure diminishes, but in consequence of the fall of temperature the saturation vapor pressure diminishes about five times as rapidly, and the air eventually becomes saturated.

3 Thermodynamic reference processes

3.7 Change of state of saturated air during expansion and compression: saturated reference processes, and saturation potential temperature θ_s

In moist air which has risen above its saturation level, or in which the pressure is below that at which the dry adiabatic expansion produces saturation, another reference process is chosen, taking into account the latent heat of condensation or evaporation involved in a change of pressure. In the *saturated adiabatic process* the air is considered to be always just saturated, and to contain liquid water in a concentration corresponding to the difference between the existing saturation mixing ratio and that occurring when the air first became saturated. This process has the advantage of being thermodynamically reversible, and the disadvantage that this initial state becomes another variable. To avoid this difficulty the more convenient *pseudo-adiabatic* process is specified, in which condensate is removed, with its heat content, or (during compression) sufficient liquid is introduced at the air temperature and evaporated to maintain a state of saturation.

An approximate expression for either process, neglecting terms containing the mixing ratios of the water vapor or condensate except that including the latent heat, is

$$c_p dT + L_v dr_s - RT dp/p = 0 \quad (3.25)$$

Assuming that the change of pressure with height is the hydrostatic change in an atmosphere with the same temperature distribution, that $r_s = \varepsilon e_s/p$, and using the Clausius-Clapeyron relation to express dr_s in terms of the other variables, Eq. 3.25 can be transformed into

$$-dT/dz = \frac{(g/c_p)(1 + ar_s)}{1 + abr_s} \quad (3.26)$$

where

$$a = \frac{L_v}{RT} \quad \text{and} \quad b = \frac{\varepsilon L_v}{c_p T}$$

This is an approximate expression for the *saturated adiabatic* or the *saturated pseudo-adiabatic lapse rate*.

The evaluation of either process is complicated by the variation of the specific heat of moist air with pressure and temperature, and of the latent heat with temperature. In practice the processes are defined by the following relations:

For the saturated *adiabatic* with respect to liquid water or ice:

$$\frac{(c_p + rc)dT}{T} + d\left(\frac{r_s L_v}{T}\right) - \frac{R d(p - e_s)}{(p - e_s)} \quad (3.27)$$
$$= 0$$
$$r_c = r - r_s$$

where r_s may be specified with respect to liquid water

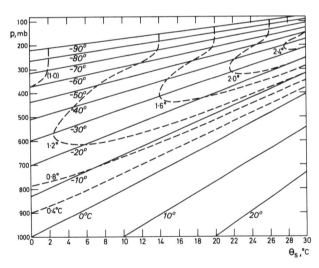

Fig. 3.8 The state of air in the *standard pseudo-adiabatic process* (solid line with respect to liquid water) as a function of pressure p (mb) and the temperature at 1000 mb (i.e., the saturation potential temperature θ_s, in °C). Over the range of conditions shown, the isotherms in this diagram are very nearly straight lines. The dashed lines show the temperature increase (°C) when saturation with respect to ice is assumed at temperatures below 0°C (according to reference 9 at temperatures down to -50°C and thereafter obtained approximately by extrapolation).

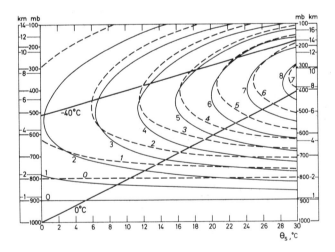

Fig. 3.9 The amount of vapor removed from saturated air after pseudo-adiabatic expansion to a pressure p mb, expressed as a concentration (g m^{-3}) at the final pressure and temperature, as a function of θ_s and for initial pressures of 900 mb (solid lines) and 800 mb (dashed lines). This amount is approximately that condensed during adiabatic expansion. The diagram contains two isotherms, and at each side approximate height scales for expansion by ascent in atmospheres of uniform θ_s of 0 and 30°C.

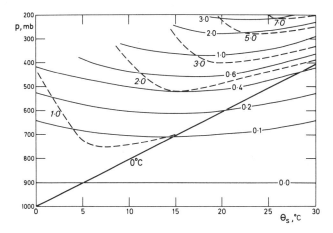

Fig. 3.10 The amount (°C) by which the temperature in the standard pseudo-adiabatic process is exceeded in the (liquid-)water-saturation adiabatic process (solid lines) and in the ice-saturation adiabatic process (dashed lines) as a function of pressure p and saturation potential temperature θ_s, for the same initial state at 900 mb (in the last process freezing of condensed water occurs at 0°C) (9).

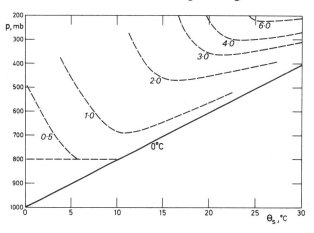

Fig. 3.11 The amount (°C) by which the temperature in the standard pseudo-adiabatic process is exceeded in the ice-saturation adiabatic process (dashed lines) as a function of the pressure p and saturation potential temperature θ_s, for the same initial state at 800 mb (in the latter process freezing of condensed water occurs at 0°C) (9).

or to ice, c is the specific heat of the condensed liquid or ice, and r, the total mixing ratio of vapor and condensed phase, has to be specified and is constant.

For the saturated *pseudo-adiabatic* with respect to liquid water or to ice:

$$\frac{(c_p + r_s c)dT}{T} + d\left(\frac{r_s L_v}{T}\right) - \frac{R d(p - e_s)}{(p - e_s)} = 0 \qquad (3.28)$$

$$r_c = 0$$

with values of r_s, c, L_v, and e_s appropriate either to liquid water or to ice.

The evaluation of the change of state due to a finite change of pressure requires the specification of both initial pressure and temperature, and in the saturated adiabatic process also of the total mixing ratio r. For convenience, therefore, the pseudo-adiabatic (with respect to liquid water, even at temperatures below 0°C) is adopted as the standard reference process, and the progressive changes of state as pressure is reduced are obtained numerically for a range of initial temperatures at the standard pressure of 1000 mb, and are displayed graphically. Just as a potential temperature θ is defined which is invariant in the dry adiabatic reference process,

so a saturation potential temperature θ_s can be defined as invariant in the saturated pseudo-adiabatic process with respect to liquid water, and equal to the temperature attained at the standard pressure of 1000 mb. Figure 3.8 displays states in this reference process as a function of pressure and θ_s. Figure 3.9 shows the amount of liquid water removed during expansion from initial pressures of 900 and 800 mb (representing a typical range of pressure at the saturation level of air near the ground), expressed as a concentration at the final pressure, and as a function of θ_s. This is practically the concentration of water condensed in the saturated adiabatic process and is the greatest that could arise in a cloud of small particles formed by ascent of air in the atmosphere. The diagram shows that for an ascent of several kilometers from levels near the surface this concentration has a maximum of several grams per cubic meter.

The temperature changes associated with large pressure changes in the standard pseudo-adiabatic reference process are appreciably smaller than those in processes in which the condensate is assumed to be ice at temperatures below 0°C or those which are reversible (9). The differences amount to a few degrees Celsius (Figs. 3.10, 3.11).

3.8 Aerological diagrams

The routine radiosonde observations represent the state of the atmosphere in terms of p, T, and u, along paths which in the study of large-scale motion systems are regarded as verticals through the sounding stations. The interpretation and discussion of the observations are made easier by their display on an *aerological diagram* which also shows the changes of state during the standard dry adiabatic and saturated pseudo-adiabatic reference processes. Abscissa, ordinate, and scale may be chosen in a variety of ways, which determine the shape of isopleths of p, T, r_s, and θ and θ_s (the dry-adiabatics and the saturated pseudo-adiabatics, respectively). It is helpful if as many as possible of the isopleths are straight or nearly so, if those of θ and θ_s obviously diverge (so that phase change of water has a readily apparent effect), and, it is generally thought, if the area within a closed curve representing a cycle of changes of state in dry air is everywhere directly proportional to work done during the cycle (mainly because the heights of certain standard pressure surfaces are then easily obtainable from a sounding). The most popular diagrams are the emagram (T-log p), the tephigram (T-entropy, or $-\log \theta$), and a T-log p diagram with skew axes on which the shape of the isopleths closely resembles that on the tephigram. Those used in this book are the tephigram (Appendix 3.1) and the $\theta_s - p$ diagram.

3.9 Wet-bulb potential temperature θ_w

Observations of the relative humidity are converted to values of dew point before they are entered in the aerological diagram, and then indicate mixing ratio r by reference to the isopleths of saturation mixing ratio r_s. During the adiabatic compression or expansion of air r is conserved; the point representing the dew point travels along an isopleth of r_s and that representing the temperature travels along an isopleth of θ; the air becomes saturated where these isopleths intersect, at a temperature T_c (Fig. 3.12). During compression from this state to a considerably higher pressure, the temperature becomes T_a or T_s according to whether the dry adiabatic or saturated pseudo-adiabatic process is used. By the introduction and evaporation of liquid water at the new constant pressure the air whose temperature is T_a can be saturated and chilled to its thermodynamic wet-bulb temperature T_{wt}, which differs from T_s because the evaporation of water proceeds in the one process at the temperature T_{wt} and in the other at temperatures between T_c and T_s. However, even under the extreme conditions of high temperature and a value of $(T_a - T_c)$ which is a few tens of degrees Celsius, corresponding to the large vertical displacement of a few kilometers in the atmosphere, the difference between T_s and T_{wt} is only about 0.2°C, while T_{wt} itself differs from the psychrometric wet-bulb temperature T_w by less than 0.1°C (Table 3.2). Thus the following propositions are valid to a good approximation:

1. The dry adiabatic through the dry-bulb temperature, the isopleth of r_s through the dew point, and the saturated pseudo-adiabatic through the (thermodynamic or psychrometric) wet-bulb temperature all meet in a point, corresponding to a state of saturation (Normand's theorem).

2. Since during the isobaric evaporation into air of liquid water at about the wet-bulb temperature the thermodynamic wet-bulb temperature remains constant, then during compression or expansion by a process which is adiabatic, or which departs from adiabatic only by the introduction and evaporation of liquid water at about the wet-bulb temperature, the point representing this temperature travels along a saturated pseudo-adiabatic. Consequently the corresponding value of θ_s defines a *wet-bulb potential temperature* θ_w which is invariant in such processes.

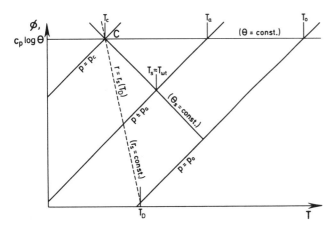

Fig. 3.12 On an aerological diagram (e.g., this schematic tephigram) the changes of state of air during adiabatic expansion from an initial state (p_0, T_0) are such that the point representing the temperature travels along an isopleth of θ (a dry adiabatic), and the point representing the dew point travels along an isopleth of saturation mixing ratio; these isopleths intersect at a point C (p_c, T_c) where the air becomes saturated. It is shown in the text that at any intermediate state (p_a, T_a) the thermodynamic wet-bulb temperature T_{wt} is given rather accurately as the temperature T_s on the isopleth of θ_s (the wet adiabatic) through the point C (Normand's theorem).

3.10 Comparison of theoretical and practical wet-bulb temperatures

Table 3.2 compares, for some typical and extreme values of p, T, and relative humidity, the wet-bulb temperatures as indicated by screen and psychrometric thermometers and by simple theory (Sections 2.4, 2.5) and as defined thermodynamically.

It appears that the simple convection theory accurately predicts the psychrometric wet-bulb temperature, and that the simple diffusion theory, also taking into account the radiative heat flux from the surroundings to the thermometer bulb, accurately predicts the somewhat higher temperature indicated by the less well-ventilated bulb in the Stevenson screen. The Meteorological Office's tables for use with screen thermometers (10) indicate that in warm, calm weather the poor ventilation of the wet bulb may result in its temperature reaching as much as 2°C above the psychrometric wet-bulb temperature.

The differences between the psychrometric wet-bulb temperature, the thermodynamic wet-bulb temperature, and the wet-bulb temperature inferred from Normand's theorem reach as much as 0.2°C only at high temperatures and low relative humidities, so that for most purposes they need not be distinguished.

Table 3.2 Comparison of theoretical and practical wet-bulb temperatures

U (%)	T_{wd} (diffusion)	T_{wdr} (diffusion and radiation)	T_{ws} (screen)	T_{wc} (convection)	T_w (pyschrometric)	T_{wt} (thermodynamic)	T_s (aerological)
			$p = 1000$ mb, $T = 20°$C				
25	9.1	11.4	11.0	10.0	10.2	10.1	9.9
50	13.1	14.5	14.2	13.8	13.8	13.9	13.8
65	15.4	16.2	16.1	15.8	15.8	15.9	15.8
75	16.8	17.3	17.2	17.0	17.1	17.1	17.1
97	19.7	19.7	19.7	19.7	19.7	19.7	19.7
			$p = 1000$ mb, $T = 40°$C				
20	20.8	24.0	23.1	21.9	22.2	22.3	22.1
40	26.9	28.8	28.3	27.7	28.1	28.3	28.1
			$p = 500$ mb, $T = -10°$C				
5	−16.0	−14.5	—	−15.3	—	−15.4	−15.5
50	−13.1	−12.3	—	−12.7	—	−12.6	−12.7
90	−10.6	−10.4	—	−10.5	—	−10.5	−10.5

Key: At some extreme and typical values of p, T, and relative humidity U, the following kinds of wet-bulb temperature are listed:

T_{wd} = the wet-bulb temperature according to the diffusion theory
T_{wdr} = the wet-bulb temperature according to the diffusion theory adjusted to allow for heat transfer by radiation, assuming the bulb and its surroundings to behave as black bodies
T_{ws} = the wet-bulb temperature indicated by a screen thermometer not specially ventilated, as read from hygrometric tables for use with thermometers in a Stevenson screen (10)
T_{wc} = the wet-bulb temperature according to the convection theory
T_w = the psychrometric wet-bulb temperature, as read from psychrometric tables (11)
T_{wt} = the thermodynamic wet-bulb temperature
T_s = the wet-bulb temperature as read from a tephigram using Normand's theorem

Appendix 3.1 The tephigram

A tephigram has temperature T as abscissa and entropy $\phi \, (= c_p \log_e \theta + \text{const})$ as ordinate, but the vertical scale is usually expressed only implicitly as a logarithmic one of the potential temperature θ. When convenient scales and ranges of T and $\log_e \theta$ have been determined the tephigram is constructed by adding isobars, isopleths of saturation mixing ratio, and saturated pseudo-adiabatics, whose form is indicated in the section drawn in Fig. 3.13.

The isobars are gently curved lines, concave toward high pressure (and temperature), which slope up from left to right. Their positions are given by Eq. 3.17; on working diagrams they are usually drawn at intervals of 10 mb from 1050 to 100 mb, and at intervals of 1 mb

3 Thermodynamic reference processes

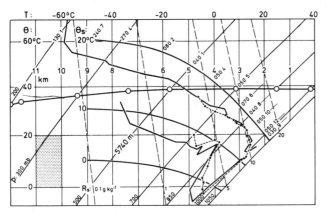

Fig. 3.13 Soundings represented on a tephigram. Isobars are labeled in mb and isopleths of saturation mixing ratio in g kg^{-1}. A nearly horizontal line is drawn through open circles representing the reported heights of standard pressure levels (e.g., 5740 m for the 500-mb level), using the indicated scale of height (km). This line allows the approximate height of any pressure level to be read. The soundings represent the vertical distributions of temperature and dew point, and winds (°C, m sec^{-1}) are entered at the appropriate levels. The solid lines and winds represent the radiosonde sounding made at Hemsby at about 1130 GMT, 19 June 1957, and the dashed lines an aircraft sounding made over the North Sea about 70 km upwind of the coast near Hemsby at about 1315 GMT on the same day. (The positions of the soundings are marked H and A, respectively, on Fig. 7.21, which shows the similar low-level flow of the previous afternoon.) The soundings indicate a close correspondence between the observations of the unusually dry layer at low levels, made with a wet-bulb thermometer on the aircraft and a hair hygrometer on the radiosonde (Appendix 2.3). They also show that the convection over the ground inland during sunshine had affected only a shallow layer a short distance from the coast, producing a warmed super-adiabatic layer and apparently a slight cooling immediately above (as might be expected if the kinetic energy of the convective circulations is transformed into potential energy near their tops; Chapter 7).

from 100 to about 50 mb, but on those reproduced in this text they are drawn only for some of the standard values 1000, 850, 700, 500, 300, 200, and 100 mb, for which height above sea level (in geopotential meters) is specified in routine reports of soundings.

Horizontal lines drawn at convenient intervals through the points in which isotherms intersect the isobar for 1000 mb define potential temperature θ and are usually called dry adiabatics; four are drawn in Fig. 3.13.

Isopleths of saturation mixing ratio r_s (usually with respect to ice at temperatures below $-40°C$) are nearly straight lines, which slope steeply up from right to left; their positions are determined by the relation

$$r_s = \frac{0.622 \, e_s}{p - e_s}$$

where $e_s(T)$ is the saturation vapor pressure (Section 2.3). They are drawn as dashed lines and in this text are labeled with values of r_s in grams per kilogram. (Proceeding upward along an isotherm r_s is almost doubled as p is halved, since $e_s \ll p$; e.g., in Fig. 3.13, at 4°C $r_s \approx 5$ g kg^{-1} at 1000 mb and $r_s \approx 10$ g kg^{-1} at 500 mb.)

Saturated pseudo-adiabatics are also drawn at convenient values through temperatures on the isobar for 1000 mb which define values of θ_s; three are included in Fig. 3.13. These curves are given by Eq. 3.28, assuming saturation with respect to liquid water throughout, and are usually terminated on the isotherm for $-40°C$, where they are practically parallel to the dry adiabatics.

It is shown in texts on meteorological thermodynamics that the work done on a unit mass of air which undergoes a closed cycle of changes of state is proportional to the area enclosed on the tephigram by the curve representing the cycle; thus an area on the tephigram is proportional to an amount of energy. The proportionality on a particular diagram is given by considering the rectangular area enclosed by two isotherms and two adiabatics, such as that stippled in Fig. 3.13, bounded by $\theta_1 = 273°K$ and $\theta_2 = 293°K$, and of width $\Delta T = 10°K$. The corresponding energy is $c_p \Delta T \log_e(293/273) \approx 0.71$ J g^{-1}.

A possible disadvantage of representing a sounding (usually given by values of T, T_d or U or r and wind at particular pressure levels) on a tephigram is that height is not an obvious coordinate. However, the heights of the standard pressure levels are included in reports of routine soundings, or may readily be obtained when a sounding has been plotted on a working tephigram. A relation between pressure and height which is sufficiently accurate for most purposes may then be entered on the tephigram (as shown in Fig. 3.13) by using the temperature scale as a height scale, making an interval of 10°C correspond to 1 km, and joining by straight lines points on isobars representing the heights of the standard pressure levels. On many of the soundings illustrated in this text the resulting line has not been included, but the method has been used to make marks against the temperature curve to indicate height at intervals of 2 km, and to mark the pressure levels appropriate to winds which are sometimes reported at particular heights (rather than pressure levels).

Appendix 3.2   The rate of condensation in ascending saturated air

Table 3.3 is an extract of a more extensive one in the *Smithsonian Meteorological Tables* (7, p. 324), which specifies the rate of condensation R of water vapor, expressed as a precipitation rate, in a layer of air 100 m thick (considered to be of uniform density and temperature) ascending (pseudo-) adiabatically at a steady speed.

Table 3.3 The rate of condensation R (units of 10^{-4} mm hr^{-1}) of precipitable water in a slab of saturated air 100 m thick, ascending adiabatically at 1 cm sec^{-1}, as a function of pressure p, approximate height z, and temperature T

z (km)	p (mb)	T(°C) 20	10	0	−20	−40
0	1,000	94	80	60	23	5
1	900	88	75	58	23	5
3	700		64	51	22	5
6	500			43	20	5
9	300				17	5

References

1. Goody, R.M. 1964. *Atmospheric Radiation*, pt. 1. *Theoretical Basis*. Clarendon Press, Oxford.
2. Saunders, P.M. 1964. Sea smoke and steam fog, *Q. J. R. Meteorol. Soc.*, **90**, 156–65.
3. Downie, C.S., and Silverman, B.A. 1960. Jet aircraft condensation trails, in *Handbook of Geophysics*, rev. ed. Macmillan, New York.
4. Pilié, R.J., and Jiusto, J.E. 1958. A laboratory study of contrails, *J. Meteorol.*, **15**, 149–54.
5. Taylor, J.H., and Church, J.F. 1966. *The Ice Fog Problem at Eielson AFB, Alaska*. Air Force Survey in Geophysics 176, Office of Aerospace Research, U.S.A.F., Bedford, Mass.
6. Oliver, V.J., and Oliver, M.B. 1949. Ice fogs in the interior of Alaska, *Bull. Am. Meteorol. Soc.*, **30**, 23–26; Benson, C.S. 1965. *Ice Fog: Low Temperature Air Pollution*. UAG R-173, University of Alaska, College.
7. List, R.J. 1958. *Smithsonian Meteorological Tables*, 6th rev. ed. Smithsonian Institution, Washington, D.C.
8. Dufour, L. 1965. Sur la température thermodynamique du thermomètre mouillé, *Acad. r. Belg. Bull. Climatol. Sci.*, 5th ser., **51**, 298–317.
9. Saunders, P.M. 1957. The thermodynamics of saturated air: a contribution to the classical theory, *Q. J. R. Meteorol. Soc.*, **83**, 342–50.
10. Meteorological Office, 1948. *Hygrometric Tables*. M.O.265, H.M.S.O., London.
11. Marvin, C.F. 1941. *Psychrometric Tables*. U.S. Weather Bureau, Washington, D.C.
12. aufm Kampe, H.J. 1942. *Kondensation und Sublimation in der oberen Troposphäre*, Forschungsbericht 1491, Deutsche Luftfahrtforsch., Berlin.
13. Lyons, W.A., and Pease, S.R. 1972. "Steam Devils" over Lake Michigan during a January arctic outbreak, *Mon. Weather Rev.*, **100**, 235–37.

4 Atmospheric aerosol

4.1 Sources of atmospheric trace substances

In addition to certain gases which are present in concentrations known or presumed to be practically constant (such as argon, helium, neon, and methane), the atmosphere contains other substances in the form of gases, vapors, or particulates, which occur in small and variable concentrations but have great biological and meteorological importance. Of these, water vapor and the gases ozone and carbon dioxide are meteorologically important in the heat economy, while some present mainly in the troposphere are intimately involved in the processes of the condensation and precipitation of water. The latter substances enter the atmosphere from the surface as particles, vapors, or gases of three principal kinds:

Particles of size mainly $> 10^{-5}$ cm, which are produced by the fragmentation of surface materials as sprays over the ocean and as dusts over arid lands and deserts.

Particles of size mainly $< 10^{-5}$ cm, produced in smokes by combustion and vaporization in natural and artificial fires.

Gases and vapors produced by fires, volcanic exhalations, or the metabolism and decay of organic material.

Sea spray

To become airborne, particles of surface material must be projected to levels where the upward speeds in the small-scale turbulent motions exceed their settling speeds. Generally the mechanics of the projection process allow the particles to reach heights of several centimeters above the surface, where in moderately fresh winds the turbulent upward speeds are several centimeters per second, the magnitude of the settling speed of a particle of unit density which has a radius of about 20 μ. Particles larger than this are unlikely to travel more than a few meters before settling back (except in strong winds), and accordingly this is generally an upper limit to the size of the particles which enter the atmosphere directly from the surface.

Droplets of sea spray enter the atmosphere during the breaking of waves, which occurs on coasts in all weathers and over the open sea when the wind speed exceeds about 6 m sec^{-1}. After a wave has broken, bubbles of trapped air rise to the surface; hardly any have a diameter of less than about 100 μ, smaller ones probably disappearing as the air they contain dissolves in the sea water. After a while each bubble breaks; the rupture of the thin upper film scatters into the air about 100 small droplets whose diameters are less than about 1 μ, and whose content of sea salt ranges from about 10^{-18} to rather more than 10^{-14} g. After a bubble has broken the depression formed by its lower surface fills and produces a jet of sea water extending vertically upward; this breaks into several larger droplets, containing up to 10^{-9} to 10^{-8} g of sea salt, which are projected to heights of several centimeters. The sea salt introduced into a shallow surface layer in these ways is carried by eddy diffusion and convection into a deeper layer of the atmosphere, at a rate estimated to be about 10^9 tons yr^{-1} (2).

Dust

Erosion by mechanical, chemical, and physical processes reduces surface materials to particles of various sizes; those of diameter more than about 100 μ can be regarded as *sand*, to be distinguished from the smaller *dust* particles which can be raised by the wind, mainly in arid regions.

Sufficiently prolonged mechanical grinding of minerals leads to a characteristic size distribution of the small particles, with a minimum diameter of about 10^{-5} cm (3), and this is generally regarded as the lower size limit of natural dust particles.

A layer of fine dust with a smooth surface is not easily disturbed by a strong wind, but the particles can become airborne in eddies where the surface is irregular or ridged, especially in the presence of sand grains which are only briefly thrown into the air, moving along by bouncing (or "saltation") and disturbing the dust layer by repeated impacts (4).

Evidence such as the loess deposits in parts of China and the composition of clays on the ocean beds downwind of deserts indicates that large quantities of dust are transported by the atmosphere. However, the source regions are small and most of the dust is raised only intermittently when there are strong winds; the average rate of introduction of dust into the atmosphere, almost all in low latitudes, is probably less than 10^8 tons yr^{-1}. [In the Caribbean during the summer months the concentration near the ground of Saharan dust particles of diameter greater than about 2 μ has an average value of about 2 μg m^{-3}, and over two-day periods is occasionally ten times as great and exceeds the average concentration of sea salt (48). Occasionally Saharan dust is transported into middle latitudes, arriving there in the middle troposphere and reaching the ground in rain which sometimes leaves typically red-brown deposits. Recently a cloud of such dust was visible on routine satellite pictures (49), and about 10^6 tons reached the ground in rain over England and Wales.] Such dusts consist predominantly of silicates derived from the weathering of the most common minerals—quartz, feldspar, and mica—and their particles are about the most efficient natural ice nucleants (50; Chapter 5).

It has been suggested (5) that salt crusts on arid soils might crack under intense sunshine and emit particles which could become airborne. Although some experiments appear to have confirmed the existence of such a process, there is not yet any good evidence that even locally it is meteorologically important.

Solid particles occasionally are spectacularly injected into the atmosphere by volcanic eruptions. Ordinarily the clouds of condensed water and ash over active volcanoes are confined to the troposphere and are then probably an unimportant source of solid particles, but the more violent paroxysmal eruptions (occurring perhaps once in a decade) produce clouds which rise above the tropopause and introduce great quantities of fine ash into the lower stratosphere. Thus the eruptions of Krakatoa in 1883 produced clouds which even before the final great explosions towered to heights estimated as 27–34 km (39). In more recent times volcanic clouds reached 18 km over Sakura-jima (Kyushu) in 1914 (6), and 35–45 km over Bezymianny in Kamchatka on 30 March 1956. In Kamchatka the mass of material ejected during the explosions was estimated from the change in the shape of the volcano to have been about 10^9 tons, of which only a small fraction may have been borne away by the winds in the form of particles, but which nevertheless would be sufficient to produce concentrations of order 1 cm^{-3} or more in a deep layer over vast areas. Still more recently volcanic particles from the eruption of Mt. Agung (Bali) on 17 March 1963 were collected some months later at a height of 20 km over Australia, and found to have mean diameters of less than 1 μ and concentrations of up to about 10^{-1} cm^{-3} (8).

Smokes

Smokes of very small and numerous particles are produced by the rapid quenching and dispersion of vapors when the supply of air to fires is inadequate for the complete combustion of the fuel, especially if it consists of the hydrocarbons consumed in most natural and artificial fires. The chilling accompanying dilution with air produces vapor pressures greatly exceeding the saturation vapor pressures and leads to condensation upon molecular aggregates (by the process of *homogeneous nucleation*) and to characteristically very high concentrations of particles ($>10^6$ cm^{-3}).

Accordingly, near fires and, for example, within cities where most places are within an hour's wind drift from a smoke source the total concentration of airborne particles near the ground is typically about 10^6 cm^{-3}, but well away from such sources it decreases toward the more representative value of about 10^3 cm^{-3} consistent with Table 2.2.

The most abundant component of aerosol material produced by combustion is SO_2, which in the atmosphere is soon oxidized and appears in aerosol particles as sulfate radical SO_4. From the annual consumption of coal and crude oil it is estimated that SO_4 is produced in the atmosphere by artificial fires at the rate of about 10^8 tons yr^{-1} (1), mostly in the northern hemisphere. The contribution from natural combustion (principally forest fires ignited by lightning) is probably at least an order of magnitude smaller, and can be regarded as a minor part of the source of trace substance provided by the ordinary decay of organic matter.

Gases and vapors

Volcanic emanations provide considerable quantities of H_2S, SO_2, and other gases, at a total rate which probably exceeds 10^7 tons yr^{-1}. However, more important sources of gases and vapors are the metabolism and decay of organisms (mainly on land, but also in the

4 Atmospheric aerosol

surface layers of seas and lakes); in particular, from these sources H_2S and NH_3 enter the atmosphere at rates estimated to be about 10^8 tons yr^{-1} [volatile organic vapors, mainly terpenes which give the air odors characteristic of various kinds of vegetation, may enter at a comparable rate (9)].

The magnitude of these important sources of trace substance is related to the solar constant. During the course of the year about half of the solar energy arriving in the outer atmosphere reaches the surface and is available for photosynthesis of organic material, but the efficiency of the process is only about 10% and it utilizes a part of the spectrum containing only about half of the received energy; moreover, the growing season occupies only a part of the year, and a large fraction of the incident energy is used in respiration. Considering these factors and that the energy of composition of dry organic matter is about 5×10^3 cal g^{-1}, the maximum annual rate of production of organic matter on a globe covered with vegetation amounts to about 10^{12} tons yr^{-1}. The actual total production is estimated to be about 10^{11} tons yr^{-1}, mostly on land (10). Since the proportion by weight of sulfur in organic matter is less than 10^{-3} (2), the magnitude of the provision of SO_4 to the atmosphere by the decay of organic material at the same rate is likely to be about 10^8 tons yr^{-1}. A similar magnitude can in the same way be anticipated for the rate of provision of the NH_4 radical. Thus the rate of use of solar energy by organic material implies that these two important radicals enter the atmosphere at a rate of order 10^8 tons yr^{-1}. Almost all the incident energy is used to evaporate water, whose rate of provision is from similar considerations over six orders of magnitude greater, consistent with the mean global annual rainfall of about 100 cm (equivalent to 5×10^{14} tons yr^{-1}).

Especially at high humidities and in bright sunshine, chemical reactions occur between the trace gases and vapors. The reactions proceed partly on the surfaces of particles already present, but also are a significant if not the most important source of very small particles.

The average mass concentration of aerosol particles in the lower troposphere

The average mass concentration of S and NH_3 near the ground is several times greater in the gaseous than in the particulate form, but considering the various source rates just estimated it is evident that over both land and ocean material for aerosol particles is provided at the total rate of about 10^9 tons yr^{-1}, sufficient to give an average mass concentration in the lowermost 3 km of about 40 μg m^{-3} after 1 week. Since this is about the average value observed away from prolific sources (1), one week is evidently the average "residence time" spent in the atmosphere before return to the surface, and is about the same as the average residence time of water vapor (corresponding to the mean global rainfall of 100 cm yr^{-1} and a mean precipitable water vapor content of about 2 cm; the mean concentration of water vapor in the lowest 3 km is about 10 g m^{-3}, nearly six orders of magnitude greater than that of the aerosol particles).

4.2 The size distribution of tropospheric aerosol particles

In the atmosphere electrically charged molecular clusters with diameters of about 10^{-7} cm ("small ions") are continually produced by cosmic radiation and the decay of radioactive trace gases, and aggregates of about this size are also formed during chemical reactions involving other trace gases. Very soon after their formation they become attached to the much larger particles introduced into the atmosphere as smokes or fragmentation products, whose radii extend over the great range from about 10^{-6} to about 10^{-3} cm. The processes of supply and removal and of aggregation by Brownian motion lead to characteristic size distributions in which the concentration of particles within a given size range decreases rapidly as the average radius increases above a value near 10^{-5} cm.

In the Aitken counter a sudden large expansion (limited to avoid affecting the small ions, whose average concentration is a few hundred per cubic centimeter) condenses water upon and renders visible particles of initial radius greater than about 5×10^{-7} cm. By its construction, which limits the volume of air sampled to a few cubic centimeters, the counter samples only the smallest particles of radius little more than 10^{-5} cm (the *Aitken nuclei*), but since the concentration of larger particles is comparatively insignificant, it indicates the total concentration N of particles. The size distribution of the Aitken nuclei is explored by examining their diffusivity, or the distribution of the mobility in electrical fields of those which are charged. Somewhat larger particles can be extracted from the air, by thermal, electrical, or mechanical precipitation, for examination under the electron or optical microscope, or their concentration and size distribution may be inferred from their light-scattering properties. The largest particles, present in concentrations as small as 1 m^{-3}, can be measured after settling upon slides exposed for long periods.

Although the concentrations of particles found by such techniques are obtained as histograms, it is usual to generalize them into continuous size-distribution spectra on diagrams in which $n(r)\,dr$, the concentration of particles with radii between r and $(r + dr)$, or N_r, the concentration of all particles of radius larger than r, is plotted against r. Because both N_r and r range over several orders of magnitude it is convenient to use logarithmic scales, as in Fig. 4.1. This shows simplified size-distribution spectra characteristic of particular meteorological situations. The uppermost curve represents a spectrum typical of air near the ground and not far from prolific sources. At radii less than about 10^{-5} cm the curve is dashed because it represents approximately a spectrum which is there rather irregular, tending to contain groups of particles in discrete size ranges (11), perhaps from particular local sources. At radii greater than about 10^{-5} cm, on the other hand, the spectrum appears more continuous: approximately

$$n(r)\,dr = c\,d(\log r)/r^\beta,$$
$$10^{-5} \lesssim r \lesssim 2 \times 10^{-3} \text{ cm} \quad (4.1)$$

In nucleus-rich air a characteristic value of about 3 is found for the exponent β, not only in the upper part of the relevant size range by the counting of samples of particles, but also in the lower part ($8 \times 10^{-6} < r < 6 \times 10^{-5}$ cm) by studies (12) of the optical properties of haze: the extinction, as a function of wavelength, and the scattering, as a function of wavelength and the angle between the incident and the scattered light. For this value of β equal intervals of $(\log r)$ represent the same volume or mass concentration. In the cleaner air commonly found over the ocean or on mountaintops the optical evidence suggests smaller typical values of β (1).

The observed establishment of the size distribution characteristic of the aerosol over land within a few days of the replacement of nucleus-rich by clean maritime air suggests that the spectrum with a value of $\beta \approx 3$ represents a quasi-steady balance between the process of Brownian aggregation, by which material enters the lower end of the relevant range of size ($r \approx 10^{-5}$ cm) and moves toward higher values of particle radius, and processes by which material is removed from the air. (Over the sea the form of the spectrum may be distorted by the comparative absence of sources of the smallest particles.) At the upper end of the size range ($r \gtrsim 10^{-3}$ cm) settling under gravity upon the earth's surface is the principal process, other than precipitation in rain, by which particles are removed: near the lower end the settling speed is negligible, but measurements of the ratio between particle concentration near the surface and deposition rate indicate that removal

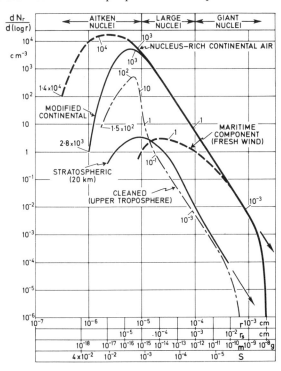

Fig. 4.1 Aerosol size-distribution spectra characteristic of various meteorological situations. The numbers beside the curves represent the total concentration N_r cm^{-3} of particles with radii greater than r. Approximate scales of the equilibrium radius r_s in saturated air, the mass m of water-soluble substance, and the maximum equilibrium supersaturation S are included.

The uppermost curve is typical of the nucleus-rich aerosol of the planetary boundary layer not far from prolific ground sources of nuclei. For radii between about 10^{-5} and 10^{-3} cm this curve obeys the law $dN/d(\log r) = 5 \times 10^{-9}/r^3$. At radii considerably less than 10^{-5} cm the curve is dashed to indicate that it is a smoothed representation of a distribution which usually has irregular peaks.

During drifting for some days away from the principal sources of small particles, aggregation changes the size distribution to that marked *modified continental*; a further change to that marked *cleaned* occurs when the air passes through a raincloud into the free atmosphere. Over the oceans in fresh winds the distribution of sea-salt particles marked *maritime component* is added (dashed curve).

The size-distribution curves near the base of the diagram are curved to indicate an upper limit of 10–20 μ for the radius of aerosol particles, since it was thought that this would be imposed by the rather large corresponding settling speed of several centimeters per second. Recent observations, however, have shown that at least over the continents in the summer season the curves extend with approximately the same slope of -3 (Eq. 4.1) to particle radii of 100 μ or even more, as shown by the arrows (51). The nature of these particles has not been established, nor has the process responsible for their consistent presence when the fall speeds of the largest are about 10 km/3 hr. They are apparently mainly solid and their surfaces only locally hygroscopic (52); they are perhaps introduced into the upper troposphere locally by deep cumulus or cumulonimbus convection and transported considerable distances before settling into the boundary layer.

by impaction under Brownian motion is an important process. In either case, the rate of removal depends on the rate of diffusion of particles toward the surface by small-scale air motions. In moderate winds the effective deposition speed of particles with radii even as small as 10^{-5} cm is observed to be of order 10^{-1} cm sec^{-1} (13), and only at radii of more than about 10^{-3} cm is it approximately the settling speed in still air (of order 1 cm sec^{-1}). A mean deposition speed of 10^{-1} cm sec^{-1} and the observed average mass concentration of 40 µg m^{-3} (1) implies a mean global deposition rate of the same magnitude as the estimated source rate (10^9 tons yr^{-1}); the general effectiveness of this process of removal is comparable with that of the removal by incorporation into rain water. The rate of cleansing of the atmosphere, and the typical general concentration in the lower troposphere of aerosol substance of surface origin (ordinary natural and artificial pollution) is therefore appreciably affected by rainfall, but probably not sensitively dependent upon it: in the complete absence of rain the concentration would increase by less than an order of magnitude.

Theoretical and numerical studies (14) of the coagulation of aerosol particles, initially uniform or with various size distributions, show that a size spectrum may tend toward a definite self-preserving form; one attempt to infer this form, considering settling under gravity as the only process removing particles, suggested a value of $\beta = 7/2$ in the lower part of the range of particle radii. Although this is not altogether inconsistent with the value usually observed, the characteristic form of the observed size distribution is almost certainly also dependent upon the statistics of the processes of supply and removal of the particles (15).

After nucleus-rich air has traveled for a few days away from the principal particle sources, aggregation by Brownian motion reduces the concentration of the smaller particles, shifts the peak in the size-distribution spectrum nearer to a radius of 10^{-5} cm, and reduces the total concentration from over 10^4 cm^{-3} to a few thousand per cubic centimeter (consistent with Tables 2.2 and 2.3), as shown by the curve in Fig. 4.1 labeled modified continental. Within a few further days the air is likely to be subject to the cloud- and precipitation-forming processes which accompany its ascent in a large-scale circulation, and which reduce the concentration of particles of radius less than 10^{-5} cm by about an order of magnitude, and even more efficiently remove larger particles, as indicated by the curve labeled cleaned. This spectrum, with a total concentration of particles which is typically about 100 cm^{-3} but occasionally as little as 10 cm^{-3}, is characteristic of the upper troposphere, and thus also of the planetary boundary layer where air from the free atmosphere enters it away from the principal continental particle sources [most commonly, therefore, in airstreams in or from high latitudes (16)—where the air is renowned for its transparency in fair weather]. Over the ocean in fresh winds, however, sea spray adds or restores large particles, as shown by the dashed curve marked oceanic.

4.3 The nomenclature of tropospheric aerosol particles

It is convenient to classify aerosol particles according to size into the following groups:

$r \lesssim 10^{-7}$ cm: These particles are short-lived and have no direct importance in atmospheric condensation processes but include the *small* ions, of not much more than molecular size, which are important in atmospheric electrical processes.

$5 \times 10^{-7} \lesssim r < 10^{-5}$: *Aitken* nuclei (which by the attachment to small ions may become large ions of very much smaller mobility), whose motion is predominantly a Brownian motion effective in causing aggregation.

$10^{-5} < r < 10^{-4}$ cm: *Large nuclei*, manifest as haze by their effectiveness in scattering or attenuating light, whose Brownian and settling motion is unimportant but which are always involved in cloud formation.

$r > 10^{-4}$ cm: *Giant nuclei*, which although occurring in comparatively small concentrations may play an important part in precipitation formation.

The description "nuclei" refers to the ability of the particles to nucleate the condensation of water vapor at modest supersaturations; however, only a proportion of the Aitken nuclei become involved in atmospheric cloud formation.

4.4 The composition of tropospheric aerosol particles

In nucleus-rich air the Aitken nuclei compose only a small part of the total mass of aerosol particles (most of which is made up of approximately equal contributions from the large and the giant nuclei), and their chemical composition is difficult to determine. However, they appear to be predominantly of continental origin, and can be presumed to have a composition similar to that of the larger continental nuclei, in which the principal radicals in the water-soluble material are NH_4 and SO_4, in proportions rather greater in SO_4 than that corresponding to ammonium sulfate, $(NH_4)_2SO_4$, with lesser amounts of Cl and NO_3. The water-soluble component is associated with similar masses of organic substances, and insoluble solids derived by aggregation with dust and ash particles.

It seems that the ocean is not a prolific source of Aitken nuclei. The principal component of the large and giant nuclei produced by sea spray is, of course, NaCl, but the proportion of Cl to Na gradually becomes reduced as the aerosol ages, probably by chemical reaction with trace gases of continental origin. Just as the aerosol substances characteristically produced over land are frequently carried long distances out to sea before they are mostly removed in precipitating cloud systems (the remainder provide most of the content of Aitken nuclei generally observed over the oceans), so also sea-spray particles are often carried far into the interior of the continents.

4.5 Extraterrestrial sources of aerosol particles

In addition to the planets and their satellites the solar system contains many smaller bodies and finely divided matter, most of which is derived from the asteroids and the tails of comets, and whose accretion by the earth is a significant source of trace materials in the stratosphere.

The particles arrive in the upper atmosphere with speeds of order 10 km sec^{-1}, and those with radii of more than several microns are heated to temperatures above $1000°K$ and melt or boil. Those of radius larger than about $100\,\mu$ become sufficiently incandescent to be visible at night as shooting stars or meteors. On the average several *sporadic meteors* can be seen each hour, but for a few successive nights at intervals throughout the year the earth intersects the orbits of concentrations of meteoric material, and the meteors appear in showers, radiating from a particular point in the sky. Many meteors, especially the fainter ones, seem to have a low density and to disintegrate under aerodynamic pressure and heating, so that they probably consist of loosely aggregated mineral particles, perhaps held together by ices of such substances as H_2O, CO_2, and NH_3.

Meteorites, micrometeorites, and interplanetary dust

The great majority of meteors evaporate completely at heights between about 60 and 100 km. Some, however, are large enough to reach the ground as *meteorites*, and are then found to be of two principal kinds, those consisting mainly of nickel-iron alloys and the much more numerous *stony* meteorites consisting mainly of silicates. Rarely, very large meteorites reach the earth's surface (with impact speeds of order 1 km sec^{-1}), producing tremendous explosions and leaving craters up to about 1 km across. In recent times the two most famous fell in Russia, one on the Sikhote-Alin mountains north of Vladivostok on 12 February 1947 (18), the other the Great Siberian Meteorite on 30 June 1908 (17), which produced an explosion that was recorded on barographs in England, and its debris in the atmosphere caused anomalous twilights over much of Europe.

Such great meteorites and those meteors large enough to be seen in daylight leave behind in the atmosphere trails of smoke with a pale blue tinge which shows that the predominant radius of the particles is less than 10^{-5} cm. The earth accretes material in the form of meteors at an average rate of about 2×10^2 tons yr^{-1} (19), a large fraction of which appears in the atmosphere as smoke particles. This rate is small compared with the rate of accretion of *micrometeorites*, those interplanetary particles that are too small to become melted on entering the atmosphere but still may be a significant stratospheric source of Aitken nuclei. Thus if the particles are assumed to have a residence time in the stratosphere of one year, which is that observed for the radioactive debris of nuclear bombs, their average concentration in the layer between 20 and 100 km (which contains a total volume of 4×10^{25} cm^3) is about $2 \times 10^{-9}/r^3$ cm^{-3}, where r is the mean radius. If r is taken as 10^{-6} cm, the concentration implied is of order 10^{-1} cm^{-3}; experimental simulation of meteor evaporation (20) suggests that a smaller value of r may be more appropriate, implying a concentration considerably greater than 1 cm^{-3}, but not too high to be maintained against aggregation by Brownian motion. It is clear that the evaporation of meteors cannot be a significant source of particles with a radius of considerably more than 10^{-6} cm. On the other hand, the micrometeoric material near the earth's orbit probably contains no particles with radii smaller than about 3×10^{-5} cm, because the radiation pressure of sunlight is sufficient to expel them from the solar system. Larger particles spiral toward the sun under the Poynting-Robertson and other

4 Atmospheric aerosol

momentum transfer effects (21) at a cosmically rapid rate, so that to maintain their supply a plausible source rate of order 1 ton sec^{-1} from comets is required.

This material is concentrated close to the plane of the planetary orbits, and by scattering sunlight produces the *zodiacal light*, which is described later in the chapter.

4.6 Stratospheric aerosol particles

Aerosol particles have been examined in the stratosphere at heights up to about 20 km by collection during aircraft flights, at heights up to nearly 30 km by sampling and by sunlight-scattering measurements made with balloon-borne instruments, and up to about 60 km by measurements from the ground of the twilight intensity at low elevations above the horizon. The particles are not always uniformly distributed, concentrations sometimes occurring in thin layers.

The direct measurements (1) show that in periods undisturbed by violent volcanic eruptions the concentration of Aitken nuclei decreases above the tropopause, to become about 1 cm^{-3} at 20 km and thereafter to diminish more slowly with height, so that their total concentration in unit *mass* of air, N_g, is virtually independent of height, as might be expected under simple sedimentation from above.

The large nuclei, however, have a minimum concentration near the tropopause and a maximum, of about 10^{-1} cm^{-3}, between about 17 and 25 km (Fig. 4.2). They are composed mainly of ammonium sulfate and persulfate with some free sulfuric acid (53), probably formed on Aitken nuclei by photochemical oxidation of the trace gases SO_2 and H_2S, derived from the troposphere (the exchange of aerosol and trace gases between troposphere and lower stratosphere is effected by the large-scale motions discussed in Section 9.5). The size-distribution spectrum of the Aitken and the large nuclei has been assumed to be continuous and of the kind characteristic of tropospheric aerosol in interpretations of observations of light scattering (22, 23), but there is some evidence that the spectrum of the large nuclei has a peak at about 3×10^{-5} cm, and that there are comparatively few nuclei in the size range from about 5×10^{-6} to 10^{-5} cm (24).

Measurements of twilight intensity imply that there is a significant concentration of particles at heights up to at least 60 km, where the value of N_g is close to that at 30 km. Direct evidence of particles in the mesosphere has been sought by retrieving, and examining under the electron microscope, slides exposed during rocket flights. The first samples were taken in northern Sweden as part of a study of noctilucent clouds (25), which occasionally appear during the high latitude summer at a height of about 80 km. On one occasion, when there were no noctilucent clouds, few particles were obtained: their concentration was inferred to be of order 10^{-3} cm^{-3}, or 10^5 g^{-1}, not much different from the value of 10^4 g^{-1} for the Aitken nuclei at 25 km. On a second occasion, when noctilucent clouds were present, many more particles were obtained. Their radii ranged between about 3×10^{-6} and 7×10^{-5} cm, and the total concentration reached the extraordinarily high if not improbable value of order 1 cm^{-3}, or 10^8 g^{-1}. The photographs showed the larger particles to have halos, indicating that they had possessed volatile coats, thought to be composed of ice. In more recent samplings the collectors certainly passed through a noctilucent cloud on one occasion (over Canada in 1968), and may have done so on a few other occasions, but the solid particles collected, with diameters mainly in the range 5×10^{-6} to 2×10^{-5} cm, did not show any signs of having possessed volatile coats (54). Their total average concentration along the whole sampled path was of order 10^5 g^{-1}, but could have been nearly two orders of magnitude greater had they mostly been collected from a layer of cloud only 1 km deep. It seems more

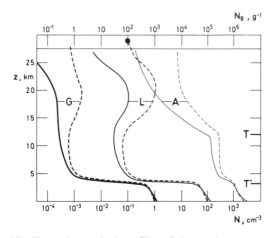

Fig. 4.2 Tentative typical profiles of the total concentration of Aitken (A), large (L; radius $>0.15\ \mu$), and giant (G; radius $>1\ \mu$) nuclei as a function of height above the ground in middle latitudes. The total number density N cm^{-3} is shown by the solid lines and the total number mixing ratio N_g g^{-1} by the dashed lines. The point in the upper panel represents the mixing ratio of the large nuclei at 50–65 km, as determined from twilight measurements (22). T marks the assumed level of the tropopause, and T' that of the top of a sharply defined planetary boundary layer.

probable that the samples were representative of a normal concentration of particles in the high mesosphere; more extensive and detailed exploration will be necessary to establish the nature of the particles of noctilucent clouds.

4.7 The equilibrium of solution droplets

At high relative humidities the hygroscopic components of aerosol particles dissolve in absorbed water, and the particles are *solution droplets* containing the insoluble components. At a particular relative humidity a solution droplet has an equilibrium radius r_e, determined principally by the mass m of dissolved substance, which can be estimated as follows.

Over a curved surface of pure water with radius of curvature r the equilibrium vapor pressure e' exceeds the saturated vapor pressure e_s:

$$\log_e(e'/e_s) = 2\gamma M/\rho_w R^* T r \quad (4.2)$$

where γ, ρ_w, and M are the surface tension, density, and molecular weight of water, respectively, and R^* is the universal gas constant.

If $(\Delta e'/e_s) = (e' - e_s)/e_s$ is small (say $\lesssim 0.2$), then Eq. 4.2 may be written as

$$\Delta e'/e_s = a/r$$
$$a \approx 10^{-7} \text{ cm at ordinary temperatures} \quad (4.3)$$

On the other hand the equilibrium vapor pressure e'' over a solution is less than the saturated vapor pressure (over the pure liquid) in accordance with Raoult's law:

$$\frac{\Delta e''}{e_s} = \frac{e'' - e_s}{e_s} = \frac{-iM'}{iM' + M} \quad (4.4)$$

where M' and M are, respectively, the number of moles of solute and water, and the empirical van't Hoff factor i is introduced to take into account the dissociation of electrolytes in solution. In extremely dilute solutions i is approximately the number of ions corresponding to one molecule of the solute: 2 for NaCl and 3 for $(NH_4)_2SO_4$, for example; for the moderate dilution appropriate to solution nuclei under high relative humidities i is somewhat less: about 1.9 for NaCl and between about 1.9 and 2.2 for $(NH_4)_2SO_4$, for example (26). Thus for spherical solution droplets of radius r, containing a mass m of solute, at high relative humidities Eq. 4.4 can be written in the approximate form

$$\Delta e''/e_s = -b/r^3 \quad (4.5)$$

where $b \approx m/7$ cm^3 where the solute is NaCl
$b \approx m/14$ cm^3 when the solute is $(NH_4)_2SO_4$
and m is in grams.

Since the precise value of b for particular nuclei is usually unimportant in the discussion of the properties of a population of nuclei of variable composition, in which m extends over several orders of magnitude, for simplicity we assume $b \approx m/10$ for all nuclei and combine (4.3) and (4.5) into

$$\frac{\Delta e_r}{e_s} = \frac{\Delta e' + \Delta e''}{e_s} = \frac{a}{r} - \frac{b}{r^3} \quad (4.6)$$

where $\Delta e_r = (e_r - e_s)$ is the difference between the equilibrium vapor pressure e_r over a solution droplet of radius r and the saturated vapor pressure. If $\Delta e = e - e_s$, where e is the vapor pressure in the surroundings, the condition for equilibrium is

$$\Delta e = \Delta e_r$$

or

$$\left.\begin{array}{l}\Delta e/e_s = a/r - b/r^3 \\ a = 10^{-7} \text{ cm} \\ b = 10^{-1} \text{ m cm}^3\end{array}\right\} \quad (4.7)$$

The equilibrium radius r_s of a solution droplet in saturated air is therefore

$$r_s = (b/a)^{1/2} = 10^3 m^{1/2} \quad (4.8)$$

At the high relative humidities (70–90%) typical of air near the surface, the equilibrium radius r_0 is determined by values of $\Delta e/e_s$ between -0.3 and -0.1. If the mass of solute is considerably greater than the observed minimum significant value of about 10^{-18} g, r_0 is smaller than r_s by a factor of more than 2 and the last term dominates the right of Eq. 4.7. It is therefore a convenient simplification to regard the radius of a solution droplet in equilibrium at ordinary relative humidities as simply

$$r_0 = m^{1/3} \quad (4.9)$$

This radius is not particularly sensitive to the exact value of the relative humidity if it is less than the value of 90% at which it is most appropriate (indeed, because of the presence of considerable insoluble components in the continental aerosol particles, the observed ratio of their equilibrium sizes at relative humidities of 95 and 40% is only about 1.3). Accordingly, Eq. 4.9 and 4.8 have been used to provide scales of m and r_s over most of the size range covered in the diagram of the typical

4 Atmospheric aerosol

size-distribution spectra observed at ordinary relative humidities (Fig. 4.1).

Conditions for the equilibrium of solution droplets

Figure 4.3 presents Eq. 4.7 in graphical form. Evidently at all subsaturations there is a single equilibrium radius of a solution droplet, while for a range of supersaturations there are two possible equilibrium radii. At the smaller, and at subsaturations, the equilibrium is stable, whereas at the larger the equilibrium is unstable; further, there is maximum value S of the supersaturation at which there can be an equilibrium, which is marginally unstable, and at which the droplet has a radius R. At any supersaturation greater than S there is no equilibrium radius and if it is maintained, the solution droplet grows indefinitely.

The value of R, obtained by differentiating Eq. 4.6 with respect to r, is

$$R = (3b/a)^{1/2} = 10^3 (3m)^{1/2} \qquad (4.10)$$

and the corresponding value S of the supersaturation ratio $s_r = \Delta e_r/e_s$ is

$$S = (2a/3)(a/3b)^{1/2} = 2a/3R \qquad (4.11)$$

(The surface area of a solution droplet with the radius R is thus three times its area when it is in equilibrium in a saturated atmosphere.)

By expressing r and $s_r = \Delta e_r/e_s$ (or the corresponding ratio $\Delta \rho_r/\rho_s$) as the fractions k and h of R and S, Eq. 4.7 can be put in nondimensional form, and is represented in Fig. 4.3 by a single curve valid for all solution nuclei. If the supersaturation s of the environment is similarly expressed as a fraction g of S, the condition for equilibrium becomes

$$gS = hS = \frac{a}{kR} - \frac{b}{k^3 R^3}$$

or

$$g \times \frac{2a}{3R} = \frac{k^2 R^2 a - b}{k^3 R^3}$$

or

$$2gk^3 = 3k^2 - 1, \qquad g = \frac{s}{S}, \qquad k = \frac{r}{R} \qquad (4.12)$$

Table 4.1 Approximate values of the equilibrium radius r_s in saturated air, the maximum stable equilibrium radius R, and the corresponding supersaturation ratio $S = (e - e_s)/e_s$, as a function of the mass m of solute in a solution droplet

Nuclei	m (g)	r_s (cm)	R (cm)	S (%)
Aitken ↑	10^{-18}	5×10^{-7}	1×10^{-6}	4
↕	10^{-15}	3×10^{-5}	6×10^{-5}	1×10^{-1}
Large	10^{-12}	1×10^{-3}	2×10^{-3}	4×10^{-3}
↕				
Giant	10^{-9}	3×10^{-2}	6×10^{-2}	1×10^{-4}
↓	10^{-8}	1×10^{-1}	2×10^{-1}	4×10^{-5}

Values of R and S are given as functions of m in Table 4.1. It appears that for the giant nuclei r_s and R are very large and S exceedingly small; we shall see that the equilibrium radii of such nuclei at vapor pressures near the saturated vapor pressure have no direct significance. The principal value of the equilibrium functions is that they may be assumed to provide the vapor density at the surface of a solution droplet, given its radius and the mass of solute, and hence in a specified environment the gradient of vapor density which determines the rate of growth or evaporation of the solution droplet, and they will be used in this way in Chapter 5.

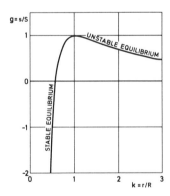

Fig. 4.3 Generalized form of the relation between the equilibrium radius r of a dilute solution nucleus and the supersaturation ratio $s = (e - e_s)/e_s$ of the surroundings, in terms of the ratios r/R and s/S, where R and S are the maximum radius and supersaturation for equilibrium. The curve is valid for all dilute nuclei.

4.8 The scattering of light by atmospheric aerosol; visual range

Many interesting and beautiful phenomena to be seen in the sky arise from the scattering of sunlight by the molecules of the air and by the particles of the atmospheric aerosol, of which the most effective in nucleus-rich air are the large nuclei whose radii vary from less to more than the wavelengths of visible light.

Visibility and visual range

The highly variable scattering due to aerosol particles disturbs the polarization pattern (27) and, more noticeably, it pales the blue of the sky and reduces the distance at which an object can be perceived and recognized by eye. This distance is important in many kinds of activity. It varies considerably according to the conditions of illumination and from one observer to another, but generally can be related to the routine meteorological observation of the visibility, the greatest horizontal distance at which the contrast between an object and its surroundings just exceeds that to which the eye is sensitive. To reduce the subjectivity of this observation, it is made upon large, dark objects seen against the horizon sky, and a note is made if it is markedly less in particular directions.

The contrast c is defined as the ratio $(I_S - I_B)/I_S$, where I_B and I_S are, respectively, the intensities of the light received from an object and from the neighboring part of the sky. In the simplest circumstances we consider the body to be black, that is, $I_B = 0$, when it is near at hand. If now it is removed to some distance s, it appears brighter because of the scattering toward the eye of light from intervening particles. It is assumed that these particles are nonabsorbent and that the concentration $n(r)dr$ of those with radii between r and $(r + dr)$, their size distribution, and the total flux density E of the light incident upon each of them (which may be from the sun, the earth, or other particles) are all uniform along a horizontal path to the object and beyond it. Then, considering the particles in an elementary volume dV bounded by a cone extending along the path, and by normal faces separated by a small distance ds, which subtend a small solid angle $d\omega$ at the eye, the intensity of the light (i.e., the flux of energy per unit solid angle), from Eq. 2.40, is

$$dI'_B = \frac{E\,dV}{s^2\,d\omega}\int f(m, x, \theta, \phi, \lambda)n(r)\,dr \quad (4.13)$$

$$= [E\int f(m, x, \theta, \phi, \lambda)n(r)\,dr]\,ds \quad (4.14)$$

The integral is clearly a very complicated function, but under the assumed conditions the whole of the term in brackets is a constant along the path, and can be assumed to be proportional to the total scattering cross section $\int C_S n(r)\,dr$ of the particles in a unit volume, which by Eq. 2.43 and 2.44 was written as a total scattering or attenuation coefficient τ for nonabsorbent particles. Hence

$$dI'_B = a\tau\,ds \quad (4.15)$$

where a is the unknown constant of proportionality. However, the scattered light is attenuated before reaching the eye with intensity dI_B by further scattering from the particles along the intervening path of length s:

$$dI_B = a\tau e^{-\tau s}\,ds$$

Thus

$$I_B = a\int_0^s \tau e^{-\tau s}\,ds = a(1 - e^{-\tau s}) \quad (4.16)$$

Similarly, if the path is directed toward the neighboring sky, we obtain

$$I_S = a\int_0^\infty \tau e^{-\tau s}\,ds = a$$

whence

$$c = \frac{I_S - I_B}{I_S} = e^{-\tau s} \quad (4.17)$$

Strictly,

$$c = e^{-\tau(\lambda)s} \quad (4.18)$$

Now the smallest value c_R of the contrast c which can be perceived varies from person to person, and with other factors such as the state of adaptation of the eye, the apparent size, the color, the sharpness of the boundaries of the object observed, and other stimuli in the field of view. Under the ordinary conditions of viewing toward the horizon in daytime, the average value of c_R for objects subtending more than 1 min of arc to the eye is generally assumed to be about 0.02, although in tests with different observers outdoors the individual assessments of the visibility implied a mean value of about 0.03 and a scatter of values between 0.01 and 0.1 or even more (29). If the conventional value of 0.02 is accepted, then from Eq. 4.18, the corresponding *visual range* V_R is

$$V_R = 3.9/\tau(\lambda) \quad (4.19)$$

a definite quantity which can be measured at night with a light projector and receiver.

Although this expression for the visual range has been obtained for particularly simple circumstances, it can be shown that if the viewed body is black, it remains valid even if the aerosol particles are partly absorbent, and Eq. 4.17 has been found to express fairly satisfactorily measurements of the variation of contrast with range made with various kinds of target exposed outdoors (29).

Generally the attenuation coefficient $\tau(\lambda)$ has components due to the following:

1. Scattering by the air molecules, $\tau_0(\lambda)$, which is given by Eq. 2.49. Considering the wavelength $\lambda = 0.55\,\mu$ for which the eye is most sensitive and ordinary conditions

4   Atmospheric aerosol

at the ground, $\tau_0 \approx 1.2 \times 10^{-2}$ km^{-1} (τ_0 decreases with height in proportion to the air density).

2. Selective absorption by air molecules.
3. Scattering by the aerosol particles, τ_p.
4. Absorption by the aerosol particles.

The second component and the fourth component [except in the immediate vicinity of smoky atmospheres (28) and in hazes of dust raised from arid ground] are generally negligible compared with the other two, so that the attenuation coefficient can be regarded as the sum ($\tau_0 + \tau_p$).

Visual range near the ground

Within the planetary boundary layer the component associated with the aerosol particles, given by Eq. 2.45, is usually considerably greater than τ_0. For the aerosol particle size distribution specified in Fig. 4.1 as characteristic of nucleus-rich continental air at ordinary relative humidities $\tau_p \approx 12 \times 10^{-2}$ km^{-1}, of which about 0.7 is due to the large nuclei of radius between 0.2 and 0.7 μ: comparatively small contributions are made by the Aitken nuclei because of their small scattering efficiency, and by the giant nuclei because of their small concentration. The corresponding visual range is therefore given as $3.9/1.2 \times 10^{-2} \approx 30$ km. As the relative humidity approaches 100% the radii of the aerosol particles increase. Those of the optically effective large nuclei, which are shown in Chapter 5 to correspond closely to the equilibrium radii, increase considerably and the visual range can be anticipated to be reduced substantially. For example, accepting the relation (4.9) for the radius at ordinary relative humidities, it can readily be found from Eq. 4.7 that for all the optically effective nuclei, the radius is greater at a relative humidity of 95% by a factor of 1.3, corresponding to a reduction in the visual range by a factor of about 1.7. As the relative humidity more closely approaches 100% the equilibrium radius of the larger nuclei increases relatively more: the ratio of the equilibrium radii at 100 and at 90% is about 3 for nuclei containing 10^{-15} g of solute and about 10 for those containing 10^{-12} g of solute, so that a large reduction in the visual range to about 1 km could be anticipated just before the formation of fog. However, conditions are rarely so uniform in the ground layer that this result would be obvious.

The visual range of about 30 km inferred as typical in nucleus-rich air over the continents may seem from experience to be too great, probably because most routine observations are made in cities or at airfields in their immediate neighborhood, close to prolific artificial nucleus sources, and often where there are few suitable objects to observe at ranges beyond several kilometers. Also, when the visual range varies with direction, observers report its least value. Nevertheless, statistics show that over most of the central United States, for example, the visual range is between 20 and 40 km in about half of all routine observations, even though occasions of fog and precipitation are not excluded (30).

Visual range in strongly polluted air

In and downwind of cities and other prolific sources of artificial nuclei the concentration of nuclei close to the ground is considerably greater than that considered above as generally representative of the nucleus-rich continental air in the planetary boundary layer. In such situations the scattering coefficient in the air close to the ground commonly reaches 1 km^{-1}, and absorption has been observed to increase the extinction coefficient by a factor of 2 or more (28). This absorption was restricted to the immediate vicinity of the artificial nucleus sources, and was probably associated with rather large particles of soot and ash produced by inefficient combustion. The haze plumes which can often be seen to extend far downwind of industrial regions are produced simply by the intensified scatter from the artificial nuclei.

Visual range in "cleaned" air

In the nucleus spectrum specified in Fig. 4.1 as representative of the cleaned air of the upper troposphere, the concentrations of the large nuclei are more than one order of magnitude less than in the nucleus-rich continental air, and accordingly in such air the scattering coefficient of the aerosol particles is only a fraction of the molecular coefficient (as actually measured from aircraft; 31), so that effectively the latter alone determines the visual range. In such clean air the visual range is over 300 km, and it is difficult to avoid greatly underestimating the distance of mountains and cumulonimbus tops seen from an aircraft. The limitation on the distance to which such clouds can be seen is set as much by the earth's curvature and the height of the clouds and observer above the top of the hazy boundary layer as by the visual range. In the clean air in which the molecular scattering is a substantial fraction of the whole the clarity of distant clouds or other objects in photographs is improved if infrared filters and films sensitive in the infrared are used. The effective wavelength of the light is then usually in the range 0.8–0.9 μ, for which the molecular scattering coefficient is less than that in the range of visible wavelengths by nearly an order of magnitude.

A transparency almost as great as that normal in the upper troposphere is found in those regions, especially in high latitudes and over or close to the ocean in light winds, where the cleaned air settles back into the boundary layer but is not rapidly supplied with fresh nuclei.

Even in moderately fresh winds the population of sea-spray particles added over the ocean, represented by the spectrum in Fig. 4.1 marked maritime component, has a scattering coefficient (10^{-2} km^{-1}) only about equal to the molecular scattering coefficient, on account of the relatively sparse concentrations of the smaller of the large nuclei. Thus in such maritime aerosol the visual range can be anticipated to be well over 100 km, and noticeably greater than in the interior of the continents, except perhaps over deserts. In settled weather the maritime component of the aerosol may be less, or even absent. The bases of cumulus can be seen clearly at elevations down to 0° (at a distance of about 100 km or more; Fig. 2.14), or to even less from elevated places. Dense cirrus and cumulonimbus tops can easily be detected at elevations as small as 0.5°, corresponding to distances of about 300 km. Then in fair weather an observer with an unobstructed horizon can survey clouds over an area which is so great that he can examine the behavior of many individual clouds, use perspective to establish their typical three-dimensional structure and arrangement, and recognize the influence of various ground features such as lakes, coasts, and mountains. Moreover, he can gain a direct appreciation of the evolution of the large-scale cloud systems which is denied to the observer in an industrial region, and which still usefully supplements the insight recently provided by observations from geosynchronous satellites.

Slant visual range

The theory of visual range in circumstances less simple than those previously assumed is very complicated (29). In particular, in the theory of the slant visual range at which objects on the ground can be distinguished from an aircraft flying above the boundary layer, or objects aloft can be identified from the ground, it has to be considered that the aerosol properties vary substantially along the line of sight, and in sunshine the angular dependence of their scattering functions has to be taken into account. The more intense forward scattering of the aerosol particles (Tables 2.8, 2.9) causes the slant visual range to be least in the direction upsun, especially when the sun is at a low elevation (32). It is responsible for a colorless aureole around the sun, in which the increase in the brightness of the sky from a radius of 30° to within a few degrees of the sun exceeds a factor of 10, even in cleaned air (measurement of the ratio by instruments on free balloons has provided estimates of aerosol particle concentration at heights up to about 20 km; 23). It is also strikingly manifest by the brilliance of the top of the hazy planetary boundary layer viewed toward the sun when it is at a low elevation (Pl. 4.1); in the opposite direction the considerable backscatter from illuminated particles is shown by the distinct shadow thrown on the haze top by a mountain peak (Pl. 4.2).

Other circumstances in which the simple theory of visual range is not applicable are those in which the illumination is not uniform along and beyond the line of sight, for example, after sunset, when objects seen toward the sun become clearer, silhouetted against the twilight sky. The great increase in the scattering when the line of sight becomes illuminated by direct sunshine, after being in the shadow of dense cloud, is often obvious from the striking loss of clarity in cloud details when the sun emerges from behind a receding layer of cloud.

Appendix 4.1 Twilight phenomena

When the sun is near the horizon in fair weather the colors which appear in the sky and illuminate haze and clouds present a fascinating spectacle. Its discussion and even description have been neglected in scientific English literature. The most striking aspects are here briefly described as they occur around sunset in a cloudless sky.

As the sun approaches the horizon the silvery aureole becomes increasingly golden, and the sky immediately above the horizon below the sun also acquires a warm yellow tinge, while above the opposite horizon the *countertwilight* begins, a shallow segment becoming orange and changing upward through yellow and green to the ordinary blue of the sky several degrees above the horizon. The colors of the countertwilight remain much weaker than those near the sun because of the lower intensity of the backward scattering from the aerosol particles. Because of the great refraction of the sunlight at the shorter wavelengths, especially when the refraction is abnormally great, the sun has a narrow green upper rim, which near sunset may become an obvious green segment and at sunset occasionally is exaggerated into the green flash (33, 34). The outline of the sun when it is close to the horizon may become distorted or even horizontally divided (by the "blind strip") when there is a low-level inversion of lapse rate.

Variations in refractive index associated with the detailed stratification of the atmosphere may produce in the outline of the sun one or two major horizontal indentations, or many fine horizontal notches, whose

4 Atmospheric aerosol

elevation remains fixed while the sun sinks. Near its upper and lower rim there may be small traveling distortions showing the presence of irregularities of refractive index (Appendix 2.6), moving with the wind at some level which is often shown to be well above the surface by the steadiness of the image of distant objects and low clouds near the horizon.

Civil twilight; the purple light

As the sun sinks out of sight, the colored *horizon band* develops, extending on either side along the horizon in a very long, shallow arch, its color changing upward from orange to a yellow and then a greenish tint. Over a radius of from 20 to 30° from the sun the horizon band is at first swamped by the sinking aureole, whose upper part extends to as much as 40° above the horizon as a translucent blue-white glow, the *clear shine*, whose indefinite border is sometimes marked by a very faint brown-grey ring several degrees across, the *brown ring*.

When the sun is nearly 2° below the horizon the golden aureole has disappeared, leaving an uninterrupted horizon band whose color now changes upward from a brownish red to orange and then yellow. Above the eastern horizon the dull grey-blue *dark segment*, sometimes called the earth's shadow, has risen, and immediately above it the narrow red, orange, orange-yellow, greenish-blue, and finally grey-blue bands of the countertwilight become more pronounced.

As the sun descends toward 3° below the horizon the clear shine sinks a little and tends to assume a green tinge, but rather suddenly and dramatically a zone at its upper border, some 20–30° above the horizon, begins to develop a rosy red glow.

This soon spreads and intensifies into the entrancing *purple light*, which measurements show, contrary to impression, arises not from a rose-red glow of increasing intensity but only from an increasing proportion of red in the generally fading twilight. When the sun is between 3 and 4° below the horizon the purple light reaches its maximum development, appearing brightest about 20° above the horizon. Mountains and isolated tall clouds near the horizon sometimes throw shadows into the purple light, which appear as pale green crepuscular rays (35). Later the purple light begins to sink into the remnants of the clear shine. In the meantime in the opposite part of the sky the dark segment has been rising more rapidly than the sun has been sinking, now reaching 5–8° above the horizon, and seems to have been extinguishing the colors of the countertwilight.

As the sun sinks beyond 5° below the horizon, the purple light fades and collapses, but seems to cause a recrudescence of the horizon band, which had been fading but now glows with fresh intensity, shining first orange and later, as the purple light sinks completely

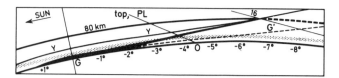

Fig. 4.4 Passage of sunlight through the atmosphere during twilight. The sunlight grazes the earth's surface at the point G, and the angular depression of the sun below the horizon is marked in degrees at the appropriate points along the earth's surface. The dark zone is that over which the sunlight is reddened, and below the line YY it is tinged yellow. The stippling shows the layer in the lower stratosphere in which there is a concentration of large nuclei. An observer at O sees the top of the purple light where the particles in the uppermost part of the layer are still in red sunlight. If there are large obstructions near the point G, such as mountains and tall clouds, they throw shadows through the layer so that the purple light appears in diverging rays, between which the sky has a blue color more or less tinged with green (35). The line GG' marks the position the grazing ray of the sunlight would occupy if it were not for extinction during passage through the atmosphere, and that marked 16 shows the horizontal ray from an observer where the sun's depression is 16° (at greater depressions it is not possible to see clouds at 80 km). The diagram is not drawn to scale.

into it, a fiery red. The countertwilight fades away, and the border of the dark segment becomes hardly distinguishable as its elevation rises beyond 15–20°.

Most of the twilight phenomena can be explained qualitatively in a fairly satisfactory way by the theory of simple scattering (36); in particular, from the occurrence of the purple light it was long ago inferred that there must be in the lower stratosphere a layer containing a concentration of particles of the size of the large nuclei. These scatter downward, at angles within about 40° of the incident beam, sunlight which has been strongly reddened during its passage through the troposphere (Fig. 4.4). The height of the layer was estimated (36) as about 25 km and the optically dominant radius of the scattering particles as about 0.5–0.8 μ, values close to those which have since been observed directly.

Astronomical twilight; the zodiacal light

As the sun's depression increases beyond 6° and civil twilight ends, the horizon band fades into a shallow orange segment, with yellow and green tinges above, beneath a faint blue-white glow at first reaching up to an elevation of about 20°, apparently to the border of the dark segment, which seems to have passed rapidly and imperceptibly through the zenith. Nevertheless, the sky above this border, where the upper stratosphere is still in direct sunlight, is for some time still not as dark as the true night sky, and indeed in the summer in high latitudes may appear abnormally bright, especially on occasions when the noctilucent clouds of the high mesosphere appear. The twilight glow continues to fade and its upper border to descend more rapidly than the sun, but it does not disappear below the horizon until the sun's depression exceeds about 16°, and astronomical twilight ends.

In the rapidly increasing darkness the *zodiacal light* may now be discerned extending upward from a region on the horizon, near but not exactly over the position of the sun, in a tapering cone with a noticeable lean, its axis lying approximately along the ecliptic and extending to some 30° or more above the horizon. It is most clearly seen in open country, well away from the glare of artificial lights, on moonless nights, and in low latitudes, where the plane of the ecliptic (and the belt of constellations forming the zodiac) rises high above the horizon. In the best viewing conditions the zodiacal light can be seen to extend in a very feeble band all along the zodiac. It is composed of sunlight scattered from interplanetary dust distributed mainly in the plane of the planetary orbits. From the polarization of the light, which merges continuously into the sun's outer corona, it is inferred that some of it is due to scattering from free electrons, while its increase of brightness with decreasing angular distance from the sun implies that the scattering from the particles is mainly forward; most of the light at angles within about 5° of the sun is due to particles with radii of several microns, and that in the zodiacal light can be explained by size distributions of smaller particles similar to that characteristic of the large condensation nuclei (37, 23).

From satellite and other observations it appears possible that the concentration of the interplanetary dust is greater in the vicinity of the earth, increasing by perhaps an order of magnitude within about 300 km of the earth's surface. A local concentration, or enhanced scatter in a backward direction, has been proposed to explain the *counterglow*, a very weak and diffuse patch of light in the penumbra of the earth, visible in the night sky almost diametrically opposite the sun, but there is some evidence of gaseous emission, and a controversy over a particulate or gaseous origin has still to be resolved. Remarkably, very faint glows have recently been detected near the antisolar point in directions approximately corresponding to the libration points which form equilateral triangles with the moon and the earth, where concentrations of interplanetary dust might be expected (38).

Abnormal twilight phenomena

The clouds from forest fires and from most volcanic eruptions remain within the troposphere. In the upper troposphere the small particles may be carried far downwind and recognized as layers of haze. For example, haze from forest fires in Alberta, Canada (September 1950; 37), and in Newfoundland (August 1961) has been observed in Europe, and has given the sun's disc a blue color. However, the particles of such haze do not remain in the troposphere in appreciable concentration for many days, and are likely to be nearly all removed within a week or two by aggregation and by incorporation into precipitation.

In contrast, as mentioned earlier, the clouds produced by the most violent volcanic explosions, and by large nuclear bombs, rise through the tropopause and deposit large quantities of small particles in the lower stratosphere. These may substantially augment the normal concentration of large nuclei there, even after diffusing over a whole or both hemispheres during a year or longer, and thereby spectacularly enhance the twilight phenomena, especially the purple light and the red glow following its collapse into the horizon band. The most prolonged period of such intensified twilights which has been recorded was that following the eruptions of the volcano Krakatoa in 1883, which began in May and culminated in several tremendous explosions during the paroxysmal phase on 26 and 27 August.

The quantity of material ejected during these explosions was estimated to be about 50 km^3, or about 10^{11} tons, two orders of magnitude greater than during the recent eruption in Kamchatka mentioned earlier in the chapter. A large fraction of this fell within 50 km, and another fraction fell as a fine dust at much greater distances, but perhaps as much as a tenth was carried away as a dust cloud in the lower stratosphere, at heights up to about 30 km. The cloud moved westward at an average speed of over 30 m sec^{-1}, recognizably making two circuits of the globe in low latitudes during which it took three or four days to pass most places, corresponding to a length of about 10,000 km. Under this cloud the sun was noticed to be unusually dim, and near its edges to be colored, in some places green, in others blue, in other copper.

With the spread of the cloud, twilight became prolonged and its colors, especially the reds, greatly enhanced. After about a week the haze layer was observed at Honolulu, by the Rev. Bishop, as "a continuous sheet at a height far above the highest cirrus, a slight wavy ripple being noticeable in its structure." He also described the diffraction phenomenon which has come to be known as *Bishop's ring*, a broad red or brownish- or purple-red ring around the sun at a radius of between about 15 and 30°. From this angle a radius of between about 0.5 and 1 μ has been inferred for the particles, but a low value of β in the size distribution of the large nuclei which scatter efficiently at this angle may have the same result (37).

The dust, at first visible even in the daytime, spread latitudinally until within about a month it occupied the whole belt within the tropics, and soon after reached middle latitudes in the southern hemisphere, which was in its winter season. There seems to have been a pause in the progression of the border of the dust cloud into the northern hemisphere, until late in November, when the dust and its accompanying twilight glows spread quickly eastward across the United States and Europe (39). Here after some weeks it became invisible during full daylight, but appeared as a diffuse cloud about sunrise and sunset, sometimes in bands of differing density, but without any detail except occasional striations across the line of sight in the bright part of the sky when the sun was a little below the horizon, with hardly any apparent motion. The dust and Bishop's ring, though gradually fading, are said to have outlasted the unusual brilliance of the twilights, but all became hardly perceptible toward the end of 1885.

Abnormal twilights and the paintings of J.M.W. Turner

Optical phenomena of the kind that followed the eruption of Krakatoa had of course been noticed before, but never so thoroughly documented (40). In particular, another great eruption in the East Indies, that of Tomboro in Sumbawa during April 1815, probably produced amounts of stratospheric dust comparable with those from the explosions on Krakatoa. The eruptions were audible at distances up to about 1500 km, compared with about 3000 km from Krakatoa, but the total amount of material ejected was thought to be greater at Tomboro. Enhanced twilights began in England a few months later, just as after the explosions on Krakatoa, and were the most remarkable recorded up to that time. In subsequent years unusual sunsets continued to be noted in Britain, perhaps more because of the interest aroused by the earlier displays than because of fresh eruptions, although in 1821 and 1822 there were violent eruptions at Vesuvius and in Iceland and Java. In 1831 clouds over a submarine volcano off Sicily reached a height of at least 20 km, and may have been responsible for a dense haze which subsequently spread over the whole hemisphere. It was variously reported that the sun was so dimmed that it became visible only at elevations above 15°, that it could be observed all day with the unprotected eye, and that the sky was never dark at midnight: small print could be read outdoors, even when the sun was 19° below the horizon. Again fiery red sunsets were noted, for example, in Germany for as long as one and a half hours after sunset. It seems likely that the extraordinary sunsets of this period stimulated the renowned studies of clouds and skyscapes by J.M.W. Turner (1775–1851), and were the inspiration of his glowing paintings of the rising and setting sun; according to the critic Ruskin it was not until after about 1820 that he became the first artist to paint clouds *scarlet*, and the first to represent not only sunshine, but the true colors of sunshine, which many thought were fancifully exaggerated in his paintings.

Recent abnormal twilights

In recent years the most notable twilight disturbances in Europe followed the eruptions of Mt. Spurr, in Alaska, on 9 July 1953, of Bezymianny in Kamchatka on 30 March 1956, and of the Agung volcano in Bali on 17 March 1963.

Toward the end of July 1953 abnormally high layers of dense haze were observed over England above aircraft flying at heights between about 12 and 14 km. The haze was most easily visible at low elevations, and hardly at all when looking overhead. On a special flight the layer was found to be only about 150 m thick, with a top at about 15 km (41; Pl. 4.3). Other reports in the following two months estimated a similar height and thickness, and mentioned that the haze was sometimes concentrated in very long streaks and that the undersides of the layers were slightly "waved" or "rolled."

These streaks, and especially the undulations, made it possible to see the haze layer from the ground at elevations below about 30°, but only when the sun was near the horizon; then the path of the sunlight through the layer varies greatly according to position with respect to the crests (Fig. 4.5). Under favorable conditions around sunset the haze undulations were visible (always with their length approximately normal to the line of sight) from Dunstable, in Bedfordshire, throughout August and intermittently until the end of September, but not subsequently. The wavelength of the undulations varied somewhat with position in the sky, but assuming an average height of 15 km varied between 1.4 and 4.2 km in 17 measurements on 5 separate occasions, with an average value of 2.6 km. A height rather lower than that of the layer of large nuclei normally responsible for the purple light was suggested by the observation that usually the undulations were best seen in the clear shine, disappearing as the purple light began, but later reappearing and becoming tinged with the rose color of the lower part of the purple light before again becoming invisible as the purple light sank below about 10°.

Fig. 4.5 Diagram illustrating the great variation in the length of the path of sunlight through a gently undulating haze layer when the sun is at a low elevation.

The subsequent red twilight glows were slightly more intense than normal.

The undulations moved in position only very slowly, if at all; their largest displacement seen through a theodolite over an interval of 2 min corresponded to a speed of less than 5 m sec^{-1}. Another striking feature was a persistent tendency for the undulations to be more prominent, as in Pl. 4.4, over a narrow region with a well-defined farther edge (the upwind edge), usually in a position near the Cotswold hills, which are frequently responsible for the formation of orographic wave clouds, and even of cirrus clouds in the high troposphere.

The general appearance and properties of these haze undulations, including their dimensions and slowness of displacement, and their association with enhanced twilight glows, correspond exactly to those previously described for clouds called *twilight cirrus* or *ultracirrus* (42), although it was recognized that they were not like cirrus and probably were not clouds in the ordinary sense. *Stratospheric haze streaks* or *waves* seem better descriptions.

Haze thick enough to be mistaken by inexperienced observers for ordinary cloud seems to have reached England within four days of the eruption of Bezymianny, and even a month later the haze waves, although not always visible on apparently favorable occasions, were at other times more marked than after the Alaskan eruption.

Great intensification and prolongation of the red twilight glow occurred all over the southern hemisphere in low latitudes within about a month of the eruption of the Agung volcano in 1963, and persisted for about a year. The spread of the haze into the northern hemisphere was much slower; the unusually intense twilights began in the southern United States (43) and in India (44) in September, and in Europe in November, and lasted there at least until the following February, with pronounced stratospheric haze waves visible on favorable occasions.

Still more recently, abnormal twilights and stratospheric hazes were visible in England during late November and December 1964, following the detonation of a nuclear bomb in China. On its first arrival over England the dust layer was unusually dense (Pl. 4.5).

The influence of volcanic dust veils on solar radiation

Haze layers in the lower stratosphere which are sufficiently dense to dim the sun noticeably throughout the day, as observed over very large areas for a month or more after major eruptions, naturally disturb the radiative economy of the atmosphere. Prolonged disturbances of this kind in periods of exceptional volcanic activity have been suggested as causes of climatic variation and even of the ice ages (45, 46). There is no persuasive evidence of any significant change in the character of the general atmospheric circulation following the explosion of Krakatoa and the more recent violent eruptions, but there are measurements of appreciable reductions in the intensity of the direct solar radiation received at the ground in middle latitudes in the northern hemisphere, amounting to about 10% or more of the average during the periods of about two years following the eruptions of Krakatoa in 1883, Mt. Pelée (Martinique) in 1902, and Katmai (Alaska) in 1912.

For some months after the Agung eruption in 1963 a similar reduction of the direct solar radiation in south Africa was completely compensated by an increase in the diffuse radiation from the sky (rather than partially, as might have been anticipated). For this very recent occurrence routine observations of wind and temperature are for the first time available from the lower stratosphere, and those from Ascension Island (8°S, in the Atlantic) suggest that for two years following the eruption the monthly mean temperatures at the 50-mb level (about 21 km) were about 3°C above normal, with some variation in the mean zonal wind, which was not inconsistent (47). These disturbances were not apparent at the 100- and 30-mb levels, and so might be attributed to increased absorption of sunshine in a haze layer confined to the lower stratosphere. It appears that individual violent eruptions have a readily detectable effect on the passage of solar energy through the lower stratosphere over at least one hemisphere for up to a year or two, apart from the energetically trivial optical effects during twilight. Any possible repercussions on the general atmospheric circulation are evidently too small to be assessed directly from the observations but could be studied in a theoretical experiment.

References

1. Junge, C.E. 1963. *Air Chemistry and Radioactivity*. Academic Press, London. This is the most comprehensive account of the properties of atmospheric aerosol and is largely responsible for giving the subject a coherent structure.
2. Eriksson, E. 1959. The yearly circulation of chloride and

sulphur in nature; meteorological, geochemical and pedological implications, Part I, *Tellus*, **11**, 375–403. Also 1960, Part II, *Tellus*, **12**, 63–109.
3. Green, H.L., and Lane, W.R. 1964. *Particulate Clouds*. E. and F. N. Spon, London.
4. Bagnold, R.A. 1954. *The Physics of Blown Sand and Desert Dunes*. Methuen, London.
5. Twomey, S. 1960. On the nature and origin of natural cloud nuclei, *Bull. Obs. Puy de Dôme*, 1–19.
6. Koto, B. 1916. The great eruption of Sakura-jima in 1914, *J. College Sci.* (Imperial University, Tokyo), **38**.
7. Gorshkov, G.S. 1959. Gigantic eruption of the volcano Bezymianny, *Bull. Vulcan.*, **20**, 77–109.
8. Mossop, S.C. 1964. Volcanic dust collected at an altitude of 20 km, *Nature*, **203**, 824–27.
9. Went, F.W. 1966. On the nature of Aitken condensation nuclei, *Tellus*, **18**, 549–56.
10. Smith, W.R. 1965. The planetary food potential, *Ann. N.Y. Acad. Sci.*, **118**, 645–718.
11. Twomey, S., and Severynse, G.T. 1964. Size distributions of natural aerosols below 0.1 micron, *J. Atmos. Sci.*, **21**, 558–64.
12. Bullrich, K. 1964. Scattered radiation in the atmosphere and the natural aerosol, *Adv. Geophys.*, **10**, 101–260.
13. Chamberlain, A.C. 1966. Transport by lycopodium spores and other small particles to rough surfaces, *Proc. Roy. Soc.*, A, **296**, 45–70.
14. Friedlander, S.K. 1960. Similarity considerations for a particle-size spectrum of a coagulating, sedimenting aerosol, *J. Meteorol.*, **17**, 479–83; Hidy, G.M. 1965. On the theory of the coagulation of non-interacting particles in Brownian motion, *J. Colloid Sci.*, **20**, 123–44.
15. Junge, C.E. 1969. Comments on "Concentration and size distribution measurements of atmospheric aerosols and a test of the theory of self-preserving size distributions," *J. Atmos. Sci.*, **26**, 603–8.
16. Fenn, R.W. 1960. Measurements of the concentration and size distribution of particles in the Arctic air of Greenland, *J. Geophys. Res.*, **65**, 3371–76; Kumai, M. 1965. The properties of marine air and marine fog at Barrow, Alaska, *Proc. Int. Conf. Cloud Phys.*, 24 May–1 June 1965, Tokyo and Sapporo, Japan, 52–55.
17. Whipple, F.J.W. 1930. The great Siberian meteorite and the waves, seismic and aerial, which it produced, *Q. J. R. Meteorol. Soc.*, **56**, 287–301; 1934. On phenomena related to the great Siberian meteorite, ibid., **60**, 505–13.
18. Fessenkov, V.G. 1949. The mass of the atmospheric residue of the Sikhote-Alin meteorite, *Dok. Akad. Nauk. SSSR*, **66** (translation T 133 R, Defence Sci. Inf. Service, Canada, 1954).
19. Lovell, A.C.B. 1957. Geophysical aspects of meteors, in *Encyclopedia of Physics*, **48**, Geophysics II, Springer-Verlag, Berlin, pp. 427–54.
20. Rosinski, J., and Snow, R.H. 1961. Secondary particulate matter from meteor vapors, *J. Meteorol.*, **18**, 736–45.
21. Wyatt, S.T., and Whipple, F.L. 1950. The Poynting-Robertson effect on meteor orbits, *Astrophys. J.*, **111**, 134–41.
22. Volz, F.E., and Goody, R.M. 1962. The intensity of the twilight and upper atmospheric dust, *J. Atmos. Sci.*, **19**, 385–406.
23. Newkirk, G., and Eddy, J.A. 1964. Light scattering by particles in the upper atmosphere, *J. Atmos. Sci.*, **21**, 35–60.
24. Friend, J.P. 1966. Properties of the stratospheric aerosol, *Tellus*, **18**, 465–73.
25. Hemenway, C.L., Fullum, E.F., Skrivanek, R.A., Sobermann, R.K. and Witt, G. 1964. Electron microscope studies of noctilucent cloud particles, *Tellus*, **16**, 96–102.
26. Low, R.D.H. 1969. A generalised equation for the solution effect in droplet growth, *J. Atmos. Sci.*, **26**, 608–11, 1345–46.
27. Sekera, Z. 1951. Polarisation of skylight, in *Compendium of Meteorology*, American Meteorological Society, Boston, pp. 79–90.
28. Waldram, J.M. 1945. Measurement of the photometric properties of the upper atmosphere, *Q. J. R. Meteorol. Soc.*, **71**, 319–36.
29. Middleton, W.E.K. 1952. *Vision through the Atmosphere*. University of Toronto Press, Toronto.
30. Eldridge, R.J. 1966. Climatic visibilities of the United States, *J. Appl. Meteorol.*, **5**, 277–82.
31. Packer, D.M., and Lock, C. 1951. The brightness and polarisation of the daylight sky at altitudes of 18,000 to 38,000 ft above sea level, *J. Opt. Soc. Am.*, **41**, 473–78.
32. Condron, T.P. 1960. Visibility, in *Handbook of Geophysics*, rev. ed. Macmillan, New York.
33. Minnaert, M. 1954. *The Nature of Light and Colour in the Open Air*. Dover, New York.
34. O'Connell, D.J.K. 1958. *The Green Flash and Other Low Sun Phenomena*. Interscience, New York.
35. *Weather*. 1967. **22**. The issue for June has a color photograph on the cover of a purple light with crepuscular rays.
36. Gruner, P. 1943. Dämmerungserscheinungen, in *Handbuch der Geophysik*, vol. 8. Gebrüder Borntrâger, Berlin-Zehlendorf, pp. 432–526.
37. van de Hulst, H.C. 1957. *Light Scattering by Small Particles*. Chapman and Hall, London.
38. Pskovski, Y.P. 1962. *Dust in Circumterrestrial Space*. Translation T 383 R, Def. Sci. Inf. Service, Canada.
39. Wexler, H. 1951. Spread of the Krakatoa volcanic dust cloud as related to the high-level circulation, *Bull. Am. Meteorol. Soc.*, **32**, 48–51.
40. Symons, G.J., ed. 1888. The eruption of Krakatoa and subsequent phenomena, in *Report of the Krakatoa Committee of the Royal Society*. Trübner and Co., London.
41. Jacobs, L. 1954. Dust cloud in the stratosphere, *Meteorol. Mag.*, **83**, 115–19.
42. Wolf, M. 1916. Über die höchsten Dämmerungswolken, *Meteorol. Ztg.*, **33**, 517–19.
43. Volz, F.E. 1965. Note on the global variation of stratospheric turbidity since the eruption of Agung Volcano, *Tellus*, **17**, 513–15.
44. Shah, G.M. 1969. Enhanced twilight glow caused by the volcanic eruption on Bali Island in March and September, 1963, *Tellus*, **21**, 636–40.
45. Wexler, H. 1956. Variations in insolation, general circulation and climate, *Tellus*, **8**, 480–94.
46. Budyko, M.I. 1969. The effect of solar radiation variations on the climate of the Earth, *Tellus*, **21**, 611–19. For a chronology of major volcanic eruptions since A.D. 1500, see Lamb, H.H. 1970. Volcanic dust in the atmosphere, with a chronology and assessment of its meteorological significance, *Phil. Trans. R. Soc.*, London, **266**, 425–533.
47. Ebdon, R.A. 1967. Possible effects of volcanic dust on stratospheric temperatures and winds, *Weather*, **22**, 245–49.
48. Prospero, J.M. 1968. Atmospheric dust studies on Bar-

bados, *Bull. Am. Meteorol. Soc.*, **49**, 645–52.
49. Stevenson, C.M. 1969. The dust fall and severe storms of 1 July 1968, *Weather*, **24**, 126–32.
50. Mason, B.J. 1960. Ice-nucleating properties of clay minerals and stony meteorites, *Q. J. R. Meteorol. Soc.*, **86**, 552–56.
51. Jaenicke, R., and Junge, C. 1967. Studien zur oberen Grenzgrösse des natürlichen Aerosoles, *Beitr. Phys. Atmos.*, **20**, 129–43.
52. Vonnegut, B., Blanchard, D.C., and Cudrey, R.A. 1969. Structure and modification of clouds and fogs, *Project Themis, Annual Summary Report 1*. The Research Foundation of State University of New York, Albany.
53. Bigg, E.K., Ono, A., and Thompson, W.J. 1970. Aerosols at altitudes between 20 and 37 km, *Tellus*, **22**, 550–63; Bigg, E.K., Kviz, Z., and Thompson, W.J. 1971. Electron microscope photographs of extraterrestrial particles, *Tellus*, **23**, 247–59.
54. Farlow, N.H., Ferry, G.V., and Blanchard, M.B. 1970. Examination of surfaces exposed to a noctilucent cloud, August 1, 1968, *J. Geophys. Res.*, **75**, 6736–50; Fechtig, H., and Feuerstein, M. 1970. Particle collection results from a rocket flight on August 1, 1968, *J. Geophys. Res.*, **33**, 6751–57.

5 The formation, growth, and evaporation of cloud particles

5.1 The growth of solution droplets

When the relative humidity increases in air containing a solution droplet, vapor condenses on the droplet; latent heat is liberated, and during its growth its surface becomes warmer than the surroundings, thus accompanying the diffusion of vapor toward the droplet there is a diffusion of heat away from its surface, predominantly into the surroundings. The rate of growth can be calculated as a function of the droplet radius and hygroscopicity, and of the vapor density in the surroundings. Usually the difference in the temperatures of the droplet surface and the surroundings is sufficiently small to allow the use of the Clausius-Clapeyron equation in its finite difference form (2.7), and the heat flux into the droplet can be neglected. In the following discussion it is further assumed that $m \geq 10^{-16}$g ($r_s \geq 10^{-5}$ cm), so that the relations (4.6) to (4.8) can be used.

The vapor density ρ_v remote from a droplet is expressed in terms of a supersaturation ratio s (which need not be positive):

$$\rho_v = \rho_{ST}(1 + s) \quad (5.1)$$

where T is the air temperature remote from the droplet, while the vapor density ρ'_v at the surface of a droplet of temperature $T' = T + \Delta T$ is

$$\rho'_v = \rho_{ST'}(1 + s_r) \quad (5.2)$$

where the term s_r is due to the combined effects of the surface curvature and the hygroscopicity, and is assumed to be small (less than about 0.2), so that from Eq. 4.6,

$$s_r = a/r - b/r^3 \quad (5.3)$$

If further, ΔT is small (not more than about 2°C), then to a good approximation, from Eq. 2.7,

$$\rho_{ST'} = \rho_{ST}(1 + f_1 \Delta T) \quad (5.4)$$

$$f_1 = \frac{L_v - R_v T}{R_v T^2}$$

Combining Eq. 5.1, 5.2, and 5.4 gives

$$(\rho_v - \rho'_v) = \rho_{ST}(1 + s) \\ - \rho_{ST}(1 + f_1 \Delta T)(1 + s_r) \quad (5.5)$$

The equations usually employed to describe the steady molecular fluxes of water and heat to and from the droplet surface are similar to Eq. 2.17 and 2.18:

$$dM/dt = 4\pi r^2 \rho_w \, dr/dt = 4\pi r D (\rho_v - \rho'_v) \quad (5.6)$$

and

$$H = L \, dM/dt = 4\pi r K \Delta T \quad (5.7)$$

where D is the coefficient of diffusion of water vapor in air and $K = k\rho_a c_p$ is the coefficient of thermal diffusion in air.

These equations rest on several assumptions:

1. Transfers of heat and moisture are by steady-state molecular diffusion under ordinary diffusion coefficients.
2. $L \, dM/dt \gg cM \, dT/dt$.
3. The vapor density at the droplet surface is that under which the droplet would persist in equilibrium.
4. There is no disturbance to the field of vapor density by neighboring droplets or by the motion of the droplet.

None of these assumptions is strictly valid. However, more rigorous treatments using solutions of the diffusion equations with time-variable boundary conditions lead to essentially the same final equations, while study of the molecular kinetics near the droplet surface indicates that errors resulting from the use of an ordinary diffusion coefficient become appreciable only when the droplet radius is so small as to be comparable with the mean free path of the diffusing molecules, and therefore usually considerably less than about 1 μ. Although an adjusted coefficient would seem necessary in considering the growth of Aitken and even large nuclei, in the principal problem of the evolution of a population of nuclei whose initial size and hygroscopicity are not accurately specified and range over orders of magnitude there is little justification for such a refinement. The assumption concerning the vapor density at the droplet surfaces appears to be sufficiently accurate for the small vapor density gradients and growth rates ordinarily encountered in the atmosphere. Further, the concentration of the droplets in the atmosphere is such that their separation is at least about 100 times greater than their radii, even when they have grown to the size of cloud droplets, so that the assumption that the growth of neighboring droplets is independent and that there is a vapor density characteristic of an "environment" seems reasonable. On the other hand, the last assumption, that the droplet has no motion relative to the air, is valid only if the droplet radius is less than about 40 μ; for larger droplets an empirical *ventilation coefficient*, which is a function of the Reynolds number, has to be introduced into the diffusion equations (as discussed in Section 5.8). The problems mentioned are discussed in reference 4.

From Eq. 5.6 and 5.7,

$$\Delta T = \frac{DL(\rho_v - \rho_v')}{K} \quad (5.8)$$

and substitution into (5.5) gives

$$(\rho_v - \rho_v') = \frac{\rho_{ST}(s - s_r)}{1 + Df_2\rho_{ST}(1 + s_r)} \quad (5.9)$$

$$f_2 = \frac{Lf_1}{K} = \frac{L_v(L_v - R_vT)}{KR_vT^2}$$

The growth equation (5.6) can be written

$$\rho_w r\, dr/dt = \frac{D\rho_{ST}(s - s_r)}{1 + Df_2\rho_{ST}(1 + s_r)} \quad (5.10)$$

For practically pure water droplets of radius greater

Table 5.1 Variation with temperature and pressure of heat-economy factor $f_3(p, T) = 1 + Df_2(T)\rho_{ST} = 1 + DL_v(L_v - R_vT)\rho_{ST}/KR_vT^2$

Temp. (°C)	D (cm^2 sec^{-1})	$Df_2(T)\rho_{ST}$	$f_3(p, T)$
	Pressure 900 mb		
20	0.286	2.77	3.77
15	0.277	2.10	3.10
10	0.268	1.55	2.55
5	0.259	1.18	2.18
0	0.251	0.87	1.87
	Pressure 800 mb		
10	0.301	1.74	2.74
5	0.291	1.32	2.32
0	0.282	0.98	1.98
−5	0.224	0.71	1.71
−10	0.264	0.50	1.50
	Pressure 300 mb[a]		
−30	0.610	0.26	1.26
−35	0.589	0.16	1.16
−40	0.567	0.10	1.10
−45	0.543	0.06	1.06
−50	0.520	0.03	1.03

[a]$Df_2(T)\rho_{ST}$ and $f_3(p, T)$ with respect to ice.

than about 0.1 μ, $s_r \ll 1$ and the equation can be written

$$\frac{\rho_w r\, dr}{dt} = \frac{D\rho_{ST}(s - s_r)}{f_3(p, T)} \quad (5.10a)$$

where

$$f_3(p, T) = 1 + Df_2\rho_{ST}$$

can be regarded as a correction factor to be applied after making the convenient but incorrect assumption that the surface temperature of a droplet is the same as the general air temperature. The correction factor varies mainly with temperature and is about 3 at ordinary temperatures in the low troposphere, but decreases to become only a little greater than unity at the low temperatures of the upper troposphere, as seen in Table 5.1.

Given s_r from Eq. 5.3, Eq. 5.10 allows the growth of a solution droplet to be calculated without reference to its heat economy.

5.2   The deviation of solution droplets from equilibrium

The magnitude of the lag behind its equilibrium size of a growing solution droplet can be estimated by considering its growth after being suddenly plunged into saturated air with its equilibrium size at an ordinary humidity. Then if the radius r is expressed as the fraction x of the equilibrium radius r_s in saturated air and with

5 The formation, growth, and evaporation of cloud particles

$s = 0$, we have

$$\left(\frac{aD\rho_{ST}}{r_s^3 \rho_w}\right)dt = \frac{x^4}{1-x^2}\left\{1 + Df_2\rho_{ST}\left[1 - \frac{a(1-x^2)}{x^3 r_s}\right]\right\}dx \quad (5.11)$$

or, upon integration,

$$\left(\frac{aD\rho_{ST}}{r_s^3 \rho_w}\right)t = \left[(1 + Df_2\rho_{ST})\left(\tfrac{1}{2}\log_e\frac{1+x}{1-x} - x - \frac{x^3}{3}\right) - x^2\left(\frac{Df_2\rho_{ST}}{2}\right)\frac{a}{r_s}\right]_{x_0}^{x_1} \quad (5.12)$$

where t is the time required for the radius of the solution droplet to increase from the fraction x_0 to the fraction x_1 of its equilibrium radius r_s.

Now from Eq. 4.7 and 4.8, at ordinary humidities to a good approximation for solution droplets containing a mass of solute $m \geq 10^{-16}$ g,

$$r_0 = m^{1/3}, \quad r_s = 10^3 m^{1/2}, \quad \text{and} \quad x_0 = r_0/r_s < 0.5$$

For this value of x_0 the value of the expression on the right of Eq. 5.12 becomes negligible compared with its value for $x_1 \approx 1$, while for the latter condition the value of the second term becomes negligible compared with the first (the term $Df_2\rho_{ST}$ is always of order 1, and if $x \approx 1$ the logarithmic term dominates, while a/r_s has a maximum value of $10^{-7}/10^{-5}$, or 10^{-2}). Approximately, therefore,

$$\left(\frac{aD\rho_{ST}}{r_s^3 \rho_w}\right)t = (1 + Df_2\rho_{ST})\left(\tfrac{1}{2}\log_e\frac{1+x_1}{1-x_1} - x_1 - \frac{x_1^3}{3}\right) \quad (5.13)$$

and for $x_1 = 0.94$, in particular,

Table 5.2 Time required for the radius r_0 of solution droplets in equilibrium at ordinary humidities to reach 94% of the new equilibrium radius r_s when plunged into saturated air (at 900 mb, 20°C)

Nuclei	r_0 (cm)	r_s (cm)	t
Aitken ↑↓	5×10^{-6}	10^{-5}	10^{-2} sec
	10^{-5}	3×10^{-5}	0.5 sec
Large ↑↓	2×10^{-5}	10^{-4}	0.25 min
	5×10^{-5}	3×10^{-4}	8 min
	10^{-4}	10^{-3}	4 hr
Giant ↑↓	2×10^{-4}	3×10^{-3}	5 days
	5×10^{-4}	10^{-2}	5 mon
	10^{-3}	3×10^{-2}	13 yr

Table 5.3 Increase in radius of giant nuclei after immersion in saturated air (at 900 mb, 20°C) for 10^3 sec (initial radius at ordinary humidities r_0; final radius $r_1 = x_1 r_s$, where r_s is the equilibrium radius in saturated air; mass of solute m)

m (g)	x_1	r_0 (μ)	r_1 (μ)
10^{-12}	0.8	1	8
10^{-11}	0.45	2	14
10^{-10}	0.23	5	23
10^{-9}	0.12	10	32
10^{-8}	0.06	21	58

$$t = 2(1 + Df_2\rho_{ST})\frac{r_s^3 \rho_w}{aD\rho_{ST}} \quad (5.14)$$

At a temperature of 20°C, $L_v = 586$ cal g^{-1}, $K = 6.14 \times 10^{-5}$ cal cm^{-1} sec^{-1} °K^{-1}, and $\rho_{ST} = 17.3 \times 10^{-6}$ g cm^{-3}, and if also the pressure is 900 mb, $D = 0.285$ cm^2 sec^{-1}; accordingly, $Df_2\rho_{ST} = 2.8$ and $\rho_w/aD\rho_{ST} = 2 \times 10^{12}$ sec cm^{-3}. Then

$$t \approx 15 \times 10^{12} r_s^3 \quad (5.15)$$

and the values in Table 5.2 are obtained.

Considering that in meteorological processes air is brought to a state of saturation over periods which mostly vary from a few minutes up to about an hour, the implication of Table 5.2 is that as the air approaches saturation only the Aitken nuclei are always very near to their equilibrium size.

This is more realistically illustrated by calculating the increase in the size of the large and giant nuclei after they have been immersed in saturated air for a period of 1000 sec, corresponding to an ascent of 1 km in the atmosphere at a speed of 1 m sec^{-1}. The results are given in Table 5.3 for nuclei of mass $> 10^{-12}$ g, for which

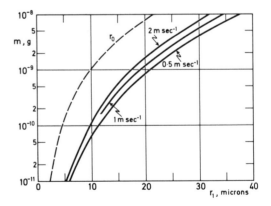

Fig. 5.1 Calculated radius r_1 of sea-salt particles of mass m, upon reaching the saturation level after adiabatic ascent at speeds of 0.5, 1.0, and 2.0 m sec^{-1} from a level (about 900 m lower) where the temperature was 25°C and the relative humidity 65% (1). The dashed curve shows the equilibrium radius r_0 at ordinary humidities, which is assumed in this text.

both x_0 and x_1 are small fractions, and the lower limit of the integral in Eq. 5.12 can be neglected. The second expression within the integral also can be neglected, while the first becomes to a good approximation $(1 + Df_2\rho_{ST})x^5/5$, so that the solution becomes simply

$$x_1^5 = 5aD\rho_{ST}t/r_s^3\rho_w(1 + Df_2\rho_{ST}) \quad (5.16)$$

or for the assumed conditions

$$x_1^5 = 6.6 \times 10^{-22}m^{-3/2}t$$

so that x_1 decreases by a factor of 2 when m increases by a factor of 10.

The growth of the nuclei would, of course, be much less during an actual ascent of air at 1 m sec^{-1} over about 1 km, as seen in Fig. 5.1, giving the results of numerical computations for nuclei composed of sea salt (1).

5.3 The formation of a cloud from a population of solution droplets

The behavior of a population of solution droplets in air which ascends, and in which the vapor pressure approaches and eventually exceeds the saturated vapor pressure, is conveniently examined by considering the vapor and droplets associated with a unit mass of dry air, and assuming adiabatic motion in surroundings of the same temperature and pressure. If the speed of ascent of the air approaches 1 m sec^{-1}, then up to the stage at which the air becomes saturated the growth of the droplets removes very little water vapor: not much more than about 10^{-3} g m^{-3}, compared with the vapor density of about 10 g m^{-3}. Their growth therefore has a negligible effect on the change of relative humidity or temperature below the saturation level. In air recently polluted with smoke the growth of the Aitken nuclei near the saturation level may substantially increase the numbers of solution droplets with radii near 0.25 μ, which effectively scatter visible light (Section 4.8) so that their presence is manifest by haze. In cleaner masses the transparency or appearance of the air does not change noticeably until much more abundant condensation commences very soon after the saturation level is reached: cloud bases are well defined at the saturation level.

During adiabatic rise above this level the solution droplets grow more rapidly and a remarkable transformation of their size distribution occurs in a short period following the attainment of saturation. The saturation mixing ratio r_s is approximately

$$r_s = \varepsilon e_s p$$

thus

$$dr_s/r_s = de_s/e_s - dp/p$$

and from the Clausius-Clapeyron and hydrostatic relations

$$dr_s/r_s = L_v dT/R_v T^2 + g dz/RT \quad (5.17)$$

If the mixing ratio of the water vapor is r_v, then the expression for the conservation of energy is similar to Eq. 3.25:

$$c_p dT + g dz + L_v dr_v = 0 \quad (5.18)$$

Since the supersaturation s is defined by $r_v = (1 + s)r_s$,

$$ds = \frac{dr_v}{r_s} - \frac{(1 + s)dr_s}{r_s} \quad (5.19)$$

Combining Eq. 5.17 to 5.19, and writing $U = dz/dt$ for the rate of ascent of air, gives

$$\frac{ds}{dt} = \frac{Ug(1 + s)(\varepsilon L_v/c_p T - 1)}{RT}$$
$$+ \left[\frac{1}{r_s} + \frac{L_v^2(1 + s)}{c_p R_v T^2}\right]\frac{dr_v}{dt} \quad (5.20)$$

Now

$$\frac{dr_v}{dt} = -\sum N_m \cdot \frac{dM}{dt}$$

or, from Eq. 5.10,

$$\frac{dr_v}{dt} = -\sum N_m \cdot \frac{4\pi r D\rho_{ST}(s - s_r)}{1 + Df_2\rho_{ST}(1 + s_r)} \quad (5.21)$$

where N_m is the concentration (g^{-1}) of solution droplets whose mass of solute and radius lie within small ranges about the average values m and r.

The simultaneous solution of Eq. 5.20 and 5.21 can be obtained numerically, and for a given initial size-distribution spectrum of the solution droplets at the saturation level (at $s = 0, t = 0$) provides the subsequent evolution of the spectrum and of the supersaturation s as a function primarily of the rate of ascent U and the temperature at the saturation level (2–4). A particularly interesting feature of the results is that in nucleus-rich air the supersaturation reaches a maximum of less than about 1%, at a level about a decameter above the saturation level; subsequently it decreases, and the particles become separated into two distinct groups, one containing small solution droplets whose maximum stable supersaturation S was not exceeded, and which remain with equilibrium radii of 10^{-1} μ or less (*inactivated nuclei*), and another containing larger solution droplets whose minimum radius becomes a few microns.

5 The formation, growth, and evaporation of cloud particles

Some insight into this characteristic behavior can be gained by considering the growth of a population of solution droplets assumed to consist of several discrete groups of uniform size and hygroscopicity (solute mass m, concentration N_m), representing approximately those in nucleus-rich continental air (Fig. 4.1), as listed in the first columns of Table 5.4. At the saturation level the solution droplets in each group have the radius r_0 specified in Table 5.4, which for the giant nuclei is that indicated in Fig. 5.1, and for the smaller nuclei is the equilibrium radius r_s.

Anticipating that the values of s, s_r, and s' of interest are very small ($\ll 1$), the relevant equations 5.10, 5.20, and 5.21 are simplified to

$$\left.\begin{array}{l} \rho_w r\, dr = \dfrac{D\rho_{ST}(s - s_r)dt}{f_3} \\[6pt] f_3(p, T) = 1 + Df_2\rho_{ST} \\[4pt] \qquad\quad = 1 + \dfrac{DL_v(L_v - R_v T)\rho_{ST}}{KR_v T^2} \end{array}\right\} \quad (5.22)$$

$$\dfrac{ds}{dt} = A - B \tag{5.23}$$

$$\left.\begin{array}{l} = Uf_4 + f_5\dfrac{dr_v}{dt} \\[6pt] f_4(T) = \dfrac{g(\varepsilon L_v/c_p T - 1)}{RT} \\[8pt] f_5(p, T) = \left(\dfrac{1}{r_s} + \dfrac{\varepsilon L_v^{\,2}}{c_p R T^2}\right) \end{array}\right\} \quad (5.24)$$

and

$$\left.\begin{array}{l} f_5\dfrac{dr_v}{dt} = -f_5 f_6 \sum N_m r(s - s_r) \\[6pt] f_6(p, T) = \dfrac{4\pi D\rho_{ST}}{f_3} \end{array}\right\} \quad (5.25)$$

Some values of the term f_3 have already been listed in Table 5.1. The terms f_4, f_5, and $f_5 f_6$ vary little over the general range of conditions of concern; thus f_4 decreases from 6.4×10^{-6} cm^{-1} at $-10°C$ to 4.9×10^{-6} cm^{-1} at $20°C$, and at pressures between 900 and 800 mb and temperatures between -10 and $20°C$ $f_5 f_6$ varies by less than 3% from a mean value of 3.4×10^{-3}. Further, f_4 and f_5 vary by less than about 1% over a possible range of conditions near a specified height (and temperature) for the saturation level, which as previously will be assumed to be at a pressure of 900 mb and a temperature of 20°C.

Now we shall consider the growth of each group of the solution droplets from the radius r_0 to the radius r' under the supersaturation s' which would arise during adiabatic ascent of air *without condensation* at the steady speed of 1 m sec^{-1}, which is typical at the bases of ordinary convection clouds. In a natural process accompanied by condensation it will appear that such a supersaturation could be maintained only by a violent upward acceleration, and that in the realistic circumstance of very small acceleration $ds/dt \approx ds'/dt = A$ at $t = 0$, that subsequently the condensation upon the solution droplets, whose effect is expressed in the term B of Eq. 5.23, causes s and r to become less than the values s' and r', and indeed very soon leads to the term B becoming equal to and then greater than the term A,

Table 5.4 Approximate trend of growth of a population of discrete groups of solution nuclei (of mass m of dissolved substance, concentration N_m, maximum stable equilibrium radius R, saturation equilibrium r_s, initial radius at saturation level r_0, equilibrium supersaturation $s_{r(0)}$ at the radius r_0, and maximum equilibrium supersaturation S) under a supersaturation s' which steadily increases with the time t at a rate A corresponding to adiabatic ascent at 1 m sec^{-1} without condensation. The radius r' is given at several time t after passing the saturation level; B' represents the rate at which condensation on each group of nuclei tends to reduce the supersaturation: in a natural process the supersaturation ceases to increase when $\Sigma(B'/A)$ reaches the value unity

Nuclei	m (g)	N_m (g^{-1})	R (μ)	r_s (μ)	r_0 (μ)	S	$s_{r(0)}$	$t = 0.4$ sec $s' = 2 \times 10^{-4}$ r' (μ)	B'/A	$t = 2$ sec $s' = 10^{-3}$ r' (μ)	B'/A	$t = 3$ sec $s' = 1.5 \times 10^{-3}$ r' (μ)	B'/A	$t = 4$ sec $s' = 2 \times 10^{-3}$ r' (μ)	B'/A
↑	10^{-9}	2	550	—	20	—	-1.3×10^{-2}	20.	$<10^{-3}$	20	$<10^{-3}$	20	$<10^{-3}$	20	$<10^{-3}$
Giant	10^{-10}	20	170	—	10	—	-1.0×10^{-2}	10	10^{-3}	10	10^{-3}	10	10^{-3}	10	2×10^{-3}
↓	10^{-11}	2×10^2	55	—	6	—	-4.4×10^{-3}	6	4×10^{-3}	6	4×10^{-3}	6	4×10^{-3}	6	5×10^{-3}
↑	10^{-12}	2×10^3	17	10	4	4×10^{-5}	-1.3×10^{-3}	4	6×10^{-3}	4	1×10^{-2}	4	1×10^{-2}	4	2×10^{-2}
Large	10^{-13}	2×10^4	5.5	3	3	1.2×10^{-4}	0	3	5×10^{-3}	3	3×10^{-2}	3	5×10^{-2}	3	8×10^{-2}
↓	10^{-14}	2×10^5	1.7	1	1	4×10^{-4}	0	1	—	1	7×10^{-2}	1	0.2	1	0.3
↑	10^{-15}	1×10^6	0.5	0.3	0.3	1.2×10^{-3}	0	0.3	—	0.4	—	0.5	0.1	0.8	0.6
Aitken	10^{-16}	2×10^6	0.2	0.1	0.1	4×10^{-3}	0	0.1	—	0.1	—	0.1	—	0.1	—
↓							$\Sigma(B'/A)$:		0.02		0.1		0.4		1.0

so that s soon reaches a maximum value and subsequently decreases.

Under all the assumed conditions we have $f_3 = 3.8$, $U = 10^2$ cm sec^{-1}, $f_4 = 4.9 \times 10^{-6}$ cm^{-1}, $f_5 = 211$, $D\rho_{ST} = 4.9 \times 10^{-6}$ g cm^{-1} sec^{-1}, $f_6 = 1.65 \times 10^{-5}$ g cm^{-1} sec^{-1}, and $f_5 f_6 \approx 3.4 \times 10^{-3}$; then numerically the equations become

$$r' dr' = 1.3 \times 10^{-6}(s' - s_{r'})dt \tag{5.26}$$

$$ds'/dt = A = 4.9 \times 10^{-4} \tag{5.27}$$

$$211 dr_v/dt = -3.4 \times 10^{-3} \Sigma N_m r'(s' - s_{r'})$$
$$= -\Sigma B' \tag{5.28}$$

and provide the approximate values of r', B'/A, and $\Sigma B'/A$ as a function of time which are listed in Table 5.4.

Several features of the *natural* process with the same population of solution droplets and rate of ascent of air can be anticipated from inspection of the table, considering that $s \approx s'$ and $r \approx r'$ while $\Sigma(B'/A) \ll 1$, and that s increases with time at least until $\Sigma(B'/A) \approx 1$, but must diminish once $\Sigma(B'/A)$ appreciably exceeds 1. In the short period of interest:

1. The *giant nuclei*, with initial radii between several and about 20 μ, hardly change in size and cannot interfere with the rise of the supersaturation s.
2. The bigger of the *large nuclei* also grow only slowly, with radii of a few microns.
3. Of the *other nuclei* some (the *activated nuclei*) attain radii of 1 μ or more very soon after the supersaturation appreciably exceeds their maximum equilibrium supersaturation S [this behavior of the activated nuclei is otherwise evident if an effective supersaturation $(s - s_r) \approx S$ is used in Eq. 5.26: after 4 sec the radius becomes at least $(10^{-5} S)^{1/2}$, that is, the original radius is already 1 μ or more, or very soon after activation becomes so, for all values of m which need be considered].
4. The supersaturation s can be anticipated to reach at least that value of s' for which $\Sigma(B'/A) = 1$ (since $B < \Sigma B'$), but by virtue of the previous conclusion cannot much exceed that value of S for which

$$3.4 \times 10^{-3} N_m r S \approx 4.9 \times 10^{-6} U$$

where r is of order 1 μ [since then within seconds $(B/A) > 1$ and from Eq. 5.23 s must decrease], or

$$N_m S \approx 15 U \tag{5.29}$$

which in this example, with $U = 10^2$ cm sec^{-1}, is satisfied for $S \approx 10^{-3}$; correspondingly, it appears from the table that s is likely to exceed 1.5×10^{-3}, but not 2×10^{-3}.
5. After attaining its maximum s_{max} the supersaturation s diminishes, and all those nuclei for which $S > s_{max}$ (here the Aitken nuclei with $m \leq 10^{-16}$ g) remain as submicroscopic particles, while the activated nuclei grow into cloud droplets of radius at least some microns.

In this example their concentration is large, about 10^3 cm^{-3}.

The magnitude of the maximum supersaturation attained, and hence of the concentration of cloud droplets produced, can be estimated for a continuous nucleus size spectrum by neglecting the presence of the giant nuclei and observing that for all activated nuclei the droplet radius is of order 1 μ and the effective supersaturation $(s - s_r)$ is about S_1, the maximum equilibrium supersaturation of the smallest activated nuclei, whose radius at ordinary relative humidities is r_1. The nucleus spectrum is supposed to be characterized by the relations

$$dN_m = c \, d(\log r)/r^\beta = c \, dr/2.3 r^{\beta+1}, \quad r > r_1 \tag{5.30}$$

while from Eq. 4.8 to 4.10

$$S_1 = 2a/3R = 2 \times 10^{-10}/3(3r_1^3)^{1/2} \tag{5.31}$$

After activation the nuclei are assumed all to have a radius r' of 1 μ and to be growing under an effective supersaturation $(s - s_{r'}) = S_1$ (it will appear that it is not necessary to specify these values precisely); then in Eq. 5.23

$$ds/dt = A - B \tag{5.23}$$

the terms on the right are

$$A = 4.9 \times 10^{-6} U \tag{5.32}$$

and

$$B \approx 3.4 \times 10^{-3} \int r' S_1 \, dN_m \tag{5.33}$$

$$\approx \frac{3.4 \times 10^{-7} \times 2 \times 10^{-10}}{3(3r_1^3)^{1/2}} \int_{r_1}^{r_2} dN_m$$

$$\approx 5.7 \times 10^{-18} c r_1^{-3/2} \int_{r_1}^{r_2} dr/r^{\beta+1} \tag{5.34}$$

where $r_2 \gg r_1$ is the radius of the largest nucleus considered; thus

$$B \approx (5.7 \times 10^{-18} c/\beta) r_1^{-(\beta+3/2)} \tag{5.35}$$

Introducing this value for B into Eq. 5.23 and writing $ds/dt = 0$ gives

$$r_1^{2\beta+3} \approx (1.2 \times 10^{-12} c/\beta U)^2 \tag{5.36}$$

which may be regarded as generally valid, considering the previously noted minor variation with the temperature and pressure at the saturation level of the numerical coefficients in Eq. 5.31 and 5.32.

In nucleus-rich continental air $c \approx 5 \times 10^{-9}$ and $\beta \approx 3$ over a large part of the nucleus spectrum (Fig. 4.1), and for these values and an ascent speed U of 1 m sec^{-1} Eq. 5.36 becomes

$$r_1^9 \approx 0.4 \times 10^{-45}$$

5   The formation, growth, and evaporation of cloud particles

so that
$$r_1 \approx 10^{-5} \text{ cm}$$
a value lying just within that part of the nucleus spectrum for which c and β have about the assumed values, and closely corresponding to that already anticipated from inspection of Table 5.4. A similar method (5) of estimating s_{max}, r_1, and the total concentration of activated nuclei,
$$N'_g = \int_{r_1}^{r_2} dN_m \text{ g}^{-1}$$
leads to practically the same result.

Equation 5.36 shows that the value of r_1 is not very sensitive to the values of the parameters c, β, and U. In the nucleus-rich continental air for which β is about 3, variation in the updraft speed U by a factor of 10 produces an inverse change in r_1 by a factor of $10^{2/9} = 1.7$, and a direct change in N'_g by a factor of $10^{6/9} = 4.6$. If the parameter c, representing the nucleus concentrations, is increased by a factor of 10, then N'_g *decreases* by a factor of 4.6.

The result of a decrease in the nucleus concentrations is less easy to foresee, for r_1 then decreases below 10^{-5} cm and enters a part of the size-distribution spectrum where β changes rapidly, as seen in Fig. 4.1. However, it is apparent from that diagram that in "cleaned" air, which is often found in the planetary boundary layer over the ocean, the *total* concentration of nuclei is of order 10^2 cm^{-3} or less, corresponding to a value of N_g of 10^5 or less. If this value is used for N_m in Eq. 5.29 with an ascent speed U of 1 m sec^{-1}, it is implied that the supersaturation s must reach a value of about 1.5×10^{-2}, which by reference to the scale of S at the foot of the diagram shows that practically *all* of the nuclei must be activated.

Accordingly, during the production of clouds by adiabatic ascent the number concentration N' of the cloud droplets formed is determined by the kind of aerosol particle population present and the updraft speed. When the latter is not very different from 1 m sec^{-1} N' is about 10^3 cm^{-3} in nucleus-rich continental air, and about 10^2 cm^{-3} or less in cleaned air away from prolific sources of fresh Aitken nuclei. In the nucleus-rich air a maximum supersaturation of order 10^{-3} is attained within about a decameter of the saturation level; in the cleaned air the maximum is about 10^{-2} and is attained within a few decameters of the saturation level.

5.4   The supersaturation well above the saturation level

If the ascent speed of the air does not change, then following the activation of the smallest nuclei the supersaturation diminishes and the cloud droplets become more nearly uniform in size: the giant nuclei continue to grow only very slowly, while the hygroscopicity (represented by the term s_r) of the smaller activated nuclei becomes negligible, so that all grow under the same effective supersaturation, and therefore at a rate inversely proportional to their radius (Eq. 5.22). Accordingly, at levels more than several decameters above the saturation level in clouds formed by adiabatic ascent the bulk of the condensed water is contained in droplets with a well-defined mean radius $\bar{r}$, determined by the mixing ratio r_c of condensed water and the concentration N'_g of the cloud droplets:
$$4\pi \bar{r}^3 \rho_w N'_g / 3 = r_c \quad (5.37)$$

Several hundred meters above a saturation level at which the pressure is 900 mb, the concentration of condensed water is about 1 g m^{-3} (Fig. 3.10), that is, r_c is about 1 g kg^{-1} or 10^{-3}. In clouds formed in nucleus-rich continental air this implies a mean droplet radius $\bar{r}$ of about 4, 6, and 10 μ for ascent speeds near the saturation level of 10 m sec^{-1}, 1 m sec^{-1}, and 10 cm sec^{-1}, respectively. In clouds formed in cleaned air the mean radius is about 13 μ or more. In the presence of some droplets of radius greater than about 20 μ, like those provided in very small concentrations by the giant nuclei, the condition for the efficient onset of the process of coalescence by differential settling ($\bar{r} >$ about 9 μ; Chapter 2) is therefore always fulfilled in the latter clouds at levels more than a few hundred meters above the saturation level, but not necessarily at any level in the former, since even if they are very deep, r_c may have a maximum value of less than 10 g kg^{-1}, and N'_g may be too great to provide the required minimum value of $\bar{r}$.

Well above the saturation level it can be assumed that the supersaturation s'' is small and does not change rapidly with height. Then the rate of condensation is $dr_c/dt = -dr_s/dt$ and can be estimated using the approximate relations (3.25) and $r_s = \varepsilon e_s/p$. Together with the Clausius-Clapeyron (2.3) and hydrostatic relations, these give
$$\frac{dr_s}{dz} = \frac{-gr_s(L_v - c_pT/\varepsilon)}{L_v^2 r_s + R_v c_p T^2} \quad (5.38)$$

Values of dr_s/dz are given for selected temperatures and pressures in Table 5.5.

Now Eq. 5.25, simplified by neglecting the term s_r and by writing $\sum N_m r$ as $N'_g \bar{r}$, and r_v as r_s, gives

$$\frac{dr_c}{dt} = \frac{-U\, dr_s}{dz}$$

$$= f_6 N'_g \bar{r} s''$$

whence

$$s'' = -\frac{U\, dr_s/dz}{f_6 N'_g \bar{r}}$$

$$= \frac{Ug(L_v - c_p T/\varepsilon)(1 + Df_2 \rho_{ST})}{4\pi N'_g \bar{r} D\rho_a (L_v^2 r_s + R_v c_p T^2)} \quad (5.39)$$

In this expression for the supersaturation s'' well above the saturation level, N'_g and $\bar{r}$ are determined by conditions very near the saturation level, while of the other terms all but D and r_s are functions of temperature only (the product $D\rho_a$ which occurs is constant). Indeed, to a sufficient accuracy for present purposes we may write

$$s'' = f_7 U/N'_g \bar{r} \quad (5.40)$$

where f_7 is a function principally of temperature; for adiabatic expansion with $\theta_s = 20°C$ it has the values listed in Table 5.6. These can be regarded as applicable at the appropriate temperatures during adiabatic expansion with any value of θ_s between 16 and 24°C, since even at these extremes the errors incurred are less than 10% at high pressures, and still less at low pressures (and temperatures) when $Df_2 \rho_{ST}$ and $L_v^2 r_s/R_v c_p T$ become small fractions.

Table 5.6 includes approximate values of the mean radius $\bar{r}$ of the cloud droplets produced during expansion with $\theta_s = 20°C$, obtained from Eq. 5.37 on the assumptions that the pressure at the saturation level is 900 mb and that the concentration of droplets formed near it has the value of 10^6 g^{-1} already inferred as appropriate in nucleus-rich continental air ascending at 1 m sec^{-1}. If the speed of ascent remains 1 m sec^{-1}, the supersaturation s''_1 at levels well above the saturation level de-

Table 5.5 Variation of $-10^8\, dr_s/dz$ cm^{-1} with pressure p and temperature T. Values in parentheses are appropriate to saturation with respect to ice

T	p(mb)			
(°C)	900	700	500	300
20	2.3	2.5	—	—
10	1.9	2.1	2.3	—
0	1.4	1.6	1.9	—
−10	0.9	1.1	1.4	—
−20	—	0.64	0.87 (0.70)	1.18 (1.06)
−30	—	0.32	0.43 (0.37)	0.68 (0.57)
−40	—	—	0.19 (0.15)	0.30 (0.24)
−50	—	—	—	0.12 (0.08)

Table 5.6 Variation with temperature, during adiabatic expansion during ascent with $\theta_s = 20°C$ (pressure at saturation level 900 mb), of estimated values of the following functions:

$$f_7 = \frac{g(L_v - c_p T/\varepsilon)(1 + Df_2 \rho_{ST})}{4\pi D\rho_a(L_v^2 r_s + R_v c_p T^2)}$$

r_c = the mixing ratio of condensed water
$\bar{r}$ = the mean droplet radius when the concentration of droplets is 10^6 g^{-1} (corresponding to an ascent speed U' at the saturation level of about 1 m sec^{-1})
s''_1 = the supersaturation for a constant ascent speed of 1 m sec^{-1}
U''_1 = the ascent speed required to produce a supersaturation s'' equal to the maximum attained near the saturation level when the ascent speed U' there is 1 m sec^{-1}
U''_5 = the corresponding ascent speed when $U' = 5$ m sec^{-1}

It is assumed that the condensation occurs in nucleus-rich continental air in which the size distribution parameters of the aerosol particles (the condensation nuclei) are $c = 5 \times 10^{-9}$ and $\beta = 3$, with a total concentration of about 10^7 g^{-1} (Fig. 4.1)

p (mb)	T (°C)	$10^3 f_7$ (sec g^{-1})	$10^3 r_c$	$\bar{r}$ (μ)	$10^4 s''_1$	U''_1 (m sec^{-1})	U''_5 (m sec^{-1})
877	15	1.4	—	—	—	—	—
770	10	1.5	3.0	8.9	1.7	8	29
680	5	1.6	4.9	10.5	1.6	9	30
606	0	1.7	6.5	11.6	1.5	9	32
542	−5	1.8	8.0	12.4	1.5	10	33
488	−10	1.9	9.2	13.0	1.5	10	33
402	−20	2.2	10.9	13.8	1.6	10	33
337	−30	2.5	11.9	14.2	1.7	8	28
286	−40	2.8	12.5	14.4	1.9	7	25
207	−50	3.3	12.7	14.5	2.3	6	21

creases from the maximum of nearly 2×10^{-3} attained during the activation of the nuclei just above the saturation level, to a value which at all levels is only about one tenth as great. If, therefore, there is no great increase in the speed of ascent, the concentration $N'_g \text{ g}^{-1}$ of cloud droplets produced in an adiabatic expansion is determined within seconds of the attainment of the state of saturation; subsequently no other droplets are formed.

However, if the speed of ascent increases sufficiently, to exceed some value U'' which depends upon the speed U' near the saturation level, the supersaturation s'' exceeds the maximum s' attained near the saturation level and a fresh activation of nuclei and formation of cloud droplets occurs. In the nucleus-rich air Eq. 5.30 and 5.36 lead to

$$N'_g = \frac{c}{2.3\beta r_1^\beta} = 7.2 \times \frac{10^{-10}}{r_1^3} = 4.6 \times 10^4\, U'^{2/3} \quad (5.41)$$

and

$$s' = 2 \times \frac{10^{-10}}{3(3r_1^3)^{1/2}} = 3.0 \times 10^4\, U'^{1/3} \quad (5.42)$$

while from Eq. 5.40

5 The formation, growth, and evaporation of cloud particles

$$s'' = \frac{f_7 U}{N'_g \bar{r}}$$

and from Eq. 5.37 and 5.41

$$\bar{r} = 1.7 \times \frac{10^{-2} r_c^{1/3}}{U'^{2/9}}$$

Thus

$$s'' = \frac{f_7 U}{800 \, r_c^{1/3} \, U'^{4/9}} \quad (5.43)$$

and

$$\frac{s''}{s'} = \frac{f_7 U}{0.24 \, r_c^{1/3} \, U'^{7/9}} \quad (5.44)$$

Accordingly that value U'' of U for which $s'' = s'$ is given by

$$U'' = \frac{0.24 \, r_c^{1/3} \, U'^{7/9}}{f_7} \quad (5.45)$$

Thus U'' is not quite proportional to U', the ratio U''/U' decreasing as U' increases, from about 10 for $U' = 1$ m sec^{-1} to about 6 for $U' = 5$ m sec^{-1}, as shown in Table 5.6. The ratio is still further reduced if the droplet concentration N'_g diminishes as a result of coalescence or removal by precipitation.

5.5 Comparison of the contributions of condensation and coalescence to the rate of growth of an individual droplet

The previous discussion has shown that in clouds formed by adiabatic ascent at a speed not very different from 1 m sec^{-1} the characteristics of the size spectra of the condensation nuclei are such that the concentration of cloud droplets is about 10^3 cm^{-3} in clouds produced in nucleus-rich air and about 10^2 cm^{-3} in clouds produced in cleaned air. Consistent with these concentrations the bulk of the water condensed at levels more than about 100 m above the saturation level is contained in droplets of a mean radius of several microns, to be regarded as characteristic of natural clouds. The supersaturation, which is proportional to the ascent speed, is of order 10^{-4} and even in extreme circumstances is less than 10^{-2}, or 1%, although the effective supersaturation over giant nuclei of great hygroscopicity may have this magnitude (Table 5.3). Thus the succession of physical states during the expansion of moist air in the standard thermodynamic reference processes of adiabatic or pseudo-adiabatic ascent, which assume zero supersaturation, is not appreciably altered by the microphysics of condensation upon natural concentrations of droplet nuclei, except in the few decameters above the saturation level. It will be shown that the deviations are much greater in clouds of ice particles, whose concentration may be several orders of magnitude smaller.

The difference ΔT between the surface temperature of the droplets and the general air temperature is given by Eq. 5.8 and 5.9 as

$$\Delta T = \frac{DL\rho_{ST}(s - s_r)}{K[1 + Df_2\rho_{ST}(1 + s_r)]} \quad (5.46)$$

which, under the extreme effective supersaturation $(s - s_r)$ of 10^{-2} that may occur in droplet clouds, has a maximum value at high temperatures and pressures. For example, at 900 mb and 20°C, with the values of D and $(1 + Df_2\rho_{ST})$ given in Table 5.1,

$$\Delta T \approx \frac{10^{-2} DL\rho_{ST}}{K(1 + Df_2\rho_{ST})} \approx 0.13°C \quad (5.47)$$

The smallness of this extreme value justifies the use of the Clausius-Clapeyron relation (5.4) in the derivation of the growth equations of the cloud droplets (but demands a large value of the correction factor $f_3 = 3.8$ to take into account their heat economy; Table 5.1).

It is interesting now to consider those few cloud droplets which by virtue of their great hygroscopicity are considerably larger than the mean size, whose growth may therefore occur by coalescence as well as by condensation, and to compare the effectiveness of these two processes. The excess of temperature ΔT over the air temperature is for most of the cloud droplets independent of their radius, but the droplets of great hygroscopicity grow under a larger effective supersaturation $(s - s_r)$ with a correspondingly greater value of ΔT; consequently their coalescence with the smaller lowers their surface temperature and increases the rate of condensation upon them. Overlooking this effect, Fig. 5.2 indicates the growth rates by condensation and coalescence in the cloud formed by the adiabatic expansion of nucleus-rich air (at an ascent speed of 1 m sec^{-1}). This diagram gives the rates of growth by condensation of droplets of radius 20 and 30 μ. It appears that in the few hundred meters above the saturation level only the condensation is important, but at about 1 km above, where the bulk of the cloud water is held

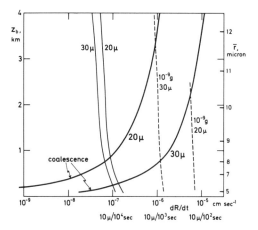

Fig. 5.2 Comparison of the rates of growth of droplets by condensation and by coalescence, as a function of the height z_b above the base of a cloud (pressure 900 mb, temperature 16°C) formed by adiabatic expansion of nucleus-rich air with $\theta_s = 20°C$. The mean droplet radius $\bar{r}$, corresponding to an ascent speed of 1 m sec^{-1} at the cloud base, is shown on the right.

The dashed curves show the rate of growth by condensation upon droplets of radius 20 and 30 μ, formed upon giant nuclei containing a mass $m = 10^{-9}$ g of solute; the growth rate is practically independent of the speed of ascent of the air. The thin solid curves are for pure droplets of the same radii and for an ascent speed of 1 m sec^{-1}; their growth rates are proportional to the local speed of ascent of the air. The thick solid lines show the rates of growth by coalescence for droplets of the same radii; their growth rates depend mainly on the concentration and size of the smaller cloud droplets.

in droplets of radius about 9 μ (which are efficiently collected), coalescence makes the dominant contribution to the growth of droplets of radius about 30 μ, and it must be overwhelmingly more important at greater radii. Practically the same result arises in the cloud formed in cleaned air, except that the level at which the mean radius of the cloud droplets first exceeds about 9 μ, and above which coalescence is the dominant process for the droplets of radius 30 μ or more, is only about 300 above the saturation level. In clouds whose concentration of condensed water is considerably less than the adiabatic values the droplet radius at which coalescence predominates is raised somewhat above 30 μ, provided efficient collection is insured by a mean droplet radius of about 9 μ or more.

5.6 The formation of clouds by isobaric mixing

Almost all atmospheric clouds are formed by the ascent of air, and in their study the most convenient reference process is adiabatic expansion, as already discussed. Some exceptional clouds, which were noted in Chapter 3, can be attributed to the mixing of moist air masses of differing temperature. An appropriate reference process is more difficult to devise, since it requires a specification of the characteristics of the mixing process, which probably produces a supersaturation that fluctuates in space and time over the whole range of values up to the maximum possible under the given temperature difference.

In those clouds produced rather suddenly, for example, the steam fogs formed where very cold air reaches open water, the temperature difference may be as large as 10°C, and correspondingly the supersaturation may be found from the graphical methods described in Chapter 3 to be of the order of 10^{-2} for mixing ratios m_1/m_2 of the moist air masses between 1/10 and 10/1, and to have a maximum of order 10^{-1}. These values are equivalent to those produced in adiabatic expansion by ascent at speeds of tens of meters per second and can be anticipated to be accompanied by the activation of practically all of the condensation nuclei present. Similarly, the supersaturation under which aircraft condensation trails form can be as high as 10^{-1}. On the other hand, within clouds which form as a result of slow and practically isobaric cooling by radiation or during the travel of air toward colder regions, the supersaturation and the droplet concentration can be anticipated to be comparatively small.

Mixing processes are of more concern in the dilution and evaporation of cloud formed by ascent. Their effects are often easily recognizable in the observed properties of clouds, as will be discussed later, but they are not well understood.

5.7 The evaporation of drops; precipitation

Just as during the formation and growth of cloud droplets in ascending air, so also during their evaporation when cloudy air descends or is diluted by mixing with clear air, the microphysical processes have no significant influence on the appropriate thermodynamic reference processes, provided that the bulk of the cloud water

5   The formation, growth, and evaporation of cloud particles

is held in droplets of radius of order microns. The evolution of considerably larger drops, such as those produced by the aggregation of cloud droplets, is more complicated for the following reasons:

1. The fall speeds of larger drops become comparable with the vertical components of the air motion, so that they no longer follow the air motion. The conditions which they experience are determined by their trajectories, and may be significantly different from the conditions following the streamlines of the air in which they originated. In particular, the drops may fall from their parent cloud and evaporate in the unsaturated air beside or beneath it. Such particles, whose paths are significantly different from the streamlines or trajectories of the air motion, can be defined as *precipitation*, although in ordinary usage the term is restricted to cloud water which reaches the earth's surface.

2. Just as the liberation of latent heat during condensation is a heat source important in concentrating and intensifying the upward branches of atmospheric circulations, occasionally to such a degree that they are called *updrafts*, so the heat sink associated with the evaporation of precipitation may have a comparable importance in the descending branches and in the production of *downdrafts*. However, whereas even in intense updrafts the vapor supersaturation and consequently the temperature excesses of the cloud particles are so small that they can be neglected in the thermodynamics of the air motion, this simplification cannot be made in the study of evaporation and air motion in downdrafts. For example, if the mixing ratio of the precipitation has the same magnitude at that of the condensed water, then whatever the general stability of the atmosphere may be, the production of downdrafts with speeds comparable with those of the updrafts requires rates of evaporation comparable with the rates of condensation in the updrafts. Then the quasi-steady supersaturation (negative in the downdrafts) is, from Eq. 5.39 inversely proportional to the term $N'_g \bar{r}$, which, from Eq. 5.37, is proportional to $1/\bar{r}^2$; hence if the supersaturation is of order 10^{-4} in clouds with $\bar{r}$ about $10\,\mu$, the subsaturation in downdrafts containing raindrops of diameter about 1 mm must be of order 10^{-1} and far too large to be neglected in any consideration. Correspondingly, the difference between the temperatures of the drops and the surrounding air, which, from Eq. 5.46, is proportional to the supersaturation and, from Eq. 5.47, is about 0.1°C for a supersaturation of 10^{-2}, may considerably exceed 1°C in the downdrafts, and is too large to permit the use of the ordinary Clausius-Clapeyron relation in the derivation of the equations for the evaporation of the drops. We shall discuss later how these complications can be taken into account.

3. With drops of considerable fall speed (Re > 1; Section 2.6), it is necessary to introduce empirical ventilation coefficients into the equations (5.6, 5.7) for the diffusion of heat and vapor to and from the droplet surfaces, which are strictly valid only for stationary droplets. These ventilation coefficients are also required in the study of the growth of hailstones by the accretion of supercooled droplets.

5.8   The ventilation coefficients in the transfer equations

The gradients of temperature and vapor density are intensified in front of a moving particle, and the transfers of heat and vapor to or from the particle as a whole are increased. Theory does not provide a general expression for the transfers, but indicates (6) that the diffusion equations should be modified to become

$$dM/dt = 4\pi r D C_v (\rho_v - \rho'_v) \qquad (5.48)$$

and

$$dH/dt = 4\pi r K C'_v \Delta T \qquad (5.49)$$

in which the *ventilation coefficients* C_v and C'_v have the form

$$C_v = 1 + a\,\mathrm{Sc}^{1/3}\,\mathrm{Re}^{1/2}$$

and

$$C'_v = 1 + a\,\mathrm{Pr}^{1/3}\,\mathrm{Re}^{1/2}$$

where a = const
   Sc = the Schmidt number = $\eta/\rho_a D$
   Pr = the Prandtl number = $\eta c_p / K$

Over the relevant range of meteorological conditions $\mathrm{Sc}^{1/3}$ and $\mathrm{Pr}^{1/3}$ have practically constant values of 0.85 and 0.89, respectively, so that

$$C_v = 1 + 0.85a\,\mathrm{Re}^{1/2} \qquad (5.50)$$
$$C'_v = 1 + 0.89a\,\mathrm{Re}^{1/2} \qquad (5.51)$$

It appears from experiment that the value of a is not quite constant; however, observations on the evaporation of water drops over the range of Re from about 2 to 800 show that the value of a is within about 10% of 0.30 [greater discrepancies have been noted at values of Re close to 1 (7), at which the velocity coefficients are comparatively unimportant]. At still higher Re, experiments on the heat and mass transfer from spherical bodies (including some on the evaporation from moist porous spheres and on the heat transfer from smooth solid spheres in air currents; 8) show that if the ventilation coefficients are expressed in the same general form, then the value of a increases to about 0.35 at $Re = 2 \times 10^4$ and to about 0.4 as Re approaches 10^5 (9). Approximately the same value was found for the combined transfer coefficient from melting ice spheres in the range of Re from about 10^4 to 4×10^4, while the corresponding value for ice spheroids increased to about 0.45 as the ratio of the minor to the major axes decreased to 0.4 (10). From the observation that the drag coefficient of artificial hailstones in this range of Re is considerably greater than that of smooth spheres (Chapter 2) a similarly high value of a might be anticipated for large and characteristically somewhat irregular or knobbly natural hailstones. In a more comprehensive series of experiments on the melting of artificial hailstones held fixed in a wind tunnel at values of Re between about 10^4 and 3×10^5, the value of a for smooth spheres increased from about 0.35 at $Re = 10^4$ to about 0.5 at $Re = 3 \times 10^5$, and the corresponding value for knobbly spheroids closely resembling natural hailstones increased to about 0.5 at $Re = 6 \times 10^4$ and thereafter more rapidly to as much as 1.2 at $Re = 3 \times 10^5$, using for the radius r and the surface area the values appropriate to a sphere of the same mass and of density 0.90 g cm^{-3} (Fig. 5.3; 11).

Considering the uncertainties usually encountered in specifying the shape, density, and fall speed of precipitation particles, as well as conditions in the surrounding air, it is a reasonable simplification to write for water droplets

$$C_v = C_v' = 1 + 0.25(\text{Re})^{1/2}, \quad 1 < \text{Re} < 200 \quad (5.52)$$

and for raindrops and small hailstones

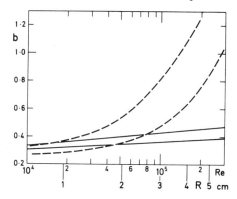

Fig. 5.3 The variation of the factor b in the ventilation coefficient of Eq. 5.54 with the Reynolds number Re, according to experiments with smooth ice spheres and artificial hailstones in a wind tunnel (adapted from reference 11). The observed values were scattered uniformly between the limits drawn as straight lines (for smooth ice spheres) and as dashed curves (for knobbly, artificially grown hailstones). An approximate scale of the radius or the equivalent spherical radius R of hailstones is included.

$$C_v = C_v' = 0.3(\text{Re})^{1/2}, \quad 200 < \text{Re} < 10^4 \quad (5.53)$$

while for larger hailstones

$$C_v = C_v' \approx b(\text{Re})^{1/2}, \quad 10^4 < \text{Re} < 3 \times 10^5 \quad (5.54)$$

where the value of the coefficient b depends on some unknown factors such as hailstone shape, surface temperature, and attitude or tumbling motion, but broadly increases over this range of Re from about 0.3 to as much as 1.2 and may be obtained approximately by reference to Fig. 5.3.

Approximate values of the ventilation coefficient are given in Table 5.7 for both water drops and hailstones; they reach about 10 for large raindrops and up to a few hundred for very large hailstones.

Table 5.7 Approximate values of the ventilation coefficients C_v, C_v' appropriate to the transfer of heat and vapor to and from water drops in the middle troposphere (see Table 2.5) and hailstones of density 0.9 g cm^{-3} in the upper troposphere (drag coefficient of large stones about 0.6; see Fig. 2.2)

Drop radius (μ)	Fall speed, V(cm sec^{-1})	Re	C_v, C_v'	Hailstone radius (mm)	Fall speed, V(m sec^{-1})	10^3 Re	C_v, C_v'
50	20	1	1.2	3	15	3	16
100	90	10	1.8	6	21	9	32
300	250	60	2.9	10	27	20	50
700	600	400	6	20	38	50	100
1,200	850	1,000	9	40	53	150	170–300

5 The formation, growth, and evaporation of cloud particles

5.9 The temperature of evaporating water drops

The surface temperature of water drops evaporating in unsaturated air may be anticipated to approximate to the wet-bulb temperature, often several degrees Celsius below the air temperature. In order to estimate the deviation from the wet-bulb temperature, the radiative heat flux can be neglected as small in comparison with the fluxes of sensible and latent heat (which is certainly justifiable, as in the theory of the psychrometer, when the drop is ventilated at a speed of more than about 1 m sec^{-1}), and the transfer of heat within the drop can be considered to be so great that the whole drop has practically the same temperature as its surface. Then if the drop has radius r and surface temperature $(T_w + \Delta T)$ while falling at a speed V through air of temperature T, wet-bulb temperature T_w and vapor density ρ_v, equating the change in the heat capacity of the drop during time dt to the difference between the supply of sensible and the loss of latent heat to the surroundings gives

$$\frac{4\pi r^3 \rho_w c d(T_w + \Delta T)}{3}$$

$$= \frac{4\pi r^3 \rho_w c d(\Delta T)}{3}$$

$$= 4\pi r C_v \{K(T - T_w - \Delta T) - L_v D[\rho_{S(T_w + \Delta T)} - \rho_v]\} dt \quad (5.55)$$

Now if ΔT does not exceed about 2°C, the Clausius-Clapeyron relation (2.7) gives

$$\rho_{S(T_w + \Delta T)} = \rho_{S(T_w)} + f(T_w)\Delta T$$

where

$$f(T_w) = \frac{\rho_{S(T_w)}(L_v - R_v T)}{R_v T^2}$$

Upon substitution in Eq. 5.55 we have

$$\frac{(r^2 \rho_w c/3C_v)d(\Delta T)}{dt}$$

$$= K(T - T_w) - L_v D[\rho_{S(T_w)} - \rho_v] - K\Delta T - L_v D f(T_w)\Delta T$$

$$= -[K + L_v D f(T_w)]\Delta T$$

since, according to psychrometric theory (e.g., see the derivation of Eq. 2.19), to a sufficient accuracy

$$K(T - T_w) = L_v D[\rho_{S(T_w)} - \rho_v]$$

Thus

$$\frac{d(\Delta T)}{\Delta T} = \frac{-3C_v[K + L_v D f(T_w)]dt}{c\rho_w r^2}$$

$$= \frac{-dt}{\lambda}$$

where λ, the lag coefficient, is the time during which an initial value of ΔT is reduced by the factor e:

$$\lambda = \frac{c\rho_w r^2}{3C_v[K + L_v D\rho_{S(T_w)}(L_v - R_v T)/R_v T^2]} \quad (5.56)$$

According to this expression λ is strongly dependent on the drop radius (which over the range of size between cloud droplets and raindrops varies by over two orders of magnitude), and to a lesser extent increases with decrease of the wet-bulb temperature on account of the presence of the term $\rho_{S(T_w)}$ in the denominator. Table 5.8 gives values of λ for the drop sizes and typical mid-tropospheric conditions specified in Table 2.5.

In a still atmosphere with a constant lapse rate dT_w/dz of wet-bulb temperature, the difference between the local wet-bulb temperature and the surface temperature of a freely falling drop tends to approach a steady value $\Delta T'$ given by

$$\Delta T' = V\lambda dT_w/dz \quad (5.57)$$

The approximate values of $\Delta T'$ which are listed in Table 5.8 are based on the average values of λ specified and a typical magnitude of (dT_w/dz), which over deep layers is limited to about the lapse rate of temperature (dT/dz). It is evident that for all but large raindrops the departure of the surface temperatures from the wet-bulb temperature is sufficiently small to be neglected. The departure can be important for the large raindrops if the air is nearly saturated, for then their rate of evaporation is substantially reduced; their surface temperatures may even fall below the dew point, so that there is condensation upon them. However, even in intense rains the largest drops contribute only a small fraction

Table 5.8 Lag coefficient λ of water drops (ventilated at their fall speeds) at an air pressure and temperature of 700 mb and 10°C, at various values of the wet-bulb temperature T_w and relative humidity R.H., and the approximate value of the steady difference $\Delta T'$ of their surface temperature from the local wet-bulb temperature when falling through an atmosphere with a lapse rate of wet-bulb temperature $dT_w/dz = -5°C$ km^{-1}

| Drop radius (μ) | λ (sec) | | | ΔT(°C) $(dT_w/dz =$ $-5°C$ km^{-1}) |
	$T_w = 8°C$ R.H. = 50%	$T_w = 5°C$ R.H. = 50%	$T_w = 0°C$ R.H. = 20%	
46	3.4×10^{-2}	3.8×10^{-2}	4.6×10^{-2}	10^{-4}
113	0.14	0.16	0.19	10^{-2}
480	1.0	1.1	1.4	3×10^{-2}
1,200	3.2	3.7	4.4	0.2
2,750	8.4	9.8	11.8	0.6

of the total mass of water evaporated, and when it is not the behavior of individual drops but the downward acceleration of the air produced by the associated cooling which is of concern, then it is a reasonable simplification to assume that the drop surfaces have the wet-bulb temperature.

5.10 The distance of fall of drops before complete evaporation; the distinction among cloud droplets, drizzle drops, and raindrops

At moderately high temperatures the magnitude of the difference between the wet-bulb temperature and the dew point in unsaturated air is several degrees Celsius, corresponding to a value of about -3×10^{-6} g cm^{-3} for the difference of vapor density $[\rho_v - \rho_{S(T_w)}]$ under which the drops which fall from a cloud evaporate. Substituting this value for the term $(\rho_v - \rho_v')$ in Eq. 5.6 and introducing the ventilation coefficient C_v gives

$$r\rho_w dr/dt = -3 \times 10^{-6} C_v D$$

or, for a drop falling through still air,

$$dz = Vr\rho_w dr/3 \times 10^{-6} C_v D \quad (5.58)$$

where z is distance fallen during evaporation. Then the distance z' fallen during complete evaporation is

$$z' = (10^6 \rho_w/3D) \int_0^r (Vr/C_v) dr$$

where D, an average value of the diffusion coefficient, can be taken as 0.33 cm^2 sec^{-1}, so that

$$z' = 10^6 \int_0^r (Vr/C_v) dr \quad (5.59)$$

where z' and r are in centimeters and V is in centimeters per second.

The distance z' is very strongly dependent on the drop radius. To simplify the integral we consider drops in three size ranges, for which the following approximations are used for the terms V and C_v without introducing serious errors:

$$r < 4 \times 10^{-3} \quad V = 10^6 r^2 \quad C_v = 1$$
$$4 \times 10^{-3} < r < 2 \times 10^{-2} \quad V = 8 \times 10^3 r \quad C_v \approx 2$$
$$2 \times 10^{-2} < r < 10^{-1} \quad V = 8 \times 10^3 r \quad C_v \approx 0.3(\mathrm{Re})^{1/2}$$
$$\approx 10^2 r$$

When Eq. 5.59 is integrated with these relations it is found that before evaporating completely in unsaturated air (a) drops of radius <40 μ falls a distance of less than 1 m, which is approximately proportional to the fourth power of the radius; (b) drops of radius > 200 μ falls a distance of more than 100 m, which is approximately proportional to the cube of the radius; and (c) drops of radius >500 μ (diameter >1 mm) fall a distance of more than 1 km, which is approximately proportional to the square of the radius.

Accordingly, a radius of about 200 μ can be regarded as separating cloud *droplets* from *precipitation*, since drops of somewhat larger size may just reach the ground (as *drizzle drops*) from very low clouds in damp weather, and also have fall speeds (of considerably more than 1 m sec^{-1}) comparable with the vertical air speeds found in clouds. Further, a radius of about 500 μ, which permits drops to reach the ground from clouds with bases as high as 1 km or more, may be regarded as about the lower size limit of *raindrops*.

5.11 The formation of ice particles

Homogeneous nucleation of ice particles

In the presence of aerosol particles in their ordinary concentrations, the change of phase of atmospheric water from the vapor to the liquid in cooling air (which begins even before the state of saturation is reached) proceeds under supersaturations which are very small. However, in damp air containing no foreign particles condensation does not occur until a very large supersaturation is reached. Unless measures are taken to remove them, too, the condensation begins upon the small ions when the supersaturation s reaches a value between about 3 and 3.5, consistent with the magnitude obtained for the equilibrium radii of droplets of the size of small ions indicated by Eq. 4.2 or 4.3. If the ions are removed, the condensation begins only when the supersaturation ratio reaches an even higher value, exceeding the equilibrium value for the larger of the clusters of water molecules continually and briefly forming within the vapor, whose frequency of occurrence as a function of their size and the general temperature can be obtained from the laws of statistical thermodynamics (12). Such a

5 The formation, growth, and evaporation of cloud particles

condensation is described as *homogeneous*, in contrast to the *heterogeneous* condensation upon nuclei consisting of particles of other substances. In the expressions derived for the threshold supersaturation at which the homogeneous condensation proceeds at an observable rate there is uncertainty about the magnitude of some of the terms, whose significance is familiar enough on macroscopic scales, but the theory reasonably agrees with the results of experiments to determine the threshold and its variation with the temperature, and shows that the homogeneous condensation has the following characteristics:

1. There is a sharply defined supersaturation ratio at which the condensation proceeds sufficiently rapidly to produce a visible cloud, since the rate at which nuclei are produced in a volume of appropriate order (say 1 cm³) increases extremely rapidly with the supersaturation (e.g., by about eight orders of magnitude as s increases from 4 to 5, and by a further four as it increases from 5 to the threshold value of about 6 at temperatures near 0°C).

2. If the threshold value is even slightly exceeded, the concentration of effective nuclei is very large, greater than 10^6 cm^{-3}, and this is the order of magnitude of the concentration of cloud particles observed when a cloud is formed.

The theory suggests that at temperatures not much below 0°C the cloud particles formed are liquid, as confirmed by experiment, but that at some temperature below about $-65°C$ the phase transition may be direct from vapor to ice. Some experimental evidence for this exists but is difficult to obtain and confirm because a homogeneous freezing of liquid water occurs at a considerably higher temperature.

The theory for the homogeneous nucleation of the ice phase in liquid water is similar to that for the nucleation from the vapor, and has the same quantitative uncertainties, but likewise indicates that the phase transition in a given sample has a certain probability, which increases with the volume of the sample, the degree of supercooling, and the period during which it is held in the supercooled state. The last, in processes of the durations common in the atmosphere and the laboratory, causes less significant variation than the much more variable volume and temperature. Their influence is illustrated in Fig. 5.4, which summarizes numerous experiments on samples of water freed with varying thoroughness from the effects of internal impurities and surface phenomena. From this diagram it can be seen, first, that the particles of radii about 1 μ or more, which can be examined optically in the laboratory during the formation of clouds, are composed of ice if the lowest temperature reached is less than about $-40°C$. Second, homogeneous nucleation of atmo-

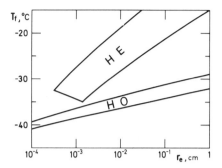

Fig. 5.4 Freezing temperatures T_f of drops of water of radius r_e, or of volumes whose equivalent spherical radius is r_e (21). The observations fall into two groups enclosed by the areas marked *HE* and *HO*; the freezing in the former was evidently by hetereogeneous nucleation within the liquid or at its supports, in spite of careful attempts at purification (e.g., by repeated distillation). The form of the variation of T_f with r_e in the latter, in which the most elaborate precautions were taken to exclude heterogeneous nucleation, is accurately represented by the theory of homogeneous nucleation.

spheric cloud droplets is not important unless the temperature falls to some value between -30 and $-40°C$.

Heterogeneous nucleation of ice particles

In the presence of suitable solid particles, ice crystals may be nucleated directly from the vapor at supersaturations with respect to ice which correspond to very large subsaturations with respect to liquid water. Apart from ice itself, however, particles of this kind are virtually absent from the atmosphere. There all solid particles soon acquire hygroscopic material, if not complete hygroscopic skins, and at high humidities they become coated with solution or immersed in solution droplets and act as *freezing nuclei* at low enough temperatures. If the freezing occurs at humidities intermediate between those representing saturation over ice and over liquid, then the crystals formed may grow to visible size, and an ice cloud may develop, without saturation with respect to liquid water ever being attained. The nuclei may then appear to have acted as *sublimation nuclei*.

It is important to meteorologists to discover how the concentration of ice crystals formed in clouds depends on the kind of aerosol present and the meteorological parameters, of which the temperature and the humidity can be anticipated to be the most important. But the nucleation of the ice phase is so complicated that from all the welter of experimental data and atmospheric observation (12, 13) it is still not possible to draw firm conclusions. It seems that the effectiveness of natural aerosol particles as ice nuclei is strongly influenced *physically* not only by their chemical nature, but also by their surface structure and condition, and contamination with organic or hygroscopic material, and *meteoro-*

logically by whether they become activated as condensation nuclei or otherwise incorporated into cloud droplets, and perhaps by the duration of their exposure to high humidities, or previously to low humidities or low temperatures.

The experimental information consists partly of direct observations of the nucleation of ice by individual particles or by aerosols of particular substances, some of which occur in the atmosphere, and partly of observations on the concentrations of ice crystals which arise in artificial clouds produced by chilling samples of atmospheric aerosol.

Theory indicates, and experiments of the first kind confirm, that the important ice nuclei are solid, insoluble particles with radii greater than about 10^{-6} cm. At sufficiently low temperatures and high supersaturations (with respect to ice) they may behave as sublimation nuclei, and at higher temperatures as freezing nuclei, since during the chilling of air a droplet cloud forms before the attainment of the supersaturation at which they can act as sublimation nuclei. In given samples of mineral dusts the various particles are active as freezing nuclei over a range of temperatures; a significant proportion (about 1 in 10^4 or more) is effective at temperatures above about $-15°C$ in dusts prepared from many silicaceous minerals and volcanic ash, and even above $-10°C$ in those composed of some common clays and micas. Much more efficient nucleating materials have been discovered which do not exist naturally in the atmosphere, notably silver iodide AgI, lead iodide PbI_2, and cupric sulfide CuS; for example, silver iodide particles behave as sublimation nuclei at temperatures below about $-12°C$ and as freezing nuclei at temperatures up to about $-4°C$.

The conditions of temperature and supersaturation in experiments involving the production of clouds are not readily controllable, and in particular those which prevail during the formation of ordinary natural clouds cannot be reproduced faithfully. Comparatively very great supersaturations arise in clouds formed by the mixing of warm vapor and chilled air, as already noted in Section 5.6, and can hardly be avoided during the important phase of cloud formation when condensation is produced alternatively in the laboratory by expansion. This is because the transfer of heat from the walls of chambers of convenient size is sufficient to prevent the required degree of cooling unless the expansion is extremely rapid, equivalent to a rate of ascent in the atmosphere of 100 m sec^{-1} or more. Thus in clouds produced by both methods practically all of the aerosol particles are likely to be activated as condensation nuclei, possibly with a marked effect on their behavior as ice nuclei. Further, the supersaturation and the concentration of liquid water produced in the cloud may vary widely from one experiment or apparatus to another,

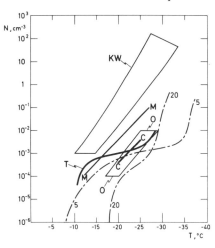

Fig. 5.5 Variation of ice crystal concentration N with temperature T in artificial clouds, according to various observers.

Values observed in fogs produced by injecting hot steam into a cold chamber of room size lie within the area marked KW(17); those observed in Oxfordshire, England, in an expansion chamber of volume 5 liters, lie within the area marked O.

Average values observed with conventional portable expansion chambers are shown by the lines marked C (representative of the clean air found generally in the southern hemisphere and at some places in the northern hemisphere; 18) and M (for summer days at a rural site in Missouri; 16). The curve marked T represents values obtained with a mixing chamber at tropical Australian stations (19).

The dashed curves marked 5 and 20 are the average values in clouds formed in a large expansion chamber in a suburb of Prague, for rates of expansion at the time of cloud formation equivalent to ascent speeds in the atmosphere of 5 and 20 m sec^{-1}, respectively (14).

and the concentration may sometimes be very much greater than is typical of natural clouds, while in laboratory clouds which are maintained for more than a few seconds there may be an important loss of droplets and embedded material to the walls or base of the chamber, regardless of whether the air is kept stirred. Until recently, arrangements for counting the ice crystals in laboratory clouds were also unsatisfactory, mostly relying on a twinkling in dark-field illumination to identify the crystals, and a subjective estimate of their separation to assess their concentration. Considering the difficulties, and the almost complete disregard of possible local sources of efficient nuclei or of trace vapors or trace gases with inhibiting effects, and the generally poor appreciation of the kind of aerosol from which the samples were taken, it is not surprising that the results of the many series of experiments undertaken in different places have not so far provided consistent indications of the nature, distribution, and properties of the atmospheric ice nuclei. Some of the results are summarized in Fig. 5.5, from which it appears that the concentration of crystals observed at a particular

5   The formation, growth, and evaporation of cloud particles

temperature varies, even in the average values quoted by different observers, by two orders of magnitude or even more.

Observations of ice crystal concentration in artificial clouds produced in samples of atmospheric aerosol

Observations of particular interest shown in Fig. 5.5 include the oldest (14), obtained in a suburb of Prague, with a specially large cloud chamber (of volume about 2 m^3) in which by elaborate precautions virtually adiabatic expansions were achieved at rates corresponding to ascent speeds in the atmosphere as low as 5 m sec^{-1}. The curves drawn show the substantial influence of the expansion rate at the time of cloud formation; the subsequent variation of expansion rate was unimportant, so that the influence was evidently exerted during the selection of aerosol particles for activation as droplet nuclei. It was also reported that with difficulty it was possible to halt the expansion at temperatures between the frost point and the dew point, and then to observe the formation of ice crystals in about the same concentration that arose when an expansion was continued and a droplet cloud produced. The curve for the lowest ascent speed may still be the most representative of the natural ice-particle formation in clouds produced in continental nucleus-rich air. This curve and its companion for a higher ascent speed are the only curves in the diagram for clouds in which a definite and practically constant supersaturation was produced during their formation. Some experiments (15) in which the degree of expansion in a conventional cloud chamber was varied indicated that when the supersaturation ratio produced was about 2.6 the crystal concentrations observed at −24°C were about the same as those in a cloud produced by mixing, while if the expansion ratio was decreased and the supersaturation ratio was about 1.8 the crystal concentration decreased by as much as an order of magnitude.

It might be anticipated that the solid particles which constitute the ice nuclei active at temperatures above about −30°C are derived from land surfaces, and that their concentration in nucleus-rich continental air would be markedly greater than in the cleaned air found in the upper troposphere and within the boundary layer over the oceans, particularly as investigations have shown that most of the ice nuclei are associated with the large condensation nuclei. Consistently, the concentration of the ice nuclei near the ground is observed to decrease markedly during prolonged rain, and an afternoon minimum in a diurnal variation may be interpreted as a dilution as small-scale convection increases the depth of the boundary layer. However, there is little direct evidence in the form of comparative measurements in different places of the concentration of crystals formed in an artificial cloud at a particular temperature (usually −20°C), aircraft observations of its variation with height, and the nucleating efficiency of particles filtered from the atmosphere during aircraft and balloon flights (13). At various places throughout the southern hemisphere, and in the northern hemisphere in high latitudes and in maritime regions (in the Caribbean and at Hawaii), the reported average ice crystal concentration at −20°C is less than 1 liter^{-1}, smaller by about one order of magnitude than that observed in a number of continental places in the northern hemisphere. Flight observations to 15,000 ft above the ground in Arizona and South Africa show little if any variation in the concentrations of ice crystals formed at temperatures between −25 and −30°C, while others over Australia with a different apparatus found concentrations above the tropopause, at heights of about 44,000 ft, to be about the same as at 25,000 ft, and more than an order of magnitude greater than the average at the ground. Particles filtered from the lower stratosphere (at heights between 15 and 21 km) in the same region were found in concentrations varying by more than an order of magnitude, but on the average represented the equivalent of rather more than 10^{-1} liter^{-1} active at −15°C, consistent with the values of about 10 liter^{-1}, active at lower temperatures, previously observed during flights above the tropopause (Fig. 5.5). Since such concentrations are larger than the average at the ground, the more so when expressed as number mixing ratios, they have been regarded as evidence for a high-level or even extraterrestrial source of ice nuclei. Nevertheless, they represent only a small fraction of the large condensation nuclei, and a still smaller fraction of the Aitken nuclei which originate predominantly near the ground, and other interpretations cannot be excluded. Possibly, for example, the effectiveness as ice nuclei of the more numerous solid particles in the nucleus-rich continental air is diminished by the presence of other trace substances which are virtually absent in the free atmosphere. Certainly there are observations which suggest that the solid particles characteristic of desert air, which is practically free of industrial and organic contamination, are particularly effective as ice nuclei. Thus abnormally high concentrations of about 100 liter^{-1} of nuclei active at −20°C have been found to occur in Japan in air carrying loess dust from northern China, and in southern France and eastern Spain on some occasions when apparently air arrives from the Sahara, even though it must usually be at heights of 3 km or more and unlikely to influence surface observations unless reached by the small-scale convection overland during the day.

Remarkably, the central parts of almost all of several hundred snow crystals collected upon the Greenland ice cap were found to contain solid particles which under the electron microscope could be measured and identified by electron diffraction as composed of the common clay minerals. Similar particles were found in snow crystals collected in Japan and the northern United States. Although other smaller particles were observed, and interpreted as Aitken nuclei acquired by aggregation during the growth of the crystals, and although the mineral particles may have been collected in the same way, it is a plausible though questionable inference that the latter were the nuclei responsible for the formation of the crystals (13).

The meteorological significance of the observed concentrations of ice nuclei

Although conditions in the experimental clouds differ significantly from those in natural clouds, the observations described probably provide within an order of magnitude the concentration of ice crystals likely to form within a natural droplet cloud, and indicate that it is a function mainly of the temperature. The concentration due to heterogeneous nucleation is very much smaller than the droplet concentration which is typically of order 10^2 cm^{-3}; it increases with decrease of temperature from about 10^{-5} cm^{-3} at $-10°C$ to 10^{-3} cm^{-3} at $-20°C$ and 10^{-1} cm^{-3} at $-35°C$. At temperatures below about $-35°C$ the homogeneous freezing of cloud droplets becomes the dominant process of ice particle nucleation, and at temperatures below about $-40°C$ ensures the freezing of all activated condensation nuclei in clouds of ordinary duration (more than about 1 min).

At such low temperatures it might be anticipated that the homogeneous freezing of the watery parts of the hygroscopic condensation nuclei would produce abundant nuclei upon which crystals could grow at any supersaturation with respect to ice. However, at temperatures below $-40°C$ the relative humidity corresponding to saturation over ice is less than 70%, and even at moderate supersaturations with respect to ice the concentration of the solute in the watery parts of the condensation nuclei may be high enough to depress by $10°C$ or even more the temperature at which the solid components of the nuclei act as freezing nuclei, or the temperature of a homogeneous freezing (21). Accordingly, even at temperatures as low as $-50°C$ the condensation nuclei are unlikely to be effective as ice nuclei until an increase in the relative humidity above about 90% with respect to the liquid produces a substantial dilution of their watery parts. At temperatures above $-40°C$, on the other hand, the concentration of ice crystals observed in the experimental clouds is so low that generally, even were they effective in the atmosphere as sublimation nuclei, they could not prevent the attainment of saturation with respect to liquid water and the formation of a supercooled droplet cloud in persistently ascending air. This is demonstrated in the following section after consideration of the rate of growth of ice crystals. It therefore happens that except at very low temperatures ice crystals are rarely observed to form independently of droplet clouds.

5.12 The growth of ice crystals by condensation

The form of ice crystals

Ice crystals occur in two principal forms: as hexagonal prisms whose growth is mainly on the prism faces, producing thin *plates*, or mainly on the base faces, producing tall *columns* or prisms (12). There are many minor modifications of these basic habits; for example, long imperfect prisms known as needles, prisms capped by plates, prisms terminating in pyramids at one or both ends, prisms which are hollow, having cavities tapering inward from one or both ends, plates which have extensions at the corners (sector plates), and thin, star-shaped crystals whose six arms may have complicated fernlike branches (*dendrites*). From laboratory studies (22; results are summarized in Fig. 5.6) it appears that the basic crystal form is determined by temperature, and becomes more "imperfect" or complicated as the supersaturation (and rate of growth) is increased.

Observations of the form of natural ice crystals are generally consistent with these results, although they are difficult to interpret, especially at temperatures not much below $0°C$, because of the considerable range of height and conditions through which the crystals may have fallen before they are caught. Ground and flight observations (23) have shown that plates and stellar forms predominate at temperatures above about $-20°C$, and that in the upper troposphere, at temperatures below about $-30°C$, the crystals are columnar. In isolated high clouds consisting of trails of ice particles (cirrus fallstreaks) the crystals collected during flights were prisms whose length was up to about 0.5 mm and about six times greater than their breadth. They had large open cavities, so that their mean density must have been considerably less than that of solid ice, and they were

5 The formation, growth, and evaporation of cloud particles

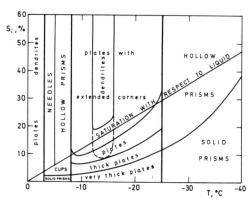

Fig. 5.6 The dependence of the form of ice crystals grown in the laboratory upon the temperature T and supersaturation s_i with respect to ice (22). The principal transitions from columnar (prismatic) to platelike forms occur at temperatures of about -3, -8, and $-25°C$; increasing supersaturation affects secondary features, leading to incomplete (hollow) prisms, and to plates thinner by comparison with their diameters and to extensions at their corners, eventually of the fernlike kind called dendritic. In the atmosphere the minimum supersaturation s_i at which the dendritic growth occurs is somewhat less than that representing saturation over liquid water (as indicated by the dashed extension to the growth region), perhaps because of the enhancement of the vapor density gradient near the crystals due to their falling motion.

often grouped in clusters of several, all radiating from a common center (because these fragile structures easily smashed on impact with the collecting slide, most if not all of the crystals may have been clustered before they were collected). The crystals of more widespread and aged ice clouds were smaller and more compact, with blunted corners and no or only small cavities (Pl. 5.1).

The thin hexagonal plates collected at higher temperatures had diameters up to about 0.5 mm and were less than 50 μ thick. Similar crystals grown in laboratory clouds have been observed to have a thickness of between one third and one sixth of the diameter for diameters up to about 0.2 mm (24), and an almost constant thickness of about 40 μ for diameters between 0.4 and 1 mm (25).

The fall speeds of ice crystals

The fall speeds of thin platelike and long columnar ice crystals are probably given satisfactorily by calculations using the empirical drag coefficients for thin discs and long cylinders found in Fig. 2.3.

The values listed in Table 5.9 for hexagonal plates in the middle troposphere are based on a density of 0.9 g cm^{-3}, and a corner-to-corner diameter 20 times greater than the thickness (which is probably too large for the smaller diameters, at which the fall speed is unimportant, and too small for diameters of 1 mm and more, for which fall speeds are also given for a constant thickness of 40 μ). Table 5.10 gives calculated fall speeds in the upper troposphere for single long prisms of mean

Table 5.9 Fall speeds V in the middle troposphere ($p = 475$ mb, $T = -15°C$, $\theta_s = 18°C$, $\rho_a = 0.64 \times 10^{-3}$) of thin hexagonal ice crystals of density 0.9 g cm^{-3} and center-to-corner dimension $r = 20b$, where $2b$ is the thickness (calculated from the drag coefficients for thin discs given in Fig. 2.3)

Re	$r\ (\mu)$	V (cm sec^{-1})	$C_v = 1 + 0.25(Re)^{1/2}$
1	115	12	1.2
2	150	19	1.3
3	170	25	1.4
5	220	33	1.6
10	300	48	1.8
20	400	70	2.1
40	600	100	2.6
60	700	120	2.9
100	960	150	3.5
For constant thickness of 40 μ			
20	400	70	2.1
40	700	80	2.6
60	980	86	2.9
100	1,500	93	3.5

Table 5.10 Fall speeds V in the high troposphere ($p = 300$ mb, $T = -42°C$, $\theta_s = 18°C$, $\rho_a = 0.45 \times 10^{-3}$ g cm^{-3}) of ice crystals in the form of prisms with cavities reducing the mean density to 0.7 g cm^{-3}, and of length $2r$ six times the thickness $2a$ (calculated from the drag coefficients of extremely long cylinders given in Fig. 2.3, and obtained from their fall speeds V_∞ by the empirical formula $V = V_\infty / [1 + (1.2/Re^{0.6})(a/r)]$ suggested in reference 26. The dimension in the Reynolds number is $2a$)

Re	Length $2r$ (μ)	V (cm sec^{-1})
0.06	140	4
0.1	180	6
0.2	230	11
0.3	270	16
0.6	370	25
1	460	38

density 0.7 g cm^{-3} (considerably less than that of ice, to take some account of the presence of large cavities) and of lengths 6 times their breadths, assumed to be falling with their lengths horizontal. The actual fall speeds of the crystals might be somewhat greater, considering that at the low values of the Reynolds number implied they might fall stably in other attitudes or they might be clustered.

The growth of ice crystals by condensation

The field of vapor density around an isolated particle growing by steady-state diffusion and the field of electrical potential around a charged conductor of the same shape satisfy the same equations (27). It is therefore convenient to write Eq. 5.6 for the growth of a sphere as

$$dM/dt = 4\pi CD(\rho_v - \rho'_v). \quad (5.60)$$

where C is the electrostatic capacitance of a conductor of the same shape and size as the particle being considered.

For the sphere and for spheroids of radius or semi-major axis r, C has the following values:

(a) Sphere:

$$C = r$$

(b) Prolate spheroid of minor semi-axis a:

$$C = A/\log_e[(r + A)/a] \quad (5.61)$$

where

$$A = (r^2 - a^2)^{1/2}$$

(c) Oblate spheroid of minor semi-axis b:

$$C = re/\sin^{-1} e \quad (5.62)$$

where

$$e = (1 - b^2/r^2)^{1/2}$$

As the asphericity of a particle becomes greater, so C becomes a smaller fraction of r. From (b), for example, for prolate spheroids which are long and thin:

$$\begin{aligned} &\text{if } r = 2a, & &C \approx r/1.5 \\ &\text{if } r = 6a, & &C \approx r/2.5 \end{aligned} \quad (5.63)$$

and

$$\text{if } r = 10a, \quad C \approx r/3$$

while for thin oblate spheroids

$$\text{if } r = 2b, \quad C \approx r/1.2$$

and

$$\text{if } r = 10b, \quad C \approx r/(\pi/2) \quad (5.64)$$

the last being the limiting value of C for the extremely thin disc. These formulas, together with the growth equation (5.60), indicate that under otherwise similar conditions the rate of condensation upon crystals of the same major dimension is less the more aspherical the crystal. On the other hand, for crystals of the same *volume* the rate of condensation is somewhat greater for markedly aspherical shapes. For a thin platelike ice crystal, for example, the ratio of the diameter $2r$ to the thickness $2b$ is about 10; applying the formula of (c), it appears that the rate of condensation on the flat spheroid which such a crystal resembles exceeds that on the sphere of the same volume by a factor of about 1.5. Similarly, from the formula of (b) it appears that for a long thin prolate spheroid with $(2r/a) = 12$, resembling a long columnar crystal with a density of 0.7 g cm^{-3}, the rate of condensation is greater than that on a sphere of the same mass and a density of 0.9 g cm^{-3} by a factor of about 1.6. Accordingly, under a given supersaturation the condensation progresses rather more rapidly if the crystals have a markedly aspherical shape, but it will hardly be justifiable to try to take this into account in any calculation in the absence of precise information about the concentration, shape, and density of the crystals and about the appropriate ventilation coefficients when the Reynolds number of the crystals exceeds 1. It has been found (24) that the limiting value $C = 2r/\pi$ is reasonably accurate (to better than about 10%) for thin hexagonal plates, and even for thin six-armed and dendritic plates, if the radius r used is that of a circular disc having the same surface area (for the complete hexagonal plate of side or center-to-corner dimension c, $r = 0.91c$). If the plate has an appreciable thickness C is increased approximately in the ratio of the total surface area to the area of the upper and lower faces. Similarly the (two-dimensional) limiting value $C = r/\log_e(2r/a)$ is reasonably accurate for long prismatic columns whose length is at least several times greater than their thickness.

The growth of ice crystals on sublimation nuclei

Accepting the thin plate with $(r/b) = 10$ and density 0.9 g cm^{-3} and the hollow column with $(2r/a) = 12$ and mean density 0.7 g cm^{-3} as the forms of ice crystals which grow in the temperature ranges characteristic of the middle and the upper troposphere, respectively, their rates of growth in steadily ascending air can be calculated on the supposition that they form on sublimation nuclei which are active as soon as supersaturation with respect to ice develops. Since the crystals may acquire a considerable fall speed, a convection coefficient C_v must be introduced into the growth equation (5.60), to give

$$dM/dt = 4\pi CC_v D(\rho_v - \rho'_v) \quad (5.65)$$

The coefficient C_v can be assumed to have the form and magnitude appropriate to spheres of the same Reynolds number, as given by Eq. 5.52: a few observations on the evaporation of platelike crystals (28) indicate $C_v \approx 1$ at Re = 1, and $C_v \approx 4.4$ at Re = 100, approximately equal to the values given by that equation. However, as the values of Re which will be of concern in the growth of platelike crystals are considerably less than 100, and in the growth of columnar crystals not ever much more than 1, the coefficient C_v will be taken as unity in the order-of-magnitude calculations that follow.

Then, taking the vapor density ρ'_v at the surface of the crystals to be the saturated vapor density over ice *at the temperature of the environment*, by the same derivation which led to Eq. 5.10, the growth equation can be written

5 The formation, growth, and evaporation of cloud particles

$$dM/dt = 4\pi CD\,\Delta\rho/f_3(p, T) \tag{5.66}$$

where the function $f_3(p, T)$ is as defined in Table 5.1, and

$$\Delta\rho = \rho - \rho_{ST}$$

Choosing as before (Tables 5.9 and 5.10) the pressures and temperatures of 475 mb and $-15°C$ and of 300 mb and $-42°C$ as representative, respectively, of the conditions under which the platelike crystals grow in the middle troposphere and the columnar crystals grow in the upper troposphere, the following values of D and $[1 + f_3(p, T)]$ can be taken as constant during the small vertical displacements of air to be considered:

at 475 mb, $-15°C$:

$$D = 0.43 \text{ cm}^2 \text{ sec}^{-1}, \quad f_3(p, T) = 1.6$$

at 300 mb, $-42°C$:

$$D = 0.56 \text{ cm}^2 \text{ sec}^{-1}, \quad f_3(p, T) = 1.1$$

(the latter from Table 5.1).

Further, the assumed dimensions and density of the crystals provide the following relations:

for the platelike crystals:

$$C = \frac{0.91r}{(\pi/2)}$$

where r is the center-to-corner dimension, and

$$r = 1.3 M^{1/3}$$

with the appropriate values of D and $f_3(p, T)$, these give numerically (in cgs units) the growth equation

$$M^{-1/3}\,dM = 4\pi \times 1.3 \times \frac{2 \times 0.91}{\pi} \times \frac{0.43}{1.6}\,\Delta\rho\,dt$$

$$= 2.5\,\Delta\rho\,dt \tag{5.67}$$

and for the columnar crystals:

$$C = r/2.5$$

where r is the half-length, and

$$r = 2.2 M^{1/3}$$

with the appropriate values of D and $f_3(p, T)$, these give numerically the growth equation

$$M^{-1/3}\,dM = 4\pi \times \frac{2.2}{2.5} \times \frac{0.56}{1.1}\,\Delta\rho\,dt$$

$$= 5.6\,\Delta\rho\,dt \tag{5.68}$$

We now anticipate that the concentration N' cm^{-3} of the crystals is so small that during the steady ascent of the air containing them their mass concentration $N'M$ g cm^{-3} remains at least an order of magnitude less than the difference $\Delta\rho'$ between the saturated vapor densities over liquid water and over ice, so that the air cools at practically the dry adiabatic rate until the dew point is reached and a droplet cloud forms, although the crystals began their growth when the temperature fell below the frost point. Then if during the ascent the vapor density difference $\Delta\rho$ is considered to increase steadily with time to reach the value $\Delta\rho'$, leading to an increase of crystal mass to the value M', we must have

$$N'M' < 10^{-1}\Delta\rho' \tag{5.69}$$

Now for the conditions under which the platelike crystals grow, if the air is saturated with respect to ice at a temperature of $-14.0°C$ the dew point is $-15.7°C$, and in practically dry adiabatic ascent saturation with respect to liquid water is reached after an ascent of 1.7/8 km (Eq. 3.24), at a temperature of $-16.1°C$, for which the difference $\Delta\rho'$ between the saturated vapor densities over liquid water and over ice is 0.21 g m^{-3}. Thus during an ascent at speed U cm sec^{-1} we assume

$$\Delta\rho = \frac{8 \times 0.21 \times 10^{-6} Ut}{1.7 \times 10^5}$$

$$= 10^{-11} Ut \tag{5.70}$$

and substitution into Eq. 5.67 leads to the solution

$$M'^{2/3} = \frac{2}{3} \times 2.5 \times 10^{-11} \times \frac{(1.7 \times 10^5)^2}{2 \times 64 U}$$

$$M' = 2.3 \times \frac{10^{-4}}{U^{3/2}} \tag{5.71}$$

and

$$r' \approx 8 \times \frac{10^{-2}}{U^{1/2}} \tag{5.72}$$

where r' is the half-diameter of the crystals at the level of droplet-cloud formation. Accordingly, at this level the diameter of the crystals is about 160 μ if the speed of ascent is 1 m sec^{-1}, and about 0.5 mm if it is only 10 cm sec^{-1}. Correspondingly, from Eq. 5.69,

$$N' < 2.1 \times \frac{10^{-8} U^{3/2}}{2.3} \times 10^{-4}$$

$$< 9 \times 10^{-5} U^{3/2} \tag{5.73}$$

that is, even if the speed of ascent is as little as 10 cm sec^{-1}, a droplet cloud does form if

$$N' < 3 \times 10^{-3} \text{ cm}^{-3}$$

Since the concentration of ice nuclei observed to be active at about the temperature concerned ($-15°C$) is always less than 10^{-3} cm^{-3}, it is confirmed that even if they were all effective as sublimation nuclei they could not prevent the formation of a droplet cloud in a steady slow ascent of air.

In the high troposphere, at lower temperatures, there

is a greater distance between the levels at which ascending air becomes saturated first with respect to ice and then with respect to liquid water. For example, if the frost point is $-40°C$, corresponding to a dew point of $-43.8°C$, in practically adiabatic ascent saturation with respect to liquid water is attained after a further rise of $3.8/8$ km (Eq. 3.24) or 480 m, at a temperature of $-44.8°C$, for which the difference $\Delta\rho'$ between the saturated vapor densities over liquid water and over ice is 0.038 g m^{-3}. Corresponding to Eq. 5.70,

$$\Delta\rho = 8 \times 10^{-13} Ut \quad (5.74)$$

and substitution into Eq. 5.68 leads to

$$M' = 2 \times \frac{10^{-4}}{U^{3/2}} \quad (5.75)$$

$$r' = 1.3 \times \frac{10^{-1}}{U^{1/2}} \quad (5.76)$$

where r' is the half-length of the crystals at the level of droplet-cloud formation. Accordingly, at this level the length of the crystals is about 130 μ if the speed of ascent is 1 m sec^{-1} and about 400 μ if it is only 10 cm sec^{-1}. Correspondingly, from Eq. 5.69,

$$N' < 3.8 \times \frac{10^{-9} U^{3/2}}{2} \times 10^{-4}$$

$$< 2 \times 10^{-5} U^{3/2} \quad (5.77)$$

that is, if the speed of ascent is 1 m sec^{-1}, a droplet cloud must form if

$$N' < 2 \times 10^{-2}$$

However, since the concentrations of ice nuclei observed during the formation of experimental droplet clouds (Fig. 5.5) considerably exceed 10^{-2} cm^{-3} at temperatures below about $-30°C$, and since the bulk of the crystal growth is accomplished late in the envisaged ascent beyond the level of ice saturation, it seems likely that even if these nuclei acted only at high supersaturations with respect to ice, the growth of crystals upon them would probably prevent the subsequent attainment of saturation with respect to liquid water, and the formation of a droplet cloud, at all ordinary ascent speeds. These results, as will be mentioned in the discussion of cloud-forming processes, are in accord with general experience that at temperatures down to about $-30°C$ ice clouds are not observed to form independently of droplet clouds, whereas in clouds forming at lower temperatures it is unusual to detect the presence of droplets, except in the particular circumstances that the temperature is not much below $-30°C$ and the speed of ascent of air associated with the cloud formation is at least several meters per second (at temperatures below about $-40°C$ the duration of any droplets which do form must be severely restricted by homogeneous nucleation of the ice phase). The rate of radiative cooling overnight of air close to the ground has the magnitude of only 1°C hr^{-1}, and is therefore equivalent to an adiabatic ascent at a speed of order only 1 cm sec^{-1}, and may become even slower or cease before the temperature falls to the dew point. It is then occasionally observed at temperatures as low as about $-20°C$ that small ice crystals appear in the air in very low concentrations (about 10^{-3} cm^{-3}); they seem to form independently of any droplet fog or cloud, although in inhabited regions it may be difficult to be sure that they did not arise by the freezing of some of the droplets in a mixing cloud formed in the exhaust from the combustion of fuel some distance away, whose remaining droplets gradually evaporated as the cloud drifted and diffused away from its source. The crystals may be hardly detectable other than by the halo phenomena (especially the sun pillar) they cause or by their glinting in sunshine, which accounts for the name "diamond dust."

The growth of ice crystals at saturation with respect to liquid water

Ice crystals very frequently arise in a droplet cloud, where their growth proceeds under the maximum supersaturation attainable at a given temperature. This supersaturation, with respect to ice, corresponds accurately to a state of saturation with respect to liquid water. Thus in the growth equation (5.66) the term $\Delta\rho$ is constant and equal to $\Delta\rho'$, the difference between the saturated vapor densities with respect to liquid water and with respect to ice. For the growth of platelike crystals in a mid-tropospheric cloud and of columnar crystals in the high troposphere, the pressures can again be assumed to be 475 and 300 mb, respectively, so that the values of the terms in the appropriate growth equations are as listed in Table 5.11.

From the first part of Table 5.11 it appears that over a large range of temperature the term $D\Delta\rho/f_3(p, T)$ in the growth equation has the approximately constant magnitude of 5×10^{-8} g cm^{-1} sec^{-1}; by using this value and introducing a convection coefficient (equal to that for a sphere at the same Reynolds number) to avoid underestimating the growth rate, the equation can be solved to give the diameter, mass, and distance settled relative to the air, as functions of time following the crystal nucleation, with the results shown in Table 5.12.

Whether crystals have the opportunity in the atmosphere to grow in an environment saturated with respect to the liquid for the periods of 1000 sec or more considered in Table 5.12 depends on the properties of the cloud in which they occur, and in particular on its thickness and the ascent speed of the air within it. Since, however, thicknesses of about 1 km and ascent speeds of several centimeters per second or more are

Table 5.11 Values of $\Delta\rho'$, the difference between the saturated vapor densities over liquid water and over ice, and of the terms D and $f_3(p, T)$ in the growth equations for ice crystals, as functions of the temperature T

T (°C)	$10^8 \Delta\rho'$ (g cm^{-3})	D (cm^2 sec^{-1})	$f_3(p, T)$	$10^8 D \Delta\rho'/f_3(p, T)$ (g cm^{-1} sec^{-1})
At pressure p = 475 mb				
−5	16	0.46	2.45	3.0
−10	22	0.44	1.97	4.9
−12	22	0.44	1.82	5.4
−14	22	0.43	1.70	5.6
−16	21	0.42	1.59	5.7
−18	20	0.42	1.50	5.7
−20	19	0.41	1.42	5.5
−25	15	0.40	1.27	4.8
At pressure p = 300 mb				
−30	11	0.61	1.26	5.6
−35	8.2	0.59	1.16	4.2
−40	5.6	0.57	1.10	2.9
−45	3.7	0.54	1.06	1.9
−50	2.3	0.52	1.03	1.2

Table 5.12 The approximate growth rate of a platelike ice crystal in an environment saturated with respect to liquid water at temperatures between about −10 and −25°C ($D\Delta\rho'/f_3$ assumed constant = 5×10^{-8} g cm^{-1}; see Table 5.11). Values of the center-to-corner radius r, the mass M, and the radius a of a drop of the same mass, the Reynolds number Re and the convection coefficient C_v, the fall speed V and the distance S settled through the air, are all given as functions of the time t following the nucleation of the crystal. For values of $r > 200 \mu$ the crystal thickness is assumed to be 40μ

t (sec)	r (μ)	$10^6 M$ (g)	a (μ)	Re	C_v	V (cm sec^{-1})	S (m)
160	100	0.4	45	1	1.25	15	10
320	140	1.3	70	3	1.4	30	50
560	200	3.8	95	7	1.7	45	150
700	250	6	110	10	1.8	55	200
1,100	420	15	160	20	2.1	70	500
1,700	700	45	220	40	2.6	80	900
2,300	1,000	95	280	60	3.0	85	1,400
3,000	1,500	210	370	100	3.5	90	2,100

not uncommon, and because the concentration of the crystals at the temperatures considered is often so low that their mass concentration is small compared with $\Delta\rho'$ until they have reached millimetric size, it can be anticipated that the growth of the crystals frequently is maintained over such periods. Accordingly, unlike liquid cloud particles, by condensation alone they may readily attain the size of precipitation particles, and it is often possible to see them as faint trails beneath droplet clouds whose duration is more than about 20 min and whose temperature is within the range between about −10 and −20°C. At higher temperatures any crystals produced are usually too sparse to be detectable or to have any influence upon the cloud behavior, while at somewhat lower temperatures they are sufficiently numerous eventually to reduce the vapor pressure below that representing saturation with respect to the liquid, so that the droplets evaporate and the cloud becomes completely transformed (glaciated) into an ice cloud. Most of the particles may remain small, but nevertheless, they often have fall speeds sufficiently great to produce the appearance of diffuse trails or *fallstreaks* beneath the originally denser parts of the parent droplet cloud.

At the temperatures near and below −40°C ice crystals usually form and grow without the appearance of droplet clouds (Pl. 6.1.), and therefore under a supersaturation which is less than that corresponding to saturation with respect to liquid water but which may nevertheless be a high supersaturation with respect to ice. The magnitude of the maximum growth rate of the crystals at temperatures near −40°C can be assessed by using in the growth equation a value of about 2×10^{-8} g cm^{-1} sec^{-1} for the term $D\Delta\rho/f_3(p, T)$ (as indicated by the values in the lower part of Table 5.11). For single columnar crystals of the kind previously considered the solution of the equation then provides the values listed in Table 5.13. As previously remarked, such crystals probably occur in clusters of several individuals, and the fall speeds of the larger clusters and the distances which they settle through the air (and also the growth rates, because of the omission of a ventilation coefficient) may be somewhat greater than those indicated in the table. At temperatures considerably below −40°C the concentrations of effective ice nuclei are probably too great (of order 10^{-1} cm^{-3} or more) to permit the maintenance of a large supersaturation or to allow the crystal dimensions to reach more than about 100 μ.

Table 5.13 The approximate growth rate of a single columnar ice crystal in the high troposphere at a supersaturation near that representing saturation with respect to liquid water ($D\Delta\rho/f_3$ assumed constant = 2×10^{-8} g cm^{-1} sec^{-1}; see Table 5.11). The pressure is assumed to be 300 mb and the temperature about −40°C; values are listed of the crystal length $2r$, its mass M, its fall speed V, and the distance S settled through the air, as functions of the time t following the nucleation of the crystal

t (sec)	$2r$ (μ)	$10^6 M$ (g)	V (cm sec^{-1})	S (m)
70	150	0.04	4	2
200	250	0.2	13	15
450	370	0.6	25	60
840	500	1.4	40	200

5.13 The growth of ice particles by accretion

As exemplified in Table 5.12, within only about 2 min of its formation in a cloud of supercooled droplets, an ice crystal attains a fall speed exceeding 10 cm sec^{-1}, substantially greater than that of most of the cloud droplets. Consequently it may quickly become large enough to collide with the droplets and continue its growth predominantly by *accretion*. Before this transition is examined it is advantageous to consider the factors that control the temperature and the density of the rime ice which is deposited on a body during the accretion of supercooled droplets.

The temperature of rime ice

The latent heat of fusion liberated during the freezing of supercooled water accreted by a body raises the temperature of the deposited rime ice above the general air temperature. Except in some laboratory conditions in which the interior of the body is cooled, the heat conducted into it is negligible compared with that transferred into the air, so that by equating the latter to the latent heat of the accreted water the mean surface temperature of the rime can be inferred. If the surface temperature is raised sufficiently, the transfer of sensible heat is supplemented by a transfer of latent heat due to evaporation (which does not materially reduce the rate of increase of the mass of rime). Experiments in which rime has been deposited on cylinders, rotated to preserve a layer of ice of virtually uniform thickness and temperature (28), and theoretical study of the physical processes accompanying the accretion (29) have confirmed that the surface temperature is satisfactorily obtained by using the appropriate empirically determined transfer coefficients. The experiments with rimed cylinders have a direct practical application in problems concerned with icing of the leading edges of the wings and other fuselage and engine surfaces on aircraft, but in meteorology the principal interest in rime ice temperature arises in the study of the growth of large hailstones. Their shape is sufficiently nearly spherical, and the Reynolds number is sufficiently large, to justify the use of the simplified expressions (5.53) and (5.54) for the transfer coefficients, considering that other important factors such as fall speed, efficiency of catch, and cloud water concentration cannot be accurately specified. Surface irregularities, the attitude of fall, the occurrence of rotation, oscillation or tumbling, and the presence of ice particles among the accreted droplets may significantly affect the details of the energy transfers and the precise form, size, and internal structure of large hailstones, but they have been shown to have no important bearing on the magnitude of the transfers or on the rate of mass increase of the hailstones in particular meteorological conditions (30).

In a dense cloud a hailstone collects and freezes water at a rate sufficient to raise its surface temperature above the general air temperature by as much as 10°C, or even more. Indeed, even at air temperatures as low as $-40°C$, the surface temperature of a hailstone may reach 0°C; the rate of heat transfer into the surroundings, and hence also the rate of freezing of the collected water, then attains a maximum value. If the concentration of cloud water is such that water is collected at a greater rate, unfrozen liquid accumulates, and the surface of the hailstone becomes wet. The minimum concentration m' (g cm^{-3}) of cloud droplets whose accretion just raises the surface temperature to 0°C depends upon the efficiency E with which they are caught, the size and fall speed of the hailstone, and the temperature and pressure of the air. It is found by equating the rate of release H_L of the heat of fusion of all the collected liquid to the rate of transfer H of sensible and latent heat from the hailstone into its surroundings. If the hailstone is a sphere of radius R and fall speed V (compared with which the fall speed of the cloud droplets is negligible), then

$$H_L = \pi R^2 E V m'(L_f - c\Delta T) \quad (5.78)$$

(part of the latent heat of fusion L_f at 0°C is considered to warm the liquid, whose specific heat is c, by $\Delta T°C$ from the air temperature to 0°C before it freezes), while from Eq. 5.48 and 5.49

$$H = 4\pi R[KC'_v \Delta T + L_v D C_v \Delta\rho] \quad (5.79)$$

where $\Delta\rho$ is the difference between the saturated vapor densities over liquid water at 0°C and at the air temperature (the latter representing sufficiently accurately the vapor density in the cloud).

Writing the convection coefficients in the general form

$$C'_v = C_v = b(\text{Re})^{1/2} \quad (5.54)$$

and

$$H_L = H$$

leads to

$$Em' = \frac{4b(\text{Re})^{1/2}(K\Delta T + L_v D \Delta\rho)}{RV(L_f - c\Delta T)} \quad (5.80)$$

The solution of this equation requires the specification of the size and mean density of the hailstone, and the density and temperature of the air. To illustrate its implications the density and drag coefficient of the hailstone may be given the constant values of 0.9 g cm^{-3} and 0.55, respectively, and the air density and tem-

5 The formation, growth, and evaporation of cloud particles

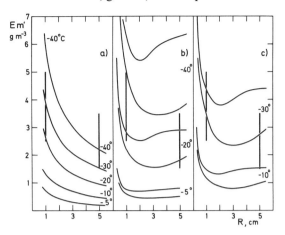

Fig. 5.7 The effective concentration Em' of supercooled cloud water, in a cloud with $\theta_s \approx 20°C$, in which the surface temperature of a hailstone (of density 0.9 g cm^{-3} and drag coefficient 0.55) is raised by the freezing of accreted water just to $0°C$, as a function of the air temperature ($0°C$) and the radius or equivalent spherical radius R of the hailstone (adapted from reference 11). The values used for the ventilation coefficient factor b are those indicated in Fig. 5.3: the average values for smooth ice spheres (diagram a), and the extremes for artificially grown hailstones (diagrams b and c). Considering that the concentration m and mean radius r of the supercooled cloud droplets in hailstorm clouds are likely to vary between about 4 and 6 g m^{-3} (Fig. 3.9) and between about 10 and 15 μ, respectively, the range of values likely for Em when $R = 1$ and $R = 5$ cm is indicated by the thick vertical lines (the values for the smaller hailstones are the larger, because their efficiency of catch E is greater; Fig. 2.7). When $Em > Em'$ part of the water accreted by a hailstone remains unfrozen.

perature assumed to correspond to a value of $\theta_s = 20°C$ which is typical of the air in clouds which produce large hailstones. The quantity (Em') is then a function of hailstone size and the air temperature only, as illustrated in Fig. 5.7. This diagram also shows the concentrations of cloud water corresponding to adiabatic ascent from cloud bases at the 800- and 900-mb levels (from Fig. 3.9).

Under these conditions the bulk of the supercooled cloud water can be anticipated to be in droplets of radius greater than 10 μ. The efficiency of catch E is then greater than 0.85 for the smallest hailstone radius considered, but may be only about 0.5 for the largest (Fig. 2.7). Accordingly, it is clear from the diagram that even very small hailstones are likely to have wet surfaces at air temperatures several degrees below $0°C$, while if the cloud water concentration is a large fraction of the value corresponding to adiabatic ascent, then large hailstones may be wet even at air temperatures as low as -30 to $-40°C$. Although many large hailstones have a spheroidal rather than spherical shape, consideration of the appropriate empirical drag and transfer coefficients shows that virtually the same values of (Em') are obtained if the dimension R is chosen to be the major semi-axis of the hailstone (31).

The density and structure of rime ice

When the surface of rime ice is wet or its temperature is close to $0°C$ the collected drops tend to coalesce before freezing and the density of the rime is nearly that of pure ice. On the other hand, when the mean surface temperature is well below $0°C$ the individual droplets freeze rapidly upon impact and may be loosely packed, with large air enclosures, so that the rime is friable and may have a density as low as 0.1 g cm^{-3}. The appearance of the rime is closely related to its density (as seen in Table 5.14).

From experiments in which rime was deposited on cylinders exposed in a wind tunnel installed in a large cold room (32), it appears that rime density depends in a complicated way on the mean surface temperature T_s of the rime, and on the radius r and impact speed V_0 of the droplets (the latter may be considerably less than the air speed V when the efficiency of catch is low). The first parameters affect the rate of freezing of individual droplets, and thus the likelihood that successively arriving droplets coalesce, and the last influences the spread (or even splashing) of the droplets, and perhaps the closeness of their packing when their freezing is rapid. The glazed rimes contain bubbles trapped when some of the small percentage by volume of air dissolved in the droplets and driven out of solution during freezing is trapped, while the much greater proportion of air in the rimes of lower density is enclosed during the irregular deposition of the droplets. In the experiments it was found that the rime density was practically independent of the surface temperature T_s when it fell below about $-20°C$; at higher values, up to about $-5°C$, the density ρ_i was given approximate-

Table 5.14 Density ρ_i of the seven principal recognized kinds of rime ice

Kind of rime	Approximate density ρ_i (g cm^{-3})
Glazed rimes	
Spongy rime (contains unfrozen liquid and some air bubbles)	$\gtrsim 0.92$
Clear ice (contains virtually no air)	0.92
Transparent ice (contains some air bubbles)	0.91
Milky ice (contains many small air bubbles)	0.88
Ordinary rimes	
Opaque rime (white and compact but readily crumbled)	0.4–0.8
Kernel rime (irregular surface)	0.3–0.6
Feathery rime (loose, fragile structure)	< 0.1–0.3

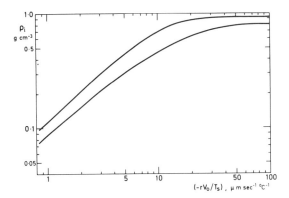

Fig. 5.8 An experimentally determined relation between the density of rime ice and the term $(-rV_0/T_s)$, where r is the mean-volume radius of the accreted droplets (μ), V_0 is their impact speed (m sec^{-1}), and T_s is the mean surface temperature of the rime (°C). The great majority of the numerous measured values for T_s of -20°C and above lie between the two lines drawn in the diagram (32). As the value of the term $(-rV_0/T_s)$ decreases below 1 μ m sec^{-1} °C^{-1} the rime density probably becomes approximately constant at a value somewhat less than 0.1 g cm^{-3}.

ly, as shown in Fig. 5.8, by the relation

$$\rho_i = a(-rV_0/T_s)^{3/4} \qquad (5.81)$$

where T_s is in degrees Celsius and the constant of proportionality a is 0.11 if r is expressed in microns and V in meters per second and is 3.6 if their units are centimeters and centimeters per second. This formula, or the more general relation illustrated in Fig. 5.8, appears to be the best available guide to the density of the accretions on hailstones grown in atmospheric clouds.

In the same series of experiments it was found that during the growth of rime in clouds containing ice particles as well as supercooled droplets, the ice particles were not accreted in appreciable quantities unless the surface of the rime became wet. They could then be accreted at a rate comparable to the rate of accumulation of unfrozen water, and at air temperatures as high as -2 to -1°C could cause the retention of this liquid, which otherwise moved to the rear of the cylinder and was shed into its wake as drops of diameter between about 1 and 2 mm.

At lower air temperatures unfrozen liquid is retained in the rime even when ice particles are present in the air in only extremely small mass concentration, probably because it is held in the spaces between dendritic crystals which grow rapidly through the film of accreted water (33). The existence of a limit to the rate at which the liquid water accreted by a hailstone can be frozen does not therefore imply a corresponding limit to the rate of increase in the mass of the hailstone. Rime containing a considerable proportion of unfrozen water is often referred to as "spongy" ice because, although it is generally sufficiently strong mechanically not to be broken away or distorted by the air flow, it is easily compressible and the liquid in laboratory specimens can be squeezed out between finger and thumb.

The formation of soft hail

The details of the transformation of a crystal of regular form into a small hailstone, after the onset of the accretion process, are very difficult to follow, depending upon uncertain and complicated matters such as the size and concentration of the cloud droplets, and especially the size, shape, and altitude and speed of fall of the ice particle (which control the efficiency of catch of the particle and the rate of accretion), and how they are modified during the course of the riming.

However, the magnitude of the time required for the transformation can be assessed with the help of some simplifying assumptions. In particular, in the virtual absence of any conclusive experimental or theoretical evidence about the efficiency of catch of a disclike collector at the low values of the Reynolds number involved (between about 1 and 50), it will be assumed to be equal to that of a sphere of the same radius and fall speed.* Intuitively it can be anticipated to be somewhat greater, but it can also be anticipated that the discrepancy will not affect the inference that the accretion process, just as the coalescence process among droplets, does not proceed at a significant rate unless most of the cloud water is held in droplets of radii at least several microns and the collecting particle has a fall speed of some tens of centimeters per second. If the cloud droplets are smaller, as may often happen in a cloud of low mass concentration (produced by a small upward displacement of air, or diluted by mixing processes), an ice crystal may continue its growth into a snow crystal of millimetric size without any riming. On the other hand, in denser clouds of larger droplets the riming and transformation into a soft hailstone begins when the crystal is still small and proceeds rapidly, so that the condensation soon becomes negligible. This may be illustrated by considering the growth of a platelike crystal of center-to-corner dimension r five times greater than its thickness (the form previously regarded as likely to arise at temperatures between about -8 and -25°C). The conditions in the parent cloud are taken to be those assumed in the derivation of Table 5.12,

---

*Experiments (34) on the collection of small spores and pollen particles by paper discs at Reynolds numbers in the range 25–150 indicated small but finite values of efficiency of catch E, of order 10^{-2}, for values of the parameter K (Eq. 2.32) of order 10^{-2}. Although according to the theoretical relation (2.33) E should be zero for K less than about 1.2, these measured values of E are not inconsistent with the experimental values for spherical collectors which provided the basis for Fig. 2.7.

but in addition droplets of radius 10 μ are supposed to be present in mass concentrations of 1 or 4 g m^{-3}, corresponding, respectively, to the moderate or dense concentrations of cloud water which may occur in the upper parts of clouds produced by large upward displacements of air.

First, Fig. 2.3 shows that in the range of Reynolds numbers concerned the drag coefficients of the thin disc and of the sphere are about the same at a particular value of the Reynolds number. Accordingly, from inspection of Eq. 2.28 written in the forms

$$\pi r^2 \cdot \frac{1}{2} \rho_a V^2 \cdot C_D = \pi r^2 \cdot \frac{r}{5} \cdot 0.9 g$$

(plate, density 0.9 g cm^{-3})

$$\pi r^2 \cdot \frac{1}{2} \rho_a V^2 \cdot C_D = \pi r^2 \cdot \frac{4r}{3} \cdot \delta g$$

(sphere, density δ)

it appears that the fall speeds are the same if the radii are the same and if the density of the sphere is

$$\delta = 3 \times 0.9/4 \times 5$$
$$= 0.13 \text{ g cm}^{-3}$$

Now from Fig. 5.8 it appears that the accretion of droplets of radius about 10 μ at impact speeds of a few tens of centimeters per second and particle surface temperatures in the range -10 to $-20°$C (corresponding to values of the expression $-rV_0/T_s$ of somewhat less than 1 μm sec^{-1} °C^{-1}) happens to be likely to produce a rime deposit of similar or even somewhat lower density. Thus the efficiency of catch and rate of mass accretion by the ice particle throughout the course of the riming can reasonably be regarded as always about that upon a sphere of the same radius and of mean density 0.13 g cm^{-3}. [Observations of the mean density

Table 5.15 Approximate rates of addition of mass dM/dt by condensation and by accretion to an ice particle of radius r, Reynolds number Re, and fall speed V, originally in the form of a thin hexagonal plate. The growth is in a cloud at a temperature between about -8 and $-25°$C, containing droplets of radius 10 μ in a concentration m of 1 or 4 g m^{-3}, and is estimated on the basis of the assumptions specified in the text

			$10^8 \, dM/dt$ (g sec^{-1})		
r (μ)	Re	V (cm sec^{-1})	Condensation	Accretion $m = 1$ g m^{-3}	Accretion $m = 4$ g m^{-3}
80	0.5	9	0.3	—	0.1
110	1	13	0.5	0.1	0.2
140	2	21	0.7	0.3	1.2
180	4	32	1.1	0.9	3.8
220	6	44	1.4	2.8	11
260	10	55	1.9	4.5	18

Table 5.16 Rate of increase of radius r and fall speed V of an ice particle in the moderately dense and the dense clouds of supercooled droplets specified in Table 5.15. The growth is at first by condensation upon a thin hexagonal plate, and subsequently by accretion of the droplets, producing a soft hailstone of density about 0.1 g cm^{-3}. Also listed are values of the distance S which the particle falls, through clear air of relative humidity 50%, before its evaporation is complete

		Time t to reach radius r (sec)		
r (μ)	V (cm sec^{-1})	Moderately dense cloud	Dense cloud	S (m)
120	15	240	240	10
200	40	560	370	40
360	80	940	460	200
500	120	1,100	500	500

of soft hailstones of radii less than about 4 mm grown in cold-weather showers have given values varying from about 0.08 to 0.2 g cm^{-3} (35).] The possible contributions of condensation and accretion to the rate of mass increase of the ice particle are then readily evaluated as a function of its radius, and are presented in Table 5.15. In the compilation of the table the rate of condensation has been estimated assuming the convection coefficient appropriate to a spherical particle.

It appears from the table that the contribution to the growth by accretion exceeds that due to condensation when the particle radius reaches about 120 μ in the dense cloud and about 200 μ in the moderately dense cloud. Thereafter, as the particle becomes larger accretion rapidly becomes the dominant process. Thus the increase of size with time can reasonably be assessed by ignoring the accretion at smaller values of the radii than those specified above, and by ignoring the condensation at larger values, with the results seen in Table 5.16.

The evaporation of ice particles

The clear air near droplet clouds is supersaturated with respect to ice, so that ice particles which grow sufficiently large to fall out of the clouds do not immediately evaporate. Those which fall from extensive droplet clouds may continue their growth, in layers several hundred meters deep. However, away from isolated droplet clouds which form in or penetrate into the upper troposphere (cumulus tops), the relative humidity more often has the low values typical of the clear air above the planetary boundary layer, and the ice particles which fall from such clouds soon begin to evaporate under subsaturations (with respect to ice) of about the same magnitude as the maximum supersaturations, representing saturation with respect to the liquid, under which they have grown, such as were assumed in the derivation of Tables 5.12 and 5.13. Accordingly, the values in these tables giving the trend of size and of distance settled relative to the air can be regarded, with a change of sign, as representative also of that which occurs during evaporation. It appears that crystals whose mass is a few micrograms, whose fall speed is about 0.5 m sec^{-1}, and which in the high troposphere are about the largest observed, may settle a few hundred meters before their evaporation is complete, while the corresponding distance for snow crystals of millimetric size may be as much as a kilometer.

If the evaporation of small, soft hailstones is considered under a relative humidity of 50% with respect to liquid water, and at a temperature of $-20°C$ in the middle troposphere, the distance S fallen before complete evaporation is found to vary with the particle size in the manner shown in Table 5.16. According to the criteria (tentatively proposed in Sections 5.7 and 5.10) for distinguishing precipitation from cloud particles—the fall speed should be comparable with the upward air speed associated with the cloud (here about 1 m sec^{-1}) and that the distance fallen through clear air during evaporation should be at least a few hundred meters—the soft hailstone with a radius of more than about 500 μ can be regarded as a precipitation particle. According to the other entries in the table, this size is not attained, even in a dense cloud, until nearly 10 min after the nucleation of the ice crystal, perhaps by the freezing of a small cloud droplet. In the moderately dense cloud this period is almost 20 min. These long intervals might be reduced by assuming a different shape for the ice crystal during its early growth by condensation, and a higher efficiency of catch during the later growth by accretion, but not reasonably by a factor as great as 2.

5.14 The multiplication of ice particles by fragmentation

Observations made in some kinds of natural clouds have shown that at temperatures not far below 0°C the concentrations of ice particles encountered may be much greater than would be expected from the concentrations of ice nuclei normally effective at the minimum temperatures within the clouds (36). The discrepancy tends to be more marked as these temperatures approach 0°C, and may reach three orders of magnitude at temperatures between about -5 and $-10°C$. The possibility cannot yet be excluded that conventional methods of measuring concentrations of ice nuclei greatly underestimate the concentration of those which may be effective in some natural conditions, but it seems much more likely that nucleated particles may multiply by some fragmentation processes. Laboratory studies have confirmed the existence of such processes, although they have not yet established their effectiveness quantitatively, and their importance is strongly suggested by observations discussed in Chapter 8, which show that in cumuliform clouds the ice phase often develops extensively at temperatures not much below 0°C, but only after the coalescence of cloud droplets has advanced to the stage of shower formation and significant concentrations of large drops (some of which freeze) have been produced. There are several recognized processes which lead to the fragmentation of ice particles.

First, the frail extensions of dendritic ice crystals may be broken away during collisions between the crystals, or even simply by the flow of air over their surfaces such as must occur in the atmosphere while they are settling. This kind of fragmentation has been observed from frost deposits on the surfaces of laboratory apparatus, and care has to be taken to prevent it in cloud chambers used to determine the concentrations of active ice nuclei. Nothing is known about its effectiveness in the atmosphere, where the crystals most liable to fragment are those with the feathery arms whose thickness, even when their diameter is several millimeters, is only about 10 μ (37). These crystals grow only in the narrow range of temperature between about -12 and $-16°C$, and only if the supersaturation is sufficiently high (Fig. 5.6). Even so, by such fragmentation during their settling they might provide within a deep layer a source of fresh crystals as important as that provided by sparse ice nuclei entering it in a slowly ascending airstream.

Second, in clouds which contain supercooled droplets fragments of ice may be produced during the freezing of the droplets. A thin surface layer of a droplet, from which the latent heat of fusion is readily transferred into the surroundings, may freeze rapidly and completely to form a shell of ice while the warmed interior is still liquid. The shell may subsequently be ruptured with the ejection of small ice fragments, under the stresses imposed by the more gradual progress of the internal freezing and liberation of dissolved gas. In the laboratory such shattering has been observed during the freezing of drops usually either suspended in cold air and artificially nucleated, or allowed to fall through air with a large temperature gradient. The behavior of such drops depends markedly on the experimental conditions, especially the degree of ventilation and the kind and concentration of the originally dissolved gases. Some experiments in which freezing droplets have been found to shatter frequently and to be a prolific source of ice fragments have almost certainly been affected by the presence of CO_2 (commonly used in solid form as a cooling or nucleating agent), which at temperatures below 0°C is nearly two orders of magnitude more soluble in water than is air (38). The liability of freezing droplets to shatter during free fall in air is likely to depend on whether they are nucleated at their surfaces and on their size; probably shattering occurs only if they are larger than ordinary cloud droplets. No investigations yet reported can be regarded as giving reliable indications of shattering behavior during natural freezing. The most relevant appear to be some in which droplets with radii in the range between about 10 and 75 μ were observed during free fall in air with a vertical temperature gradient not greater than 0.5°C cm^{-1} (39). The droplets froze at air temperatures between about -20 and $-32°C$ after nucleation by particles naturally present in the water, or at temperatures near $-8°C$ when the droplets were of water into which particles of silver iodide had been introduced. No droplets of radius less than 25 μ were observed to shatter during freezing. On the other hand, at least 5% of the larger droplets shattered during freezing, irrespective of the temperature, since this was the proportion of all those, frozen or unfrozen, which were observed to shatter while passing through a limited field of view. Visually a few large fragments of ice could be seen to be ejected during a shattering, but nucleation in a tray of supercooled sugar solution placed below and to one side of the region of freezing showed that many smaller particles were also ejected. No measurement or even estimate was made of the total number of fragments produced by a shattering, so that it is not possible to assess the significance of the process, but evidently in a population of droplets of this size range the concentration of ice crystals arising during their freezing might be at least several times greater than the concentration of effective nuclei. Nothing is known of the behavior of still larger drops during freezing in free fall, but it is possible that they are less liable to shatter, since experiments on the freezing of large suspended drops indicate that shattering may not occur unless the freezing proceeds symmetrically inward, which may be unlikely when a drop is sufficiently large (Re $\gg$ 1) to be ventilated asymmetrically during free fall.

It has long been suspected that a more prolific production of ice fragments may accompany the freezing of supercooled droplets during riming. Experiments in which an artificial hailstone was exposed to drafts of supercooled cloud established that ice fragments were produced during the impaction and freezing of the droplets, unless the surface of the hailstone was wet, or unless the radius of the droplets was less than about 15 μ (40). The average number of ice fragments produced for each impacting droplet varied little, having the rather high magnitude of 10. Although the validity of these experiments has been questioned, this result is consistent with experience in the previously mentioned experiments (32) to study the density of rime ice, in which within some minutes of the onset of riming the concentration of ice particles in the cold-room cloud increased by several orders of magnitude, reaching values of about 10^{-1} cm^{-3}, even at air temperatures as high as $-5°C$. Rime was deposited not only on objects exposed inside the small wind tunnel but also on parts of its inner surfaces, and therefore over areas probably of order 10 cm^2. The cloud contained droplets of mean radius about 30 μ in concentrations of about 10 cm^{-3}, so that inside the tunnel in a draft of a few meters per second they were collected at the rate of about 5×10^4 sec^{-1}. This is equivalent to about 10^7 in a few minutes, after which interval the number of ice particles in the whole room, of volume about 10^3 m^3, was of order 10^8.

Examination of the rime deposits under a microscope showed that some of the frozen droplets possessed narrow spicules up to several hundred microns long. These are produced by the extrusion of liquid through cracks in a frozen surface layer. Other frozen droplets appeared to have only the stumps of spicules which had been broken away by the draft or by the impact of other droplets. Yet other structures suggested that large drops, of radius about 100 μ or more, had splashed upon impact and scattered frozen residues over the rime surface and perhaps also into the air. Rime of low density and needlelike protrusions are particularly fragile, and pieces may be broken away readily by the impact of even small droplets or even merely by a draft. Recent experiments (41) suggest that the mechanical rupture of needlelike crystals growing (by condensation) upon rime in a narrow range of surface temperatures near $-5°C$ (Fig. 5.6) may be the most significant of all the ice particle multiplication processes.

Assuming that during the growth of hail as many as 10 ice fragments are on the average produced for each accreted droplet of radius more than about 15 μ, this source of ice particles is clearly likely to become much more important than the primary ice nuclei if the following conditions are fulfilled: (a) the concentration of ice nuclei is less than the concentration of droplets of this size; and (b) a substantial fraction of the droplets are accreted by the hail. Suppose, for example, small hailstones of low mean density and radius R of about 2 mm, and of fall speed V about 10 m sec^{-1} (Table 2.5), are present in a concentration N_h of about 10 m^{-3} (corresponding to the small mass concentration of

10^{-1} g m^{-3}). If they are accreting droplets of radius greater than 15 μ in the concentration N_d of only 1 cm^{-3}, then the rate of production F of ice fragments is

$$F \approx 10\pi R^2 \, EVN_h N_d \, \text{cm}^{-3} \, \text{sec}^{-1} \quad (5.82)$$

$$\approx 10^{-2} \, \text{cm}^{-3} \, \text{sec}^{-1}$$

This exceeds the rate of provision of ice particles by the flux of ice nuclei in air ascending at a speed of order 1 m sec^{-1} at all temperatures down to about $-15°$C (Fig. 5.5). Even at low temperatures the fragmentation may beome the more important source because once under way it is progressive until the supply of supercooled droplets is greatly depleted. Collisions between the ice fragments and cloud drops tend to increase the concentration of the hail, and hence the total rate of droplet accretion and fragmentation. The probable trend in the composition of a cloud cannot be established without consideration also of the pattern and endurance of the air motion. However, it is clear that in a cloud sufficiently thick and persistent to produce some frozen precipitation, but in which the temperature is nowhere much lower than about $-15°$C, the concentration of ice particles may become very much greater than would be anticipated by supposing that they were formed only upon the primary ice nuclei. [It is interesting that irregular small ice fragments have been observed in concentrations of order as large as 1 cm^{-3} at temperatures between about -10 and $-15°$C on the summit of a small Japanese mountain, inside winter clouds which produce falls of snow crystals and small soft hail (42). However, since riming occurs also on the mountain slopes, such fragments may not all be derived from the precipitation particles in the air.]

References

1. Keith, C.H., and Arons, A.B. 1954. The growth of sea-salt particles by condensation of atmospheric water vapour, *J. Meteorol.*, **11**, 173–84.
2. Squires, P. 1952. The growth of cloud drops by condensation. I. General characteristics; II. The formation of large cloud drops, *Austr. J. Sci. Res.*, A, **5**, 59–86; 473–99.
3. Mordy, W. 1959. Computations of the growth by condensation of a population of cloud droplets, *Tellus*, **11**, 16–44.
4. Neiburger, M., and Chien, C.W. 1960. Computations of the growth of cloud drops by condensation using an electronic digital computer, in *Physics of Precipitation*. Geophysics Monograph no. 5, American Geophysical Union, Washington, D.C., pp. 191–210.
5. Twomey, S. 1959. The nuclei of natural cloud formation, Part II. The supersaturation in natural clouds and the variation of cloud droplet concentration, *Geofis. Pur. Appl.*, **43**, 243–49.
6. Ranz, W.E., and Marshall, W.R. 1952. Evaporation from drops, *Chem. Eng. Prog.*, **48**, 141–46; 173–80.
7. Kinzer, G.D., and Gunn, R. 1951. The evaporation, temperature and thermal relaxation-time of freely falling water-drops, *J. Meteorol.*, **8**, 71–83.
8. Karve, C.S. 1942. Evaporation from water drops and wet spherical surfaces, *Proc. Indian Acad. Sci.*, **16**, 2, A, 103; Kramers, H. 1946. Heat transfer from spheres to flowing media, *Physica*, **12**, 61.
9. Macklin, W.C. 1964. Comments on the "General heat and mass exchange of spherical hailstones," *J. Atmos. Sci.*, **21**, 227–28.
10. Macklin, W.C. 1963. Heat transfer from hailstones, *Q. J. R. Meteorol. Soc.*, **89**, 360–69.
11. Macklin, W.C. 1968. Heat transfer from artificial hailstones, *Q. J. R. Meteoral. Soc.*, **94**, 93–98.
12. Mason, B.J. 1971. *The Physics of Clouds*, 2d d. Clarendon Press, Oxford.
13. Mossop, S.C. 1963. Atmospheric ice nuclei, *Z. angew. Math. Phys.*, **14**, 456–86.
14. Findeisen, W., and Schulz, G. 1944. Experimentelle Untersuchungen über die atmosphärische Eisteilchenbildung I, *Forsch. Erfahr. Reichsamt Wetterdienst*, A, **27**; reprinted in Schulz, G. 1947. *Die Arbeiten und Forschungsergebnisse der Wolkenforschungsstelle des Reichsamts für Wetterdienst in Prag*, Berlin Wetterdienst.
15. Soulage, G. 1961. Origins, concentrations and meteorological importance of atmospheric freezing nuclei, *Nubila*, **4**, 43–67.
16. Bourquard, A.D. 1963. Ice nucleus concentrations at the ground, *J. Atmos. Sci.*, **20**, 386–91.
17. aufm Kampe, H.J., and Weickmann, H.K. 1951. The effectiveness of natural and artificial aerosols as freezing nuclei, *J. Meteorol.*, **8**, 283–88.
18. Bigg, E.K. 1960. Summary of measurements of ice nucleus concentrations, *Bull. Obs. Puy de Dôme*, 89–98.
19. Bigg, E.K. 1961. Natural atmospheric ice nuclei, *Sci. Prog.* **49**, 458–75.
20. Palmer, H.P. 1949. Natural ice particle nuclei, *Q. J. R. Meteorol. Soc.*, **75**, 17–22.
21. Hoffer, T.E. 1961. A laboratory investigation of droplet freezing, *J. Meteorol.*, **18**, 766–78.
22. Mason, B.J., Bryant, G.W., and van de Heuvel, A.P. 1963. The growth habits and surface structure of ice crystals, *Phil. Mag.*, **8**, 505–26.
23. Magono, C., and Woo Lee, C. 1966. Meteorological classification of natural snow crystals, *J. Fac. Sci., Hokkaido Univ.*, ser. 7, **2**, 321–35 (proposes and illustrates profusely a detailed classification of the various regular, irregular, and rimed forms of ice particles collected at the ground); Weickmann, H.K. 1945. Formen und Bildung atmosphärischer Eiskristalle, *Beitr. Phys. freien Atmos.*, **28**, 12–52 (contains photographs of various forms of ice crystals collected during flights up to high levels in the troposphere).
24. Reynolds, S.E. 1952. Ice-crystal growth, *J. Meteorol.*, **9**, 36–40.
25. Schaefer, V.J. 1947. Properties of particles of snow and the electrical effects they produce in storms, *Trans. Am. Geophys. Un.*, **28**, 387–614.
26. Jayaweera, K.O.L.F. 1965. The behaviour of small clusters

of bodies falling in a viscous fluid, Ph.D. Thesis, Imperial College, University of London.
27. McDonald, J.E. 1963. Use of electrostatic analogy in studies of ice crystal growth, *Z. angew. Math. Phys.*, **14**, 610–20.
28. Macklin, W.C. 1961. Accretion in mixed clouds, *Q. J. R. Meteorol. Soc.*, **87**, 413–24.
29. Macklin, W.C., and Payne, G.S. 1967. A theoretical study of the ice accretion process, *Q. J. R. Meteorol. Soc.*, **93**, 195–213.
30. Macklin, W.C. 1964. Factors affecting the heat transfer from hailstones, *Q. J. R. Meteorol. Soc.*, **90**, 84–90.
31. Macklin, W.C. 1963. Heat transfer from hailstones, *Q. J. R. Meteorol. Soc.*, **89**, 360–69.
32. Macklin, W.C. 1962. The density and structure of ice formed by accretion, *Q. J. R. Meteorol. Soc.*, **88**, 30–50.
33. Macklin, W.C., and Ryan, B.F. 1965. The structure of ice grown in bulk supercooled water, *J. Atmos. Sci.*, **22**, 452–59. See also Carras, J.N., and Macklin, W.C. 1973. The shedding of accreted water during hailstone growth, *Q. J. R. Meteorol. Soc.*, **99**, 639–48.
34. Starr, J.R., and Mason, B.J. 1966. The capture of airborne particles by water drops and simulated snow crystals, *Q. J. R. Meteorol. Soc.*, **92**, 490–99.
35. Browning, K.A., Ludlam, F.H., and Macklin, W.C. 1963. The density and structure of hailstones, *Q. J. R. Meteorol. Soc.*, **89**, 75–84.
36. Hobbs, P.V. 1969. Ice multiplication in clouds, *J. Atmos. Sci.*, **26**, 315–18; Burrows, D.A., and Robertson, C.E. 1969. Comments on "Ice multiplication in clouds," *J. Atmos. Sci.*, **26**, 1340–41; Mossop, S.C., Cottis, R.E., and Bartlett, B.M. 1972. Ice crystal concentration in cumulus and stratocumulus clouds, *Q. J. R. Meteorol. Soc.*, **98**, 105–23.
37. Nakaya, U. 1954. *Snow Crystals*. Harvard University Press, Cambridge, Mass.
38. Dye, J.E., and Hobbs, P.V. 1966. Effect of carbon dioxide on the shattering of freezing water drops, *Nature*, **209**, 464–66; 1968. The influence of environmental parameters on the freezing and fragmentation of suspended water drops, *J. Atmos. Sci.*, **25**, 82–96; Pena, J.A., Pena R.G. de, and Hosler, C.L. 1969. Freezing of water droplets in equilibrium with different gases, *J. Atmos. Sci.*, **26**, 309–14.
39. Hobbs, P.V., and Alkezweeny, A.J. 1968. The fragmentation of freezing water droplets in free fall, *J. Atmos. Sci.*, **25**, 881–88.
40. Latham, J., and Mason, B.J. 1961. Generation of electric charge associated with the formation of soft hail in thunderclouds, *Proc. Roy. Soc. A*, **260**, 537–49.
41. Hallett, J., and Mossop, S.C. 1974. Production of secondary ice particles during the riming process, *Nature*, **249**, 26–28.
42. Magono, C., et al. 1962. Investigation on the growth and distribution of natural snow crystals by the use of observation points distributed vertically, III, *J. Fac. Sci., Hokkaido Univ.*, ser. 7, **1**, 373–91.

6 Atmospheric motion systems

6.1 The identification of motion systems

The kaleidoscopic phenomena manifest in the atmosphere are studied scientifically as patterns in the fields of the atmospheric variables of state. It is natural to identify particular patterns, recognized primarily from the distribution of air velocity as *motion systems*, and to classify them into a number of kinds (Fig. 6.1), each of which has its own characteristic dimensions and distributions of the other physical variables. The difficulty of making such a classification precise, arising principally because atmospheric motion is *turbulent* (i.e., unsteady or containing simultaneously systems having a wide range of scales), is a fundamental one which has not been given sufficient discussion.

Some motion systems contain condensed water in the form of clouds and rains, so that their presence is shown simply by the shapes of individual clouds or by the arrangement of populations of clouds; for example, an individual circulation of the ordinary kind of convection may be marked by a cumulus cloud which occupies its upper part, while the location of the great vortices of cyclones is made apparent in a survey from a satellite by accompanying concentric spiral bands of clouds. Each motion system is distinguishable from others throughout a limited period during which it has a characteristic evolution. Some systems are virtually stationary, evidently remaining closely associated with some topographical feature, while others travel, often for distances several times greater than their characteristic horizontal dimension. Thus a thundercloud may develop in an afternoon and travel for some hundreds of kilometers before dissolving in the night, or a cyclone may form off the east coast of North America and move across the North Atlantic before decaying beyond recognition over Northern Europe several days later.

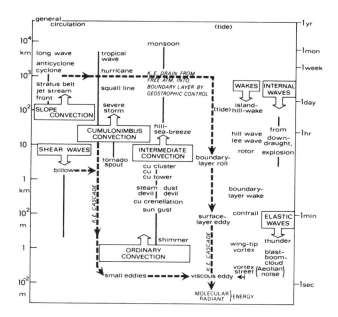

Fig. 6.1 Atmospheric motion systems arranged according to characteristic horizontal dimension, with an indication of approximate corresponding time scale. Kinetic energy is generated in the atmosphere on a great range of scales, especially effectively on those above about 1000 km, and is removed in a boundary layer by a "cascade" through eddy motions leading to a viscous dissipation (Fig. 6.2), and in the atmosphere above probably mainly by a consequent systematic mean flow toward higher pressures, although partly also by a similar cascade in accompanying phenomena of smaller scale. Many of the terms in the figure are defined in the text.

113

6   Atmospheric motion systems

The physical description of motion systems

It is not possible to explain in a simple and satisfying manner that the observed mean condition of the atmosphere and the properties of its individual motion systems are consistent with the familiar physical laws, assuming only a few fundamental physical parameters such as the earth's composition, geometry, and distance from the sun. It might seem feasible that with only such data and with powerful computational aids calculations could be made of the evolution of an irradiated atmosphere supposed originally in a number of realistic or simple physical states (e.g., one in which it is dry, of uniform temperature, and at rest), which would show that after some time the "model" atmospheres reproduced observed mean states, contained realistic motion systems, and revealed their mechanisms. Remarkable progress has been achieved in this kind of numerical study of the atmosphere, which is the basis of scientific weather prediction; in particular, mean states and motion systems have been calculated which closely resemble those observed. Even so, the correspondence has been to some extent imposed by arbitrarily incorporating important observed terrestrial properties, such as fixed climatological fields of radiative energy fluxes, or of surface temperature. Otherwise the calculations would become so excessively complicated and laborious as to be impracticable. Moreover, progressive reduction of the number of observed parameters obviously demands simultaneous consideration of the ocean, and eventually of the chemical as well as physical properties of the earth as a whole. Further, the results of involved numerical calculations, even with experiments in which the introduced parameters are varied one at a time, do not necessarily lead to improved insight into mechanism, compared with more elementary studies using drastically simplified models and having recourse to parameters whose values are based upon observation rather than inference. It therefore appears inevitable that any physical description and discussion of atmospheric behavior is guided by appeal to some observed conditions.

6.2 The scales of motion systems

Thus theoretical studies of the atmosphere employ mathematical models, consisting of equations (and boundary conditions) simplified as far as is consistent with the aim of relating quantitatively various properties of observed systems, or two successive observed states, as well as making clear the dependence of the relation upon physical laws. Mathematical analysis and observational measurement and interpretation provide mutual stimulus: the data concerning what are regarded as the significant features inspire drastic but profitable simplification of the models, and the analysis alters the physical interpretation of the data or directs attention to the need for improved or new kinds of data. Most motion systems are essentially three-dimensional in space and all have an evolution in time, but their analysis in detail as four-dimensional phenomena can generally be pursued only by numerical methods. However, valuable insight into their physical mechanism is often obtained from models in which general analytical solutions can be obtained by omitting one, two, or even three dimensions. For example, a persistent phenomenon (in the sense that particles pass through its prominent part in a time short compared with its duration) may be modeled by assuming a *steady state* in which terms containing $\partial/\partial t$ are omitted, thus introducing great simplification at the cost of relinquishing consideration of how the phenomenon appears or decays (and perhaps with less obvious penalties). Similarly, simplification may be achieved by assuming one horizontal space dimension to be irrelevant (e.g., presuming $\partial/\partial y = 0$ where y is distance across a flow). Another much exploited technique is the *perturbation method*, in which one of a few steady states given by simplified equations of motion is subject to a small disturbance, or perturbation, whose properties are examined in a new solution obtained by presuming that the squares and products of the small changes of the variables of state (including the motion) are negligible. In this way the stability of the initial flow to small disturbances, and the early structure of new motion systems which may develop from such disturbances (considered always present in a real flow), may be established.

An appropriate simplification of the general governing equations can often be made when the scale of the phenomenon under consideration is known. Then some terms may be omitted as insignificant; the effects of phenomena of other scales, always present in the atmosphere, can be excluded or, if they are important, represented in some simple manner not demanding a complete specification (e.g., diffusion effected by systems of smaller scale may be expressed by some transfer coefficient empirically related to the gradient of a property on the scale being considered).

In this way the concept of scale has a special importance in theory. Similarly, it enters the technique and analysis of observation. Intuitively it is evident that the description of the essential features of a particular system requires measurements at points separated in space by distances of magnitude an order less than that of the extent of the system, and in time by intervals of magnitude an order less than that of the duration of the system. Three kinds of difficulty arise in making the measurements.

First, the instruments should ideally sample volumes consistent with their separation, with a response time consistent with the time interval required of the observations. Generally the instruments in use have dimensions small with respect to a man, and in the atmosphere they are therefore sensitive to systems on an enormous range of larger scales up to the greatest encountered. It then frequently happens that an instrument is strongly affected by a system of smaller scale than that under consideration, so that its reading does not fit smoothly into those of the other instruments, and it is said to be unrepresentative. Accordingly, the analyst who uses the observations to define a pattern must continually decide that some are *unrepresentative* and either reject them or adjust them in some manner.

Second, the number of measurements needed to define satisfactorily a single motion system is very large. If several variables of state (velocity, temperature, pressure, concentration of water vapor, and concentration and particle size distribution of condensed phases of water) are measured at places separated by one tenth of the dimension of a system at ten intervals during its evolution, then the total number of measurements is of order (10 elements) $\times 10^3 \times 10$, or 10^5. If, as is usually convenient, the observations are made from fixed places, it has to be considered that the system usually appears at an unpredictable place and time and during its evolution travels over a distance typically an order of magnitude greater than its horizontal dimension. Moreover, upon examination it is hard if not impossible to define the bounds of systems in space and time; most seem to be *open systems*, intimately linked with the remainder of the atmosphere, if only tenuously with its remoter parts, so that the explanation of their behavior often seems to demand information from well outside their apparent borders. These complications increase the needed number of measurements to at least the order of 10^7.

Third, under the limitations of the conditions in which they are used, the instruments may be inaccurate. Thus far economy has dictated that widespread samplings of the free atmosphere be made with small balloons carrying light and cheap instruments. Balloon-borne thermometers are subject to variable radiation errors which increase with height to become disturbing in the upper troposphere and serious in the stratosphere. More important, measurements of the state of atmospheric water are inadequate. The balloon-borne hygrometers now used are unreliable in the upper troposphere and useless in the stratosphere, while instruments to measure the properties of condensed water are generally of such size or complexity that they can be used only if carried on aircraft, so that their use is severely restricted and even so is generally subject to difficulties of calibration at the high air speeds which are usual.

6.3 The development of the theory of motion systems

For the reasons just outlined unambiguous observations have been obtained only for systems of the largest scales in the extratropical continental regions of the northern hemisphere. There the economic value of locating these systems, and of producing weather forecasts for periods of a day or two ahead simply (until recently) by extrapolating their behavior, has brought into existence a network of radiosonde stations which routinely sound the atmosphere twice a day up to heights of about 25 km. Over large areas the average separation of these stations is about 400 km, and their observations therefore reasonably well define the structure of motion systems, such as cyclones, whose horizontal dimension is at least several times greater, and whose general properties are not immediately affected by local phase changes of water, although they depend on the general stratification which is related to the prevalence of condensation and precipitation in the atmosphere.

These observations have been an essential help in the achievement of a theory of the motion systems as large as or larger than cyclones. Presumably mainly because of the lack of observations sufficiently abundant to define the evolution of smaller systems, and partly because of a natural preoccupation of most theorists with the comparatively plentiful data of immediate economic value, the theory of all systems of smaller scale, and particularly of those in which the phase changes of water play an important or even essential role, is fragmentary and poorly developed. Not until the scale of the systems is reduced to less than a few meters and the required measurements fall within the compass of an individual investigator does one again encounter comparatively advanced theory, and then only for the statistical properties of the motion systems close to the earth's surface, which are an important link in the transfer of energy from solar radiation into the atmosphere.

It can hardly be anticipated that rapid progress will be made in the theory of motion systems of intermediate scale, simply because of the expense of the observations which seem necessary to provide a sufficient stimulus. The cost of the measurements needed to define the evolution of a single cyclone can be regarded as of order 1% of the annual budget of the official weather services over an area the size of North America or Europe, that is, about £100,000. This great expense is not any the less if the system under study is smaller than a cyclone. On the contrary the program of observations is likely to be more costly for the following reasons: (*a*) the program is not a routine which has become efficient through long use, but requires the assembly of a number of specialists at some favorable location; (*b*) the smaller dimensions and duration of the system are likely to demand more responsive and elaborate instruments, carried by aircraft rather than small, free balloons; (*c*) the deployment and the coordination of the observers becomes more difficult and may require sophisticated facilities such as radar.

Theories of cloud formation and rain production have mostly been based on very crude representations of the associated motion system, ranging from upward air currents presumed uniform in space and time to patterns of streamlines in two or three dimensions and unchanging in time. These latter, the most realistic constructed until very recently, may have been consistent (kinematically) with the conservation of mass, but have not been shown to be entirely consistent (dynamically) with the other conservation laws.

The crudity of the representations is not a serious matter in the study of the *formation* of cloud particles, when the relevant physical dimensions are less than 1 cm and accordingly also less than those of the smallest of the motion systems, but the inability to specify the motion systems in detail soon intrudes as an obstacle in the study of the *evolution* of cloud particles and of course in the explanation of the behavior of a cloud as a whole. The advance of the studies of clouds and rains from a *cloud physics*, consisting of the application of the laws governing the behavior of cloud particles to a kinematical description of the parent motion system, toward a *cloud dynamics* must proceed eventually by the same method used to establish and refine the theory of the large-scale motion systems. In this method a simplified set of equations expressing all the conservation laws, together with appropriate boundary conditions, is solved numerically to predict a future from a specified initial state. The success of the prediction, and therefore of the theory, is judged by comparison with observed evolutions of systems.

It seems intrinsically more difficult to construct theories of the motion systems which have scales smaller than that of cyclones, partly on account of the difficulty of obtaining the required observations, and also because the structure and behavior of the systems may be strongly modified by or even essentially dependent upon the phase changes of water. These are comparatively unimportant in the evolution of individual large-scale systems and have only recently and tentatively been included explicitly in the theories on which are based the numerical techniques of weather prediction now replacing the qualitative extrapolation techniques formerly used to produce forecasts.

Nevertheless, as attempts are made to refine these theories in order to include rainfall in forecasts, in order to extend their useful period from two or three days to as much as two or three weeks, and in order to forecast weather in tropical regions (where clouds seem to play a more important role in even the large-scale systems), so more account has to be taken of the energy transformations and fluxes accompanying the small-scale systems and the associated phase changes of water. A main objective of the study of the small-scale systems, which may be attained more readily than satisfactory theories of their own evolution, is the establishment of simple but sufficiently accurate ways to represent them in the formulation of the theory of the large-scale systems.

6.4 The classification of motion systems

The kinetic energy of atmospheric motion systems in general is generated and maintained against frictional dissipation by the conversion of gravitational potential energy provided by heat sources and sinks. It often appears that the *generation* of what are recognized as individual systems is not *immediately* related to the heat sources and sinks. For example, the tropospheric motion systems of very large scale derive their energy from the latitudinal temperature gradient and are continually acting to reduce it, while radiative fluxes act to restore it. However, the action of these fluxes appears to be widespread and sluggish, with a time constant which is considerably longer than that which characterizes the *initial* rate of growth of the motion systems which continually develop here and there. Thus the calculation of the radiative equilibrium of the troposphere (Section 1.2 and Fig. 1.1) indicates that a disturbance from the equilibrium temperature is halved by radiative fluxes in a period of about 20 days. In contrast, the initial rate of growth of the large-scale motion systems typically represents a doubling of their intensity in a period of about $\frac{1}{2}$–2 days. Moreover, the estimated mean rate at which tropospheric kinetic energy is dissipated by friction is about 2×10^5 erg g^{-1} day^{-1} (1), corresponding to a halving period of about 4 days. Accordingly, the conditions for the generation of large-scale motion systems may be studied with the simplifying assumptions that the motions are adiabatic and frictionless.

The possible existence of these and other kinds of systems, their dimensions, and their initial growth rates can be established analytically by examining the stability to a small perturbation of some atmospheric equilibrium state, for example, a state of rest or of a steady horizontal zonal motion with a uniform meridional temperature gradient. For mathematical convenience the applied perturbation is given a sinusoidal form, and the motion systems which are found to develop are at least in their early stages wavelike. Three fundamental mechanically distinct classes of motion system are recognized, to be regarded as extreme cases of more general motions which are a combination of classes:

Elastic waves or sonic disturbances, including sound waves: particles are displaced parallel to the direction of wave propagation, with transformations between kinetic and internal energy. The disturbances move with about the speed of sound and are refracted by gradients of temperature and wind.

Gravity waves, including ordinary convection: particles are displaced up and down in waves which propagate transversely, with transformations between kinetic and gravitational potential (together with internal) energy, and sometimes preexisting kinetic energy. The waves have a characteristic frequency depending on the degree of stability of the atmosphere to vertical displacement, and partly upon their structure, which for wavelengths of a few kilometers corresponds to a speed of a few tens of meters per second.

Quasi-geostrophic waves: these waves develop in a large-scale flow and are essentially affected by the rotation of the earth. The particle displacements are almost horizontal. Transformations occur between kinetic energy, gravitational potential (together with internal) energy, and preexisting kinetic energy; in the extreme case of *Rossby waves* the motion is horizontal: no gravitational potential or internal energy is involved, and the original kinetic energy is only redistributed.

Motion systems (apart from sonic disturbances) which produce clouds or other obvious phenomena generally move with about the mean velocity of the wind in the layer which they occupy, or with the velocity of some external agency (such as an aircraft). Particular examples of the latter kind, called *forced phenomena*, are stationary disturbances over or near mountains, which may produce the *wave clouds* described later (Pl. 6.4), or disturbances over traveling convective circulations, which may produce similar clouds (pileus; e.g., Pl. 7.3b). Tidal oscillations can be included in this class of motions. Motion systems of the former kind are often thought of as carried by a current, but careful examination of all systems shows that there is always a flow of air through the pattern or phenomenon regarded as representing a motion system.

The main atmospheric phenomena corresponding to waves of the three principal classes and their combinations are listed according to the magnitude of their horizontal dimension in Fig. 6.1, together with some processes of very small scale, of interest in cloud physics. Some of the names which are used are explained in the following paragraphs.

The *elastic waves* of the first principal class have very little direct importance in the atmosphere; those which are not audible are hardly ever noticed except near explosions. An explosion is produced by very rapid local generation of heat or gas, and is accompanied by outward-traveling elastic waves. In some places the pressure variation in a wave may be comparable with or exceed the normal atmospheric pressure, so that it is recognized as a *shock wave*. Near an exploding nuclear bomb the compression is sufficient to heat the air to incandescence; farther away, and sometimes near a chemical explosion, the expansion is sufficient to produce a (*blast*) cloud in the form of a rapidly expanding, thin, hemispherical shell. Clouds may also appear in the shock waves produced by supersonic aircraft (*boom clouds*). As an elastic wave progresses an outward displacement of air is effected by which the atmosphere accommodates the introduced or expanded gas; in particular those spreading from a heat source of meteorological intensity permit the expansion of the heated air, and the decrease in its density rather than increase in its pressure, which is the first observed effect of the heating (2). The intensity of these waves is so small that they are not detected.

Among the phenomena produced by the *gravity waves*, composing the second principal class, are various kinds of convection which draw upon gravitational potential energy to effect transfers of heat in the troposphere from low to high levels and from low to high latitudes. *Ordinary* or *small-scale* convection consists of circulations whose horizontal and vertical dimensions are comparable, in which the ascent appears concentrated into *plumes* or *thermals*. The kinetic energy is provided by the interchange of particles in the vertical, with a consequent release of gravitational potential energy (at least in the lower part of the layer occupied, buoyancy is generated by an upward displacement of a small volume of air, so that the stratification is said to be statically unstable). The convection transfers energy up from near the earth's surface, partly as sensible and partly as latent heat. When condensation of water vapor occurs in the ascending air of the circulations, typically a population of separate pyramidal *cumulus* (*heap*) clouds appears; the flat base of a cumulus has a diameter about equal to the height of its tops above the base level, usually only 1 or 2 km. The circulations contain motions on a range of lesser scales down to about 1 cm (below which the energy of all atmospheric motions is dissipated increasingly effectively by molecular viscosity), manifest on cloud borders as towers, crenellations, and smaller irregularities. Cumulus represent the prevalent form of cloud, especially in low latitudes.

Cumulus of ordinary convection occasionally reach up into the middle or upper troposphere and produce showers; they are then called *cumulonimbus* (*raining heap clouds*). Because the evaporation of rain may greatly assist the descent of air and then appears often to produce a notable increase in the intensity and scale of the whole circulations it is convenient to distinguish *cumulonimbus convection* from cumulus convection.

A convective circulation arises also in the presence of a horizontal (strictly an isobaric) gradient of temperature, for then the interchange of particles separated sufficiently in the horizontal, as well as vertically, diminishes the gravitational potential energy, even though an interchange in any vertical cannot, so that the stratification is described as statically stable. Ordinary convection introduces horizontal temperature gradients into the atmosphere on a variety of scales corresponding to the distances over which the intensity of a heat source at the earth's surface is variable. On a range of scales up to 1000 km or more, such variations are principally topographical, due to the different thermal properties of land and sea, or low and high ground. The *intermediate-scale convection* which arises is familiarly manifest near the surface by sea and mountain breezes.

Another kind of gravity wave which commonly occurs in stably stratified airstreams is that which causes a vertical oscillation about an equilibrium level, induced by flow over an obstacle. Such waves are sometimes spectacularly shown by a series of regularly spaced *wave clouds* in the lee of mountains, formed by condensation of water vapor in the wave crests. Other gravity waves, which also take their energy from the kinetic energy of the motion on a larger scale, form in flows characterized by a pronounced *shear*, the vector change of velocity with distance. In the atmosphere the strong shears are usually in the vertical, and are especially prevalent immediately above the earth's surface, where friction reduces the air velocity to zero. The gravity waves which arise when the shear passes some threshold value develop secondary instabilities and motion systems in a range of smaller scales, or *eddies*, through which kinetic energy of the mean flow is passed, eventually to be dissipated by viscous forces and transformed into internal energy. Two kinds of waves are recognized, one essentially dependent on the viscosity of the fluid and the other on the existence of a point of inflection in a profile of the wind speed in a vertical plane. Those of the second kind which develop against the resistance to vertical displacements implied by a statically stable stratification are ascribed to a *Kelvin-Helmholtz instability*. They sometimes produce evanescent billow clouds well above the ground, and are probably the main cause of the bumpiness in clear air which is occasionally experienced by aircraft at high levels; they may also contribute significantly to the dissipation of the kinetic energy of the atmosphere, but the eddies predominantly responsible for this dissipation are those in the virtually ever-present turbulent boundary layer.

The third principal class of wave is the *quasi-geostrophic wave*, mainly of large horizontal scale, whose time scale is a large fraction of a day or longer, so that the motion is strongly influenced by the earth's rotation and streamlines and isobars become nearly parallel. The most important kinds of such waves are the *long waves* and *cyclone waves* of middle latitudes, which appear on maps of the flow in the middle and upper troposphere as wavelike distortions of a general flow from west to east, accompanied by *troughs* of low pressure and *ridges* of high pressure, and on surface weather maps as vortices in the flow, accompanied by concentric isobars around centers of low pressure (*depressions* or *cyclones*) and high pressure (*anticyclones*). Examined in other ways, which more clearly indicate the vertical component of the air flow, the cyclone waves can be seen as the *slope convection* which works in the meridional temperature gradient to transfer heat up through the extratropical troposphere and from low into high latitudes. Again the general stratification is stable, but gravitational potential energy is provided for the generation of the kinetic energy of the waves by the interchange of particles separated not only in the vertical but by a sufficiently great distance in the horizontal. The slope of the paths along which most of the particles move in this convection is typically only about 1 in 1000. Within the developing waves the vertical and horizontal motions intensify in comparatively narrow strips called *frontal zones*, or simply *fronts*, accompanied by cloud and rain belts, and *jet streams* containing upper tropospheric wind speeds as high as 100 m sec^{-1}.

In low latitudes the static stability is less and cumulonimbus, sometimes organized in systems of large scale, are the most evident and prevalent form of deep convection. The quasi-geostrophic motion systems are less intense than in middle latitudes and their structure and frequency are uncertain; they appear often to occur combined with cumulonimbus convection, as in the *easterly waves* which move from east to west in the trade wind regions. These occasionally intensify into the most intense of terrestrial storms, the *tropical cyclones* (*hurricanes* or *typhoons*), which are essentially dependent upon heat extracted from the ocean.

The time scale of motion systems

Since, with the exception of the elastic and tidal waves, the speed of propagation of the atmospheric motion systems does not exceed the speed of some wind on their scale, on all scales down to about 10 km it has the magnitude of 10 m sec^{-1}, then diminishes with scale to 1 m sec^{-1} at about 100 m and to 1 cm sec^{-1} at a few centimeters. The typical *relative velocity* with which particles traverse a motion system is about the same, and so the time taken to traverse the system diminishes steadily with scale from about 1 week in the largest to about 1 sec in the smallest. Accordingly, it is possible in Fig. 6.1 to add a scale of time which corresponds to that of horizontal dimension; except in those few systems such as lee waves in which the motion of a particle is oscillatory, the duration of an individual system does not much exceed the time taken by some particle to traverse it completely, and so the time scale is also an approximate indication of the typical duration of the systems.

Our own perception processes have their characteristic time constant, amounting to some fraction of a second, so that our appreciation of a visibly evolving phenomenon is best achieved when it lasts not much less or much more than several seconds. This is exploited in the cinematograph, whose individual still frames succeed each other sufficiently quickly to give the illusion of continuous motion, and in high-speed and time-lapse cine photography, which allows the perception of processes with very short and very long time constants, respectively. Time-lapse photography of clouds is particularly illuminating; the space scale of those in the ordinary field of view is 1 km or rather more, and their time scale correspondingly a considerable fraction of an hour (Fig. 6.1), so that their evolution is best seen when frames naturally exposed at intervals of a few seconds are projected at an ordinary rate (corresponding to an increase in their apparent rate of change by a factor of about 60). When pictures taken throughout a day at longer intervals are projected at an ordinary rate the evolution of the individual clouds disappears and is replaced by an impression of the evolution of cloud systems and their diurnal variation. Films of this kind taken from a satellite stationary relative to the earth provide insight into the behavior of the large-scale cloud systems, especially in low latitudes. It is difficult to gain this appreciation of the evolution of the large-scale motion systems from the several maps of synchronous or "synoptic" states placed side by side or superimposed in conventional ways, so that in those used for illustration it is desirable to display as much information on each map as the risk of confusion will allow, and to employ special techniques, including the use of distinctive colors, to improve the clarity and to draw attention to significant features.

The dimensions of atmospheric motion systems are ultimately limited by the size of the earth, but their time scale can be extended by entering the realm of *climate*, statistics of meteorological observations. The simplest and most commonly used statistic is the arithmetical average, which usually has a regular diurnal and annual variation, but otherwise fluctuates irregularly and does not settle down to a definite value as the period considered is increased. In the study of individual motion systems it is useful to be aware of the mean fields of the meteorological elements which in sequence they compose, and in which the more extreme occur sporadically. These mean fields are illustrated in climatological handbooks, usually in the form of global or hemispheric distributions of variables, especially winds, averaged over a season, a year, or many years. They represent the main features of the *general circulation*, which will be considered later. Since our concern is with individual motion systems, and especially those that produce clouds and rains, and not with their statistics, we may regard the general circulation as sufficiently well defined by observations from a few successive seasons, or even only one, although variations in the character of the seasons which seem large to ordinary people as well as climatologists occur from one year to another.

It is demanded of modern meteorological theory that it grasp the interdependence of all the phenomena listed in Fig. 6.1, each of which modifies the energetics of the atmosphere, although some play a trivial part. The aim of the study of clouds and rains from this general point of view is to establish the influence of the phase changes of water on gravity wave phenomena, especially the forms of convection, and on the lesser of the quasi-geostrophic wave phenomena, on the range of scales between about 1 and 1000 km. This also involves the consideration of the microphysical processes which affect cloud particles, on scales less than about 1 cm

6 Atmospheric motion systems

(which has so far been the preoccupation of cloud physics).

The kinetic energy of motion systems

Patterns recognized as representing a particular kind of motion system are difficult to define exactly; their shapes are not simple, no two are quite alike, and their spacing is irregular, so that their boundaries and dimensions are indefinite. At least in the early stages of their development, however, individual motion systems generally appear as a series of wavelike distortions of a more extensive flow, and it is customary to analyze their scale and intensity by representing them along some axis as the resultant of a (Fourier) series of sinusoidal waves superimposed upon a mean uniform flow. In this way, for example, spectral distributions of u', v', and w', fluctuations in the horizontal components u and v, and the vertical component w of the mean velocity, may be obtained showing the contributions $s(k)dk$ made to the energies $\overline{u'^2}$, $\overline{v'^2}$, and $\overline{w'^2}$ by components of wavenumber $k (= 2\pi/\lambda$, where λ is wavelength) between k and $(k + dk)$. The function $s(k)$ is called the *spectral density function*. The advantage of this kind of analysis is that it is exact, and the advantage of employing the sinusoidal function is that both it and its derivative have a clear and simple distinction between amplitude (intensity) and scale, but since real patterns are only approximately sinusoidal the form of the spectral distribution may not have a clear physical interpretation, and in particular it may not correspond well with what might be anticipated from mere visual inspection of the flow. However, satisfactory theories have been established for the spectral form of some kinds of turbulent motion, and it is instructive to display its probable structure over the whole range of the scales of motion which may occur in the troposphere (Fig. 6.2). We consider the scale to be represented by horizontal dimension; since this varies over nearly ten orders of magnitude it is convenient to use the logarithm of the wavelength λ as abscissa and the function $ks(k)$ as ordinate. Then if the ordinate scale is linear, the kinetic energy within a given range of wavelengths is proportional to the area under the curve showing the form of the spectrum. However, since the function $ks(k)$ also varies over several orders of magnitude, in Fig. 6.2 it too is shown on a logarithmic scale. This scale is marked in units of $kE(k)$, the contribution to the "eddy" kinetic energy $(\overline{u'^2} + \overline{v'^2})/2$ of horizontal motion (of 1 g of air) contained in motions of scales which lie within a unit interval of the natural logarithm of the wavenumber or wavelength; the scale therefore indicates by reference to the curves in the diagram the magnitude of the components u', v' of the deviation from the mean velocity of the flow, as a function of horizontal scale.

On the other hand, a second pair of (dashed) curves is drawn for w^2 (not $w^2/2$), where w is vertical velocity (the mean always being regarded as zero), in order that where at the smaller scales the motions are approximately isotropic, with $u' \approx v' \approx w$, this property is manifest simply by coincidence of the appropriate curves.

The principal pairs of curves are drawn to represent the mean annual conditions in the northern hemisphere within the planetary boundary layer, in which the eddy kinetic energy is predominantly associated with the motion systems of ordinary convection, and also within the remaining part of the troposphere, the free atmosphere (Section 1.3). In addition, several other curves are included to indicate the contributions made by various motion systems discussed in the text. In respect of some kinds of motion system which are only locally or intermittently present, a distinction is made between the spectrum of kinetic energy within regions occupied by the motion systems (dashed lines) and their contribution to the space-time tropospheric mean value (solid lines). Most of the curves are estimates, generally consistent with the data given in the references listed in the figure legend. The data are nearly all derived from observations made virtually simultaneously, but some represent a time series of observations made at a particular place, and for these a propagation speed of 10 m sec^{-1} has been assumed to give the approximate equivalence of time and space scales indicated by the periods entered along the horizontal axis.

Several features of the diagram should be noted:

1. The overwhelming concentration of the tropospheric eddy kinetic energy is in the horizontal motion of the large-scale (*trough-ridge*) systems. The total mean eddy kinetic energy implied by the diagram is about $(12 \text{ m sec}^{-1})^2/2 \text{ g}^{-1}$, approximately equal to the kinetic energy of the mean (zonal) flow. Most of the tropospheric eddy kinetic energy is contained in the *long waves* of which between two and four span the hemisphere. *Cyclones* (of hemispheric wavenumber between about 6 and 10) make a substantial contribution to the eddy kinetic energy, and can be shown to be the dominant source of the kinetic energy of the motion on other scales (including the mean zonal flow), transforming potential energy represented by the latitudinal temperature gradient into kinetic energy during slope convection (see, e.g., reference 3, and Chapter 9).

2. The eddy kinetic energy present in the troposphere has a secondary maximum at a scale of a few kilometers due to the *ordinary* convection in the boundary layer. The form of the spectra in the interval of scales between a few kilometers and a few hundred kilometers can be estimated only tentatively because of the limited resolution in space and time of routine aerological soundings and the inadequacy or sparsity of other kinds

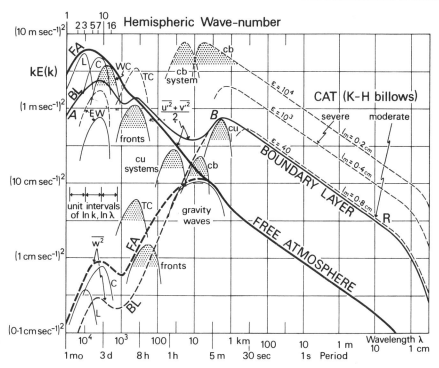

Fig. 6.2 Estimated forms of the spectral distribution of the horizontal eddy kinetic energy $kE(k)$ and of the spectral energy function $ks(k)$ of the vertical component w of the eddy velocities, as a function of horizontal wavelength $\lambda = 2\pi/k$, for the mean annual conditions over the northern hemisphere in the convective boundary layer (BL) during moderately vigorous convection, and in the remainder of the troposphere (the free atmosphere, FA). The ordinate scale gives the energy of unit mass (1 g) contained within unit interval of the natural logarithm of the wavenumber k or the wavelength λ (the width of such intervals is indicated in the left center). Along the abscissa a scale of period has been drawn, related to wavelength by a constant translation speed of 10 m sec^{-1} at the longer wavelengths, and used to incorporate in the diagram the results of some analyses of observations made at a fixed place.

The distributions within regions occupied by individual phenomena which are local and intermittent are indicated by dashed lines, and their contributions to the averages by continuous lines, some labeled with letters having the following meanings:

- L = long waves (large-scale trough-ridge systems)
- C = cyclones
- WC = wave cyclones formed in frontal zones
- TC = tropical cyclones
- EW = tropical easterly waves
- cb = cumulonimbus
- cu = cumulus
- CAT = clear air turbulence in the free atmosphere, originating at least mainly as K-H
- K-H = Kelvin-Helmholtz billows
- ε = rate of dissipation of eddy kinetic energy in 1 g of air (cm^2 sec^{-3})
- l_m = the lower limiting scale of turbulence in the inertial subrange

Those phenomena whose properties are strongly dependent on the condensation of water vapor are distinguished by a stippling.

The distribution for the boundary layer between the points A and B is from an analysis of winds measured at a height of about 100 m above Long Island, New York (8), and its form is consistent with observations at other places in the United States at about the same height (8) and also there and in other regions nearer the ground (9). Those for large scales in the free atmosphere are broadly consistent with analyses made for various tropospheric levels and latitudes (10), often as a function of hemispheric wavenumber, here drawn to correspond to the scale of wavelength in latitude 40°.

The distributions drawn for the free atmosphere at wavelengths between about 1000 and 1 km are estimates not based on any comprehensive data, while those drawn for wavelengths of about 1 km or less are based on analyses of measurements made during flights through thunderstorms (cb) and cumulus (cu) over the United States (12), and through clear air turbulence in the upper troposphere (13). The analyses have been smoothed into the straight lines of slope $-2/3$ at wavelengths down to a limiting scale l_m in accordance with the theory of the inertial subrange of homogeneous isotropic turbulence. The point marked R indicates that the position and slope of the curve corresponding to clear air turbulence of moderate intensity in the upper troposphere is approximately confirmed by observations of radar echoes from associated fluctuations of refractive index (15).

of observations, and because of the consequent unsatisfactory knowledge of the kinds and properties of the motion systems which exist with these scales. Probably in the free atmosphere there are major contributions from the *fronts* and *jet streams* of middle latitudes, on scales of order 100 km, and lesser contributions on smaller scales from gravity waves (principally *mountain* and *lee waves* and *wake eddies* due to flow over topographical features), or from convective agitation of the boundary layer. Within the boundary layer the energy is generated in *intermediate-scale circulations* (associated with topographical features) and other systems whose structure has not been defined satisfactorily, including some whose presence is manifest by groupings of cumulus and cumulonimbus.

3. In the free atmosphere energy is intermittently and locally introduced on scales mainly of order 100 m, and rarely as large as a few kilometers, by transfer from the large-scale flow during the development of Kelvin-Helmholtz *billows* in shallow layers of pronounced wind shear, usually at the top of the boundary layer or in the upper troposphere. They are the predominant cause of bumpy flight conditions experienced by commercial aircraft (sensitive to motions of wavelength between a few tens of meters and a few kilometers), and since they are only rarely accompanied by cloud formation they have been described and studied as *clear-air turbulence*.

4. Most billows rapidly become turbulent, generating motions of successively smaller scale through which the eddy kinetic energy is transferred until at scales of about 1 cm it is *dissipated* (transformed into internal energy) by molecular viscosity. The kinetic energy of the shear-dominated motions near the ground in the boundary layer and much of that generated in its convective circulations is similarly subject to a *cascade* through a range of smaller scales and eventual dissipation. In these circumstances the form of the spectral density function $s(k)$ has been found to be close to that predicted for an extensive range of scales (the inertial subrange) in steadily maintained isotropic turbulence in a neutrally stratified fluid. Then $s(k)$ is dependent only upon k and upon the rate ε at which the kinetic energy is dissipated below a limiting scale l_m (or introduced above an upper limiting scale l_o by an organized motion system):

$$s(k) = a\varepsilon^{2/3} k^{-5/3} \quad (6.1)$$

where a is a constant whose value is about 0.5 along the direction of the mean flow and larger by a factor of 4/3 in the lateral and vertical directions (4). Accordingly, in Fig. 6.2 the relevant spectra have been extended toward the limiting small scale l_m as straight lines with a slope of $-2/3$. [The wavelength l_m (cm) is a function of the coefficient of kinematic viscosity v (cm^2 sec^{-1}):

$$l_m \approx 6(v^3/\varepsilon)^{1/4} \quad (6.2)$$

Since in typical conditions $v^{3/4}$ varies only from about 0.3 in the lower troposphere to 0.4 in the high troposphere we may write

$$l_m \approx 2/\varepsilon^{1/4} \quad (6.3)$$

implying that l_m is approximately a function of ε only.]

This form of the spectra has been shown by aircraft measurements to be approximately correct, and about the same for the vertical and the horizontal components of the air motion, in the range of wavelengths between about 20 m and 1 km. These measurements have been made in clear-air turbulence, in cumulus, and in cumulonimbus. Moreover, the form and position of the spectrum for clear-air turbulence of moderate intensity is supported in the region of centimetric wavelengths by measurements of the intensity of the radar echoes which can be received from the associated fluctuations of refractive index (5). [In the upper troposphere, these fluctuations are effectively due only to accompanying variations of temperature. In a typical mean vertical gradient of refractive index the fluctuations associated with moderate clear air turbulence ($\varepsilon \approx 10^2$ cm^2 sec^{-3}) correspond to a total backscatter cross section per unit volume, η (Section 2.8), of about 10^{-16} cm^{-1}. An echo is then just detectable at ranges of a few tens of kilometers by the most sensitive (high-power, narrow-beam) radars using wavelengths of about 10 cm. In the lower troposphere where the pronounced gradients and fluctuations of refractive index are dominated and magnified by variations of vapor density, the reflectivities of turbulent regions are commonly greater, and in particular echoes from the upper boundaries of convective circulations (Section 7.6 and Pl. 7.1) may be detectable by such radars at ranges up to about 100 km.] The position of the spectral curve for the average horizontal eddy kinetic energy of the free atmosphere at small wavelengths was estimated on the assumption that shallow layers of weak or moderate clear-air turbulence are present much of the time, as suggested by observations of stellar scintillation (Appendix 2.6). It is consistent with a mean upper tropospheric value of $\varepsilon \approx 0.1$ cm^2 sec^{-3} inferred from observations of the diffusion of smoke puffs and atomic bomb debris (6), and implies that nearly all of the dissipation of tropospheric kinetic energy occurs in the convective boundary layer (this dissipation is responsible for the removal of kinetic energy also from layers above, through the mechanism of geostrophic control discussed in Section 9.2). Assuming that this layer occupies about 0.7 of the area of the hemisphere, the average total dissipation rate indicated is $\varepsilon \approx 20$ cm^2 sec^{-3}, consistent with the estimate from the energy balance of $\varepsilon = 2.3$ (-1.2) erg cm^{-2} sec^{-1} (7), or ≈ 4 cm^2 sec^{-3}, if the average depth of the layer is about 200 mb (2 km).

5. The properties—frequency, dimension, and intensity—of all motion systems are in some degree dependent on the presence of water in the atmosphere. In many, however, its influence is indirect, expressed only by the stratification which is imposed by the radiative properties of water vapor and by its continual condensation and precipitation. In others the condensation and precipitation play a more immediate and important part. As examples of the former kind there are the large-scale long waves and cyclones, the gravity waves due to flow over topographical features, and the small-scale Kelvin-Helmholtz billows and turbulent motions which include the *wave cyclones* that appear in cloudy frontal zones, the tropical cyclones, cumulonimbus, and cumulus other than the smallest. The distinction between the two kinds is not sharp, but the principal motion systems of the latter kind have horizontal dimensions between about 2 and 2000 km, and their contributions to the energy spectra in Fig. 6.2 are marked by a stippling. The discussion of these systems occupies the major part of this text.

6.5 The classification and nomenclature of clouds

It is not possible to discuss things without naming them, and the choice of names requires care and understanding if undesirable connotations are to be avoided. The diligent observation of clouds began toward the end of the eighteenth century, well before sound bases had been laid on which to develop meteorology as a science, and indeed not long after it could be written that the winds were among the most mysterious of all natural phenomena ("Men of genius who rifled Nature by the Torchlight of Reason even to her very nudities, have been run aground in this Channel; the Wind has blown out the candle of Reason, and left them all in the dark"; 16). During the nineteenth century numbers of enthusiastic cloud observers devised classifications of clouds, based on appearance; although they were primarily morphological, these classifications usually contained some rudimentary ideas of the physical nature of clouds. As meteorology developed internationally in the last half of the century the need for a universally accepted nomenclature eventually led to an agreed International Classification which perforce was still morphological, and indeed desirably so since thereby it could be correctly used by the many observers with no special training. In recent decades the bases for a classification according to physical principles have emerged, but the old nomenclature has persisted, partly because the necessity for a change has been avoided by the use in forecasting practice of codes of numbers for reporting the state of the sky and the weather. [The codes and advice on their use are given in the *International Cloud Atlas* (21) and other official publications, e.g., reference 23.] They were so well devised to represent the important physical features of the large-scale motion systems (at least outside the tropical regions; 17) that no need for any modification has become apparent. Moreover, the study of cloud physics, which has expanded so much in recent decades, was at first concerned mainly with the microphysics of the particles, rather than with the associated motion systems, and workers have mostly been content to use the conventional cloud nomenclature. Some dynamicists (18, 24) have sharply criticized recent elaborations of the nomenclature which indeed have virtually no meteorological significance, but a physical classification which combines the appealing merits of clarity, economy, and comprehensiveness has not yet been proposed. Any which is advanced should take account not only of the forms of individual clouds but also of their organization on larger scales, which has become apparent from satellite pictures (and has revealed large-scale motions, especially within the tropics, which are still not understood).

Morphological classifications; the International Classification

The naturalist Lamarck is usually credited with being the first to publish a classification of cloud forms. In a first version of 1802 he named the following:

   Nuages en balayures (clouds in sweeps)
   Nuages pommelés (dappled clouds)
   Nuages groupés (heaped clouds)
   Nuages en voile (sheet clouds)
   Nuages attroupés (clouds in flocks)

To these he soon added several other sensible and immediately interpretable definitions, but his work seems to have been without effect, partly because of his use of names not generally understood, and perhaps partly because it appeared in a publication discredited because it contained weather forecasts based upon astrological suppositions.

In contrast, a classification published by Luke Howard in England at about the same time (1803) in *Tilloch's Philosophical Magazine* (and reproduced in books and encyclopedias in later years) used Latin names, was soon adopted by many scientists, and had a profound influence on painters (especially John Constable) and poets (especially Goethe, who wrote odes in honor of Howard and his cloud forms, and did much to recommend his views among philosophers throughout Europe). Howard's classification remains the basis of the present internationally accepted classification. He recognized three principal kinds of cloud:

> 1. *Cirrus:* parallel, flexuous, or diverging fibers.
> 2. *Cumulus:* convex or conical heaps, increasing upward from a horizontal base.
> 3. *Stratus:* a widely extended, continuous, horizontal sheet, increasing from below.

The last two kinds correspond to the clouds associated with what we have already distinguished as the two principal kinds of atmospheric motion system: the small-scale circulation of ordinary convection, and the nearly horizontal circulation of the large-scale slope convection. These must therefore be the two basic kinds of cloud also in a physical classification.

The distinctiveness of the first kind of cloud, striking from a morphological point of view, is less fundamental dynamically, arising from certain peculiarities of isolated high clouds of ice crystals. They form only locally in air which may generally be appreciably supersaturated with respect to ice. Often the ice crystals have such small concentrations that they become much larger than cloud droplets (Section 5.12), with fall speeds reaching a large fraction of 1 m sec^{-1} (Table 5.13), so that the clouds become virtually trails of precipitation (fallstreaks; Pl. 6.1), with a fibrous texture unlike that of ordinary droplet clouds. Moreover, at the low temperatures characteristic of high levels the evaporation of ice crystals which fall into unsaturated air usually proceeds more slowly than the evaporation of cloud droplets in the lower troposphere, because the difference of the vapor density between the crystals and their surroundings is limited by the small magnitude of the saturated vapor density, as well as because of the larger size of the particles. Consequently, even when trails are poorly developed, or the ice clouds have patterns imposed by other kinds of small-scale motion, their details have diffuse edges, unlike the sharply defined if ragged edges of droplet clouds. Thus the introduction of a third principal kind of cloud into a morphological classification can be justified also on a physical if not a dynamical basis (however, the reasons for the distinctiveness of the cirrus were not understood by Howard, who attributed it to electrical properties).

In addition to the three principal kinds of clouds, Howard defined others by compound names indicating transition or association:

> 4. Cirro-cumulus: small, well-defined roundish masses, in close horizontal arrangement;
> 5. Cirro-stratus: horizontal or slightly inclined masses, attenuated towards a part or whole of their circumference, bent downward, or undulated, separate, or in groups of small clouds having these characters;
> 6. Cumulo-stratus: the cirrostratus blended with the cumulus. [It seems that by this name, which is no longer used, Howard did not mean the shower- or thunder-cloud which has spreading fibrous tops and which much later was defined as the cumulonimbus; rather, he regarded the latter cloud as a small-scale variety of his last kind, the]
> 7. Cumulo-cirro-stratus, or nimbus: the rain cloud. A cloud or system of clouds from which rain is falling. It is a horizontal sheet, above which the cirrus spreads, while the cumulus enters it laterally and from beneath.

Throughout the nineteenth century various modifications and extensions of Howard's classification were proposed, besides some completely novel schemes. However, gradually it became generally agreed that Howard's classification should remain as a basis, and if possible be arranged also to indicate the height of a cloud, since the various forms seemed to be most prevalent in particular ranges of height; if so, widespread observations of the kind and angular velocity of clouds clearly had a great potential value in revealing the forms of the atmospheric circulation, then not accessible to direct observation. Toward the end of the century the leading enthusiasts (especially Abercromby in England and Hildebrandsson at Uppsala in Sweden) made concerted efforts to devise a classification which could be universally accepted. Abercromby collected cloud photographs from many places and made two voyages around the world to ascertain that the same principal cloud forms occurred everywhere, and at Uppsala and some other observatories extensive series of measurements of cloud height were undertaken, first by double-theodolite triangulation and later by photogrammetry, in an attempt to establish the ranges of height within which the various forms occurred. In discussing the first measurements made at Uppsala Abercromby stated (in 1887):

> The most important result is that clouds are not distributed promiscuously at all heights, but have a decided tendency to form at three definite levels, whose mean summer level at Uppsala is: Low clouds ... 2–6,000 ft; middle clouds ... 12.9–15,000 ft; high clouds ... 20–27,000 ft. It would be premature to speculate on the physical significance of this fact, but we find the same definite layers in the tropics, and no future nomenclature will be satisfactory which does not take the idea of these levels into account.

In a later review of measurements made also at Berlin, Storlien (63°N, 12°E), and Blue Hill (Mass.), Clayton (19) found two zones of maximum frequency, one between 1 and 2 km, and another (less marked and more diffuse) between 7 and 10 km. He concluded that "the common division of clouds into lower clouds and upper clouds has a natural basis; but there appears to be no basis of this kind, as was thought by Abercromby, for the division of the clouds into three levels, upper, middle, and lower clouds." [The former division might be anticipated, corresponding to one between the low clouds of ordinary convection, the most prevalent of all clouds, and the high clouds of slope and deep ordinary convection, which spread far from the rain systems and showers over which they form, and are therefore also frequent in regions of fair weather. A reason for a third region of maximum frequency in the middle troposphere is still not clear. It may be a feature of occasions of small fractional cloud cover over land (suitable for measurements of cloud height from the ground), when there is or recently has been cumulus convection to a considerably greater height than is general over the ocean, producing a layer of high relative humidity in the lower middle troposphere. Other kinds of motion systems, such as intermediate-scale circulations, orographic and shearing-instability waves, may then locally produce clouds confined to such levels. They might be expected to occur frequently over land in middle latitudes in summer, near coasts in winter when there has been deep cumulus convection over the sea, and throughout the year over moist lands and islands in low latitudes.]

However, Abercromby and Hildebrandsson had by this time proposed a new classification which was based on Howard's scheme but grouped clouds into four categories according to the levels they occupied (the cumulus and thundery clouds which could have great vertical extension forming a separate fourth class). The compound cumulostratus of Howard for the raining heap cloud or thundercloud was replaced by the name cumulonimbus, extensive rainy layer clouds were called nimbus, and extensive low clouds with a gross dappled or rolled structure were called stratocumulus; these three names had all been suggested by earlier writers. Cirrocumulus was reserved for a cloud of small dapples (but placed in the category of middle-level clouds), and the prefix alto- was introduced to distinguish the principal forms of middle clouds: altocumulus, consisting of large dapples or balls, and altostratus, a thick grey layer with indistinct or irregular details.

The classification of Abercromby and Hildebrandsson was recommended by an International Meteorological Conference held in 1891, and it soon became official. Subsequent criticisms were mainly to the effect that it should recognize more detail, and these were taken into account when a revision was undertaken by an international commission in 1926. Their work resulted in the publication of a large cloud atlas, illustrating a classification which, in an effort to preserve the main features of one which by that time had become familiar and in widespread use, described the previous categories as *genera*, which were subdivided into numbers of *species* defining visual attributes of lesser importance.

The International Classification was further elab-

Table 6.1 Some useful terms from the International Classification (18)

High clouds (with characteristic fibrous or silken textures, and often optical phenomena, which indicate a composition of ice crystals; Section 9.6)
    Cirrus: separate tufts or trails of cloud
    Cirrostratus: an extensive layer of optically thin cloud
Middle-level clouds
    Altocumulus: a layer or patch of cloud, usually shallow, with a dappled or wavy pattern (composed predominantly of droplets)
    Altostratus: a layer or patch of cloud with no clear internal pattern (preferably reserved for an extensive and often thick or opaque layer with a fibrous texture which indicates the composition to be predominantly of ice; Pl. 6.2)
Low clouds
    Stratus: a shallow layer or patch of cloud, with no clear detail when seen from below, but sometimes with a dappled and wavy pattern when seen from above
    Stratocumulus: an extensive layer of cloud with a gross dappled or rolled pattern (seen from above or below), usually representing shallow cumulus which have increased in amount to form a complete cover with a uniform base level
Clouds which may have marked vertical extension
    Cumulus: separate heap clouds, sometimes very shallow and ragged, sometimes shallow and tabloid, and at other times clustered and towering
    Cumulonimbus: the tall shower clouds or thunderclouds whose tops become frozen and tend to become persistent and spread horizontally as ice clouds

Species
    Uncinus: cirrus in the shape of a comma, often with a longer tail, whose top usually has the form of an upright tuft
    Castellanus: clouds whose upper parts sprout into cumuliform columns, often from a common base, which have a "miniature" appearance compared with ordinary cumulus
    Lenticularis: clouds with the shape of lenses or almonds, especially when seen from the side; they are often separate or isolated, with well-defined edges, but sometimes occupy a large part of the sky
Supplementary features
    Mamma: protuberances hanging like udders from the lower surface of a cloud
    Virga (fallstreaks): trails of precipitation beneath a cloud, not reaching the ground
Accessory clouds
    Pileus: a small and shallow arched cap or hood of smooth texture which forms above cumulus or cumulonimbus
Special clouds
    Noctilucent clouds: tenuous clouds in the high stratosphere, seen occasionally in high latitudes during the summer, and only when the sun is sufficiently far below the horizon for the whole troposphere to be in the earth's shadow
    Nacreous clouds: shallow clouds in the lower stratosphere, usually of lenticular form, seen rarely near mountains in high latitudes during the winter. They are strikingly illuminated against a twilight sky when the sun is just below the horizon, and are then often distinguished by brilliant iridescent coloring

6 Atmospheric motion systems

orated by another commission in the years 1947–53, and defined and illustrated in a new atlas, published in 1956. The concept of genera was retained, the species were increased and modified, and *varieties, supplementary features, mother clouds,* and *accessory clouds* were added. By this thorough and devoted work a comprehensive morphological classification was constructed, which if used slavishly for description is extraordinarily unwieldy and uninformative (18) compared, for example, with what can now be achieved easily by the provision of a photograph, especially if in color, or (in large-scale analysis) by the specification of a number in the international synoptic codes. Nevertheless, some terms of the classification are so conveniently brief and in such general use that they naturally appear in any discussion of clouds. Table 6.1 list some of these with their generally accepted meanings.

Physical classifications

After about 1920 the development by the Norwegian school of their frontal theory of cyclones, and concept of air masses, led to several proposed physical, or genetic, classifications of clouds (22). These differed mainly in the extent to which they endeavored to include detail representing small-scale or infrequent processes; in essence they were very similar, recognizing four principal mechanisms which lead to cloud formation by cooling air:

1. Large-scale ascent in slope convection, concentrated into frontal zones, and producing layer clouds at all levels.
2. Local ascent in ordinary convection over warm surfaces, producing cumulus and cumulonimbus.
3. Turbulent heat transfer from airstreams moving over cool surfaces, perhaps aided by radiative cooling, producing widespread low-level layer clouds.
4. Local ascent due to flow over irregular terrain, producing lenticular clouds (or over cumulus towers, producing pileus).

Further, it was recognized that internal convection due to radiative fluxes produced the dappling or banding characteristic of most shallow layer clouds, and an ordinary convection following the formation of cloud in unstably stratified layers produced castellanus, distinguished from cumulus by having no direct connection with the surface underneath.

The four principal mechanisms constitute an adequate basis for a physical classification and do not call for any

Table 6.2 Terms either not included or not given appropriate emphasis in the International Classification

---

Rows: long parallel lines of small cumulus, aligned in the direction of the wind (often called *streets* by sailplane pilots; Section 7.6). The clouds are sometimes joined together at their base level, and rarely have a poorly developed cumuliform structure, so that the clouds have the appearance of longitudinal rolls

Longitudinal rolls: commonly develop in layers of low stratus; they likewise show the presence of vortices with axes lying along the wind direction, which probably derive their energy from the kinetic energy of the mean wind (Pl. 6.3; Section 7.6)

Clusters: aggregates of cumulus clouds, often joined together at the base, of which the central members are tallest (Section 7.6); they indicate the presence of intermediate-scale circulations associated with surface features (such as hills or coasts)

Shelves: shallow layer clouds produced by the horizontal spreading of cumulus tops (Section 7.10), formed in the same way as the anvil clouds of cumulonimbus, but in the absence of abundant precipitation retaining a well-defined and almost level undersurface. In the International Classification they are regarded as parts of cumulus mother clouds, and are described as stratocumulus cumulogenitus or altocumulus cumulogenitus

Scud: fragmentary clouds below the general level of the base of neighboring cumulus or cumulonimbus, formed over sea-breeze fronts (Section 7.10) or squall fronts (Section 7.3)

Billows: groups of several wavelike clouds which grow rapidly before "breaking," becoming distorted and frayed with much irregular detail. They are due to a shearing (Kelvin-Helmholtz) instability in a shallow layer

Shreds and puffs: small ragged clouds which occur singly or in groups of several individuals spaced at fairly regular intervals, and soon evaporate or, at high levels, soon become diffuse and tenuous; the puff clouds have a more pronounced vertical development.

They are probably due to Kelvin-Helmholtz instability, since similar clouds sometimes appear beneath billow clouds. They occur at all levels, and sometimes are the first clouds to form during the development of a population of cumulus clouds in fresh winds; in the International Classification they are then cumulus of the species fractus, a term restricted to cumulus and stratus, and applicable also to small ragged cumulus in ordinary convective circulations which barely reach the condensation level

Hill-wave clouds: lenticular clouds formed over hills. The term hill-wave refers to the disturbance in an airstream over or near a hill, which is most pronounced over long ridges lying across the direction of the air flow. In addition to the hill wave there may be a series of waves in the lee of the hill (Pl. 6.4), producing in the middle or lower troposphere

Lee-wave clouds: a regular series of separate and smooth-edged stationary lenticular clouds composed wholly or predominantly of droplets. These clouds are often greatly elongated across the wind direction (because they are most readily produced by a long ridge lying across the airstream)

Rotor clouds: over the ground beneath the crest of a lee wave of large amplitude there may be a closed circulation, or rotor. Its upper part sometimes contains a rotor cloud which is turbulent, has ragged edges and an arched top, and appears to rotate (Pl. 6.4)

Plume clouds: a name given to clouds which trail horizontally a long distance downwind of a source, and sometimes applied to anvil clouds, but preferably reserved for long trails of cirrus which form in hill waves (Pl. 6.4)

Sea smokes and steam fogs: fogs, usually shallow, which form in very cold air flowing over warm water (Sections 3.2 and 7.5)

revision of the familiar nomenclature of the International Classification.

However, additional nomenclature is needed for some clouds or cloud details associated with various kinds of small-scale motion system, and for some large-scale cloud systems, for example, those tropical systems in which cumulonimbus convection and large-scale motion (weak by comparison with the slope convection of middle latitudes) may be essentially in operation together. English words, or their literal equivalents in other languages, are already in widespread use for both purposes. For example, cumulonimbus organized into a long belt are described as a *line squall*, while other disturbances such as tropical depressions and cyclones are regarded as large-scale cumulonimbus systems. The inner ring of cumulonimbus in the tropical cyclone is called the *eye wall*, a shallow central mound of low clouds in the eye is the *hub cloud*, and the extensive spread of anvil cirrus over the cyclone is called the cirrus *shield*. Other words are used to refer to individual clouds or their details which are associated with particular parts of familiar motion systems, or with small-scale motion systems which have only recently been recognized or studied. Thus the intense cumulonimbus has features such as the *dome*, the *anvil*, the *arch cloud*, and the *funnel cloud* (all discussed in Chapter 8); the last three are included in the International Classification as the supplementary features *incus*, *arcus*, and *tuba*, but the ordinary names seem more natural to use in a verbal or written discussion.

Some other terms have been used with definite meanings in this text, and in other publications, or in atlases of cloud pictures arranged according to the processes which produce them (24–26). Table 6.2 lists some of these terms.

References

1. Smagorinsky, J., Manabe, S., and Holloway, J.L. 1965. Numerical results of a nine-level general circulation model of the atmosphere, *Mon. Weather Rev.*, **93**, 727–68.
2. Scorer, R.S. 1952. Sonic and advective disturbances, *Q. J. R. Meteorol. Soc.*, **78**, 76–81.
3. Saltzmann, B., and Teweles, S. 1964. Further statistics on the exchange of kinetic energy between harmonic components of the atmospheric flow, *Tellus*, **16**, 432–35.
4. Lumley, J.L., and Panofsky, H.A. 1964. *The Structure of Atmospheric Turbulence*. Interscience, New York.
5. Hardy, K.R., Atlas, D., and Glover, K.M. 1966. Multiwavelength backscatter from the clear atmosphere, *J. Geophys. Res.*, **71**, 1537–52.
6. Wilkins, E.M. 1963. Decay rates for turbulent energy throughout the atmosphere, *J. Atmos. Sci.*, **20**, 473–76.
7. Oort, A.H. 1964. On estimates of the atmospheric energy cycle, *Mon. Weather Rev.*, **92**, 483–93.
8. Van der Hoven, I. 1957. Power spectrum of horizontal wind speed in the frequency range from 0.0007 to 900 cycles per hour, *J. Meteorol.*, **14**, 160–64.
9. Oort, A.H., and Taylor, A. 1969. On the kinetic energy spectrum near the ground, *Mon. Weather Rev.*, **97**, 623–36; Hwang, H.J. 1970. Power density spectrum of surface wind speed on Palmyra Island [6°N, 162°W], *Mon. Weather Rev.*, **98**, 70–74.
10. See, for example, Kao, S.-K., and Wendell, L.L. 1970. The kinetic energy of the large-scale atmospheric motion in Wavenumber-Frequency Space: I. Northern Hemisphere, *J. Atmos. Sci.*, **27**, 359–75; Nitta, T. 1970. Statistical study of tropospheric wave disturbances in the tropical Pacific region, *J. Meteorol. Soc. Japan*, **48**, 47–59.
11. Julian, P.R., Washington, W.M., Hembree, L., and Ridley, C. 1970. On the spectral distribution of large-scale atmospheric kinetic energy, *J. Atmos. Sci.*, **27**, 376–87, and included references.
12. Steiner, R., and Rhyne, R.H. 1962. *Some measured characteristics of severe storm turbulence*. National Severe Storms Project Report 10, U.S. Weather Bureau, Washington, D.C.
13. Reiter, E.R., and Burns, A. 1966. The structure of clear air turbulence derived from "TOPCAT" aircraft measurements, *J. Atmos. Sci.*, **23**, 206–12.
14. Pinus, N.Z., Reiter, E.R., Shur, G.N., and Vinnichenko, N.K. 1967. Power spectra of turbulence in the free atmosphere, *Tellus*, **19**, 206–13.
15. Atlas, D., Hardy, K.R., and Naito, K. 1966. Optmizing the radar detection of clear air turbulence, *J. Appl. Meteorol.*, **5**, 450–60.
16. Defoe, Daniel. *The Storm; a collection of the most remarkable casualties and disasters which happened in the late dreadful Tempest, both by sea and land, on Friday the 25th November, 1703*, 2d ed., London; quoted in Ludlam, F.H. 1966. *The Cyclone Problem: A History of Models of the Cyclonic Storm*. Inaugural Lectures, Imperial College of Science and Technology, University of London, pp. 19–49.
17. Bergeron, T. 1928 Über die dreidimensionale Verknüpfende Wetteranalyse. Part I: Prinzipielle Einführung in das Problem der Luftmassen—und Frontenbildung. *Geofys. Publ.*, **5**, no. 6; see also Part IV in Godske, C.L., Bergeron, T., Bjerknes, J., and Bundgaard, R.C. 1957. *Dynamic Meteorology and Weather Forecasting*. American Meteorological Society, Boston.
18. Scorer, R.S. 1963. Cloud nomenclature, *Q. J. R. Meteorol. Soc.*, **89**, 248–53.

For discussions of the historical development of cloud classifications references 19–22 can be consulted.

19. Clayton, H.H. 1896. Discussions of the cloud observations; Chapter I, Historical sketch of cloud nomenclature, *Ann. Astr. Obsy. Harvard College*, **30**, pt. IV, Cambridge, Mass.
20. Besson, L. 1923. Aperçu historique sur la classification

des nuages, *Mem. de l'Office Nat. Met. de France*, no. 2, Paris.
21. *International Cloud Atlas.* 1956. Preface to the 1939 edition, vol. 1, World Meteorological Organisation, Paris.
22. Howell, W.E. 1951. The classification of cloud forms, in *Compendium of Meteorology.* American Meteorological Society, Boston, pp. 1161–66.
23. *Observer's Handbook.* 1956. 2d ed. Meteorological Office, H.M.S.O., London.
24. Scorer, R.S., and Wexler, H. 1967. *Cloud Studies in Colour.* Pergamon Press, Oxford.
25. Ludlam, F.H., and Scorer, R.S. 1957. *Cloud Study.* John Murray, London.
26. Scorer, R.S. 1972. *Clouds of the World.* David and Charles, Newton Abbot (Great Britain). [An extraordinarily comprehensive collection of cloud pictures (mostly in color) and interpretative comment.]

7 Cumulus convection

More than half of the energy of solar radiation absorbed at the earth's surface is transferred into the troposphere, as both sensible and latent heat, by motions with a range of small scales. The local rate of the energy transfer varies greatly, and its sign is occasionally reversed when the earth's surface is a heat sink, as over the ocean during the flow of air toward cooler water, or over land on a clear night. The small-scale motions are described collectively as *convection*, a word customarily reserved by the meteorologist to imply that heat is transferred vertically; *advection* is used to mean transport by a uniform or mean flow, usually in the horizontal. There is said to be *free convection* when the virtual potential temperature in a layer decreases with height and small vertical displacements generate buoyancy forces tending to accelerate the motion—so that the stratification is described as *unstable*; and *forced convection* occurs when there are no buoyancy forces or they oppose the motion—that is, when there is no available potential energy and the kinetic energy of the small-scale motions must be drawn from that of the mean flow. The convection described in this chapter generally occupies only the low troposphere, in contrast with deep cumulonimbus convection, and with the *slope convection* of large horizontal scale, which is prevalent throughout the extratropical troposphere. In the slope convection available potential energy and upward heat transfer depend on the interchange of air not only in the vertical, but also in the horizontal (Chapter 6).

Convection occurs intermittently over land, mainly in sunshine, and persistently by day and night in those airstreams of the large-scale slope convection which flow over the ocean toward warmer water. It plays an essential part in introducing into the troposphere the horizontal temperature gradients upon which the large-scale motion systems depend. Its study is therefore concerned not only with the properties of its individual small-scale motion systems, but also with its statistical properties as a diffusive process which is part of the mechanics of the large-scale motion systems.

These properties are well established only at levels very close to the earth's surface, where in the very confused motion when there is a general wind it seems hardly possible or profitable to identify individual motion systems. At higher levels, however, it seems unlikely that measurements will become so abundant or amenable to analysis that satisfactory statistical representations of the convection will be established without recognition of the form and arrangement of individual systems. Moreover, this kind of knowledge is needed to understand the liability of clouds to produce rain, or to develop into cumulonimbus, either naturally or by artificial modification. The chapter therefore begins with a review of some statistical properties of flow near the earth's surface, and of the characteristic features of the stratifications that accompany ordinary convection. It will appear that in the energetics of the large-scale motion systems upward transfers of heat and water vapor can conveniently be considered separately over land and over sea, and that the occasional downward transfers during forced convection are comparatively trivial. The remainder of the chapter consists of a discussion of the properties of individual cumulus, the most prevalent form of cloud.

7 Cumulus convection

7.1 General aspects of transfers of heat and water vapor between the atmosphere and the underlying medium

Underneath the land surface the vertical diffusion of heat proceeds by molecular conduction at a rate q proportional to the vertical temperature gradient:

$$q = -K\, \partial T/\partial z \qquad (7.1)$$

where K is the coefficient of molecular conductivity of heat (cal cm^{-1} °K^{-1}) and z is distance from the surface. Also

$$\frac{\partial T}{\partial t} = \frac{\partial}{\partial z}\left(\frac{K}{c\rho}\frac{\partial T}{\partial z}\right)$$

$$= \frac{\partial}{\partial z}\left(k\frac{\partial T}{\partial z}\right) \qquad (7.2)$$

where c, ρ, and k are, respectively, the specific heat, density, and thermometric conductivity (cm^2 sec^{-1}) of the medium.

In the atmosphere and ocean the diffusion can be expressed by a similar equation if k is replaced by k_E, an *eddy transfer coefficient*, whose value is determined empirically; it is found to be orders of magnitude larger than the molecular coefficient, because of the mobility of the medium, and to vary greatly according to the nature of the flow. Whereas, for example, the molecular coefficient in air (previously used in Eq. 2.18) is under ordinary conditions about 0.2 cm^2 sec^{-1}, in the atmosphere it is found that in moderate winds the magnitude of k_E some distance from the earth's surface varies from about 10^3 cm^2 sec^{-1} when the air is warmer than the earth's surface, to between about 10^5 and 10^7 cm^2 sec^{-1} when the air is being heated. In the compressible atmosphere it is presumed that T should be replaced by an appropriate potential temperature, and that in an unsaturated atmosphere the transfer of sensible heat can be associated with the departure of the temperature gradient from the dry adiabatic lapse rate:

$$\frac{\partial \theta}{\partial t} = \frac{\partial}{\partial z}\left[k_E\left(\frac{\partial T}{\partial z} + \Gamma\right)\right] = \frac{\partial}{\partial z}\left(k_E \frac{\partial \theta}{\partial z}\right) \qquad (7.3)$$

Insofar as the atmosphere and the underlying medium may each be characterized by a single value of the transfer coefficient and variations of pressure in the atmosphere are unimportant, the changes of temperature in response to some imposed variation of temperature, or to some imposed heat source, at their interface, are evidently the same in each medium if distance z from the interface is expressed in a scale proportional to $k^{1/2}$ (or $k_E^{1/2}$) (1). Since the heat capacity of a unit volume is ρc (ρc_p in the atmosphere), the total heat capacity of layers between the interface and levels of equal temperature variation is proportional to $\rho c k^{1/2}$ (or $\rho c_p k_e^{1/2}$), the *conductive heat capacity* of the medium. In any circumstances in which a heat transfer occurs between the two media, or into each from a source at their interface, the significant physical properties of a medium are the transfer coefficient k (or k_E) and the conductive capacity, for which some typical values are listed in Table 7.1.

The table shows that the values of k_E in the atmosphere (appropriate to levels above a very shallow surface layer of negligible heat capacity) are much greater than those characteristic of land and ocean; they thus imply that during heat transfers the depth of the layer affected is much greater in the atmosphere than in the underlying media. For example, the depth of the layer chilled in an airstream which travels across a cold surface for a few days is about 1 km, and the depth of the layer heated by convection from a warm surface generally exceeds 1 km, whereas the depth of the layer affected in the underlying medium is less than about 1 m.

The circumstances in which heat is extensively diffused into or from the atmosphere may be divided into four main categories, according to whether the heat is provided by the absorption of sunshine at the earth's surface or is exchanged with the underlying medium as a result of large latitudinal displacements of air, and according to whether the air is over the ocean or over land.

In clear sea water only about two thirds of the energy of sunshine is absorbed in the uppermost meter. Apart from this notable transparency, however, the large value of the conductive heat capacity of wind- or current-disturbed sea water (compared with that of the air) results in a small diurnal variation of surface temperature (reaching 1°C only in calm, fine weather), and in only a small fraction of the absorbed energy immediately entering the atmosphere. On the other

Table 7.1 Thermal diffusivity and conductive capacity of the atmosphere and various underlying media (2)

Medium	k (or k_E) (cm^2 sec^{-1})	Conductive capacity (cal, cgs units)
Still air	2×10^{-1}	10^{-4}
Airstreams		
Over cooler surface	10^3	10^{-2}
Subject to weak convection over warmer surface	10^5	10^{-1}
Subject to intense convection over warmer surface	10^7	1
Dry ground	10^{-3}	10^{-2}
Moist soil	5×10^{-3}	5×10^{-2}
Still water	10^{-3}	4×10^{-2}
Disturbed water	10^{-2}	10

Table 7.2 Heat transfer between air and underlying medium during 24 hr, with a wind of 10 m sec^{-1} and an initial temperature difference of 10°C (2)

Medium	Heat transferred (cal cm^{-2} day^{-1})
Air chilled over dry ground	<30
Air warmed over dry ground	30
Air warmed over wet soil	100
Air warmed over ocean	240 ($k_E = 10^5$ cm^2 sec^{-1})
	580 ($k_E = 10^7$ cm^2 sec^{-1})
Air chilled over ocean	30 ($k_E = 10^3$ cm^2 sec^{-1})

Table 7.3 Typical magnitudes of the transfers into or from the atmosphere of sensible heat and (in parentheses) of the sum of the sensible and latent heat (cal cm^{-2} day^{-1}) in the principal meteorological circumstances

	During advection toward surfaces which are:				In sunny weather over ground which is:	
	Cooler		Warmer			
	Over land	Over sea	Over land	Over sea	Moist	Arid
Middle latitudes	<30	30	<30	300 (600)	50 (300)	250 (300) (in summer)
Low latitudes	—	<10	—	20 (300)	50 (300)	300 (<350)

hand, land media have a comparatively small conductive heat capacity, so that during sunshine the energy absorbed at the surface passes predominantly into the atmosphere.

With some simplifying assumptions the diffusion equations may readily be solved (2) for the circumstance that an airstream of initially uniform potential temperature differing from that of an underlying medium by 10°C flows over it for 24 hr at a speed of 10 m sec^{-1}, to estimate the typical magnitude of heat transfers into or from atmospheric airstreams. Some results are reported in Table 7.2.

Considering that the mean annual upward energy flux from the earth's surface is about 200 cal cm^{-2} day^{-1}, corresponding to that required by the mean radiative energy loss of the troposphere (equivalent to a rate of cooling of about 1°C day^{-1}), Table 7.2 shows that the energy exchanges between the earth's surface and the atmosphere due to advection are small except in airstreams which are warmed over the ocean. However, in this circumstance the comparatively great conductive heat capacity of the water is such that its own temperature hardly changes.

The other circumstance in which significant energy transfer occurs is intense sunshine over land, when the comparatively small conductive heat capacity of the ground insures that nearly all the net radiation at the surface is transferred into the atmosphere. According to observation and indirect measurements (3), over land in middle latitudes during fine summer weather, and in low latitudes throughout the year, the flux into the lower troposphere amounts to about 300 cal cm^{-2} day^{-1}, of which a large fraction (except over arid ground) is in the form of latent heat (3, 4). Also over the sea it is observed that in the cool airstreams of middle latitudes the transfers of energy in the forms of latent and of sensible heat are comparable in magnitude, while in low latitudes where the gradients of sea surface temperature are small latent heat predominates. Table 7.3 can be constructed to show the magnitudes of the energy transfers in the principal meteorological circumstances. It demonstrates, in combination with Table 7.1, that in the study of the individual large-scale atmospheric motion systems to a good approximation the land can be regarded as a perfect thermal insulator and the ocean as a perfect thermal conductor of unlimited heat capacity, whose surface temperature does not change.

It appears also that in such a study it may not be necessary to consider an appropriate transfer coefficient in order to estimate the diffusive heat transfer, for in the circumstance of most importance, when the air is heated from below, the circulations of the convection which arises have a time scale of a fraction of an hour (Fig. 6.1), much smaller than the time scale of the large motion systems or the period during which the heat is introduced, which even over land is a large part of a day. Hence when the convection is widespread it is generally in a layer with a characteristic distribution of potential temperature (suitably defined if clouds are present), which represents nearly a neutral stratification for the kind of convection present (clear air, cumulus, or cumulonimbus convection). In this stratification the small horizontal temperature differences that arise between the ascending and descending branches of the circulations are sufficient, since they imply vertical velocities of order 1 m sec^{-1}, to provide a heat flux of the required magnitude (Chapter 1).

The convection layer has a well-defined top where the potential temperature is about equal to that set by the temperature at the earth's surface; the temperature everywhere within the layer is determined approximately by this temperature and the characteristic stratification. Over the sea, then (with some reservations discussed later), the modification of an airstream is governed by the change of sea surface temperature which it experiences (Fig. 7.1). In a layer which is heated over land in sunshine, on the other hand, it is the energy supplied which controls the temperature changes. However, the

7 Cumulus convection

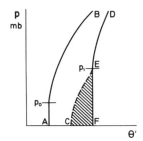

Fig. 7.1 The change of stratification due to convection over the ocean or during sunshine over land. It is assumed that the convection is accompanied by a practically uniform potential temperature θ' (suitably defined, and equal to θ when the air remains unsaturated). An initial stratification represented by the curve AB would, during a certain period, be transformed by advection and radiation into that represented by the curve CD.

Over the ocean the convection is associated with flow toward warmer water, and the curve CD is appropriate to a place downwind where the sea surface temperature corresponds to a higher surface value of θ' represented by the point F. The stratification produced there is then given by the curve FED—the top of the layer affected by convection has risen from the level p_0 to the level p_1. Over land the convection is produced by sunshine, and the surface temperature and the part FE of the curve are approximately determined by the condition that the heat represented by the net radiation at the surface, less that used for the evaporation of water, is passed into the atmosphere (it is related to the shaded area CEF).

depths of the layer occupied by convection, and in the former circumstance also the total heat transfer, are affected by the radiative heat loss, and especially by large-scale vertical motion. The rate of rise of temperature in the lower troposphere in airstreams flowing over the ocean, and during sunshine over land, is commonly $10°C$ day^{-1}; the radiative heat loss is generally equivalent to a rate of cooling of about $1°C$ day^{-1}, whereas in the middle troposphere the rate of change of temperature due to a typical vertical component of the large-scale flow is about $5°C$ day^{-1} (corresponding to a vertical velocity of 1 cm sec^{-1} in a lapse rate equal to half of the dry adiabatic). Since cumulus convection occurs predominantly in airstreams which flow toward lower latitudes and are subject therefore to large-scale descent, the vertical motion generally tends to confine the convection to the low troposphere, and to restrict the flux of sensible heat from the ocean.

7.2 The simple parcel theory of convection

Whereas in the cellular convection described in Section 1.3 the fluid is about equally divided into regions of ascent and descent, a striking feature of laboratory convection in deeper fluids heated from below, and in the cumulus convection which occurs in the atmosphere, is the tendency for the ascending currents to be more localized and hence more intense than the descending currents. Clouds formed in the ascending currents are commonly separated by large clear spaces. Because showers, lightning, and hail often form in these clouds, attention has been concentrated upon the conditions in the ascending rather than the descending air. The first theories have been concerned not so much with the transfers effected by convection, but with the depth of the layer affected. The convection has been represented as the ascent of limited volumes of buoyant air within a quiescent environment of much greater horizontal extent, whose unchanging stratification can be regarded as the product of some motion system of comparatively great space and time scale. This approach was encouraged by a practical problem: Given a morning sounding of the tropospheric stratification, forecast the thickness of clouds (and hence the probability of showers and thunderstorms) which develop over land as a result of sunshine.

According to the simple parcel theory convection begins in a layer, usually over the surface, in which the lapse rate exceeds the dry adiabatic. It extends into an upper layer in which the lapse rate is less than the dry adiabatic (Fig. 7.2). In the lower layer a small mass of air (a *parcel*) displaced upward becomes warmer and

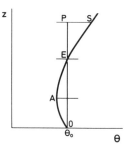

Fig. 7.2 Convection in unsaturated air according to the simple parcel theory. In an atmosphere whose distribution of potential temperature in the vertical is shown by the curve OS, a parcel ascends adiabatically from the surface (along the path OP) with an acceleration and vertical velocity which reach maxima at the levels A and E (the *equilibrium level*), respectively, and first comes to rest at the level P. On an aerological diagram in which a closed path includes an area proportional to work done the areas OAE and ESP are equal.

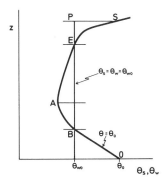

Fig. 7.3a The simple parcel theory of cumulus or cumulonimbus convection, represented on a z-θ_s, θ_w diagram; the atmospheric stratification is shown by the curve *OBAES*. Air ascends adiabatically from near the surface with the potential temperature θ_0 to become saturated at the level B, and thereafter with the saturated potential temperature corresponding to its wet-bulb potential temperature θ_{w0} near the surface. It acquires a maximum buoyancy near the level A and passes the equilibrium level E with a maximum velocity before coming to rest at the level P.

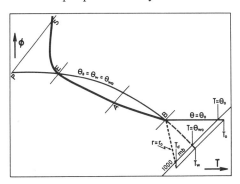

Fig. 7.3b The simple parcel theory of cumulus or cumulonimbus convection, represented on a tephigram; the dry- and wet-bulb temperatures T_0 and T_w near the surface correspond to the potential and wet-bulb potential temperatures θ_0 and θ_{w0}, respectively, and the mixing ratio there is r_0. The areas *BAE* and *ESP* are equal, if the small area between the dry adiabatic with potential temperature θ_0 and the curve representing the state of the environment below the cloud base is ignored, as in the diagram. In both (a) and (b) the difference between air temperature and virtual temperature or cloud virtual temperature is neglected.

less dense than its surroundings, and therefore subject to an upward buoyancy force; the temperature in the ascending currents of ordinary convection is given by the assumption that they contain parcels which rise adiabatically without disturbing their surroundings or the prescribed general stratification. The greatest accelerations and vertical displacements arise for those parcels displaced from the base of the layer, where the potential temperature θ in the lower layer has a maximum value.

If at a particular level z the densities and temperatures (strictly, the virtual temperatures) of the parcel and its surroundings are, respectively, ρ', ρ, T', and T, and if w is the vertical velocity of the parcel, then assuming that the pressure of the parcel is the same as that in its surroundings at the same level,

$$\frac{dw}{dt} = \frac{g(\rho - \rho')}{\rho'} = \frac{g(T' - T)}{T} \qquad (7.4)$$

The acceleration is upward while the parcel is warmer than its surroundings, and becomes zero at an *equilibrium level E* where its temperature becomes equal to that in the surroundings and a maximum vertical velocity is attained; above this level the parcel becomes colder than its surroundings and decelerates until it reaches some level P. [Clearly it may retrace its path in a series of oscillations about the equilibrium level (Fig. 7.2), but these are not considered.]

In layers of unsaturated air subject to ordinary convection the observed lapse rate is nearly the dry adiabatic except very close to the ground, but on occasions of cumulus convection the lapse rate in the clear air between clouds is often obviously intermediate between the dry and the saturated adiabatic lapse rates. If the clouds are supposed to contain air which has risen from close to the ground following the saturated adiabatic (or pseudo-adiabatic) process the levels E and P are clearly defined (Fig. 7.3). Given a sounding the theory can be tested by comparing the level of the cloud tops and the level P. Generally it is found that cumulonimbus tops are well below the level P, while in a population of cumulus only a few individual clouds reach considerably above the level E at which the theory predicts a maximum vertical velocity. Further, measurements of properties inside clouds give values which are only fractions of those corresponding to adiabatic ascent from the cloud base. Evidently and not surprisingly, the postulates of the simple parcel theory that conditions in the ascending currents can be considered without regard of the remainder of the circulations, and in particular that the ascent is from close to the surface and adiabatic or pseudo-adiabatic, are not satisfactory. In spite of this, weather forecasters often consider the anticipated height of the level E, together with the large-scale weather situation, when making a subjective assessment of the probable depth and intensity of afternoon convection, for want of a better method of relating the convection to the anticipated stratification. Moreover, in the succeeding discussion it remains convenient to refer continually to the dry and pseudo-saturated adiabatic processes, and to infer the departures from these processes which are implied by the observed distributions of potential temperature (θ, θ_w, and θ_s) and mixing ratio (r).

7 Cumulus convection

7.3 Fluxes and gradients near the surface

The vertical gradient of potential temperature θ in a deep layer of unsaturated air subject to ordinary convection has been anticipated to be very small. Close to the surface, however, the restriction upon the magnitude of the vertical components of the air motion demands that large gradients of θ and mixing ratio r accompany the typical upward fluxes of heat and water vapor.

Measurements of the transfers of momentum, heat, and water vapor between the earth's surface and the atmosphere as well as theory of flux rates (2, 5, 6) refer almost entirely to conditions near the surface, in the lowest few decameters, which are the most readily accessible for detailed observation. At suitable sites within this shallow layer the fluxes can be assumed independent of position; the vertical component of the wind can be neglected compared with its fluctuation w', and accurate measurements of the vertical fluxes can be made by correlating with w' the fluctuations there about suitable mean values (u, T, and r) of the horizontal wind speed, the temperature, and the mixing ratio. The period over which to determine mean values seems appropriately to be the fraction of an hour which is large compared with the period of intense convective fluctuation but small compared with that of diurnal changes or the passage of large-scale motion systems. There has been a preoccupation with the derivation of relations between the mean vertical gradients and fluxes, but these are simultaneously determined as part of the evolution of processes of greater scale, which have not yet been given comparable attention. However, simplified relations are quoted here to complete the description of the stratification of convective layers, and of some features of convective clouds, such as the height of their bases.

Near the surface radiative heat transfer may be important, and within about 1 cm the fluxes proceed under molecular diffusivity, circumstances which complicate a comprehensive theory of convective transfers. However, this complication will be overlooked because the decreases of θ and r upward through this very shallow layer are only small fractions of those of principal interest, fundamentally related to the height of cloud formation, which occur in the lowest few decameters. (Moreover, the near equality of the molecular diffusivities and the small contribution of the radiative fluxes in an airstream led to no serious error when the same simplification was made in the analogous problem of the fluxes of heat and moisture near a wet-bulb thermometer, in Chapter 3.)

The fluxes of momentum, heat, and water vapor can be related to the gradients of the mean values of the wind speed, potential temperature, and mixing ratio by turbulent transfer coefficients k_M, k_H, and k_W:

$$\tau = \rho k_M \partial u / \partial z \quad (7.5)$$

$$H = -c_p \rho k_H \partial \theta / \partial z \quad (7.6)$$

$$E = -\rho k_W \partial r / \partial z \quad (7.7)$$

where τ is the downward flux of momentum. By assuming constancy of flux, and that buoyancy forces have a negligible influence upon the motions, it is found that

$$\partial u / \partial z = u_* / k'z \quad (7.8)$$

where u_*, the *friction velocity*, is defined by

$$u_*^2 = \tau / \rho \quad (7.9)$$

and k' is a constant whose value is about 0.4. Upon integration this gives a logarithmic vertical profile of u, applicable above some level $z = z_0$ known as the

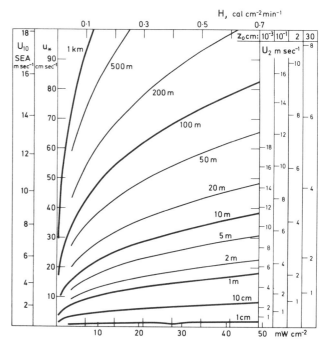

Fig. 7.4 The variation of the parameter $-L$ with the friction velocity u_* and the heat flux H. (H near the surface during moderately vigorous convection over the sea, and over land in summer sunshine, is typically about 0.3 cal cm^{-2} min^{-1}, but it reaches extremes of about 0.7 cal cm^{-2} min^{-1} in invasions of cool air over the oceans and in intense sunshine over deserts.)

A scale on the left gives U_{10}, the approximate wind speed (m sec^{-1}) at a height of 10 m above the sea surface, computed from a roughness length z_0 given by Eq. 7.11 and a value of the constant $a = 17$. Scales on the right give U_2, the approximate wind speed (m sec^{-1}) at a height of 2 m above the land surface, for several values of the roughness length z_0: 10^{-3}, 10^{-1}, 2, and 30 cm, corresponding, respectively, to smooth mud flats, short mown grass land or natural prairie, open grassland, and bush-covered land (over which the wind is appropriate to a somewhat higher level clear of the bushes). All the scales are strictly valid only for neutral conditions, but are nearly correct when $-L$ is large compared with the heights considered.

roughness length, which is empirically related to the kind of surface:

$$u/u_* = (1/k')\log_e(z/z_0) \qquad (7.10)$$

The roughness length z_0 varies from about 10^{-3} m over smooth flat surfaces to 10^{-1} m or more over vegetated surfaces, generally being about one tenth of the height of the vegetation.

Over the ocean z_0 varies with the degree of disturbance of the surface, and hence with the wind; it appears to be given approximately by the relation

$$z_0 = u_*^2/ag \qquad (7.11)$$

where a is a constant, estimates of whose value vary between about 13 and 20. In fresh winds z_0 is typically about 0.1 cm; in strong winds it reaches about 0.3 cm.

In neutral stratifications observations in the lowest decameters show that the profiles of θ and r are similarly logarithmic and imply the approximate equality there of the coefficients k_M, k_H, and k_W.

When convection is in progress and buoyancy forces are significant the profiles of the properties depart appreciably from the logarithmic form and the relations between their mean gradients and fluxes become more complicated (5), involving another length scale L (in addition to z_0) defined by

$$L = -u_*^3 c_p \rho T/k'gH$$

[The dimensionless number (z/L) is related to the Richardson number Ri by

$$\mathrm{Ri} = \frac{(g/\theta)\partial\theta/\partial z}{(\partial u/\partial z)^2}$$

which arises in discussions of motion systems whose kinetic energy may be derived partly from that of the mean motion because of the presence of a shear in the vertical, and in which accelerations under buoyancy forces may arise; when k_M and k_H are equal, $\mathrm{Ri} = (z/L)$. The particular form in which the Richardson number has been expressed above can be regarded as the ratio of the kinetic energy $(\Delta z)^2(g\partial\theta/\partial z)2\theta$ which may be generated under a buoyancy during a vertical displacement Δz, and that $(\Delta z\,\partial u/\partial z)^2/2$ which may be derived from the shear of the mean wind over the same distance z.]

When the heat flux is large and the wind light, as may often happen over land, the magnitude of L is small compared with the height of screen level; on the other hand, in moderate or fresh winds with a typical heat flux L is 1–2 dkm (Fig. 7.4).

7.4 The estimation of convective transfers; the condensation level

The measurements supporting the validity of the general relations have required special instrumentation at carefully chosen exposures. They have little direct value in the inference or prediction of representative values of the fluxes over large areas during the modification of airstreams over land and sea. These fluxes, and the associated gradients, are determined by large-scale conditions, mainly the flow and general stratification of the air, the temperature of the water over sea, and the net radiation at the ground and its moistness over land. For most purposes the modification may be sufficiently well estimated by simple means, given conditions at a single height a few meters above the surface, supplemented by information on surface temperature over the ocean and net radiation over land.

Over the ocean the important transfers of heat and water vapor occur when winds are fresh or strong. Then the wind speed, air temperature, and dew point routinely measured on board ships are observed at a height which is comparable with the magnitude of L and are not especially sensitive to the precise level of measurement. The transfer equations may be written

$$\tau = \rho C_D u^2$$

$$H = c_p \rho C_D u \Delta T \qquad (7.12)$$

$$E = \rho C_D u \Delta r \qquad (7.13)$$

where ΔT and Δr are, respectively, the differences between the values of temperature and mixing ratio at the ship's deck level and at the sea surface (where the air is assumed to be saturated); C_D is a drag coefficient defined by

$$C_D = \tau/\rho u^2 = (u_*/u)^2$$

and depending, but not sensitively, upon the particular level where the wind speed u is measured. An empirical relation between C_D and u can be used: at a height of several meters (appropriate to the routine ship observations) C_D appears to increase from about 10^{-3} in light winds to over 2×10^{-3} in strong winds. In the derivation of the relevant values in Table 7.2 a constant value $C_D = 2.5 \times 10^{-3}$, appropriate to the stronger winds, was chosen. Smaller constant average values have been used in the estimation of climatic values of the fluxes, for example, 2×10^{-3} (8), and 1.4×10^{-3} for conditions in the Caribbean Sea (9; appropriate to a mean speed of about 6 m sec^{-1} in the trade winds there).

Over land the evaluation of general fluxes from routine

observations is more difficult because they are made only at screen level and there are large local and diurnal variations. Climatic values over large regions are usually assessed from the estimated total net radiation at the ground, and the total evaporation (regarded as the difference between precipitation and runoff; 8). The partitioning of the net radiation between transfers into the atmosphere of sensible and of latent heat is strongly dependent upon the water content of the soil and the behavior of the vegetation, and over small areas and short periods, such as a day, can be determined only with special instrumentation. Where the soil is moist or there is abundant vegetation the upward flux of sensible heat is usually only a fraction of the flux of latent heat. However, during prolonged sunny and rainless weather the soil may eventually become so dry that the plants wilt and evaporation suddenly becomes small. Variations from day to day in the ratio H/LE between about 0.4 and 3 have been measured during a period in which occasional summer rains moistened dry pasture land (4); over the included month the total precipitation was only 0.8 cm, while the water evaporated was equivalent to over 6 cm, drawn mainly from depths below about 1 m (reached by the roots of the clumps of coarse grass).

The Bowen ratio H/LE

The Bowen ratio H/LE between fluxes of sensible and latent heat is

$$\frac{H}{LE} = \frac{c_p k_H \partial\theta/\partial z}{L k_W \partial r/\partial z}$$

where the gradients $\partial\theta/\partial z$ and $\partial r/\partial z$ are measured near the surface. Presuming that $k_H = k_W$ even very close to the surface, a maximum value of the Bowen ratio over a water surface is given approximately by

$$[(L/c_p)(\partial r_s/\partial T)_{T_0}]^{-1}$$

where r_s is the saturation mixing ratio and T_0 is the temperature of the surface (r cannot exceed r_s and if it is less, then E is increased).

The ratio is readily evaluated using the Clausius-Clapeyron relation (2.3), and varies with temperature as shown in Table 7.4.

Similarly, it may reasonably be assumed that at levels well away from as well as near the surface

$$\frac{\partial r/\partial t}{\partial T/\partial t} \approx \frac{c_p E}{H}$$

$$> \frac{\partial r_s/\partial t}{\partial T/\partial t}$$

$$> \left(\frac{\partial r_s}{\partial T}\right)_{T_0}$$

Table 7.4 Variation of $(H/LE)_{max}$ over a water surface of temperature T_0

T_0 (°C)	$(H/LE)_{max}$
0	1.5
5	1.1
10	0.8
15	0.6
20	0.5
25	0.4

since aloft the temperature T and hence $\partial r_s/\partial T$ have values smaller than at the surface. Accordingly, over a water surface convection in a layer of limited depth brings it toward a state of saturation, and indeed a final equilibrium state can be regarded as one in which θ_w (or θ_s) in the layer is everywhere equal to the value at the surface, and the mixing ratio of vapor and condensed water is constant and equal to the saturation mixing ratio at the surface. This state is practically never observed because the convection is never sufficiently prolonged. Generally it is continually extending upward into the dry air of the free atmosphere (although large-scale subsidence may prevent the top of the convection layer from rising relative to the ground). Cumulus clouds which form in the upper part of a layer of convection are usually isolated, separated by clear spaces in which the air has low relative humidity.

Over the ocean in middle latitudes the transfer of sensible heat into air which streams rapidly toward regions of higher sea surface temperature can be large (Table 7.3). In the Northwest Atlantic, for example, the change of sea surface temperature along the trajectory of air traveling at 15 m sec^{-1} for one day into lower latitudes may reach about 12°C. If a layer 300 mb deep is modified by the convection alone, with an *average* temperature rise of several degrees in one day, the heat flux at the surface is

$$H \approx 300\, c_p \times 5 \text{ cal cm}^{-2} \text{ day}^{-1}$$

$$\approx 0.25 \text{ cal cm}^{-2} \text{ min}^{-1} \approx 4 \times 10^{-3} \text{ cal cm}^{-2} \text{ sec}^{-1}$$

a value attained over land only in the middle of a day of strong sunshine. Such large fluxes imply values of several degrees Celsius, and occasionally as much as 10°C, for the difference ΔT between the temperature at the sea surface and at ship's deck level. For example, if the wind speed is u, then for a layer 300 mb deep moving over an average sea surface temperature gradient of about 1°C (100 km)$^{-1}$ implied above, Eq. 7.12 gives

$$10^{-7} \times 300\, c_p u/2 = c_p \rho C_D u \Delta T$$

or

$$1.5 \times 10^{-5} = \rho C_D \Delta T$$

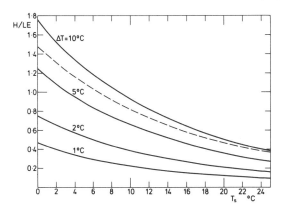

Fig. 7.5 Variation of the Bowen ratio (H/LE) between the transfers into the atmosphere of sensible and latent heat from an ocean of surface temperature T_s, when the relative humidity at ship's deck level is 80%, for various values of ΔT, the difference between T_s and the temperature at ship's deck level (solid lines). The dashed line shows the approximate maximum value of the Bowen ratio over water.

from which it appears that ΔT over the open sea may depend upon u only through the drag coefficient C_D, which in fresh winds is typically about 2×10^{-3}. For this value ΔT is about 7°C. Near the Gulf Stream the gradient of sea surface temperature and the values of ΔT to be anticipated are about twice as great.

In air subject to convection over the oceans it is observed that the relative humidity at ship's deck level is usually between 75 and 80%, implying a value of the Bowen ratio (H/LE) which varies with ΔT and T_s, the temperature of the sea surface, as shown in Fig. 7.5. This diagram includes the maximum value of the Bowen ratio as estimated in Table 7.4 from Eq. 7.25, and shows that the ratio can be as high as 1.0 only over cold water, and must always be a small fraction over warm water.

Transfers inferred from synoptic data

Transfers of heat and water vapor into the atmosphere can in principle be estimated from routine soundings over land, which also show changes of stratification along the air trajectories; however, it is usually difficult to make these estimates accurate.

In studies of changes due to cloudless convection it has been shown that the adiabatic layer produced, with the approximately constant potential temperature θ_a, extends to a height considerably above the level at which θ_a is observed in the unmodified air before the onset of the convection (9, 10). The convection is accompanied by the cooling of the air in the upper part of the layer which it modifies (cf. Fig. 3.13), and some of the warming in the lower part has to be attributed to the redistribution of potential temperature throughout the whole layer. Such cooling (perhaps only relative to a general warming produced by large-scale subsidence) is probably characteristic of all convection, with or without cloud, consisting of unsteady circulations. It can be regarded as due to the expenditure of some of the kinetic energy (acquired under buoyancy forces in the lower part of the layer of convection) of masses of ascending air where they reach upward into stably stratified air and become relatively cool, together with a turbulent mixing which results in a limited subsequent sinking of greater masses. (Some kinetic energy may be lost also by the spread of gravity waves away from the intruding air masses.) In cloudy convection the cooling may be enhanced by the evaporation of water condensed at lower levels; on the other hand, if showers develop, the precipitation of water upon the surface and its partial evaporation at various levels may also alter the vertical distribution of the change of potential temperature. The scheme for this change illustrated in Fig. 7.1 is therefore evidently too simple. Other difficulties in estimating the heat economy of a layer of convection arise, especially if it is deep and occupied by cumulus clouds. In particular it is difficult to assess the net radiative loss of energy by long-wave radiation, which in a deep layer over land may during 24 hr be as much as half of the solar radiation absorbed at the ground during the daylight hours. (In one study of the development of an adiabatic layer, over England during 8 hr of a cloudless summer day, transfers of sensible heat upward from the ground and downward through the top of the layer, and a net radiative gain of heat in the layer—the absorption of solar radiation by the air and the aerosol particles exceeding the net loss by long-wave radiation—appeared to be all of the same magnitude; 10.)

In the study of the transfers of heat and water into the atmosphere over the ocean some simplification in the interpretation of observations arises from the steadier and more uniform meteorological conditions. However, the sparseness of the soundings makes it hardly possible to take satisfactory account of the large-scale vertical motion, and the almost invariable presence of clouds and also of showers adds other difficulties.

In an investigation into the modification of air in the broad trade wind current in the North Pacific covering the route from San Francisco to the Hawaiian Islands, the large-scale vertical motion was obtained by assuming that the variation of the mean wind in the vertical planes normal to a mean trajectory was negligible compared with that in the vertical along it (8, p. 162). The conditions along the trajectory are represented in Fig. 7.6, which clearly shows a large-scale sink of air, leading it from the middle troposphere into the cloud layer; at a height of 3 km at the beginning of the trajectory the mean speed of descent was about 0.5 cm sec^{-1}. From this and other studies it appears that in the trade winds the typical magnitudes of the transfers of sensible and

7 Cumulus convection

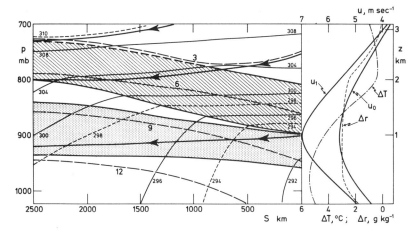

Fig. 7.6 Modification of air in the trade winds of the North Pacific, according to mean data for the period July to October 1945, along a trajectory on the southeast flank of the subtropical anticyclone, extending a distance S from 32°N 136°W to the Hawaiian Islands (21°N, 158°W) (adapted from the original source quoted in reference 8, p. 162).

On the left of the diagram are shown three streamlines of the mean flow, and isopleths of potential temperature θ (solid lines) and mixing ratio r (dashed lines), labeled, respectively, in °K and g kg^{-1}. The cumulus cloud layer as defined in this text, extending upward from the level of the cloud bases to the level where the decreasing lapse rate first becomes equal to the saturated adiabatic, is stippled. The cumulus tower layer extends upward into the hatched layer, characterized in the original reference as the inversion layer, in which the mean lapse rate abruptly becomes much smaller. In the lower part of this layer, at distances S up to about 900 km, there is an inversion (temperature increasing with height). The top of the layer is arbitrarily drawn at a level where the lapse rate increases again, which is well marked only on the right of the diagram, where it is drawn as a solid rather than as a dashed line.

To the right of the diagram the changes ΔT and Δr of mean temperature and mixing ratio along the trajectory are drawn as functions of height, together with the profiles u_0, u_1 of the mean wind speed u at its beginning and end.

latent heat are about 20 and 200 cal cm^{-2} day^{-1}, so that the Bowen ratio (H/LE) is only about 0.1. The difficulty of assessing the heat economy of the layer of convection is illustrated by the magnitude of the estimated net radiative heat loss, at about 100 cal cm^{-2} day^{-1}.

In the middle latitude airstreams subject to convection over the ocean the large-scale vertical velocity is usually downward and commonly greater, but is more variable. Where the flow returns toward higher latitudes on the eastern side of pronounced troughs it may be very small or even directed upward. Very large rates of transfer of sensible heat which have sometimes been inferred from changes in the stratification along trajectories, for example, 800–1500 cal cm^{-2} day^{-1} over the Northeast Atlantic (11) and 1000 cal cm^{-2} day^{-1} over the sea between Japan and the neighboring continent (8, p. 261), are probably too great because of the neglect or underestimate of the large-scale descent. Estimation of the accompanying fluxes of latent heat is hampered by the difficulty of taking into account the precipitation of water from shower clouds which, over the sea, is not measured. The implied values of the Bowen ratio are probably overestimated on both accounts, and characteristically seem to be too high (considering the maximum values specified in Table 7.4): between 3 and 5, and 2.3, in the two studies quoted, even though in the latter the precipitation was in the form of snow and estimated without serious error from observations at coastal stations. In this case observations made on ships in the Japan Sea showed Bowen ratios decreasing from 1.4 to 0.6 as the temperature of the sea surface rose from about 3 to 15°C (and T decreased from 14 to 10°C), as might be anticipated from Fig. 7.5.

The height of cumulus bases

If there are no, or only very shallow, clouds formed in the ascending air of the convective circulations, the layer of almost adiabatic lapse rate extends up to a shallow transition layer of marked stability which separates it from the unaffected free atmosphere above. When more substantial clouds are formed the release of latent heat affects the stratification of the layer which they occupy, and the thickness of the adiabatic layer is limited by the level of the cloud bases.

Under the usual condition that there is little decrease of mixing ratio with height in the adiabatic layer, the cumulus bases are approximately at the saturation level, or *lifting condensation level* (where air ascending adiabatically becomes saturated), of the air at the top of the superadiabatic layer. This is appreciably above the saturation level of air at screen level over land (4 ft, or 1.2 m above the ground), which according to Eq. 3.24 is given by the observed depression ΔT_D of the dew point below the air temperature. For example, in 100 daytime aircraft soundings on occasions of cumulus convection

7.4 The estimation of convective transfers; the condensation level

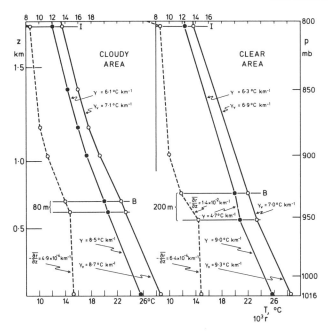

Fig. 7.7 Average stratification over the Caribbean Sea during cumulus convection in fresh trade winds, in areas where the clouds were clustered, and also in comparatively cloud-free areas (from numerous aircraft soundings made during the period 10–28 April 1946; 25). Average values of the lapse rate of mean mixing ratio $\partial \bar{r}/\partial z$, mean temperature γ, and mean virtual temperature γ_v are included. The transition layer below the level B of the cumulus bases is indicated, and also the base of the trade wind inversion I. In the mean the sea temperature was 0.4°C above the dry bulb, and 4.2°C above the wet-bulb temperatures recorded on board a research vessel in the locality of the sounding.

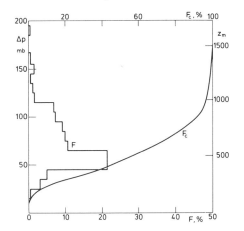

Fig. 7.8 Frequency distribution of cumulus base heights in 264 special observations made by British meteorological reconnaissance aircraft over the Northeast Atlantic during the period September 1956–December 1958. F is the percentage frequency with which the reported difference Δp between the pressures at sea level and at cloud base fell within intervals of 10 mb, and z is an approximate corresponding scale of height. F_c is a smoothed curve representing the cumulative percentage frequency as a function of height.

over eastern England and Northern Ireland the mean pressure observed at the level of the cloud bases was 902 mb, a pressure about 22 mb less and a height about 200 m greater than the values calculated from the appropriate (unventilated) screen thermometer observations (similar differences are indicated in Fig. 7.13 and 7.16); the differences did not change noticeably over the observed range of the height of the cloud bases, from about 500 to about 1500 m (12). Similarly, during a study of summer cumulus in central Sweden the height of their bases (determined by theodolite triangulation, and varying between about 1 and 2 km) was during the greater part of the day consistently about 300 m above the level calculated from observations made with an Assman psychrometer 2 m above the ground (13). Considerably greater differences develop over land in the late afternoon when the cumulus typically persist for a while after the screen temperature has begun to fall and when it is no longer representative of θ in the adiabatic layer (Fig. 7.11). Nevertheless, if this behavior is taken into account, an estimate of the level of cumulus bases can evidently be made from screen temperatures which probably is considerably better than the visual estimate of even an experienced observer with no special aids.

Cumulus base level over the sea

The average level of the cumulus bases over the tropical oceans, where advection over the small gradients of sea surface temperature produces only very small fluxes of sensible heat, is several hundred meters (Fig. 7.6, 7.7), and is consistent with the assumption that the mean value of the upward flux of water vapor near the surface has the same magnitude as the average rate of precipitation, that is, about 120 cm yr^{-1}, or 4×10^{-6} g cm^{-2} sec^{-1}. Thus in the Caribbean, the average rate of evaporation has been estimated (8, p. 152) to be a little greater: about 4.5×10^{-6} g cm^{-2} sec^{-1}. Here, where the mean wind speed near the surface is about 6 m sec^{-1}, the appropriate value of the drag coefficient C_D is about 1.5×10^{-3} and the mean value of the difference Δr between the mixing ratios at the sea surface and ship's deck level is given by

$$4.5 \times 10^{-6} = 1.2 \times 10^{-3} \times 1.5 \times 10^{-3} \times 6 \times 10^2 \overline{\Delta r}$$

or

$$\overline{\Delta r} = 4.2 \times 10^{-3}$$

Assuming that the temperature at ship's deck level is practically that at the sea surface this value of Δr at a temperature of 25°C implies a difference ΔT_D between the air temperature and dew point of about 4°C and an adiabatic lifting condensation level of only 450 m

7 Cumulus convection

(Eq. 3.24), a little below the average level of cumulus bases observed in this region. Evidently a similar height for the cumulus bases away from the rain systems can be anticipated generally over the tropical oceans, where the average upward flux of sensible heat at the surface is small, but the average rate of evaporation is large.

Since the height of cumulus bases is closely related to the depression ΔT_D of the dew point at screen level, over the oceans it is more generally a function of the relative magnitude of the differences Δr and ΔT in Eq. 7.12 and 7.13, that is, of the Bowen ratio (H/LE). It is likely to be sensitive to the large-scale vertical motion, in which descent, by warming and drying of the middle troposphere, tends to reduce ΔT and increase Δr, and thereby to increase ΔT_D and the height of the cloud bases; conversely large-scale ascent tends to lower the cloud bases. The large-scale vertical velocities attain greater values in the slope convection of the middle latitudes than in the tropics, and are probably mainly responsible for an apparently greater variability there in the height of cumulus bases over the oceans. In both regions the most frequent height is several hundred meters, and on account of the higher values of the Bowen ratio (H/LE) reached over the cooler waters it may even be somewhat lower in the middle latitudes; in the Northeast Atlantic it appears to be about 450 m (Fig. 7.8).

The large-scale vertical motion may be anticipated to have a strong influence also upon the height to which the cloudy convection penetrates. The sinking motion typical of the west flanks of troughs, where the flow is toward warmer water, tends to lower the cloud tops; on the other hand, on the east flanks, where the flow becomes almost parallel to the sea surface temperatures, the large-scale vertical motion is small and may even be upward, so that it tends to prolong the heat transfer from the surface and to maintain or even raise the level of the cloud tops. Such an influence is indicated in the situation over the North Atlantic illustrated in Fig. 7.9, where a value of ΔT of a few degrees Celsius was maintained over a large area on the east flank of a deep trough. Showers from convection clouds in the part of a cool flow out of the Davis Strait which reached as far south as about the 50th parallel continued, and may even have intensified, after it had turned poleward as a southwesterly current, persisting even as far north as the North Cape of Norway in latitude 70°N.

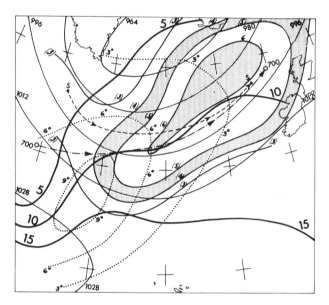

Fig. 7.9 Some properties of convection over the North Atlantic, 17–18 December 1951, indicating the influence of large-scale vertical motion. For 1200 GMT on 18 December, isopleths are drawn of (*a*) sea surface temperature (heavy lines labeled in °C); (*b*) sea-level isobars at intervals of 16 mb (thin lines, labeled in mb); (*c*) difference ΔT of temperature between the sea surface and ship's deck level (dotted lines, at intervals of 3°C); (*d*) the height reached by cumulus tops, based partly on radiosonde soundings and reports by reconnaissance aircraft, but mainly on numerous reports by commercial aircraft (thin lines, labeled with italic figures in brackets, at intervals of 1 km, stippled over two intervals).

Two "trajectories" *SS* and *700–700* show paths of air at the surface and at the 700-mb level (about 3 km) during the previous 24 hours. The temperature difference ΔT reaches a maximum of over 9°C approximately where the air flows most rapidly over the largest gradient of sea temperature. Values of several degrees persist where the air subsequently flows almost along the isotherms in the sea surface, on the east side of the pressure trough. On this side the cloud tops reach their maximum height of over 6 km. On the other side large-scale subsidence probably plays an important part in limiting the convection to about the lowest 4 km, whereas farther east the great reduction in, or even reversal of the large-scale vertical velocity probably supports the maintenance of a large flux of heat from the sea and a deep cloud layer.

Cumulus base level over the land

During convection over land the air adjacent to the surface is generally not saturated, and the level of cumulus bases, particularly in arid regions, is typically higher than over the ocean (Fig. 7.10). Except far inland there is also a diurnal variation in which the base level rises during the morning and reaches a maximum in the afternoon, at about the time of the maximum temperature at screen level. Where air which has been subject to convection over the ocean flows inland, the clouds which form first in the morning appear at about the height of several hundred meters corresponding to the base level previously found over the sea, but subsequently ΔT_d, the depression of the dew point at screen level, increases as the screen temperature rises markedly with only small change of dew point (if the moist layer over the sea is shallow and surmounted by very dry

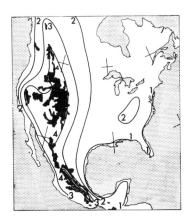

Fig. 7.10 Average height of cumulus bases (in kilometers above sea level) over North America in mid-afternoon in July, inferred from mean daily maximum screen temperature and the associated relative humidity. The average height is several hundred meters over the ocean, but increases to 1 km or more a short distance inland. The intersections of the meridians 70, 90, and 110°W and the 30th and 50th parallels are marked, and ground above 2 km is shown in black.

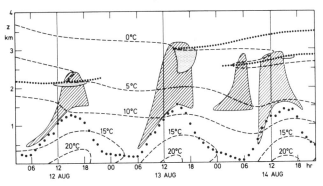

Fig. 7.11 Diurnal variations in cumulus near Cranfield, central England, on three successive summer days (12–14 August 1952) constructed from observations by theodolite from the ground, from flights in sailplanes, and from soundings by aircraft and tethered balloons. The hatched and stippled regions are those occupied, respectively, by cumulus and by shelf clouds formed by the spreading of their tops. Dashed lines are isotherms labeled in °C, and rows of dots mark the bases of inversions. The large black circles show the variation in the mean lifting condensation level of the air at screen level, from observations at five neighboring airfields (under the conditions prevailing in the afternoons the wet-bulb thermometers in the screens, which were not specially ventilated, may indicate temperatures which are about 0.5°C too high, and therefore lifting condensation levels of the air at their level which are 100 m or rather more too high; Section 3.10).

air, the dew point at screen level may even decrease substantially as the air flows inland and the convection extends upward into the dry layer). Consequently, in airstreams from the ocean the level of the cumulus bases inland rises rapidly during the morning by as much as 1 km or more, as seen in Fig. 7.11, representing typical summer conditions in central England. After mid-afternoon the base level remains at about the maximum height attained, or it may lower a little in the late afternoon and evening, when the clouds persist for some time after the fall of temperature at screen level has made it no longer representative of θ in the adiabatic layer. By the evening the lifting condensation level corresponding to the screen temperatures has fallen well below the level of the cloud bases. Well inland, remote from coasts, the level of the cloud bases has a similar but smaller diurnal variation.

Large-scale vertical motion influences the level of cumulus bases over land in the same sense as over the sea, but over thermally irregular terrain topographically

On the first two days a decreasing southwesterly wind brought inland air which had been subject to cumulus convection over the ocean west of the British Isles; by the third day the winds had fallen light and variable near the surface, but in the early hours of that day castellanus clouds formed in air which had been subject to cumulus convection over northern France during the previous afternoon.

induced circulations of intermediate scale have effects which are commonly at least as great and more obvious, manifest by noticeable local variations in the height of the cumulus bases and in the degree of their vertical development. The most striking examples occur near sea-breeze fronts (e.g., see Pl. 7.14–7.17), which may also be responsible for local persistence of cumulus into the late evening.

7.5 The stratification during cumulus convection

The superadiabatic layer

At heights which are only a small fraction of $-L$ the kinetic energy of the small-scale motion systems is drawn predominantly from that of the mean flow and the effects of buoyancy forces are insignificant. There gradients are indistinguishable from those found in neutral stratifications; synchronous recordings of temperature and wind show continual irregular fluctuations of temperature and of the vertical component w' of the wind (Fig. 7.12), of which the major have a frequency of order 1 sec^{-1}. The fluctuations appear disordered,

7 Cumulus convection

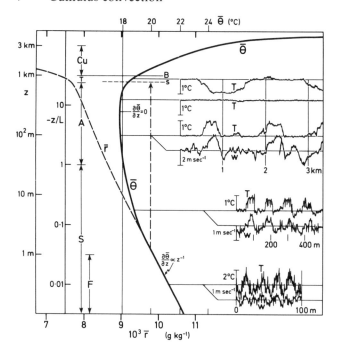

Fig. 7.12 Typical vertical distributions of mean values of potential temperature $\bar{\theta}$ and mixing ratio $\bar{r}$ during cumulus convection, and associated horizontal variations of temperature T and vertical component w of air velocity drawn as a function of the scaled height $-(z/L)$. In the layer of *forced convection* F (where buoyancy forces are not significant) $\partial\bar{\theta}/\partial z = z^\beta$ with $\beta = -1$, and the profile of mixing ratio $\bar{r}$, suitably scaled, is identical to that of $\bar{\theta}$. The level $z = -L$ is arbitrarily chosen to define the top of the *superadiabatic layer* S; above it is the adiabatic layer A in which θ is almost constant. Near the middle of the adiabatic layer $\partial\bar{\theta}/\partial z$ becomes zero, and in its upper part both $\bar{\theta}$ and $\bar{\theta}_v$ increase slightly with height, while $\bar{r}$ continues to decrease. In a shallow *transition layer* T below the layer Cu occupied by cumulus clouds the magnitudes of $\partial\bar{\theta}/\partial z$ and $\partial\bar{r}/\partial z$ become substantially greater. Scales of $\bar{\theta}$, $\bar{r}$, and z have been added to make the profiles representative of convection over a dry land surface in a moderate wind during the middle of a day of intense sunshine (upward flux of sensible and latent heat each equal to 0.30 cal cm^{-2} min^{-1} = 21 m W cm^{-2}, corresponding to a rate of evaporation E of about 10^{-5} g cm^{-2} sec^{-1}, $u_* = 0.5$ m sec^{-1} and $-L = 35$ m). The height of the cumulus bases is drawn as $B = 1$ km; in the cloud bases $\theta \approx 18°C$ and $\theta_w \approx 13°C$; these values are 200 m lower then the saturation level s of the air at screen level, and 2 and 3°C lower than $\bar{\theta}$ and $\bar{\theta}_w$ there.

but have a correlation consistent with the upward heat flux.

At greater scaled heights, particularly near and for some distance above the level $z = -L$, the records have a different structure, which indicates the presence of motion systems with scales comparable with the magnitude of L. There are again rapid fluctuations in the horizontal and vertical components of velocity, but during sharply defined "disturbed periods" of a fraction of a minute the temperature fluctuates irregularly about a value distinctly greater than one more steadily maintained during intervening "quiescent periods," whose duration is typically two or three times longer. It appears that in the disturbed periods there is buoyant ascent of air, some of which has risen almost adiabatically from near the surface, while in the thermally quiescent periods there is more widespread comparatively slow descent from the layer in which the lapse rate is very nearly the adiabatic (which will be called the *adiabatic layer*, in contrast with the *superadiabatic layer* immediately above the surface, in which the lapse rate significantly exceeds the adiabatic).

From the later arrival of the warm disturbed periods at the higher of a vertical array of thermometers it has been inferred that there are plumes of rising air tilted forward and traveling with the speed of the wind at a very low level (2). An alternative interpretation is that plumes are arranged in long rows lying nearly along the wind, with a sideways tilt and a small lateral velocity which carries them past an array of thermometers; a width about equal to their height would then not be inconsistent with the much greater dimension suggested by their recorded duration and the total local mean wind speed. If visible, such rows might resemble (on a reduced scale) lines of cumulus, reaching upward to about the level $z = -L$; they are perhaps manifest in steam fogs, as in Pl. 3.1. Their dimensions, including vertical extent and horizontal separation, might be expected to increase downwind (with the suppression or disappearance of some), until a set of neighboring rows leaves the surface in the ascending part of a circulation on the scale of the depth of the entire layer of convection, corresponding to the updraft below a cumulus. If so, associated systematic variations in lateral wind components (which have been less well studied, particularly when their variability over several minutes is large), and in the dimensions and orientation of the rows, might be expected on scales of a few kilometers, or over periods of several minutes or more at a particular place. [On the occasion of Pl. 3.1 the wind speeds between 10 m and 1 km above the lake were 10–15 m sec^{-1}. A layer about 150 mb—1.5 km—deep was warmed by about 5°C during passage across the lake. The heat flux from the surface can be estimated, presuming that this warming occurred over a path of 100 km, or by using the relation (7.12), to have been 0.5 ± 0.2 cal cm^{-2} min^{-1} corresponding (Fig. 7.4) to a value of $-L$ of about 20 m. This is about the typical vertical extent of the long turreted plumes of steam fog in the plate, which are separated horizontally by rather greater distances; here and there, separated by distances of about 200 m, the fog is several tens of meters deep, while in places 1 km or more apart wisps reach up to several hundred meters, or perhaps even to the cumulus at about 1 km. The orientation of the fog rows clearly varies by 10° or more from place to place, and there are comparatively clear spaces or regions of very shallow fog which may mark the regions of the most pronounced downward parts of the cumulus circulations. The implication of these patterns in a single picture is uncertain, but they provide some support for the view that circulations are generated on the scale of the depth of the whole layer of convection, and on that of the superadiabatic layer, nearly two orders of magnitude less, but not on intermediate scales.]

Dust and steam devils

Narrow rotating columns of cloud have been observed to reach from steam fog up into the bases of cumulus, recalling the dust devils of arid lands in intense sunshine. Generally no rotation is detected in the air ascending from low levels into and within cumulus. Evidently sufficient vorticity to concentrate in the air converging into the updrafts is present only when the low-level wind flow is disturbed near and downwind of irregularities of terrain over land or near coasts; even then vortices become evident only when there is dust or steam fog to raise (in more favorable special circumstances their rotation is sufficiently intense to produce a considerable pressure fall near their axes with the formation of funnel clouds, as in the waterspouts and tornadoes discussed in Section 8.9).

Dust devils last only a few minutes; they occur in light winds during intense sunshine over dusty ground (14–16). In some places they are common; for example, in Arizona several tens have been observed each summer day over an area of about 500 km^2, and the frequency over particular areas of 1 km^2 in the lee of small hills may be nearly as great (16). The wind speed near the ground in a dust devil reaches 10 m sec^{-1} or more within a few meters of its axis (the air rotating in either sense), where the air temperature is about 10°C higher and the air pressure about 3 mb lower than in the surroundings. (These magnitudes are consistent with the theory of whirlwinds mentioned in reference 8, p. 205, according to which an important requirement for the formation of a dust devil is an extreme vertical gradient of temperature near the surface.)

The adiabatic layer

Most of the observations made in the study of convection above the lowest few decameters have been made using single aircraft. The difficulty of measuring representative mean values is especially great during the day over land, not only because of topographically induced irregularities but because prolonged flights at each of several levels are needed, during which significant general changes may occur. It has been confirmed that the gradients of the mean values of both the actual and

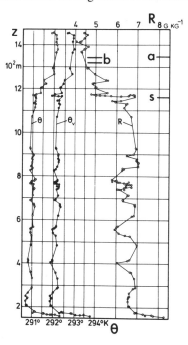

Fig. 7.13 Sounding made inland over southern England during an ascent in a towed sailplane by Miss Woodward, from 1140 to 1153 GMT, 18 July 1961. Potential temperature θ, virtual potential temperature θ_v, and mixing ratio r are shown as functions of height z above sea level (in hundreds of meters). The lowest level of observation is at about screen level. The surface wind was northerly, 3 m sec^{-1}; cumulus were observed with bases between the levels marked b; their tops were at about 1500 m. The adiabatic lifting condensation level of the air at screen level is marked s; that of the moister air in the adiabatic layer ($\theta = 291°$, $R = 6.9$ g kg^{-1}) is marked a.

the virtual potential temperature well above the ground are very small and change sign at a height of a few hundred meters. Thus the *superadiabatic layer*, in which there are pronounced upward decreases of θ and r from their surface values, can be distinguished from an (almost) *adiabatic layer*, below any cumulus clouds present, in which θ and often also r have nearly constant values. In any but extreme conditions these values, which are of interest in the study of the cloud properties, are about equal to those at the level $z = -L$ which accordingly will be regarded as defining the top of the superadiabatic layer.

The adiabatic layer extends up to the bases of cumulus clouds, and between the clouds usually to within about 200 or 300 m of their base level. The mean mixing ratio $\bar{r}$ decreases upward, but often by less than 1 g kg^{-1} over the whole layer. The mean virtual potential temperature increases slightly with height throughout most of the layer (Fig. 7.7, 7.13). [Occasionally a large upward decrease in $\bar{r}$ has been observed (17), perhaps when there has been insufficient time to respond completely to a sudden change of surface condition, as in the middle of the day near a coast, where air from the ocean has entered land over which considerably higher values of both temperature and mixing ratio have been established near the surface.]

In the lower part of the adiabatic layer the records obtained on an aircraft flying horizontally (18, 19) show the same features as those from fixed instruments in the upper part of the superadiabatic layer: sharp-edged pulses of higher temperatures and increased mixing ratios mark the regions of ascending air, in which the mean upward velocities reach a few meters per second. At least in the lowest part of the adiabatic layer, powered aircraft experience about as much high-frequency bumpiness (corresponding to significant changes of vertical air velocity on horizontal scales of about 100 m) between the regions of ascending air as within them. It has been inferred that small-scale motion systems are present everywhere, but it is difficult to distinguish a bumpiness due to flight through oscillatory motions (gravity waves) transmitted upward from lower levels. Higher in the adiabatic layer it is the experience of pilots in sailplanes, whose lower air speed makes them sensitive to motions of still smaller scale, that they are present only within and near the edges of the ascending air, there producing tingling sensations and a distinctive wind noise, which are not encountered elsewhere. It is probable, therefore, that intense, small-scale motions which effect mixing within the largest convective circulations are confined to the superadiabatic layer and to the regions where air ascends and where it begins to spread horizontally at the top of the layer of convection.

Although the mean upward speed of ascending air is maintained, its excess temperature diminishes with height, and in the lower part of the adiabatic layer it becomes difficult to recognize, although there still may be an excess of virtual temperature. Toward the top of the layer the regions of ascending air are appreciably cooler but still moister than their surroundings; however, the virtual temperature is distinctly lower than the mean value, so that the buoyancy has been reversed.

From the spacing of shallow cumulus which sometimes form above the adiabatic layer, and other evidence, it clearly contains circulations whose horizontal scale is two or three times its depth, much greater than that of the circulations which are recognized in the upper part of the superadiabatic layer. Probably there are distinct and widely separated scales of motion, but it is difficult to obtain the kinds of observation which would establish this and the form of the larger circulations, because the methods generally obtain information only from points or from traverses made by aircraft. The observations have therefore neither sufficient abundance nor accuracy to define the whole field of motion, and moreover are usually obtained over land where their interpretation is hindered by the unknown but often strong influence of irregularities introduced by the terrain and traveling

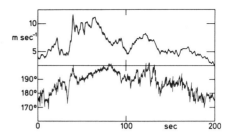

Fig. 7.14 Typical form of "sun-gust" in a moderate wind during sunny weather with convection in progress (22). A sudden increase in wind speed (upper trace) is associated with a veer of wind and a slight fall in temperature. For a while the fluctuations of wind direction (lower trace) decrease; the wind speed is irregularly maintained before decreasing gradually during a period of several minutes before the arrival of another gust, corresponding to a spacing of a few kilometers.

cloud shadows. Near the ground evidence for organization on the larger scale is provided by the occurrence of pronounced gusts of wind, at intervals representing a spacing in the wind direction which is several kilometers. These intermittent gusts, which are so characteristic of warm sunny weather over land that they have been called sun-gusts (*Sonnenboen*, 21), have the typical structure seen in Fig. 7.14 (22).

In the gusts an increase of wind speed by several meters per second during a few seconds is accompanied by a slight fall of temperature. The speed is irregularly maintained for a few tens of seconds and then gradually decreases while the temperature slowly recovers. If the general wind speed is only several meters per second, there may be a complete lull before the arrival of the succeeding gust. In observations of the behavior of soaring birds and insects in an arid part of India (23) it was noticed that for a while after a gust dragonflies a few meters above the ground were unable to fly without wing-flapping, indicating settling air. On the other hand, during the onset of the gust, birds at a higher level could begin soaring and after drifting downwind for a few kilometers could reach heights of up to 1 km. Sometimes gusts were seen to originate, suddenly raising a thin curtain of dust extending 10–50 m across the wind and several meters high. In these observations such dust curtains frequently could be followed for a few kilometers (24). However, in studies over England it was found that an individual gust could not be traced downwind from one anemometer to another even if they were only 100 m apart (22): evidently it is an unsteady group of gusts which can be traced for several kilometers, moving with about the mean wind speed in the adiabatic layer. The air in and behind the gusts has descended from near the top of the superadiabatic layer, and locally where it spreads out close to the ground a more stable stratification is temporarily substituted. The displaced warm air must rise away from the ground equally irregularly, but higher up may form a more

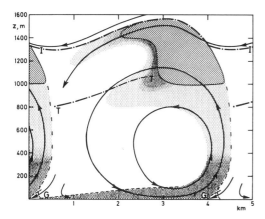

Fig. 7.15 Tentative scheme of mean relative flow in a vertical section through a layer of cumulus convection (the circulations have a variable persistence and spacing and the streamlines drawn cannot be regarded as trajectories). The superadiabatic layer is lifted off the surface over the groups of gusts at the positions G, supplying buoyant air for updrafts beneath the shaded cumulus clouds, which are considerably stronger than the downdrafts in the intervening spaces. The general wind is from right to left and if it has a speed of more than a few meters per second prevents a reversal of the surface wind direction near the gusts. Motions on a range of smaller scales in the clear air are most intense where stippling is added: in and near the superadiabatic layer, the stronger updrafts, and the clouds. They are associated with diffusive transfers, in particular across the shallow markedly stable layer whose top I is the top of the layer modified by convection (between clouds there is another shallow stable layer, the transition layer, with base marked T).

Unless the surface temperature is rising rapidly, descent into the superadiabatic layer of air from well above the level of the cloud bases cannot occur within one cell of the horizontal dimension indicated, and the section may better represent the motion transverse to a row of cumulus and the general wind direction.

persistently ascending column. Soaring birds usually form a cluster, suggesting that only a limited volume of air rises, consistent with the group of gusts which marks its source having a life of only a few minutes. Sometimes, however, the birds form a column several hundred meters tall, and it is the experience of the sailplane pilot that it is sometimes possible to join another seen soaring at a greater height by flying to a position beneath him and locating an updraft, showing that a traveling region of ascent has a duration of at least several minutes. Most evidence, however, indicates the marked unsteadiness which is typical of nearly all convection. A scheme of the mean circulation in the vertical which is implied is sketched in Fig. 7.15.

The transition layer

Especially over irregular terrain the circulations that develop in the adiabatic layer carry modified air intermittently to a variable height, whence a wind shear progressively spreads it across the layer of convection. The top of the adiabatic layer may only locally be sharply defined, away from clouds by a shallow *transition layer* of marked stability which separates it from the cumulus layer. Much of the air entering cloud bases rises through the upper part of the adiabatic layer with an almost constant potential temperature and mixing ratio (and a maximum upward speed which over land in summer sunshine is usually between 3 and 5 m sec^{-1}). Away from clouds the transition layer extends from about the level of the cloud bases to about 200 m below. In it the mixing ratio decreases upward by 1–3 g kg^{-1} and θ increases by about 1°C (Fig. 7.7, 7.13). The air settling into this layer from the clear spaces between the clouds eventually acquires the considerably higher mixing ratio which characterizes the adiabatic layer by mixing motions whose scale and mechanics have not been studied. The nearly constant mixing ratio found generally below the transition layer suggests that a pronounced spreading of the ascending currents occurs near its base, as also indicated by the large proportion of the sky near this level which is found on horizontal flights to be occupied by the moister regions (20). Consistently, sailplane pilots find upward air speeds sufficient to carry them into the clouds occur over only a small fraction of the area of their bases (usually on the upwind side).

Low in the adiabatic layer, where the stratification is slightly unstable and buoyancy is retained during the mixing between ascending air and its surroundings, the diameters of the regions of rising air can be anticipated to increase with height. On the other hand, in the upper part of the adiabatic layer, and especially in the transition layer, the mixture (and eventually all the ascending air) acquires a reversed buoyancy. This causes a progressive reduction in the diameter of the regions of rising air, or to their "erosion."

Assuming virtually adiabatic ascent through the shallow transition layer, the air which reaches its condensation level must there be as much as 1°C colder than its surroundings. If the clouds are not more than a few hundred meters tall, its deceleration is marked, so that soaring sailplanes (whose sinking speed relative to the air is about 1 m sec^{-1}), although easily rising to just beneath the clouds, are unable to enter them. Such clouds are wholly cooler than their surroundings.

The cumulus layer and the cumulus tower layer

The layer occupied by cumulus has a distinctive stratification. Some of the air within clouds more than a few hundred meters tall regains a positive buoyancy corresponding to a cloud virtual temperature exceeding the virtual temperature of the surroundings, and the upward velocities of 3–5 m sec^{-1} found below the cloud bases are maintained or recovered, and locally even somewhat exceeded, in the cloud interiors.

7 Cumulus convection

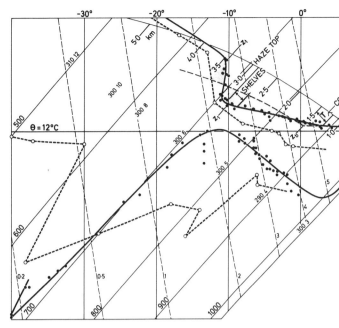

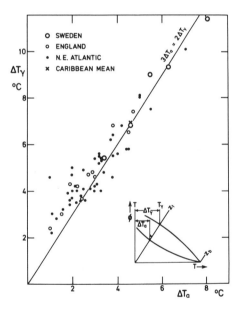

Fig. 7.16 The stratification near Dunstable, central England, in mid-afternoon on 15 June 1956. The dots are individual measurements of temperature and frost point (converted to dew point at temperatures below $-10°C$) made by an aircraft of the British Meteorological Research Flight during a spiral ascent. Lines have been drawn arbitrarily through these points to represent mean conditions, and extended to screen level on the basis of a sounding with a tethered balloon made at midday, and to levels above 3.5 km on the basis of routine radiosonde ascents. Marks indicate the heights of the cumulus bases z_b, of the cumulus layer as defined in the text z_1, and of the tallest cloud tops z_t. Also marked are the transition layer T, the top of the haze in the convection layer, and the layer occupied by extensive shelf clouds. Winds are entered in degrees and meters per second.

Heavy dashed lines joining significant points marked by open circles are from the sounding made at Liverpool at 0300 GMT, which is representative of the stratification in the west-northwesterly stream of maritime air from the Northeast Atlantic before it entered England. The convection over the sea had not reached above about 1.5 km.

Fig. 7.17 The mean stratification of the cumulus layer, given by some soundings on occasions of simultaneous cloud observations, shown by the relation between ΔT_a and ΔT_γ, where $\Delta T_a = T_A - T$, $\Delta T_\gamma = T_\gamma - T$, and T_A, T, and T_γ are respectively, the temperatures observed at the top of the cumulus layer, and those produced there by dry and by saturated adiabatic ascent from its base (as indicated in the inset tephigram).

The small black circles and the small open and large open circles represent, respectively, individual observations by British meteorological reconnaissance aircraft over the Northeast Atlantic, and observations made during field studies inland in central England and in central Sweden. The cross represents a mean value in cloudy areas of the trade winds over the Caribbean (from Fig. 7.7).

Air which rises through the cloud bases generally has the potential temperature θ and mixing ratio r which characterize the adiabatic layer, and therefore has θ and θ_w commonly between 1 and 2°C less than their screen-level values. The near constancy of potential temperature and mixing ratio in the adiabatic layer results in the cloud bases usually appearing remarkably level and uniform in height (within a few tens of meters; the microphysics of the condensation processes insures that the base level can be defined visually within as little as about 2 dkm, except in very polluted air).

In the lower part of the layer occupied by cumulus, the lapse rate between the clouds (and the mean lapse rate inside them, since the horizontal temperature differences can be anticipated to have the small values consistent with the magnitudes of the vertical velocities and upward heat flux) considerably exceeds the saturated adiabatic but is less than the dry adiabatic. It diminishes with height, and eventually becomes equal to the saturated (pseudo-) adiabatic, at a level which can conveniently be regarded as defining the top of a *cumulus layer*, since usually the great majority of the clouds are confined below it. Above this level the lapse rate frequently is markedly less than the saturated adiabatic (and there may even be an inversion), so that cloud air which rises beyond it quickly loses buoyancy and it is not far below the level reached by the tops of the tallest clouds (Fig. 7.16). On other occasions the lapse rate becomes only a little less than the saturated adiabatic, so that the buoyancy is only gradually reduced and reversed, and the peak heights reached by some clouds may be as much as 2 km or even more above the top of the cumulus layer as it has been defined. It is therefore necessary to distinguish a further layer, the *cumulus tower layer*, which is another kind of transition layer, separating the layer of convection from the free atmosphere above.

From a number of occasions when soundings and simultaneous cloud observations in some detail have been made, it appears that the mean lapse rate in the cumulus layer usually has about the same relation to the saturated adiabatic: at the top of the layer the temperature is about $2\Delta T/3$ higher than that corresponding to dry adiabatic ascent from the bottom of the layer (the level of the cloud bases), where ΔT is the difference in the temperatures produced at the top of the layer by dry and by saturated adiabatic ascents (Fig. 7.17). Thus the *mean* lapse rate in the layer is nearly halfway between the dry and the saturated adiabatic lapse rates.

7.6 The arrangement of convective circulations

Convective circulations shown by radar; mantle echoes

The intense, small-scale motions in the superadiabatic layer draw kinetic energy from both buoyancy and the large-scale flow near the ground. The adiabatic and cumulus layers are also strongly turbulent in and near the regions of vigorous ascent. The circulations penetrate some distance into the very stable stratification above the layer of convection, and a considerable part of their kinetic energy may be transmitted upward and away in the form of gravity waves. However, it is probable that most of their kinetic energy is dissipated in small-scale motions in the superadiabatic layer, within regions of ascending air, and in the regions where it spreads out beneath the transition layer or the top of the cumulus layer. The kinetic energy is transferred to motions of successively smaller scale, until eventually it is transformed into heat by molecular viscosity, or converted into potential energy during the associated mixing with the stable air and the gradual upward extension of the layer of convection.

The small-scale motions are accompanied by fluctuations of potential temperature and water vapor density, throughout a whole range of scales from an upper limit of order 100 m to a lower limit of order 1 cm set by the intensity of the motions and the molecular viscosity. There are corresponding fluctuations in refractive index, the most pronounced occurring in regions where there are large mean gradients of water vapor density, that is, in the transition layer or at the sides and tops of cumulus in the low troposphere. (A change in temperature of 1°C alters the refractive index by between 1 and 2 N units, whereas a change in mixing ratio of 10^{-3}—about equally probable in warm air—alters it by about 6 N units; Chapter 2). Consequently, it is the radiation scattered back from these regions which is most readily detected. A radar is sensitive to variations of refractive index averaged over a distance of half of the wavelength in the direction of the beam. The theory of the intensity of echoes received by centimetric radars from such variations in a region of turbulent motion is based on the reasonable assumption that the motions of such small scale lie within the inertial subrange of the similarity theory of turbulence, in which fluctuations are homogeneous and isotropic, with properties determined by the average rate of dissipation of turbulent kinetic energy (26). Theory and experience show that echoes can be detected only by radars of unusually high power, using large antennas (narrow beams) and long wavelengths—several centimeters or more, above the limiting scale at which the intensity of the turbulent motions is substantially reduced by molecular viscosity. The radar reflectivity η of turbulent air is then hardly dependent upon the wavelength λ, distinguishing this kind of echo from that due to backscattering from small particles, for which η is proportional to λ^{-4}. The echoes from turbulent air are weaker than precipitation echoes displayed on conventional weather radars, having the intensity of those in which the radar reflectivity factor Z is about 1 $mm^6 m^{-3}$, or less. Those from cumulus were first noticed during systematic search for precipitation echoes forming in individual growing clouds held under visual survey at very short ranges (Pl. 8.4). When it was established by double-theodolite survey that echoes could be observed from the sides and tops of the cumulus they were called *mantle echoes*, because they cloaked the clouds and it was desirable to distinguish them from precipitation echoes and the *angel echoes* commonly received from insects and birds. [On one occasion when echoes were observed with a radar using a wavelength of 3.2 cm the radar reflectivity was shown to be too great to be associated with fluctuations of refractive index, and was attributed to abnormal and widespread concentrations (of order 10^{-3} m^{-3}) of airborne ants, whose presence near the ground at the radar site was a nuisance (28). It does not seem possible that passive airborne insects should after some time be distributed other than uniformly in the layer of convection, but if purposeful predatory birds exploit and become concentrated in the regions of ascending air, as commonly happens in sea-breeze fronts, they can make a dominant contribution to the radar echoes.]

Examples of echoes associated with turbulent convection are shown in Pl. 7.1. The echoes on an RHI display usually appear as shallow domes which outline the upper parts of the circulations, often at short ranges fusing into a continuous undulating band of echo which locally extends down farther toward the ground (Pl. 7.1*g*). On PPI displays made with the radar scanning at a small elevation angle so that the beam intersects the upper parts of the domes, the echoes appear as roughly circular or oval rings, irregularly distributed on occasions of light winds or little wind shear, but otherwise usually in rows aligned approximately in the direction of the wind (27; Pl. 7.1*h, j*). The whole pattern moves with the wind, but usually is not steadily maintained: detailed observations have shown the development of individual circulations, during which domes rise and expand radially at speeds of about 1 m sec^{-1} before their tops reach a maximum height; subsequently the tops subside slowly while the domes separate from lower echo and continue to broaden before gradually weakening and eventually disappearing after 10–20 min (28). The rings seen on the PPI displays have rather irregular outlines and have a range of sizes, perhaps partly because the domes are intersected by the radar beam at a variable height or at a variable stage in their development, but generally the separation of adjacent circulations appears to be about twice the height reached by the tops of the domes and more nearly three times as great as the average height of the layer of convection.

It is likely that unless the surface temperature is rising rapidly (by at least a few degrees Celsius per hour), the patterns of adjacent circulations with all dimensions only a few kilometers are replaced by rows when cumulus develop which are more than about 1 km tall. This is because much of the air that leaves such clouds enters the clear spaces of the cumulus layer with a potential temperature as much as 2°C higher than that which characterizes the adiabatic layer. It cannot then return to near the surface within the period of about an hour which is set by a horizontal dimension of a few kilometers and a relative horizontal velocity of about 1 m sec^{-1}. Rather, typical rates of radiative heat loss or of afternoon surface temperature change over land, implying that the return requires a horizontal (relative) path of at least tens of kilometers, more readily provided by descent in the spaces between long rows of clouds. The tentative scheme of the circulation in the vertical which is given in Fig. 7.15 may then better represent the projection upon a section across one of a pair of rows along the general direction of motion.

Cumulus rows

Almost always over the ocean (30), and often over land (particularly early and late in the day, perhaps because then the topographical influences are least effective; 31), cumulus are assembled into long parallel rows lying approximately along the wind (Pl. 7.7–7.10). The rows are not always regularly arranged, but the clear spaces separating them are usually between one and two times as wide as the rows, and the distance between the axes of the rows is between two and three times the depth of the layer of convection. Usually there are individual clouds in a row, separated by distances about equal to the diameter of a cloud at its base, showing that two distinct scales of motion are present. Occasionally, however, clouds in a row, especially the larger, are joined together more or less continuously near their base level. A row may then mark a continuous strip of updraft which sailplane pilots can exploit to make fast cross-country flights of up to several hundred kilometers.

When a row first forms several small clouds often appear nearly simultaneously along its length, and it may extend by the addition of other individuals at either end, showing that a circulation previously existed in the dry layer independently of cloud formation, which probably resulted from its intensification or deepening. The presence in the adiabatic layer of such circulations, longitudinal "roll vortices," can be inferred from a variety of evidence, including helical paths followed by constant-volume balloons (32), variations of wind observed near the surface (33), especially as cloud rows pass sideways overhead (34), and the common tendency for the tops of small members of a row of clouds to have an obvious relative motion across the row. It is not uncommon for a row to consist of separate elongated clouds which increase in size toward the center of the row (Pl. 9.2*b*). The relative motion across the row seems more likely to be that in the upper part of a roll vortex than to represent an unusual large-scale shear of wind near the top of the layer of convection. Rows of small clouds have been observed to move as a whole more slowly than the individual clouds within them, and to have one end which extends upwind, sometimes even relative to the ground. In contrast, a long row which forms over an obviously favorable topographical feature has one end fixed, extends steadily downwind of the source while developing, and may be isolated.

Rows of cumulus characteristically occur in winds which are at least moderately fresh. A corresponding organization of the updrafts in the adiabatic layer during convection near coasts is shown by soaring gulls (35): in light winds they soar in clusters by circling in updrafts which evidently are narrow and carry the birds downwind, whereas in fresh winds they are able to soar while flying steadily upwind, and are sometimes distributed in narrow vertical sheets lying along the wind direction.

Cumulus convection over the sea generally requires air to be moved briskly toward warmer water. Accordingly, cumulus in rows are widespread over the oceans and are probably the most prevalent form of atmospheric cloud. Usually some rows are better developed than others, and frequently they waver in direction and bow or fork (Pl. 7.9), sometimes to form a cellular pattern of larger horizontal scale, discussed later.

On the basis of observations over New England (36) it was suggested that rows occur only when the curvature of the vertical wind profile in the layer of convection exceeds some value, at which the tendency of air displaced vertically to preserve its vorticity suppresses circulations in vertical planes along the direction of the flow. The required profile is characteristic of situations in which there is a gradient of mean temperature across the flow such that a thermal wind (Section 9.2) is opposed to the geostrophic wind; the wind speed then often reaches a maximum in the layer of convection. (It increases with height in its lower part because of friction over the surface and, because of the large-scale variation of geostrophic wind associated with a temperature gradient across the flow, decreases with height in its upper part. If, as usual in convection over the sea, the large-scale flow is toward generally warmer regions, the thermal wind has also a component across the flow which implies that the geostrophic wind backs with height. On the other hand, in the convection layer the effect of friction over the ground is to deflect the air near the surface toward low pressure, and thus ordinarily to produce a marked veer with height. If this also is opposed by a component of the thermal wind, the flow throughout the layer of convection may be almost uniform in direction.)

However, although in special observations in Florida maxima of wind speed and uniformity of wind direction in a layer of convection were frequent, no definite relation could be found between the occurrence of cumulus in rows and any property of the wind profile (31; in the period of observation days of arrangement into rows, or rows with some bands of larger clouds, were about as frequent as days when the cumulus showed no obvious pattern). Since all the winds were observed during routine soundings at stations on the coast they may not have been representative of the region inland where the cumulus which were studied developed during the day; perhaps for the same reason the rows were often aligned at a large angle to the reported wind directions, especially in the lighter winds (sometimes the degree of organization of the clouds into rows, and the alignment of the rows, both vary from one part of the peninsula to another). Moreover, in a subsequent study balloons released into fields of developing cumulus indicated winds which varied with time of day, and with position relative to the clouds (by up to $\pm 50°$ in direction and ± 3 m sec^{-1} on some days). Evidently the local winds associated with the cumulus circulations have speeds comparable with those of the general wind if the latter is only moderately strong, so that representative values are difficult to determine. In the earlier study it was found that when the presumed mean wind speed exceeded about 5 m sec^{-1} the cloud rows were on the average almost parallel to the mean wind direction at the surface, and backed from it by about 15° at the level of the cloud bases.

In aircraft observations of cumulus in the trade winds, where the flux of sensible heat is small, upward motion of air below clouds is difficult to detect, being much weaker than is typical over land in sunshine, and bumpiness in flight near the level of cloud bases is not significantly greater below clouds than in regions well away from them (8). Frequently, moreover, the cumulus in such regions develop narrow towers whose shape, like that of castellanus towers (Appendix 8.8), implies that the pronounced updrafts develop in clouds formed by a much slower ascent of air. It appears that the speed of ascent of air into clouds over the ocean is probably substantially less than the few meters per second typical of ordinary convection over land.

The slow ascent below the cumulus and their arrangement in rows may be due to an instability characteristic of a boundary layer in a rotating fluid, which generates long vortices with horizontal axes aligned approximately along the direction of the flow. The presence of instabilities of this kind has been demonstrated in laboratory experiments with steady flows of neutrally stratified layers of water and air over rotating plates (37, 38). In some experiments with water, for example, the vortices have been made visible by injecting dye, and found to be aligned in a direction backed from that of the free flow, on the average by about 14°.

The generation of such vortices depends on the nature of the flow and the stratification in the layer affected by friction. In particular it requires the presence of an

7 Cumulus convection

inflection in the profile of the speed of the mean flow projected upon a vertical plane. In the simple theory of steady large-scale flow under friction it is assumed that there is a balance between constant pressure gradient, Coriolis forces due to the earth's rotation, and frictional forces associated with a constant coefficient of momentum transfer k_M (Eq. 7.5). The flow is then represented by a vector whose length varies exponentially with height and whose reference angle varies linearly with height. Close to the surface the flow is directed across the isobars toward low pressure, at an angle of up to 45° (depending on the choice of a lower boundary condition); the flow turns continuously with height, and first becomes parallel to the isobars, with nearly the geostrophic speed, at a height $D = \pi(2k_M/f)^{1/2}$, where f is the Coriolis parameter. In the atmosphere a similar variation of wind with height is often found with a value of D of about 1 km (implying that in middle latitudes an appropriate average value for k_M is about 5×10^4 cm^2 sec^{-1}). The theoretical vertical profile of the component of the wind speed normal to the geostrophic wind has a point of inflection at a height of about $D/2$, and for a sufficiently nearly neutral stratification (small mean value of the Richardson number Ri) the flow can be anticipated to be unstable, permitting the growth of disturbances in the form of roll vortices which draw their kinetic energy from that of the mean flow.

Perturbation analyses and numerical calculations show that in a neutrally stratified fluid with a geostrophic speed U, two-dimensional vortices may arise when the Reynolds number, defined as $\text{Re} \equiv UD/k_M$, exceeds a threshold value of about 10^2, and that the vortices may be stationary or slow-moving with the velocity of flow at about the level of inflection on the vertical profile, and that they should be aligned at an angle of about 15° from the direction of the geostrophic wind (39). In the calculations quasi-steady roll vortices developed, and Fig. 7.18 shows flow patterns, and some trajectories, in an example thought to represent about the most intense which might occur in the atmosphere. The maximum variations in the horizontal wind speeds are over 30% of the geostrophic wind speed, but the maximum vertical velocities are only about 4% of it, that is, only about 0.5 m sec^{-1}. The vertical displacements of air have a maximum near the middle of the layer of height D, and amount to about 100 m near its top.

The relevance of these particular studies to the atmosphere depends on the validity of using the simple model of geostrophic flow under friction with a constant eddy transfer coefficient, and on the effect of a stratification which in general is not quite neutral. In fresh winds the critical value of the Reynolds number may always be exceeded if a coefficient of the magnitude specified above is appropriate. On the other hand, the circulations may be suppressed by an even slightly stable stratification (Ri > 0.03; 39), while during vigorous convection associated with more than a small flux of sensible heat the smaller-scale motion of the kind normal over land in the middle of the day, which draws its energy mainly from buoyancy, must be expected also to be present and to be dominant. When present the roll vortices are probably confined to the layer below the clouds, where they may contribute significantly to fluxes through the adiabatic layer and the near-uniformity of mixing ratio within it. They may be important, not for the formation of clouds, but for their arrangement. However, they probably play a part in the maintenance of shallow cumulus in airstreams returning poleward ahead of cold fronts (Pl. 9.2a, b), and may be responsible for the formation of continuous long rolls of cloud which occasionally are seen over middle-latitude coastal waters in shallow layers of air which are cooled slightly after leaving the land (Pl. 6.3), and for the apparent patchiness of many sea fogs.

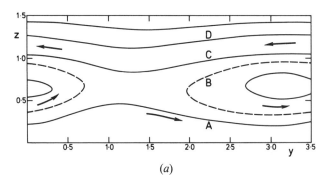
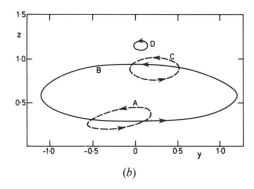

Fig. 7.18 Calculated patterns of flow in a boundary layer circulation, from Faller and Kaylor (39). (a) *Streamlines* of the flow in a vertical plane approximately normal to the direction of the geostrophic wind and across the length of the roll vortices; (b) projections upon the vertical plane of *trajectories* in the moving roll vortices of particles lying on the streamlines marked A, B, C, and D in (a). The trajectory D lies very nearly along the geostrophic wind. Vertical and horizontal distances z and y are marked in units equal to the characteristic vertical dimension D of the boundary layer.

Large cellular patterns of cumulus

Over the sea a row pattern often develops large cells in which the ratio between the diameter and the depth of the layer of convection is of order 10, with an average value of about 30, corresponding to a cell diameter of about 50 km (40). Two kinds of such cells have been distinguished, open and closed, according to whether they are mainly open spaces whose boundaries are delineated by narrow belts or groups of clouds, or are cloudy patches outlined by narrow clear spaces. It has been thought that in the former kind the upward motion is near the edges of the cells, while in the latter it is near their centers, with sink at the edges. Circulations in either sense might be anticipated to produce eventually a complete overcast at the top of the layer, so that from above they would be indistinguishable, and an alternative explanation is simply that in the closed cells the convection has progressed to the stage of saturating the air at the top of the cumulus layer, while in the open cells it has not.

A close inspection of pictures having a high resolution shows that the closed cells consist mostly of the shelf cloud of cumulus convection. Locally within the shelves it may be possible to see the protruding towers of the parent cumulus, which have a higher albedo (Pl. 7.11, 7.12). Sometimes isolated small cumulus can be seen from above as comparatively bright clouds, through thin parts of a shelf cloud. Most routinely transmitted satellite pictures have had a horizontal resolution of a few kilometers, sufficient to show only the clusters of clouds which contain large cumulus, or extensive shelves. Accordingly, it is the closed cells which have been most prominent in these pictures.

If the closed cells consist mainly of shelf cloud, they might be expected to be characteristic of a shallow layer of convection with a small Bowen ratio. Another favorable circumstance may be small values of the general wind speed, and therefore of the wind shear near the top of the layer of convection (which favors mixing with the dry free atmosphere above). Consistently, closed cells are seen mainly over those parts of the ocean where the upward flux of sensible heat is small. Even when the air is eventually moved toward cooler water, as most often happens near the west coasts of the continents, extensive shelf cloud and convection down to the surface may be maintained by a radiative heat loss. Generally convection within the shelf cloud itself produces a pattern in it with a horizontal scale comparable with its thickness.

It is often noticeable from the ground that the shelf clouds of neighbouring cumulus clusters spread toward each other and yet remain distinct, separated by a narrow clear space. Similarly, when a large cumulus grows over land and reaches into a layer of shelf cloud which has persisted from the previous day, or arrived from another region, the spreading shelf which it produces is usually bordered by a narrow ring of clear air.

Estimates of the diameters of closed cells, from satellite pictures, and of the depth of the layer of convection, from ship or island radiosonde soundings thought to be representative of the convection, suggest that the deeper layers have the larger cells, but do not clearly show a direct proportionality. The very large value of the ratio between the diameter and the depth of the cells has not yet been explained satisfactorily. In the classical theory of steady cellular convection (Chapter 1) the ratio is about 3, a result confirmed experimentally, and exhibited also in the spacing of cumulus and of cumulus rows. A larger ratio can be obtained by introducing into the theory transfer coefficients for heat and momentum which are larger in the horizontal than the vertical, by at least one order of magnitude (41). However, there is no useful theory of the required anisotropy of the transfer coefficients, nor any independent observational evidence of it. Other suggested explanations involve the influence of rotation of the flow, and the control of the ratio between the horizontal areas of ascending cloudy air and descending clear air (between clouds) by a variable stable stratification in the clear air (42). It seems more probable that the increase of horizontal scale is due to the intervention of downdrafts from some cumulus large enough to produce showers (which over the sea demands a thickness of only about 2 km; Table 8.7). The cool downdrafts spread out over the surface and temporarily interrupt convection sufficiently deep to produce cloud. The scale is then rather to be regarded as a minimum associated with cumulonimbus.

Intermediate-scale circulations associated with topographic features

Local variations in surface roughness, albedo, and emissivity, in conductive heat capacity and moisture content, and in height and angle of inclination to the solar beam produce near the surface complicated patterns in the distribution of virtual potential temperature θ_v. These are most marked in light winds and alter with change in the general wind and the time of day. The variations in θ_v are introduced into the adiabatic layer, generating *intermediate-scale circulations* (Chapter 6) on a great variety of scales, in which air ascends in and above regions of high θ_v, and subsides over neighboring regions. The subsidence over such regions causes an increase of θ_v in and above the upper part of the adiabatic layer (where the stratification is stable); consequently the air ascending there in the local small-scale circulations loses its buoyancy more rapidly, and does not reach the levels attained elsewhere: the adi-

7 Cumulus convection

abatic layer is shallower, and in its circulations the ascending air is less likely to become saturated. Large variations of θ_v near the surface, such as those associated with hills which rise well above a surrounding plain, or with a boundary between land and water, produce pronounced intermediate-scale circulations manifest by the familiar mountain and lake or sea breezes. Aloft their presence may be strikingly shown by the clustering of cumulus, often obviously over the high ground or in a belt near the edge of the lake or sea breeze, and frequently by the complete absence of cumulus in neighboring regions.

The circulations which have been most carefully observed and studied are those associated with sea breezes, which often have a dominating influence in determining regions of thunderstorm formation even over large land masses, such as western Europe, which have an extensive seaboard. When the large-scale wind is light the sea breeze on a sunny day is observed to begin early in the day near the coast; subsequently the circulation increases in depth and horizontal extent, and its intensity reaches a maximum in mid-afternoon at about the time of the afternoon maximum of surface temperature inland. The landward flow at low levels attains a speed of as much as 10 m sec^{-1}, and has above it a deeper and weaker return flow toward the sea which extends up to a height of 1–2 km (generally to about the level of cumulus bases inland, above which the warming produced by the convection is very small). The circulation persists into the evening, and the flow near the surface from the direction of the sea may penetrate inland as far as 50 km from the coast in middle latitudes, and considerably more in lower latitudes. When the large-scale wind is moderately strong and onshore the sea-breeze circulation is weaker and may be difficult to detect; on the other hand, in a fresh offshore wind the circulation may be intense, but does not extend so far inland.

A land breeze which develops overnight is a much less intense circulation since the cooling of the land surface is communicated to only a very shallow layer of air. Nevertheless, in favorable locations, such as bays which enter land masses (43) or straits, it may be responsible for the development of offshore cumulus or even thunderstorms during the night or early hours of the morning.

The general features of the sea-breeze circulation have been reproduced in numerical models (e.g., 44). They also show slight sinking motion inland of the zone of pronounced ascent (a result supported by the experience of sailplane pilots), adiabatic warming and increased stability in the air which sinks over the sea, and the tendency to produce a front at the leading edge of the flow from the sea, over which the upward vertical motion is concentrated, especially when the general

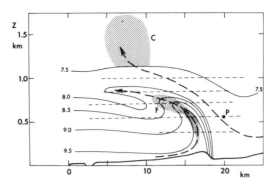

Fig. 7.19 Distribution of cloud and mixing ratio (isopleths labeled in g kg^{-1}) in a vertical plane lying from north to south through the south coast of England, and inferred streamlines (from ground and flight observations, along the paths shown as dashed lines, by J.E. Simpson). Distance inland from the coast and height z are indicated in km.

The stippled areas marked C and F are the regions occupied, respectively, by a main band of cumulus, formed by the ascent of air from landward of the sea-breeze front, and lower ragged clouds formed by the ascent of cooler and moister air from the seaward side of the front. These clouds are shown in Pl. 7.16, a view to the south from the position marked P. The finer stipple shows a region of haze, probably due to a high concentration of aerosol particles acquired during the flow of the air across a coastal town.

wind is offshore. The sea-breeze front and the region of pronounced vertical motion over it are well-known to sailplane pilots. Frequently their position is conspicuously marked by a belt of large cumulus when there are only scattered small clouds farther inland, and perhaps none at all to seaward (Pl. 7.14, 7.15; Fig. 7.19–22). The sea-breeze front usually moves inland at a speed which is fairly regular when measured over hourly intervals, and only about half the speed of the sea breeze behind it. At the ground a major part of the transition between the conditions typical of the

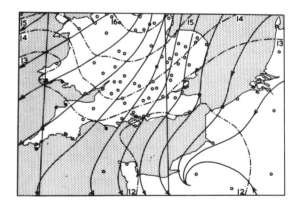

Fig. 7.20 Streamlines of the surface wind and sea level isobars (dot-dash lines labeled in mb above 1000 mb) over England, 0600 GMT, 18 June 1957. The positions of reporting stations are shown by circles. Later sea-breeze fronts developed, shown in Fig. 7.21.

Fig. 7.21 The flow near the surface at 1200 GMT, 18 June 1957. The number of observations available was nearly twice as great as at 0600 GMT. (The points marked H and A show the positions of the soundings of Fig. 3.13 made on the following day.)

inland air and of the cooler sea breeze often is concentrated into a horizontal distance of about 1 km. The transition zone aloft is also rather narrow (Fig. 7.19); the sailplane pilot usually finds that the width of the zone in which he can soar without interruption is little more than 1 km wide and sharply bounded by a turbulent region on its seaward side. The depression of the dew point near the surface is usually less in the sea-breeze air than in the air inland of the front, and the ascent of some of the sea-breeze air is commonly shown by the formation of cloud as much as 0.5 km below the base level of the main belt of cumulus over the front. This lower cloud is usually in ragged and tenuous patches here and there along the front, often appearing

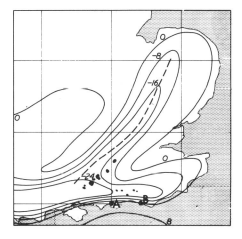

Fig. 7.22 Isopleths of the horizontal divergence of the surface winds at 1200 GMT, 18 June 1957 (labeled in units of 10^{-5} sec^{-1}), measured over areas of about 10^3 km^2. The small black areas represent radar echoes at 1415 GMT, from the first cumulonimbus of the day. Plate 7.14 is a panoramic view looking landward from the position over the coast marked A, at 1400 GMT, and Pl. 7.15 is a view west-northwest from over the coast at the position marked B, at 1422 GMT.

The arrangement of convective circulations 7.6

as an irregular fringe below the bases of the higher clouds (Pl. 7.16) but occasionally thickening into clouds which look like ordinary cumulus (Pl. 7.17).

The zone of pronounced updraft at the sea-breeze front is exploited not only by sailplane pilots but also by birds, particularly swifts, which feed on insects carried aloft in high concentrations from close to the ground. Over England in summer an average concentration of swifts at the front between about 0.5 and 1 km above the ground has been estimated from observation in a sailplane to be about 3×10^{-6} m^{-3} (45), and appreciably higher values are likely to occur locally, sufficient to be detected at moderate ranges by powerful centimetric radars, so that their displays show the positions of the fronts by lines of echo (Pl. 7.1).

Sea-breeze circulations, like land breezes, are more pronounced where there are hills near the coast, since they increase the horizontal differences of potential temperature between land and sea, and over broad promontories. The effects upon cumulus distribution are very obvious near mountainous islands in the trade winds. Overnight the clouds are usually massed over the windward slopes, while the lee slopes are clear; with the development of the sea-breeze circulation during the day, cloud-free clear rings develop around the islands, extending out to sea a distance comparable with the width of most islands, and up to between 50 and 100 km from large islands (46; Pl. 7.18). Similar but sometimes even broader clear zones form around larger land masses, but may be due to the upwelling of relatively cool water near the coasts (47). In the east of the oceans, where the stratification of the air above the cumulus layer is very stable, the island cumulus remain small, with the unfortunate effect that no showers form to augment the scanty rainfall (48), but in a more favorable location in the large-scale circulation the cumulus may often be sufficiently large to produce abundant rainfall, as on Hawaii (49), while in the west of the oceans, for example, over the Caribbean islands, it is common for thunderstorms to develop during the afternoon. Intermediate-scale circulations which produce large cumulus or cumulonimbus may be intensified by the release of latent heat. Where they are variable in space and time they greatly complicate the interpretation of the stratification observed in a place downwind, where air modified by convection may arrive at a variety of levels above the top of the local layer of convection.

Sometimes clouds first appear just downwind of islands, and evidently their formation is not solely due to the islands behaving as a heat source, but is associated with a disturbance which it introduces into the field of motion, perhaps associated with a change of friction at the surface. Such cumulus may persist over the sea to form a line as much as 100 km long (Pl. 7.19). Cloud lines are particularly persistent if convection is in

7 Cumulus convection

progress also over the sea, and may show by their curvature *wake eddies* in the lee of islands. Such wake eddies are occasionally also visible (even in the routine satellite pictures) in extensive cumulus and shelf clouds formed over the sea, downwind of islands or capes of height comparable with the depth of the layer of convection (Pl. 7.20). The eddies appear analogous to those in the vortex-streets formed by the vortices shed from obstacles in laboratory flows at high values of the Reynolds number, quantitatively if the molecular viscosity is replaced by an eddy viscosity with the magnitude of about 10^7 cm^2 sec^{-1} for the transfer coefficient (50). Unlike the intermediate-scale circulations produced by irregular surface heating, the wake eddies draw their kinetic energy from that of the mean flow. They may persist very far downstream, for example, over 500 km (corresponding to a lifetime of about 12 hr) beyond the island of Madeira (width about 40 km). [In the lee of this island Sir Francis Chichester on his global voyage experienced sequences of lulls and periods of variable wind which made the handling of his yacht very troublesome for several hours (51, p. 32). In the times of commercial sailing ships the waters in the lee of large islands were avoided because they were notoriously liable to such disturbances.] Wake eddies of even larger scale are observed in satellite pictures to extend from Iceland as far as Ireland. Similar low-level eddies are probably not recognized when there are no clouds to make them visible.

7.7 Experimental forms of convection

A general theory of convection in an atmosphere heated from below should relate initial conditions and the rate of heating to the depth of the layer of convection which develops, its stratification, and the distributions of velocity, temperature, and water substance (in the vapor and condensed phases) in the circulations which develop in it. Mainly because of the turbulence of variable intensity which characterizes the circulations, the construction of such a theory is extremely difficult, and so far only tentative progress has been made. For the purpose of introducing convection into the modeling of the large-scale atmospheric motion, very simple representations based on an assumed stratification in the layer of convection have so far seemed fairly satisfactory, except in the tropics where the (cumulonimbus) convection frequently occupies the whole depth of the troposphere and has a less clearly understood relation to the kinds of large-scale motion which occur there. The need of a detailed theory has therefore only recently seemed pressing, as attempts are made to improve and extend numerical prediction.

On the other hand, for two or three decades there have been intensive studies of the properties of individual convection clouds, because of the importance of the phenomena associated with the larger (e.g., rain and hail) and the prospects of modifying them artificially. Because of the small duration of many clouds which are well separated with sharply defined boundaries it is reasonable to regard the clouds and their "updrafts" as distinct from a practically inert and unchanging "environment" which they pass through. This point of view may have been encouraged by the simplicity of the parcel theory of convection, and by the introduction of the graphic description "penetrative convection" to distinguish the rise of cumulus towers from "cumulative convection," by which was meant large-scale slope convection (52). Although such a concept can hardly be maintained for long, and indeed is difficult to define, considering that air evidently passes through the boundaries of cumulus, just as of clouds in general, it can be closely represented in experiments with fluids in the laboratory. Since these experiments give some insight into the manner in which air does pass through cloud boundaries, and have provided some concepts and terminology used in interpreting cloud behavior, their description here precedes a discussion of the observed properties of clouds, awareness of which will be necessary for the construction of more realistic theories.

In the laboratory the kinds of motion which arise when buoyant fluid moves through neutral or stably stratified surroundings are conveniently studied by introducing into a liquid a source of a miscible liquid of different density. The motions which develop have characteristic forms, and can be assumed to be similar to those which occur in the compressible atmosphere, providing that the Reynolds number is sufficiently large for turbulent motion to develop, that the local differences of density are small compared with the density of either fluid, and that attention is given to the region away from the neighborhood of the source, where the mechanical action of the device for liberating the buoyant fluid usually causes some unrepresentative motion. Two principal forms of motion have been studied, the conical *plume* which develops over a maintained source, and the *thermal* which develops from a virtually instantaneous source. In both the motion in and near the buoyant fluid becomes turbulent, and it mixes with the surrounding fluid, so that if it is marked with a cloud of small particles or dye, the cloud increases in size with distance away from the source, and becomes more tenuous. It is the similar mixing which is responsible for the large departures from adiabatic changes of state that are observed inside convection clouds.

The buoyant plume

Both the plume and the thermal in neutrally stratified surroundings are modeled with the simple assumptions that there is a steady mean motion, that distributions of mean velocity and mean buoyancy in horizontal sections have a similar form depending only on the height above (or below) the source (the only fundamental linear dimension), and that the small-scale mixing motions are characterized by velocities which at a particular height are proportional to a mean vertical velocity there (say that on the axis of the plume). In the application of the assumptions it is necessary to prescribe the form of the horizontal profiles of mean velocity and buoyancy force; for a simple illustration they can be supposed uniform across a plume rising through an incompressible liquid of constant density ρ from a constant point source of weight deficiency, equivalent to a constant point source of heat in a neutrally stratified compressible atmosphere (Fig. 7.23). If w and ρ' are the mean vertical velocity and the mean density of the plume at the height z and the radius r from the axis, then considering the nondivergence of the flow and the conservation of momentum and mass

$$\frac{d}{dz}(\pi r^2 w) = \alpha \cdot 2\pi r w$$

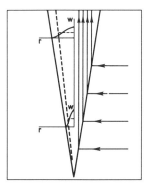

Fig. 7.23 The axially symmetric buoyant plume in surroundings with a neutral stratification. On the right of the axis are streamlines of the mean flow, and on the left are profiles of the mean vertical velocity w at two levels. The solid lines show the approximately Gaussian profile which is observed, in a plume forming a cone with a semi-angle of about $9°$, and the dashed lines show the hypothetical uniform profile and narrower plume (with the same upward flux of density deficiency) which is considered in the text.

$$\frac{d}{dz}(\pi r^2 w \rho') = \pi r^2 g(\rho - \rho')$$

$$\pi r^2 w g(\rho - \rho') = \text{const}$$

where α is the constant ratio of the inflow velocity at the edge of the plume to the mean vertical velocity within it. If the densities within and outside the plume differ only slightly, the density ρ' on the left of the second equation can be replaced by the constant density ρ without significant error, and the equations can be rewritten as

$$\frac{d}{dz}(r^2 w) = 2\alpha r w \qquad (7.14)$$

$$\frac{d}{dz}(r^2 w^2) = \frac{r^2 g(\rho - \rho')}{\rho}$$

$$= \frac{Q}{w} \qquad (7.15)$$

where

$$Q = \text{const} = \frac{r^2 w g(\rho - \rho')}{\rho}$$

The solutions to these equations can be written

$$r = az$$

$$w = \left(\frac{b}{a}\right) z^{-1/3}$$

$$\frac{g(\rho - \rho')}{\rho} = \left(\frac{Q}{ab}\right) z^{-5/3} \qquad (7.16)$$

where $5a = 6\alpha$, $10b^3 = 9\alpha Q$, relations whose form can easily be derived by dimensional analysis.

Similarly, solutions for the properties of the plume can be found for more realistic horizontal profiles of velocity and buoyancy, and for ascent through stably stratified surroundings, and show satisfactory agreement with experiment (53). If a plume is marked with a dye, its edge is seen instantaneously to have irregular, sharply defined protuberances associated with individual small-scale motions or eddies; however, if a picture is made with a time exposure over a sufficiently long period, the edge appears diffuse and has to be defined arbitrarily. It is found experimentally that the form of the horizontal profiles of mean properties across the plume is closely similar to a normal distribution curve centered on the axis. If the edge of the plume is taken to be at a radius where the mean value of the concentration of dye, the vertical velocity, and the density deficiency in the plume are $1/e$ of their axial values, then in neutrally stratified surroundings the semi-angle of the cone of the plume is about $6\frac{1}{2}°$ ($5/6\alpha = 9$); using the larger radius at which the fractional mean values are considerably smaller, and which is a more obvious boundary to the eye, the semi-angle is about $9°$

7 Cumulus convection

($5/6\alpha \approx 6$). About the same value is appropriate for an equivalent plume with uniform profiles which has the same flux of density deficiency.

The buoyant thermal

Experimental thermals are conveniently produced by the sudden release into a tank of water of a volume of solution slightly denser than water, marked with a fine precipitate or dye, and suspended just below the free surface. Soon after its release the marked liquid assumes a characteristic configuration, sinking as an expanding knobbly volume. In surroundings of constant density (neutral stratification) the sides of the thermal sweep out a cone with a vertex near the point of release (Pl. 7.2).

Again the experiments confirm the expectation that in neutrally stratified surroundings the size of the thermal, defined as the half-diameter r of its widest part, increases linearly with the distance traveled, so that if the height z of the cap is measured from the vertex of the cone which the thermal describes (the *virtual origin*),

$$z = nr \qquad (7.17)$$

while the value of n appears to be constant for any one thermal, it has been found to vary from one to another between 3 and 5, with a mean value of about 4.0 (54). These variations appeared to arise from small differences in the manner of release, which was by a mechanical arrangement not finely controllable and whose operation and subsequent presence always produced some disturbance in the liquids. The volume V of the thermal, which retains a well-defined outline, is given by

$$V = mr^3 \qquad (7.18)$$

where the constant m also varies from thermal to thermal about a mean value of about 3.0. The speed of travel w_t of the thermal, conveniently defined as the rate of advance of its most readily identifiable feature, the point where its front intersects its axis, has to be a function only of the mean buoyancy force upon the thermal and its size, and must therefore on dimensional grounds be given by an expression of the form

$$w_t^2 = C^2 rg(\rho - \rho')/\rho \qquad (7.19)$$

where the constant C^2 is a so-called Froude number, a ratio between kinetic and available potential energy. Experimentally it has been found (54) that $C \approx 1.2$ (with some evidence that a more nearly constant parameter is $Cn^{1/2} \approx 2.4$).

The characteristic form of the thermal is shown in Fig. 7.24 (55) and Pl. 7.2. The mean motion in its vicinity is rather like that in a vortex ring, but a prominent feature of a marked thermal is the continual development of

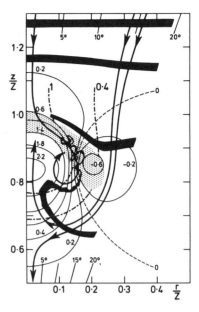

Fig. 7.24 Properties of the axially symmetric thermal in surroundings with a neutral stratification (adapted from reference 55). The distribution of mean vertical velocity w is shown by isopleths of w/w_t (thin lines), where w_t is the upward speed of the thermal cap, against coordinates of z/Z and r/Z, where z is height, r is distance from the axis, and Z is the height above the virtual origin. The region within the cap of the thermal in which vigorous small-scale motions and mixing occur is stippled. Away from this region the flow is comparatively smooth.

The heavy lines are streamlines of the relative flow, showing that fluid lying within a cone of semi-angle about $15°$ (with vertex at the virtual origin) becomes incorporated into the thermal. The dark bands show successive positions of a strip of fluid lying in the path of the thermal.

Considering only the mean flow, particles cannot fall out of the thermal unless their fall speeds exceed some value depending on their position. The dot-dash lines labeled 1 and 0.4, in boldface figures, show the boundaries within which particles of fall speeds less than w_t and $0.4 w_t$, respectively, remain circulating within the thermal.

protuberances over an almost hemispherical leading cap. The protuberances are like small thermals, and as they grow exterior fluid is captured between them and incorporated into the main thermal. The mean circulation carries the mixed material down the sides of the cap and eventually toward the axis of the thermal on its rear side; the vigorous small-scale circulations which produce the protuberances and the mixing are confined to a shallow zone near the cap of the thermal, and elsewhere, especially near the axis in its rear, the flow is comparatively smooth. In the middle of the thermal the vertical velocities are about twice the rate of its advance, so that the fluid in the interior is rapidly brought into the mixing zone, and exterior fluid incorporated there returns from the rear during a period in which the thermal advances by hardly more than a diameter. In a dyed thermal the maximum concentration of the dye, and therefore the greatest local buoyancy, is found in a ring surrounding the region of maximum upward velocity; the maximum local buoyancy is about twice as great as the mean value over the whole volume of the thermal.

The thermal appears as a device which by a combination of scales of motion executes an efficient mixing between the buoyant fluid and its surroundings, while retaining a sharply defined outline to its leading surface. Nearly all of the exterior fluid which is incorporated into the visible thermal enters through the leading surface, with a mean velocity which is everywhere about $w_t/3$. The dissipation of kinetic energy by the small-scale motions apparently results in only about half of the potential energy released by the travel of a thermal appearing as kinetic energy of mean motion.

The starting plume

If a source of buoyancy is maintained for a short period it produces a cone-shaped thermal, or *starting plume*. After the extinction of the source the buoyant fluid eventually acquires the configuration of the ordinary thermal, but for a while it has properties intermediate between those of the plume and the thermal. Near the leading part the motion is like that in the thermal, but it is supplied with fresh buoyant fluid from the plumelike part in the rear as well as exterior fluid from the front. From experiment it appears that in neutral surroundings the two contributions are about equal, and that the rate of spread of the front part of the marked starting plume is considerably less than that of the thermal (53). The rate of fractional change of volume with height is therefore also considerably smaller than for the thermal; whereas for the latter

$$(1/V)dV/dz = 3/nr, \qquad n \approx 4 \qquad (7.20)$$

for the starting plume

$$(1/V)dV/dz = 3/nr, \qquad n \approx 6.5 \qquad (7.21)$$

while the effective fractional rate of dilution with the surroundings may correspond to a value only about half as great.

Buoyant ascent in stably stratified surroundings; erosion

During the ascent of buoyant fluid through surroundings of neutral stability the mixture with exterior fluid acquires buoyancy and also ascends; if, however, the surroundings are stably stratified at least part of the mixture loses buoyancy and some remains behind, so that the ascending volume is said to be eroded. A marked volume leaves a trail (Pl. 7.2b), so that superficially the whole more closely resembles a cumulus cloud than a plume or a thermal in neutral surroundings. The head may shrink, and eventually it ceases to ascend. No systematic studies appear to have been made of behavior in these more complicated circumstances, in which the motion of a thermal at some level can be anticipated to depend not only on its size and mean buoyancy, and the stability of the surroundings, but also on the velocity with which it arrives. Studies of the behavior of thermals during their passage across an interface from one liquid of uniform density into another of greater density will be considered in the discussion of cumulonimbus convection.

Thermals with increasing total buoyancy

In an attempt to simulate the natural circumstance that the liberation of latent heat in an ascending cloud may increase its buoyancy, experiments have been made (56) with thermals in which chemical action during mixing with the surrounding fluid produced many small gas bubbles; these were retained in the thermals and caused their total buoyancy to increase with time. Whereas in the experiments with thermals in neutral surroundings which were previously described their total buoyancy remained constant and their speed was inversely proportional to the distance traveled, with the new arrangement some of the thermals produced and chosen for study had either an approximately constant ascent speed or a constant acceleration. The diameter of these thermals also increased linearly with height, but the mean of the constant n in the relation (7.17) was more than the value of 4 found for ordinary thermals: it was nearly 4.5 for the former kind, and 5.0 for the accelerating thermals, without any clear dependence on the acceleration.

The mushroom clouds of nuclear bombs: atmospheric thermals

When a nuclear bomb is exploded in the atmosphere the air in its vicinity is heated to incandescence, and a very rapidly growing *ball of fire* is produced. In a period of order seconds this reaches a maximum diameter of up to about 1 km and subsequently fades, while a shock wave of rapidly diminishing intensity travels several kilometers outward. Over an interval of a few kilometers well beyond the fireball the rise of pressure at the front of the shock wave is followed by a rarefaction and temperature fall which may be sufficient to cause condensation in air of high relative humidity. A quickly expanding thin spherical shell of *blast cloud* then appears. (Similar but much smaller and even more evanescent clouds can sometimes be seen around the explosions of ordinary bombs falling upon moist terrain.)

7 Cumulus convection

The heated air rises and it cools, mainly by radiation at first and later by mixing and expansion. The rate of rise attains a value of some tens of meters per second, and after an explosion near the earth's surface a height of 3 km is reached within about 1 min. Subsequently the rate of rise diminishes and the cooling becomes sufficient for a cloud to form by the condensation of water vapor. The cloud makes visible a large thermal, which grows in size as it rises, so that in the high troposphere its diameter is a few kilometers (Pl. 7.3). Over land a shallow layer of air above the surface is warmed and filled with dust and smoke, and this rises up toward the thermal in a column which is dark with this debris. Over the sea the stem of the mushroom cloud so produced is usually drawn mainly from the adiabatic layer of the cumulus convection; the stem is a white cloud which then has practically the same base level as any cumulus which are present. The lower part of the stem, the fluted sheaths which may form around it where higher damp layers of air are drawn into the circulation of the thermal, and the cap (pileus) clouds which often form above the head of the mushroom cloud, all have a regular outline indicating the smooth nature of the flow in the outer parts of the circulation, in contrast to the crenellated, "boiling" appearance of the head and the upper part of the stem, where the flow is turbulent (Pl. 7.3).

7.8 Observations of the properties of cumulus clouds

Visual observations; properties of cumulus populations

Individual cumulus can be identified and watched throughout a lifetime which is short compared with the duration of a cloud population. No systematic studies seem to have been made, but the typical lifetime of a cloud appears to increase with the size attained, from a few minutes for the smallest to about 20 min for those about 1 km deep, and an hour or perhaps rather more for those a few kilometers deep. Visual studies of the clouds from the ground are best made with sequences of still photographs, from which acceptably accurate measurements can be made of cloud dimensions (Appendix 7.1), supplemented by time-lapse films. Some of the most informative observations have been made by photographic surveys from aircraft, in which successive pictures taken at intervals of several seconds can be used for photogrammetry.

Plate 7.4 is a picture from a series of photo-reconnaissances made to study the populations of cumulus which form over the almost featureless terrain of Florida during the daytime; the flights were made on 19 summer days (57). The area included in the foreground and middle distance amounts to about 2000 km^2, and in this example and often in other situations, especially over the ocean, it is obvious that the clouds form a rather uniform population with fairly well-defined properties. In this study, for example, it was found that the size distribution of the clouds could be assessed satisfactorily if the area over which they were examined was as large as about 200 km^2, thereafter hardly changing as the area was increased to about 2000 km^2 (58). The clouds tapered upward from level bases whose outlines in a photograph taken looking vertically downward were irregular but generally roughly circular, so that an equivalent circular diameter D for a base could readily be estimated, and the smaller clouds were sufficiently well separated to be distinct, so that an areal number concentration could be found as a function of size. The horizontal displacement of cloud shadows could be measured, so that the height of the cloud bases could be determined. The height t of cloud tops above their bases was found from photographs like Pl. 7.4, taken looking obliquely downward.

The smallest value of the diameter D which could be measured was as little as a few decameters, set by the conditions of illumination and the photographic techniques. However, the analysis could be made because the number of small clouds did not increase indefinitely as their size decreased: evidently there was a definite spacing of circulations in the adiabatic layer which had developed to a stage at which clouds were just produced in their upper parts. (The number of clouds counted, however, almost certainly exceeded the number of distinct circulations, because many near the edges of larger clouds could be regarded as only irregularities.) A different kind of difficulty arose predominantly in the middle of the day and the afternoon, when some of the clouds tended to form clusters; if they remained distinct, although with separations smaller than were general, the sizing and counting procedure remained definite, but sometimes the clouds fused at the base or were separated only by cracks in what appeared to be a larger entity, so that the procedure had to involve subjective judgments. This tendency for the larger clouds to appear in clusters is common (see, e.g., Pl. 7.6). The clusters probably mark intermediate-scale circulations, which in Florida can be expected to be weak, especially in the early part of the day (the clouds examined were in localities carefully selected to have terrain of the most uniform character). Consequently, the clusters there were not so numerous or pronounced as to make the analysis impracticable. Its results were similar to those obtained in a few similar studies made over other southern states and northeastern states of the United States and over Brazil.

It appeared that in typical populations the areal concentration of small clouds ($D < 0.3$ km), expressed as the number in an area of 100 km^2, was over 10^2 in morning hours when the maximum diameter was less than 2 km, decreasing in mid-afternoon to about 10^2 when the maximum diameter was about 4 km, and to about 10 when the maximum diameter was about 10 km (and the population included some shower clouds). The areal number concentration of the clouds whose diameter D fell within a small size interval decreased approximately exponentially with increase of D, rather closely in the morning hours, when the clouds were more regularly spaced. In the afternoon, even over the almost uniform terrain, the form of the size spectrum was appreciably distorted by clustering: there were relatively fewer clouds of intermediate size and relatively more of nearly the largest size, which were mostly constituents of the clusters, around whose fringes were very large concentrations of the smallest clouds. Considering the method of counting, the number concentrations mentioned are not inconsistent with the small clouds of the mornings being associated with individual circulations in the adiabatic layer, spaced at intervals rather greater than the depth of the layer; in the later hours of the day many of the small clouds seem to have been associated with larger clouds or with intermediate-scale circulations leading to clusters containing larger clouds.

Some other results of the analysis over Florida are summarized in Fig. 7.25. This diagram shows a diurnal variation in the height of the cumulus bases typical in flows inland from the ocean; in the middle of the day the thickness of the clouds of moderate size was about equal to or greater than the diameter of their bases, while early and later in the day such clouds were distinctly flatter. The proportion α of a large horizontal area covered by clouds in the middle of the day was between a third and a half, and on some occasions was even more; probably the fraction of the area at the level of the cloud bases which was occupied by rising air was less, considerably so if only upward velocities exceeding 1 m sec^{-1} are considered.

The shape of individual cumulus; cumulus towers

The impulsive character of cumulus convection is strikingly shown by the "towers" of clouds more than about 1 km deep, each of which in turn composes the cloud summits before it sinks aside and is replaced. Towers generally rise on the upwind or upshear side of the cloud, and sink and evaporate on the other side, where the cloud base is less definite, as in Pl. 7.5. In this example the successive towers are in the later stages of their evolution well separated, but on other occasions individual towers are less easily distinguishable, and rarely a large cloud grows by a progressive swelling without any individual towers becoming identifiable.

The form of ascending towers is just like that of experimental thermals (Pl. 7.2): the outline of the surface is sharply defined, its leading part is roughly hemispherical, and it is covered with protuberances whose dimensions range down to a few meters and perhaps less, so that the tower has the cauliflower appearance usually mentioned in definitions of the species cumulus *congestus* (swelling cumulus). The protuberances continually emerge from the cap of the cloud and subside down its sides (Fig. 7.26).

After reaching a peak height a tower subsides; the protuberances become less pronounced or disappear, and the cloud edges less sharp, and eventually ragged and tenuous before the tower evaporates completely, or disappears behind other cloud, at some height not much below the peak height. While a tower is dissolving there is not usually any suggestion that the evaporation proceeds steadily from the outside inward; rather, a stage is often reached at which there are only fragments scattered over almost the whole of the volume previously occupied, which disappear almost simultaneously. It then seems that the descent, rather than mixing with the surroundings, has played the major part in causing the evaporation of a cloud inside which the liquid water concentration was not far from uniform.

From studies in central Sweden (59) it appeared that if attention was concentrated on well-defined towers whose diameters could be measured without much

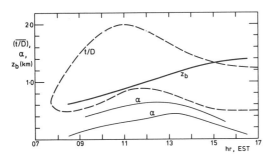

Fig. 7.25 The typical diurnal variation of some properties of cumulus populations over Florida: the height z_b of the cloud bases, the total horizontal fraction α of areal cover (thin solid curves, of which the upper shows the trend on unusually cloudy days), and the average value of t/D, where t is vertical thickness of clouds of base diameter (D) about half the maximum diameter in the population: the dashed line is the envelope within which points representing 100 population samples were nearly uniformly scattered (58).

7 Cumulus convection

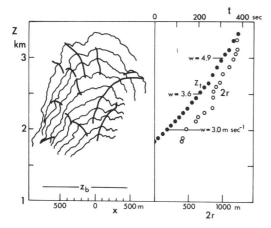

Fig. 7.26 The rise and broadening of a cumulus tower (in central Sweden, 10 August 1955; 59), inferred from time-lapse film and other observations. The left diagram shows the successive outlines of the tower (at intervals of 37 sec) and the diffluent paths of some prominent nodules (heavy lines). The approximate extent of the well-defined part of the cloud base and its height are shown by the line z_b.

The right diagram shows the height z_t of the top of the tower as a function of time t, and its approximate diameter $2r$ as a function of height. The rate of rise of the top increases from about 3 to about 5 m sec^{-1}; there are short-period variations in the rate of rise associated with the intermittent emergence of protuberances from the upper part of the tower.

ambiguity soon after they had emerged from the bulk or flanks of the parent cloud, the half-diameter r varied approximately linearly with height z of the tower summit. The average relation from 10 days of observations on 167 towers observed at various heights over the range from about 1 to 11 km above the ground (and therefore including towers rising from shower clouds as well as cumulus) was

$$(z - z_0) = n'r, \qquad n' = 5 \pm 0.5 \qquad (7.22)$$

where z_0 was the height of a virtual origin which generally was about half the height z_b of the cloud bases above the ground (z_b varied between 500 and 2200 m). This is similar to the relation obtained for experimental thermals in neutrally stratified surroundings, in which the corresponding coefficient n varied from about 3 to 5 with a mean value of 4.0.

Similar observations have been made (60) of the change of size during their evolution of individual small cumulus over a pronounced cloud source, the San Francisco Peaks in Arizona, which rise to just over 1 km above the general level of the neighboring terrain (about 2500 m above sea level). Almost all of the clouds studied increased in diameter as their tops rose, until at some level the rate of rise greatly decreased; subsequently the diameter began to decrease, the cloud outline became less definite, and the cloud base became ragged, and eventually the top of the cloud sank and

it evaporated. Again the cloud diameter during the broadening phase was an approximately linear function of height, with a coefficient n' having a mean value of 4, indicating a virtual origin at some level between the cloud base (at 5300 m) and the mountain top.

The two series of observations described show both that individual cloud towers ascending into clear air broaden with height for a while like experimental thermals, and that the diameter of emergent towers measured over a great range of heights shows about the same linear relation between height and diameter. This kind of relation is qualitatively supported by general experience, and can be recognized in many pictures of large clouds which show towers at a range of heights. In particular, when convection develops in a cloud formed in the middle troposphere (by a large-scale motion system), the cloud towers produced are characteristically much narrower than cumulus towers reaching the same heights from low levels. Their "miniature" appearance indicates that the convection does not extend down to the surface, and the observer reports them as the species *castellanus* (Appendix 8.8). However, it has sometimes not been possible to find any consistent relation between the diameter and height of cumulus towers (61), while some clouds of a population, and occasionally all, do not produce distinct towers. Among clouds over the tropical oceans there is some evidence (62; cloud pictures in other references) for an approximately linear relation, but with a larger value of n' (about 8) than specified in Eq. 7.22; that is, at a given height above the surface the cloud towers are considerably narrower.

The rate of rise of cumulus towers

It is usually found that after the top of a tower has emerged several hundred meters from a large cumulus into the surrounding clear air (i.e., within little more than a minute of its first appearance), its ascent speed decreases, and the ascent ceases after the tower has risen a distance of up to 1–2 km, about equal to its diameter upon emergence. In the studies made over Sweden (13, 59), the rates of rise were measured as soon as possible after the recognition of the towers, but when plotted as a function of height above the cloud base were found to show a large scatter. Figure 7.27 combines all the measurements made during three successive days when the stratifications (Fig. 8.19) and the properties of the convection clouds (Figs. 8.8, 8.17) were very similar; it appears that the maximum rate of ascent attained by towers whose parent clouds were not shower clouds was nearly 6 m sec^{-1}, at a height between 2 and 3 km above the level of the cloud bases.

160

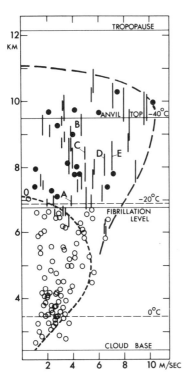

Fig. 7.27 The rate of rise of the tops of towers emerging from cumulus and shower clouds (open and full circles, respectively, determined over intervals of about 200 m from theodolite measurements; and vertical lines, showing mean rate of rise over the intervals indicated, determined from time-lapse films). The values have a large scatter, but seem to have maxima lying within two well-defined curves drawn as dashed lines, according to whether showers were produced by the parent clouds. Since the showers were recognized visually there is some uncertainty about their occurrence in those clouds which barely reached a "fibrillation level" (whose significance is explained in Chapter 8 where the diagram is discussed again), which seemed to separate cumulus from taller clouds in which showers developed. The observations were made on three successive summer days of similar cloud behavior in central Sweden (10–12 July 1955; further discussed in references 13 and 59). The observation marked O refers to the first fibrillating tower of the group pictured in Pl. 8.5; those marked $A-E$ refer to the subsequent glaciated towers identified on the plate.

Flight observations of cumulus properties

A number of studies have been made with aircraft of the internal properties of cumulus clouds: the temperature, the refractive index, the air velocity (usually as inferred from accelerations experienced by the aircraft), the particle nature and size distribution, and the concentration of condensed water. It has not usually been possible to attempt the measurement of all these properties simultaneously and continuously during traverses of the clouds, and most instrumental techniques have had serious limitations (63). Moreover, usually only a single aircraft has been employed, which cannot make more than a few traverses of a selected cloud throughout its existence, and which does not permit the airborne observer a continuous view of its evolution, or even sometimes to be certain that the successive traverses have been made in the same cloud. These circumstances are troublesome because the cumulus properties are very dependent upon position inside it and the stage in its evolution. Ideally several aircraft should be deployed under a controller in the air or on the ground, who can keep the cloud under continuous photographic and radar survey, but such an operation demands elaborate organization and is very expensive (Section 6.3).

In the interpretation of the available data it is desirable to have some model of typical cumulus structure and evolution in mind, which can be based on visual experience, even if it becomes apparent that it is not representative of all circumstances and may even eventually require drastic modification, supplementation, or replacement. The assumption used here is that in the almost universal presence of some shear of wind in the vertical, large cumulus are asymmetrical and consist of an assemblage of a few towers resembling thermals or starting plumes, which continually rise on the upshear flanks of the cloud as a whole, and subside, spread, and evaporate on the downshear flanks, mainly in a layer some distance below a peak level reached occasionally by only the extreme tops of the cloud. This model is illustrated by Pl. 7.5 and 7.6 and Fig. 7.28. It will be assumed that within rising cloud towers the single most important coordinate determining mean cloud properties is the height above the base level. In the figure the individual cloud towers are clearly separated, especially on the downshear flanks, but sometimes when there is a lesser wind shear or apparently a steadier circulation it may be difficult to distinguish towers. [Sometimes, too, the evaporation of the cloud on its flanks proceeds more slowly, and a narrowing "shelf" (of *stratocumulus cumulogenitus*) extends away from the summits. Such shelves may spread over areas several times greater than those covered by the parent cumulus, or by amalgamating they may produce an almost complete overcast over a large region. The shelves frequently survive their parent clouds, for a while indicating their mode of formation by their oval shapes, and after some time acquire a dappling on a scale comparable with their thickness—usually only a fraction of a kilometer—with the onset of a slow convection produced by the radiative heat fluxes. Seen from just above, the tops of the shelf clouds usually are clearly all at about the same level, above which the tallest cumulus towers intermittently rise by up to about 1 km, as in Pl. 7.6.]

7 Cumulus convection

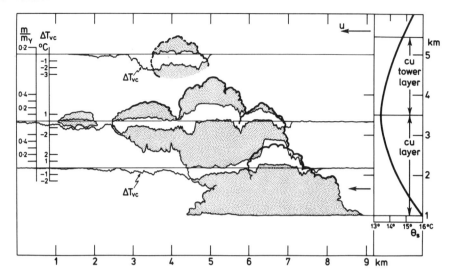

Fig. 7.28 Model of the structure of large cumulus in the presence of some shear in the vertical of the wind u, or when grown in a wind over a persistent source at the ground. The cloud is regarded as composed of several towers, which rise in succession to reach a peak level and form the cloud summits, before subsiding and evaporating on the downwind or downshear side of the cloud. Profiles of the excess ΔT_{vc} of the cloud virtual temperature over the general virtual temperature of the clear surroundings (ΔT_{vc} is proportional to the buoyancy of the cloud air), and of the cloud water concentration m (expressed as a fraction of the value m_y corresponding to adiabatic ascent from the cloud base) are drawn along three horizontal traverses. The uppermost of these intersects a cloud tower at the highest levels occasionally reached.

Also given are approximate scales of height and horizontal distance, and, on the right, a typical vertical profile of θ_s in the clear air above the level of the cloud bases and well away from the cumulus, defining the *cumulus layer* and the *cumulus tower layer*. When shelf cloud is extensive the profile has a sharper minimum at about the level of its tops, which then mark the top of the cumulus layer.

Cloud temperature, liquid water concentration, and buoyancy

Figure 7.29 is a typical set (64) of the instrumental records of temperature, vapor mixing ratio, and concentration of cloud water, which are obtained from horizontal flights through the tops of large cumulus. In this example the air temperature was recorded by an electrical resistance thermometer (housed in a vortex tube designed to prevent cloud particles striking the element), with a relative accuracy of about 0.2°C and a response time of about 0.01 sec, corresponding to about 1 m of flight path. The water vapor pressure, or mixing ratio, was inferred from a simultaneous and about equally sensitive and responsive recording of the refractive index of the air (measured in a duct inside the aircraft), whose variations in practically horizontal flights at temperatures near and above 0°C are due mainly to changes of vapor pressure, changes of temperature and air pressure making only minor contri-

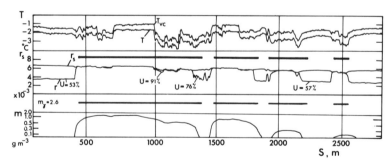

Fig. 7.29 Records of temperature T, mixing ratio r, and liquid water concentration m during a practically horizontal traverse of large cumulus cloud at 525 mb (about 5.2 km above sea level), near Tucson, Arizona, at 1430 MST, 19 July 1955 (64). Also shown are the inferred values of the cloud virtual temperature T_{vc}, the saturation mixing ratio r_s, and the liquid water concentration m corresponding to adiabatic ascent from the cloud base (at 630 mb, about 3.9 km). Some values of the relative humidity U in the clear air are indicated.

The heavy horizontal lines show the intervals (of horizontal distance S) during which the aircraft was in cloud. The cloud was on approach a sharply defined and probably rising tower, entered only about 300 m below its top.

butions, as noted previously. The concentration of cloud water was recorded by an instrument which measures the electrical resistance of a moving strip of paper after it has been exposed behind a slit to the airstream and wetted by the cloud droplets. The instrument had a response time of about 0.5 sec (corresponding to about 50 m of flight path), during which period it sampled about 5 cm^3 of cloud air; it was considered to give the concentration of water in the form of cloud droplets to within about 25%. The cloud virtual temperature T_{vc} was inferred from these records and is also shown in the figure.

It can be seen that the aircraft passed through a principal cloud tower about 1 km in diameter, in which the air near its edges was not everywhere quite saturated, and subsequently through three further and successively smaller cloudy regions of diminishing cloud water concentration, the latter pair of which were wholly colder and denser than their surroundings. These can be regarded as subsiding and evaporating towers of the kind illustrated in Pl. 7.5 and drawn in Fig. 7.28. Characteristically, the boundaries of the towers appear to be more sharply defined on one side, probably the upwind or upshear side on which lay the principal tower of the cloud as a whole. In this tower the air in an inner third had a uniform temperature about $\frac{1}{3}$°C above that in the surrounding clear air, and a buoyancy corresponding to about the same excess of virtual temperature. In two zones inside the cloud edges, on the other hand, the presence of small-scale motions and mixing with the clear air is shown by the marked fluctuations of temperature about a mean value *below* that in the clear surroundings, corresponding to a deficit of cloud virtual temperature and hence to a negative buoyancy, which is especially pronounced on the far side of the cloud tower. The concentration of cloud water reached a maximum of just over 1 g m^{-3}, about half of the value corresponding to the adiabatic ascent of air from the cloud base.

An analysis of the records from numerous cloud traverses at several levels showed that the regions of low and irregularly fluctuating temperature within the cloud boundaries were often very narrow, especially in the cloud towers: altogether more than half were less than 100 m across. At all levels it was common to find in the cloud interiors small regions in which the refractive index was very nearly that at the same level well away from the clouds.

The general range of temperatures measured in clouds and the average dew points observed in the surrounding clear air are represented on a tephigram in Fig. 7.30, together with the local radiosonde sounding. The larger clouds examined by the aircraft were clustered over high ground, and although the two kinds of measurement cannot be compared with much confidence, the

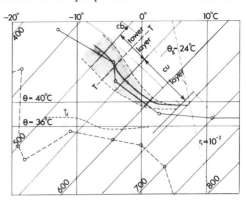

Fig. 7.30 The stratification (shown on a tephigram) on the occasion of the aircraft traverse through cumulus tops represented in Fig. 7.29 (64). [Isobars are labeled in hundreds of mb; two dry adiabatics for $\theta = 36$ and 40°C are drawn, and the isopleth of saturation mixing ratio $r_s = 10^{-2}$. The saturated (pseudo-) adiabatic $\theta_s = 24$°C and another through the maximum temperature encountered at the mean level of the cumulus bases are also included.] The clouds studied were over mountains in Arizona which rise about 1 km above the general level of the plain (about 1 km above sea level), with peaks to about 2 km above. The clouds were small and scattered away from the high ground, but large and clustered over it, with a mean base level of about 3.7 km (642 mb) and towers reaching about 5.6 km (500 mb).

The distributions of temperature and dew point according to the radiosonde released from Tucson at 1400 MST are shown by the continuous and dashed straight lines joining the open circles. The winds up to the 450-mb level were light easterly. Arbitrarily smoothed distributions of temperature and dew point T_d encountered by the aircraft away from clouds are drawn as continuous and dashed curves; inside clouds temperatures measured fell within the stippled region. It will be seen that over the high ground the temperatures and dew points in and below the cumulus layer were somewhat higher than those indicated over the lower ground by the radiosonde, consistent with the presence of an intermediate-scale circulation.

During the afternoon large amounts of shelf cloud formed over the high ground, at the top of the cumulus layer (550 mb), and some cumulonimbus (thunderclouds) with tops at about 12 km (about 200 mb). The full circle labeled *cb* shows the high temperature encountered in the middle of a traverse of one of these cumulonimbus, suggesting the practically adiabatic ascent of air from a cloud base considerably lower and warmer than the mean cumulus base.

somewhat higher average clear air temperatures and dew points found by the aircraft are consistent with the presence of an intermediate-scale circulation producing ascent of air over the high ground and descent over the plain, where the radiosonde sounding was made.

Another interesting set of records from a cloud traverse reproduced in Fig. 7.31 was obtained (64) on an occasion when the top of the cumulus layer was surmounted by a pronounced inversion of the lapse rate and very dry air. The flight through a cloud top in the inversion layer shows it to be markedly colder

7 Cumulus convection

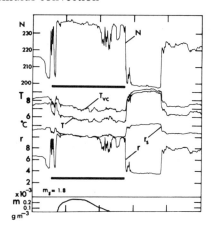

Fig. 7.31 Records of the refractive modulus N of the air at microwave frequencies, the temperature T, mixing ratio r, and liquid water concentration m during a horizontal flight about 200 m below the peak top of a cumulus near Boston, Massachusetts (64). The heavy horizontal lines show the intervals during which the aircraft was in cloud. Also shown are values of the cloud virtual temperature T_{vc}, the saturation mixing ratio r_s, and the liquid water concentration m corresponding to adiabatic ascent from the cloud base, at 812 mb (about 2.0 km).

than the clear air, and, particularly on one side, to have a neighboring and probably descending shell of warm and very dry air. The traverse was made only 300 m below the peak top of the cloud, and the cloud water concentration is very low; the cloud tower was probably in a late stage of its evolution—others encountered contained concentrations up to about half that corresponding to adiabatic ascent from the cloud base. In these moderately large cumulus the strong influence of mixing with clear air upon the cloud properties is shown by the peculiar distribution of cloud temperatures with height (Fig. 7.32): above the inversion the highest cloud temperatures appear to be the result of mixing with the surrounding warm air, rather than of more nearly adiabatic ascent from the cloud base. It also appears that the warm dry air encountered just outside the boundaries of the cumulus towers had probably not descended more than about 200 m. Shelf clouds were observed near the larger clouds; among the cloud clusters the clear air in the upper part of the cumulus layer had a generally high relative humidity, and the top of the cumulus layer appeared as a well-defined haze top.

In the records illustrated in Fig. 7.29 and 7.31 the sluggishness of the response of the instrument used to measure the cloud water concentration is indicated by the smooth form of its record. When the responsiveness is improved fluctuations inside the cloud suggested by noticeable variation of visibility become evident, and the concentration of cloud water may have an obvious correlation with the temperature in the

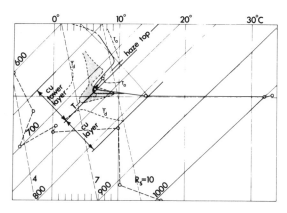

Fig. 7.32 A representation on a tephigram of the stratification on the occasion of Fig. 7.31, when moderately large cumulus towers rose above a marked inversion at the top of the cumulus layer (64). The clouds were observed over uniform ground inland, about 40 km northwest of the local radiosonde station during the period 1400–1600 EDT.

The distributions of temperature and dew point according to the radiosonde released at 1400 EDT are shown by the solid and dashed straight lines joining the open circles. The winds up to the 600-mb level were light westerly. Arbitrarily smoothed distributions of temperature and dew point $\bar{T}_d$ encountered by the aircraft away from clouds are drawn as solid and dashed curves; inside clouds the temperatures measured fell within the stippled region. The dot-dash lines labeled T_a show the average maximum clear air temperature measured just outside the cloud boundaries.

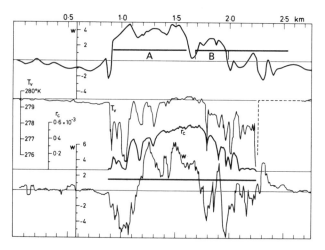

Fig. 7.33 Examples of vertical air speeds inferred from accelerometer records obtained during horizontal traverses of cumulus clouds in aircraft, at levels about 1 km above the cloud base. The lower traces (63) include concentration of liquid water and virtual temperature, from a traverse about 1250 m above the cloud base and 250 m below the cloud top. The upper trace was obtained from a traverse at a height of 1520 m through a cumulus with a base at about 700 m in the trade winds over the Caribbean Sea; the top of the major tower intersected was at about 2 km (72). The heavy horizontal lines indicate flight in cloud; on the upper trace the segment A marks a major cloud tower, and the segment B a minor tower.

expected sense that the warmer air has the higher concentration (63; Fig. 7.33), at least in the cumulus layer and for a small distance into the cumulus tower layer. When droplets are sampled from volumes of about 10 cm³ (over a path length of about 1 m) at intervals of about 200 m on a horizontal traverse of a cloud, the droplet concentration and the total water concentration are found to vary together by a factor of between about 3 and 5, excluding a zone often less than 200 m wide at the cloud edge (corresponding to the zone of strong temperature fluctuation) in which the values of both may be comparatively small (65). In many traversed clouds there are regions in the interior where the cloud water concentration is very low, presumably corresponding to the spaces between the volumes which are or will become the cloud towers. The general features of the records of temperature and cloud water concentration obtained during traverses of large cumulus can therefore be accommodated in the model of Fig. 7.28, which represents the cloud as composed of a few volumes in which the motion is at least qualitatively similar to that in experimental thermals. Most of the available observations suggest that in clouds which are not in a late stage of their evolution horizontal traverses intersect one or two such volumes, within which it appears reasonable to specify mean values of temperature and cloud water concentration, which provide a significant measure of buoyancy, and maximum values which may be greater by a factor of two or more but which usually are encountered over very small regions, not more than some tens of meters across. For application in problems of cloud and precipitation growth it certainly seems desirable in assessing mean values to exclude from consideration regions of low temperature and very small cloud water concentration, which often occupy a considerable fraction of the traversed cloud and even all of it in the late stages of its evolution. It is also reasonable to relate these mean values to height above the level of the cloud base, which would be the only significant parameter if the air in a cloud rose adiabatically from its base.

Figure 7.34 summarizes information on cloud water concentration from several sources which appear to be among the most reliable (65–71), representing as a function of height above cloud base the ratio of $\bar{m}/m$ and $\bar{m}_{max}/m$, where $\bar{m}$, m_{max}, and m are, respectively, the observed "mean" and maximum concentrations, and that corresponding to adiabatic ascent from the cloud base. Near the cloud base the latter ratio may exceed unity, probably on account of the inaccuracy of the measurement of the concentration of cloud water, or of the height above the cloud base. Otherwise all the observations show that the ratio is less than unity and decreases rapidly with the height above cloud base.

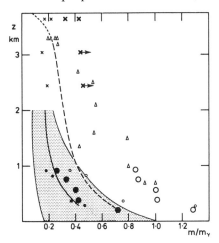

Fig. 7.34 Observed variation with height z above the cloud base of the concentration of liquid water m in cumulus clouds, expressed as a fraction of the concentration m_γ corresponding to adiabatic ascent from the cloud base.

The stippled zone and the central line show, respectively, the approximate range and average values in clouds observed near the eastern coast of Australia, using for m a value representative of the higher measured concentrations (65). The sensor gave values consistent with those obtained by observing local distributions of cloud droplet size in small volumes (66). The dashed line represents the mean concentration observed in cumulus congestus near Leningrad (67). The larger full and open circles are average values using, respectively, the "mean" and maximum concentrations observed in several tens of clouds over the Caribbean Sea (68). The small and large crosses are "mean" and maximum values observed in large cumulus over England, using an instrument which could not measure concentrations exceeding 1.8 g m⁻³ (69). The smaller full and open circles are values using the "mean" and maximum concentrations observed in 17 small clouds on 4 days over the central United States (70). The triangles represent maximum values observed in an allied investigation over the central United States (adapted from reference 71).

Considering that the very low values represented by the left of the stippled area were characteristic of all the clouds examined and perhaps therefore of a late stage in their evolution, and that in the other investigations the mean values were apparently taken over the whole cloud traverse, a representative trend for the values appropriate to the growing parts of large cumulus is probably that indicated by the dashed line representing the Russian observations. This indicates that the ratio $\bar{m}/m_\gamma$ is only about 0.4 at a height of 1 km above the cloud base, and at considerably greater heights falls to a value of little more than 0.2. The mean concentration of droplets has been observed to decrease with height at a roughly similar or even greater rate (65): it is, of course, subject to reduction not only by the admixture of clear air from levels above the cloud base, where the droplets are formed, but also by coalescence.

7 Cumulus convection

The vertical velocity of air in cumulus

According to the experience of sailplane pilots the ascent speed of the air in large cumulus is usually 3–4 m sec^{-1} at cloud base and increases to about twice as much 2–3 km higher. The measurement of vertical air speeds of this magnitude from the accelerations experienced by powered aircraft during horizontal traverses is difficult, but values of 4–6 m sec^{-1} are inferred at heights about 1 km above the level of the cloud base (63, 72) and as much as 15 m sec^{-1} at 3–4 km above (73). In a summer investigation in the foothills of the northern Caucasus measurements were made by radar of the ascent speed of "no-lift" balloons (of nearly zero buoyancy) which entered cumulus after being released from the ground beneath (74). It was estimated that the errors incurred in interpreting the ascent speed of the balloons as the vertical air speeds were less than 30% for speeds of a few meters per second and less than 20% for speeds of about 10 m sec^{-1}. The distribution of the averaged ascent speeds with the height above cloud base in small and large cumulus is seen in Fig. 7.35. From the diagram it appears that the vertical air speed in clouds increased with height above cloud base at the rate of about 4 m sec^{-1} km^{-1}, but after reaching a maximum decreased in their upper parts. Since the level of maximum vertical speed was higher in the larger clouds, it probably corresponds approximately to the height of the top of the cumulus layer, above which the buoyancy of rising cloud air is soon lost and then reversed.

This result, and experience with aircraft, is consistent with the previously quoted visual observations of the

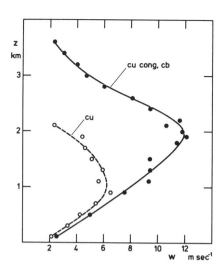

Fig. 7.35 Distribution with height z above cloud base of the average ascent speeds w of no-lift balloons inside cumulus clouds (74). Open circles represent observations from 33 cumulus and cumulus mediocris (small clouds); full circles represent observations from 16 cumulus congestus and cumulonimbus (large clouds).

maximum rates of rise of cumulus tower tops, which generally increased with height above cloud base at a rate of about 2 m sec^{-1} km^{-1}, considering that if the motions in the cloud towers are similar to those in thermals, the speed of ascent of air in their interior should be about twice the rate of rise of the tower tops. (After entering the upper part of a thermal, however, a no-lift balloon or sailplane can be anticipated thereafter to ascend at a lesser mean speed, about equal to that of the thermal top.)

7.9 The interpretation of observed cumulus properties

It is clear from both the visible behavior of cumulus and the instrumental observation of their internal properties that the cloud air is subject to pronounced mixing with its surroundings. It is particularly interesting that on many occasions the diameter of towers which are growing or which are emerging from the general bulk of the parent clouds is approximately a linear function of height, with a constant of proportionality comparable with those found for experimental thermals or the heads of starting plumes in neutral stratifications. The experiments give some insight into the mechanism of the mixing, showing that a rising cloud tower incorporates clear air lying ahead of it, while the observations imply a rate of mixing which may be used, given a sounding of the clear air, to calculate the average internal properties of the tower as a function of height above the observed level of the cloud base, assuming horizontal (isobaric) mixing.

To illustrate the kind of calculations which may be made, and the typical result, we refer to some observations of cumulus convection on three successive days in central Sweden. On each day the behavior of the well-scattered clouds was similar, and the several routine radiosonde soundings made over the whole period differed little among themselves; the mean sounding, seen in Fig. 7.36, is taken to represent conditions in the clear air. From a study (59) of 53 cloud towers, at heights between 2.9 and 10.9 km, it appeared that the height z and the half-diameter r of an emergent tower were well related by the equation

$$z - z_0 = n'r \qquad (7.22)$$

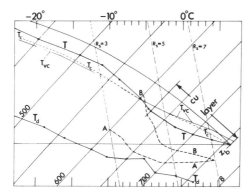

Fig. 7.36 Mean conditions during three successive similar days of cumulus convection over central Sweden (near Östersund, 10–12 July 1955), represented on a tephigram. The mean distributions of temperature T and dew point T_d in the clear air, obtained from five routine radiosonde soundings, are shown by the solid lines joining full circles. The mean cloud base was at 2.5 km above sea level (the level z_b, at 760 mb); dry and saturated adiabatics are drawn through the point representing the clear air conditions at the level of the cloud base.

The level of the virtual origin of the cloud towers was at about 1.5 km. Line AA shows the distribution of clear air dew point required to maintain a state of saturation in air rising from the cloud base level and incorporating sufficient clear air for its volume to increase proportionally to the cube of the height above the virtual origin. Similarly, line BB shows the distribution of dew points required in the same circumstances to maintain a concentration of cloud water equal to one quarter of that corresponding to adiabatic ascent from the cloud base.

If the clear air incorporated is first saturated at its observed temperature, the temperature of the cloud air is shown by the dashed line T_c; the solid line T_{vc} shows by comparison with that marked T the difference between the virtual temperatures in the cloud air (taking into account the presence of condensed water) and in the clear air with its observed dew point T_d. This difference is proportional to the buoyancy of the cloud air, which is at first positive but becomes negative a small distance above the top of the cumulus layer. The tops of the observed cumulus towers occasionally reached above the 450-mb level (at the extreme left of the diagram).

with values of the constants n' and z_0, the height of a virtual origin, estimated on each day, respectively, as $n' = 5.7 \pm 0.7$, 4.8 ± 0.5, 4.8 ± 0.4, and $z_0 = 1.35 \pm 0.25$, 1.7 ± 0.2, and 1.8 ± 0.35 km. The mean value of the latter is about 1.5 km, and the mean height of the cloud base level was 2.5 km. A calculation of the average mixing ratio, as a function of height, in a mass of air which is saturated at the height of the cloud base, and which then while rising incorporates sufficient clear air for its volume to be proportional to the cube of its height above the level of the virtual origin, gives the remarkable result that the state of saturation can persist for only a few hundred meters above the level of the cloud base. This arises because of the dryness of the surrounding clear air, corresponding to a dew point depression of about 5°C, a common value.

Another calculation may be made to obtain the distribution with height of the minimum possible mixing ratio of water vapor in the clear air which will preserve an expanding cloud tower (assuming it to have a temperature no higher than that of the clear air and to contain a negligible concentration of condensed water). The distribution is shown in Fig. 7.36 by the line AA, and it requires dew points considerably higher than those observed. Yet another distribution, shown by the line BB, is obtained by assuming that the cloud contains a concentration of condensed water which is one quarter of that corresponding to adiabatic ascent from the level of the cloud base, and which Fig. 7.34 shows to be a minimum plausible value. The implied relative humidities in the clear air are, of course, still higher, and more strongly emphasize that the observed relation between the height and diameter of the cloud towers cannot be reconciled with the assumption that they mix with the clear air during rise from the level of the cloud base.

Since most cumulus towers apparently emerge from a larger parent cloud, and subsequently do not rise far through clear air, an alternative assumption is that they reach the levels at which they are first observed after rising through saturated air composed of the incompletely evaporated residues of previous cloud towers. These can be supposed to have acquired approximately the temperature (or virtual temperature) of the clear air by vertical displacement, and to contain negligible concentrations of condensed water. Then no evaporation, but only dilution, occurs during the rise of a cloud tower, and the calculated distribution with height of the mean concentration of condensed water is that shown in Fig. 7.37. In the diagram it is expressed also as the fraction f' of the value corresponding to adiabatic ascent, and compared with the fraction f according to the following simple theory.

It is assumed that adiabatic ascent of air saturated at the level of the cloud base produces a temperature excess ΔT_γ above that of the clear air and a concentration $r_{c\gamma}$ of condensed water which both increase linearly with distance above the height z_b of the cloud base (where their values are zero). If the mass M (rather than the volume) of a cloud tower is proportional to the cube of its height z, and all heights are measured above the virtual origin, then

$$\Delta T_\gamma = a(z - z_b)$$
$$r_{c\gamma} = b(z - z_b) \quad (7.23)$$
$$M = cz^3$$

where a, b, and c are constants.

If, at the height z_1, $\overline{\Delta T}$ is the difference between the mean temperature of the cloud tower and the temperature of the clear air, then assuming that the ascent

7 Cumulus convection

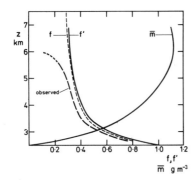

Fig. 7.37 The mean concentration $\bar{m}$ of condensed water in rising cloud towers as a function of height z, according to the calculation made for the conditions represented in Fig. 7.36, assuming that the incorporated clear air is first saturated at its observed temperature (the cloud base height is 2.5 km).

The curves f, f' show values of the ratio $\bar{m}/m_\gamma$, where m_γ is the concentration of water condensed in adiabatic ascent from the level of the cloud base, according to the calculation mentioned (f'), and alternatively according to the simple theory leading to Eq. 7.25 (f).

The thicker dashed curve (observed) is that included in Fig. 7.34; it is considered to provide representative values of the ratio $\bar{m}/m_\gamma$ in large cumulus.

consists of small adiabatic vertical displacements followed by horizontal mixing (without evaporation), and that at every level all the cloud air continues to rise, we have approximately

$$cz_1^3 \overline{\Delta T} = \int_{z_b}^{z_1} a(z_1 - z) \cdot 3cz^2 \, dz + cz_b^3 \, a(z_1 - z_b)$$

or

$$= 3\left[\frac{z_1 z^3}{3} - \frac{z^4}{4}\right] + z_b^3(z_1 - z_b) \quad (7.24)$$

This leads to

$$\frac{\overline{\Delta T}}{a(z_1 - z_b)} = \frac{\overline{\Delta T}}{\Delta T_\gamma} = f = \frac{z_1^3 + z_1^2 z_b + z_b^2 z_1 + z_b^3}{4 z_1^3}$$

or

$$f = \frac{1 + h + h^2 + h^3}{4} \quad (7.25)$$

where $h = z_b/z_1$. The same value of f is obtained for the ratio between the mean concentration of the cloud water and that produced by adiabatic ascent from the level of the cloud base. Accordingly, for a virtual origin about halfway between the surface and a cloud base level at about 2 km, f decreases rapidly with height, becoming a small fraction not much above this level: only 0.37 at 2 km above, thereafter gradually approaching the value 0.25.

From Fig. 7.37 it appears that the value f' of this ratio calculated for the conditions in Fig. 7.36 is close to the corresponding theoretical value f. Moreover, the curve representing its variation with height is remarkably similar to the dashed curve transferred from Fig. 7.34, which appeared to be representative of the mean observed value in growing cumulus. This correspondence seems to give the simple theory some plausibility and suggests that the inferred values of the mean cloud temperature and cloud virtual temperature (included in Fig. 7.36) may also be about correct. Values of the latter can be associated with the observed diameter and maximum rate of rise of cloud towers and introduced into the expression (7.19) for the rate of rise of experimental thermals

$$w_t^2 = C^2 rg(\rho - \rho')/\rho \quad (7.19)$$

written as

$$w_t^2 = C_1^2 rg(T_{vc} - T_v)/T \quad (7.26)$$

(where T_{vc} and T_v are, respectively, the virtual temperatures of the cloudy and the clear air) to obtain a value of the constant C_1^2 for comparison with the value of $C^2 \approx 1.2$ found for experimental thermals. For example, from Fig. 7.36 at 3.3 km $(T_{vc} - T_v) = 1°K$, $T = 275°K$, and $r = (z - z_0)/n' \approx 1.8/5$ km, while from Fig. 7.27 $w_t \approx 4$ m sec^{-1}; accordingly

$$C_1^2 = 16 \times 10^4 \times 275 \times 5/1.8 \times 10^5 \times 980$$
$$= 1.2$$

just equal to experimental value. This agreement is, of course, partly fortuitous, for in the non-neutral stratification the relation (7.26) is not strictly applicable and, for example, could not be used at all in the present conditions above 4.5 km; above this level the inferred buoyancy of a cloud tower becomes negative, but the ascent speed appeared to be about at its maximum and some towers rose a further 2 km or more (Fig. 7.27). Nevertheless, the agreement in the region where the buoyancy is changing only slowly with height adds to the evidence that the cloud towers behave like thermals.

In the calculation of the internal properties of growing cloud towers whose diameters increase linearly with height the constant of proportionality enters only indirectly in that its observed value determines a height for the virtual origin. If, as usual, this is intermediate between the surface and the height of the cloud bases, realistic values for the internal properties of the towers can be obtained only by assuming that they rise through air which is saturated, rather than in the low relative humidity commonly observed in the clear air well away from clouds. The implication of this result is that tall clouds can develop only in restricted regions where the relative humidity is unrepresentatively high; such a region is usually marked by a cluster of clouds and due to the persistent presence of the ascending part of an intermediate-scale circulation. These circulations, associated with irregularities of terrain, obviously ar-range and locally intensify cumulus convection over land, and also over the sea the existence of circulations with similar horizontal scales of tens of kilometers is often indicated by patterns in cumulus populations. In their presence the relative humidity of the clear air between the clustered clouds is observed to be distinctly greater than elsewhere, as shown in Fig. 7.30 and 7.32. Consistently, although the cloud towers which reach the greater heights are broader than others, the general shape of a large cumulus, or of the cluster within which the tallest clouds occur, is one which tapers upward. The cloud towers that reach highest are generally those whose ascent is for the greatest part through air which has been modified by previous clouds or by slow ascent on the intermediate scale, and which may even still contain incompletely evaporated cloud residues, especially near the level of the cloud bases. Whereas the nuclear bomb clouds produced by sudden enormous sources of buoyancy rise through clear air to levels far above the saturation level, just like great thermals (Pl. 7.3), in natural convection a cloud tower is rarely seen in isolation, and then only briefly in a late stage of its evolution (an example can be seen near the middle of Pl. 7.14). Generally a cloud tower is seen to emerge from a parent cloud or cluster of clouds of much greater horizontal extent; when observing from a distance on the ground, it is not possible to appreciate that the individual clouds of a cluster may be horizontally separated, or to be sure that a particular tower protrudes from lower cloud rather than rising into view from a position beyond it, but it remains probable that in the cumulus layer it ascended through saturated or nearly saturated air. Late in its ascent, when it is clearly visible, a cloud tower unquestionably rises through unsaturated air, whose relative humidity is likely to have the low values representative of levels above the cumulus layer, but significantly the tower then usually is observed to ascend a distance only about equal to its diameter when it first emerged into view. Most growing clouds produce several towers which reach successively greater heights, and occasionally when the growth is steadier a cloud broadens as its renewed tops reach higher levels, so that it is reasonable to think that the air ascending in its central parts is progressively less affected by mixing with clear air.

Evidently a local source of buoyant air with a given intensity, or, from an alternative point of view, an intermediate-scale circulation of given intensity, can maintain a cumulus of a certain size. An increase in the intensity can be anticipated to cause the cloud to build, and the internal properties of its towers to approach more nearly those corresponding to adiabatic ascent. The implication of Fig. 7.34, however, is that the intensity usually associated with even the largest clouds in typical cumulus populations is such that inside the clouds the departure from adiabatic conditions is large and easily observable. (It seems that in the more intense convection associated with cumulonimbus, discussed in Chapter 8, there is a close approach to adiabatic conditions in parts of the clouds, even at heights far above the cloud base. Some evidence of this difference appears already in the aircraft observations presented in Fig. 7.30.)

7.10 Theories of cumulus convection

According to the previous discussion small cumulus probably modify very little the general characteristics of the convection in the unsaturated air below them. On the other hand, in populations of clouds large enough for the phase changes of water to become significant, intermediate-scale circulations (associated mainly with shower development or with irregularities of terrain) and large-scale vertical motion intervene to exert a dominating influence on the spacing, size, and properties of the larger of the clouds. It is therefore necessary to be cautious in appealing to observation during the construction and testing of theories of cumulus convection. Existing studies are of four kinds: perturbation theories, theories supposing the existence of steady circulations, elaborations of the simple parcel theory, and numerical modeling employing arbitrary mixing laws. None is satisfactory; all have obvious deficiencies, but all have interesting and informative features.

Perturbation theories

In perturbation theory the possibility is examined that a stationary pattern of flow in a fluid, or a state of rest, can be sustained against small imposed disturbances. The fluid is said to be *stable* if the disturbances lead only to small oscillations about the initial state (which in real fluids are damped and eventually disappear), and *unstable* if at least some tend to amplify in such a way that the departure from the initial state becomes progressively greater and it is not restored. In obtaining the equations which govern infinitesimal departures of the physical variables of state from their initial values, terms involving the products of the departures or their squares (and higher powers) can be neglected, so that the equations are *linearized* and made more tractable. In general the initial rate of amplification of perturbations in an unstable fluid is found to depend on their scale, and on intuitive and other grounds the scale of a per-

7 Cumulus convection

turbation for which the rate of amplification has a maximum value is regarded as that which will eventually become dominant, and that to be compared with the scale characteristic of the motions observed to develop in real fluids. Further analysis leads to a structure for the developing disturbance and the transports of heat and momentum which it effects, which are likewise anticipated to be characteristic also of the disturbance when it attains finite amplitudes.

Such *linear perturbation theory* has provided useful insight into the (billow) motions which arise in sheared flows, and into the large-scale slope convection (in which a simple initial state is that of geostrophic zonal flow with a uniform thermal wind). In each case it has indicated a realistic dominant (horizontal) dimension and a realistic structure. It has been less successful in the study of the convection which arises in a deep layer of fluid following heating from below (or cooling from above) and the production of an unstable stratification. In such a layer exponentially amplifying small perturbations of the layer grow at essentially the same rate $(gB)^{1/2}$, where $B = -(g/\theta)\partial\theta/\partial z$. Diffusion inhibits the growth of the perturbations of smallest scale; however, if a magnitude for the wavelength L of the largest perturbations which are likely to be seriously affected is estimated by comparing the rate of dissipation of kinetic energy due to the introduction of a coefficient of diffusion of momentum k_M with the rate of generation of kinetic energy corresponding to the growth rate of perturbations, the result obtained is $L^2 \approx k_M/50(gB)^{1/2}$. For typical conditions ($\partial\theta/\partial z$ of order $0.1°C\ km^{-1}$) L is of order 1 cm if the molecular diffusion coefficient is used, and still only of order 10 m if a large value for an eddy diffusion coefficient is used. This result seems more likely to be relevant to the small-scale motions which can be seen to develop on the caps of thermals and of cumulus towers than to the observed circulations on the scale of the depth of the adiabatic and the cumulus layers.

Wind shear, whose importance is again associated with the value of a Richardson number, also can have a disrupting effect on the motions of small scale. For $Ri \approx -2$ the unstable perturbation aligned across the wind shear has a dominant wavelength $L \approx 4h$, where h is the depth of the unstable layer. This result seems more promising, but the magnitude of the appropriate Richardson number in cumulus convection is probably at least ten times greater, and the influence of the shear is then much reduced. Moreover, if the shear is in one direction, only the preferred mode of motion is in longitudinal rolls (aligned along the shear), and the shear does not then influence the perturbation mechanics.

It therefore seems that the perturbation theory as conventionally formulated is incomplete, lacking a shear or an eddy diffusion which is variable in space (and determined or controlled by the convection which develops), or that the processes by which a dominant wavelength is selected are essentially nonlinear.

Steady-state theories; mixing between cloudy and clear air; shelf clouds

Since the depth of the cumulus layer, the associated stratification, and the fractional area occupied by cloud hardly change during the life of a cumulus, especially over the sea, it might be thought that the detailed behavior of the individual cloud, and its unsteady nature, need not be considered in modeling these general aspects of cumulus convection and relating them to a heat transport into or through the cumulus layer.

In one model which has been examined (75) the layer is considered to consist of an array of cells of uniform width, and of depth d, each of which contains an axially symmetric cloudy updraft, occupying a fraction S of the horizontal area and surrounded by sinking air in which there is no evaporation. The lapse rate in the ascending air is made greater than the wet adiabatic by the introduction of a parameter which corresponds to a rate of mixing with the surrounding air (of the magnitude that occurs with plumes or thermals). A solution is then obtained for a steady circulation in each cell, containing relations between S, the mean ascent speed, the mean (potential) temperature difference between ascending and descending air (together implying a mean upward heat flux H), and a constant vertical gradient of potential temperature (or mean lapse rate γ), the last being a function of S, H, a, and d. It is found that with S not much different from 0.1, d between 1 and 3 km, and $a \approx 2d$, γ has a fairly definite minimum value for a given value of temperature T at the bottom of the layer and of H. These latter parameters are accordingly regarded as determining a preferred mode of convection, which for a large range of H (between about 100 and 1000 cal cm^{-2} day^{-1}) is characterized by the following values of d, S, and γ:

with $T = 15°C$:
$\quad d = 1\text{–}2\ km, \quad S = 0.08\text{–}0.12,$
and $\quad \gamma = 8.1\text{–}8.6°C\ km^{-1}$

with $T = 20°C$:
$\quad d = 2\text{–}3\ km, \quad S = 0.03\text{–}0.06,$
and $\quad \gamma = 5.5\text{–}6.1°C\ km^{-1}$

On the basis of these results it is speculated that uniform heating cannot be responsible for vigorous cumulus convection in which S and d have substantially larger values.

Although the results appear to be realistic (in particular, the inferred mean lapse rates are a little nearer to the wet adiabatic than to the dry adiabatic, as usually found in the cumulus layer), the model has some unsatisfactory features, principally the lack of consideration of water substance and its phase changes, which can be anticipated to be an essential part of cumulus convection. Thus whereas some processes can be imagined to provide upward heat fluxes at the base and top of the layer to allow the assumed continuous upward transport through it, no such process can be envisaged for the associated transfer of water substance. Nor is it obvious that cumulus convection occurs only as a response to a requirement for a transport of sensible heat; in convection associated with warming at the surface the maintenance of characteristic lapse rates suggests that the divergence of the heat flux in the vertical is about the same at all heights, but it is not clear how the heat flux itself varies with height, particularly in and near the base and top of the cumulus layer. It has already been remarked that continued convection in a confined layer over a moist surface can be expected to lead toward a state of complete cloud cover and eventually of general saturation, so that the observed duration of populations of scattered cumulus with a characteristic mean stratification of temperature, and mixing ratio, should perhaps be regarded as essentially related to the continuous incorporation of air with little vapor content, either by the large-scale subsidence of air into the cumulus layer or by the progression upward of its top.

It is therefore likely that a satisfactory theory will specify the mean mass transfers of air and of water vapor accompanying the cumulus convection. It is interesting to consider the mass transfers of air associated with the cumulus updrafts and the downdrafts produced by chilling where condensed water is evaporated during mixing between the cumulus and the clear air. It was shown in Chapter 3 that an isobaric mixing just sufficient to evaporate all the condensed cloud water, of specific humidity q, produces a temperature $\Delta T'$ above that of the clear air which is given by

$$\Delta T' = \frac{\Delta T - \Delta T_0}{1 + m} \quad (3.6)$$

where ΔT is the initial temperature excess of the cloudy air, m is the mass of clear air mixed with unit mass of the cloudy air, and

$$\Delta T_0 = \frac{q}{L_v/c_p} \quad (7.27)$$

is the temperature excess of the cloudy air for which the temperature of the saturated mixture is equal to the temperature of the clear air. If q has the value q_γ corresponding to adiabatic ascent from the cloud base, then ΔT_0 will be written as $(\Delta T_0)_\gamma$. (Since the densities of the mixture and of the clear air will be compared, strictly it is the virtual temperatures which are concerned. These, as shown in Chapter 3, may differ significantly from air temperatures which are well above 0°C; however, here this complication will be overlooked, since it does not materially affect the qualitative conclusions to be drawn.)

We consider the ascent of a unit mass of air from the cloud base, first adiabatically, acquiring a temperature excess $\Delta T = \Delta T_\gamma$, before isobaric mixing with clear air just evaporates the condensed water. If the lapse rate in the clear air is the dry adiabatic, then

$$\Delta T = \Delta T_\gamma = \Delta T_0 = (\Delta T_0)_\gamma$$

and in *any* mixture in which the cloud has evaporated the temperature is that of the clear air.

However, if the lapse rate is that typical of a cumulus layer,

$$\Delta T = \Delta T_\gamma \approx (\Delta T_0)_\gamma/3$$

while the appropriate value for ΔT_0 remains $(\Delta T_0)_\gamma$; thus the temperature of such a mixture is always less than the clear air temperature, and is a minimum in the just-saturated mixture of unit mass of cloud air and the mass m of clear air. At heights of 1–2 km above the cloud base $(\Delta T_0)_\gamma$ in warm weather reaches values of 5–10°C, and accordingly ΔT even in adiabatic ascent to such levels reaches only about 1–3°C. For such values, and still more for the smaller values arising from non-adiabatic ascent, the value of m is dependent practically only upon q and the relative humidity of the clear air (Eq. 3.9). For a relative humidity $u = 0.8$, which according to Fig. 7.7 is an average value in the cumulus layer of the trade winds, typical values of m in the range of heights mentioned are between about 2 and 4, and accordingly the chilling of the clear air by the evaporation of cloud during mixing is about 1°C; in convection over land the relative humidity in the cumulus layer may often be lower (e.q., between 0.6 and 0.7 in Fig. 7.37), and the chilling correspondingly somewhat greater.

It is difficult to specify more realistic conditions in which cloud air is subject to some mixing already during its ascent, and which before the evaporation of its liquid water by mixing has a temperature more nearly that of the clear air, and a concentration of liquid water substantially less than the adiabatic value. However, considering that Eq. 7.25, which was derived on the basis of some simple assumptions, indicated a value for the ratio f between the concentration of liquid water representative of the cloud interior and the adiabatic value which is about that observed, we may use it to suppose that both the temperature excess ΔT and the concentration of the condensed water are the same fraction f of their adiabatic values ΔT_γ and q_γ. Then the sign of the numerator in the expression for $\Delta T'$ (Eq. 3.6) is not changed, for

$$\Delta T = f \Delta T_\gamma \approx f (\Delta T_0)_\gamma/3$$

7 Cumulus convection

and the appropriate value for ΔT_0 is now $f(\Delta T_0)_\gamma$. Moreover, since the value of m is proportional to q, that is, also to ΔT_0, the magnitude of $\Delta T'$ is hardly changed: it will still usually be about or more than 1°C in the upper part of the cumulus layer. Thus following the evaporation of the cloud the saturated air sinks, and in a cumulus layer with a lapse rate which has the typical relation to the dry adiabatic (less by about 4°C km^{-1} in warm weather) it can be expected to reach equilibrium at a level about 250 m lower if the motion is adiabatic, and rather less if mixing with the clear air continues. Consistently, details on the outside of a cumulus, which always appear to be sinking relative to a rising top, have little vertical motion when they begin to lose their identity, and may be observed to descend a little before they disappear (as shown in Fig. 7.26; in the cumulus tower layer, however, where all the air in a tower may become much cooler than the clear air before its ascent ceases, the following subsidence is correspondingly more pronounced, as illustrated in Pl. 7.5).

At every level in the cumulus layer the mass of air which is chilled by the evaporation of cloud during mixing of course exceeds the mass of cloud which evaporates after reaching that level from below. However, the evaporated cloud can sink only a small distance, while of the cloudy air which ascends through the lower levels, some reaches every higher level up to the top of the cumulus layer before it is evaporated. Considering for a short period an individual cloud occupying the whole layer together with a volume of surrounding air which it is supposed could be identified as containing water evaporated from the cloud, it appears that in the lower part of the volume there is a net upward mass flux of air, while in the upper part of the volume there is a net downward mass flux (this clearly must be true near the top of the volume). In the parts of a cumulus layer between individual clouds and their residues, compensating mass transfers can be expected, tending to increase the lapse rate and possibly to change the vertical profile of the general wind across the cumulus population (76). Since the mass transfers depend on the lapse rate and the distribution of vapor mixing ratio in the cumulus layer it is evident that these are not independent, but their relation cannot be examined without a complete specification of the processes affecting the convection.

It is remarkable, however, that if cloudy air is considered to ascend adiabatically (or with mixing) to some level in a cumulus layer with an observed stratification (as given by Fig. 7.7 and 7.36, e.g.), or into the much drier cumulus tower layer, before the cloud is evaporated by mixing with the clear air and the saturated mixture sinks to an equilibrium level, the vapor mixing ratio in the mixture is usually there found to be somewhat greater than that observed in the clear air (when this decreases with height the difference is reduced if mixing is considered to continue during the sinking). The approximate agreement tends to confirm that an individual cumulus might be modeled on the assumption that it has a negligible effect on a quasi-steady stratification, while the sign of the discrepancy suggests that a quasi-steady stratification with scattered clouds can be maintained only if the net upward flux of water vapor into the cumulus layer is accompanied by an upward extension of the layer into drier air or by a horizontal divergence of water vapor associated with large-scale subsidence.

There are two other interesting aspects of the stratification associated with cumulus convection. First, the cumulus layer is often surmounted by a shallow layer in which there is a pronounced inversion (as shown in Fig. 7.16). Such inversions are commonly regarded as strongly impeding vertical transfers, but it must be recognized that the inversions are produced by the extension of the convection into potentially warm air, just as the inversions previously described which are observed above layers of dry convection. It is indeed obvious from the usual large upward decrease of vapor mixing ratio in an inversion layer (well defined by several points in Fig. 7.16) that it is a transition layer in which the air has been partially modified by the convection. It has been shown theoretically and illustrated by observations with balloon-borne radiometers (77) that the large decreases of mixing ratio, which can be expected when there is pronounced large-scale subsidence in the troposphere, lead to an abnormally large rate of radiative loss of energy in the transition layer, which tends to maintain or intensify the inversion and to increase the lapse rate below it.

Second, the radiative loss of energy is still more intensified if shelf clouds near the top of the cumulus layer become extensive. It is not clear under what conditions they develop; they are most common over the ocean, where the mixing ratio and relative humidity of the air in the cumulus layer may rise at a significant rate when the large-scale subsidence and the heat flux from the surface diminish or cease (usually when the airstream eventually turns into a more nearly zonal direction), but an upward flux of water vapor from the surface continues. The presence of shelf clouds seems to imply an unusually small rate of exchange between the moist top of the cumulus layer and the comparatively very dry free atmosphere. Since the cumulus layer is a layer of small wind shear, a marked wind shear is often concentrated in the cumulus tower layer and may materially assist the exchange by the Kelvin-Helmholtz instability described in Chapter 9. Shelf clouds might therefore be a symptom of moderate or light general winds in which such marked shear is absent.

Once the shelf clouds have begun to form the associated intensified radiative cooling may quickly lead to their more extensive development. (Over England cumulus which form on a bright morning may soon reach the top of a layer recently subject to cumulus convection over the sea, and then sometimes produce shelf clouds which quickly spread into an overcast and make it a dull day. On these occasions the wind is usually moderate, with little or no northerly component —as might be anticipated from the discussion above— and the air near the top of the cumulus layer is usually already moist with a high relative humidity, and lies beneath an inversion or layer of small lapse rate. The ascent of the airstream as it passes over hills, or intermediate-scale circulations which cluster the cumulus over hills, may play an important part in saturating the air near the top of the cumulus layer, in which the renewed convection early in the day is not sufficiently vigorous to penetrate far into the drier air above, and therefore in leading to the formation of extensive shelf cloud. While the shelf clouds are developing it is sometimes possible to see clouds forming at their level independently of cumulus, clearly indicating some ascent on a larger scale. After the development of extensive shelf cloud the radiative fluxes tend to maintain or even thicken it, while the shading of the ground from sunshine may prevent a more vigorous convection which might lead to its dissipation by more intensive convective mixing with the drier air above. However, an impressive change of weather from a promising morning to a dull day depends upon marginal meteorological trends difficult to identify, still more to forecast.) The shelf clouds typically spread from the tallest clouds, in which air reaches the top of the cumulus layer with some buoyancy. Sometimes their formation seems to coincide with the production of showers; a relation is plausible, for a reduction of liquid cloud water leads to reduced chilling and sinking of air during the evaporation of cloud by mixing, and therefore may produce a higher relative humidity in the mixture containing the evaporated cloud. In particular, according to the previous discussion based on Eq. 3.6 and 7.27, if as much as two thirds of the liquid water concentration in buoyant cloud air is precipitated before the air is subject to mixing with clear air, then the mixture is not chilled below the clear air temperature but retains some buoyancy. The consequent reduction of the ordinarily pronounced temperature gradient and shear near the cloud edges can be expected to reduce the intensity of the small-scale mixing motions and the readiness with which they lead to the evaporation of cloud. The onset of precipitation in a large cloud may therefore result in the air of its principal updrafts spreading and remaining visible as a rather tenuous shelf cloud near the top of the cumulus layer.

Modifications of the simple parcel theory

As already indicated, the simple parcel theory can be modified by providing for an interchange of mass between the followed volume and its surroundings, assuming some rate of exchange based on the laboratory experiments with plumes or thermals (made with incompressible, and mostly neutrally stratified, fluids). The resulting theory can be applied, with empirical adjustments, to predict the mean internal properties, rate of rise, and peak heights of clouds in prescribed or observed stratifications, as in an extensive series of observations made in recent years over the Caribbean Sea and Florida (78).

In this work the theory is based on the experimental result (53) that the same form of equation can be written for the vertical acceleration of a buoyant volume, and therefore presumably a cloud tower, whether it behaves as a plume or a thermal:

$$\frac{dw}{dt} = \frac{d(w^2/2)}{dz}$$

$$= \frac{gB(z)}{1+\gamma} - \frac{3(3K_2/4 + C_D)w^2}{r(z)} \quad (7.28)$$

where γ = a *virtual mass coefficient*, which can be considered to arise because the available potential energy is used to set in motion not only the volume followed, but also surrounding fluid

K_2 = an *entrainment coefficient*, which expresses the exchange of mass between the volume (of diameter $2r$) and its surroundings

C_D = drag coefficient

An appropriate value for K_2 is based on a theoretical value for a fractional rate of dilution of the mass M of rising fluid with the surrounding fluid (78):

$$\frac{(1/M)dM}{dz} = \frac{9K_2}{32r}$$

and the value indicated by the experimental result for the "starting plume," from Eq. 7.21, appropriately modified, is

$$\frac{(1/M)dM}{dz} \approx \frac{0.2}{r} \quad (7.29)$$

according to which $K_2 \approx 0.7$. This entrainment coefficient of course also affects the temperature and liquid water content of the rising volume, and from some observations of their mean values in cloud towers traversed by aircraft in the middle troposphere it was decided that for application it was reasonable to assume $K_2 = 0.65$, $C_D = 0$, and $\gamma = 0.5$ (a value obtained by experiment).

Most of the work was directed at finding cloud towers

with temperatures in the range from about -4 to $-20°C$, which contain a high proportion of supercooled cloud water, thoroughly freezing some by a massive injection of artificial ice nuclei and thereby substantially increasing the buoyancy, and then comparing the observed subsequent peak heights attained by treated and untreated clouds with theoretical estimates. The aircraft therefore entered cloud towers at levels between about 5 and 8 km, where the half-diameter r varied between 500 and 1300 m (79). The variation of the tower diameter $2r$ with height during the previous growth was not known. However, it was found that if it was assumed to have been constant, then according to the theory the clouds could always have grown to the flight level from the observed level of the cloud bases (where the buoyancy was assumed zero and the speed of rise 1 m sec^{-1}) in the stratification given by a representative sounding. (This is not surprising, considering that in these situations the cumulus layer was rather deep, the lapse rate exceeding the saturated adiabatic up to a height usually between 4 and 5 km.) Moreover, in the subsequent growth of the clouds into the upper troposphere their properties are not very sensitive to a presumed rate of entrainment, because of the limitation placed by the low temperatures on the concentration of cloud water which can be evaporated. Consequently the tower diameter was presumed to remain constant, even though on some occasions it obviously changed. The effect of this assumption on the computed subsequent growth was less important than the magnitude of r, and still less important than the stratification in the upper troposphere and the assessed magnitude of the buoyancy $[B = (T_{vc} - T_v)/T_v$, where T_{vc} is the cloud virtual temperature (Section 2.6) and T_v is the virtual temperature of the clear air at the same level]. In maritime airstreams clouds of the size studied are invariably shower clouds, and in the first formulation of the theory it was arbitrarily assumed that half of the concentration of cloud water was removed by precipitation as it was condensed.

The prevalent stratification was of the kind described in Chapter 8 as typical of present or recent ordinary cumulonimbus convection, in which a layer of small stability surmounts the cumulus layer before the lapse rate in the upper troposphere again becomes nearly equal to or even slightly greater than the saturated adiabatic. In these circumstances most cloud towers do not rise far above the top of the cumulus layer, but a small proportion, perhaps those of sufficient size, may retain some buoyancy, or regain it by the precipitation or freezing of cloud water, and may then ascend a further few kilometers into the high troposphere (Section 8.2).

The untreated cloud towers which were studied ascended only about 1 km or less after the time of initial observation, reaching peak heights between about 7 and 8 km (with one exception to be mentioned later), generally as predicted by the theory. In the treated cloud towers it was assumed that latent heat was released near the $-6°C$ level by the freezing of all the cloud water and the lowering of the vapor pressure to saturation with respect to ice (usually leading to a warming of between 1.5 and 2.0°C), and that during subsequent ascent saturation with respect to ice was maintained. Some of these towers grew little higher, if at all, but seven of the total of fourteen rose a further 3–5 km to reach a peak height of about 11 km; sometimes the rise was spectacular (80), occurring in about 15 min subsequent to the treatment, following a previous similar period of little and irregular growth. In all examples the peak height was approximately as predicted by the theory including the freezing of the cloud water, thus apparently substantiating the theory as well as demonstrating a real artificial modification of the cloud behavior.

However, the exceptional untreated tower mentioned above also rose by nearly 3 km to a height of 10 km, as predicted by the theory for the frozen clouds, suggesting that in this example, at least, the development was associated with a natural freezing of the cloud particles. The progress of the natural development of the ice phase in the kind of maritime clouds studied has not yet been established: it is believed that the ice particles do not form in significant concentrations in freshly rising towers until the temperature of the cloud air has fallen below $-20°C$, but on the other hand it is known that concentrations much higher than would be expected from measurements of ice nucleus concentration occur in clouds which rise barely above the 0°C level, especially following the production of rain (as will be discussed at the end of Section 8.1). Moreover, at least in clouds over land, a spectacular growth surge soon after the formation of rain in clouds with supercooled tops is a common phenomenon, as described in Section 8.2. The evidence for an artificial cloud modification, while strong, therefore needs further examination. Our concern here, however, is only to illustrate the kind of use which has been made of a more elaborate formulation of a parcel theory. Recently in the investigations described it has been even further extended by parameterization of the processes of conversion of condensed water into precipitation, with the aim of obtaining more realistic concentrations of cloud water and of precipitation (for comparison with observations of radar reflectivity).

Evidently the experimental information on the properties of buoyant plumes and thermals can provide a modified parcel theory which can usefully be applied to the study of individual cloud towers. However, this theory has the inherent disadvantage of retaining the concept of an environment of predetermined properties and, even if used only for diagnosis, of depending upon some information about the local rate of generation of buoyancy or the local upward flux of buoyant air, the equivalent of the source strength Q (Eq. 7.15) of the plume, and of the initial buoyancy and size of the thermal. Thus if the rate of exchange of mass between the rising air and its surroundings is expressed as $(1/M)dM/dz$ and is assessed from observations of the height reached by cloud tops, the results vary from very small values (<0.1 km^{-1}) for very large clouds, implying hardly significant mixing, to large values (≈ 1 km^{-1}) for very small clouds. The value implied by the observations of the expansion with height of towers (regarded as thermals) represented by Eq. 7.22, for example, is about $0.6/r$, or about 2 km^{-1} near the level of the cloud base and about 0.3 km^{-1} in the high troposphere. No studies have yet been made of the causes of this great apparent variation. In general, no presumed mixing law has any prognostic value: the apparent success of the modified parcel theory in the Caribbean work described depends upon the observation of a value of r in the middle troposphere and the application of the theory in the upper troposphere. In the kind of stratification observed there in the upper troposphere the buoyancy of a cloud tower, and its vertical displacement, are sensitive no less to the assumed degree of mixing between cloud and environment than to the initial value of the buoyancy, to the presumed appropriate reference process for the rise of cloud air, and to the stratification of the environment. Since the uncertainties of the environment, due on the one hand to the poorly understood physical processes of change of phase and precipitation of cloud water, and on the other hand to inaccuracies and unrepresentativity of the available radiosonde soundings, correspond to possible errors in the estimation of the buoyancy which have the magnitude of the buoyancy itself, it is remarkable that the theory appears to have been as useful as the observations suggest. It is possible that the apparent degree of success is partly a fortuitous result of previous natural cumulonimbus convection, in which some clouds whose tops reached heights between 7 and 8 km (temperature about $-20°$C) rose a further few kilometers as a result of an increase of buoyancy due to an extensive natural freezing, and produced a stratification which insured a similar development of somewhat smaller clouds in whose tops the freezing was artificially induced.

The numerical modeling of cumulus convection

Increasing effort is being given to the study of cumulus convection by numerical integration of equations expressing the conservation of momentum, matter, and energy, using tentative diffusion laws and chosen simple initial and boundary conditions. The governing equations are usually also simplified to make them more easily tractable, and so that their solutions contain only the kind of motion system required [e.g., by assuming that the fractional range of potential temperature is small, and that the time scale of the motions is that characteristic of gravity waves, of order minutes (81), which excludes the elastic (acoustic) waves of comparatively small time scale and the large-scale motions subject to Coriolis accelerations].

The resulting equations have been solved numerically to show how air with an unstable stratification overturns after being subject to a small disturbance; in this way the growth of a thermal can be simulated. The detail in the results depends strongly on the manner in which diffusion coefficients representing motions of scale smaller than the grid-spacing are handled; even if they are removed, finite difference approximations smooth the distributions of the properties of the flow, simulating a moderately effective diffusion process. However, solutions have been obtained in which features of the pattern of motion change almost uniformly with height above a virtual origin, just as in an experimental thermal; in particular, the rate of rise of the foremost part of the thermal is soon very close to that inferred from the similarity theory using experimentally determined constants. For example, in two such calculations relating to a dry atmosphere (82) the values of the parameters n (the broadening coefficient in Eq. 7.17), C (relating the velocity w_t of the thermal top to the mean buoyancy in Eq. 7.19), and w_{max}/w_t (where w_{max} is the maximum vertical velocity inside the thermal) were very close to average experimental values (Table 7.5). Accordingly, the numerical method was extended to include the cloudy thermal in an atmosphere with a lapse rate exceeding the saturated adiabatic lapse rate (83), using a constant coefficient of diffusion. It is difficult to estimate a realistic value for this coefficient, but it can plausibly be regarded as the product of a length scale and a velocity characteristic of the turbulent motion, say 50 m and 2 m sec^{-1} for a small cumulus, corresponding to 10^6 cm^2 sec^{-1} (accelerometer records obtained on flights through cumulus have indicated an average value of about 7×10^5 cm^2 sec^{-1}; 84). Values of 4×10^5 or 4×10^4 cm^2 sec^{-1} were used, with an initial lapse rate of 7.2°C

Table 7.5 Comparison of the values of the parameters n (Eq. 7.17), C (Eq. 7.19), and w_{max}/w_t in two numerical experiments on simulated thermals (82), and the average experimental values

Parameter	Experiment 1	Experiment 2	Laboratory
n	4.0	4.9	4.0
C	1.4	1.3	1.2
w_{max}/w_t	2.3	2.3	2.3

km^{-1} (considerably greater than the saturated adiabatic lapse rate). The results gave an exponentially increasing rate of upward growth of a cloud soon after its formation with unrealistically large temperature excesses and vertical velocities, suggesting that the effect of the diffusion terms was too small.

In general, in these and similar calculations the clouds produced do not retain the compact shape of a thermal steadily broadening with height, but tend to acquire the shape of a mushroom with a head surmounting a columnar stem. They emphasize the importance of the diffusion processes, whose magnitude and distribution influence not only the shape but also the internal properties of clouds. They also emphasize a mechanism other than frictional dissipation by which clouds may lose kinetic energy, namely the excitation of gravity waves in surroundings which are stably stratified. The vertical displacement of air in these waves is in general less than 100 m even near the clouds; however, their existence in the atmosphere is occasionally manifest by a disturbance in a cloud layer approached or penetrated by a rapidly rising cumulus top or by the formation of a cap cloud (pileus) above it (Pl. 7.3*b*). The calculations described and others which include circulations of intermediate scale (85) represent only tentative first steps toward a satisfactory treatment of the motion systems and clouds that arise in an extensive airstream moving over a surface which is a source of heat and moisture, and liable to contain circulations on a great range of scales. A first aim must be the establishment, in conjunction with better observations, of magnitudes and laws for the small-scale diffusion processes.

The requirements of a theory of cumulus convection

An improved theory of cumulus convection is needed for two principal applications. First, a simple but adequate representation is required of the laws governing the transfer of momentum, heat, and water vapor from the earth's surface into the atmosphere by convection, and their distribution in the vertical, for inclusion in models of atmospheric motion used in the numerical prediction of large-scale weather. For this purpose it may be sufficient to assess transfers from assumed values of transfer coefficients and the differences between conditions in the lowest layer of the model atmosphere and at the surface [given over land by energy balances involving various empirical relations (90, 91), and over the sea by a constant surface temperature]. If, further, it is assumed that the convection produces a characteristic stratification in layers above the lowermost, the height to which the convection extends (this from the stratification of temperature alone) and the degree of modification follow. Procedures of this kind have already been introduced into large-scale models which have the necessary vertical resolution (e.g., 86). In addition to expressions for the transfers and stratification associated with cumulus convection, relations must be sought between properties of the large-scale flow and other features of the cloud populations, particularly the fractional area of cloud cover which determines general albedo and the radiative exchanges which are important over long periods.

Second, a detailed theory is required to relate the stratification (including the wind distribution) to the properties of individual convection clouds, especially their vertical dimensions, updraft speeds, and concentrations of condensed water. As discussed in Chapter 8, these are important in the study of the development of showers, and hence also of thunderstorms and other violent local weather. The unsteadiness of the convection on cloud scales, manifest by the ascent of limited volumes of air behaving like thermals, recalls the unsteadiness of laboratory convection in deep fluids (53, p. 226) but equally has not yet been satisfactorily explained. The aircraft observations described in this chapter suggest that a cloud tower (probably its most important component) can be regarded as consisting of a strongly turbulent transition shell and an interior within which properties have almost uniform mean values (varying with height above cloud base). It is not known whether the frequency of this structure is due to a bias toward observing newly rising cloud towers, or how it could result from the mixing motions which occur on a great range of scales (the assumption sometimes made that mixing with clear air produces everywhere cloud properties which vary only with height may obscure mechanisms important in the evolution of droplet spectra).

Extensive further development of numerical models of cumulus, with more comprehensive boundary conditions and laws of small-scale (turbulent) diffusion, can be expected eventually to lead to the required theory. Together with improved observational studies, they will establish the properties of those circulations which occupy the whole of the convection layer, in which, in contrast to those motions effecting transfers close to the surface, it is possible to discern some kinds of organization. The improvement to the required degree of both the numerical models and the observations is likely to be slow because of the complication, and of the technical difficulty of obtaining data which are at once comprehensive, accurate, and representative.

It may be possible to obtain useful insight into the large-scale problem by regarding the convective circulations as separable in scale, those near the surface being expressed statistically in the usual way, and others which occupy deeper layers consisting partly of restricted volumes of air which are subject to mixing while rising through an environment and partly of some circulations of much greater extent and persistence whose presence results in a local mean vertical motion. The ascending volumes which settle into equilibrium near the top of the layer of convection become part of the environment, which sinks during the ascent of fresh volumes; in the presence of circulations of intermediate or large scale this sinking is locally changed in magnitude and perhaps even in sign.

If the ascending volumes and the environment are considered to have horizontally uniform properties, and the intensity of the mixing between them is prescribed, it is possible to examine the heat and moisture economy of a cumulus layer, and to show that it is continually destabilized by the cloudy convection, the environment being cooled in its upper part and warmed in its lower part (76, 87). The layer therefore must extend upward unless there is large-scale subsidence.

If some proportion of the kinetic energy of ascending volumes, generated under buoyancy in the lower part of the layer of convection, is considered to be consumed by transformation into potential energy under reversed buoyancy in the upper, stably stratified part of the layer (rather than dispersed by gravity waves or dissipated in turbulent motions), then the relation between the stratification of the layer and its depth can be examined, most easily for convection in which clouds do not form (9). Such an analysis has been made of the stratification in a layer of convection maintained over a cool surface by radiative heat loss from a complete cover of low cloud, leading in particular to a theory for the intensity of the inversion which develops at the level of the cloud top (88).

This kind of analysis can be extended to relate the stratification of a whole layer of cumulus convection, in respect of θ (89) and r, its evolution in time, the properties of the ascending volumes (including vertical velocity and concentration of condensed water), and the large-scale conditions. It may be advantageous to introduce scales such as S' and S_m to refer to the size of the ascending volumes and to the intensity of the mixing between the volumes and the environment (e.g., the height interval over which the mass of a volume increases by a chosen factor, say e). From the previous discussion it appears that cumulus sufficiently tall to contain air which regains buoyancy (after rising through the stable upper part of the adiabatic layer and transition layer) are characterized by the appearance of towers whose behavior invites identification with the separate volumes of the model. Accordingly, an appropriate value for a radius S' seems to be a small fraction—about 0.20 (Eq. 7.22)—of the height above the surface (or even a constant related to the mean height of the cumulus layer). The variation of S' with height and the assumption that an ascending volume is horizontally homogeneous imply a definite value of S_m; for example, if S' varies linearly with height, as found with thermals ascending through a neutral environment and in some observations of cumulus towers, then neglecting the variation of air density with height,

$$S_m = M\,dz/dM = S'/0.6 \approx z/3$$

On the other hand, in the modification of the simple parcel theory based on the observations of cloud towers over the Caribbean Sea and Florida it was found necessary to assume the lesser rate of dilution implied by Eq. 7.29:

$$S_m \approx S'/0.2$$

In the analysis of those observations S' was taken to have a constant value given by the size of a cloud tower when it was measured in the middle troposphere. This is inconsistent with the assumption of horizontal homogeneity, but could be regarded as effectively recognizing that a rising cloud tower leaves behind a proportion of its material while the remainder is still subject to a rate of dilution expressed by S_m. Evidently further study is required of the validity of the use of variables such as S' and S_m and their consistency with a physical model of the cumulus convection. However, it appears that with a suitable choice of values (even constants) it may be possible to formulate a set of equations which express the economy of heat and water in the several strata of the whole convection layer, and by their simultaneous solution to relate the stratification and its evolution to the large-scale conditions. That these last exert a dominating control on the convection is evident from the following considerations.

In the layer below cloud the presence of the fixed lower boundary as the heat source is responsible for the shallowness of the markedly unstably stratified layer which is the obvious source of buoyancy; in the cloudy layer the unstably stratified part occupies a much larger fraction, usually more than half, of the whole. Below cloud the magnitude of the sensible heat flux and the vertical velocity of an ascending volume generated in adiabatic ascent by a buoyancy corresponding to a temperature excess of magnitude 1°C together imply the reasonable magnitudes $w_a \approx 1$ m sec^{-1} and $\alpha \approx 0.1$, respectively, for the upward speed of the ascending volumes and the fractional horizontal area α which they occupy. In the cloudy layer the depth over which the ascending volumes may accelerate is greater, and vertical velocities of several meters per second are known to be attained, but the magnitude of the excess of cloud virtual temperature can be anticipated not to exceed 1°C and is generally observed to be less (Fig. 7.29). Thus the general lapse rate is close to that followed by an ascending cloudy volume subject to mixing with the environment. This is determined mainly by the scales S' and S_m, and by the difference between the saturation mixing ratio and the mixing ratio in the environment.

An important feature of the convection is that except in the shallow layer close to the ground the mean lapse rate is less than the dry adiabatic, that is, $\partial\bar{\theta}/\partial z$ is positive. Consequently a steady rise in the local temperature of the environment, and in the mean temperature of the whole layer when the fraction α is small, can be associated directly with the mean vertical velocity $w_d = -\alpha w_a/(1 - \alpha)$ of the environment (in response to the continual ascent of individual volumes from near the surface), rather than to turbulent diffusion of the excess of potential temperature from the ascending volumes.

This association makes plain the significant role in the development of cumulus convection played by a large-scale vertical velocity w, since as shown below it commonly has the same magnitude as w_d and its effect is simply additive: $\partial\bar{\theta}/\partial t = -(w + w_d)\partial\bar{\theta}/\partial z$, where $\bar{\theta}$ is the mean potential temperature at a particular level. Cumulus convection in middle latitudes is prevalent in airstreams moving into lower latitudes, with large-scale subsidence, or returning poleward after recent such excursions, when there may be little subsidence or even upward large-scale motion. In the former kind of airstream w may reach several centimeters per second in the lower middle troposphere, but probably not as much as 1 cm sec^{-1} within about 1 km of the surface, as exemplified by analyses of some major southward excursions of air over the United States (Table 7.6). At such low levels it may be attributable to departures from geostrophic equilibrium (Chapter 9) associated with friction near the surface or acceleration in moving systems.

The magnitude of w_d below cumulus may be as much as several centimeters per second if the earlier estimates of w_a (≈ 1 m sec^{-1}) and α (≈ 0.1) are accepted. In the cumulus layer both α and w_d are probably substantially less. The magnitude of w_d there can be estimated from the observed rates of change of potential temperature on occasions when w is thought to be small (if large-scale subsidence contributes to an observed warming but is neglected, then w_d will be overestimated). However, assuming that the stratification characteristic of the whole layer of convection is maintained, $\partial\bar{\theta}/\partial t$ in the cumulus layer can be anticipated to be approximately that near the surface. In fresh winds over the oceans this can be as large as 10°C day^{-1}, or 10^{-4} °C sec^{-1} in middle latitudes (in low latitudes it is much less). Over land on sunny days the corresponding rate of rise of temperature may exceed 1°C hr^{-1}, or 2×10^{-4} °C sec^{-1}. However, the rate of increase of $\bar{\theta}$ in the cumulus layer is less because of the accompanying rise in the height of cloud bases, and the value of about 10^{-4} °C sec^{-1} is probably representative of vigorous cumulus convection over both land and sea. Since $\partial\bar{\theta}/\partial z$ in the cumulus layer is about 3°C km^{-1} in warm and 2°C km^{-1} in cool weather, it follows that w_d is not more than a few centimeters per second in middle latitudes, and less than 1 cm sec^{-1} over the sea in low latitudes. Since the observed values of the speed of ascent of air inside cumulus are a few meters per second in both situations, the implied values of α are about and considerably less than 10^{-2}, respectively, much smaller than in the layer below cloud base because of the comparatively great stability. (The small inferred values of fractional cloud cover are appropriate to levels in the middle of a cumulus layer, and to areas occupied by ascending air, and are not to be compared with the fractional cover at the level of the cloud bases, which usually is much greater and may include areas occupied by air barely entering the cumulus layer before returning to the dry layer.)

Evidently the large-scale vertical velocity w has the same magnitude as w_d, and normally the same sign. Until the theory is established the effect cannot be discussed quantitatively, but it can be anticipated to provide a satisfactory explanation of the strong control exerted over cumulus development by the large-scale vertical motion: intense subsidence brings potentially warm and dry air toward the surface, makes the convection layer shallow, and raises the level of the cloud bases or even prevents cloud formation. In regions of only slight subsidence or large-scale ascent the level of the cloud bases is comparatively low, the amount of cloud large, and the layer of convection deep, readily extending upward into the middle or high troposphere (usually with the complication of transformation into cumulonimbus covection, discussed in Chapter 8, in which the cloud water no longer moves with the air but falls out, and by its partial evaporation tends to intensify the convection).

The close relation between the height to which the convection extends and the large-scale vertical motion is familiar to weather forecasters. Until recently their assessment of the large-scale vertical motion and its

Table 7.6 Mean vertical velocity w (over areas of order 10^6 km^2) during four occasions of major excursions of tropospheric air south of 45°N (92, p. 302)

Pressure level (mb)	Approximate height (km)	$-w$ (cm sec^{-1})			
		1	2	3	4
400	7	1.4	4.0	0.4	2.6
500	5.5	2.0	4.8	1.9	4.8
600	4	2.2	4.2	2.3	3.8
700	3	1.6	3.1	1.9	2.1
800	2	1.0	1.7	1.2	1.2
900	1	0.4	0.7	0.8	0.5

effect was necessarily indirect and qualitative, based on experience of weather according to position in the patterns of the large-scale weather systems, and supported by studies finding a marked correlation between the depth of the layer of convection and characteristics of the flow in the lower troposphere which are indicators of the recent vertical motion. These indicators have been simply the curvature of the sea-level isobars (12), and more reliably the vertical component of the relative vorticity in the lower troposphere (93). Similarly, cloud reconnaissance flights over the northern Pacific Ocean established a close relation between the height of the tops of convection clouds and the apparent rate of change of the vorticity in the flow in the high troposphere relative to the motion systems (30; convection reaching into the high troposphere over the ocean is associated with high-level horizontal divergence or low-level convergence, implying mean upward tropospheric motion and increase along the relative flow of anticyclonic vorticity at high levels and of cyclonic vorticity at low levels). Figure 7.10 presented a situation over the ocean in which it appeared that the depth of the layer of convection (and the heat flux at the surface) was governed more by the large-scale vertical motion than by the rate of change of sea surface temperature following the low-level flow. That this should be, and that the wind shear in the layer of convection is approximately the thermal wind (Chapter 9) set by the large-scale motion, is not inconsistent with the concept of a characteristic stratification, for this can be accommodated with an independent pattern of surface temperature by a variable $\Delta\theta$ (controlling the heat flux) in the shallow superadiabatic layer.

Appendix 7.1   Cloud mensuration

In most studies of clouds which require a knowledge of their dimensions it is the height of details which is required, such as the height of the bases and tops of cumulus, and estimates liable to errors of 100 m or even more are acceptable. Often cloud photographs can be scaled when some dimension, such as the range of a known detail, has been determined by radar or by a laser rangefinder. A suitable laser is normally used with a pulse repetition rate of not more than about $1\ \text{sec}^{-1}$, but even so can be used as a "lidar" (in the manner of a radar) to give very accurate cloud ranges even in daylight at distances of up to about 50 km; especially if used in conjunction with conventional photography it therefore has a great potential value in the study of clouds, whose exploitation has only recently begun.

In the absence of aids such as radar and lidar, cloud mensuration requires triangulation with theodolites or photogrammetry. Experienced observers using theodolites and in telephonic communication can measure the height of selected details in cumulus tops with a maximum likely error of between 100 and 200 m over a wide field of view at distances up to about 80 km (13). Photogrammetry has the advantage of allowing the measurement of many more cloud details, but usually over a more restricted field of view, and requires elaborate precautions if a similar accuracy is to be achieved (94).

In many studies a lesser degree of accuracy is still useful, and further recommends the use of photography as a valuable part of the records of all cloud observations. For example, the heights of cumulus tops can be determined, with an accuracy (maximum likely error about $\pm 10\%$) acceptable in large-scale analysis, from time-lapse films taken on reconnaissance flights, even when narrow- guage (16 mm) film is used (30). Similarly useful estimates of cumulus dimensions and velocities can be made from time-lapse films taken on the ground when a theodolite has been used to establish the position of the horizon in the field of view (95), even if the linear scale is introduced only by the height of the cloud bases estimated from screen-level observations, or from the angular elevations of the bases of clouds evidently clustered over a ground feature at a known distance.

References

1. Sutton, O.G. 1953. *Micrometeorology*. McGraw-Hill, London.
2. Priestley, C.H.B. 1959. *Turbulent Transfer in the Lower Atmosphere*. University of Chicago Press, Chicago.
3. Monteith, J.L., and Sceicz, G. 1961. The radiation balance of bare soil and vegetation, *Q. J. R. Meteorol. Soc.*, **87**, 159–70; Sellers, W.E. 1965. *Physical Climatology*. University of Chicago Press, Chicago.
4. Dyer, A.J. 1961. Measurements of evaporation and heat transfer in the lower atmosphere by an automatic eddy-correlation technique, *Q. J. R. Meteorol. Soc.*, **87**, 401–12.
5. Brown, R.A. 1974. *Analytical Methods in Planetary Boundary Layer Modelling*. Adam Hilger, London.
6. Webb, E.K. 1965. Aerial microclimate, *Met. Monogr.*, **6**, 27–58.
7. Budydo, M.I. 1956. *The Heat Balance of the Earth's*

7 Cumulus convection

Surface. Gidromet. izdatel'stvo, Leningrad; translation published by Office of Technical Services, U.S. Department of Commerce PB 131692, Washington, D.C.

8. Malkus, J.S. 1962. Interchange of properties between sea and air: large-scale interactions, in *The Sea*, vol. 1. Interscience, London, ch. 4.
9. Ball, F.K. 1960. Control of inversion height by surface heating, *Q. J. R. Meterol. Soc.*, **86**, 483–94; discussion, ibid., **88**, 102–5.
10. Zobel, R.F. 1966. Temperature and humidity changes in the lowest few thousand feet of atmosphere, *Q. J. R. Meteorol. Soc.*, **92**, 196–209.
11. Craddock, J.M. 1951. The warming of arctic air masses over the eastern North Atlantic, *Q. J. R. Meteorol. Soc.*, **77**, 355–64.
12. Petterssen, S., Knighting, E., Janes, R.W., and Herlofson, N. 1945. Convection in theory and practice, *Geofys. Publ.* (Oslo), **16**, no. 10.
13. Ludlam, F.H., and Saunders, P.M. 1956. Shower formation in large cumulus, *Tellus*, **8**, 424–42.
14. Ives, R.L. 1947. Behaviour of dust devils, *Bull. Am. Meteorol. Soc.*, **28**, 168–74.
15. Williams, N.R. 1948. Development of dust-whirls and similar small-scale vortices, *Bull. Am. Meteorol. Soc.*, **29**, 106–17.
16. Sinclair, P.C. 1964. Some preliminary dust-devil measurements, *Mon. Weather Rev.*, **92**, 363–67.
17. Warner, J. 1963. Observations relating to theoretical models of a thermal, *J. Atmos. Sci.*, **20**, 546–50.
18. Warner, J., and Telford, J.W. 1967. Convection below cloud base, *J. Atmos. Sci.*, **24**, 374–82.
19. James, D.G. 1953. Fluctuations of temperature below cumulus clouds, *Q. J. R. Meteorol. Soc.*, **79**, 425–28.
20. Grant, D.R. 1965. Some aspects of convection as measured from aircraft, *Q. J. R. Meteorol. Soc.*, **91**, 268–81.
21. Albrecht, F. 1942. Die thermische Konvektion in der freien Atmosphäre und ihr Bedeutung für den Wärmeumsatz zwischen Erdoberfläche und Luft, *Wiss. Abhand.*, **9**, no. 5 (Deutsches Reich Reichsamt für Wetterdienst, Berlin).
22. Durst, C.S., and Giblett, M.A. 1932. The structure of wind over level country, *Geophys. Mem.*, **6**, no. 54.
23. Scorer, R.S. 1954. The nature of convection as revealed by soaring birds and dragonflies, *Q. J. R. Meteorol. Soc.*, **80**, 68–77.
24. Hankin, E.H. 1921. On dust-raising winds and descending currents, *Mem. Indian Met. Dept.*, **22**, pt. 6, 567–73.
25. Bunker, A.F., Haurwitz, B., Malkus, J.S., and Stommel, H. 1949. Vertical distribution of temperature and humidity over the Caribbean Sea. *Papers Phys. Oceanog. Met., M.I.T. and Woods Hole Oceanog. Inst.*, **11**, no. 1.
26. Atlas, D., Hardy, K.R., and Naito, K. 1966. Optimizing the radar detection of clear air turbulence, *J. Appl. Meteorol.*, **5**, 450–60.
27. Atlas, D., and Hardy, K.R. 1966. Radar analysis of the clear atmosphere: angels, *Proc. 15th Gen. Ass. URSI*, 5–15 Sept., Munich, Germany, 401–69.
28. Konrad, T.G. 1968. The alignment of clear-air convective cells, *Proc. Int. Conf. Cloud Physics*, 26–30 Aug., Toronto, Canada, 539–43.
29. Hardy, K.R., and Ottersten, H. 1969. Radar investigation of convective patterns in the clear atmosphere, *J. Atmos. Sci.*, **26**, 666–72.
30. Malkus, J.S., and Riehl, H. 1964. *Cloud Structure and Distributions over the Tropical Pacific Ocean*. University of California Press, Berkeley.
31. Plank, V.G. 1966. Wind conditions in situations of patternform and non-patternform cumulus convection, *Tellus*, **18**, 1–12.
32. Angell, J.K., Pack, D.H., and Dickson, C.R. 1968. A Lagrangian study of helical circulation in the planetary boundary layer, *J. Atmos. Sci.*, **25**, 707–17; Angell, J.K. 1972. A comparison of circulations in transverse and longitudinal planes in an unstable planetary boundary layer, *J. Atmos. Sci.*, **29**, 1252–61.
33. LeMone, M.A. 1973. The structure and dynamics of horizontal roll vortices in the planetary boundary layer, *J. Atmos. Sci.*, **30**, 1077–91.
34. Conover, J.H. 1959. *Cloud Patterns and Related Air Motions Derived by Photography*. Final Report, Contract AF 19 (604)-1589, Blue Hill Met. Obsy., Harvard Univ., Milton, Mass.
35. Woodcock, A.H. 1940. Convection and soaring over the open ocean, *J. Marine Res.*, **3**, 248–53.
36. Kuettner, J. 1959. The band structure of the atmosphere, *Tellus*, **11**, 267–94; 1971. Cloud bands in the earth's atmosphere, *Tellus*, **23**, 404–26.
37. Faller, A.J. 1965. Large eddies in the atmospheric boundary layer and their possible role in the formation of cloud rows, *J. Atmos. Sci.*, **22**, 176–84.
38. Tatro, P.R., and Mollo-Christensen, E.L. 1967. Experiments on Ekman layer instability, *J. Fluid Mech.*, **28**, 531–44.
39. Lilly, D.K. 1966. On the instability of Ekman boundary flow, *J. Atmos. Sci.*, **23**, 481–94; Faller, A.J., and Kaylor, R.E. 1966. A numerical study of the instability of the laminar Ekman boundary layer, *J. Atmos. Sci.*, **23**, 466–80.
40. Hubert, L.F. 1966. *Mesoscale Cellular Convection*. Met. Satellite Lab. Rep. 37, U.S. Environmental Science Services Administration, Washington, D.C.
41. Ray, D. 1965. Cellular convection with nonisotropic eddies, *Tellus*, **17**, 434–39.
42. Kuo, H.L. 1965. Further studies of the properties of cellular convection in a conditionally unstable atmosphere, *Tellus*, **17**, 413–33.
43. Neumann, J. 1951. Land breezes and nocturnal thunderstorms, *J. Meteorol.*, **8**, 60–67.
44. Estoque, M.A. 1962. The sea breeze as a function of the prevailing synoptic situation, *J. Atmos. Sci.*, **19**, 244–50.
45. Simpson, J.E. 1967. Aerial and radar observations of some sea-breeze fronts, *Weather*, **22**, 306–16.
46. Malkus, J.S. 1955. The effects of a large island upon the trade-wind stream, *Q. J. R. Meteorol. Soc.*, **81**, 538–50.
47. Raghavan, K. 1969. Satellite evidence of sea-air interactions during the Indian monsoon, *Mon. Weather Rev.*, **97**, 905–8.
48. Garcia-Prieto, P.R., Ludlam, F.H., and Saunders, P.M. 1960. The possibility of artificially increasing rainfall on Tenerife in the Canary Islands, *Weather*, **15**, 39–51.
49. Mordy, W. 1957. Geographic and climatic notes on Project Shower area, *Tellus*, **9**, 472–74.
50. Chopra, K.P., and Hubert, L.F. 1965. Mesoscale eddies in the wake of islands, *J. Atmos. Sci.*, **22**, 652–57.
51. Chichester, F. 1967. *Gipsy Moth Circles the World*. Hodder and Stoughton, London.
52. Shaw, Sir Napier. 1930. *Manual of Meteorology*, vol. 3. Cambridge University Press, Cambridge, p. 307.
53. Turner, J.S. 1973. *Buoyancy Effects in Fluids*. Cambridge University Press, Cambridge.
54. Scorer, R.S. 1957. Experiments on convection of isolated masses of buoyant fluid, *J. Fluid Mech.*, **2**, 583–94. These experiments are also discussed in Scorer, R.S. 1958.

Natural Aerodynamics. Pergamon Press, London, ch. 7.
55. Woodward, W. 1959. The motion in and around isolated thermals, *Q. J. R. Meteorol. Soc.*, **85**, 144–51.
56. Turner, J.S. 1963. Model experiments relating to thermals with increasing buoyancy, *Q. J. R. Meteorol. Soc.*, **89**, 62–74.
57. Plank, V.G. 1965. *The Cumulus and Meteorological Events of the Florida Peninsula during a Particular Summertime Period*. Environmental Research Paper 151, Air Force Cambridge Research Center, Bedford, Mass.
58. Plank, V.G. 1969. The size distribution of cumulus clouds in representative Florida populations, *J. Appl. Meteorol.*, **8**, 46–67.
59. Saunders, P.M. 1961. An observational study of cumulus, *J. Meteorol.*, **18**, 451–67.
60. Glass, M., and Carlson, T.N. 1963. The growth characteristics of small cumulus clouds, *J. Atmos. Sci.*, **20**, 397–406.
61. Orville, H.D. 1965. A photogrammetric study of the initiation of cumulus clouds over mountainous terrain, *J. Atmos. Sci.*, **22**, 700–709.
62. Malkus, J.S. 1960. Recent developments in studies of penetrative convection and an application to hurricane cumulonimbus towers, in *Cumulus Dynamics*. Pergamon Press, London, pp. 65–84.
63. Telford, J.W., and Warner, J. 1962. On the measurement from an aircraft of buoyancy and vertical air velocity in cloud, *J. Atmos. Sci.*, **19**, 415–23.
64. Cunningham, R.M., Plank, V.G., and Campen, C.M. 1956. *Cloud Refractive Index Studies*. Geophysics Research Paper 51, Geophysics Research Directorate, Air Force Cambridge Research Center, Bedford, Mass.
65. Squires, P. 1958. The spatial variation of liquid water and droplet concentration in cumuli, *Tellus*, **10**, 372–80.
66. Warner, J. 1955. The water content of cumuliform cloud, *Tellus*, **7**, 449–57; Warner, J., and Squires, P. 1958. Liquid water content and the adiabatic model of cumulus development, *Tellus*, **10**, 390–94.
67. Chuvaev, A.P., and Kryukova, G.T. (1954) quoted (p. 102) in Khrgian, A. Kh., Ed. 1961. *Cloud Physics*. Translated from the Russian by the Israel Program for Scientific Translations, and published by the Office of Technical Services, U.S. Department of Commerce, Washington, D.C.
68. Ackerman, B. 1959. The variability of water contents in tropical cumuli, *J. Meteorol.*, **16**, 191–98.
69. Day, G.J. 1956. *Further Observations of Large Cumuliform Clouds by the Meteorological Research Flight*. M.R.P. 980, Meteorological Office, London (a paper of the Meteorological Research Committee, London, available from the Library of the Meteorological Office).
70. Battan, L. J., and Reitan, C.H. 1957. Droplet size measurements in convective clouds, in *Artificial Stimulation of Rain*. Pergamon Press, London, pp. 184–91.
71. Draginis, M. 1958. Liquid water within convective clouds, *J. Meteorol.*, **15**, 481–85.
72. Malkus, J.S. 1954. Some results of a trade-cumulus cloud investigation, *J. Meteorol.*, **11**, 220–37.
73. Byers, H.R., and Braham, R.R. 1949. *The Thunderstorm*. U.S. Department of Commerce, Washington, D.C., p. 20.
74. Sulakvelidze, G.K., Bibilashvili, N.Sh., and Lapcheva, V.F. 1965. *Formation of Precipitation and Modification of Hail Processes*. Translated from the Russian by the Israel Program for Scientific Translations, 1967, and published by the U.S. Dept. of Commerce, Clearinghouse for Federal Scientific and Technical Information, Springfield, Va., p. 48.
75. Asai, T. 1968. Cellular cumulus convection in a moist atmospheric layer heated below, *J. Meteorol. Soc. Japan*, **46**, 301–6.
76. Fraser, A.B. 1968. The white box: The mean mechanics of the cumulus cycle, *Q. J. R. Meteorol. Soc.*, **94**, 71–87.
77. Staley, D.O. 1965. Radiative cooling in the vicinity of inversions and the tropopause, *Q. J. R. Meteorol. Soc.*, **91**, 282–301.
78. Simpson, J.S., and Wiggert, V. 1969. Models of precipitating cumulus towers, *Mon. Weather Rev.*, **97**, 471–89.
79. Simpson, J., Brier, G.W., and Simpson, R.H. 1967. Stormfury Cumulus Seeding Experiment 1965: Statistical analysis and main results, *J. Appl. Meteorol.*, **24**, 508–21.
80. Simpson, J. 1967. An experimental approach to cumulus clouds and hurricanes, *Weather*, **22**, 95–114.
81. Ogura, Y., and Phillips, N.A. 1962. Scale analysis of deep and shallow convection in the atmosphere, *J. Atmos. Sci.*, **19**, 173–79.
82. Ogura, Y. 1962. Convection of isolated masses of a buoyant fluid: a numerical calculation, *J. Atmos. Sci.*, **19**, 492–502.
83. Ogura, Y. 1963. The evolution of a moist convective element in a shallow conditionally unstable atmosphere: A numerical calculation, *J. Atmos. Sci.*, **20**, 407–24.
84. Matvejev, L.T. 1964. On the formation and development of layer clouds, *Tellus*, **16**, 139–46.
85. Orville, H.D. 1965. A numerical study of the initiation of cumulus clouds over mountainous terrain, *J. Atmos. Sci.*, **22**, 684–99.
86. Gadd, A.J., and Keers, J.F. 1970. Surface exchanges of sensible and latent heat in a 10-level model atmosphere, *Q. J. R. Meteorol. Soc.*, **96**, 297–308.
87. Haman, K. 1969. On the influence of convective clouds on the large scale stratification, *Tellus*, **21**, 40–53.
88. Lilly, D.K. 1968. Models of cloud-topped mixed layers under a strong inversion, *Q. J. R. Meteorol. Soc.*, **94**, 292–309.
89. Betts, A.K. 1973. Non-precipitating cumulus convection and its parameterisation, *Q. J. R. Meteorol. Soc.*, **99**, 178–96.
90. Kondrat'yev, K. Ya. 1956. *Radiative Heat Exchange in the Atmosphere*. Pergamon Press, Oxford, 1965.
91. Monteith, J.L., and Szeicz, G. 1962. Radiative temperature and heat balance of natural surfaces, *Q. J. R. Meteorol. Soc.*, **88**, 496–507; Stanhill, G., Hofstede, G.F., and Kalma, J.D. 1966. Radiation balance of natural and agricultural vegetation, *Q. J. R. Meteorol. Soc.*, **92**, 128–40; Lumb, F.E. 1964. The influence of cloud on hourly amounts of total solar radiation at the sea surface, *Q. J. R. Meteorol. Soc.*, **90**, 43–56.
92. Palmén, E., and Newton, C.W. 1969. *Atmospheric Circulation Systems*. Academic Press, New York.
93. Crutcher, H.L., Hunter, J.C., Sanders, R.A., and Price, S. 1950. Forecasting the heights of cumulus cloud tops on the Washington-Bermuda airways route, *Bull. Am. Meteorol. Soc.*, **31**, 1–7.
94. Kassander, A.R., and Sims, L.L. 1957. Cloud photogrammetry with ground-located K-17 aerial cameras, *J. Meteorol.*, **14**, 43–49.
95. Saunders, P.M. 1963. Simple sky photogrammetry, *Weather*, **18**, 8–11.
96. Scientific and Technical Information Division, Office of Technology Utilization. 1967. Earth photographs from Gemini III, IV and V, National Aeronautics and Space Administration, Washington, D.C.
97. Lowman, P.D. 1968. *Space Panorama*. Weltflugbild Reinhold A. Müller, Zurich (69 plates).

8 Cumulonimbus convection

In this chapter, concerned with convection in which showers occur, we begin by considering the evolution of the size distribution of the droplets in cumuliform clouds and of the associated radar reflectivity (because indirect observation by radar is an invaluable help in this and other studies). In particular a relation is sought between the stage at which drops large enough to fall from the clouds are first present in significant concentrations (loosely described as the stage of shower "formation") and the size of growing cumulus.

Following the onset of a shower, universally recognized as defining the transformation of a cumulus into a cumulonimbus, the convection often becomes markedly more intense and the individual clouds larger and more persistent, although still impulsive. The cumulonimbus convection has its own characteristic stratification and will be regarded as distinct from cumulus convection. Moreover, in the presence of a suitable wind shear in the vertical it becomes rather steady, with a circulation containing not only an updraft but a pronounced neighboring downdraft passing through the region of precipitation. Accordingly, *ordinary* cumulonimbus convection is further distinguished from an *organized* form, which causes particularly severe local storms when the available potential energy is large. An intense organized storm is described, and such storms are shown to be common during some seasons in several parts of the world.

The properties of severe local storms and the conditions under which they occur are considered with reference to observations made in the United States and western Europe, and the remainder of the chapter is given to discussion of some phenomena which accompany them: whirlwinds, downdrafts, hail, and intense rains.

8.1 The formation of showers

The size distribution of cumulus droplets

In Section 5.3 the properties of the population of cloud droplets produced by condensation (without aggregation) in air ascending adiabatically were shown to be determined by the initial vapor density, by the kind of aerosol particle population present, by the speed of ascent very near the saturation level, and by the height above the saturation level. In particular, in ascent at the few meters per second characteristic of the updrafts in cumulus bases, the concentration of the cloud droplets is determined in the first few decameters above the saturation level. Unless the speed of ascent increases with height at rates (specified in the last rows of Table 5.6) much exceeding the value of about 2 m sec^{-1} km^{-1} typical of large cumulus, or unless the concentration of cloud droplets is greatly reduced by aggregation or precipitation, no further small aerosol particles grow into cloud droplets at higher levels in the cloud. Since a few decameters is a small distance compared with the dimensions of cumulus, it might be anticipated that the observed properties of droplets near the cloud bases correspond closely to those calculated on the assumption of adiabatic ascent. This appears to have been confirmed for the total number concentration N by sampling the aerosol below cumulus bases, computing a corresponding cloud droplet concentration assuming adiabatic ascent at 3 m sec^{-1}, and comparing it with a mean value of N measured at levels not more than about 300 m above their bases (1). N varied from cloud to cloud, with values as low as 50 cm^{-3} over the sea, but with an average value of 10^3 cm^{-3} inland. According to general experience N near cumulus bases well inland is usually several hundred per cubic centimeter (2–5), whereas in cumulus formed in oceanic air it is usually less than 100 cm^{-3} (3, 5).

The evolution of the concentration and size distribution of the cloud droplets beyond the earliest stages is very much complicated by the subsequent distinct departure from adiabatic conditions. The associated variability of the size spectra encountered within a cloud makes them difficult to interpret without some reference to at least a kinematical model of the air flow in the cloud.

The processes which affect the evolution of the size distribution of cumulus droplets are represented schematically in Fig. 8.1, which is based on Fig. 7.27. The regions of the cloud in which the dominant processes are the formation F, aggregation into precipitation P, and evaporation E of the cloud particles are considered to be the base, the interior, and one side of the cloud, respectively. Rain probably is generated in those parts of the cloud which contain the stronger updrafts and the larger concentrations of cloud water; these may not be well represented by summaries or averages of observations made indiscriminately throughout a cloud and including data from other and perhaps more extensive parts where the concentration of condensed water may have been trivial or much reduced by evaporation. In the cloud base the updraft speed w varies somewhat with place and time, and hence already introduces some variability in N_g and in a conveniently defined mean droplet size $\bar{r}$. (According to Section 5.3 an increase in w by a factor of about 3 may cause an increase in N_g by a factor of 2 in nucleus-rich air, and hence a decrease in $\bar{r}$ by a factor of about 1.3. In clean air the alterations would probably be trivial, since if w has the magnitude of 1 m sec^{-1}, practically all of the aerosol particles become cloud droplets irrespective of its actual value.) Above the cloud base aggregation A and the introduction of clear air, or the reintroduction of partly evaporated cloud, accompanied or succeeded by mixing M, are likely to reduce N substantially and to extend the range of droplet sizes. (Clear air surrounding a cumulus usually has a lower concentration of condensation nuclei than air which rises through the base, and is several hundred meters below its saturation level. However, some air

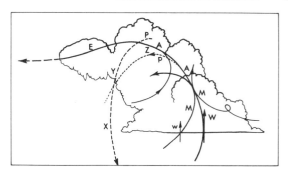

Fig 8.1 Schematic representation of processes significant in the evolution of the particle size distribution spectrum in cumulus, based on the cumulus model of Fig. 7.27. Vapor begins to condense at the cloud base C, quickly developing a size distribution of droplets which depends on aerosol properties, temperature, and the vertical velocity W of the air (which may vary appreciably in position or time). The regions in which the cloud properties are most significant for the development of rain are those containing the high concentrations of condensed water, through which a schematic air trajectory is drawn (and not those where evaporation E is dominant). Along this path the cloud droplets grow by condensation and aggregation A, with a size distribution substantially affected also by mixing M with air from other parts of the cloud or from outside it. Eventually particles develop with fall speeds of more than about 1 m sec^{-1}, which can be regarded as precipitation, since their paths now differ significantly from the air trajectories. Some precipitation p leaves the air being followed, while some P arrives from other places. Some falls out of the side or base of the cloud, and some may evaporate before reaching the surface. The complete particle spectrum is very variable according to position in the cloud model; raindrops occur in some places X alone, and in others Y with only small cloud droplets, while in other places Z there may be a practically continuous range of sizes between those of cloud droplets and of large raindrops.

entering the sides of a cloud in the circulations of the nodules may rise such a distance before there is appreciable mixing with air already containing droplets, so that condensation occurs on fresh nuclei, perhaps thus maintaining high in the cloud droplets of smaller size than would occur in adiabatic ascent.) Observations made in regions where evaporation is dominant should evidently not be combined with those of the regions richer in condensed water: mean properties obtained in this way are not likely to be useful in the construction of a theory of rain formation.

A further difficulty in the interpretation of the observations arises when the aggregation A has progressed to the stage at which considerable concentrations of cloud particles have fall speeds of about 1 m sec^{-1} or more, so that their paths differ significantly from air trajectories and they are appropriately regarded as precipitation. In some places precipitation removes condensed water from the air, while in others it arrives from elsewhere. Following the development of precipitation which leaves the side or base of the cloud, the significance of a particle

8   Cumulonimbus convection

spectrum is even more obviously impossible to appreciate without reference to some kinematical model: in some places there may be only raindrops, perhaps of nearly uniform size, in others only raindrops and sparse small cloud droplets, and in yet others a practically continuous spectrum of intermediate sizes.

The most notable feature of the size spectra produced by adiabatic ascent is that most of the condensed water is contained in an overwhelming majority of droplets of nearly the same size, whose radius is therefore given approximately by partitioning the total concentration of condensed water uniformly among the total concentration of cloud droplets. This is illustrated by the dashed curves in Fig. 8.2, which show the cumulative concentrations of an assumed spectrum of nuclei, first at a relative humidity of 75% and then after lifting to the saturation

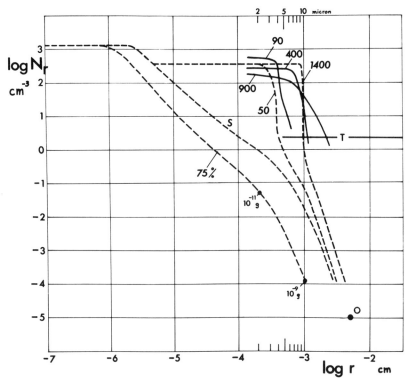

Fig. 8.2 Calculated and observed size-distribution spectra of cumulus droplets; the curves show the concentration N_r cm^{-3} of all droplets of radius larger than r cm, on logarithmic scales.

The dashed curves marked 75% and S show, respectively, assumed size distributions of sea-salt nuclei, in equilibrium at a relative humidity of 75% and on arrival at the saturation level during the adiabatic ascent of air at 1 m sec^{-1}. At the radii indicated on the first curve the giant nuclei have masses of 10^{-11} and 10^{-9} g.

The calculated size distribution of droplets at a height of 50 m above the saturation level after continued adiabatic ascent at this speed is represented by the dashed curve marked 50 (6). It shows a typical total droplet concentration of a few hundred per cubic centimeter; further condensation during adiabatic ascent does not change the total concentration (in g^{-1}) and for some distance hardly alters the concentration (in cm^{-3}). At a height of 50 m above the saturation level the upper part of the curve representing the spectrum is already steep, implying a predominant droplet radius of a few microns, and this characteristic feature of spectra produced by adiabatic ascent becomes more pronounced farther above the saturation level, as shown by the dashed curve marked 1400, representing the spectrum 1400 m above the saturation level. There the predominant droplet radius is nearly 10 μ; it is rather accurately given by dividing the concentration of condensed water by the total concentration of cloud droplets.

The solid curves are droplet spectra observed in cumulus (7); they terminate near the threshold concentration T because of the limitation of the instrumental technique. The curves marked 90 and 400 represent spectra observed at 90 and 400 m above cloud bases (Samples 105 and 113 of Table 8.1) and have the characteristic shape (and concentration of condensed water) of those produced by adiabatic ascent. More generally the observed spectra contain smaller concentrations of condensed water and droplets, and a broader range of droplet sizes, as illustrated by the curve marked 900 (Sample 140 of Table 8.1), and cannot be produced by the adiabatic ascent of air.

Below the threshold concentration T the curves representing the spectra near the cloud base probably more nearly coincide with the dashed curves produced by the growth of the giant condensation nuclei, which is determined by the duration of their exposure to saturated air rather than by the nature of the ascent. This expectation is supported by the position of the point O, which shows a typical concentration of large droplets in small cumulus in the same region, measured by a different technique (9).

level, and the corresponding spectra of cloud droplets calculated assuming further adiabatic ascent at 1 m sec^{-1} to levels 50 m (6) and 1400 m above the saturation level. During cloud formation the total concentration of cloud droplets produced is 450 cm^{-3}; at 50 m above the saturation level the concentration of condensed water is about 0.1 g m^{-3}, corresponding to a mean droplet radius of nearly 4μ. This can be seen to be about the predominant calculated radius: droplets with radii greater than 5μ are present in concentrations of less than 1 cm^{-3}. Continued adiabatic ascent with droplet growth by condensation steepens the size-distribution curve still more; 1400 m above the saturation level the predominant radius is nearly $10\,\mu$, and the concentration of droplets of radius more than $10\,\mu$ is only 1 cm^{-3}.

In striking contrast, fundamental for the rapid progress of aggregation, individual and mean droplet spectra observed in cumulus are represented by curves which in general have much less steep slopes, indicating a broader size distribution and a less obvious mean droplet size. Sometimes the spectra are distinctly bimodal, suggesting the mixture of cloudy air from two dominant kinds of source region, perhaps the deep interior of the cloud and the outer transition shell often conspicuous on horizontal traverses made to measure temperature and concentration of liquid water (Section 7.8 and Fig. 7.29). The spectra found by successive sampling even at the same level in one cloud are usually reported as very variable. In many publications such samples are given only after they have been summarized into mean values whose physical significance is obscure. One exception provides 150 individual spectra (7) from flights into cumulus over southern England on 10 separate occasions (8), some of which are listed in Table 8.1 to illustrate the different kinds encountered. During these flights samples were taken repeatedly at several levels during horizontal traverses of clouds, using a device by which a glass slide coated with a layer of magnesium oxide was briefly exposed to the airstream beside the aircraft. Cloud droplets left impact craters in the oxide coat whose diameters were found by experiment to be greater than the droplet diameters by a factor of about 1.4; the collection efficiency of the slide was thought to be practically unity for droplet radii greater than about 2.7 μ, and this value was assumed to be valid for all sizes of counted droplets, probably resulting in some underestimate of the number of droplets in the smallest of the size categories adopted (radii less than 3.2 μ). The area of slide examined to obtain each spectrum, the exposure time (corresponding to about 1 m of flight path), and the air speed were such that a volume examined was between 2 and 3 cm^3, so

Table 8.1 Individual size distributions and total number concentrations N at height z above the cloud bases, and values of the total concentration m of liquid water and of the concentration m_γ corresponding to adiabatic ascent from the cloud base, derived from data in reference 7. (a) Narrow size distributions in which N is large and $m \approx m_\gamma$; (b) N is small and m is small compared with m_γ; (c) size distribution is broad, N has intermediate values, and m is a considerable fraction of m_γ; (d) size distribution is broad and $m > m_\gamma$.

Sample number	Height (m) of cloud: Bases	Tops	z	1.3– 3.2	3.2– 4.5	4.5– 6.3	6.3– 9.0	9.0– 12.3	12.3– 18.0	18– 26	26– 36	36– 52	52– 73	N (cm^{-3})	m (g m^{-3})	m_γ (g m^{-3})	m/m_γ
a																	
102	600	1,500	90	111	171	10								292	0.05	0.15	
105	600	1,500	90	87	464	35	(4)							590	0.13	0.15	1
113	600	1,500	400		13	24	186	57	(1)					281	0.65	0.65	1
b																	
111	600	1,500	400	(1)	(5)	12	12		(0.5)					30	0.03	0.65	
124	1,000	2,700	1,200	(6)	(7)	(5)	(4)	12	(6)	(1)	(0.5)			41	0.25	2.0	
143	1,200	2,700	900	29	27	19	(6)	(4)						85	0.06	1.7	
c																	
112	600	1,500	400	9	45	43	79	40	12	(1)				229	0.60	0.65	0.9
140	1,200	2,700	900	20	24	24	55	41	12	(5)	(1)			182	0.95	1.7	0.6
144	1,200	2,700	900	25	24	22	145	66	(3)					285	0.66	1.7	0.4
146	1,200	2,700	1,200	18	44	70	70	14	(2)	(1)	(0.5)	(0.5)		220	0.51	2.0	0.4
148	1,200	2,700	1,500	17	13	19	36	17	12	(3)	(1)	(0.5)		119	0.70	2.3	0.3
d																	
133	1,200	2,700	270	(4)	10	17	62	36	53	11	(1)		(0.5)	195	1.6	0.5	
145	1,200	2,700	1,200	12	16	9	8	10	13	11	15	(4)	(1)	99	5.0	2.0	

that the threshold concentration above which the droplets were likely to be fairly sampled, as in similar impaction techniques, is several per cubic centimeter. Note that if $n(r)\,dr$ is the mean concentration of randomly distributed droplets of radius between r and $(r + dr)$, the probability P_i that in sampling a volume x the number of impacted particles will be i is $P_i = e^{-xn}(xn)^i/i!$, an expression closely corresponding to the normal distribution $(2\pi xn)^{-1/2}\exp[-(i - xn)^2/2xn]$. Thus it appears that a sample of at least 20 particles is required to provide, with a probability P of 90%, a value of particle concentration which lies within the range $n/2^{1/2}$ and $2^{1/2}n$, where n is the mean concentration. Accordingly, in Table 8.1 the entered values of concentrations of 7 cm^{-3} or less are enclosed in parentheses to indicate that they are not derived from a sufficiently large sample to be representative of the mean concentrations within the criteria stated. The lower useful limit to the concentration in which droplets can be detected is marked by the line T in Fig. 8.2.

Although the technique is subject to some inaccuracies, there is no clear evidence of any serious defects, but the droplet spectra obtained show several interesting features not all readily explicable. These are illustrated by the selected spectra specified in Table 8.1. The first three represent spectra occasionally observed at heights of up to a few hundred meters above the cloud bases, in which the narrowness of the size distribution of the droplets, their high total number concentration (several hundred per cubic centimeter), and the concentration of liquid water are such that they almost certainly indicate the adiabatic ascent of air from the cloud bases, and thus provide some confidence in the technique of observation. The first of these spectra is shown in Fig. 8.2 by the curve marked 90; it clearly has the character of the dashed curves representing spectra calculated assuming adiabatic ascent, but unlike those, its lower limit is restricted by the threshold of sensitivity (near curve T). However, that the observed spectrum actually is prolonged approximately along the dashed curves is indicated by the full circle labeled O, which represents a typical observation a few hundred meters above the base of small cumulus in the same region (9), made by a different impaction technique (using silver foil mounted on a screen, with which a larger volume can be sampled because only droplets of radius larger than about 50 μ leave an impression).

Another kind of spectrum represents a very small total concentration of liquid water, presumably from the tenuous regions found near cloud edges and commonly also well inside clouds. In such spectra the total number concentration of the droplets is only several tens per cubic centimeter, having been reduced by evaporation, probably combined with dilution during mixing with clear air, and perhaps also by some previous aggregation. Examples are given in the second group of three entries in Table 8.1.

A third kind of spectrum contains a total concentration of liquid water m which is a considerable fraction of the adiabatic value m_y, an intermediate total number concentration of about 200 cm^{-3}, and a broad size distribution, the radius of the largest droplets present in the threshold concentration of a few per cubic centimeter entering the range between about 20 and 30 μ. The five examples from two flights given in Table 8.1 illustrate that in these frequently encountered spectra the ratio m/m_y tends to decrease with height and is consistent with the mean values obtained with instruments which measure the average values of m over distance of at least some decameters (Fig. 7.34).

In some spectra, exemplified by the last pair of Table 8.1, the radii of the largest droplets extend up to values between about 40 and 70 μ, and the concentration of liquid water considerably exceeds the adiabatic value. Since the fall speeds of the larger droplets are less than about 0.5 m sec^{-1}, and they can barely be regarded when sampled as having a motion significantly different from the air motion, the adiabatic concentration of liquid water is surprising. Some such spectra are from levels near the cloud base, where this concentration may have been underestimated through inaccuracy in assessing the saturation level (from observations of the height of the cloud base), but the broad nature of the recorded size distributions at such low levels is itself remarkable and shows that this cannot be a complete explanation. Again, although in other spectra from higher levels a substantial part of the difference between the observed and adiabatic liquid water concentrations is present in large droplets whose concentrations are so small that they may not have been fairly sampled, it seems that the observations cannot on this account be regarded as merely unrepresentative. If they are not due to some other kind of error, for example, the shattering of rain on the slide-holder, then at least locally in the cloud the liquid water concentration reaches, or exceeds slightly, the adiabatic value. Other series of samplings give some support to this conclusion. In one sampling, using a technique of impaction on soot-covered slides (10), it was found that in a very small fraction of the cloud interior the inferred liquid water concentration was about equally likely to reach the adiabatic value or to fall to zero. Similarly, in observations over New Jersey and Florida (11) of droplets caught and photographed on oil-covered slides the inferred liquid water concentrations at levels between 1500 and 2000 m above cumulus bases occasionally reached or exceeded the adiabatic values (3–4 g m^{-3}). In the droplet spectra found in these studies the radii of the largest droplets extended up to about 80 μ. If the occasional high water concentrations are to be attributed to the settling into the sampled air of the larger droplets from some other region, their fall speeds, combined with settling times of order 100 sec, imply that significant variations in water concentration occur over distances of less than 100 m. These are probably beyond the resolution of instruments which measure the water concentration directly, but they are suggested by rapid large fluctuations in visual range during cloud traverses, which transmissometer records (11) show to have a space scale of some decameters.

Thus the droplet spectra with abnormally large values of the total liquid water concentration can provisionally be regarded as representing only local variations of the third and prevalent kind, whose principal characteristic is the breadth of the size distribution. A particularly interesting feature is the concentration of droplets with radii larger than about 20 μ, which commonly reaches a value as high as 1 cm^{-3}. According to the discussion in Chapter 5, the small relative fall speeds and collision efficiencies of droplets with radii smaller than about 30 μ imply that aggregation plays an insignificant part in their growth; Fig. 5.2 shows that in a liquid water concentration of about 2 g m^{-3} in the form of uniform droplets of radius about 8 μ (near the level 1 km above the base of the model cloud), the average rate of growth by aggregation of a droplet of radius between 20 and 30 μ is between 10 $\mu/10^4$ sec and 10 $\mu/10^3$ sec (it will later be shown that a very small proportion of droplets may grow several times faster than the average rate, but this statistical result does not explain the high concentrations noted for droplets with radii of a little more than 20 μ). Since 10^3 sec can be regarded as an upper limit to the period available for growth in the lower parts of a cumulus, it seems very unlikely that aggregation can contribute appreciably to the production of droplets with radii up to between 20 and 30 μ.

On the other hand, in a model cloud in which only condensation occurs, if the concentration of droplets is some hundreds per cubic centimeter, the prevalent droplet radius can hardly exceed to 10 μ, and only the giant condensation nuclei can attain radii as large as 20–30 μ. Their hygroscopicity is so large that their growth rate is virtually independent of the supersaturation in the cloud (as illustrated in the upper rows of Table 5.4). In almost any circumstances, after growth by condensation in the cloud for a period of magnitude 10^3 sec, they compose a droplet spectrum like that shown by the lower part of the dashed line labeled 1400 in Fig. 8.2 (corresponding to a growth period of 1400 sec). This spectrum contains droplets of radius larger than 20 μ in concentrations of not more than about 10^{-2} cm^{-3}, at least two orders of magnitude smaller than those observed in the broad droplet spectra, represented in Fig. 8.2 by the continuous curve marked 900 (sample 140, from Table 8.1c).

Giant condensation nuclei therefore cannot be the source of the droplets of radius between about 20 and 30 μ which are observed in the broad size-distribution spectra. Their presence has special interest because of their importance as evidence of the bridging of the two regimes in which condensation on the one hand and coalescence on the other are dominant. If it is accepted that they are produced by condensation only, in conjunction with some particular (nonadiabatic) kind of air motion, it seems necessary to suppose that locally in the updrafts the concentration of droplets is not more than a few tens per cubic centimeter (i.e., not more than about a tenth of the values of a few hundreds per cubic centimeter, which seem to be typical near the bases of the clouds under discussion). Such comparatively low values can be anticipated near cloud edges, and perhaps also locally in a cloud interior, where mixing with introduced clear air has led to dilution and incomplete evaporation. (They are present in some of the reported droplet samples; Table 8.1b.) If air from such regions is drawn back into the cloud and lifted several hundred meters before there is significant further mixing, concentrations of about 1 g m^{-3} water are condensed on the sparse droplets (considerations of the kind in the last part of Section 5.4, which led to Table 5.6, suggest that if the original mean radius of the droplets is several microns and their concentration not much less than one tenth of that produced in the updrafts at the cloud base, then condensation nuclei also present are probably not activated and the condensation proceeds upon the existing droplets only). Thus small droplets in concentrations of order 10 cm^{-3}, which may be characteristic of an appreciable fraction of the cloud volume, can by motions of scale somewhat smaller than that of the cloud itself attain radii of 20–30 μ locally in the cloud interior. There, after the air containing them has been mixed with large proportions of the surrounding cloud, they could be more widely distributed among smaller droplets in the commonly observed concentrations of the magnitude 1 cm^{-3}. A more careful examination of the suggested mechanism for the production of the large droplets, based on more extensive and reliable sampling (related to some kinematical model of the cloud), would be necessary to confirm whether it is the essentially nonadiabatic process which must be sought. At present, the plausibility of some such process and of the individual observed droplet spectra which have been quoted suggest that within a few hundred meters of a cumulus base some large droplets of a size (radius 30 μ or more) and concentration (approaching 1 cm^{-3}) sufficient to promote the aggregation of the cloud droplets at a significant rate are readily produced when the aerosol is such that droplet concentrations of a few hundred per cubic centimeter arise at the cloud base. They should be produced still more readily in "cleaned" or maritime air (in which the latter concentrations are less than 100 cm^{-3}), but perhaps

8 Cumulonimbus convection

hardly at all in the nucleus-rich air found in and downwind of prolific nucleus sources (in which the latter concentrations approach 10^3 cm^{-3}).

Stages in the evolution of the droplet spectra found in large cumulus are further illustrated in Fig. 8.3, in which the scales of Fig. 8.2 have been extended to allow the inclusion of precipitation particles. The sloping straight line represents a particle spectrum which contains a liquid water concentration of 1 g m^{-3} in each unit interval of ($\log r$). The corresponding total concentration is a few grams per cubic meter, the magnitude of the concentrations typical of both cloud droplets (which enter the clouds in updrafts of a few meters per second) and precipitation particles (which generally leave the cloud at speeds not much greater). The diagram also contains typical spectra of aerosol particles, and again indicates the sparsity of the large cloud droplets which can be produced by the giant condensation nuclei alone during the period of about 10^3 sec provided inside cumulus clouds.

In the upper part of the diagram is a droplet spectrum not much different from those of group c in Table 8.1, which are represented by sample 140 and the curve 900 of Fig. 8.2. It is a mean spectrum (3), regarded also as representative, from the upper parts (probably between 2 and 3 km above the cloud bases) of large cumulus in the central United States which did not produce rain [as judged by the absence of echoes on the aircraft radar (12), implying a radar reflectivity factor $Z < 10$ mm^6 m^{-3}]. Because the volume of individual samples was about 8 cm^3, the mean spectrum can be extended downward on the diagram to a concentration of 10^{-1} cm^{-3}, at which the droplet radius is almost 30 μ. The curve has been prolonged to pass between the points B and C, which represent a range of concentrations commonly found by the silver foil technique for droplets of radius greater than 50 μ in large cumulus over England (9). It is further extended to a point D, representing an estimated concentration and radius for still larger drops in clouds in maritime air over southern England on three successive days when the cloud thickness was about 1500 m (Fig. 7.12); none of the clouds examined produced a shower (13). The complete curve is drawn to emphasize the evidence that cumulus not producing showers often contain drops large enough to be regarded as precipitation particles, but in very small concentrations.

A second composite spectrum entered as a dashed curve was obtained in the upper parts of oceanic cumulus; again the sampling was partly by collection upon slides (3) and partly, in another series of flights, by the silver foil technique (14). The total number concentration of droplets has the small value typical of clouds formed in clean air. As might therefore be anticipated from the previous discussion, the concentrations of droplets with radii in the range 20–30 μ is considerably greater than

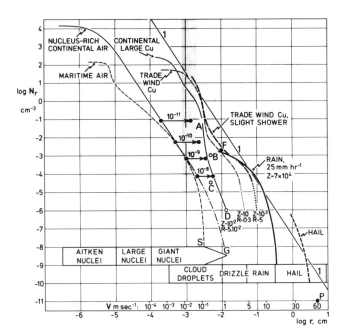

Fig. 8.3 Typical observed size distribution spectra of particles in precipitation, in cumulus, and in aerosol (left-hand pair of curves, extended down to the point S representative of the sea-salt nuclei in fresh winds over the ocean, and the point G which may be representative of the largest particles in nucleus-rich continental air). The arrows marked 10^{-11} to 10^{-8} show the growth of salt nuclei of these masses (in g) which occurs by condensation after immersion in saturated air for 10^3 sec.

The cloud droplet spectrum labeled Continental Large Cumulus is a mean spectrum from the upper parts of clouds in the central United States (3); it does not differ significantly from the broad spectra of Table 8.1c, obtained in smaller clouds over England. Below the point A, the threshold concentration of the instrumental technique, the curve is extended to pass between the points B and C, indicating the typical range of concentrations of larger droplets observed over England by the silver foil technique (9), and to the point D (13), to give a spectrum probably representative of the upper parts of large clouds not producing showers.

The thin dashed curve is a mean spectrum for the upper parts of trade wind cumulus about 2 km thick (3, 14). The thick dashed curve is an individual spectrum obtained at a height of nearly 2 km in a trade wind cumulus producing a slight shower (15).

The minimum concentration detectable in the techniques used to obtain the latter two spectra is more than 10^{-5} cm^{-3}; both curves have been extended downward arbitrarily to a concentration of 10^{-6} cm^{-3} without significantly affecting the approximate values entered for the radar reflectivity factor Z (mm^6 m^{-3}) and the rainfall rate R (mm hr^{-1}). The point F marks a concentration of uniform particles which is estimated in the text as just sufficient to be visible in a fibrillation trail.

The solid thick curve represents the average spectrum of rain sampled at the ground in moderately intense showers (16).

Inferred concentrations aloft for the largest hailstones in two severe storms in England (19) are shown by the remaining curve, and for very large hailstones which fell at Potter, Nebraska, in 1928 (20) by the point P.

in clouds examined over land. The lower part of the curve represents a stage in the advance toward the kind of spectrum which contains rain, during which the curve composing the spectrum extends downward on the diagram and shifts to the right. A still more advanced stage is shown by another spectrum, from the upper part of a trade wind cumulus, which produced a slight shower (15). In its lower part this spectrum is almost coincident with the upper part of an average spectrum of raindrops at the ground in England when the rainfall rate is 25 mm hr^{-1} (16), typical of a moderately intense shower.

In general, individual and mean droplet spectra obtained in and near cumulus clouds are intermediate between those approximating to the steep curve 90 of Fig. 8.2 (extended down through the point O even in small cumulus if the technique of observation is sufficiently sensitive) and those approximating to the curves in Fig. 8.3 representing the shower cloud and the rain at the ground. Locally in cumulus the particles from certain size ranges may be much depleted or even absent (e.g., at Y in Fig. 8.1), and the droplet spectrum may not be continuous or may have more than one radius at which dN/dr has a maximum value.

Accordingly, it is only the mean values of large numbers of individual spectra which provide smooth size-distribution curves that can be described reasonably well by simple expressions. Such formulas are sometimes useful in the estimation of radar reflectivity or optical properties, for example. One is

$$a - F = \exp\left[-(r/a)^c\right] \qquad (8.1)$$

where F is the fraction of the liquid water concentration contained in droplets of radius less than r, and a and c are constants (17). Another is

$$n(r) = ar^2 \exp\left[-br\right] \qquad (8.2)$$

where $n(r)dr$ is the concentration of droplets with radii in the interval between r and $(r + dr)$, a is a constant,

$$b^{-1} = (1/3N)\int_0^\infty rn(r)\,dr$$

and

$$N = \int_0^\infty n(r)\,dr$$

is the total concentration of droplets (18). However, these and more complicated expressions have no obvious physical significance, and while for some purposes it may be convenient to use the simplest, most meteorological discussions are better based on individual observed or assumed distributions.

The size distribution and radar reflectivity factor Z of raindrops

Many observations have been made, mostly at the ground, of the size distribution of the drops in various kinds of rain, and these have been used to obtain empirical relations of great practical value between the rainfall intensity R, liquid water concentration M, and radar reflectivity factor Z, where

$$Z = \int_0^\infty n(D)D^6\,dD$$

[usually expressed in mm hr^{-1}, g m^{-3}, and mm^6 m^{-3}, respectively (21)]. The results of sampling volumes of order 1 m^3 in individual rains show considerable variability (22), but a simple size-distribution law which is reasonably accurate for raindrops with diameters larger than about 1 mm, especially as an average over the much larger volumes and time intervals of minutes or longer that are involved in the measurement of rainfall by radar, is the exponential relation

$$n(D) = N_0 e^{-\Lambda D} \qquad (8.3)$$

where D = drop diameter (cm)
$n(D)dD$ = concentration (cm^{-3}) of drops with diameters between D and $(D + dD)$
N_0 = 0.08 cm^{-4}

and

$$\Lambda = 3.67/D_0 = 41\,R^{-0.21}\,\text{cm}^{-1} \qquad (8.4)$$

where D_0 is the volume median diameter (half of the drops have smaller volumes) and R is expressed in mm hr^{-1}. The concentration N_D of all drops with diameters greater than D is

$$N_D = (N_0/\Lambda)e^{-\Lambda D} = (N_0/\Lambda)10^{-\Lambda D\,2.3} \qquad (8.5)$$

A little over 4/5 of the mass concentration M is held in drops with diameters in the range between $D_0/2$ and $2D_0$.

Other empirical relations derived directly from the observations providing this size distribution law are

$$Z = aR^b = 2.4 \times 10^4 M^{1.82} \qquad (8.6)$$

where

$$a = 200,\quad b = 1.6,\quad \text{and}\quad M = 0.072R^{0.88} \qquad (8.7)$$

where Z, R, and M are expressed in units of mm^6 m^{-3}, mm hr^{-1}, and g m^{-3}, respectively (Fig. 8.4).

The values of the coefficients a and b quoted for the relation (8.6) were originally derived for widespread rains, and subsequently other values have been proposed as better averages for shower and thunderstorm rains, for example, for thunderstorms in Illinois (21):

$$a = 486, b = 1.37$$

8 Cumulonimbus convection

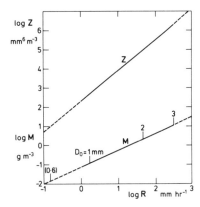

Fig. 8.4 Variation of radar reflectivity factor Z and liquid water concentration M with the rainfall intensity R, according to relation (8.6). The lines are dashed where they are extrapolated beyond the range of rainfall intensities in the observations upon which the relation is based. Some values of median raindrop volume diameter D_0 are entered beside line M.

and from observations made in Illinois, Florida, Japan, and Germany (23):

$a = 450, b = 1.46$ for thunderstorms
$a = 300, b = 1.37$ for showers
$a = 205, b = 1.48$ for widespread rains

However, these several pairs of values used with rainfall intensities within the large range 1–100 mm hr^{-1} lead to values of log Z which differ from those implied by (8.6) by less than about 0.2 (2 db), which is hardly within the resolution of a radar even in favorable circumstances.

The expressions (8.3) and (8.4) generally lead to overestimates of the concentrations of the smaller raindrops (those with diameters less than about 2 and 1 mm, respectively, in showers of moderate and slight intensity), which are those whose size is most affected by evaporation during fall toward the ground. On the other hand, the expressions lead to great underestimation of concentrations if they are applied in the size range of cloud droplets ($D < 10^{-2}$ cm). The expressions evidently represent a law which is approximately valid for a population of drops that has arisen by the coalescence of cloud droplets under gravity. Computations indicate that over a considerable range of shower rain intensities the implied raindrop population removes cloud water at about the typical rate at which it is provided by condensation, and that initial size spectra which are exponential but with slopes steeper than the value $-\Lambda$ indicated by the law, or which have no slope at all, are quickly brought toward it during fall through and beneath cloud.

Visual aspects of shower formation

During the diurnal evolution of cumulus convection overland it frequently appears that showers fall only from those clouds whose tops reach some particular level, or, perhaps, some particular height above the level of the cloud bases. This condition might be anticipated on the basis that the relevant cloud properties (e.g., condensed water concentration and updraft speed) of those parts of the cloud in which the rain is generated are principally a function of height above the base, and, further, that the generation of rain demands a minimum period of evolution of the droplet spectrum which is provided only when the cloud depth is sufficiently great (Fig. 8.5).

The required cloud depth often seems to be well defined on any one day, but it varies substantially with the meteorological situation. It is smallest in maritime air, often less than 2 km, so that if the surface temperature is about 20°C or more, the cloud tops may be below the 0°C level. Inland in middle latitudes the required depth is usually considerably more and the cloud bases are higher, so that the temperature of the cloud tops is well below 0°C, while in high latitudes or over arid regions the clouds may be wholly above the 0°C level, and in these circumstances the ice phase may appear in the cloud water and affect the development of the particle size spectrum.

The first visible signs of shower development in large

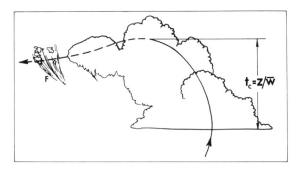

Fig. 8.5 Schematic representation of the concept of the time available for the generation of rain in a cumulus. The line through the model cloud represents an idealized (smoothed) trajectory of air through the cloud; it is drawn continuously through the regions of strong updrafts and greatest concentrations of condensed water, and dashed where evaporation is dominant. Cloud droplets are formed at the cloud base and their size distribution evolves to produce increasing concentrations of the larger drops, and eventually, if a sufficient time t_r is available, significant concentrations of drops of diameters about 1 mm, so that the cloud can be said to contain rain which will eventually reach the ground. The period t_r is dependent primarily upon the initial droplet size distribution and the small-scale processes of condensation and aggregation which cause its evolution, and the production of the rain demands that $t_r < t_c$, where t_c is an available time dependent on the large-scale air motion in the cloud; it is approximately $z/\bar{w}$, where z is the height above the cloud base attained by the volumes with high concentrations of condensed water, and $\bar{w}$ their mean rate of rise. If t_r is somewhat less than t_c, trails of smaller drops may be seen among the dissolving cloud summits, producing the phenomenon of fibrillation (F).

cumulus can often be detected by observing the dissolution of cloud towers. Commonly a tower can be seen to rise 1–2 km into clear air, and to evaporate completely during a smaller subsequent subsidence, so that it does not meanwhile become obscured by other parts of the parent cloud. (Since the mixing with the surrounding air occurs predominantly while the tower is rising, this behavior demonstrates its content of condensed water to have been only a fraction of that corresponding to adiabatic ascent.) Moreover, as previously remarked, the evaporation of the tower does not usually proceed steadily from the outside; rather, a stage is reached in which only a few shreds of cloud remain throughout the volume previously occupied by the tower, as illustrated in Pl. 8.1. Although these shreds have a ragged texture, their edges are well defined, and they finally disappear quickly without much subsidence, showing that the droplets of which they are composed are small. However, if a cloud tower has reached sufficiently high, among these residues there are faint diffuse trails which are narrower than the original tower, and evidently are composed of particles large enough to have significant fall speeds through the air (about 1 m sec^{-1} or more); they usually survive the shred clouds by 2–3, so that the whole period of their evaporation is about 5 min. This phenomenon is called *fibrillation* (Pl. 8.1 and 8.3).

The size of the particles in these trails can be estimated by assuming that the evaporation occurs in surroundings of mean relative humidity about 90%. Then by introducing values of $D = 0.4$ cm^2 sec^{-1} and $[\rho_v - \rho_{S(T_w)}] \approx 0.1$ g m^{-3} ($\rho_{ST} \approx 2$ g m^{-3}) into Eq. 5.6 and 5.58 as appropriate to the middle troposphere, a period of 5 min for complete evaporation and a corresponding fall path of 300 m are found to imply an original droplet radius of about 10^{-2} cm, for which the fall speed is about 1 m sec^{-1}. It therefore seems reasonable to suppose that the fibrillation streaks contain particles of about this size, in concentrations just sufficient to make them visible.

The concentration which a cloud of such particles must have if they are to be visible depends not only on their nature and the dimensions of the cloud, but also on the conditions under which they are illuminated and viewed. However, a magnitude can be estimated, if they are assumed to be liquid spheres, to have a narrow size distribution about an average radius r_p of 10^{-2} cm, and to be present with a total concentration N at a distance D of about 60 km, in a volume whose dimension L along the line of sight is a fraction of the original diameter of the parent cumulus tower, say about 1 km. If the line of sight passes through air in and below the cumulus layer, all in sunshine, it can be supposed that the brightness of the sky in the vicinity of the cloud is due practically entirely to the scattering of sunlight by the intervening aerosol particles (of radius r) in the lower troposphere, effectively over a path length of about $D/2$. It has already been remarked in Chapter 4 that the optically effective aerosol particles are the large nuclei ($r > 10^{-5}$ cm); inspection of Table 4.2 shows that the average value of the scattering function appropriate to these particles (which determines the intensity of the incident radiation scattered in particular directions) is about equal to that for the optically very large particles of radius $r_p \approx 10^{-2}$ cm, regardless of the scattering angle (in the present context this varies with the angle between the line of sight and the sun). Hence the brightness of the cloud of precipitation, and of the neighboring sky (determined by the integrated radiation scattered toward the eye by the particles within a small solid angle), is proportional simply to

$$\frac{D}{2} \int n(r)r^2 \, dr + LNr_p^2 \quad \text{and} \quad \frac{D}{2} \int n(r)r^2 \, dr$$

respectively. In nucleus-rich continental air

$$\pi \int Q(x) n(r) r^2 \, dr \approx 10^{-6} \text{ cm}^{-1}$$

(and in cleaner air may be less by a factor of up to about 5), where an average value of $Q(x)$ is about 2 (Fig. 4.4). Then, assuming that the cloud of precipitation is just visible if its brightness is a small percentage above that of the background sky,

$$2 \times 10^{-2} D \int n(r) r^2 \, dr \approx 10^{-8} \frac{D}{\pi} \approx LNr_p^2$$

and

$$N \approx 10^{-8} \frac{D}{\pi} Lr_p^2 \text{ cm}^{-3}$$

Thus, inserting the suggested values $D = 60$ km, $L = 1$ km, and $r_p = 10^{-2}$ cm,

$$Nr_p^2 \approx 2 \times 10^{-7} \text{ cm}^{-1}$$

and

$$N \approx 2 \times 10^{-3} \text{ cm}^{-3}$$

A point F has been included in Fig. 8.3, corresponding to this estimate of the concentration of particles in the first detectable fibrillation trails. It lies between the drop size-distribution curves for which the radar reflectivity factor Z has the values 10 and 10^3 mm^6 m^{-3}. If in these distributions only those drops with radii greater than 10^{-2} cm are considered, the corresponding values of $\int n(r)r^2 dr$, to be compared with that of 2×10^{-7} cm^{-1} estimated above for Nr_p^2, are, respectively, about 3×10^{-7} and 2×10^{-6} cm^{-1}. In the second distribution even those drops of radius greater than about 7×10^{-2} cm, whose concentration is only about 10^{-5} cm^{-3}, provide a contribution of 3×10^{-7} cm^{-1}. Moreover, since during several minutes such drops would descend a few kilometers through clear air and so lengthen the fibrillation trails much beyond the extent usually first observed, it seems that the fibrillation is usually just detectable when the spectrum of droplets in the cloud towers has developed to some stage intermediate between the two distributions, implying a radar reflectivity factor Z of about 10^2 and a precipitation rate of about 1 mm hr^{-1} (Fig. 8.4).

In cloud towers which reach well above the 0°C level a high proportion of the particles composing the fibrillation trails are likely to be frozen. Assuming, for example, that freezing nuclei are collected during the aggregation of cloud droplets in an average concentration of 1 g m^{-3}, drops of radii 10^{-2} cm have each swept a volume of about 4 cm^3 and a high proportion are likely to be frozen if their temperature has fallen as low as about $-30°C$ (Fig. 5.4); the corresponding temperature for drops of radius 7×10^{-2} cm is about $-15°C$. The brightness and persistence of well-developed fibrillation trails of frozen particles are likely to be substantially greater than postulated for those which first become detectable by eye.

Shower formation as observed by radar

Radar echoes from the particles of cumulus clouds provide valuable evidence not only of the cloud locations (and thereby, in conjunction with visual observations, of their dimensions and rates of growth), but of the evolution of the size distributions and phase of the cloud particles.

Most of the radars so far used in meteorological studies have a sensitivity which allows the detection of an echo whose intensity corresponds to a radar reflectivity factor Z of about $10^{-1} \text{ mm}^6 \text{ m}^{-3}$ at a range of 10 km, assuming that the reflecting particles fill the beam (strictly, a volume with the dimensions of the beam widths and the pulse length). However, although very powerful modern centimetric radars exist which have a sensitivity about 100 times greater, with beam widths of only a fraction of a degree, and which would be very suitable for the study of the early stages of shower formation, their use for this purpose has not so far been reported. The conventional centimetric radars have been exploited mainly to survey the atmosphere over large horizontal areas, and their displays at very short ranges have often been dimmed to reduce the glare of intense echoes from neighboring ground features (which otherwise interferes with the examination of more distant targets), so that their useful working ranges have been several tens of kilometers. Moreover, their beam widths, at least in one direction, have been a few degrees, so that the interesting parts of cumulus in the early stages of shower formation have at such ranges not completely filled their beams. For these reasons their effective sensitivity has been appreciably reduced, and the minimum detectable echo intensity has corresponded to a value of Z usually in the range $1-10 \text{ mm}^6 \text{ m}^{-3}$. Consequently the only description yet provided of the nature of the radar echo in the very early stages of shower formation refers to records on a time-height display of a vertically pointing millimetric radar (wavelength 1.25 cm), observing large cumulus at ranges of only a few kilometers with a minimum detectable echo intensity corresponding to values of Z between about 10^{-2} and $10^{-1} \text{ mm}^6 \text{ m}^{-3}$ (25).

In these observations, made inland over Massachusetts, no echoes were received from small cumulus. Echoes were first detected in the upper parts of large cumulus (about 1500 m or more deep); they appeared with irregular heights and shapes, but with a tendency to have tufted tops (probably corresponding to individual cloud towers) and trailing bases. In a more developed stage, when the maximum values of Z exceeded $1 \text{ mm}^6 \text{ m}^{-3}$, some echoes were received from regions beyond the edges of the clouds visible overhead, and the echo trails extended down to below the cloud bases, or even to the ground.

Coordinated radar and visual observations (including double-theodolite ranging) of cumulus at distances often less than 20 km have been made inland over England, using a 10-cm radar with which a detectable echo implied a minimum value of Z of about $1 \text{ mm}^6 \text{ m}^{-3}$. It was found that echoes from the cloud particles frequently developed within cumulus only about 2000 m deep and extended down to the ground, although no fibrillation was seen in the cloud towers. In cumulus of this size towers do not project much above the general level of the tops and often subside out of sight before their evaporation can be observed. The only visual sign of rain was an occasional slightly diffuse appearance of the cloud base. The observed value of Z indicated a rainfall rate of only about 0.1 mm hr^{-1} (Fig. 8.4), or about one raindrop on 1 m^2 of ground each second.

The first detectable echo in larger clouds is about 1 km wide and 1–2 km tall, with its top a few hundred meters below the top of the visible cloud. (The columnar shape often said to be characteristic of shower echoes is due mainly to the compressed horizontal scale of the ordinary display, which shows heights up to about 15 km but ranges up to about 150 km.) Sometimes the echo fades and disappears, perhaps after a brief intensification and descent, but in growing clouds the echo intensifies rapidly, typically at the rate of several decibels per minute. It broadens somewhat and lengthens, its top remaining some hundreds of meters below the visible cloud top (and therefore fairly accurately indicating its rate of rise); its base lowers at several meters per second and after several minutes reaches the ground (26; Pl. 8.2). In experience in Arizona (27) and in England, the majority of first definite echoes, and their parent clouds, have tops which do not rise more than about 1 km during the few minutes following their recognition, and subsequently sink. Accordingly, if a radar is operated in an automatic scanning cycle designed to survey a whole hemisphere of sky over a period of several minutes (3 min is usually about the minimum practicable), the intensities and the heights of the tops and bases of the echoes which first appear in its records are more variable on any one day than when attention is concentrated on individual clouds.

The rate of rise of an echo top is approximately the rate of rise of the top of the visible cloud, and the descent of an echo top is an indication of the subsidence of the cloud top. On the other hand, the rate of descent of the base of an echo is not readily related to any vertical velocity of the air or of the particles: there may previously have been an echo, too faint to be detected, which extended down to the cloud base or even to the ground, so that the apparent descent of the echo should perhaps be interpreted rather as the intensification of progressively lower parts to the degree at which they can be observed.

When echoes are first observed from large clouds they may appear as narrowly separated columns corresponding to individual towers. Occasionally, in growing clouds, they appear for a short while to be forked, with the shape of an inverted V (an example is seen in Pl. 8.2). This probably implies that the larger precipitation particles grown in the upper parts of a persistent updraft are displaced into a sheath around it (the motion system of a thermal would favor this kind of distribution). The internal column relatively free of them is likely to have a diameter less than about half that typical of a cloud tower at the same height z, that is, less than about $z/5$ (Eq. 7.22), or about 1 km at a height of 5 km. The clear core of the echo is therefore not likely to be observed at ranges exceeding $1/\theta$ km, where θ is the nominal horizontal beam width (in radians) of the radar, and then only when the azimuth of the axis of the radar beam is carefully adjusted, as otherwise echo from just beyond the sides of the clear column must be displayed at the same range as its center. For the radar whose RHI display is seen in Pl. 8.2, $\theta \approx 3.5 \times 10^{-2}$, consistent with the maximum range of about 30 km at which forked echoes could be distinguished.

The variation on individual days of the minimum thickness of shower-producing cumulus

The dimensions of cumulus which do not produce radar echoes can readily be determined by double-theodolite ranging, the small nodules on their sharply outlined summits providing ideal sighting targets for experienced observers in telephonic communication (28). There are no such definite details in the cloud bases, but their heights are satisfactorily measured from their angular elevation, on the assumption that they have the same range as the cloud tops. Similarly satisfactory estimates can be made by observing the angular elevation of cloud bases near showers whose range is determined by radar (although care has to be taken to recognize that near intense showers the cloud base is often somewhat below the general level of the cumulus bases). Thus with a conventional radar and only one theodolite sufficiently accurate estimates can be made of the summit levels attained during the growth and decay of cumulus even if only some produce radar echoes.

During the course of a summer day of good visibility over land, observations by these means frequently can be made on several tens of individual large cumulus at the suitable working ranges between about 15 and 75 km (especially from near a coast or lake shore, because an intermediate-scale circulation there tends to remove small clouds which may hide the tops of distant large clouds). Generally on a day of scattered showers the

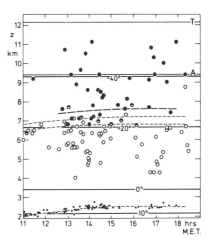

Fig. 8.6 Observations of peak heights reached by the towers of convection clouds, and of their fibrillation or glaciation, during three successive summer days of similar convection in central Sweden. The observations are based on height determinations by the double-theodolite method (28).

At the foot of the diagram the dots represent observations of the heights of cloud bases; it increased from about 2.0 km early in the day to about 2.5 km in the afternoon. Open circles show the peak heights reached by cumulus towers which evaporated without fibrillation, and the full circles the heights reached by towers which clearly showed fibrillation or became glaciated. The half-filled circles refer to towers which showed slight or uncertain signs of fibrillation.

The determinations of the peak heights of the cloud towers are likely to be accurate within ± 200 m. It appears that only those towers which reached or surpassed the narrow range of heights indicated by the thin dashed lines became fibrillated, while those towers which rose above the thick dashed line a few hundred meters higher became glaciated.

Also entered are the approximate heights of the levels at which the clear air temperature was 10, 0, -20, and $-40°C$, of the tropopause T, and of the level A of the upper surface of the cumulonimbus anvils (θ_s was about $18°C$ at both this last level and the level of the cloud bases in mid-afternoon).

8 Cumulonimbus convection

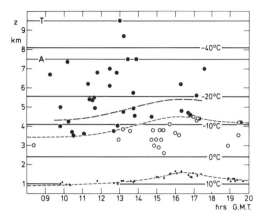

Fig. 8.7 Observations of the peak heights reached by the towers of convection clouds, and of their production of a detectable radar echo (radar reflectivity factor Z about $10 \text{ mm}^6 \text{ m}^{-3}$) or glaciation, during a day of convection over southeastern England (2 July 1956).

At the foot of the diagram the dots represent observations of the heights of cloud bases, which increased from about 1.0 km early in the day to about 1.6 km in the afternoon. Above, the open circles, dots, and full circles show, respectively, the peak heights reached by cloud towers in which no radar echo was detected, in which a radar echo was detected, and which showed both a radar echo and glaciation.

The determinations of the peak heights of the cloud towers were made by both double-theodolite and radar ranging, and are likely to be accurate within ± 200 m. It appears that only those towers which exceeded the heights indicated by the thin dashed line produced detectable radar echoes, while those which reached higher than the levels marked by the thick dashed line also became glaciated. In this and the previous figure there is some evidence that all the dashed lines, including that representing the level of the cloud base, have similar trends.

Also entered are the approximate heights of the levels at which the clear air temperature in the early afternoon was 10, 0, -20, and $-40°$C, of the tropopause T, and of the level A of the upper surface of the cumulonimbus anvils (in mid-afternoon θ_s was about 14.4°C at this last level and 14.2°C at the cloud base level).

early clouds are too small to produce rain, but eventually some become sufficiently large. According to experience in special studies made in Sweden (28) and subsequently in England and Italy, if attention is concentrated successively on growing clouds of promising size, the observations obtained can be about equally divided between those on clouds which eventually give some sign of shower development and those on clouds which do not. It appeared from these studies that the sign was recognized if the cloud summits surpassed a well-defined level z_s on any one day. Examples of observations which demonstrate this result are presented in Fig. 8.6, which includes all the observations made on three successive days of similar cumulus convection, in which the fibrillation of the cloud towers was taken as the sign of shower formation, and in Fig. 8.7, which refers to a single day over England when the criterion was the detection of a radar echo. The determinations of the cloud heights were liable to errors of up to about 200 m, but it is unlikely that the level of the extreme tops of a cumulus tower can usefully be specified with any greater accuracy. However, it was usually possible to specify a level z_s to within a few hundred meters, and even to discern some evidence that it has a diurnal trend in which the variation of height is about equal to the variation of several hundred meters observed in the level of the cloud bases (implying that the cloud thickness, rather than the height reached by its tops, is indeed a parameter of great significance in the evolution of the size spectrum of its particles).

Studies in other regions of the heights of the radar echoes first detected in cumulus clouds, some of whose results are summarized in Appendix 8.2, suggest that a level z_s is not well defined. Typically, for example, on any one day the heights of the tops of the echoes when first detected have been spread over an interval of about 3 km. However, a large part of the spread may have been due to the long period of a scanning cycle, a considerable variation with range in the effective sensitivity of the radar, a diurnal variation (see, e.g., Fig. 8.7), and by differences in the character of the clouds within the area surveyed (e.g., the cumulus convection may be intensified locally over a mountain, over a lake- or sea-breeze front, or, as will be discussed later, near an intense shower). Nevertheless, the average height reported for the echoes first observed on any one day generally has considerable variations from one occasion to another, presumably associated with changes in the large-scale characteristics of the airstreams.

The variation of the minimum thickness of shower clouds with the large-scale meteorological situation

Some typical characteristics of the clouds observed during special studies in Europe are listed in Appendix 8.1, including the minimum thickness $(z_s - z_b)$ of the cumulus associated with shower formation in the early afternoon (a time of the day when the cloud properties change only slowly, and when the stratification was observed by radiosonde and aircraft soundings). Although on any one day the thickness $(z_s - z_b)$ was definite to within a few hundred meters, it varied greatly from day to day, between a minimum value of only about 1 km and a maximum value of over 5 km.

The lowest values of the minimum thickness occur in air which can be characterized as *maritime* and which has only recently entered the Continent. This air can be regarded as having sunk in a large-scale motion system into the layer of convection over the ocean, with the low aerosol particle concentrations associated with cleaned air (Section 4.2). During travel in the fresh winds which accompany convection over the sea, large and giant condensation nuclei are abundantly replenished by sea spray (Fig. 4.1), but the total concentration of aerosol particles remains low until the air reaches the prolific land sources of nuclei, and is there modified during the course of a few days into air which can be characterized as *continental*. When maritime air enters the Continent in strong winds, then within a few hundred kilometers of the coast the minimum thickness of cumulus which produce showers is only about 1 km and the height of the cloud bases is only the several hundred meters typical over the ocean (Fig. 7.9), so that the height of the tops of shower clouds can be as little as 2 km. The minimum temperature in the clouds is then barely below 0°C even in high latitudes, and distinctly above 0°C in middle latitudes. Although there is no precise evidence, observations made over the Northeast Atlantic show that also over the open ocean showers are produced by nearly all cumulus of thickness as much as 2 km, and by a large fraction of those of thickness about 1 km (Fig. 8.8). A similar result was found over the Caribbean Sea, where individual clouds were examined at short range with airborne radar (29; Fig. 8.9 and Appendix 8.2). Its sensitivity was such that the detection of an echo implied a radar reflectivity factor Z more than about 10 mm^6 m^{-3}, corresponding to a rainfall rate of order 10^{-1} mm hr^{-1}. In these observations radar echoes were observed from few clouds of thickness less than about 1.5 km, but from all those of thickness about 3 km or more. It seems likely that if the evolution of individual clouds over the open ocean were to be examined, then just as in the studies made over land a rather definite vertical dimension associated with shower formation would be evident.

As maritime air is modified over Europe in the summer there is usually a considerable increase in both the height

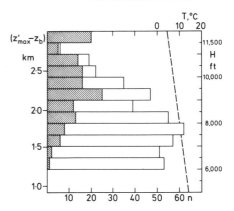

Fig. 8.9 Shower occurrence over the Caribbean Sea, as a function of the height H of the tops of cumulus encountered during 15 flights made by meteorological reconnaissance aircraft during a winter and spring (29), and of the thickness ($z'_{max} - z_b$) of the individual clouds, assuming the height z_b of their bases was 600 m.

The hatched and the clear blocks of the histogram show the number of occasions n on which radar echoes, respectively, were or were not observed when the height of the cloud tops lay in the indicated intervals of 500 ft. The minimum detectable echo intensity corresponded to a radar reflectivity factor Z of about 10 mm^6 m^{-3}. The average value of the (hardly varying) temperature T at the height H of the cloud tops is shown by the dashed line.

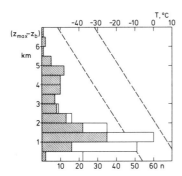

Fig. 8.8 Shower occurrence over the open ocean (the Northeast Atlantic) as a function of the maximum thickness ($z_{max} - z_b$) of convection clouds (z_{max} and z_b being, respectively, the maximum height of the cloud tops and the height of their bases); from 241 special observations made by British meteorological reconnaissance aircraft during two years.

The hatched and the clear blocks of the histogram show the number of occasions n on which showers or no showers, respectively, were recorded when the maximum cloud thickness lay within the indicated intervals of 0.5 km. The approximate minimum and maximum temperatures T at the level of the cloud tops are shown by the dashed lines.

of cumulus bases and the minimum thickness of the clouds associated with shower formation. In fresh winds from the sea this thickness has the small values of between about 1 and 2 km characteristic of the open ocean; in weaker winds its value in the afternoon is usually between 2 and 3 km, and in maritime air which has been over land for as long as 24 hr it may exceed 3 km. Sometimes the changes between one day and the next are striking. For example, in an invasion of maritime air from the Norwegian Sea into central Sweden a thickness of only about 1.1 km was observed. Overnight this air became stagnant, and on the following day it was returning slowly northward; the height of cumulus bases in the afternoon was 1.2 km, even lower than on the previous day (indicating large-scale upward motion), but the minimum cumulus thickness associated with shower formation had become about 4 km (and the height reached by the tops of the shower clouds increased from less than 3 to over 8 km). However, on the next day westerlies once more set in from the Norwegian Sea and the minimum thickness had returned to a small value (1.6 km).

On another occasion in central Sweden a shallow layer of maritime air arrived from the northwest during the morning. Well-modified air in a weak flow from the Continent farther east was by midday displaced up to a height of about 1500 m, 300 m above the level of the cumulus bases. It appeared very hazy by comparison with the air below, in which the visibility was at first more than 100 km. For about an hour the lowest parts of cumulus were seen clearly to great distances, gleaming white in the sunshine; their upper parts, which reached about 4 km, had a dusky tinge and were seen only indistinctly beyond 40 km. However, this differentiation soon disappeared, the vigorous convection evidently effectively mixing the two kinds of air, and the minimum thickness of the shower clouds which developed was 3.0 km, more typical of well-modified than of the fresh maritime air.

During a period in which special observations of shower formation were made in northern Italy, the Mediterranean was frequently invaded by maritime air from the Atlantic. Before reaching Italy the maritime air at low levels had passed over western Europe; the smallest value of the minimum cumulus thickness associated with shower formation was over 3 km, and usually it was between about 4 and 4.5 km. Such a value seems to be typical of maritime air which has become well modified after spending more than 24 hr inland. It is remarkable that the highest value (5.2 km) found in observations over southern England occurred when air at levels above about 1 km arrived from the east after having been subject to convection over central and eastern Europe on previous days. (Again, in a very small layer near the surface maritime air arrived from the North Sea, but the convection inland over southeast England during the day mixed this air thoroughly with that of Continental origin above it.)

Several series of observations made during the summer in the United States, some details of which are given in Appendix 8.2, similarly indicate that the minimum cumulus thickness associated with shower formation varies from about 3 to over 5 km. The highest values occur well inland in the central and southwestern states, and the lowest, in these data, over Texas in air probably drawn from the Gulf of Mexico (small values are also implied in the previously mentioned observations with millimetric radar, made over Massachusetts near the northeast coast).

The interpretation of the observations of shower formation

The occurrence of showers from cumulus in which the temperature is everywhere above 0°C, commonly in low latitudes and occasionally also in middle latitudes, shows that the evolution of the size spectrum of their droplets to the stage of rain formation can proceed without complications due to the development of the ice phase.

The evolutions of the droplets and of the air motions associated with large clouds are mutually dependent, since the motions are driven principally by the heat sources and sinks associated with condensation and evaporation, while, on the other hand, the motions on a range of scales modify and limit in space and time the microphysical processes of phase change and aggregation which govern the number, mass concentration, and size distribution of the cloud particles. In seeking to explain observed features of the evolution of the size distribution of the particles, and in particular the formation of showers, a beginning is made by postponing inclusion of the dynamics of the clouds, and considering the microphysical processes in the contest of some simple kinematical model of the air motions. Even so, specification of a motion system which realistically represents a cumulus, and calculation of the formation and evolution of the size distribution of the droplets within it, are complicated and difficult problems not yet treated satisfactorily.

In the simplest kind of model it is recognized that in a large cumulus the great majority of the cloud droplets acquire by condensation radii of several microns and a mass concentration of about 1 g m^{-3}, while already near the cloud base the giant condensation nuclei produce a small proportion with radii of about 30 μ (Fig. 8.3). The rate at which these collide (by differential settling under gravity) and coalesce with the smaller, providing that the latter have a mean radius of nearly 10 μ or more, is sufficient to allow them to reach the size of raindrops within the period of about 1 hr, which is the typical duration of a large cumulus (Fig. 2.8). The concentration of the giant nuclei which produce the droplets of radius about 30 μ has generally about the magnitude of that of small raindrops (10^{-4} cm^{-3}), so that their growth can be regarded as capable of leading to the production of a shower. It is supposed that the coalescence sets in at a significant rate in air which has risen only a few hundred meters from the cloud base, and that thereafter the contribution of condensation to the growth of the large droplets can be neglected (although on giant nuclei it is likely to be appreciable at first; see Fig. 5.2).

Attention is concentrated upon the growth of a single large droplet of radius R and fall speed V, among the many smaller droplets of radius r and fall speed v, which collectively compose a concentration m of cloud water. The equation which describes the mean rate of growth of the droplet is

$$4\pi R^2 \rho \, dR = \int \pi R^2 E(V - v) m(r) \, dr \, dt \qquad (8.8)$$

where ρ = density of liquid water
$E(R,r)$ = efficiency of catch (Section 2.6) of droplets of radius R among droplets of radius r
$m(r)$ = concentration of liquid water held in droplets with radii between r and $(r + dr)$

If v, even initially, is regarded as negligible compared with V, and the small droplets are considered to have uniform radii, then

$$4\rho dR/Em = dS \qquad (8.9)$$

where S is distance settled through the air. For a concentration $m = 1 \text{ g m}^{-3}$ and a final radius of 1 mm, the distance settled through the air is 4 km or more, considering that E has a maximum value of about 1.

The large droplet of initial radius R_0 can be considered to enter the base of a cloud in which the air has a constant velocity U in the vertical only; if m and r are functions of height z only, then

$$4\rho dR = E(V - v) m \, dt$$
$$dz = (U - V) dt$$

and

$$\frac{(U - V) dR}{V - v} = \frac{Em \, dz}{4\rho} \qquad (8.10)$$

The droplet introduced at the cloud base rises until $V = U$ and subsequently descends, eventually leaving the cloud with a maximum radius R'. Again neglecting v compared with V, and admitting the other reasonable approximations that V is a function only of R, and E only of z, then

$$R' - R_0 = U \int_{R_0}^{R'} dR/V \qquad (8.11)$$

The integral can be evaluated numerically after specifying V as a function of R, and is shown in Fig. 8.10 for values appropriate to the lower troposphere.

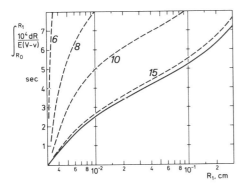

Fig. 8.10 Approximate values of the integral

$$\int_{R_0}^{R_1} dR/E(V - v)$$

(R is the radius and V the fall speed of a drop collecting smaller droplets of radius r and fall speed v, with an efficiency of catch E), displayed here for an initial radius $R_0 = 30 \, \mu$. The dashed curves show the integral for values of r of 6, 8, 10, and 15 μ, and the solid line shows the integral for $E = 1$, $v = 0$. Otherwise the values of v and V are those appropriate to 700 mb and 10°C (as in Table 2.5), and the values of E are taken from Fig. 2.6 and 2.7.

For an initial radius $R_0 = 30\mu$, R_1 is 1 mm if $U \approx 2$ m sec^{-1} and is 3 mm, about the maximum stable size, if $U \approx 4$ m sec^{-1}. These results appear to confirm that raindrops can be produced readily from droplets of the assumed initial size in cumulus containing moderately strong updrafts, but their relevance to natural shower formation is suspect in view of the implication that a smaller initial size produces a larger raindrop (arising from the unrestricted vertical dimension of the cloud), and the observation that raindrops of radius as much as 2 mm have very low concentrations (only about 10^{-6} cm^{-3}) even in intense showers (Fig. 8.3).

It is more interesting to examine the progress of the coalescence as air rises in a model cloud, for simplicity specifying U, r, and m as functions of height only, with the eventual aim of explaining the observed minimum values of the thickness of cumulus which produce rain. Further simplifications can be made to obtain some results with the minimum of complication.

First, r and v will be regarded as constants; second, the growing droplet which is followed will be considered to have a mean rate of rise w_t equal to the mean vertical velocity of a cloud tower or of some internal circulation resembling a thermal, which is reasonable while w_t is considerably greater than the fall speed V of the droplet (see the final part of the caption to Fig. 7.24). The ascent rate w_t can later be given some realistic variation with height, and, with the same approximations for E and V as before, the variables in the growth equation (8.10) can now be separated to give

$$\int_{R_0}^{R_1} dR/E(V - v) = \int_{Z_0}^{Z_1} m \, dz/4\rho w_t \qquad (8.12)$$

The integral on the left is shown in Fig. 8.10 as a function of R_1 for $R_0 = 30 \, \mu$ and several values of r, calculated for conditions appropriate to the lower middle troposphere and extracting values of E from Fig. 2.6 and 2.7. The integral on the right will be evaluated for the clouds observed in central Sweden during the afternoons of 10–12 July 1955, when observations were made which suggest appropriate values of w_t (as indicated in Fig. 7.27). The distribution of m with height above the cloud base will be taken from that drawn for the mean liquid water concentration $\bar{m}$ in Fig. 7.37, which is consistent with the generally observed trend of its ratio to the value corresponding to adiabatic ascent from the cloud base (Fig. 7.34). The corresponding distribution of the term $(m/4\rho w_t)$ is given in Table 8.2. A rather large value is inferred near the maximum height reached by the clouds, caused by the rapid decrease of the ascent speed w_t. However, it is reasonable to suppose that m also decreases rapidly in this region, and accordingly that it has no significance for the calculation of the growth of the cloud particles, which will be terminated several hundred meters below the level where $w_t = 0$.

8   Cumulonimbus convection

Table 8.2 Inferred values of the mean concentration m of cloud water in growing parts of cumulus (from Fig. 7.37), and of the ascent speed w_t appropriate to these parts and cloud towers (from Fig. 7.27), and the corresponding values of the term $m/4\rho w_t$, for the clouds observed in central Sweden in the afternoons of 10–12 July 1955

Height above cloud base (km)	m (g m^{-3})	w_t (m sec^{-1})	$(m/4\rho w_t)10^{-5}$ (sec km^{-1})
0	0	1.0	0
0.25	0.3	1.5	5.0
0.5	0.5	2.0	6.0
1.0	0.7	3.2	5.5
2.0	1.0	5.1	5.0
3.0	1.1	5.2	5.0
4.0	1.1	4.8	6.0
(4.5	1.0	2.0	12)
(4.7		0	—)

From Table 8.2 it appears that at all levels more than a few hundred meters above the cloud base the value of the term $m/4\rho w_t$ is approximately 5.5×10^{-5} sec km^{-1}; thus the value of the integral on the right of Eq. 8.12 over the interval of height between levels 500 m and 4 km above the cloud base is about $3.5 \times 5.5 \times 10^{-5}$ sec, or 2×10^{-4} sec.

Now on the occasion of these observations 4 km appeared to be about the minimum cloud thickness associated with shower formation, as indicated by the fibrillation of the cloud towers, which (from visual and radar observation) implies that drops of radius 100 μ were present in concentrations of about 10^{-3} cm^{-3}. However, from Fig. 8.12 it can be seen that according to Eq. 8.12 such drops could have been grown only from those droplets which 500 m above the cloud base had radii of nearly 40 μ, and then only if the cloud water had at all times been held in droplets of sufficient size to provide an efficiency of catch E of about the maximum possible value. Since the thickness of 4 km is one of the *highest* values thought to be associated with the production of showers by coalescence, and values of only 2 km or even less occur on occasions when the appropriate values of m and w_t are unlikely to differ much from those adopted in the present case, it seems that the coalescence proceeds substantially more rapidly than envisaged. The calculated rate could be increased somewhat by reducing the assumed values of w_t, presuming the relevant parts of the cloud to have liquid water concentrations nearer the adiabatic values (even if only locally), by appealing to condensation to increase the rate of growth of the large droplets in the early stages, or by supposing that the mixing processes in the cloud somehow considerably increase the period during which their growth proceeds. However, it will appear that the discrepancy between observation and calculation is more probably an indication that the evolution of a size distribution of droplets cannot be assessed satisfactorily by arbitrarily dividing them into two distinct groups, and examining the average rate of growth of the larger by coalescence only with the smaller.

Observed and calculated evolutions of the size distribution of cloud droplets

In the consideration of the evolution of a size distribution of cloud droplets there are some advantages in specifying the *volume* (or mass) of a droplet rather than its radius, and the *volume* of water which is held in a unit volume of air by droplets whose radii fall within chosen intervals. Figure 8.11 contains four size distributions of droplets corresponding to four typical stages in the evolution of a spectrum of cumulus droplets formed in partly modified maritime air, and containing a cloud water concentration of about 1 g m^{-3}, as described in the figure caption. The fourth distribution, which is typical of a moderate rain shower, does not include droplets of small sizes because the raindrops here are spatially well separated from the cloud droplets among which they began their growth: the complete spectrum of sizes then depends upon where the rain is sampled (e.g., whether within or outside the cloud). Since in this last stage the intensity of the rain (composed of drops with a mean fall speed of several meters per second) approximates to the flux of condensable water vapor into the cloud (several grams per cubic meter in an updraft of about 1 m sec^{-1}), the concentration of the rain water is about 1 g m^{-3}. Thus the problem of explaining the evolution of the droplet size distribution can be regarded as essentially one of explaining the transfer of a concentration of condensed water of about 1 g m^{-3} from a mean (volume) radius of about 10 μ to one of about 1 mm.

Calculations of the evolution of a size distribution of droplets subject to change by coalescence following collisions due to differential settling under gravity have so far been made only when this is the sole process of concern in the interior of a cloud and the air is motionless, and mostly also under the assumption that the cloud is sufficiently extensive for the size distribution, in the region considered, to be a function of time only. The statistical treatment of the evolution is based on the following formulation. It is assumed that the probability that the center of a droplet whose volume lies between x and $x + dx$ can be found in a unit volume of air is given by a function $n(x,t)dx$. To the first order in dx the function $n(x,t)dx$ also gives the average concentration of droplets with volumes between x and $x + dx$ at the time t (39). It is further assumed that the presence of one droplet has no influence on the probability of another being found in its neighborhood, and that the function $n(x,t)$ has a uniform distribution in space (the cloud considered has a thickness sufficient for the dis-

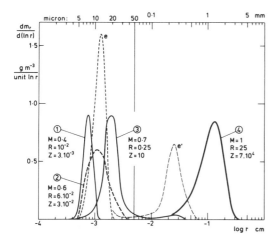

Fig. 8.11 The size distributions of cumulus droplets from smoothed observed distributions, chosen to represent four typical stages of evolution in large cumulus formed in partly modified maritime air. The distributions are represented with log r as abscissa, and with ordinate showing the mass concentration function $dm_r/d(\ln r)$, or rdm_r/dr, where dm_r is the mass concentration of water held in droplets with radii between r and $r + dr$.

The total mass concentration of cloud water is proportional to the area under the curve showing the size distribution. Values are indicated for M, the total concentration of cloud water (g m^{-3}); R, the "rainfall" intensity in still air (mm hr^{-1}); and Z, the radar reflectivity factor (mm^6 m^{-3}).

The observed size distributions, numbered 1 to 4, and an arbitrary one marked e, are as follows:

1. A size distribution of a kind sometimes found near the cloud base, whose narrowness and total cloud water concentration suggest virtually adiabatic ascent from the cloud base. It has been extended to radii greater than the maximum observed (18 μ) by including the size distribution of giant condensation nuclei represented in Fig. 8.3, after growth in a saturated environment for a few minutes.

2. A size distribution of the kind found several hundred meters above the cloud base, in which the total concentration of cloud water is a large fraction of the value corresponding to adiabatic ascent from the cloud base. It has been broadened, perhaps not so much by coalescence as by condensation during mixing processes (sample 112 of Table 8.1c, from 400 m above a cloud base, extended to include the point C on Fig. 8.3).

3. A size distribution representing the stage loosely called "shower formation," in which an appreciable fraction of cloud water has been transferred into drops of radius more than 100 μ, leading to the fibrillation of cloud towers, and the detection of a faint echo by conventional radar (based upon the average distribution in the upper parts of large trade wind cumulus not producing a visible shower, as shown in Fig. 8.3).

4. A size distribution representing an intense rain shower (as shown in Fig. 8.3). The distribution is not extended to radii of less than about 10^{-2} cm, because this part varies greatly according to the place in which the rain is sampled.

e. A size distribution of a form whose evolution by coalescence is readily calculated analytically for simple conditions and growth kernels ($M = 1$ g m^{-3} and the total droplet concentration = 239 cm^{-3}). A numerical calculation of its evolution gave the distribution e' of large droplets after 30 min (41). The distributions 1 to 4 and e' can be regarded as representative of those found successively in air rising through a large cumulus along a path drawn as a continuous line in Fig. 8.7: in a first approach to the study of shower clouds it may be possible to explain the transformations as a function only of the time or the height above the cloud base along such a path.

tribution to remain constant over the whole period envisaged).

The probability that during an interval dt a collision, assumed to be followed by coalescence, will be experienced by a particular droplet of volume x_2, radius r_2, and fall speed v_2, with a smaller droplet of volume x_1, radius r_1, and fall speed v_1, is given by the product of four terms:

1. A swept volume $\pi(r_2 + r_1)^2(v_2 - v_1)dt$.
2. The efficiency of catch E' defined not as with E in Section 2.6 (following Eq. 2.31) with respect to a swept volume of cross section πr_2^2, but with respect to one of cross section $\pi(r_2 + r_1)^2$.
3. The probability $n(x_1,t)dx_1$.
4. The probability $n(x_2,t)dx_2$.

If $P(x_2,x_1)dt$ is the probability of the collision for unit values of the last two terms (the function P is called the *kernel* and is definable for aggregation by processes other than differential settling under gravity, e.g., by Brownian motion or electrostatic attraction), then

$$P(x_2,x_1)dt = E'(x_2,x_1)\pi(r_2+r_1)^2(v_2-v_1)dt \quad (8.13)$$

It is convenient to extend the definition of P by stating that

$$P(x_2,x_1) = P(x_1,x_2), \quad x_2 \gtrless x_1 \quad (8.14)$$

Then during the same interval dt the effect of coalescences upon the concentration $n(x_2,t)$ of droplets of volume x_2, compounded of a reduction due to collisions of droplets of this size with both smaller and larger droplets, and an addition due to the coalescence of droplets of smaller volumes x_1 and $x_2 - x_1$, can be expressed by

$$\frac{\partial n(x_2,t)}{\partial t} = \frac{1}{2}\int_0^{x_2} P(x_2-x_1,x_1)n(x_2-x_1,t)n(x_1,t)dx_1$$
$$- n(x_2,t)\int_0^{\infty} P(x_2,x_1)n(x_1,t)dx_1 \quad (8.15)$$

(The factor $\frac{1}{2}$ in the first term on the right arises because when x_2 is taken as the upper limit of the integral each collision of concern is included twice.)

Analytical solutions of Eq. 8.15 have been found (39) for some particular simple kernels, for example, $P = C$ and $P = b(x_2 + x_1)$, where C and b are constants, and for two simple kinds of initial size distribution: that containing two discrete sizes, and a symmetrical or skewed Gaussian-like distribution represented by a function containing an exponential term. The particular specified forms of the kernel P have no physical basis, although with appropriate choice of the constant b the second can be made to give values very approximately those of the real kernel over a considerable range of values of r_1 and r_2 when r_2 is about 50 μ or more. The value of the analytical solutions for the evolution of the droplet spectra is in providing some insight into the important controls, and confirmation of the accuracy of numerical calculations based on more realistic and therefore more complicated collection kernels.

For the kernel of $b(x_2 + x_1)$ the analytical solution to Eq. 8.15 was first obtained for the simple initial distribution function

$$n(x, 0) = \left(\frac{N_0}{\bar{x}_0}\right) \exp(-x/\bar{x}_0)$$

where N_0 and $\bar{x}_0$ are, respectively, the initial total droplet concentration and the initial mean droplet volume. This kind of distribution is illustrated in Fig. 8.11 by curve e, representing a total mass concentration of 1 g m^{-3} and a mean droplet volume corresponding to a radius of 10 μ, for which $N_0 = 239$ cm^{-3} (in this distribution the number concentration N_r of droplets of radius greater than r is $N_0 \exp(-x/\bar{x}_0)$, or $N_0 \exp(-10^9 r^3)$).

The analytically and numerically derived evolutions of such droplet spectra (40–42) differ strikingly in some respects from those implied by considering the average growth of an individual large droplet. Growth rates both smaller and larger are given for proportions of droplets of the same size; in particular, a few grow very much more quickly (and others much more slowly). This arises from the introduction into the governing equation (8.15) of the *probabilities* $n(x_1, t)$, $n(x_2, t)$ and $n(x_2 - x_1, t)$, which may be finite within a large volume of cloud and then imply that some coalescences occur within a certain time interval, however small: droplets of all sizes are produced by coalescence up to some limit determined by the volume of the whole cloud or by some lesser volume corresponding to a droplet concentration which is regarded as the minimum that is meteorologically significant. In the latter process it is not possible to speak of the growth of individual droplets, but only of the change of the radius or radii corresponding to a specified concentration exceeding this threshold value.

From the numerical calculations using the statistical formulation, and from our previous discussion, it appears that the period of evolution to the stage corresponding to shower formation is that needed for the mass concentration of water held in droplets of radii 100 μ or more to attain a value of order 10^{-2} g m^{-3} (requiring their total concentration to be about 10^{-2} cm^{-3}). For a cloud water concentration of 1 g m^{-3}, the initial spectrum e of Fig. 8.11, and the collection kernel $b(x_2 + x_1)$ (or a more realistic kernel), this period is about 1400 sec.

The continuous model of droplet growth demands a period sufficient for individual droplets of radius about 20 μ (initially present in the required concentration of about 10^{-2} cm^{-3}) also to attain a radius of about 100 μ. This period can be estimated by evaluating Eq. 8.8 with an appropriate choice of a (constant) radius r_1 for the smaller droplets. With $r_1 \approx 12$ μ it is about twice as long as the period given by the statistical model. In the assumed kind of initial distribution, with an initial total droplet concentration of magnitude 10^2 cm^{-3}, the growth rate in the statistical model at concentrations of order 10^{-4} to 10^{-2} cm^{-3} is initially about six times greater than that in the continuous model (and at lower concentrations it is even more). This ratio decreases with time and the two rates become practically the same when $x_2 \gg x_1$. However, the implication of the comparison made is that the ratio may retain an average value as high as 2 over the period postulated as necessary for shower formation.

The curve e' in Fig. 8.11 shows a size distribution of raindrops produced soon after the stage of rain formation during the evolution of the spectrum e according to the statistical treatment. It illustrates an interesting feature of this and other calculated evolutions: in this kind of representation, of the distribution of the mass concentration of the liquid water, there is typically a minimum extending over a range of droplet radius between about 30 and 100 μ. It so happens that a similar minimum is present in the drop spectra 3 and 4 selected from Fig. 8.3 as representative of observed stages in shower development, although it is uncertain since it depends on the *curvature* of the approximate spectra drawn in that diagram. The presence of this minimum in the calculated spectra arises because of the manner in which the collection kernel increases with drop size, mainly as the drops emerge from a regime in which the fall speed is proportional to r^2 (for $r \lesssim 30$ μ) to one in which it is proportional to r (for $r \gtrsim 100$ μ; Table 2.5. The minimum gives further support for regarding a drop radius of more than 100 μ as separating cloud droplets from precipitation particles; Section 5.10). As the calculated evolution of a droplet spectrum proceeds, the curve representing the cloud droplets shrinks without much lateral spreading, while that representing the rain rapidly grows to contain an increasing proportion of the liquid water, shifts to the right, and rapidly extends to include drops of large radii (principally by mutual coalescence).

The calculations based on the statistical treatment are obviously a great advance on those considering only continuous collection, and are likely to lead to a more satisfactory explanation of the time required for shower formation and the apparent suddenness of its onset. They are also amenable to parameterization, assisting their inclusion in dynamical models of shower clouds

(42). Such models, which will need to include the processes of condensation and mixing, especially for the earliest stages of the development and broadening of droplet spectra, have not yet been constructed because of the formidable complexity.

However, consideration of the characteristic values of the (volume) mean droplet radius $\bar{r}_0 = (m/N)^{1/3}$ to be expected in rising air at a height of a few hundred meters above cumulus bases, where coalescence is first likely to set in at a significant rate, suggests that it is this property of an actual initial size distribution of droplets which most strongly affects the rapidity of its evolution toward the stage of shower formation. This radius will generally be a good measure of the size of the cloud droplets which comprise most of the condensed water. It depends mainly on the mean droplet concentration N' arising at the cloud base which, as discussed in Section 5.3, varies substantially according to the kind of aerosol from which the cloud is formed (and to a lesser degree with some other factors, more especially the speed of ascent of the air at the cloud base). It is generally between 50 and 200 cm^{-3} in clean maritime airstreams, increasing as the air is enriched with condensation nuclei during travel over land and reaching a maximum of 1000 cm^{-3} or more near prolific sources of aerosol particles. The corresponding values of the mean droplet radius decrease from about 15 μ in the cleanest air to about 5 μ in the nucleus-rich air.

The importance of its particular value within this range was already suggested in Fig. 2.8, referring to the growth by coalescence of droplets of radius R among the bulk of radius r, which implies that they cannot attain the size of raindrops within the period of less than 1 hr which is provided within cumulus unless initially R is about 30 μ or more and r is nearly 10 μ or more. It is otherwise apparent from Fig. 2.6 and 2.7 that unless R exceeds 20 μ and r about 5 μ, the efficiency of catch E (or E') is so small (less than about 10^{-2}) that the growth of the larger droplets by coalescence cannot proceed at a significant rate. When the continuous-growth model is replaced by a statistical treatment of the evolution of a size distribution of droplets these conclusions are qualitatively still valid. The sensitivity of the rate of evolution of the size distribution to the mean radius, as it varies between about 5 and 15 μ, is indicated by Fig. 8.10, which refers to the continuous-growth model. On this diagram the ordinate is, for a given concentration of cloud water held in droplets of radius r, proportional to the time taken for a droplet of initial radius 30 μ to reach a radius R_1; for a value of R_1 of 100 μ or slightly more the dashed lines show that this period has a minimum when r is about 15 μ, is nearly doubled when r is reduced to 10 μ, and increases by further substantial amounts as r is reduced still more.

The efficient onset of the coalescence process therefore requires the presence of appreciable concentrations of droplets of radius exceeding about 20 μ among a majority of radius approaching 10 μ, and the period required to reach the stage of shower formation is sensitive to the initial total droplet concentration. Probably the first requirement has less practical importance, since in most natural clouds adequate concentrations of large droplets are probably always provided by condensation, allied with a mixing process which makes it unnecessary to rely on the presence of giant condensation nuclei (as discussed earlier in this chapter, droplets of radius about 30 μ may be observed within several hundred meters of cumulus bases in concentrations of order 10^{-3} or even 10^{-2} cm^{-3}, some two orders greater than the concentrations of the giant nuclei capable of reaching this size after a few hundred seconds in the cloud). Then the large variations in the magnitude of the growth kernel associated with the changes of the efficiency of catch as $\bar{r}_0$ varies between about 5 and 15 μ are probably mainly responsible for the observed change in the minimum thickness of shower clouds as clean air moves inland from the sea and is progressively enriched with aerosol particles which act as condensation nuclei. On the basis of the observations of shower formation and the previous discussion, Fig. 8.12 has been constructed in anticipation of an improved theory, to show as functions of the thicknesses attained by the clouds over Sweden which were previously described, and to which Table 8.2 refers, the maximum values within the clouds of the following properties:

1. M, the concentration of cloud water in the form of drops of radius greater than 100 μ.
2. R, the corresponding rainfall intensity, which is a better indicator of the progress toward shower production than R', the value associated with the whole population of cloud droplets.
3. The radar reflectivity factor Z.

The diagram is drawn to imply that M becomes a large fraction of the concentration of cloud water, and R rapidly increases from insignificant to substantial values, as the peak height reached by the cloud tops increases by several hundred meters to approach and then surpass the level F at which fibrillation occurs in the summits. The observer who relies on visual evidence can readily obtain the impression that the shower formation occurs suddenly in clouds which reach a certain size. The observer with a conventional radar also finds that radar echoes associated with rain are detected only from clouds of a certain size, depending on the performance of his equipment; on the other hand, with a sufficiently sensitive radar echoes are detected even from small clouds, and a more complete impression is obtained of the rapid but progressive evolution toward shower production as the thickness of the clouds

Cumulonimbus convection

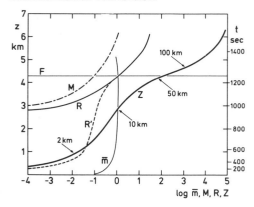

Fig. 8.12 A tentative estimation of the properties of cumulus over central Sweden on 10–12 July 1955 (see also Fig. 7.27, 7.36, and 8.6). The several curves show, as a function of height z of the tops above the bases, the maximum values within the clouds of the following properties:

M, the concentration of cloud water in the form of droplets of radius greater than 100 μ (in g m^{-3}).

R, the corresponding rainfall intensity (in mm hr^{-1}).

R', the "rainfall" intensity associated with cloud droplets of all sizes (in mm hr^{-1}).

Z, the radar reflectivity factor, in mm^6 m^{-3} [arrows indicate values which in volumes 0.5 km across provide just detectable echoes at the indicated ranges of 10–100 km from a conventional (TPS-10) radar, and at a range of about 2 km from a vertically directed millimetric radar (25). This last threshold is also appropriate to powerful and sensitive modern radars used at ranges of some tens of kilometers].

Another curve shows the mean concentration $\bar{m}$ of condensed water in air rising inside the clouds. The line F marks the level exceeded by the cloud towers which became fibrillated. The scale on the right of the diagram indicates the probable period provided for the progress of the coalescence process in air rising inside the clouds.

increases. On the right of the diagram is a time scale relating the period over which the size distribution of the cloud droplets has evolved to the cloud thickness, as implied by the model illustrated in Fig. 8.5.

With the additional assumption that in all airstreams subject to widespread cumulus convection, the rising speed of cloud air has about the same variation with height above the cloud bases, Table 8.3 can be composed as a tentative summary of the way in which the character of the aerosol controls the minimum thickness of cumulus in which showers form (by the coalescence of droplets). Typical values are included of the temperature at the height of the tops of these clouds, which over and near the oceans is well above 0°C in low latitudes and may be about or just above 0°C in middle latitudes. Elsewhere it is usually below 0°C, and well inland it is usually between about -5 and -30°C, while in high latitudes and over arid regions the temperature may be near or below 0°C even at the cumulus bases. In these circumstances the appearance of the ice phase may influence the formation of showers and certainly affects their subsequent development.

The appearance of the ice phase in cumulus

Ice particles appear in cumulus above the 0°C level, probably mostly by the freezing of droplets, in concentrations which can be expected to increase as the temperature lowers, approximately as found in experiments with cloud chambers (Fig. 5.4). The estimation of the growth of such ice particles is complicated by the variability of crystal habit with temperature (Fig. 5.5),

Table 8.3 Tentative association of the kind of airstream, the minimum thickness, and other features of shower clouds

Kind of airstream	N (cm^{-3})	$\bar{r}$ (μ)	t_s (min)	$z_s - z_b$ (km)	z_b (km)	z_s (km)	Warm airstreams T_b (°C)	T_s (°C)	Cool airstreams T_b (°C)	T_s (°C)
Clean maritime air over the sea or near coasts										
In fresh winds	50	15	10	1	0.6	1.6	18	11	7	0
Otherwise up to about	200	10	13	2	1	3	16	2	5	-10
Clean maritime air enriched	300	8	17	3	1.5	4.5	14	-5	2	-20
with nuclei after 1 or 2	to	to	to	to	to	to				
days inland	500	7	20	4	2	6	10	-15	0	-30
Continental nucleus-rich air	$>10^3$	<5	>25	>5	>2	>7	>10	<-25		

Key:
N = total concentration of droplets a few hundred meters above cloud bases (where the coalescence process effectively begins)
$\bar{r}$ = the corresponding (volume) mean radius
t_s = the minimum period required for the production of a shower, associated with a minimum cloud thickness $z_s - z_b$
z_b = the height of cloud bases
z_s = the minimum height of shower cloud tops
Typical temperatures T_b and T_s at the heights z_b and z_s are given for the cool airstreams of middle latitudes ($\theta_s \approx 10$°C) and the warm airstreams of low latitudes or continental summers ($\theta_s \approx 20$°C).

and the consequent difficulty of assessing the fall speeds and efficiencies of catch which determine the contribution of the accretion with unfrozen droplets. According to the rough estimates made in Chapter 5 (see Table 5.15), ice crystals growing among the unfrozen droplets of a cumulus are likely to develop into small, soft hailstones within several hundred seconds; if present in concentrations of order 10^{-4} cm^{-3} or more they would then comprise a considerable fraction of the cloud water and so this period can be regarded as that required for the formation of a shower by the growth of ice particles. Since ice nuclei are usually active in this concentration at temperatures below about $-10°C$ (Fig. 5.5), a cloud whose tops surpassed the level of this temperature by rather more than 1 km could be expected to produce a shower solely by the development of the ice phase, irrespective of the coalescence of cloud droplets. Moreover, even if the concentration and size of the cloud droplets were too small for their accretion to be significant, Table 5.12 indicates that after several hundred seconds in the cloud the ice crystals may by condensation alone acquire fall speeds of 0.5 m sec^{-1} or more and a size which allows them to settle several hundred meters through clear air before evaporating. Nevertheless, although in these ways showers of soft hail or trails of snow crystals may develop in small cumulus in very cold weather or in small cumuliform clouds in the middle troposphere, it appears that in the more general circumstance that the cumulus base is at about or below the 0°C level the development of frozen precipitation and the transfer of a substantial part of the cloud water into the ice phase, accompanied by the glaciation of the upper parts of the cloud, does not occur until the coalescence process has advanced to the stage of shower formation.

The glaciation of the cloud tops, which in extratropical regions frequently accompanies the development of showers, was long held to be evidence that the introduction of a small proportion of ice crystals among supercooled cloud droplets is a necessary condition for the formation of a shower. (The rapid initial growth of the crystals by condensation, to a size at which accretion of droplets could begin, was regarded as necessary to produce a broad enough size distribution of particles for their aggregation under gravity to proceed efficiently.) However, careful visual observations, particularly in low latitudes, are sufficient to throw doubt on this view, and more conclusive evidence that the development of the ice phase usually accompanies or follows the formation of showers by the coalescence of droplets, rather than the reverse, is provided by aircraft observations of cloud composition during traverses of their tops.

The most significant observations are those made in clouds sufficiently thick to produce showers but which do not extend much above the 0°C level, so that there is little doubt about the temperature at which the ice particles encountered could have formed. Such observations have been made in England (43, 44), Australia (45), and the central and southern United States (46, 47), and they can be summarized as follows:

1. During traverses at levels where the temperature is between -5 and $-10°C$ through a cloud tower whose minimum temperature is above about $-15°C$, it is difficult to detect the presence of ice particles unless a broad size distribution of droplets is present. When droplets of radius greater than 100 μ occur in concentrations approaching 10^{-2} cm^{-3} (corresponding to the stage of shower formation) ice particles of comparable size and concentration are found, although even at the minimum cloud temperatures a typical concentration of ice nuclei is only 10^{-5} cm^{-3} (Fig. 5.4).

2. On successive traverses after a cloud tower reaches the flight level the large particles are initially found to be predominantly liquid, but subsequently the proportion of the cloud water which is frozen increases and after about 15 min little liquid remains. At first the large ice particles are round with smooth surfaces, have a density close to 0.9 g cm^{-3}, and are evidently frozen drops; some are transparent, but others are opaque and appear to have acquired coats of rime ice by the accretion of droplets. Smaller ice particles are also present, with irregular shapes even when their diameters are as small as 20 μ, and the total concentration of ice particles may reach values as high as 1 cm^{-3} (46). Eventually, when little liquid water remains in the cloud, ice crystals of regular form (hexagonal plates and columns) are found.

The concentration of the ice particles, of both large and small sizes, which arise during the glaciation of cumulus tops is often some orders of magnitude greater than that of the ice nuclei active at the temperature concerned, so that evidently a process of multiplication of nucleated ice particles is at work. The irregular form of the small particles observed during the glaciation suggests that they are produced by fragmentation of rime or freezing droplets during the growth by accretion of ice particles formed upon a small proportion of the large drops which freeze (Section 5.14). The relevant laboratory data on the fragmentation cannot yet be regarded as established, but if they are accepted it is possible to calculate the progress of the glaciation and to assess how the breadth of the initial droplet size distribution affects the period required for most of the liquid water to become frozen. This has been done (48) under several simplifying assumptions; in particular the cloud air was assumed to be motionless and the rate of production of ice fragments during riming was assumed to be that found in the laboratory experiments described in reference 239.

The principal results were as follows:

8   Cumulonimbus convection

1. When the initial size distribution of the droplets is narrow (e.g., $\bar{r}_0 = 5\ \mu$ and N_r of order 10^{-4} cm^{-3} for $r \geq 40\ \mu$) the period required for the conversion of a considerable fraction (10% or more) of the cloud water into ice is long ($\gtrsim 30$ min).

2. When the initial size distribution of the droplets is broad, approximately corresponding to the stage of shower formation, this period is only several minutes and the conversion into an ice cloud is practically complete in about 10 min. After such a period small ice fragments, of diameter 10–20 μ, accumulate in large concentrations, of order 1 cm^{-3}.

3. When the rime fragmentation is excluded, then with the very broad initial size distribution of droplets the glaciation proceeds nearly as rapidly at first, but requires more than 20 min to reach virtual completion, and the concentration of small ice particles remains low.

It is not clear how such results might be modified if the laboratory evidence of the rate of ice fragmentation were revised, or if a more realistic cloud model were used. However, they appear to confirm that the observed rapid glaciations of cumulus at high temperatures, and the concentrations of ice crystals in their glaciated residues which exceed those of freezing nuclei by orders of magnitude, are dependent on the development of the cloud droplet distribution to the stage of shower formation, and on a fragmentation process active during the growth of the ice particles.

It can be expected that after the glaciation of a cumulus tower succeeding adjacent cloud towers will have high concentrations of ice particles introduced into them from its residues, so that their glaciation may proceed even more rapidly and completely. In this way the ice phase appears more extensively and at lower levels in shower clouds than if it were dependent solely on the very small concentrations of natural ice nuclei, and it may spread across a group of cloud towers with the considerable speed (several kilometers in 10 min) of the air motions which they contain.

The extensive glaciation of small shower clouds with summit-level temperatures little below 0°C is a common phenomenon in the maritime airstreams of middle latitudes. Although the first precipitation may consist mainly of rain or small hail, the eventual predominance of ice crystals in the fibrous trails of precipitation is often indicated by optical phenomena or the arrival of snow at the surface, especially on high ground. It is striking that the development of the ice phase should so readily follow shower formation in these comparatively clean airstreams, while farther inland cumulus tops reach levels where the temperature is $-20°$C or even lower without producing either showers or any evidence of glaciation. Plate 8.3 shows a typical example of the glaciated residues of a small shower cloud, beside a cumulus which just failed to attain the minimum thickness required for shower formation. Such shower clouds may produce shelf clouds which survive for some time as droplet clouds with well-defined edges, above the glaciated residues of the original cumulus. Less extensive patches or shreds of droplet cloud commonly occur at the tops of shower clouds of small or moderate size, and may be particularly prominent because of their brilliance when they are near the sun's azimuth (because, unlike the larger and perhaps crystalline particles of the precipitation trails, small droplets scatter light predominantly forward); in the previously mentioned studies of shower clouds in Sweden and England they were observed to occur at temperatures as low as about $-25°$C. They are usually the remnants of shelf clouds (Pl. 8.4), but especially at the lower temperatures occasionally form just above or at the upper fringes of glaciated residues, and grow for a few minutes before slowly evaporating.

8.2 The vertical development of cumulonimbus

The growth surge following glaciation

When showers form in cumulus towers which have risen well above the 0°C level, as typically happens during cumulus convection over land, glaciation may substantially increase their buoyancy in three ways: by the freezing of liquid cloud water and the release of latent heat of fusion, by the precipitation of cloud water, and, in the presence of a sufficient concentration of small ice particles, by the release of additional latent heat of condensation during the reduction of the vapor density from that representing saturation over liquid water toward that representing saturation over ice. The maximum contributions of these three processes to the buoyancy of a cumulus tower can be assessed as the increases in the virtual temperature due to the freezing of a concentration of about 1 g m^{-3} of cloud water, before its precipitation, followed by a reduction in the vapor density to the state of saturation over ice. In the middle troposphere at a temperature of about $-10°$C ($\rho_a = 2/3$ kg m^{-3}) these increases are, respectively, about 0.5, 0.4, and 0.3°C, or altogether over 1°C. Even though only fractions of these maximum increases are likely to be realized, a glaciation may evidently be accompanied by an increase in the virtual temperature of several tenths of a degree. If a cloud tower behaves like a thermal

204

and Eq. 7.26 is applicable, a tower of diameter 2 km and a buoyancy equivalent to an excess virtual temperature of 0.5°C may attain a rising speed of about 5 m sec^{-1}. It frequently has been observed that whereas prior to shower formation the tops of cloud rising through the cumulus tower layer decelerate as they lose buoyancy and eventually acquire a small negative buoyancy, following shower formation the tops of glaciated clouds retain ascent speeds of several meters per second in the upper part of the cumulus tower layer, and rise into the high troposphere.

The first large cumulus which develop into cumulonimbus with glaciated tops are usually in isolated groups associated with the more intense intermediate-scale circulations (such as those over mountains or sea- or lake-breeze fronts). A steady growth of large cumulus into glaciated cumulonimbus is very rarely observed. More commonly the increase in the size of the clouds as convection intensifies over land during the first part of the day is gradual, until eventually a striking surge occurs in the growth of those clouds in which shower formation and glaciation first occur. Whereas the rate of rise of the level reached by the tallest cumulus is generally only between 1 and 2 km hr^{-1}, and the onset of shower formation does not occur until near or after midday (when the cloud tops are in the middle troposphere), within the following half hour succeeding towers of the same cloud group have often reached up to near the tropopause. Usually they persist and spread, predominantly on one side, to produce the characteristic anvil of the cumulonimbus, or mature glaciated shower cloud. Examples of this striking growth surge are illustrated in Pl. 8.1 (and Fig. 8.13) and Pl. 8.5 (and Fig. 7.27). Similar behavior has been noted in a number of places and latitudes: in the previously described observations of showers made in Sweden, England, and Italy; in Iceland (49); in Malaya (50); and in Cuba (51). In Malaya, as in Sweden (Fig. 7.27), the rising speed of freshly glaciated cloud towers through the upper troposphere was usually about 10 m sec^{-1}. The precipitation and freezing of condensed water evidently can provide a significant additional source of buoyancy for cumulonimbus clouds. However, it is also likely that a downdraft produced by the associated precipitation may intensify the intermediate scale circulations and increase the strength and persistence of the updrafts even in the lower troposphere, with a consequent nearer approach to adiabatic ascent. On the occasions of widespread and only moderately intense cumulonimbus convection described, there was no obvious corresponding increase of tower diameter above the fibrillation level, and this process may have had only secondary importance. On the other hand, in the development of severe local storms, discussed later, it is dominant.

In middle latitudes the stratification of the upper troposphere is predominantly under the control of the large-scale motion systems, and when it is markedly stable there is no pronounced growth surge. On the day following the surge illustrated in Pl. 8.1, for example, shower formation occurred under almost identical conditions in the lower troposphere, in cumulus with tops at about 6 km. However, the high troposphere was warmer and glaciated cloud tops rose barely above 8 km, whereas on the previous day they had quickly reached through the tropopause to over 12 km.

The stratification associated with cumulonimbus convection

In middle latitudes the intermittency of cumulonimbus convection even overland in the summer season makes it difficult to identify a stratification typical of widespread cumulonimbus convection, corresponding to that which appears generally to be characteristic of cumulus convection. Even in low latitudes, where the cumulonimbus convection is more prevalent and plays a major part in the heat economy of the troposphere, the clouds are often clustered in traveling disturbances of greater scale, and the more severe storms seem to require particular large-scale conditions for their development. However, it seems reasonable to regard the mean summer stratification over Florida as one dominated by cumulonimbus convection, since in this season it occurs over the peninsula (and neighboring islands) almost every afternoon. The thunderstorm frequency is probably the

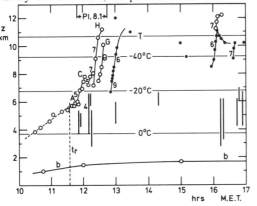

Fig. 8.13 Development of cloud towers near Verona, Italy, on 8 July 1958. Peak heights (above sea level) attained by the tops of towers are shown by open circles, and the tops of radar echoes by full circles, joined by lines to show rates of rise (in m sec^{-1}). The period marked at the top of the diagram is that during which the pictures of Pl. 8.1 were taken, showing towers of the cloud group A, the fibrillating tower C, and the glaciating tower H. The time of detection of the first radar echo is shown by the dashed line t_r, and the vertical extents of radar echoes subsequently observed at an early stage in their development are shown by vertical lines. The variation of the level of the cumulus bases is shown by the line marked b, and the heights of the tropopause T and some isotherms are also entered.

8 Cumulonimbus convection

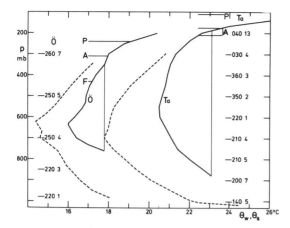

Fig. 8.14 The mean stratifications observed at Tampa (Ta), near the east coast of central Florida at 1200 GMT during July 1965, and near Östersund (Ö), central Sweden, at 1500 GMT during the three successive days of similar cumulonimbus convection, 10–12 July 1955, represented on a p-θ_w, θ_s diagram. The distributions of θ_s above the level of the cloud bases are shown by solid lines, and the distributions of θ_w above screen level are shown by dashed lines. Mean winds in tens of degrees and meters per second are entered at the appropriate levels, and are seen to be light in both situations. Against the Östersund sounding the letters F, A, and P, respectively, mark the fibrillation level, the mean level of anvil tops, and the peak height reached by cumulonimbus towers. On the Tampa sounding values thought to be typical of the height of the anvil tops, and about the extreme for the cloud tops, are indicated similarly. The value of $-(\overline{\mathrm{Ri}})$ (Section 8.5) is about 7 on the sounding from Tampa (but may be underestimated), and about 25 on that from Östersund.

highest attained anywhere in the world, and the mean tropospheric winds are light, suggesting a minimal influence of large-scale motion systems. The stratification observed is illustrated in Fig. 8.14 by the mean sounding at Tampa, near the east coast, during July 1965, when thunderstorms were reported every day either from Tampa (27 days) or from Fort Myers (about 170 km to the southeast; 29 days).

The mean sounding indicates an unusually deep cumulus layer, with its top at about 550 mb, and a lapse rate within it which is closer to the saturated (pseudo-) adiabatic than previously represented as typical of widespread cumulus convection. Thus, whereas in Fig. 7.17 a value of about 2/3 appears for the ratio $\Delta T_a/\Delta T_\gamma$, in the present mean sounding, the corresponding value is nearly 4/5, similar to that indicated on a mean sounding for the Caribbean region during the hurricane season (52). This difference may be due partly to the averaging process in constructing the mean sounding, for on individual soundings a change of lapse rate and decrease of relative humidity frequently suggests a lower level for the layer. On the mean sounding the minimum of θ_w at 700 mb is probably a better indication of the mean level of the top of the cumulus layer.

In the upper troposphere the mean lapse rate is close to but distinctly less than the saturated adiabatic until the tropopause is approached, which again because of the averaging process is less sharply defined than on many individual soundings. Matters of special interest are the average level of the top of the anvil clouds, which show where most of the air ascending from near the surface spreads out in the high troposphere, and the greatest heights which are reached by the cumulonimbus towers. No systematic studies appear to have been made of these features, but from incidental data in several references (e.g., 53–56) it seems that the height of the anvil tops is usually between about 12 and 13.5 km, and that the tops of the visible clouds and their radar echoes occasionally reach or even exceed 16 km, as indicated in Fig. 8.14. Thus θ_s in the anvil tops has nearly the value found in the cloud bases, and the mean lapse rate over the entire cumulonimbus layer is close to the saturated (pseudo-) adiabatic.

Figure 8.14 shows this and other features of the stratification in the mean sounding for the previously discussed three successive days of cumulonimbus convection in central Sweden. (Plate 8.6, from a similar day of convection there, shows how the main features are often virtually obvious.) Also in the studies of showers over England and Italy θ_s in the anvil tops was within 1–2°C of the value in the cloud bases. This result, which would be anticipated on the basis of the simple parcel theory, might seem to imply that in the cumulonimbus, in contrast to the cumulus, air ascends without significant mixing with the surrounding clear air. However, it is associated with the choice of the pseudo-adiabatic process with respect to liquid water to define θ_s. In a reference process in which all the condensate is retained and frozen the corresponding temperature in the high troposphere is several degrees Celsius higher if θ_s at the cloud base is about 20°C or more (Fig. 3.10). In the upper parts of cumulonimbus the actual process may be expected to be intermediate between these extremes, so that ascent without mixing from near the 0°C level into the high troposphere should produce there temperatures which are a few degrees *higher* than implied by conserving θ_s as ordinarily defined. In the upper and especially in the high troposphere, the evaporation and chilling due to the admixture of a given proportion of clear air is about an order of magnitude smaller than in the lower troposphere, on account of the small increase of mixing ratio required to saturate the clear air. Accordingly, even if the degree of mixing with clear air experienced by clouds is the same in both upper and lower troposphere, little change of θ_s is likely in the layer of subzero temperatures extending upward from a level somewhat above the top of the cumulus layer to the level of the anvil tops, and the mean lapse rate over the whole cumulonimbus layer must seem much closer to the adiabatic lapse rate than the mean lapse rate in the cumulus layer.

8.3 The properties of ordinary cumulonimbus

Visual observations

An obvious feature of most cumulonimbus is that their growth is not steady, but impulsive: just as with cumulus, a large cloud appears to be composed of a succession of towers, which rise into view mainly on its upwind or upshear side.

The rate of rise of the individual towers can be measured using theodolites or cameras together with radar ranging, or even by a radar alone if it has adequate resolution and is used with care for this purpose. The highest values of the ascent rate are often found near the tropopause (Fig. 7.27) when the stability of the high troposphere is small.

The diameters of well-defined towers are usually approximately proportional to their heights. The proportionality $n' = (z - z_0)/r \approx 5$ (Eq. 7.22) observed in Sweden and elsewhere in Europe was about the same for both cumulus towers and glaciated cumulonimbus towers. Observations in other regions suggest different and perhaps more variable values of n'; some cumulonimbus towers near the tropopause over Florida (57), for example, imply values between about 6 and 12 if the height z_0 of the virtual origin is assumed to have been near the ground. These towers attained maximum ascent speeds near the tropopause, and rose as much as 2.5 km above it. Saunders (57) compares the extent of these penetrations into the very stably stratified (isothermal) stratosphere, above a sharply defined tropopause, with estimates based on the simple parcel theory and others based on experiments with laboratory thermals which, after descending through one layer of uniform density, penetrated a limited distance into another of uniform greater density. For the former estimates it can be assumed that the buoyancy is zero at the tropopause, that the parcel lapse rate (at the low temperature concerned) closely approximates the dry-adiabatic lapse rate ($\partial \theta/\partial z = 0$), and that the stratosphere is isothermal. It then follows that a parcel is brought to rest after a penetration P which is proportional to its vertical velocity w' at the level of the tropopause, and which can be expressed as

$$P = w'/s^{1/2} \tag{8.16}$$

where the stratospheric *stability* $s = (g/\theta)\partial\theta/\partial z$ is approximately constant.

On the other hand, an analysis of the results of the experiments with thermals provided the following approximate formulas for the maximum penetration P_t of a thermal reaching the tropopause with a diameter d and an ascent speed of w'_t:

$$P_t \approx 1.6(w_t'^2 d/s)^{1/3}, \quad 0.1 < P_t/d < 0.6$$
$$P_t \approx 2.2 w'_t/s^{1/2}, \quad 0.6 < P_t/d < 3 \tag{8.17}$$

Thus a small penetration is predicted to be a function of both the ascent speed and the diameter at the tropopause, and larger penetrations are predicted to depend only on the ascent speed. For these the value $s = 4.8 \times 10^{-4}$ sec^{-2} appropriate to the observations in Florida (and approximately valid for any isothermal lower stratosphere) implies a penetration p_t of 1 km for an ascent speed w'_t of about 10 m sec^{-1}, whereas the corresponding value w' according to the parcel theory (from Eq. 8.16) is about 22 m sec^{-1}. This discrepancy practically disappears when it is recalled that the vertical velocity of some of the fluid inside a thermal is about twice as great as the rate of rise of its cap.

The penetration according to Eq. 8.17 is not particularly sensitive to changes in the diameter d over the range of values likely to be appropriate for cloud towers. The diameters d and rising speeds w'_t of five cumulonimbus towers observed over Florida varied between 2.5 and 5 km, and between 10 and 25 m sec^{-1}, respectively; the corresponding ratio w'_t/P varied from 9.5 to 11 m sec^{-1} km^{-1}, about equal to the value of 10 m sec^{-1} according to Eq. 8.17.

Another relevant observation is that of the penetration of a tower into the stratosphere over northern Italy on 21 July 1958 (Appendix 8.1). The rate of rise of the top of its radar echo at the level of the tropopause (at 11.8 km), where its diameter was 5 km, was about 27 m sec^{-1}. The peak height reached by the radar echo was 14.3 km, representing a penetration into the stratosphere of 2.5 km. Also on this occasion the stability s was 4.8×10^{-4} sec^{-2}, and the penetration according to Eq. 8.17 is 2.7 km, remarkably close to the actual value.

After reaching peak heights cumulonimbus towers subside before spreading and amalgamating into a persistent shelf cloud. When the winds are very light throughout the troposphere the shelf cloud may spread out symmetrically on all sides. Generally, however, the shelf cloud becomes greatly extended on one side, approximately in the direction of the relative or actual wind at its level (often the two are hardly distinguishable). Away from the cloud summits, composed of recently risen towers, the upper surface of the shelf cloud is practically horizontal; probably the thickness of the layer of air which spreads out after ascending in the main part of the cloud quickly shrinks to less than about 1 km, but usually this thickness is not obvious because the base of the shelf cloud is made indistinct by the precipitation which falls from it. Near the cloud towers the precipitation is intense and composed of large particles, so that there fibrous trails extend down to the ground, but at increasing distances it is composed of smaller and smaller particles, which evaporate at ever

8　Cumulonimbus convection

higher levels, so that the shelf cloud appears to taper outward and is aptly called an *anvil cloud*. The persistence of the updrafts is often so great and the evaporation of dense cloud at the low temperatures of the high troposphere is so slow that the anvil cloud is a normal and extensive feature, sometimes hundreds of kilometers long. The mamma that often develop on its underside reveal small-scale convection produced by the cooling which accompanies evaporation of the precipitation. Poorly developed and smaller mamma, with ragged surfaces, are sometimes seen on the underside of the shelf clouds of cumulus near the principal cloud towers.

Observations from aircraft

Observations of the internal properties of cumulonimbus from aircraft, which might be expected to provide the most definite and useful information on which to base a dynamic theory of the cloud structure, have so far proved less stimulating than visual and radar studies. This can be attributed partly to the difficulty and expense of obtaining with aircraft information sufficiently accurate and abundant to define the structure of the whole cloud and its evolution, partly to the lack of satisfactory instrumentation for measuring the cloud properties from a fast-moving platform, and perhaps also partly to the deterrent effect of the hazards of flight through the more intense cumulonimbus. In contrast, visual and photographic studies of the outer surface of the clouds provide at comparatively little expense and trouble information which although indirect has good resolution in two space dimensions and time. Suitable radar is an even more prolific source of data which are indirect, refer only to the precipitation, and have less resolution, but nevertheless have the great advantage of presentation in all three space dimensions and in time. Plans for the use of pilotless aircraft, or for putting instruments into the clouds by dropping them from above or by introducing balloons into the updraft at low levels seen promising, but if fulfilled, they will still have the disadvantage that the amount of information obtained will be limited and difficult to place within the framework of a good model of the cloud. The most rewarding, even if the most expensive information, will be gained by exploiting all these techniques simultaneously using several aircraft, making not only explorations of cloud interiors but also photographic surveys of clouds under simultaneous radar examination, and measuring the wind field at several levels in the clear air surrounding them.

Few systematic programs of aircraft observation in the interior of cumulonimbus have been attempted. None has provided entirely reliable information on the distribution of cloud temperature or of cloud water, and the most valuable measurements obtained are those of the dimensions and magnitudes of the vertical component of the air motion encountered during horizontal traverses of the clouds. Even so, the derivation of the vertical air speeds from the recorded behavior of aircraft is subject to uncertainties due to the complication of the reaction of the aircraft and their pilots in the rapidly fluctuating air motion experienced during flight at high speeds. Moreover, the aircraft have generally been directed into the cumulonimbus as it appears on the PPI display of a radar, often deliberately to avoid the region of the most echo (believed also to be the most hazardous), and the observations have to be interpreted according to position with respect to the radar echo, and not, as would be desirable, with respect to some model of cumulonimbus structure like that suggested for cumulus in Fig. 7.28. The most interesting observations are those which indicate whether a model of the kind shown in that figure, with suitably increased scales, can also represent cumulonimbus convection.

The most extensive series of aircraft observations were made in Florida and Ohio in 1946 and 1947 (58), and recently in the midwestern United States where thunderstorms are infrequent but sometimes very severe (see, e.g., 59). The analysis of the first series recognized the typical cumulonimbus as a complex composed of several intense parts, or cells, each of which had an evolution through three principal stages. In the first, the *cumulus stage*, the pronounced vertical velocities in the cell were upward, constituting an updraft whose average speed was assessed as the average rate of upward displacement of the aircraft traversing it (the aircraft was as far as possible flown with constant power and a minimum of control, and recorded control movements were used to make some adjustments to the directly indicated updraft speed). The second, *mature stage*, was defined as beginning with the arrival of rain at the ground below the cell. Its distinguishing feature was the development with the precipitation of a region of downdraft, which appeared in the middle or lower parts of the cell and then increased in horizontal and vertical extent. Early in this stage the updraft was found beside or partly over the downdraft, and attained its greatest intensity (increasing with height in the range of heights explored, between 1.5 and 7.5 km). Subsequently the region of downdraft spread, until after 15–30 min it occupied the whole lower part of the cell, marking the end of the mature stage. In the succeeding *dissipation stage* the downdraft extended to high levels and weakened, until after about 30 min the cell contained little or no vertical motion.

In the mature stage the precipitation encountered inside the clouds was usually reported as snow or rain; hail, which when small can be recognized only by the distinctive sound of the impacts on the aircraft, was reported on only a small percentage of the traverses. In the regions in which the investigations were made the frequency of hail at the ground is small: less than 1 day each year in Florida and only about 2 in Ohio.

Rain was sometimes reported even at the greatest heights flown, where the temperature was about $-15°C$. In more recent flight studies of the cumulonimbus in hurricanes and over the midwest of the United States rain has been reported at still lower temperatures, even below $-40°C$ (60), and in concentrations of 10 g m^{-3} or more (61). It seems likely that the precipitation actually consisted of spongy hailstones containing a large proportion of unfrozen water, particularly as sometimes it produced little airframe icing.

Measurements of the total concentration of condensed water at high levels (above 8 km) in severe thunderstorms over Oklahoma provided maximum values of about 10 g m^{-3}, and in one case 44 g m^{-3} (when the air crew reported frequent lightning, and the visual impression of "running into a wall of water"). In view of their great effect on the buoyancy of the cloud air such large values are likely to be local or transitory but are not improbable, since they are also implied by the high rainfall rates at the ground which are known to occur over a period of a few minutes in severe thunderstorms, as discussed in Section 8.12. [For example, rain at an average rate of 200 mm hr^{-1} during a period of 5 min is likely to be experienced about once in 5 yr at a point on the ground in the Gulf states (62), and implies a concentration of rain water in the air near the ground of at least 7 g m^{-3}, while rates of up to 1000 mm hr^{-1}, or even more, have been recorded over shorter periods.]

High values have also been measured for the concentrations of ice particles in tropical cumulonimbus (63), using a calibrated instrument in which the particles entered a tube exposed outside the aircraft and were melted before collection. The aircraft was flown through cumulonimbus anvils for periods of up to about half an hour, usually at heights between about 6 and 8 km and just avoiding the regions of intense radar echo. The concentration of ice particles was generally between about 0.5 and 3 g m^{-3}, but for a small percentage of the flying time rose to between 4 and 7 g m^{-3}. On some of these and other flights made over the Mediterranean and the British Isles the size distribution of the ice particles was estimated from their impact dents on aluminum foil, and for diameters between about 100 μ and 5 mm was found to be well represented by the relation

$$n(D) = 10^3/D^3 \qquad (8.18)$$

where $n(D)\,dD$ is the number concentration of particles (m^{-3}) of diameters between D and $D + dD$ mm (64); the total mass concentration was probably usually more than 1 g m^{-3}. The corresponding curve of cumulative concentration N_r lies somewhat above that included in Fig. 8.3 for rain with an intensity of 5 mm hr^{-1}, but the lower densities and fall speeds of the ice particles imply a smaller precipitation rate.

In the flight investigations of cumulonimbus over Florida and Ohio the average speeds of the updrafts and downdrafts were estimated primarily from the vertical displacements of the aircraft, which were flown at a constant power setting and with a minimum of pilot control. The drafts encountered were mostly between about 1.2 and 1.5 km across, the width showing little variation with height. The updrafts were rather broader and stronger than the downdrafts; their speed was usually in the range 5–10 m sec^{-1}, but sometimes exceeded 15 m sec^{-1}. Similar speeds and dimensions have been inferred from special flights made through summer thunderstorms in England; the width of the transition zone between updraft and downdraft has often been observed to be less than 1 km (65).

Flight through cumulonimbus at the air speeds of modern aircraft (about 200 m sec^{-1}) is extremely bumpy, giving the impression that the air motion is very turbulent. The problem of inferring the form of the actual air motion from the records of the aircraft behavior is difficult (and not one of immediate interest to the aircraft designer, who is usually more concerned with the statistics of the loads imposed on the aircraft). Detailed profiles of vertical, lateral, and longitudinal air speed have been constructed for traverses made at heights between about 8 and 12 km across large thunderstorms over or near Oklahoma (59), and examples from one flight are presented in Fig. 8.15. The vertical velocity w_g was determined from the basic equation

$$w_g = V\alpha - V\theta + w_a + l\,d\theta/dt$$

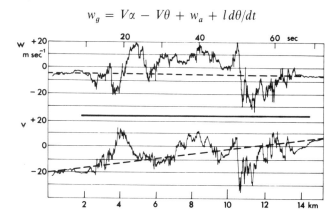

Fig. 8.15 Vertical velocity w and lateral (westward) air velocity v obtained by the analysis of measurements made during a traverse of a thunderstorm at a height of 12 km (about 2 km below the cloud tops) near Oklahoma City (59). The extent of cloud along the flight path is indicated by the thick horizontal line in the center of the figure. The true reference zeros of velocity are probably as shown by the dashed lines. The general wind was west-southwesterly, increasing from about 10 m sec^{-1} at 3 km to 40 m sec^{-1} at 12 km; diagram shows that the upward motions encountered inside the cloud were associated with weaker westerly components, and the downward motions with stronger westerly components.

where V = true air speed
α = (small) angles of incidence of the air (shown by a flow vane)
θ = pitch inclination of the aircraft
w_a = vertical velocity of the aircraft
l = distance between the flow vane and an accelerometer from the integration of whose record w_a was determined

The vertical acceleration, α, θ, and $dθ/dt$ were measured at intervals of 0.05 sec as departures from their mean values throughout the whole record of a traverse, to provide values of vertical air speed at the same intervals. The errors and differing response times of the recording instruments lead to errors of several meters per second, and the integration of the accelerometer record leads to inaccuracies which are only partly removed by taking into account the total vertical displacement of the aircraft as given by the pressure altimeter. For these reasons the variations of vertical air speed indicated are probably fairly accurate over distances of about 0.5–1 km, but the absolute values, and in particular the average speeds over greater distances (of the "drafts" rather than the "gusts"), which also have meteorological interest, are less reliable. Intrinsically the drafts produce only small loads on the aircraft, but they cause handling problems and pilot reactions which can also be hazardous and which complicate the interpretation of the aircraft's behavior. The most striking results of the investigations are, first, that the maximum derived upward velocities were large (although the aircraft probably did not pass through the most intense parts of the storms), often exceeding 15 m sec^{-1} and sometimes exceeding 40 m sec^{-1}; and second, the magnitudes of the vertical, longitudinal, and lateral components were comparable.

It is difficult to reconcile these results with the experience of sailplane pilots who have soared in cumulonimbus updrafts and found the flow to be remarkably smooth until the sailplane leaves the updraft (66). However, they also report the updrafts to be only a few hundred meters across, and accordingly may have tried to remain in parts where the upward motion is both most intense and comparatively smooth. The upward speeds encountered have increased with height and have reached values as high as 30 m sec^{-1} at a height of about 7 km. The sailplane pilots have not provided any useful information about downdrafts, because generally these have been intentionally avoided.

It is evident from the aircraft observations that there is a qualitative difference between cumulus and cumulonimbus convection: whereas in the cumulus downward motions are confined mainly to the outer shell of the cloud and produce only small vertical displacements, with the development of precipitation in the cumulonimbus a strong downdraft appears which extends down to the ground, and whose speed, horizontal dimensions, and duration in the middle and lower troposphere become comparable with those of the updraft. According to the aircraft observations the air in the downdrafts is up to about 3°C colder than the clear air at the same level, implying the evaporation of a concentration of cloud water comparable with that found in the updrafts. Large downward displacements of air to levels below the cloud base and even to the ground demand the evaporation of substantial amounts of water from the precipitation. Insofar as mixing processes and other heat sources or sinks (e.g., the melting of ice particles) are unimportant, the wet-bulb potential temperature $θ_w$ is conserved during descent.

The weather and the downdraft at the ground

The studies in which aircraft flew through cumulonimbus over Florida and Ohio also used dense networks of recording instruments at the ground to examine the weather beneath cumulonimbus; from the records of such networks maps can be prepared at intervals of a few minutes which show in detail the evolution of the wind and pressure field over an area some tens of kilometers across during the development of individual cumulonimbus cells (67–69). They show that the first measurable rain reaches the ground over an area limited to a few square kilometers, attains its greatest intensity (of order 10 mm hr^{-1}) early in the mature stage of a cell with a sharply defined boundary, and subsequently weakens and spreads over a larger area.

Large changes of wind and temperature occur near the ground beneath cumulonimbus in their mature stage, and spread well beyond the edges of the rain area. These are produced by the spreading out, mainly within the lowest kilometer, of air which descends in a downdraft. This air is separated from its surroundings by a narrow transition zone recognized at the surface as a squall front. (In routine observation the wind speed reported from stations with anemometers is generally the mean speed over an interval of 5–10 min preceding the time of observation. The frequent fluctuations, associated with eddies and convective circulations near the surface, which have a duration of less than 1 min are called *lulls* and *gusts*. It has already been mentioned that during cumulus convection groups of gusts often occur at intervals of 10 min or so, implying some organization on the scale of the layer of cumulus convection. The more pronounced organization associated with cumulonimbus convection, and in particular with its strong downdrafts, leads to marked increases of wind speed near the ground which have a duration of several minutes and are distinguished as *squalls*. Squalls contain gusts, and the wind speed usually reported for a squall is the maximum recorded by an anemometer, which generally represents an interval of only a small fraction of a minute. In the observing practice of the United States a squall is reported only if a speed of 8 m sec^{-1} or more is sustained for at least 2 min.

When the tropospheric winds are light, and cumulonimbus move only slowly, a squall front expands in a ring around the rain area, with the winds behind it directed radially outward. More commonly the rain area travels at speeds of about 10 m sec^{-1}, with a velocity close to the mean wind velocity in the lower troposphere. The downdraft then spreads predominantly forward of the storm and the squall front is well defined only on that side.

Beneath a cell entering the mature stage the squall front moves away from the rain area at a speed often greater than 10 m sec^{-1}, and sometimes obviously greater than the speed of the winds behind it, measured a few meters above the ground. In the front the wind velocity commonly changes by 10 m sec^{-1} in less than 1 min, corresponding to a horizontal distance of a few hundred meters. As the front travels farther away from the rain area it weakens and slows, but usually is still recognizable after traveling 30 km or more. It may be overtaken by a front from another cell.

Behind a squall front dry- and wet-bulb temperatures fall, and in and near the rain they reach minimum values which are commonly several degrees Celsius below those ahead of the front. At screen level the values are particularly liable to be affected by convection from the surface, but even so the decrease of wet-bulb temperature is usually sufficient to show that the air has descended from well above the ground, and sometimes from above the cloud base. For example, in Florida the minimum screen-level wet-bulb temperature during or after thunderstorms is about 21°C, corresponding to a value of θ_w (about 20°C) which on the mean sounding (Fig. 8.14) is found near the level of the cloud base. (A particular case is illustrated in Fig. 8.16.) The air in the cold outflow is often nearly saturated, but remarkably the relative humidity sometimes falls in and near intense rain; at screen level it may become as low as 70 or even 60%, so that in humid regions such as Florida it may become lower than before the rain.

Beneath cumulonimbus there are also interesting changes of surface pressure. A slight fall of pressure, amounting to less than 0.5 mb, is observed during a period of 5–15 min before the appearance of a radar echo, over an area several times wider than the maximum horizontal extent of the echo. Subsequently the pressure directly beneath the echo steadies, but in the surroundings continues to fall for a while, often more rapidly than before. Within minutes of the first measurable rain reaching the surface the downdraft becomes established and the pressure in the rain area rises by 0.5 mb or more; later the pressure rise spreads more

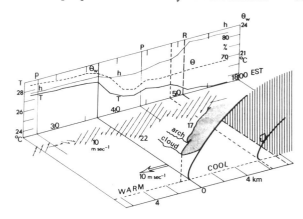

Fig. 8.16 The sequence in time of the surface wind, temperature T, relative humidity h, and wet-bulb potential temperature θ_w, during the passage of a squall over Miami on 24 May 1961, and the spatial distribution implied by the observed speed of 10 m sec^{-1} at which the gust front traveled south (adapted from reference 85). The letters at the top of the diagram show when the surface pressure began to rise (p), when the rise was a maximum (0.8 mb; P), and when the heavy rain began (R). There may have been a second gust front in the position indicated. The height of the base of the arch cloud was about 550 m.

evenly into the region occupied by the downdraft air outside the rain area, and during the dissipation stage gradually everywhere returns to its undisturbed value.

Such changes of pressure can be related to accelerations in the flow of the air near the ground, which, if sufficiently persistent, can be regarded as steady relative to the storm as a traveling motion system, and to a first approximation to be also horizontal and not seriously affected by friction. Then along a trajectory the pressure p and relative velocity v are related by Bernoulli's equation $(p + \rho v^2/2) = $ constant, given that the fractional change in the air density ρ is very small. The small pressure falls before the approach of a squall front indicate an acceleration of low-level air toward the cumulonimbus, and the pressure rise behind the squall front must accompany the acceleration of air away from the foot of the downdraft. Behind a squall front the pressure rise Δp can be expected to be approximately $\rho V^2/2$, where V is the change of wind speed during the squall, a rise of 1 mb therefore corresponding to a squall speed of about 13 m sec^{-1}. In ordinary storms the pressure rise of rather more than 0.5 mb is consistent with squall speeds of about 10 m sec^{-1}, but in more severe storms to be discussed later squall speeds may reach 40 m sec^{-1} or more, with pressure changes of several millibars. In the severe hailstorm discussed in Section 8.4, for example, at several stations experiencing a squall of 20–25 m sec^{-1}, with gusts to 30 m sec^{-1} (Table 8.5), the pressure fell by between 0.6 and 1.0 mb during about an hour before the arrival of the squall (indicating a widespread acceleration of air toward the storm), and

8 Cumulonimbus convection

behind the squall rose to between 3.2 and 4.0 mb above its undisturbed value.

Experimental studies of dense outflows

The manner in which a fluid flows into and beneath another of lesser density has been examined experimentally by releasing salty water, marked by a dye or suspension, into clear water held in a deep, long trough. The dense fluid travels along the bottom of the trough behind a head formed at its front, whose height H is about twice as great as the depth h of the following layer (Fig. 8.17; 71). The steep front of the head has columns or buttresses of diameter about $H/2$, and close to the trough bottom locally forms overhanging noses. Relative to the front, advancing at the speed c, the clear fluid is displaced upward and backward, while the dense fluid approaches from behind and rises up toward the top of the head. Small-scale disturbances resembling billows (Section 9.6) form on the front of the head, develop as they move up it, and disintegrate to produce a turbulent flow behind it, where there is some marked fluid in which the dye has become too dilute to be apparent.

At large values of Reynolds number the steady velocity c is related to a characteristic depth (h or H) and a (small) density excess $\Delta\rho$ of the dense fluid (over the density ρ of the surrounding fluid), by an expression analogous to Eq. 7.19:

$$c^2 = C^2 hg\,\Delta\rho/\rho \qquad (8.19)$$

containing a Froude number C^2. If $h/D \ll 1$, where D is the total depth of the fluid, simple theory (72) gives $C^2 \approx 2$, but C^2 is considerably less (≈ 1.5) if h/D, as in the experiments, is as large as 0.1. In their results the velocity has usually been expressed in terms of H rather than h, to give

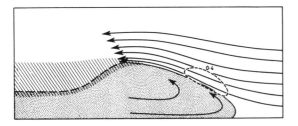

Fig. 8.17 A vertical section showing the typical form assumed by one fluid (stippled) flowing over a solid boundary into another fluid of slightly smaller density and with a neutral stratification (70). Some streamlines are drawn of the flow relative to the front of the dense fluid; a dashed line near the "head" of the dense fluid shows where the upward velocities reach 0.4 of the speed of advance of the front. The front of the dense fluid is deformed locally by smaller-scale motions, some of which amplify rapidly near the head and produce a strongly turbulent motion in the hatched area behind it.

$$c^2 \approx 0.6\,Hg\,\Delta\rho/\rho$$

which is in approximate agreement with theory if H/h is about 2.

Cold outflows in the atmosphere; haboobs and andhis

Equation 8.19 might be expected to be applicable to cumulonimbus downdraft outflows, and perhaps to shallow invasions of cool air behind sea-breeze fronts, when there is usually a deep adiabatic layer and the displaced fluid can be considered to have a neutral stratification, as in the experiments.

In some arid regions where there are large expanses of ground covered with fine sand or dust and little vegetation, the form of the cold outflows from thunderstorms early in the wet season is spectacular when they become filled with clouds of particles raised from the surface by the strong winds. The most familiar of such dust storms are the *haboob* of the Sudan (73) and the *andhi* of northwest India (74, 75). The cumulonimbus have high bases typical of nucleus-rich continental air (often about 4 km above the ground), and according to the descriptions both the visible trails of precipitation and their radar echoes become tenuous near the ground or even fail to reach it. The first dust is raised from the ground along an arc behind the squall front (in the more vigorous andhis "sometimes the dust is locally raised in great masses as if a descending current from a gigantic hose-pipe was playing on the ground"; 74). The shape of the advancing dust cloud is strikingly like that of the density current in model experiments; its advancing edge is inclined at a steep angle, and close to the ground it has lobes and overhanging noses (Pl. 8.7). The dust can be seen to have an upward and backward motion, and ahead of it the air is pushed away and lifted. From a distance a dust dome in sunshine may appear as a pale haze, but at close quarters it is usually threateningly dark with red, brown, or yellow tinges, and on arrival it may reduce the visibility to less than 30 m or even cause pitch darkness.

A sea-breeze front can also be seen to have the same characteristic shape when the sea air has passed over a coastal town or other source of pollution, so that where it rises at the front it becomes hazy (and marked also by fragmentary low clouds; 71). The average height of the head of the sea-breeze front over southern England is only about 700 m, and its speed of advance is only a few meters per second. The temperature decrease ΔT during its passage is only about 1 °C and can hardly be expected to give an accurate measure of the average density excess in the cool air, but on the assumption that it does, several occasions on which values of c, H, and ΔT were measured gave values of C^2 between about 0.4 and 1.1 (71).

Better agreement with the experimental value of C^2 has been obtained from similar measurements made in haboobs in the Sudan. Their speed of advance c near the parent cumulonimbus is usually between 10 and 15 m sec^{-1} and ΔT is several degrees Celsius; inferred values of C^2 vary between 1.1 and 1.9. A close correlation has been noticed between ΔT and the maximum gust speed V_m in haboobs at Khartoum, which has prompted an examination of the relation between these properties both in sea breezes and in more violent squall storms (71). From experiments it appears that the speed of advance c of the head of a density current is about three fourths of the speed of the steady flow near the surface beneath it, while on the basis of experience of sea-breeze fronts and haboobs it can be assumed that V_m (which is usually the speed reported for a squall) is about three fourths of the average speed of the wind in a squall, so that $V_m \approx 2c$ and $c^2 \approx V_m^2/r$. In data collected from squall storms in the midwestern United States (253), $V_m^2/4 \approx 2 \times 10^5 \Delta T$, while Eq. 8.19 gives $c^2 \approx 5\Delta Th$, again assuming that $\Delta T/T$ is representative of $\Delta \rho/\rho$ over the whole depth of the downdraft outflow. Accordingly, the implied values of h and H are about 0.4 and 1 km, respectively, which while not unreasonable seem rather small. In view of the discussion of the temperature in downdrafts in Section 8.10, it is possible that the observed large values of ΔT (mostly about 10°C and a few as much as 20°C) were appropriate to a large region well behind the storm squalls rather than to the air in the outflows from the most intense downdrafts, which are likely to produce the strongest squalls. However, in severe squall storms, which include the andhis, it will be shown that there is a more organized form of circulation whose mechanics are more complicated, so that the simple relation (8.19) is probably less relevant.

Numerical studies of cold outflows

Some numerical studies have been made of simple models of the travel of wedges of cold air into their surroundings (e.g., 76). Since the horizontal pressure gradients are not initially balanced geostrophically, the cold air subsides and spreads under the surroundings with the conversion of potential into kinetic energy; the upward and downward motions of the circulation which develops have their greatest intensity near the advancing front of the cold air. In some computations the latent heat of condensation in air ascending beyond its saturation level has been included. The available potential energy and the intensity of the vertical motion is then likely to increase, and the speed of propagation to change. The zone of upward motion (and cloud) advances into the warm air at speeds of 10–15 m sec^{-1}, consistent with observed values. Similar speeds are characteristic of downdraft outflows in more complicated numerical models of cumulonimbus evolution which have been reported recently.

The formation of cumulonimbus cells over the squall front

Over much of the area covered by the cool outflow from the downdraft the ground has been shaded or chilled and moistened during the passage of the storm, and convection from the surface has ceased. Where it continues or is renewed it is at first limited to a shallow layer, because the stratification in the cool air is stable. Consequently, with the production of the downdraft during the development of a cumulonimbus and its outflow at the ground, cumulus which may have been present in an adjacent region dissipate, and some time elapses before they can form again after the cool air has been modified by convection from the surface and by sinking as it spreads away from the region of precipitation. For example, overlooking the effect of subsidence, if the layer of cool air is 500 m deep, the provision of about 50 cal cm^{-2} is required to eliminate a mean (potential) temperature difference of 4°C between the cool air and the air above, requiring a period of some hours of sunshine over land. Over the sea, where the heat is provided by the water, the period can be estimated to be similar, in the following way.

We consider that an adiabatic layer of depth h cm is replaced by cool air in which the temperature is lower by $\Delta T(z)$°C. Originally ΔT is supposed to increase uniformly from zero at $z = h$ to a value $\Delta T_0 = x_0$ near the sea, at the top of a shallow (superadiabatic) layer of negligible heat capacity where this and subsequent values of $\Delta T = x$ are also the differences between the air and sea temperatures. The heat flux H from the sea into the cool air is assumed from the commonly used relation $H = \rho c_p C_D u x$ (Eq. 7.12), with appropriate values of the wind speed u and drag coefficient C_D. If the potential temperature of a column of modified air above the surface layer is uniform [so that $\Delta T \approx x$ over its whole depth, which is $h(x_0 - x)/x_0$, increasing as x diminishes], then the rise of temperature dx in a column and the heat flux during an interval dt give the approximate relation

$$\rho c_p C_D u x \, dt \approx \frac{\bar{\rho} c_p h (x_0 - x) \, dx}{x_0}$$

or

$$t \approx \frac{h}{C_D u}\left(\log_e \frac{x_0}{x_1} - \frac{x_0 - x_1}{x_0}\right)$$

where t is the time for x_0 to be reduced to x_1, or for the depth of the new adiabatic layer to be increased to the fraction $(x_0 - x_1)/x_0$ of the previous depth h. Thus the period required for the development of an adiabatic layer with virtually the previous potential temperature and depth is nearly $(h/C_D u)$. In a fresh wind over the sea typical values are $u = 10$ m sec^{-1}, $C_D = 2 \times 10^{-3}$, and $h = 500$ m, so that this period is several hours. Considering that the effect of subsidence is also likely to be important, a typical period for the suppression of cumulus is not likely to be more than a few hours of travel away from the region of precipitation, but with average outflow speeds relative to the cumulonimbus of several meters per second this time usually is sufficient to produce cloud-free areas several tens of kilometers across. Examples of such clear spaces can be found in Pl. 7.9, and the open cells with about this dimension which are seen in satellite pictures of convection over the ocean may all arise in this way. In airstreams with little wind shear in the vertical the edge of the cold outflow spreads out almost symmetrically from its parent cumulonimbus, which may be short-lived and soon disappear. Until the cool outflow is well modified its front may be marked by a ring of enhanced cumulus convection. On the other hand, in airstreams with more shear the downdraft spreads out predominantly away from one side of the cumulonimbus, which may appear more persistent, being continually renewed on that flank.

Similarly, over land the cold dome and the chilling of the ground by rain often removes the circulation of intermediate scale which was responsible for the clustering of cumulus and the generation of the cumulonimbus which produced the downdraft. However, it also often causes lifting near its advancing front which may be equally or even more effective in intensifying cumulus convection, so that a belt of clustered and towering large cumulus appears over the squall front just as over a lake- or sea-breeze front. Commonly, then, after the formation of a cumulonimbus succeeding cells appear over a continually renewed and perhaps lengthening squall front, and a chain or arc of cumulonimbus or large cumulus develops. [Over both land and sea (77) such cloud chains are observed to reach extreme lengths of some hundreds of kilometers.]

Usually the cumulonimbus convection continues to be impulsive, and on the PPI display of a radar with ordinary resolution (beam-width about 1°) the echo from each fresh cell is at first separate or otherwise easily distinguishable from the previous echoes. From radar data collected during the studies made in Florida and Ohio it was found that most new echoes appeared near existing ones. The relative frequencies of the appearance of new echoes were 1 more than 15 km away from existing echoes, 2 in the zone between 5 and 15 km away, increasing to 7 within 5 km of their forward and right flanks and to 11 within 5 km of each of a pair of existing echoes. On the occasion over England illustrated in Fig. 8.18 it was observed that new echoes

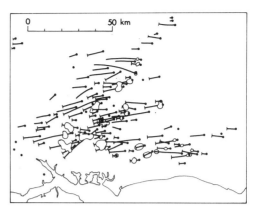

Fig. 8.18 Paths of radar echoes from individual cumulonimbus cells over southern England, between 1330 and 1530 GMT, 18 June 1957. The dots show the place of origin of the cells and the solid lines the paths of their centers, terminated by a transverse line where the echo disappeared before 1530; the open lines show the positions of the echoes at 1430 GMT. About 125 echoes from individual cells were detected after the formation of the first at 1342 GMT. Some persisted for only a few minutes, others over half an hour. Downdrafts spread out at the ground beneath the more intense cells, with weak squalls on their western sides; new cells formed mainly in the quadrants southwest and northwest of existing cells, and between about 3 and 10 km in advance of their western edges. The mean velocity of the cells was 080° 8 m sec^{-1}, about equal to the wind velocity at all levels between 1.5 and 5 km. The westernmost storms were the most intense, and they continued to spread westward during the evening.

which appeared near existing echoes were most frequently between about 3 and 10 km ahead. Such individual cells are usually visible on the display while the precipitation in them has an appreciable intensity (more than about 1 mm hr^{-1}), and they have a duration which is generally less than an hour. In the investigations made over Ohio it was found that the echoes mostly became 2–5 km across within 10–15 min of detection, and had a duration which averaged 23 min; it exceeded 45 min in only 2 of 67 cases (79). On the occasion of Fig. 8.18 it was visually obvious that in one particular cumulonimbus the updraft had practically ceased 20 min after the detection of an echo whose total duration was about 35 min; the average duration of 41 echoes from individual cells was only 24 min, corresponding to an average path length of 10 km. The rather short duration of such cell echoes facilitates the recognition of their intermittent formation on PPI displays.

Similarly, when cumulonimbus are studied visually or on the RHI display of a radar it is usually possible to recognize and follow the rise of the tops of a succession of individual towers. In the observations made over Ohio it was found that in the early stages of the development of cumulonimbus three or four successive towers reached progressively greater heights at intervals of about 17 min (58). Similar intervals separate successive neighboring towers in the cumulonimbus shown in Pl. 8.1, 8.5, and 8.13 (e.g., in Pl. 8.1, tower *K* about 14 min after *H*; in Pl. 8.5, tower *D* about 15 min after *C*; in Pl. 8.17, tower *C* about 17 min after *B*, and tower *E* about 14 min after *D*). As another example, in the storm over Italy whose penetration into the stratosphere was previously mentioned, the times at which the tops of the radar echoes from the first four principal towers reached 8 km, and the peak heights which they attained, were as follows: 2214 (8.0 km), 2231 (11.1 km), 2247 (13.8 km), and 2302 MET (14.3 km); the corresponding intervals are 17, 16, and 15 min.

These observations suggest that the impulsive growth of cumulonimbus tends to exhibit a period of a little less than 20 min. It may be significant that this corresponds approximately to the period required for the production of a downdraft with a squall front at the ground following shower formation (about 10 min; 67), and the subsequent ascent of air from over the gust front to the level of shower formation (several kilometers at an average rate of ascent of several meters per second, or about 10–15 min). Later in the development of cumulonimbus it becomes more difficult to identify successive cloud towers, and the interval between them sometimes appears to become shorter and irregular, perhaps as the squall front extends over a wider arc and produces more numerous towers whose interaction complicates the behavior.

Cumulus convection over land ceases toward sunset, but leaves a stratification in the lower troposphere in which cumulonimbus convection can be maintained into the evening or the dark hours by the formation of fresh cells in air lifted over squall fronts. In this way cumulonimbus convection which began one afternoon over southern England, in two inland belts associated with intermediate-scale circulations (Fig. 7.22), spread 200 km westward by sunset (Fig. 8.18). Groups or belts of cumulonimbus may persist even longer in cool airstreams over the sea, and they may be responsible for the features sometimes identified on synoptic charts as minor cold fronts or instability fronts.

The speed of ascent of air over the squall front

Over a squall front air ascends into cloud base at a greater speed and over a larger area than is usual below cumulus. In Fig. 7.35 the average updraft speed near the base of cumulus was indicated as between 1 and 2 m sec^{-1}, a value not inconsistent with the experience of sailplane pilots who often find greater speeds over regions as little as 200 m across. On the other hand, on the sides of thunderstorms they encounter ascent speeds of up to about 6 m sec^{-1} (66). In studies with a specially instrumented light aircraft over Colorado the average values of the ascent speed found near cumulus and cumulonimbus bases increased with the intensity of the nearby precipitation, and consistently reached values of several meters per second (Table 8.4). The observations were made ahead of cumulonimbus along flight paths extending from within 0.8 km of the middle of the rain area to 0.8 km ahead of the visible leading edge of the cloud base (a zone a few kilometers wide found to contain the strong updrafts), at levels 100–900 m below the cloud base (about 2 km above the ground). The flow experienced in the updrafts was smooth rather than turbulent; bumpy conditions were encountered mainly when entering and leaving areas of updraft.

Table 8.4 Average ($\bar{w}$) and range of values of updraft speeds encountered over Colorado near cumulus bases, and a few hundred meters below cumulonimbus bases in front of precipitation (80). The individual values were averages found over a distance of about 3 km

	Number of measurements	$\bar{w}$ (m sec^{-1})	Range of values
Cumulus	71	0.8	1.5– 5.0
Cumulonimbus			
Light precipitation	42	1.0	−1.5– 5.0
Moderate precipitation	48	2.5	−3.0– 9.5
Heavy precipitation	58	3.8	−1.0–11.5

Updrafts ahead of cumulonimbus with moderate to heavy precipitation extended over distances 10–20 km across their paths, and were generally strongest in the right front quadrant. The updrafts in this quadrant frequently persisted for rather long periods: in one, for example, speeds between about 4.5 and 5.5 m sec^{-1} were measured throughout an hour. The region of strong updrafts was nearly always marked by a dark arch cloud of a kind which familiarly accompanies squall fronts (Pl. 8.8–8.10; Fig. 8.16), whose base is often lower than the general level of the cumulus bases, probably because condensation occurs in air which has risen from near the surface with comparatively little mixing with drier air. Tattered fringes of ("scud") cloud (Pl. 8.9) always indicated strong updrafts nearby. They are similar to the ragged clouds often found below cumulus in the sea-breeze front (Pl. 7.16; Fig. 7.19), and they likewise appear where some of the comparatively moist cool air, in this case from the rain area, is lifted near the front.

The smooth, persistent flow of air into cumulonimbus indicated by arch clouds, and flight conditions near them, are consistent with observations (e.g., 81) that at least within the lower parts of the clouds the temperature and concentration of condensed water reach values closely corresponding to adiabatic ascent. The establishment of the localized cold downdraft in the cumulonimbus can be regarded as the provision of an intense intermediate-scale circulation which can move with the

8 Cumulonimbus convection

cloud. The resulting increase in scale and intensity of the updraft in the lower troposphere, together with the increase in the buoyancy of the updraft in the middle troposphere which accompanies the partial precipitation and conversion to ice of the cloud water, is responsible for the striking surge in the growth of a cumulus which is often observed during the development of a shower.

8.4 Intense cumulonimbus; severe local storms

The ordinary cumulonimbus so far described develop after the lower troposphere has been subject to widespread cumulus convection and has acquired the stratification characteristic of it. In the cumulus layer there is little variation of wind velocity with height, and often the wind shear in the upper troposphere, into which the cumulonimbus convection extends, is also small, especially over land in sunny weather.

Sometimes, however, cumulonimbus occur when there is a considerable wind shear (several meters per second per kilometer) in a deep layer, and the cumulonimbus convection then is generally more intense and persistent, the storms traveling for several hours over paths hundreds of kilometers long, accompanied almost continuously by severe surface weather. (Such storms are most frequent in the United States, where the storms which produce very strong winds—in squalls or tornadoes—or large hailstones are called severe local storms, a name now generally adopted; 96.) This behavior contrasts strikingly with the observed disruption of cumulus towers in strongly sheared layers, and the general impression that shear suppresses cumulus convection (perhaps because the shear or strong winds interfere with the development of the intermediate-scale circulations which locally favor the growth of large clouds).

The importance of the shear in cumulonimbus convection is that it causes the updraft to lean away from the vertical, or insures that its outflow in the anvil cloud spreads rapidly away from one side of the updraft. Large concentrations of precipitation which would otherwise impede the updraft then fall into clear air which, especially in the middle troposphere, appears a more favorable source for a downdraft than air in or below cloud, because of its considerably lower wet-bulb potential temperature θ_w. The possibility then arises that a cumulonimbus is organized to move at some velocity intermediate between those of the winds in the low and in the middle or upper troposphere, so that its drafts can be supplied continuously and rapidly by streams of both potentially warm and potentially cold air approaching the cumulonimbus with a large relative velocity. The greater intensity of the drafts in a steady or quasi-steady cumulonimbus compared with an ordinary cumulonimbus (in the same stratification of temperature and humidity) could be expected to arise from their larger dimensions and less effective mixing with their surroundings (at least in the lower troposphere, leading to more efficient conversion of available potential energy), and from the kinetic energy of the relative flows which supply the drafts. Figure 8.19 is a simple, idealized two-dimensional model of this kind of convective motion system.

An intense traveling storm over England

Features typical of intense traveling storms were present in one of the first of such storms to be studied intensively, with several radars and a dense network of ground observers (86). The cumulonimbus first came under radar surveillance as they approached the south coast of England from the southwest on 9 July 1959, moving almost directly toward the radar station; eventually several separate storms were identifiable, two of which attained a much greater intensity than the others. The lesser storms were groups of separate individual cells of comparatively low radar reflectivity and typically short duration (20 min or less). In contrast, new cells of the intense storms formed on their right flanks and amalgamated into large echoes within which the cells could, for a while, still be recognized as regions of high echo intensity. These echoes persisted for long periods; in particular, four adjacent cells of one storm fused into a very intense and large single echo (even though the

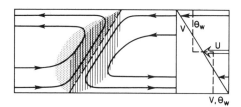

Fig. 8.19 A schematic representation of a flow through a steady-state cumulonimbus, traveling at a speed U in a layer in which the wind speed increases with height. Vertical profiles of wind speed V and wet-bulb potential temperature θ_w in the air approaching the storm are shown on the right, and on the left are streamlines of the flow relative to the storm. Condensation occurs in the stippled region in air of high θ_w, and the updraft is tilted to allow precipitation, represented by vertical hatching, to fall from the cloud into the air of low θ_w entering the rear of the storm.

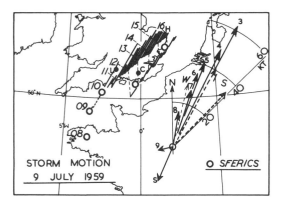

Fig. 8.20 The paths of individual cells of two severe storms (1 and 4), determined from the PPI display of a 10-cm radar in southeast England, 9 July 1959. The lines marking the boundaries between the cells, most of which merged on the display with full receiver gain, are terminated where it became impossible to distinguish the cells on displays showing only the most intense parts of the echoes. The open circles show the centers of intense "sferic" activity, and the dashed lines the positions of the radar echoes, at the hours indicated by italic numbers. Positions C, L, and H are radiosonde stations whose soundings were useful in determining the stratification near the storm. Vectors on the right show representative winds at the surface (S) and upper levels (labeled in hundreds of millibars), and the velocities of radar echoes from storms characterized as weak (w), intense (I), or severe (S).

horizontal resolution of the radar was as small as 1–3 km at the ranges involved) for about 3 hr, moving more slowly than and distinctly veered from the general winds in the troposphere (Fig. 8.20). During a phase of maximum intensity which lasted about 45 min hailstones of diameter between 3 and 4 cm arrived almost continuously at the ground beneath this large cell, and its radar echo moved steadily without any marked variation in its structure, maintaining several distinctive features which, allied to other information, permitted the general nature of the air flow in the storm to be inferred (86).

Some of these features, and the distribution of significant surface weather, are illustrated in Fig. 8.20–8.25. In its intense phase the storm produced rain at the ground over a broad swath about 100 km across (Fig. 8.24). Within this and near its right-hand edge were narrower swaths of intense rain and large hailstones: over a path about 10 km across about 15 mm of rain fell in 15 min, and in an overlapping path, also 10 km across, hailstones of diameter over 1 cm fell for about 5 min.

Over England the approach of the cumulonimbus was obscured visually by an increase of middle-level clouds and a thickening and lowering of an initially higher overcast produced by the forward spread of dense anvil cloud. The storm was heralded on its intense right flank by a low and dark arch cloud, above a squall front behind which the previously northeasterly surface wind became westerly or southwesterly, with gusts to about 30 m sec^{-1}. In the rear of the intense precipitation θ_w at

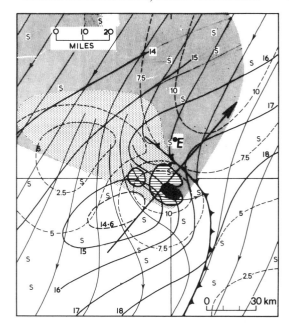

Fig. 8.21 The surface weather associated with the severe storm 1 of Fig. 8.20 and 8.22, constructed mainly from observations made at about 13h at the stations whose positions are marked by the letter S. Lines of the National Grid 100-km squares are indicated, and the position of the radar station is marked by the letter E. The air flow near the surface is shown by thin solid and dashed lines (streamlines and isotachs of mean speed, in m sec^{-1}). The position of the squall front is marked by the barbed line. The thick lines are isopleths of screen level θ_w (in °C). The area covered by the radar anvil cloud is shown in fine stipple, and that over which rain reached the ground by a coarse stipple. Over the hatched areas the radar reflectivity factor Z_e exceeds 10^4 mm^6 m^{-3} and the surface rain is probably intense; in the small heavily shaded area hail reaches the ground. The arrow shows the direction of motion of the storm.

screen level fell 2–3°C below the values prevailing before the storm, and well to the rear the surface winds returned to their former northeasterly direction, almost directly opposed to the direction of travel of the storm (Fig. 8.21). The squall front, which existed only on the right flank of the storm, trailed far behind the storm as its duration increased, and other storms, including the second intense one, formed over the front and by their own cold outflows intensified and further extended it (Fig. 8.22).

The peak level of the echo tops and the intensity of the radar echoes from the principal storm reached maxima during the intense phase in which large hail fell. Previously the radars, even at ranges over 100 km, clearly showed it to consist of separate cells. With the onset of the intense phase the speed of advance of the well-defined rear edge of the radar echo decreased from 25 to about 18 m sec^{-1}, and it was no longer possible to observe a succession of individual columns, although one radar was used continuously in a search for them. Instead, the radar echo assumed a virtually steady configuration

8   Cumulonimbus convection

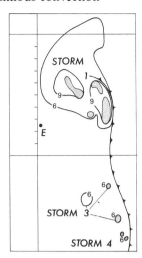

Fig. 8.22 Contours of maximum radar echo height (in km) at 14h, 9 July 1959. Parts of the 100-km squares of the National Grid and the position of the radar station (*E*) are marked. The position at the ground of the squall front extending south from the principal storm is shown by a thick line with barbs, and the stippled areas are those in which there are fresh echo cells with well-defined rising tops.

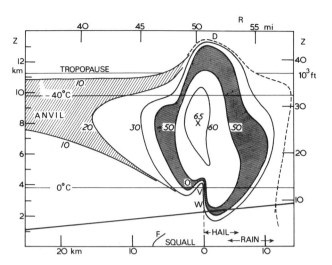

Fig. 8.23 Characteristic features of the intense phase of storm 1 (at about 12h), shown on an RHI radar display in a vertical plane through the storm in its direction of motion (from right to left, approaching the radar). The range *R* from the radar is given at the top of the diagram, and a scale of kilometers at the bottom; the vertical scale is exaggerated by a factor of about 2.

Isopleths are drawn of the (4.7 cm-) radar reflectivity factor, expressed as $10 \log Z_e$ (mm^6 m^{-3}). The letters identify the following features: *D* is the dome, almost vertically above *W*, the wall; *V* is the echo-free vault; and *O* is the forward overhang of the anvil cloud echo. *X* is the position of maximum radar echo intensity.

The quasi-steady positions of the squall front *F* and the regions of hail and rain at the ground are also shown. The dashed line at the rear of the storm shows where the echo boundary would probably have been observed in the absence of attenuation of the radar beam during its passage through the intense precipitation.

illustrated in Fig. 8.23, with the following distinctive features:

1. In the central and right-hand parts of the echo the first precipitation to reach ground observers, consisting almost solely of large hailstones, formed a *wall* with a sharply defined upright front face lying in a plane across the direction of the storm motion. The wall was practically vertically beneath . . .

2. . . . the *dome*, the highest part of the echo, whose top remained persistently about 2 km above the level of the tropopause.

3. The *anvil echo* is a diffuse echo which extended forward from the storm, becoming progressively shallower and less intense with increasing distance, and evidently corresponding to the forward part of the anvil cloud. Its top was just below the level of the tropopause, and its base descended gradually from about 7.5 km 40 km ahead of the storm, to reach the ground on the left of the wall. In a position 2–5 km ahead of the central and right-hand parts of the wall, however, it descended to a minimum height of 3–4 km, to form . . .

4. . . . a *forward overhang* whose base rose nearer to the wall, to meet it at a height of about 5 km or more, enclosing an . . .

5. . . . *echo-free vault*.

Some unsteadiness of the storm was indicated by the gustiness of the wind at the ground behind the squall front, and by small variations in the height of the tallest echo tops, in the pattern of echo intensity, and in the rainfall intensity and maximum hailstone size at the ground. Nevertheless, by assuming that the storm was a motion system in a steady state during the intense phase, throughout a period considerably longer than the time taken for air to pass through either the updrafts or the downdrafts, it is possible to infer the general form of the air flow in a vertical plane through the intense cell of the radar echo and along its direction of motion. The main constraints upon the form of the updraft are (*a*) that the supply of potentially warm air at low levels (θ_w greater than about 18°C) lay below 3 km in a layer which entered the front of the storm at relative speeds of up to nearly 30 m sec^{-1} (Fig. 8.25), (*b*) that this layer rose above the squall front ahead of the storm, and (*c*) that it produced an updraft which reached a maximum height in the dome before subsiding and spreading predominantly forward of the storm just below the level of the tropopause. The maximum upward speed attained in the updraft can be estimated as over 30 m sec^{-1} from the observed maximum penetration of the dome above the level of the tropopause (nearly 2 km), and from the size of the largest hailstones (implying fall speeds of about 35 m sec^{-1} in the upper troposphere, as shown in Fig. 2.2).

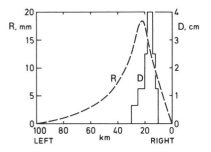

Fig. 8.24 Representative values of the precipitation R and the diameter D of the largest hailstones during the passage of the severe hailstorm of 9 July 1959 over southern England, as a function of distance from the right flank of the storm (measured normal to the direction of travel).

In addition, the position of the updraft flow in the middle troposphere behind the squall front is evidently indicated by the echo-free vault and the forward overhang, produced by the lifting of ice particles which have fallen from the anvil cloud. On the assumption that these particles were dry and spherical, with the size distribution typically found in anvil clouds, or of uniform size and a mass concentration of about 1 g m^{-3}, the observed values of the radar reflectivity factor Z_e in the overhang (up to 10^5, as shown in Fig. 8.23) imply that their diameters extended up to several millimeters, corresponding to fall speeds of up to about 10 m sec^{-1}, so that the vertical velocity of the updraft in the echo-free vault must have exceeded this value. (The inferred fall speed of the largest particles is hardly affected if it is assumed that they became wet after reentering the updraft and scattered the centimetric radiation like liquid spheres; 86.) The updraft speed probably continued to increase with height up to about 8 km, above which level the buoyancy was small or even negative, considering the weight of cloud water; above

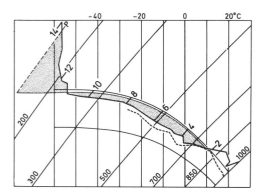

Fig. 8.25 Stratification of dry- and wet-bulb temperature for the vicinity of storm 1 (of Fig. 8.20 and 8.22) during its intense phase, shown on a tephigram. Heights above sea level are marked in kilometers, and the level P reached by the highest tops of the radar echo is indicated. The dashed line is the isopleth $\theta_s = 19.5°C$, considered representative of the potentially warm air near 2 km which supplied the storm updraft. (See also Table 8.5.)

Intense cumulonimbus; severe local storms 8.4

the tropopause there must have been a large negative buoyancy, and a rapid decrease with height in the updraft speed, as also suggested by the rapid reduction in the intensity of the radar echo, implying the falling out of the larger precipitation particles.

The general form of the downdraft flow is not as readily inferred; it seems likely to have been supplied predominantly from a layer between about 3 and 7 km which approached the storm from the rear with values of θ_w between 16 and 18°C, low enough to produce a strong downdraft at levels near and below 3 km. The air just behind the squall front, which above the ground had a forward speed greater than that of the storm, must

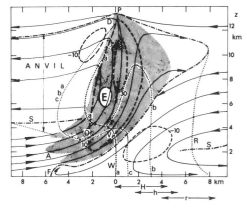

Fig. 8.26 An estimate of the flow in the storm of Fig. 8.23, in a vertical section along the path of the storm (moving from right to left in the diagram). The solid lines are relative streamlines, and dashed thicker lines are isopleths of upward component of velocity (in m sec^{-1}). Dot-dash lines marked S in the lower part of the diagram show heights at which there is no wind relative to the storm and indicate that near the storm they may differ considerably between its front and rear.

The dotted lines labeled a, b, and c are schematic trajectories, not at first distinguished, of small particles (fall speeds about 10 m sec^{-1}) which fall from the anvil cloud. They enter the foot of the updraft at about the same level, and subsequently rise and grow into hailstones (the trajectories are referred to later in the chapter, in the discussion of hailstone growth). One particle grows into a large hailstone with a fall speed of about 30 m sec^{-1}, and again falls forward of the updraft, following the path a and reaching the ground in the leading precipitation. Another, initially slightly larger, is not lifted as high and does not become as large; it falls to the rear of the updraft along the path b. The third, initially still larger, is lifted hardly at all and follows the path c. The differing trajectories show how the forward overhang O, the echo-free vault V, and the wall W of the radar echo are produced, and indicate the kind of sorting which leads to the largest hailstones reaching the ground in a zone H which lies ahead of those in which the small hail h and finally the rain r arrive (these zones are also separated in planes normal to the section). Other features marked are the squall front F, the arch cloud A above it, and the dome D with its peak P practically vertically above the wall W. The region of maximum radar reflectivity ($Z_e \approx 5 \times 10^6$ mm^6 m^{-3}) observed by radars stationed in front of the storm is shown by the clear area marked E.

8 Cumulonimbus convection

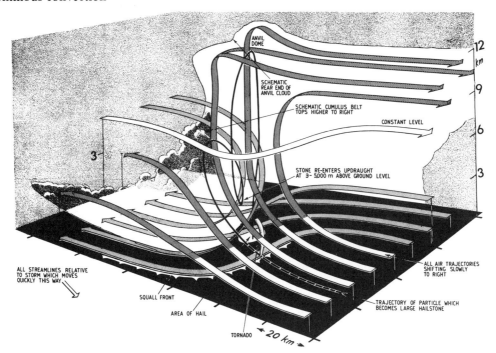

Fig. 8.27 A tentative three-dimensional model of the intense traveling storm of 9 July 1959 over southeast England (86, 96). The vertical scale is exaggerated. The relative streamlines can be regarded as trajectories and are hatched where they are in cloud or precipitation. Hail reaches the ground in the white area and rain over the stippled area. A belt of cumulus is indicated lying in a wing over the squall front on the right flank of the storm.

Much of the air which rises over the squall front forms the updraft of the cumulonimbus; a small proportion reaches a peak height in the dome before subsiding and joining the main outflow in the anvil cloud, which spreads predominantly forward of the storm.

The downdraft is drawn partly from the middle troposphere behind the storm and partly from the lower troposphere in front of it; the relative proportions of these flows is uncertain. No tornado occurred in this storm, but when one occurs it is typically near the position in which a funnel cloud is drawn, close to the squall front and the region of hail at the ground. One streamline in the upper troposphere is drawn to indicate that the general flow may be strongly disturbed near the storm.

evidently leave the plane of the section if the steady configuration is to be preserved, just as the updraft air must leave the plane near the rear lip of the anvil. It seems that the mass flux in the downdraft was substantially less than that in the updraft, as might be anticipated since the displacement of a unit mass of air from the middle troposphere to the ground requires the evaporation of nearly as much water as is condensed in a unit mass raised from low to high levels, whereas the precipitation at the ground represents a considerable fraction of the mass of water vapor condensed in the storm.

In the construction of the diagram of the flow in the chosen plane it became further evident that there was significant motion through the plane. For example, it was not possible to account for the known duration of large hailstones at the ground unless the mass flux and the dimension of the updraft aloft were about twice as great as if the flow had been simply two dimensional. Further, near the storm substantial cross-components of the wind were evident at low levels from pilot balloon data, and at higher levels from the shape of the radar echo from the anvil cloud and the observed motion of

details within it; it was estimated that at heights of several kilometers the winds near the updraft had a component toward the left flank of the storm of 10 m sec^{-1} or more. Other prominent three-dimensional aspects of the storm were the spread of the downdraft outflow behind and to the *right* of the storm, and the spread of the updraft outflow, and therefore of the anvil cloud, ahead and to the *left* of the storm (Fig. 8.21). Again, the decrease of θ_w at screen level behind the storm to values below 15°C suggests that there the air was drawn from low levels in front of and to the left of the storm (analysis of radiosonde and aircraft soundings showed that the westward decrease of θ_w at low levels, which is apparent in Fig. 8.21, did not occur in the middle troposphere near the storm).

From the general considerations outlined here the flow in the vertical plane through the intense part of the storm was inferred to be approximately as shown in Fig. 8.26; Fig. 8.27 is a corresponding three-dimensional model. It is likely that in Fig. 8.27 the components of air motion across the direction of motion of the storm have been insufficiently emphasized.

8.5 General characteristics of intense traveling storms; cumulonimbus dynamics

Partly depending on the degree of resolution in the available observations, cumulonimbus convection may be observed on a range of scales between that of the single cell in the ordinary cumulonimbus, whose updraft is a few kilometers across and lasts about half an hour, and that of the more persistent and intense storm, whose updraft extends about 20 km across its path and persists continuously if not altogether regularly for several hours. The severe storms frequently are accompanied by lesser storms, some of which may successively enter an intense phase, so that they are not readily identifiable from even modern synoptic ground observations, whose spacing is rarely less than 100 km. The earliest evidence of the extraordinary nature of the severe storms was in assembled reports which showed the narrowness and almost uninterrupted great length of swaths of crop damage, as in the notorious storm which crossed France on 13 July 1788 (87). This storm advanced northeast across France "in two columns 12 miles apart, the one on the west 10 miles broad and the other 5 miles broad, the one extending nearly 500 and the other 440 miles." The astronomer Flammarion called this "the greatest hailstorm known," and Carlyle records that "there fell, on the very edge of the harvest, the most frightful hailstorm; scattering into wild waste the Fruits of the Year, which had otherwise suffered grievously by drought." This storm was thus a significant factor among those leading to the outbreak of the French Revolution in the following year.

Early studies of intense traveling storms

Late in the last century networks of voluntary observers were used in Europe for investigations into thunderstorms and hailstorms, mostly on the initiative of enthusiastic individuals and spurred by interest in means of preventing hail damage. In this way it was established that a distinction could be made between scattered storms which affected small areas, and much less frequent intense traveling storms which could be traced over very long paths. In England, for example, it appeared that hail from an individual storm usually fell on an area a few kilometers long and less than 1 km across, and in the state of Württemberg in Germany the average farmland affected by a hailstorm was found similarly to be several kilometers in area (88). On the other hand, detailed observations made in Austria (89) showed that the length of the almost straight swaths of hail left by the intense traveling storms was often 100 km or more, and the width usually between 6 and 12 km. Occasionally they contained small areas over which no hail had fallen, while in other places they broadened temporarily with an increase in the intensity of the hail fall. These storms usually moved from between south-southwest and northwest, at an average speed of about 11 m sec^{-1} but occasionally at nearly 20 m sec^{-1}. They were shown to be moving with approximately the velocity of a middle-level current, often against the direction of the wind close to the ground (90). (These studies, based on observations by about 1000 volunteers, provided a wealth of interesting information on the behavior of the storms. In particular, it was found that hail fell in only a small part of the storm, and that the largest hail often fell only on the right-hand edge of the hail swath. Lightning rarely struck the ground, especially in the hail falls, although the discharges were frequent and the thunder typically merged into a continuous clattering din. The largest hailstones were often several centimeters in diameter and weights of up to 1 kg were reported.)

Over Austria intense traveling hailstorms occurred several times each year, but over England their frequency is only about once every few years. Nevertheless, observations of thunderstorms made in England at about the same time were sufficient to show that the more intense often traveled steadily from a southerly quarter along a straight path up to several hundred kilometers long, at speeds up to about 20 m sec^{-1}, and it was proposed that they be called *line thunderstorms* (91). Subsequently the same behavior was established for intense hailstorms in southeast England (92). Although it has long been accepted as part of a weather forecaster's task to predict the occurrence and intensity of cumulonimbus convection, his or her information on its severity has usually had to be ascertained, if at all, after the event, and the forecaster has necessarily been intent upon relating it to some property of the large-scale motion systems. Already by the beginning of the twentieth century it was established that the intense traveling storms were associated not with "strong insolation and steep lapse-rate at low levels, nor with the occurrence of a pressure minimum, but with the transition-zone between cool and warmer regions" (90). (In Austria these storms usually began in the afternoon and often persisted into the night hours; in other places, for example, southern England, they are probably more frequent in the night or early morning than in the afternoon, and therefore evidently not related to any local surface heating.)

Much later, after the development of the polar front theory of cyclones and the general adoption of the techniques of frontal analysis of synoptic weather charts, it was recognized that the intense traveling storms were associated with fronts, and they were referred to as "frontal" to distinguish them from ordinary "air-mass" or "heat" storms, although only vague ideas existed about the reasons for the association (93); even recently, when it was found in the United States that the most severe thunderstorms occurred where the winds in the middle troposphere were strong (speeds about 20 m sec^{-1} or more; 94), the result was at first regarded as a paradox, because it was then thought that the implied strong wind shear in the vertical, which must also accompany the strong horizontal temperature gradient near a front, tends to destroy organized vertical currents rather than promote them. Nevertheless, these early studies make it plain that the hailstorm of 9 July 1959, which has been described in detail, was not an abnormal phenomenon, but a typical member of a class of storms which although infrequent compared with ordinary cumulonimbus occur occasionally over southern England, and produce damaging hail over western and central Europe sufficiently often each summer to be a familiar menace to crops, especially where grapes and other fruits are grown.

The intensification of cumulonimbus convection in the presence of wind shear; cumulonimbus dynamics

That the cumulonimbus convection can be more intense in the presence of wind shear is an interesting and important phenomenon which can be illustrated by comparing three storms over southern England (Table 8.5): 18 June 1957, an occasion of small shear and obviously impulsive cumulonimbus convection (Fig. 8.18); 9 July 1959, the occasion of the intense traveling hailstorm discussed above; and 5 September 1958, when an even more intense storm, accompanied by giant hail and tornadoes, crossed southeast England (95).

On the last occasion measured weights of about 200 g confirm reported diameters of about 7 cm, while indirect evidence of still larger stones suggested equivalent spherical diameters of 8 cm. The winds were lighter with less shear in the vertical than on the occasion of the intense hailstorm of 9 July 1959, and the storm moved more slowly, but again well to the right of the direction of the general winds.

On 18 June 1957 winds were light easterly (see also Figs. 7.21, 8.18, and 8.24); the wind shear in the vertical was very small and opposed to the direction of travel of the cumulonimbus, so that their anvils tended to trail behind the lower parts of the clouds. The tallest clouds occurred in mid-afternoon with tops (observed both by aircraft and radar) generally below the tropopause at about 12 km; however, two towers briefly reached just over 13 km before subsiding rapidly, and this has been taken as the peak height attained. In contrast to the behavior of the traveling storms on the other two

Table 8.5 Approximate values of properties of the storms of 18 June 1957, 9 July 1959, and 5 September 1958 over southeast England, and of 26 May 1963 over Oklahoma

	18 June 1957	9 July 1959	5 September 1958	26 May 1963
Maximum point rainfall (cm)	4 (5 from 2 storms)	3	10	8
Maximum measured rainfall rate (mm hr^{-1})	240 (in 10 min)	110 (in 10 min)	200 (in 20 min)	—
Storm velocity (°, m sec^{-1})	080 8	230 18	230 12	290 10
Relative velocity of potentially warm air	260 5	060 15	070 15	140 20
Equivalent kinetic energy [(m sec^{-1})^2]	5^2	15^2	15^2	20^2
Magnitude of wind shear (m sec^{-1})				
From near ground to 5 km	0	30	20	22
From near ground to tropopause	7	45	30	47
Updraft duration (hr)	<0.5	3	4	6
Maximum lightning discharge rate (min^{-1})	5	15	30	—
Maximum height of cloud tops (km)	13.2^a	13.4	15	17.8
Maximum penetration of tropopause (km)	1.1	2.4	3	5.3
Maximum hailstone diameter (cm)	1	4–5	7–8	9–10
Fall speed in high troposphere (m sec^{-1})^b	25	40	50	65
Simple parcel theory				
Updraft properties				
θ_s (°C)	21	20	21	24
Maximum temperature excess (°C)	9	4	8	12
Speed at -40°C level (m sec^{-1})	67	35	55	75
Positive area [(m sec^{-1})^2]	70^2	36^2	59^2	85^2
Negative area [(m sec^{-1})^2]	35^2	44^2	60^2	95^2
Difference [(m sec^{-1})^2]	61^2	$-(25)^2$	—	$-(42)^2$
Mean Richardson number for updraft, $-(\overline{Ri})$	10^2	0.7	4	5
Maximum wind speed in squalls (m sec^{-1})	10	30	25^c	25^c

^{a} Twice attained for a few minutes only; otherwise peak heights 12.0 km.
^{b} Cf. Simple parcel theory, speed at -40°C level (below).
^{c} More near tornadoes.

occasions, the radar echoes from the individual cumulonimbus were short-lived, and the echoes from the two tallest clouds disappeared completely within about an hour. Together the two storms produced total falls of rain of between 1 and 2 in. in an area about 20 km long and 10 km across, but although high rainfall intensities were recorded (including 1.6 in. in 10 min, corresponding to 240 mm hr^{-1}), the dense network of volunteers observed remarkably little hail in these or the other storms; no definite hail swaths could be traced, and the diameter of the largest stones was about 1 cm (in five reports from scattered locations, separated by places in which no hail at all was observed).

Some of the more important properties of the storms on the three occasions are summarized in Table 8.5 (which includes information on some intensively observed storms over Oklahoma, which are discussed in Section 8.8). The table entries emphasize that on the occasion of little wind shear the local intensity of the rainfall from the principal storms was comparable with that on the other occasions, but in other respects the severity of the weather associated with the latter was markedly greater. The greater intensity of their downdrafts is indicated by the stronger squalls at the ground, and the greater speeds attained in their updrafts by the larger size of the hailstones produced, and by the increased penetration of the tropopause by the cloud tops.

The comparative intensities of the updrafts can also be considered by referring to the simple parcel theory of ordinary convection, although with strong reservations since it assumes that the motion in the drafts is frictionless and adiabatic, and that the upward acceleration of a parcel of fluid is given simply by its buoyancy. The general presence of condensed water and motions of smaller scales makes it likely that the buoyancy will be overestimated, while the uncertainty of the latter assumption is illustrated by the thermal, in which the greatest vertical velocity is not associated with the maximum local buoyancy. However, it has already been remarked that in cumulonimbus convection, in contrast to cumulus convection, the air of the outflow in the anvil cloud tops has a potential temperature closely approaching that in the cloud bases, so that the conventional pseudo-adiabatic reference process, perhaps fortuitously, may give a good measure of the temperature of the warmest part of the updraft, and it is interesting to compare the observed peak heights of the cumulonimbus with those predicted by the simple parcel theory. According to this theory (Section 7.2) the maximum speed of ascent of cloud air is attained at the equilibrium level E found on an aerological diagram from the intersection of the curve representing the state of the environment and the (pseudo-) adiabatic curve through the point representing the cloud base. (See Fig. 7.3; in the theory this is also the curve through the wet-bulb temperature near the ground. The small updraft speed which experience shows is acquired already at the cloud base, and differences between air temperature and virtual temperature, are ignored.) On a tephigram the kinetic energy at the level E is proportional to the area, described as the positive area, enclosed by these curves. Above the level E the buoyancy is reversed, and the air reaches a maximum height where the corresponding isobar and the two curves enclose an equal area known as the (upper) negative area.

It is remarkable that on some studied occasions of intense and persistent cumulonimbus over England (96) and over the United States (97), the negative area defined by the isobar given by the pressure at the observed height of the cloud top has been found to be about equal to or even distinctly greater than the positive area. Correspondingly, the observed penetration of the tropopause (usually close to the level E) has been approximately that given by Eq. 8.16. This approximate agreement between observation and the simple parcel theory has been found only when the storm has been severe and the positive area large: the area measured is not then so sensitive to the value adopted for θ_s at the cloud base and in the updraft. This value cannot usually be assessed with a likely error of less than $\pm 0.5°C$, implying a likely error of about $\pm (25 m/sec)^2$ in the positive area, when, as in Table 8.5, it is expressed in terms of an updraft speed attained near the tropopause (the corresponding error in the negative area, when the cloud tops rise a few kilometers above the tropopause, is less). In the severe storms the size of the largest hailstones suggests that the speeds implied by the simple parcel theory for the updraft at high levels are realistic, since at the top of the zone of supercooling they somewhat exceed the fall speeds of the hailstones, as would be expected from the discussion of hail growth in Section 8.11. (However, the observed heights of the domes of severe storms and the sizes of their largest hailstones imply only that the *maximum* updraft speeds have about the values given by the simple parcel theory. These are attained by only a small proportion of all the air which ascends in the updrafts, although at the cloud bases the potential temperature is nearly uniform.)

On the other hand, Table 8.5 shows that on the occasion of hardly any wind shear (18 June 1957) the negative area was only about one quarter of the positive area, even when calculated from the peak height briefly attained by the two tallest cloud towers (the other measured peak heights implied hardly any negative area). Whereas in the other storms the cloud tops maintained their peak heights and intensity for an hour or more, on this occasion the convection was clearly impulsive and cloud towers were observed at their peak heights for only a few minutes. Moreover, although the positive area was even larger than on the occasion of the tornadic hailstorm of 5 September 1958, the cumulonimbus produced only small hail, their updrafts evidently being insufficiently strong or sustained for the growth of large hailstones. The negative area was also comparatively small or absent during the period of impulsive cumulonimbus convection with little wind shear over central Sweden, which has been discussed previously (Fig. 8.14). The showers of this period contained no hail or only very small hailstones, and cloud tops did not reach the tropopause.

Accordingly, a clear distinction appears between ordinary cumulonimbus, in which the convection is obviously impulsive, and intense cumulonimbus, in which both updrafts and downdrafts are comparatively persistent and attain greater speeds, with the production of more violent weather. This difference seems to be related to the wind shear, which if sufficient results in the convection becoming more organized, with increase of scale and a more efficient conversion of potential into kinetic energy.

So far only tentative studies have been made of this more organized form of convection (e.g., 105); attempts have been made to find a steady two-dimensional flow pattern, dynamically consistent with the distribution of heating and cooling associated with the condensation of water vapor and the evaporation of precipitation. The evaporation is difficult to treat appropriately because unlike the condensation it cannot be supposed that it proceeds in air which is nearly saturated, as will be explained in Section 8.10. The evaporation is severely limited by low temperatures and by the size of precipitation particles, so that it is not possible for the flux of air in the downdraft to be closely comparable to that in the updraft, and for the overturning to be as symmetrical in the idealized model of Fig. 8.19. Theoretically difficulties also arise in accounting for the maintenance of an interface with a slope appropriate for the efficient transfer of precipitation from the updraft, unless it is across rather than along the principal flow, so that the updraft air can leave the storm on the side opposite its entry. A steady flow through the storm in only two dimensions may not be possible, and progress in the study of intense cumulonimbus will probably depend on the numerical evaluation of flows in three space dimensions. Even with some simple representation of the behavior of the water substance, not involving the development of particles with various sizes and paths, and of the small-scale motions, the problem of establishing whether a steady or quasi-steady flow can exist appears formidable. However, not until this is done will it be possible to ascertain which features of the initial stratification are the most important, and how they control the structure of cumulonimbus.

In all the formulations of the problem examined a Richardson number appears which expresses a tendency for the wind shear to organize the flow (i.e., to introduce a larger, dominant scale of motion), against a tendency for the buoyancy to disrupt it. Overlooking the probably important complication that there is a downdraft in which the buoyancy is difficult to evaluate, although it is certainly different from that present in the updraft and effective over considerably less than the whole depth of the troposphere, it is reasonable to consider the Richardson number appropriate to the updraft flow, represented as a ratio between an apparently available potential energy (a positive area in the simple parcel theory) and the kinetic energy corresponding to the vector difference of the general (relative) wind velocities at the levels of flow into and out of the updraft circulation. The relevant Richardson number is some mean value appropriate to all the air which composes the circulation, but it is probably dependent mainly on the air which has the extreme displacements, between levels near the ground and near the tropopause. A single representative value, obtained from the general stratification, and not from observations of any circulation which actually develops, may therefore be given by a number $-(\overline{\text{Ri}})$, which corresponds to the whole positive area of the simple parcel theory, and the kinetic energy of the wind shear ΔU between the level E of the theory and the level (usually near the ground) of the air of highest potential temperature; both quantities are conveniently expressed in the units of (m/sec)^2 used in Table 8.5: $-(\overline{\text{Ri}}) = (\Delta W)^2/(\Delta U)^2$, where (ΔW) is the maximum updraft speed according to the simple parcel theory. The values entered in Table 8.5 show that in the severe storms $-(\overline{\text{Ri}})$ was less than about 5, whereas on the occasion of the ordinary cumulonimbus it was as large as 10^2.

From a very limited amount of other experience it appears that either severe local storms or ordinary thunderstorms have occurred on occasions when the values of $-(\overline{\text{Ri}})$ have been between about 5 and 10, the former generally when the available potential energy was extremely great, and the latter when it was only small or moderate. On the whole, therefore, there is some support for regarding $-(\overline{\text{Ri}})$ as a significant property of a layer liable to cumulonimbus convection, and for anticipating that an organized cumulonimbus updraft and an efficient conversion of available potential energy can occur only when $-(\overline{\text{Ri}})$ is not much greater than 1. Its value is therefore given for most of the stratifications illustrated in this chapter.

Although the simple parcel theory may give a reasonable estimate of the potential energy which is available for conversion into the kinetic energy of an organized updraft, it has already been mentioned that the distribution of the kinetic energy and in particular the maximum updraft speed cannot be obtained without a more complete theory. However, an extended parcel theory can relate the properties of the inflow and the outflow of the updraft well away from the storm, where it can be assumed that the distribution of the pressure in the vertical is hydrostatic. Considering the flow along a relative streamline through a steady updraft and assuming the motion to be (pseudo-) adiabatic without any dissipation of kinetic energy, then the kinetic energy of the outflow is equal to the kinetic energy of the inflow increased by the available potential energy. This result indicates that some of the kinetic energy of the general flow is also available to the updraft. In particular, it is interesting to consider the implications in a stratification neutral in the conventional sense, but saturated at low levels and with low relative humidities in the middle or high troposphere (θ_s constant, but θ_w decreasing with height). Then according to the parcel theory (and as usual overlooking the difference between actual and cloud virtual temperature) no work need be done in lifting the saturated air and generating precipitation, while a downdraft can be produced by evaporating precipitation into the unsaturated air. In the presence of a wind shear to ensure the transfer of the precipitation, a steady organized storm can be envisaged in which the updraft would have no buoyancy but a speed of several meters per second provided by the relative speed of its inflow. More generally, the kinetic energy of an updraft may be augmented by that present in its inflow, but the contribution is unlikely to be significant when the available potential energy is large, as shown in Table 8.5, and is not sufficient to explain the discrepancies noted there between the positive and negative areas. The kinetic energy of the inflow is probably more important in rare storms in which the available potential energy is not usually large and which are not intense, but cause damaging floods by remaining almost stationary for as long as several hours. Their duration suggests that they are well organized, and in some of the examples which are described in Appendix 8.14, for which adequate data are available, it appears that the value of $-(\overline{\mathrm{Ri}})$ was approximately 1. On the occasions of such stationary storms the direction of the flow at low levels is generally opposed to that in the upper troposphere, but apparently not always directly, and their occurrence draws attention to the velocity of a storm as one of its important properties for which as yet there is no adequate theory. The velocity of a storm often alters markedly with changes in its severity: usually in the most intense phase the speed decreases, and in the northern hemisphere the direction of travel turns toward the right (in the southern hemisphere toward the left; 108). Exceptionally, as described later, the direction of motion of a storm turns in the opposite sense. Related to this deviation of the direction of motion of storms from that of the general winds is the skewness which develops in the drafts, and in particular the obvious tendency for the outflow of the downdraft to spread out from only one flank of the storm.

In Section 8.6 the occurrence of organized cumulonimbus over the world will be considered, relying mainly upon reports of the weather experienced in storms. Attention is necessarily biased toward the more intense storms, because it is hardly possible to identify weak organized storms from the ground observations at individual places, and these are generally the only data available. Of those features of the weather which are conventionally recorded, the *speed of the squall* experienced at the ground is probably the best indication of both the intensity of the downdraft and the degree of organization of the convection. Of the others, the *maximum rate of rainfall* may not be much greater in an intense traveling storm than in ordinary cumulonimbus convection, probably depending mainly on the vertical velocity of the air at the cloud base, which may not vary much, rather than the speed attained by the updraft at higher levels; moreover, evaporation in a deep and strong downdraft may significantly reduce the rain which reaches the ground. The *total rainfall* at places on the ground is affected by the duration of the rain; it may be large even on occasions of ordinary cumulonimbus when the clouds are slow-moving, or when the measuring gauge is read only once or twice a day and several storms have occurred. The occurrence of *large* or *giant hailstones* is reliable evidence of great updraft speeds, but, as will be discussed later, it seems that they are not produced in all intense storms. Moreover, they fall over small areas, and even if they do not altogether escape notice their size seldom is recorded. In most regions outside the tropics climatological data include hail frequency at well-scattered observing stations, but the values are dominated by the falls of small hailstones which may be produced by ordinary cumulonimbus. Particularly in the cool season, when small cumulonimbus often grow over the sea, the frequency of hail commonly has a maximum near coasts exposed to the cool airstreams in the rear of cyclones.

On the other hand, the better organized cumulonimbus usually has an intense downdraft and is accompanied by a stronger squall. (It will be shown later—in Section 8.10—that a high cloud base, and therefore a deep adiabatic layer, is a condition strongly favoring the production of an intense downdraft. Over arid regions, therefore, the occurrence of strong squalls near cumulonimbus may be less reliable evidence of the organization of the convection.) The squall sweeps over a larger area than may be affected by large hailstones, and accordingly is more often recorded. Because of the suddenness of their arrival such squalls are a notorious hazard to sailing ships and aircraft, and for this reason there is abundant information in the meteorological literature on their occurrence and severity, and in particular the maximum speed of the gusts in them has been commonly and accurately recorded. The difference between the wind velocity during and preceding the squall might be a more significant feature, but since the latter is usually small the speed attained in the squall itself is a useful indicator. The speed depends on the proximity to the downdraft, but in the ordinary cumulonimbus convection it is usually not more than about 10 m sec^{-1}, whereas in

8 Cumulonimbus convection

the intense traveling storms it is typically more than 15 m sec^{-1} and may reach even 50 m sec^{-1}. The squall front is usually marked by an arch cloud of great length (often spanning the whole field of view) and otherwise impressive appearance, and in the era of the square-rigged sailing ships of widespread maritime commerce became well known as a valuable warning of the approach of a violent squall (Appendix 8.4).

8.6 The distribution over the world of (intense) organized cumulonimbus

Judged by the criteria of accompanying severe squalls and arch clouds, or by the occurrence of large hailstones, the parts of the world in which organized cumulonimbus are most frequent are those seen in Fig. 8.28. Those regions in which large hailstones occur are distinguished from those in lower latitudes where rain, lightning, and squalls are severe, but large hailstones are almost unknown. The latter regions are over or close to lands where in some seasons the intertropical confluence zone is well defined and where there is then a considerable wind shear in the lower troposphere. The regions liable to severe hailstorms are also over or close to land, but are in higher latitudes where large wind shears are associated with frontal zones in the large-scale slope convection and usually extend up through the whole troposphere. Possible reasons for the usual absence of large hailstones in low-latitude storms will be considered in the discussion of hailstone growth (Section 8.11).

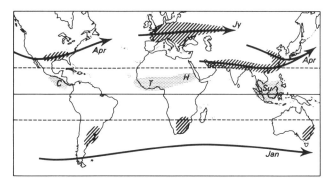

Fig. 8.28 The distribution of regions subject to severe squall storms. In low latitudes (stippled regions) the storms only very exceptionally produce large hailstones. They mostly occur where the intertropical confluence zone becomes sharply defined and accompanied by a considerable shear of wind between the low and the middle troposphere. The storms are known locally as *chubascos* (*C*), *tornadoes* or *disturbance lines* (*T*), *haboobs* (*H*), and *sumatras* (*Su*).

In middle latitudes (hatched regions) the severe storms usually produce large hailstones and sometimes the violent whirlwinds properly called tornadoes. They are closely related to the mean position of the strongest winds in the middle troposphere during the squall-storm season (broad arrows). In the southern hemisphere the storms occur in the southeastern parts of the continents during excursions northward of the strong winds from their mean position, associated with the penetration of cold fronts into their latitudes.

The squall storms of low latitudes

Although hailstones of diameter up to about 1 cm are occasionally reported from places within 20° of latitude from the equator, and even larger stones occur exceptionally (101), the characteristic feature of the intense storms in this zone is the arched squall. They are typified by the West African *disturbance lines* (102, 103), or *tornadoes* (they are not, however, accompanied by the violent whirlwinds known by this name).

These storms develop throughout the year in a belt between the coast of the Gulf of Guinea, whose latitude is about 5°N, and the intertropical confluence zone (ITCZ). This zone separates the dry east-northeasterly winds leaving the Sahara from the shallow trade winds of the southern hemisphere, which cross the equator and enter West Africa as a moist southwesterly current. It is often sufficiently well defined, even on monthly mean charts, to be called a front, and to be drawn as a line along which the screen-level dew point is about 15°C. About 300 km south of this line the mid-afternoon screen-level values of θ, θ_w, and the dew point are, respectively, about 33, 25, and 21°C, while an even lesser distance to the north the corresponding values are those typical of the central Sahara: about 42, 20, and 5°C; it is the gradient of dew point which is most marked and attracts the attention of forecasters during chart analysis. In the winter season the moist air does not penetrate far inland, but over central West Africa in the summer it reaches as far as 20°N, and over a broader inland zone covered by the moist air the storms become more extensive and severe. The moist layer a few hundred kilometers south of the ITCZ is usually between 1 and 2 km deep and it is often sharply separated from the dry easterlies aloft, but the distinction becomes less clear farther south and west, where continual cumulus and cumulonimbus convection progressively modifies the air in the upper current.

Cumulonimbus form in the afternoon over the hillier inland regions (sometimes at least as far east as the Marra mountains in the Sudan at about 25°E) and become intense in a zone which extends to within about 100 km of the ITCZ. They frequently travel west, away from their source regions, with fresh clouds forming on their flanks until a chain of cumulonimbus develops which extends roughly from north to south over several hundred kilometers. These disturbance lines may persist throughout two or three nights and days, traveling westward at a fairly uniform speed of about 15 m sec^{-1}, about equal to or even a little greater than the speed of the easterlies near the level of the outflow in the anvil clouds. Consistently, although a forward lip of anvil may extend some kilometers ahead of the columns of cloud which rise up in an almost vertical wall just behind the arch cloud, the anvils trail predominantly behind the storms, in contrast to the typical configuration in middle latitudes. A few kilometers behind the front of a squall, which is usually from between northeast and southeast, the rain begins and quickly becomes intense, 20 mm often falling in the first 15 min; it then gradually eases, falling from cloud whose base appears to be at a height of 4–5 km, and about 2 hr later the rain ceases and a light southwesterly wind has been restored. The rear edge of the upper cloud does not appear for another 1–2 hr.

In central West Africa during the summer pronounced disturbance lines affect individual places at intervals of 2–3 days; squall speeds exceeding 25 m sec^{-1} are experienced about once each month, and well inland the maximum recorded speeds reach about 40 m sec^{-1} (104). Behind a squall θ_w at screen level falls by several degrees to about the value previously found near the 700-mb level.

The haboobs of the Sudan are similar but less intense squall storms, and others occur in the East Indies and over and near the west coasts of Central America. Like the West African storms, they form over land but may travel out over the sea, and accordingly near coasts occur most often in the late hours of the day or during the night. Those in the East Indies are most intense in the Strait of Malacca, and are known locally as *sumatras*; the squall speeds are said rarely to exceed 20 m sec^{-1} (99), but in some places reach this value at least several times each year.

Figure 8.29 shows an inferred form of the relative flow in the cumulonimbus of the West African disturbance line. The flow is portrayed in the vertical plane lying in the direction of travel of the storm; the evidence that its speed of travel may somewhat exceed that of any wind in the troposphere emphasizes that there must be significant components of flow normal to this plane, and that at least some of the downdraft air may be drawn from ahead of the storm. If the disturbance line is several hundred kilometers long it must then be composed of a chain of individual updrafts with a much lesser dimension.

Intense cumulonimbus in the southern hemisphere

In the southern hemisphere the ordinary climatological data are a poor source of information about intense cumulonimbus. It is clear, however, that in South Africa (106) and Australia (107) traveling storms occasionally produce large hailstones, and rarely also tornadoes, during midsummer in the southeastern provinces (Fig. 8.28; the frequency of thunderstorms is high, but no greater than over the same continents farther north, where hail is generally rare). Near Johannesburg and Pretoria, South Africa, the average frequency of hail at a particular place is about 5 days each year; intense traveling storms which produce hailstones of diameter greater than 3 cm are relatively rare: over a selected area of 2500 km^2 about 5 occur each year, compared with about 80 hailstorms altogether (108). They produce long swaths, oriented usually toward a northerly point and to the left of the direction of the winds in the middle troposphere.

In South America the intense cumulonimbus occur mainly over Uruguay and the northeast of Argentina. The frequency of hail of any kind is said to be less than 1 or 2 days each summer at most places (109); however, this is no good indication of the absence of occasional falls of large hailstones. In the province of Mendoza near the Andes, where thunderstorms occur nearly every day in midsummer (110), the hail frequency in one area

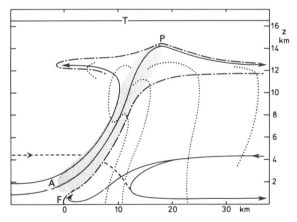

Fig. 8.29 The form of flow in the cumulonimbus of the West African disturbance line, typical of the low-latitude squall storm, in a vertical section along its path (westward, from right to left in the diagram). The vertical scale is exaggerated. The solid lines are schematic relative streamlines; a dashed line is a hypothetical streamline in a three-dimensional storm, which suggests that the downdraft, as well as the updraft, may be drawn from ahead of the storm. The region of condensation is shown by stippling; the updraft probably reaches a speed of 30 m sec^{-1} in the upper troposphere. Lines of dots show some paths of precipitation particles (all arrive at the ground as raindrops). The squall front and the arch cloud above it are marked F and A, respectively. The peak P of the cloud is shown below the level T of the tropopause; the anvil extends a short distance ahead of the storm, but trails far *behind* it, in contrast to the configuration in the intense middle-latitude storm (Fig. 8.26).

8   Cumulonimbus convection

of 4000 km² is about 30 each season (111), comparable with that quoted for South Africa. The neighborhood of the mouth of the Rio de la Plata is notorious for violent arched squalls (*turbonados*) which frequently accompany the passage of cold fronts (Appendix 8.4). The squall speed frequently reaches 35 m sec^{-1}, and extremes of 45–55 m sec^{-1} have been recorded in Montevideo.

Figure 8.28 shows that in each continent of this hemisphere the region liable to intense storms lies somewhat north of the mean summer position of the strongest mid-tropospheric winds: in all regions the storms are associated with the approach of cold fronts which displace the strong winds providing the required wind shears.

Intense cumulonimbus over Asia

Cumulonimbus convection and thunderstorms are frequent over India and Southeast Asia during both the summer monsoon and the periods of about two months which precede and follow it (in many places thunderstorms occur on several days each month, and in some, especially those over and near high ground, on more than 10 days each month). However, traveling storms accompanied by severe squalls and hail do not occur during the monsoon. They are characteristic of the periods following the monsoon, and especially of the months that precede it. They occur in a belt which extends from the Gulf of Arabia into China, and are most frequent in the north of India. Over this subcontinent the influences upon the cumulonimbus convection of motion systems covering a great range of larger scales are striking. Among these motion systems are the diurnal circulations associated with coasts and hills; passing systems of slope convection, which are evident mainly in high latitudes and the upper troposphere; circulations on the scale of India itself, a great peninsula which in the hot premonsoon period appears to produce a shallow influx of maritime air across the coast in the lee of the prevailing northwesterly winds (and hence a convergence inland, as happens on a smaller scale over Florida, where thunderstorms occur nearly every day in the summer); and ultimately the monsoon, associated with topography on the continental scale, in which the southeasterly trade wind current of the southern Indian Ocean crosses the equator and invades the whole of India as a southwesterly airstream.

In the dry regions in the west of the belt of intense storms squalls frequently raise dense clouds of dust or fine sand, and over the plains of northwest India they are the *andhis* which have already been mentioned; the name refers to the extreme darkness produced by the worst dust storms, even in the afternoon. They are often said to be accompanied by only light rain, or even no rain at all at the ground, and no thunder, but the squalls at the ground and the dust clouds spread well away from the parent cumulonimbus. (Similarly, the arch clouds and the squalls which herald the pamperos of Argentina are often observed over the sea to pass without any rain.) The occurrence of the andhis requires the spread into the arid regions of at least a shallow layer of maritime air, usually from the Arabian Sea (112) during the passage of shallow depressions eastward over northern India. Even so, the level of the cloud bases is often as high as 4 km. (When the stratification is examined the potential energy available for an updraft is usually rather small, and smaller than that apparently available for a downdraft from the middle troposphere, providing sufficient rain water could be evaporated to keep its air saturated.) However, thunderstorms and strong squalls occur even with such a high cloud base, and the radar echoes are sometimes observed to travel for at least several hours at speeds of 20–30 m sec^{-1} (113). The updrafts are probably not intense since their tops are rarely above 12 km (114) and hail hardly ever reaches the ground. The great duration of many of the squall storms is also indicated by the observation that although the squalls have a maximum frequency in the late afternoon (112) they also often occur in the night (over one quarter between 21h and 06h). The westerly squalls often attain speeds over 20 m sec^{-1}, and occasionally over 40 m sec^{-1} (e.g., in a thunderstorm at Allahabad a northwesterly squall of 45 m sec^{-1} was followed by 23 mm of rain in 15 min and a pressure rise of 8 mb; 117).

As the summer monsoon approaches storms become more intense and sometimes produce very large hailstones, not only over northwest India but also farther west near the coasts of the northern Arabian Sea and the Persian Gulf (Appendix 8.4). Such severe hail and squall storms are most frequent in Bangla Desh, formerly called Bengal (northeast India, east Pakistan, and Assam), where they are known locally as *Kal-Baisakhis* (calamities of the month of Baisakh, between April and May), and more generally as *Nor'westers*, from the usual direction of their approach and of their squalls. They also are associated with incursions of maritime air, which in this region are more frequent than in the northwest of India. Figure 8.30 shows the mean flow at 0.6 km over and near India on 9 days of April 1966 during which there was a pronounced flow at this level from the Bay of Bengal into northeast India. The moist southerlies enter in a layer less than 1 km deep, below a drier west-northwesterly airstream, which soundings at Allahabad show has an adiabatic layer nearly 4 km deep, with a potential temperature of about 40°C. The mean speed of the west-northwesterlies increases with height, to about 30 m sec^{-1} at 12 km. As the moist layer passes inland during the day it becomes subject to

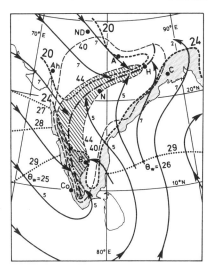

Fig. 8.30 Mean conditions near the surface over and near India during nine afternoons in April 1966 when there was a pronounced southerly flow at Calcutta (C), one of the conditions favorable for the occurrence of Nor'wester thunderstorms over northeast India. The solid thin lines are streamlines of the flow at about 0.6 km above sea level, with small figures alongside which give typical speeds (in m sec^{-1}). Inland a northwesterly flow meets a southerly flow in a convergence zone, the mean position of whose axis is marked by a heavy dot-dash line.

The thin and thick dashed lines are isopleths of the screen-level afternoon maximum values of θ and θ_w (in °C). The area with values of θ higher than 44°C, which includes the axis of the convergence zone, is hatched, and the region inland with values of θ_w over 24°C is stippled. Over the Bay of Bengal the mean value of θ_w is about 26°C. It increases to as much as 28°C just inland of the coast, but farther inland decreases as convection thickens the shallow moist layer. The dotted lines over the sea are isopleths of mean sea-surface temperature in April.

The letters beside the large dots identify the positions of the following places mentioned in the text: *A*, Allahabad; *H*, Hazaribagh; *Ah*, Ahmedabad; *N*, Nagpur; *ND*, New Delhi; and *Co*, Cochin.

convection which penetrates into the drier air above, and its depth appears to increase; for this reason, in spite of the energy introduced by sunshine, θ_w at low levels decreases below the very high value of about 26°C found over the sea. However, cumulus convection begins toward noon over the higher ground inland of the Ganges delta, with a cloud base level of about 3 km (where θ_s is usually about 24°C), and cumulonimbus frequently form in the early afternoon, especially over the Hazaribagh Range of Chota Nagpur. These hills rise to heights above 900 m and extend northeast from about 500 km west of Calcutta to Hazaribagh, about 350 km northwest of Calcutta (Fig. 8.30). Near these hills the convergence zone is frequently well defined and extends from southwest to northeast, strongly resembling the dry lines of the midwest of the United States, which are described later. Cumulonimbus form on the moist side of this zone, and often develop into squall storms which travel southeast at speeds of 20–25 m sec^{-1} to reach the Ganges delta in the late afternoon and evening. By this time their squall fronts and belts of precipitation may be some hundreds of kilometers long. After a squall the screen-level value of θ_w is usually about 21°C (118), which away from the storms is found at a height of about 3 km. The storms occur several times each month in April and May, producing squalls which typically exceed 20 m sec^{-1} and may reach 40 m sec^{-1}. Occasionally there are falls of hailstones of giant size and more rarely also tornadoes, perhaps usually associated with more isolated storms farther inland, on occasions when the cumulus convection is more widely suppressed and the available potential energy is abnormally great. The reports of giant hail and tornadoes are mainly in local newspapers, so that it is virtually impossible to compile statistics of their frequency and intensity. It is likely, however, that in the one or two months preceding the onset of the monsoon tornadic storms occur somewhere in northwest India or east Pakistan on as many as several days. Radar observations indicate that the echo tops occasionally exceed 15 km (119, 120); the tops of the most severe storms probably reach 18 km.

In this season the central and southern parts of India are also invaded by maritime air (Fig. 8.30), and near the higher ground in the south the number of days with thunderstorms in April is about 10, as great as anywhere in the north. At least on the average, the potential energy available for the production of updrafts and downdrafts seems about as great as in the northeast. Nevertheless, in this region, south of the zone of strong westerlies in the upper troposphere, hailstorms hardly ever occur at any time of the year (121), and severe squalls are also very rare, probably because the required wind shears are absent.

In the summer monsoon maritime air in the lower troposphere floods the whole country from the west or southwest (Fig. 8.31), and the strong westerly winds which occupy the upper troposphere over northern India in the premonsoon and postmonsoon periods are displaced northward, and replaced by weaker easterlies. Near Calcutta, and in particular over the hills to its west and northwest, thunderstorms are even more frequent than before the monsoon, occurring on as many as 15 days in a month. The mean stratification at Calcutta (Fig. 8.32) has a positive area which extends up to near the tropopause at 17 km, and is larger than in the premonsoon months. However, the wind shear, especially below 10 km, is comparatively weak. Consistently, the height of the tops of the radar echoes from thunderstorms more frequently exceeds 15 km during the monsoon (115, 119), and reaches extremes of 16–20 km (122, 123). (The tallest clouds tend to occur when the high troposphere is several degrees Celsius cooler

8   Cumulonimbus convection

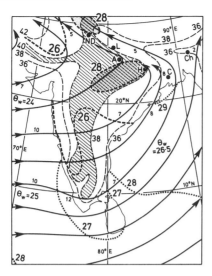

Fig. 8.31 Mean conditions near the surface over and near India during 10 afternoons in July 1967 when there was a pronounced southerly flow at Calcutta (*C*).

The solid thin lines are streamlines of the flow at about 0.6 km above sea level, with small figures alongside which give typical speeds (in m sec^{-1}).

The thin and thick dashed lines are isopleths of the screen-level afternoon maximum values of θ and θ_w (in °C). The area between the isopleths of θ for 38 and 40°C is hatched, and the regions inland where θ_w exceeds 28°C is stippled. Mean values of θ_w at screen level are indicated in some places over the sea; the dotted lines are isopleths of mean sea-surface temperature in July.

The letters beside the large dots identify the following places: *C*, Calcutta; *Ch*, Cherrapunji; *H*, Hazaribagh; *ND*, New Delhi; and *L*, Lucknow.

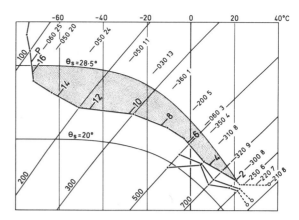

Fig. 8.32 The morning (0530 IST) sounding at Lucknow (*L* in Fig. 8.31) on 22 July 1967, adjusted at low levels (dashed lines) to include representative afternoon screen-level values of dry- and wet-bulb temperature. Winds are entered in degrees and meters per second. During the afternoon the tops of radar echoes near New Delhi were observed to reach heights of just over 16 km, about the level marked *P* (124). On this and the following 2–3 days the temperatures in the layer between about 300 and 150 mb at New Delhi and Lucknow were several degrees below the monthly averages, and a large positive area appears on the tephigram. However, the radar echoes did not reach appreciably above its top, indicating ordinary rather than organized cumulonimbus convection, consistent with the rather large value of $-(\overline{\mathrm{Ri}})$, which is about 8.

than in the monthly mean, often during the slow westward passage of troughs in the upper easterlies, so that the positive area is even larger and its top higher; an example from an occasion when the echo reached over 16 km (124) is seen in Fig. 8.32. Nevertheless, the cloud tops are barely above the level of the top of the positive area, indicating a lack of organization of the updraft, although the value of $-(\overline{\mathrm{Ri}})$ is no greater than about 8. It may be that in this kind of stratification conditions are unfavorable for the production of a strong downdraft, and that the value of $-(\overline{\mathrm{Ri}})$ obtained without reference to a downdraft is not a good indicator of the degree of organization of the storm drafts.

Generally during the monsoon the cloud tops barely reach the equilibrium level of the simple parcel theory. Hailstones, even of small size, are very rarely observed (although the rainfall intensity is often great, reaching about the 200 mm hr^{-1} implied by the observed extreme values of just over 10^6 mm^6 m^{-3} for the radar reflectivity factor Z_e; 124).

Further evidence that the intense and efficiently organized cumulonimbus are restricted to the northern regions of India during the premonsoon and postmonsoon periods is provided by the statistics of the frequency of strong squalls at the ground. If all squalls in which speeds of at least 30 mi hr^{-1} (13 m sec^{-1}) are considered, thereby including much ordinary cumulonimbus convection, then Cochin in the south of India records the most (about 20 each June, during the southwest monsoon). However, if the limit is raised to 40 mi hr^{-1} (18 m sec^{-1}), the number experienced there falls to about 10 each *year*, compared with about 7 at New Delhi and 5 at Calcutta during the premonsoon *month* of May (125, 126), although the average annual number of days with thunderstorms at Cochin—about 70—is equal to that found anywhere in the north of India. On the other hand, in August, in the middle of the summer monsoon, the total number of squalls in a 5-year period which exceeded even 30 mi hr^{-1} was only 3 at New Delhi, and 1 at Calcutta (where thunderstorms occur on about half of the days in this month). The strongest squalls occur over the whole of northern India in the premonsoon months; in the 5 years considered, the maximum speeds recorded at New Delhi and Calcutta were 30 and 33 m sec^{-1}; at Allahabad, about halfway between, it was 45 m sec^{-1}, in the month of March. Other high recorded values are 36 m sec^{-1} at Ahmedabad in May (127) and 38 m sec^{-1} at Nagpur in April (128). The positions of all these places are marked in Fig. 8.30.

It thus seems that the intense cumulonimbus convection which is characteristic of the northern regions in the premonsoon and postmonsoon periods is associated with tropospheric wind shears which are sufficient in magnitude to organize the convection efficiently; in other regions and seasons the available potential energy is often larger, but the wind shears are comparatively small. The intensity of the thunderstorms in the north has been recognized to be associated with strong winds in the upper troposphere (129; and it has been remarked that in their presence storms with small hailstones sometimes occur even when the soundings show only very small positive areas; 130), but the association has not been attributed directly to the implied large wind shears.

During the summer monsoon the tropopause over the whole of Southeast Asia and intertropical Africa is well defined at a height of about 16.5 km, where the temperature and pressure are, respectively, about $-80°C$ and 100 mb, corresponding to a value of θ_s of $28°C$. This is appreciably above the average values of θ_s which occur in the cumulonimbus bases, or indeed the daily maxima of θ_w at screen level (except in northern India). Consistently, not only are the tops of cumulonimbus generally below 16 km, but their outflows, judged by the available aircraft observations of the tops of anvil clouds and the dense cirrus found in and near regions of cumulonimbus convection, is only about 14 km (122, 132, 133). This is almost the maximum height reached by aircraft, but the cirrus seen at greater heights are reported to be shallow and tenuous. Moreover, a normal limit of about 14 km seems plausible considering that θ_s at that level has a mean value about equal to that in the cloud bases (mostly 24–25°C, but sometimes as high as 27–28°C over northern India, as in Fig. 8.32). Moreover, both on individual occasions (133) and in the mean the lapse rate decreases substantially at higher levels (from 7–8 to 3–6°C km^{-1}), which might be expected if they were more rarely reached by the convection, or more rarely occupied by extensive and persistent anvil cloud (since the associated radiative fluxes tend to increase the lapse rate in the cloud layer). Thus the tropopause in low latitudes cannot be regarded simply as the upper limit of the layer of cumulonimbus convection.

Intense cumulonimbus over Europe

Evidence indicates that over Europe the intense traveling cumulonimbus most frequently occur in midsummer in a belt extending from France across the lands bordering the Alps to as far east as Georgia, north of the Caucasus mountains. As happens elsewhere in the world, the area thus subject to hailstorms during the growing season contains many regions in which the climate also favors grain and leaf crops, and especially the soft-skinned fruits such as grapes, peaches, and pears. Consequently the hailstorms are a feared agricultural hazard and in Europe there is a long history of measures intended to prevent the fall of damaging hail. In spite of this, and even though synoptic and climatological data have been gathered for many years in this populous and prosperous area, it is difficult to assess the frequency of the more severe storms other than by reference to the occasional studies of hailstorm behavior in which dense networks of voluntary or unofficial observers have been used (principally in Austria, and more recently in France, Italy, and England).

It would be interesting and useful to have a climatology of cumulonimbus convection, specifying distribution, frequency, and intensity, with the detail which could be achieved by routine survey with meteorological radars and perhaps, after some development, with techniques of *sferic location* (i.e., location of lightning). Such a survey would identify the intermediate-scale circulations, associated with topography, which localize cumulonimbus formation, and the part played by the large-scale motion systems in controlling the occurrence, extent, and intensity of cumulonimbus convection. There is a popular tendency to regard intense cumulonimbus, because of the small size of the areas over which they are violent and therefore the great rarity with which they affect individual places, as "freak" storms, in the sense that some haphazard and unpredictable circumstance determines whether they occur. However, the intensity and duration of such a storm implies that its meteorological scale is quite large. Insofar as it is possible to assess a horizontal dimension for the entire circulation, in a storm with a duration of several hours and relative horizontal velocities of up to about 20 m sec^{-1}, it must be some hundreds of kilometers. Similarly, the local heat flux into the upper troposphere associated with an updraft speed of some tens of meters per second and a temperature excess of several degrees Celsius is 3 or 4 orders of magnitude greater than that associated with the large-scale motion systems (and the average net radiative heat sink), so that any one system could not be expected to contain more than a few persistent intense cumulonimbus. It is therefore likely that systematic survey would show that in the appropriate season the intense cumulonimbus occur regularly in the favorable part of each large-scale motion system (later shown to be near the cold front), and that recognizable variations in the large-scale situations, combined with topographical and diurnal influences, lead to predictable variations in their intensity and paths. The violent weather in them would rarely be experienced in any one locality, except perhaps in places lying near and in the prevalent direction of motion from a pronounced source (usually prominent high ground).

The conventional statistics show that at individual places in the European region liable to intense storms the average frequency of days with thunder, mostly due to ordinary cumulonimbus, generally has a maximum of between 3 and 8 days in July; on low ground the average frequency of days with hail is in the summer about one tenth as great. The ratio of days of hail to days of thunder of course varies from almost zero within the tropics to about 1 near middle-latitude coasts which face the cool airstreams in the rear of cyclones, but the value of about 1/10 seems to be characteristic of inland regions liable to intense summer storms (e.g., also in North America; 140); it may be related simply to the respective areas over which hail and thunder are experienced in a typical storm. The frequency with which thunderstorms occur somewhere within an area comparable to that covered by the forward part of a large-scale (trough-ridge) motion system is very much higher. For example, the frequency of thunderstorms somewhere over Great Britain in midsummer has an average value of about 24 days each month, and in some years reaches 30 days (141).

Information on the occurrence of the large hailstones and tornadoes which accompany the more intense storms is very sparse. In central and southeast England falls of hailstones sufficiently large to lead to newspaper reports of damage to crops or buildings have an average frequency of about $1 \text{ yr}^{-1} (10^4 \text{ km}^2)^{-1}$ (142), consistent with general experience of one or two severe traveling hailstorms each year. It appears that the average frequency in the Po Valley in northern Italy, in southern France (143), and in Austria (89) is probably about five times greater than in England. The average values may also be substantially higher in northern France, southern Germany, and Switzerland, and in combination suggest that severe traveling storms may occur somewhere in western Europe on between 10 and 20 days each summer, that is, in at least a large fraction of the more intense of the passing trough-ridge systems, which succeed each other at intervals of several days.

Tornadoes sufficiently intense to cause loss of life are almost unknown in Great Britain, but those which cause serious damage to woods or buildings (in swaths mostly only between 10 and 20 km long) occur somewhere about once every two years (144). The frequency of similar tornadoes over France and Germany in the belt liable to intense storms is probably greater (145, 146) by about the factor of 5 estimated for the relative frequency of damaging hailstorms; on the other hand, tornadoes seem to be comparatively rare in northern Italy, a region notorious for the frequency of hailstorms. Tornadoes are occasionally reported from other parts of the European belt of intense storms, for example, Hungary and southern Russia.

Less severe whirlwinds, which cause minor damage to buildings, appear to be much more frequent near the coasts of western Europe in the cool airstreams in which cumulonimbus develop over the sea, and have a maximum frequency in the winter season. In four recent years over Great Britain, tornadoes or whirlwinds which caused at least some damage to buildings were reported in newspapers on a total of 10 days in the five months May–September, and on 20 days in the five months October–February, when several incidents often were reported on the same day (147). In such winter situations the cumulonimbus tops may be as low as 6 km and the available potential energy is usually small, but there may be sufficient shear to organize the convection efficiently, so that individual storms are persistent and can be traced over long paths. When the winds above the boundary layer are strong the storms travel rapidly and their sudden squalls may cause structural damage. On 18 November 1963, for example, when damage was reported from several places along a line across southern England (147), the wind speeds in the layer between 1 and 4 km were between 30 and 40 m sec^{-1}. Because many damaging storms occur in darkness, or when people are sheltering rather than observing, there are few definite reports of funnel clouds, and unless subsequent investigation shows the destruction to be confined to a long narrow path it is impossible to be confident that it was due to a whirlwind rather than a severe squall.

Intense cumulonimbus over North America

During the summer season in the central United States the frequency of intense cumulonimbus, and of the violent and destructive phenomena which accompany them, exceeds that reported from any other part of the world. In the United States as a whole the average annual damage by hail is about $10 million to property and about $50 million to crops, amounting to rather more than that inflicted by tornadoes (which is almost wholly to property); the damage wrought mainly well inland by the intense cumulonimbus is less than but comparable with that suffered near the east and south coasts from hurricanes, which on the average now amounts to about $300 million. In the period 1916–53 the average annual number of lives lost was 230 in tornadoes (148) and 75 in hurricanes (149); in recent years deaths from both causes have been reduced by improved warning services.

It is believed that tornadoes are associated with the most intense cumulonimbus, and the techniques of forecasting them are aimed at predicting the circumstances in which there is an extreme rate of conversion of potential into kinetic energy (150). Since most thunderstorms are due to ordinary cumulonimbus, and since in the United States as elsewhere the statistics of hailstorms do not refer separately to those in which some of the hailstones are very large, the statistics of tornadoes give

The distribution over the world of (intense) organized cumulonimbus 8.6

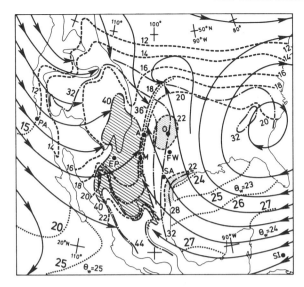

Fig. 8.33 The distribution of wind, mean potential temperature θ, and mean wet-bulb potential temperature θ_w at screen level in mid-afternoon, over and near the United States during May 1962. A smoothed ground contour for a height of 6000 ft (1800 m) is entered. The thin solid lines are streamlines of the prevailing winds, showing two principal currents entering the United States from the south and from the west, and meeting in a convergence zone whose axis is marked by a line of dots. The thick dot-dash lines over the land are isopleths of θ for 44, 40, 36 (partly drawn), 32, and 28°C; the area between the isopleths for 44 and 40°C is hatched. Dotted lines give average surface temperatures over the sea. The thick dashed lines over the land are isopleths of θ_w, labeled at intervals of 2°C; areas in which θ_w exceeds 22°C inland over the United States are stippled. Some average values of θ_w are entered over the sea.

The letters beside the large dots in this diagram and in Fig. 8.34 identify the positions of the following places mentioned in the text: *PA*, Point Arguello; *SA*, San Antonio; *EP*, El Paso; *FW*, Fort Worth; *M*, Midland; *O*, Oklahoma City; *A*, Amarillo; and *SI*, Swan Island.

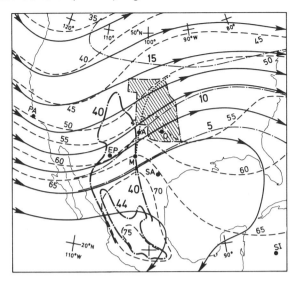

Fig. 8.34 The mean streamlines (solid) and isotachs (thin dot-dash lines, in m sec^{-1}) on the mean isentropic surface $\theta = 41°C$ over and near the United States during May 1962. The height of the surface is indicated by thick dashed lines (in tens of millibars). Isopleths of mean mid-afternoon screen-level θ for 40 and 44°C are transferred from Fig. 8.33, as is the line of dots which marks the axis of the convergence zone in the flow near the ground.

Insofar as the map represents a typical state, it implies that the air in the adiabatic layer of high potential temperature which is produced over the arid high ground of the southwestern United States (cf. Fig. 8.35) and northern Mexico ascends into the middle troposphere as it flows away northeastward across the 100th meridian, over Texas, Oklahoma, and Kansas. The upward displacement of about 100 mb in about half a day suggests typical vertical velocities of a few centimeters per second. The associated cooling of the base of the adiabatic layer is as much as 10°C, and eventually near and north of the 35th parallel removes its effectiveness as a barrier to the upward development of cumulus convection from the moist southerly flow beneath it (cf. Fig. 8.37).

The three adjoining states of Oklahoma, Kansas, and Nebraska (Table 8.6) are shown by the hatched area.

the best indication of the frequency and distribution of the most intense cumulonimbus. These show that tornadoes occur in all months of the year over the country lying east of the Rocky Mountains, but that during the winter season they are comparatively rare and virtually confined to the states bordering the Gulf of Mexico (151, 152).

In the summer the chain of high ground which is composed of the Rocky Mountains and their southward extension into the Sierra Madre of eastern Mexico divides two prevailing airstreams in the lower troposphere. To the west is the subsiding and rather dry northwesterly flow on the eastern flank of the subtropical anticyclone in the North Pacific Ocean, while to the east is the trade wind current of the North Atlantic subtropical anticyclone, which reaches the Gulf of Mexico as a distinctly warmer and moister flow and turns to enter North America as a southerly airstream. There is a marked difference in the mean depth and mixing ratio in the moist layer of ordinary convection in these two streams, related to the difference in the position of the flows with respect to the associated anticyclones and the difference in the surface temperatures of the waters over which they flow (Fig. 8.33). The stream which enters the southwest of the United States from the Pacific has a moist layer barely 1 km deep (with θ_w about 11°C); inland it is strongly heated, but because it is mainly dry and flows over arid ground the adiabatic layer there becomes deep and almost cloudless with the rather high potential temperature of about 40°C, and a θ_s in the middle troposphere of about 20°C. In the southerly airstream which enters from the Gulf of Mexico, on the other hand, the air is much moister; over the Gulf in this season the cumulus convection reaches up to nearly 3 km on the average (153), and θ_w in the lowest kilometer has a value of about 21°C which

8 Cumulonimbus convection

Table 8.6 Occurrence of intense storms in the neighboring states of Oklahoma, Kansas, and Nebraska (see Fig. 8.34) during May 1962, as indicated by reports of funnel clouds F, tornadoes T, extensive lightning damage L, and maximum reported diameter D of hailstones (154)

Date	Oklahoma D(cm)		Kansas D(cm)		Nebraska D(cm)	
3	7					
4	7					
7						T
8			8			F
9			8			
10	7					
11						L
12						$L, T?$
14					7	T
15					3	T
16		F		T	6	
17	3	T	3	T	5	T
18			5	F		T
19			6	F		
20	4	T	4	T	10	T
21			6	T		
22				T		
23	7	T				
24	10	T	13	T	10	T
25	14	T		T		
26	10	T	8	T		
27	2	T	8	T	8	F
28	5	T	4	T		T
29	4	T				T
30	9	T	10			T
31		F		T		

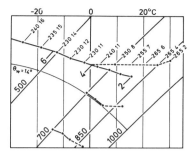

Fig. 8.35 The mean sounding for El Paso (*EP* in Fig. 8.34 and 8.35), May 1962, represented on a tephigram. The sounding is for 1200 GMT, but at low levels is adjusted to be consistent with the mean mid-afternoon sounding and screen observations (dashed lines). Heights are marked in kilometers and winds in degrees and meters per second are entered at appropriate levels. A deep (and dry) adiabatic layer with $\theta \approx 40°C$ is produced in the afternoon.

is very high compared with θ_s in the middle-tropospheric westerlies and provides a very large available potential energy. Consequently, it might be anticipated that east of the Rocky Mountains deep cumulus and cumulonimbus convection would be widespread, much as in the summer monsoon over India. Such convection is indeed common over many eastern and southern states, where the mean stratification becomes that typical of ordinary cumulonimbus convection. However, in those states which lie immediately east of the Rocky Mountains two circumstances lead to its frequent suppression, intermittently relieved by localized and violent convection; hence the weather there more nearly resembles that of the premonsoon period over northwest India.

These circumstances are conveniently discussed with reference to charts of mean conditions for a month early in the summer season, such as Figs. 8.33 and 8.34 for May 1962. This month was selected because it included an occasion of intense cumulonimbus convection which will be described in some detail later; it was a month of frequent intense storms in the midwest (Table 8.6).

The first of the significant circumstances is that the air which is heated rather than moistened during sunshine over the arid western regions (Fig. 8.35) arrives in the lee of the high ground above the moist southerly current as a dry west-southwesterly airstream characterized by a potential temperature of about 40°C, corresponding to a value of θ_s of 23–26°C at its base, which is at a height of 2–3 km over the more southern states (Fig. 8.36). On individual rather than mean soundings the transition between the two flows is often even more distinct, and marked by shallow inversion in which the temperature increases upward by a few degrees and there is a rapid veer of wind direction. The convection in the southerly current, in which θ_w is about 21°C, is thus in the south unable to penetrate the dry layer even when cumulus form. Although the depth of the layer of convection is restricted, both θ and θ_w in it tend to rise as the air flows farther north; eventually the mean

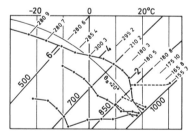

Fig. 8.36 The mean sounding for San Antonio (*SA* in Fig. 8.33 and 8.34), May 1962, represented on a tephigram. The sounding is for 1200 GMT; at low levels the dashed lines, consistent with the 2400-GMT sounding and screen observations, show the mid-afternoon conditions. Heights are marked in kilometers and winds in degrees and meters per second are entered at the appropriate levels. The cumulus convection in the afternoon is restricted to the lowest 2–3 km by the potentially warm air ($\theta \approx 42°C$) which arrives as a westerly airstream from the arid high ground of northern Mexico (cf. Fig. 8.34).

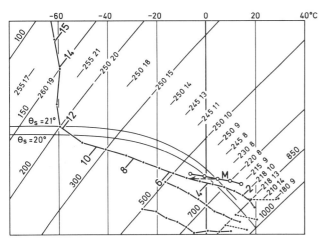

Fig. 8.37 The mean sounding for Oklahoma City (*O* in Fig. 8.33 and 8.34), May 1962, represented on a tephigram. The sounding is for 1200 GMT, but at low levels it is extended by the dashed lines, consistent with the 2400-GMT sounding and screen observations, to show the mid-afternoon conditions. Heights are marked in kilometers and winds in degrees and meters per second are entered at the appropriate levels.

The high potential temperature between 2 and 3 km here represents only a slight obstacle to deep cumulus (and cumulonimbus) convection, in contrast to the conditions farther south (cf. Fig. 8.36) and southwest. Upwind at levels above about 3 km a stronger obstacle is presented by the potentially warm air ($\theta \approx 40°C$) from the arid high lands, as shown by the mean temperature at Midland, marked *M*. However, this air ascends and cools as it approaches Oklahoma (as indicated by Fig. 8.34 and the dashed arrow along the isopleth for $\theta = 41°C$ on the present diagram), so that its effectiveness as an obstacle decreases and eventually disappears.

value of θ_w in the lowest kilometer may reach 22° or more. Moreover, as the moist air moves north the restraint of the dry layer aloft weakens, partly because of the general northward decrease in its temperature, and partly because its ascent in the large-scale flow (Fig. 8.34) leads to its cooling and a significant decrease in θ_s at its base (Fig. 8.37), and eventually it is no longer an obstacle to deep cumulus convection. This usually begins very locally in weak intermediate-scale circulations in the vicinity of the (western) edge of the moist flow, which at the surface appears as a confluence between south-southeasterly and south-southwesterly airstreams, and more strikingly as a narrow zone in which the screen-level dew point decreases westward from about 10°C to between 0 and −10°C. This transition zone is identified as a significant feature in the routine analysis of synoptic charts and has become known as the *dry line*. The position of the transition zone is strongly controlled by topography; accordingly it can be traced on climatological maps (155), and may be well defined on a map for an individual month, as in Fig. 8.33.

In the cumulonimbus convection which develops near the dry line, strong updrafts are favored by the low values of θ_s (below about 18°C) in the middle troposphere, which are associated with the large lapse rate in the dry layer aloft; moreover, strong downdrafts are also favored by this large lapse rate and the low values of θ_w (below about 16°C) associated with the dryness of its lower part (Section 8.10). Even in the mean, therefore, the stratification over the region in the lee of the Rockies contains more available potential energy for both updrafts and downdrafts than that over the eastern states.

The second circumstance contributing to the intensity of the cumulonimbus convection in the midwestern states is that during the spring and summer the troposphere over the United States remains baroclinic, although toward midsummer the zone of marked horizontal temperature gradient weakens and is displaced northward to near the 50th parallel. Accordingly considerable wind shears persist in the troposphere, and strong westerly winds at high levels, especially near the cold fronts of the cyclones which every few days appear east of the Rockies and travel mainly east or northeast across North America. In the presence of these wind shears the cumulonimbus convection, released locally in regions over and downwind of which the available potential energy is large, becomes efficiently organized and produces intense traveling storms. In the particular month for which the mean conditions have been illustrated, for example, over the three states experiencing the greatest number of tornadic storms the maximum tropospheric mean wind speeds (at about 175 mb, where $\theta_s \approx 23°C$, and therefore about the level of the updraft outflows from the most intense cumulonimbus) were nearly twice as great as in the states farther east. Also, it is possible that the wind shear in the lower troposphere is important for the organization of the cumulonimbus downdraft circulation; it is striking that the three states lie near the axis of the maximum wind speeds at 700 mb, which in the midwest is about the level of the air of lowest potential temperature (θ_w). Consistently it appears significant that as the summer season progresses the region of maximum frequency of tornadoes is displaced northward from near Missouri to Iowa, following the mean position of the strongest winds in the middle (and high) troposphere. The frequency of tornadoes diminishes in midsummer when these winds are at their weakest and farthest north, in a latitude (about 50°) infrequently reached by the potentially very warm air in the low-level southerlies. A marked tendency for tornadic storms to occur near the axis of regions of strong wind in the middle troposphere was established empirically by forecasters (96), and has recently been demonstrated in the climatological data for the states of Texas, Oklahoma, and Kansas, where it appears that especially in midsummer the strong winds may sometimes be associated with the subtropical jet stream rather than, as usual, with the frontal jet streams prevalent in middle latitudes (157).

In midsummer the area most subject to tornadoes is distinctly south of the mean position of the strong winds aloft. On the other hand, the areas of most frequent severe hailstorms have a similar but greater seasonal displacement, and in July and August extend northward into Wyoming, the Dakotas, Montana, and the Canadian province of Alberta. This trend is apparent in the conventional statistics of days of hail (140, 162), although they do not satisfactorily indicate the frequency of the more intense storms with large hailstones. Since grain crops are extensively grown and insured in the states mentioned, statistics of crop damage are another and more detailed source of information, but are affected by the susceptibility of the crops and the strength of the squalls which may accompany the storms (158); however, they too suggest that it is not until July and August that the hailstorms in the northernmost regions become so frequent that they are a serious nuisance. Thus, whereas the number of days with damaging hail has a maximum over Oklahoma in May, and over Nebraska in June, in the Dakotas the number increases abruptly from a few in May to over 20 in June and July, with a maximum in July and nearly as great a value in August over North Dakota (158, 159; Appendix 8.5; see also the special studies reported in 160–62).

Most of the intense storms develop from cumulonimbus which form in the afternoon near the dry line, or, especially in midsummer in the more northern regions, over high ground on the eastern flanks of the Rocky Mountains. Whereas cumulonimbus which form over hills and islands on occasions of little wind shear tend to remain there or to decay while drifting away, when there is a favorable wind shear the cumulonimbus intensify after their formation and travel away downwind, typically at speeds of about 15 m sec^{-1}, and persist for periods of up to several hours or even more, until they have traversed the whole of the region in which there is a significant supply of available potential energy. Accordingly, over the western plains near the formation sites the frequency of hail (and of tornadoes) has a pronounced diurnal maximum in mid-afternoon. In Denver, Colorado, for example, where these sites lie only about 50 km to the west, two thirds of the hail falls begin between 13h and 18h local time. In Alberta the mean time of hail occurrence is about 15h at a distance of 50 km from the mountains (to the southwest), and about 18h at a distance of 350 km (of all hailstorms, half travel as far as 200 km and one fifth as far as 350 km from the mountains; at still greater distances the maximum frequency might shift into the night hours if it were not dominated by the prevalence of small hail from ordinary cumulonimbus in the afternoon hours).

The diurnal variation of tornado occurrence in the various states has a similar trend (151). While in all states there is an afternoon maximum frequency, indicating the importance of local sunshine in increasing the available potential energy, it is most pronounced in those western states in which the storms originate, while in those states which lie several hundred kilometers farther east, about the distance traveled by the storms in half a day, there is a much higher proportion of occurrences during the dark hours, even though many may then escape observation.

There are large diurnal variations also in the spring and summer thunderstorm frequencies (140). Over and near the high ground in the west of the continent there is a pronounced maximum in the frequency of onset of thunderstorms in the afternoon hours, accompanied by an even more striking minimum in the night hours between local midnight and 06h. The same trend is almost as marked, especially in the summer season, near the Appalachian Mountains and the east and southeast coasts. On the other hand, in the plains states of the midwest the diurnal variation is small; over large areas less than a third of thunderstorms begin in the afternoon hours, and about as many begin during the night hours: in places near the eastern border of Nebraska the frequency of onset during the period between midnight and 06h actually exceeds that during the period between noon and 18h, and there is a corresponding nocturnal maximum in the amount of summer rainfall.

Explanations of the remarkable frequency of nocturnal thunderstorms in summer over the central plains of the United States have been sought mainly in processes which during the night would favor the *formation* of cumulonimbus or which would increase the available potential energy and lead to their intensification. Generally large-scale ascent near fronts has been regarded as the cause of cloud formation, and divergences in the net radiative fluxes or differential horizontal advection have been held responsible for changes in lapse rate well above the ground and increases in the available potential energy (163). In particular, the remarkable nocturnal intensification of the southerly flows at low levels over the midwest ahead of troughs has been regarded as causing a significant northward advection of potentially warm air near the ground during the night hours (164). During May 1962 the southerlies over Oklahoma were sufficiently persistent for the wind speed a few hundred meters above the ground to reach a maximum as high as 14 m sec^{-1} in the mean sounding for 06h at Oklahoma City (Fig. 8.37); on individual nights the wind speed in a shallow layer over the western plains often exceeds 15 m sec^{-1} in a belt a few hundred kilometers across, and reaches a maximum of 25 m sec^{-1} or more. Such strong flows have become known as "boundary layer jet streams." The diurnal variation of wind speed with a nocturnal maximum at the top of the boundary layer has been attributed to an inertial oscillation in which the wind speeds increase from afternoon values which are substantially less than the geostrophic to night values which are appreciably greater, following the cessation of ordinary convection toward sunset and the consequent great reduction in the friction which opposes the flows near the ground. However, a diurnal variation of thermal (geostrophic) wind over sloping ground plays an important part in the production of the more striking oscillations of the wind speed and the intense boundary layer jets (Appendix 8.6).

Although it has been thought that the strong nocturnal flows may favor the occurrence of thunderstorms over the central plains during the night by increasing the available potential energy, it is necessary to account not only for the high frequency of nocturnal thunderstorms there, but also for the relatively low frequency of afternoon thunderstorms. The explanation evidently involves the large-scale topography, and it has been suggested that the intermediate-scale circulations associated with the comparatively intense daytime heating and nighttime cooling over the higher ground of the Rocky and Appalachian mountains extend far over the neighboring and intervening plains, producing vertical motions sufficient to inhibit the cumulus and cumulonimbus convection during the day and to produce or intensify it during the night (163). Some support for this hypothesis was found in analyses of summer wind observations, which showed consistent mean patterns of diurnal change of horizontal divergence in the lower troposphere (of magnitude 10^{-5} sec^{-1} in 6 hr). On the other hand, it is not clear whether there are diurnal variations in the winds which can also be regarded as consistent; those that have been established and discussed seem more probably related to diurnal variations in the properties of the boundary layer (166). It appears that the simple kind of circulations envisaged play a more definite role over topography which is variable on a smaller scale, as already emphasized (sometimes causing the formation of cumulonimbus during the night over straits and lowlands between neighboring mountain ranges; Appendix 8.7).

The circulations which affect the cumulus convection over most of the central United States are of essentially larger scale. It seems that not only the principal mountain chains, but also the sloping ground of the plains states, which extends several hundred kilometers east of the Rockies, plays an important part in these circulations. Particularly in the recurrent situations favoring cumulonimbus convection, when a broad southerly airstream enters the country at low levels from the south, there are important diurnal variations in the horizontal temperature gradient and eddy viscosity in the convection layer near the ground over this sloping region, which produce the diurnal variation in the wind whose most striking feature is the nocturnal boundary layer jet. The circulations associated with the part of the air flow which is across the isobars (and therefore up or down the sloping ground) effectively extend out over the plains during the day the subsidence which would otherwise be found only near the mountains. On the other hand, during the night the katabatic downslope component of the wind near the ground, which would normally be expected, may be prevented or even reversed, and thus play an important part in increasing the potential energy available to existing cumulonimbus, or even contribute to the formation of cumulonimbus.

Over the midwest an important feature of the descent associated with daytime subsidence in the lower troposphere would be the arrival at a low level of air which had previously acquired an abnormally high potential temperature over the high ground in the west, particularly where it is arid. This air is often easily recognizable as a barrier to the ordinary convection, the more so the lower its base (as illustrated in Fig. 8.34, 8.36, and 8.37). On individual occasions its base is found at about the 700-mb level even as far east as the 90th meridian, and commonly its effectiveness as a barrier is not removed by ascent and cooling until it has reached the eastern states.

It is, nevertheless, striking that afternoon cumulonimbus convection is frequent not only in the eastern states, but also near the 100th meridian over those states containing or lying immediately downwind of high ground, storms forming over the eastern slopes or near the dry line, at the western edge of the moist southerly flow at low levels. The reasons for this behavior are not apparent in the climatological data, and will be discussed later with reference to a particular occasion. The cumulonimbus frequently become well organized and intense, with strong squalls, and travel east across the regions where deep afternoon convection is suppressed and the available potential energy is therefore large; successive storms form over the squall fronts of their predecessors and mostly on their right, or southern flanks, and a belt of storms, or a *squall line*, often develops which extends approximately from north to south over a distance of some hundreds of kilometers. Since the speed of the squall lines is typically about 15 m sec^{-1}, they pass between the 95th and 90th meridians some 12 hr after forming near the 100th meridian. Accordingly, the nocturnal maximum frequency of thunderstorm onset in this zone is probably due to the passage of squall lines formed in the afternoon farther west. Their continued progression, or similar developments over and downwind of the Appalachians, may contribute to the less marked nocturnal thunderstorm activity near the east coast beyond the 80th meridian.

The traveling cumulonimbus in the midwest, where they are intense, are observed to be accompanied by at

8 Cumulonimbus convection

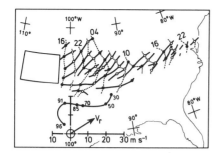

Fig. 8.38 An analysis of the tracks of individual rainstorms over the United States (adapted from reference 170). The storms formed near the eastern border of the state of Colorado, whose boundaries are marked, in the afternoon of 20 May 1949. The positions of their centers at intervals of 3 hr are shown by open or closed circles and are connected by dashed lines labeled with the hour of the day (in CST); their tracks are drawn as solid lines through positions determined from hourly rainfall records. The diagram beneath contains a hodograph showing the average winds in the neighborhood of the storms at pressure levels (in mb) and a vector showing the mean velocity V_r of the storms. The squall line composed of several storms moved eastward, individual storms shifting toward its northern end and being replaced by others continually forming at its southern end.

least moderate rains which cover areas 30–50 km or more across and last for several hours, so that they can be identified and followed as *rainstorms* on maps constructed from the published hourly rainfall amounts observed at stations with recording gauges (170). An example of the analysis of such data in Fig. 8.38 shows the travel of groups of rainstorms which formed one afternoon near the eastern border of Colorado and moved through the midwest during the night hours, reaching the east coast on the following night.

It is likely that persistent traveling cumulonimbus are responsible for nocturnal maxima of thunderstorm frequency over some other continental regions. For example, it is known that nocturnal thunderstorms are frequent in the summer over the east of Argentina, several hundred kilometers from the western slopes of the Andes between about the 30th and 40th parallels, where cumulonimbus form in the afternoons (163).

The summer monsoon over northern Mexico and the arid southwest of the United States

There is a great increase between May and July in the frequency of days of thunderstorms over and near the southern Rocky Mountains; in July it becomes almost as great as in Florida, while the frequency of storms over the Great Plains remains virtually unchanged. This feature is related to a change in the mean pattern of the flow in the lower troposphere over North America, which can be said to introduce a summer monsoon into the arid region comprising the north of Mexico and the southwestern states of Arizona and New Mexico.

In the winter and early spring this region lies beneath the axis of the strongest monthly mean winds in the middle troposphere over the United States. Especially in the winter the position of the strong westerlies is subject to considerable daily variations during the passage of troughs associated with the large-scale motion systems of middle latitudes, and in this season rains are occasionally experienced in the west of the country even as far south as the 20th parallel. In the spring these disturbances become less frequent and intense, and the weather over the southwestern inland states becomes persistently fine and dry in the subsiding flow which arrives from the eastern flank of the anticyclone in the Pacific. However, as the summer approaches, the westerlies eventually retreat well to the north, and the moist easterlies of the lower troposphere over the Caribbean Sea and the Gulf of Mexico advance to cover northern Mexico and enter Arizona and New Mexico. Before the arrival of these easterlies the air over the arid regions is too dry to allow the formation of other than traces of small cumulus, but after its arrival the mean relative humidities in the adiabatic layer increase, and although clouds which form (principally over the mountains) have the high base level of about 4 km, the mean stratification has a considerable positive area, and the clouds frequently develop into cumulonimbus in the afternoons. Usually winds and wind shears are weak and the cumulonimbus have only moderate intensity, are not well organized, and remain over the mountains where they form. Occasionally, however, they travel away and persist throughout the evening, so that the frequency with which thunder occurs at the reporting stations, mostly on low ground, depends very much on their situation with respect to the important local hill sources. Over the region as a whole there is a thunderstorm on nearly every day in the summer.

The retreat northward of the westerlies to beyond the 40th parallel and the advance into the region of the moist southeasterlies occurs in many years abruptly late in June, so that the onset of the thundery season in a particular state is often spectacularly sudden and has become a well-known climatic singularity (174; even in the mean, the rainfall at Tucson, Arizona, for example, is 5 mm in May, 7 mm in June, and 46 mm in July). It thus resembles the summer monsoon in India, and can similarly be related to changes in the circulation pattern over extensive parts of the hemisphere, but it occurs a few weeks after the retreat northward of the westerlies aloft over northern India and neighboring Asia.

8.7 Conditions for the formation of intense storms

It appears from the previous discussion that the efficient organization of cumulonimbus requires the presence of wind shear, and it may be anticipated that the most intense storms occur where there is an abnormally large available potential energy and sufficient shear between low levels and levels near the troposphere to insure a small value of $-(\overline{\mathrm{Ri}})$. Large potential energy is generally associated not with a locally abnormally low potential temperature θ_s in the middle or upper troposphere, but with a comparatively high potential temperature θ_w in the lowest 1–2 km. This arises in favorable topographical situations after the ordinary convection from the surface has for a while been confined to a shallow layer of about this depth (175). For example, in the trade winds which enter the Caribbean region in the spring the moist convection usually occupies a layer only 2–3 km deep, beneath air of high potential temperature (and low relative humidities) which has subsided from the upper troposphere; the cumulus, at whose base θ_w is about 22°C, cannot penetrate into a middle troposphere in which θ_s has about the same value. Ordinarily, where the easterly trade wind current turns northward on the western flank of a subtropical anticyclone large-scale upward motion develops and the moist layer deepens; however, where the Caribbean trade winds enter North America, to the east of the Mexican highlands, the moist current is found beneath southwesterlies whose potential temperature has been raised to a particularly high value (θ about 40°C or even more, and θ_s about 24°C or more). Moreover, subsidence in the lee of the strongly heated high ground may cause the moist layer to become even shallower, and capped by an intensified inversion which continues to confine the ordinary convection to the lowest 2–3 km (as already discussed, and illustrated in Fig. 8.33 and 8.35); during the flow inland the mean afternoon values of θ, θ_w, and θ_s at low levels tend to increase. Nevertheless, there is a latitudinal temperature gradient in the middle troposphere and near the 40th parallel the mean values of θ_s are only about 20°C, so that during the northward flow of the moist lower current the inversion and the restraint upon the cumulus convection must eventually disappear. If only these large-scale circumstances were involved, deep cumulus and cumulonimbus convection might be expected to arise continually in the lee of the Rockies near this parallel, but, as already mentioned, circulations of smaller scale intervene, and the cumulonimbus convection typically begins during the afternoon farther south, in a few places near the dry line marking the western edge of the moist flow (whose mean position is seen in Fig. 8.33). Even so, the large-scale topography, in the form of the extensive *arid* and generally *high* ground in the west of North America, over which the potential temperature θ in the adiabatic layer reaches abnormally high values, makes an essential contribution to the intensity of the cumulonimbus convection by temporarily confining the ordinary convection in its lee to a layer which is shallow and in which the mean value of θ_w therefore becomes abnormally high, and substantially greater than θ_s aloft in and downwind of the region where the cumulonimbus form. Thus after forming near the western boundary of a broad region in which the previous suppression of cumulus convection prevented the dissipation of available potential energy, and over which there is pronounced wind shear, the cumulonimbus intensify and are able to travel eastward over very long paths.

The formation of intense storms near the dry line over the United States

The circumstances in which intense cumulonimbus form near the dry line can be described in more detail for a day (26 May 1962) when severe storms developed over northern Texas, Oklahoma, Kansas, and western Missouri, and when there were many reports of tornadoes and giant hailstones, some of diameter up to about 10 cm (Table 8.6; mean conditions during this month have already been discussed). The important large-scale features are most simply illustrated by analyses of the flow in isentropic surfaces (Section 9.5) at a time (2330 GMT or 1730 CST) in the afternoon when the storms were in progress. Near the surface during the day, and also aloft over the period implied by the dimensions of the region to be considered and the horizontal wind speeds, the flow was not strictly isentropic. Further, the streamlines drawn of the flow relative to a poorly defined large-scale ridge-trough system moving eastward at about 2 m sec^{-1} can be only approximations to the trajectories followed across the entire region, but nevertheless probably represent the significant features of real trajectories. [Near the surface other complications not taken into account are diurnal variations of wind (discussed in Appendix 8.6) and of temperature in the layer subject to cumulus convection, which tend to confuse the large-scale analysis but whose effects are important in the localization of the cumulonimbus.]

Over the high ground in the west of North America on this occasion potential temperatures in the adiabatic layer of the ordinary convection everywhere reached over 40°C, and in Mexico reached 48°C. However, the air generally was dry, and even in the south the potential energy available for cloud development was small, with very little wind shear. On both eastern and western flanks of the Mexican plateau the arrival of maritime air in sea-breeze circulations can lead more readily to the

8   Cumulonimbus convection

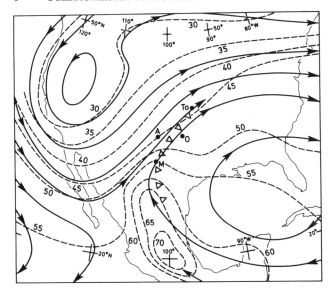

Fig. 8.39 The isentropic relative flow chart for the surface $\theta = 47°C$, based on the soundings made at 2330 GMT, 26 May 1962. The height of the surface is represented by isobars drawn as dashed lines and labeled in tens of millibars. Streamlines are drawn as continuous lines and are probably close approximations to trajectories. Over the high ground in southern Mexico θ in the adiabatic layer exceeds 47°C, and the weak flow cannot be represented. Air which is modified by ordinary convection over northern Mexico can be traced farther north (as a layer of steep lapse rate with θ about 47°) into the vicinity of the dry line near the surface [e.g., on the sounding at Midland (M in Fig. 8.43), it appears between about 560 and 350 mb. It cannot be found, however, on the sounding at Amarillo (A in Fig. 8.42), where the air at the corresponding potential temperature has not been subject to ordinary convection and is considerably drier. These neighboring stations lie on either side of the confluence zone between flows from the south and from the west].

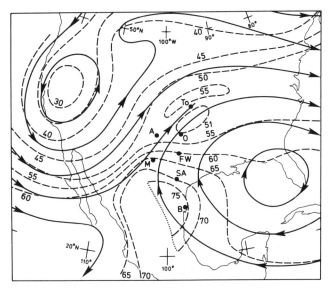

Fig. 8.40 The isentropic relative flow chart for the surface $\theta = 43°C$, at 2330 GMT, 26 May 1962, drawn in the same manner as Fig. 8.39. Again there is a marked confluence zone in the vicinity of the dry line at the surface, between flows from the south and flows from the west. Except over Mexico, in the south the flow is at levels about or somewhat above the general top of the layer of ordinary convection (as shown near the Gulf of Mexico by the sounding from Brownsville (B; Fig. 8.44). There appears to be some subsidence over the Caribbean which becomes pronounced just to the east of the Mexican highlands and leads to the appearance of an inversion above a layer of ordinary convection, which over southern Texas is constricted to a thickness of little more than 1 km (Fig. 8.44). However, as the flow continues north into Texas and toward the dry line the subsidence is replaced by upward motion and the restraint of the inversion upon the convection weakens (Fig. 8.45).

formation of cumulonimbus, as here shown by radars in western Texas, which were able to detect at extreme range tall echoes over the slopes of the Sierra Madre mountains in northeast Mexico (Fig. 8.41). However, such cumulonimbus remain near the mountains and disappear in the evenings. The potentially warm air which flows eastward above the moist current over the neighboring plains, and the lack of suitable wind shear, prevents the cumulonimbus from becoming organized and able to migrate away from the high ground.

On the isentropic chart for $\theta = 47°C$ (Fig. 8.39) it is not possible to indicate the flow over the high land in central and southern Mexico, since there the potential temperature in the adiabatic layer was raised to a higher value. Farther north, however, in the lee of the Rocky Mountains there is clearly a confluence zone between flows from the Gulf of Mexico, on the one hand, and from the Pacific and across the arid southwestern states of the United States on the other. A similar confluence in practically the same position appears in the flow on the lower isentropic surface for $\theta = 43°C$ (Fig. 8.40), which

shows a subsidence over the Caribbean becoming very pronounced where the flow reaches the lee of the Mexican highlands, followed by ascent in the confluence over western Texas. A similar pattern of vertical motion is suggested also on the isentropic surface for $\theta = 35°C$ (Fig. 8.41), which lies generally close to the top of the moist current in the Gulf of Mexico; inland this current has a western boundary at the ground marked by the dry line which separates it from the much drier airstream (with higher potential temperatures) which enters the confluence zone from the west. Thus three airstreams can be distinguished near the dry line east of the southern Rocky Mountains:

1. A stream which arrives in the middle troposphere with a steep lapse rate ($\theta \approx 47°C$; $\theta_w \approx 17°C$) after modification by ordinary convection over the Mexican highlands, referred to as "Mexican air," M.

2. A deep adiabatic layer ($\theta \approx 40°C$; $\theta_w \approx 17°C$) produced over the arid southwest during travel from the west coast, which is usually described as "continental tropical air," T_c.

240

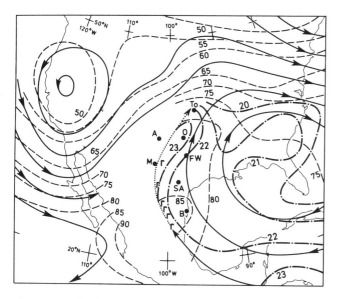

Fig. 8.41 The isentropic relative flow chart for $\theta = 35°C$, 2330 GMT, 26 May 1962, drawn in the same manner as Fig. 8.39. The dot-dash lines show the distribution of the mean wet-bulb potential temperature θ_w in the lowest kilometer of the tropical maritime air; the western edge of this flow, which lies at the dry line, and where a potential temperature of 35°C was reached close to the surface, is marked by a dotted streamline. It passes west of Oklahoma City (O) and reaches north to just west of Topeka (To). The letters r show some principal localities over which tall cumulonimbus formed during the afternoon near the southern part of the dry line, as recognized by radar at Brownsville (B), and farther north. Tall cumulonimbus and showers formed locally also near the eastern edge of high ground farther south over Mexico, at least as far south as the 19th parallel, beyond the range of the radar at Brownsville. However, only those cumulonimbus which formed north of Midland (M) appear to have moved away from their sources and persisted into the evening. These intensified into severe storms (Fig. 8.47).

3. A shallow moist layer ($\theta \approx 35°C$ or less; $\theta_w \approx 22°C$) which flows north in the lee of high ground after entering North America from the Gulf of Mexico and the Caribbean Sea; it is separated from the continental tropical air by the dry line, and is called "maritime tropical air," T_m.

The characteristics of these airstreams are illustrated in Fig. 8.42–8.46. From these diagrams it appears that throughout the middle and upper troposphere over the midwest the marked horizontal temperature gradient (indicated by the isobars on the isentropic surfaces) is characteristic of the eastern side of a large-scale trough, and might be expected to contain a frontal zone in which upward motion is concentrated where a relative flow from the southeast meets another from the northwest. At low levels the frequent and even persistent positioning of the confluence line in a southwesterly flow in the lee of the northern highlands of Mexico can be associated with the strong topographical barrier

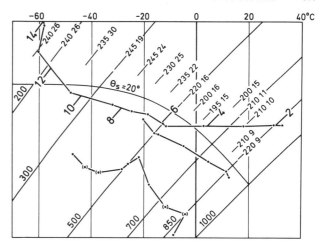

Fig. 8.42 The afternoon sounding (2330 GMT, 1730 CST) at Amarillo (A in Fig. 8.39–8.41) in northwestern Texas, on 26 May 1962, typical of the continental tropical air which flows over the arid southwest from the west. Heights are marked in kilometers and winds are entered in degrees and meters per second. A very deep dry-adiabatic layer with a potential temperature θ of a little over 42°C, which is high for this kind of airstream, extends up to nearly 6 km (the afternoon maximum of θ at screen level was 44°C). The air is generally too dry for cumulus to form, except perhaps over mountains. The relative humidity at screen level was reported as 6%; values below about 20% which are obtained from the radiosonde (from levels indicated by dew points within parentheses) are unreliable and probably overestimates. The dew point and θ_w therefore probably decrease with height throughout the layer of ordinary convection.

presented to the zonal flow by these highlands and the Rocky Mountains farther north.

Although generally the confluence line would appear as a cold front, here the peculiar circumstance is that in the low troposphere the passage of the subsiding northwesterly flow over the arid regions results in its arrival at the confluence line as a current which has a potential (and virtual potential) temperature somewhat *higher* than that in the moist flow from the southeast, so that the confluence line is recognized at the surface not as a cold front but as a dry line. The dry line itself, regarded as the western boundary of the moist southeasterly flow, extends well south along the eastern slopes of the Mexican highlands. Figure 8.40 shows that over the coastal plain in this region there is a very marked subsidence at about the level of the highlands, which is probably part of the circulation due to the intense solar heating over the neighboring high ground. It is this subsidence, rather than the simple advection aloft of air of high potential temperature, which has the important effect of suppressing the cumulus convection over the low ground into about the lowest kilometer (Fig. 8.44). Only on the flanks of the high ground does the upward flow in intermediate-scale circulations have sufficient intensity to allow deep cumulus and cumulo-

8 Cumulonimbus convection

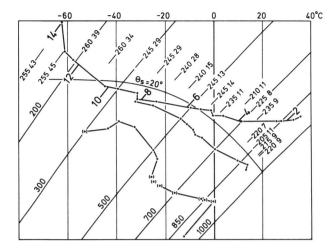

Fig. 8.43 The afternoon sounding at Midland, Texas (*M* in Fig. 8.39–8.41), on 26 May 1962. As at Amarillo, nearly 400 km farther north (Fig. 8.42), there is an adiabatic layer with a potential temperature θ of about 42°C, but here between about 4 and 8 km the air appears to have been raised to a higher potential temperature and moistened by ordinary convection over the Mexican highlands to the south. Characteristically this air has a steep lapse rate, and after ascent to these levels contains, as on this occasion, altocumulus and altocumulus-castellanus, although the sounding shows that the air is not generally saturated. At screen level in the continental tropical air the values of relative humidity and θ_w are about the same as at Amarillo (a small percentage and 17°C, respectively). In the afternoon the dry line was only just east of Midland; in the early morning the surface winds there were south-southeasterly and moist air was present.

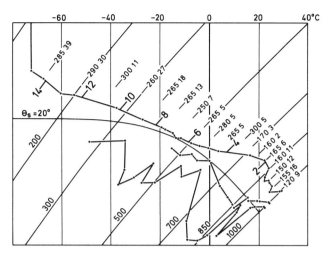

Fig. 8.44 The afternoon sounding from Brownsville, Texas (*B* in Fig. 8.40 and 8.41), 26 May 1962. Consistent with Fig. 8.40, there appears to have been pronounced subsidence in the stream of tropical maritime air entering North America as a south-southeasterly current: between the 850- and 700-mb levels θ_s has reached about 28°C, whereas θ_w in the lowest kilometer is only about 22°C, and locally the ordinary convection is confined to a layer only several hundred meters thick (small amounts of shallow cumulus at about 700 m were observed). Between about 4 and 8 km, however, the flow is from the Mexican highlands farther west, over which cumulus convection has moistened the air.

nimbus convection (Fig. 8.41); the clouds produced are unable to leave their sources, and they disappear in the evening. Accordingly, over the low ground a high wet-bulb potential temperature is preserved in a lowest moist layer, and even increases as the current flows across Texas toward the dry line farther north (Fig. 8.41). The processes responsible for the localized development of cumulonimbus very near the dry line there have not been satisfactorily identified. It is evident that there is marked ascent above the moist layer (Fig. 8.39, 8.40, and 8.45), which greatly increases the available potential energy. It must also reduce the restraint on the cumulus convection, but this cannot be removed completely unless ascent occurs also within the moist layer, or a transition layer between the moist layer and the continental tropical air above is cooled (and moistened) by the repeated intrusion of cumulus towers from below. Probably both processes are effective close to the dry line, and in the afternoon they eventually lead to the formation of clusters of tall cumulus and cumulonimbus locally where there are intermediate-scale circulations. Over the (topographically) uniform terrain these circulations are probably less intense and less easily recognized than the familiar examples over more variable surfaces. Nevertheless, on the present occasion it is

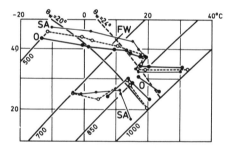

Fig. 8.45 Afternoon soundings, 26 May 1962, showing the progressive modification of the tropical maritime air passing over Texas and into Oklahoma. The soundings are from San Antonio (solid thin line), Fort Worth (dashed line) and Oklahoma City (solid thick line), *SA*, *FW*, and *O*, respectively, here and on Fig. 8.40 and 8.41. The adiabatic layer is about 1500 m deep on all soundings, and becomes warmer and moister farther inland; wet-bulb temperatures are entered only for the sounding from Oklahoma City, and show that the mean value of the wet-bulb potential temperature θ_w in the lowest kilometer there is almost 24°C (as indicated in Fig. 8.41).

In the dry air above the inversion, on the other hand, there is progressive cooling, consistent with the ascent suggested by Fig. 8.40. This leads to increasing available potential energy, and a progressive thinning of the layer with θ_s higher than θ_w in the adiabatic layer beneath, that is, to a weakening of the restraint upon the cumulus convection. A sounding made at Topeka, still farther north, shows the restraint to have disappeared (Fig. 8.46).

242

Conditions for the formation of intense storms 8.7

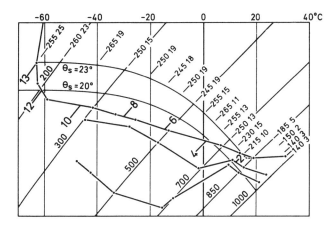

Fig. 8.46 The afternoon sounding from Topeka, Kansas (*To* in Fig. 8.40 and 8.41). The adiabatic layer has almost the same characteristics as at Oklahoma City, farther south (Fig. 8.45), but the inversion above it has disappeared and there is no longer any restraint upon deep cumulus and cumulonimbus convection.

noticeable that where the development of a cluster of clouds (*E* of Fig. 8.47 and Pl. 8.12) was observed from its earliest stages it began between two rivers, where the ground was not only higher but probably considerably drier than near the rivers, and so would be more liable to produce a higher potential temperature in the adiabatic layer. (If the dryness of the ground is important, the occurrence of storms may alter the cumulonimbus sources on subsequent days and further hinder the recognition of their relation to intermediate-scale circulations.)

Over Oklahoma and Kansas the winds in the high troposphere were not particularly strong, and the value of $-(\overline{Ri})$ indicated by the sounding for Topeka (Fig. 8.46) is as large as 10. Although the wind in the layer between 12 and 14 km was reported as over 40 m sec^{-1} only 3 hr later, implying a value of $-(\overline{Ri})$ only half as great, the general geostrophic wind at these levels was only about 26 m sec^{-1}, and the former value seems likely

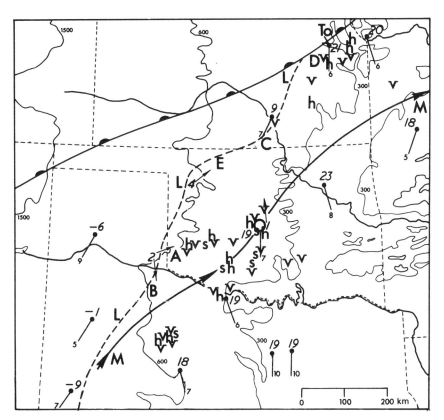

Fig. 8.47 Map for the afternoon of 26 May 1962, showing the position over Oklahoma and neighboring states (whose borders are drawn as dashed lines) of a warm front marking the northern boundary of the moist flow, the dry line *LL* (west of which no cumulus were present), and the streamline *MM* south of which Mexican air could be recognized in the middle troposphere. Ground contours are drawn for heights of 300, 600, and 1500 m, and principal rivers are included. The positions of routinely observing stations are marked (including Oklahoma City, *O*, near the center of the diagram, and Topeka, *To*, in the top right) with surface winds (speeds in m sec^{-1}) and dew points (italic figures in °C). In the moist southerly flow the dew point is generally about 20°C; west of the dry line it is less than 0°C. Reports of severe storm incidents between 1500 and 2300 CST are indicated by tornado symbols (*v*) and the letters *h* (large or giant hail) and *s* (destructive squalls). The italic figures 2 and 4, and the associated arrows, show the positions and directions of view of the pictures in Pl. 8.11 and 8.12. The letters *A* to *E* show the positions of cumulonimbus groups identified in such pictures (175).

to be representative. It may be more significant that the storms were in a belt of maximum wind speed in the lower middle troposphere. The wind shear between low levels and the dry air of the lower middle troposphere was therefore large, and may have played an important part in the organization of the downdrafts in the cumulonimbus circulations, not taken into account in the calculated simple version of the appropriate Richardson number. The steep lapse rate and extreme dryness of the lower middle troposphere in the sounding will be shown later (Section 8.10) to be important for the generation of strong downdrafts.

South of Oklahoma the wind shear in the lower troposphere was small, and the available potential energy was much less, partly on account of the lower potential temperatures attained in the adiabatic layer; moreover, in the lowest layers of the moist flow into the central states east of the dry line there appears to have been some continuation of the subsidence over the Gulf of Mexico (Fig. 8.41), producing a stratification even less favorable for cumulus convection than that just inland of the Gulf coast. Over these southern and central states it was therefore unsuitable for both the formation of cumulonimbus and the persistence of any traveling from the west after development near the dry line.

Aerial reconnaissance showed only shallow afternoon cumulus convection near Oklahoma City, not far from the dry line, but spectacular arrays of tall clouds close to the dry line, and clear skies west of it (Pl. 8.11). Isolated clusters of cumulus near the dry line north of Oklahoma City developed rapidly into groups of cumulonimbus, one of which (E) was well established only 90 min after clouds first appeared as traces of very shallow cumulus (175; Pl. 8.12).

Detailed observations of wind, temperature, and water vapor mixing ratio made on other occasions during traverses of the dry line by reconnaissance aircraft (176), flying at pressure levels between about 900 and 800 mb, have shown that the transition between the moist and dry airstreams is often remarkably sharp. For example, the change of mixing ratio between values of about 10×10^{-3} and 2×10^{-3}, which are typical of the two streams in the neighborhood of the dry line, has been found to occur over paths as short as about 0.5 km; the accompanying change of temperature, of rather more than 1°C, has generally made it difficult to determine with any confidence a difference of virtual temperature between the two streams, although it seems that by day the virtual temperature is somewhat higher in the tropical continental air up to about the 700-mb level. The shallow transition layer between the two streams consistently appears to slope upward steeply toward the east below this level, but to reverse its tilt at higher levels, where the presence of Mexican air above the moist stream may complicate the analysis.

The narrowness of the transition layer near the ground suggests horizontal convergence there; this has been confirmed by the aircraft wind observations and found to be of order 10^{-4} sec^{-1} over a belt about 20 km wide, corresponding to a mean vertical velocity of about 20 cm sec^{-1} at the 850-mb level. (The observed fields of temperature and mixing ratio in the moist air near the dry line have considerable horizontal variations on scales of tens of kilometers, which may be associated with the presence of intermediate-scale circulations.)

The ascent near the dry line, which plays an important part in the formation of cumulonimbus convection there, may be related to a frictional cross-isobar flow which is stronger in the shallower convection layer of the moist flow, and to accelerations associated with the diurnal variation of the boundary layer jet, in which the wind speed often has a maximum well inland. Analysis of the field of horizontal divergence associated with the jet, using observed winds, indicates ascent concentrated downstream of the speed maximum, where values of about 10 cm sec^{-1} are attained near the top of the moist layer, averaged over areas of (300 km)^2. General ascent may continue throughout the night and where pronounced may contribute to the persistence of storms until the following day (177).

The formation of intense storms over western Europe

Over North America, perhaps mainly on account of the barrier presented by the western mountain chain to the flow in the lower troposphere, the large-scale circulation is favorable throughout the summer half of the year for the generation of intense cumulonimbus convection in the midwest, where air which in the lowest kilometer has attained a mean value of θ_w of about 22°C over the warm waters of the Caribbean Sea is led inland and northward, to a region where θ_s in the middle troposphere has a value substantially less, so that a large amount of available potential energy is continually present.

Over western Europe the topography is not as pronounced, and the favorable combination of a high value of θ_w in the lowest kilometer and a wind shear to efficiently organize the cumulonimbus occurs more intermittently and more variably in position during the passage of large-scale motion systems. The intense cumulonimbus have long been recognized to occur typically in frontal zones on the eastern side of large-scale troughs. They form near confluence lines between distinct airstreams, but whereas over North America the confluence line is recognized as a dry line because the airstream from the west has passed over arid lands and has been raised to a high potential temperature, in western Europe it arrives after passage over the cool waters of the Northeast Atlantic and consequently the confluence line is recognized as a cold front. Moreover, in this part of the world the arid lands lie to the south, principally the great Sahara desert and some of the lands bordering the Mediterranean Sea. There is no region, corresponding to the Caribbean Sea, which is an obvious and extensive source of air with a mean value of θ_w of over 20°C in the lowest kilometer (which is required for the production of a substantial available potential energy since also over Europe θ_s in the middle troposphere has a mean value of about 18°C). Although the waters of the Mediterranean are warm, they are surrounded by lands over which the adiabatic layer reaches a considerably higher potential temperature, so that generally there is no ordinary convection over the sea, which acts as a heat sink rather than source. Study (175) shows that over western Europe the production of high values of θ_w in the lowest kilometer occurs over rather small regions in the lee of high ground, and is much more dependent on the energy introduced during the day by sunshine than over North America.

It was already mentioned early in Chapter 7 that in clear summer weather the ordinary convection due to sunshine introduces about 300 cal cm^{-2} into the atmosphere each day (Table 7.3). In a layer of given depth this represents a definite increase in the mean value of θ_w, but the depth of the layer of ordinary convection varies considerably according to the kind of surface over which it occurs. The energy is introduced mostly as latent heat over moist surfaces, but mostly as sensible heat over arid ground, so that in the latter circumstance θ near the surface and the depth of the adiabatic layer become comparatively great, while the mean increase of θ_w becomes comparatively small. Making some allowance for radiative cooling, in clear weather the mean value of θ_w can be estimated to rise by about 4°C each 24 hr in a layer about 1 km deep, but by only about 0.5°C in a layer 5 km deep. Accordingly, over Europe the mean value of θ_w in about the lowest kilometer reaches its highest values where the ordinary convection is confined to a shallow layer over moist ground. Over arid land in deep adiabatic layers θ_w at low levels does not increase much even after several days of sunshine.

In the summer a subsiding northerly airstream persistently enters northeast Africa from eastern Europe and is subject to ordinary convection as it turns westward and crosses the central Sahara. A deep adiabatic layer is produced which extends up to near the 500-mb level, in which θ is rather more than 40°C and θ_w has a mean value of about 17°C, barely more than on the coast, while the mean value of θ_w in the lowest kilometer is only about 20°C. Ahead of pronounced troughs traveling into Europe some Saharan air is drawn northward and arrives in the middle troposphere over Europe east of the confluence line and locally saturated near it, but with a value of θ_s of not more than about 18°C, which is about the seasonal mean value. This is barely below the seasonal mean afternoon maximum of θ_w at screen level inland over France and southern England, and the provision there of a large amount of available potential energy for cumulonimbus convection in frontal zones demands the production of values of θ_w in the lowest kilometer which are significantly greater. The manner in which these high values arise can be illustrated with reference to the particular occasion of the intense traveling storm of 9 July 1959 over southern England, which was discussed earlier in the chapter.

This storm was one of a series that developed on successive days in a cold front zone which reached the coastal regions of western Europe on 8 July and subsequently moved inland. The intense storm over southeast England on 9 July was succeeded by another over the English Channel on the evening of 10 July and another over Holland on the afternoon of 11 July (in which the largest hailstones had a diameter of about 6 cm; 178). The daily recurrence of the intense storms indicates the importance of each day's sunshine in producing favorable conditions, which are sufficiently well represented by analyses of the situation on the afternoon of 9 July, when the first intense storm was traveling across southeast England.

The cold front was then associated with a shallow cyclone over northern Norway (175). Near the surface a trough extended along the frontal zone into southern Spain, and in a more pronounced trough aloft the southwesterly winds reached a maximum speed of nearly 40 m sec^{-1} in the high troposphere over England (Fig. 8.20).

The distributions of θ_w and θ_s in the forward side of the trough at midday on 9 July show little difference from the monthly means except at low levels over France and southern England. The most striking feature of the distributions was that over southern France, in the lee of the Spanish highlands, θ_w at low levels reached 24°C, a value normally encountered near the Greenwich meridian only over Africa south of the intertropical convergence zone. Such high values are occasionally generated near the Alps and near the Pyrenees of northern Spain when the local ordinary convection is suppressed into a shallow layer. They reach northern France and Germany by the northward displacement of air from the previous day. The travel of such regions of high θ_w, which soundings show to be characteristic of a layer about 1 km deep, is of course difficult to follow during the night, when the screen-level observations become unrepresentatively low. That which is formed north of the Pyrenees during a day when there is a flow in the lower troposphere from a southerly quarter appears to be associated with the subsiding part of an intermediate-scale circulation, of the same kind as that which confines the ordinary convection to a shallow layer over the low ground east of the Mexican highlands,

8 Cumulonimbus convection

discussed previously. The potential temperature of the adiabatic layer over the Spanish tableland in southerly flows during midsummer is characteristically about 36°C, and a similar value is found over southern France in a layer aloft with a base at about the 850-mb level, which is produced either by advection from Spain or by subsidence of air arriving from the Mediterranean. In the former circumstance the mean value of θ_w in the layer is about 17°C, and locally rather more, and observations indicate that the air has been modified by ordinary convection over Spain; in particular the soundings then often show that the *upper* part of the layer, in the middle troposphere, has been moistened, and when it ascends in the large-scale northward flow over France bands of cumuliform clouds form in it (*castellanus*; see Appendix 8.8). On the other hand, if the layer has arrived from the Mediterranean, it is considerably drier, because over the cool sea it has not been subject to ordinary convection.

Isentropic relative flow charts show the pronounced ascents as air leaves Spain and travels north in frontal zones; consistently castellanus clouds are produced in the middle troposphere over England, and the lifting of the base of the layer, decreasing its pressure by 100 mb or more, causes θ_s in it to diminish from about 24 to only about 18°C, so that the restraint which the layer has imposed over France upon ordinary convection from below disappears.

The relative flow of the air in the lowest kilometer over France is usually considerably backed from the generally southerly direction in the lower troposphere, partly because of the friction to which the boundary layer flow is subject, perhaps also because a trough tends to appear in the lee of the Pyrenees, and sometimes because the surface pressure tends to fall most over a region of highest potential temperature θ which develops over central France, leading to the appearance of shallow depressions familiar to forecasters as "thundery lows." Accordingly, the lowest layer of high potential temperature θ_w usually has a motion with a component westward through the frontal zone toward its cool side, and ascends to produce clouds which develop into tall castellanus and cumulonimbus where the restraint upon deep convection disappears, and where there is frequently sufficient shear to organize cumulonimbus convection efficiently. The first indications of the development of thundery weather as a large-scale trough enters western Europe in summer are the sferic observations (Appendix 9.4) which locate lightning discharges in such clouds or in castellanus formed in the Spanish air, usually over the Bay of Biscay near the western coast of France. From the occurrence of the sferics over the sea, and their tendency to appear in the night or early morning (Fig. 8.20), it is evident that the first tall clouds are not produced by the heating of air over the surface in their vicinity. On days succeeding 9 July 1959 intense cumulonimbus again formed near the 50th parallel, but were displaced farther east with the frontal zone, and the region north of the Alps became a more significant source of the lowermost layer of high θ_w.

8.8 Characteristics of severe local storms

The visual appearance of intense cumulonimbus

The striking features of intense cumulonimbus which can be seen from the ground, and more readily from the air, are the great extent of the anvil cloud, the characteristic dome which reaches well above the general level of the anvil cloud, and the arch cloud over the squall front.

Plate 8.13 shows moderately intense traveling cumulonimbus as seen from the air (near the border of Colorado, on an occasion of marked wind shear). The cloud tops were persistent broad domes, with only small-scale detail, whose summits reached well above the level of forward-spreading anvil clouds. The circulation of air through the clouds appeared to be steady (179). In Pl. 8.13 the cumulonimbus anvil clouds are clearly separate, but in the regions farther south over the United States, where tornadic storms are more common, the anvil clouds of several cumulonimbus often amalgamate into widespread dense layers with almost level tops at about the height of the tropopause. Here and there these layers are penetrated briefly by towers or more persistently by domes, of more turbulent appearance; the domes may be 20 km or more across, and may protrude spectacularly several kilometers into the stratosphere (Pl. 8.14, 8.15). The scale of details on the surface of the domes, and some observations which indicate that their radiative temperature is only about 10°C below that of the neighboring cloud tops, suggest that the intense mixing between the cloud and the clear air is confined to a shell only 100–200 m thick (180), and the dome and top of the anvil cloud appear generally to retain sharply defined upper surfaces. Thus, although the larger domes penetrate well into the lower stratosphere, they are probably an inefficient means of moistening it.

The upwind edge of the anvil may spread some tens

of kilometers to the rear of the columns which mark the region of strong updraft, while on the forward side the anvil cloud spreads with approximately the speed of the general wind at its level, so that after an intense storm has been in progress for some hours the anvil cloud may be several hundred kilometers long and up to about 100 km broad (68, 181). (Plate 8.16 is a picture of a large cumulonimbus over southern Florida with an anvil cloud whose visible part is nearly 100 km long; it is taken from reference 182, which includes other pictures of even more extensive anvil clouds.)

The velocity of severe local storms

It has been customary to compare the velocity of a weather system, identified by some persistent feature, with a mean velocity of the atmosphere in its neighborhood. If the mean velocity changes only slowly, it can be used in forecasting practice to anticipate the movement of a weather system, such as a cyclone, a hurricane, or a thunderstorm.

Difficulties may arise in deciding what feature best represents the weather system, how far the system extends, and how to determine a mean wind in the surroundings. However, these difficulties are not serious when the associated motion system is only a small disturbance in a nearly uniform, more widespread flow. The identifying feature can be the simplest to observe, for example, the minimum sea-level pressure in the cyclone, the eye of the hurricane, and the most intense radar echo in the thunderstorm. The position of such a feature is sometimes found to move regularly along a smooth path, with a velocity clearly about equal to that of some mean tropospheric wind.

In most cumulonimbus rotation is very slow and there is no obvious feature by which to locate a center. Visible features, such as the top of the tallest cloud tower, are persistent (for an hour or more) only in some kinds of severe storms (which may travel hundreds of kilometers), and their position is difficult to determine and track by optical means only. However, radar echoes from cumulonimbus are detectable at ranges of up to 200 km or even more on the displays of a suitable radar, and the position of some feature of the echo on a PPI display, such as the most intense echo, is easily followed using conventional radar. Its position in the cumulonimbus may vary with elevation, with the stage in the evolution of the circulation, and with the kind and distribution of precipitation particles present; for example, early in the evolution it is in a growing isolated tower in the middle troposphere, whereas in a late stage it may be in a slowly evaporating residual anvil cloud of ice particles in the high troposphere. However, it is usually in or near the region of most intense updraft, and following it is a reasonable and almost the only practical method of assessing a velocity for a cumulonimbus. In this way it has become accepted that showers and small thunderstorms move with about the mean velocity of the winds in the troposphere, or in the layer which they occupy (184), apparently in agreement with the old view that the velocity of a thunderstorm is nearly the velocity of the atmosphere in which the bulk of the cloud is located.

A careful and extensive set of comparisons has been made between winds thought to be representative of the undisturbed environment (generally, the means from at least four separate measurements in a special net of soundings) and thunderstorm velocities, assessed from the motion of points judged to be the geometric centers of the storm echoes observed on a (full-gain) PPI radar display. Each of several tens of echoes from storms in Florida and Ohio was examined at intervals of 2 to 5 min throughout the period in which it retained a readily recognizable identity; the average duration of this period was nearly 30 min. It is not quite clear what event determined the end of the period; however, since about half of all the echoes increased in height (the mean heights of the tops of a measured proportion were about 20,000 in Ohio and about 35,000 ft in Florida), it was evidently confusion with other developing echoes at least as often as the disappearance of the echo. Nevertheless, the average short duration of the identifiable echo implies that the storms studied were distinctly impulsive in character, not intense, and were composed of short-lived individual cells, consistent with the small tropospheric wind shears observed (only about 4 m sec^{-1} in Ohio and 7 m sec^{-1} in Florida). Among the interesting results of this study were the following:

1. The average speed of the storms was less than the average of the wind speeds at all levels above 4000 ft in Ohio and 2000 ft in Florida. It appears that the relative motion (of the potentially warm air) in the lowest few thousand feet was toward the fronts of the storms, and in Ohio most often into their right flanks, with an average speed of only about 5 km.

2. The direction of motion of the storms was usually the average of the wind directions between 2000–4000 and 20,000 ft, and about equal to the wind direction at 10,000 ft.

On the whole the data could be said to support the idea that the storms moved with about the vector mean wind of the layer which they occupied, but occasional marked deviations were found. In these and other observations it has been noticed that an echo path may be irregular, often manifestly because of sudden extension or fusion with another echo on one of its sides. Thus a distinction has often been drawn between the "translation" of a storm (with some mean wind) and its "propagation."

More elaborate studies (186) have since confirmed

that small cumulonimbus echoes move approximately with the mean wind in the middle troposphere (between about 3 and 6 km), in contrast with the behavior of some large echoes (from more intense storms), which move more slowly and distinctly (frequently by as much as 30°) to the right or, less frequently, rapidly and to the left. The mean wind speeds on the occasions of severe thunderstorms may be as great as 30 m sec^{-1}, and the vector difference between the mean wind and the storm velocity may be as great as 20 m sec^{-1}. These large differences make such severe thunderstorms the only phenomena in which the forecaster finds it obviously difficult to apply the concept of a *steering current*. It may be suspected that this is an aspect of the exceptional speed of the circulation through the isolated intense cumulonimbus, in which the maximum speed of the updraft may reach or exceed 30 m sec^{-1}.

Figures 8.48–8.51 present some features of the radar echoes and surface weather observed on an occasion of severe local storms near Oklahoma City (191), providing good examples of the properties of the intense cumulonimbus of the midwest. Cumulonimbus formed in the early afternoon west and northwest of Oklahoma City in a typically favorable large-scale flow pattern, on the eastern side of a trough which was most pronounced in the low troposphere, and near a dry line advancing slowly across western Oklahoma. Individual storms could be identified by their radar echoes, which at first moved toward the northeast or east-northeast; several storms, which became more intense and persistent, developed radar echoes with the echo-free vaults and the very high domes which are characteristic of severe storms, and their paths turned to a direction south of east, that is, to the right of the direction of the general wind at any level in the troposphere (Fig. 8.48 and 8.49). During the turning the speed of travel decreased from about 15 to about 10 m sec^{-1}; thus the potentially warm air near the surface entered the storms from the south-

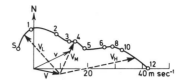

Fig. 8.49 A smoothed hodograph of the tropospheric winds observed 30 km ahead of storm G of Fig. 8.48 (heights in km; surface marked S). The vectors marked **v** (thin line) and **V** (thick line) represent the velocities of the ordinary and the intense cumulonimbus, and the dashed lines V_L, V_M, and V_H show, respectively, the velocities of the air in the low, middle, and high troposphere relative to the severe storms.

east, and the dry and potentially cold air in the middle troposphere entered from the southwest. A sounding made ahead of the severe storms showed very large available potential energy (corresponding to the positive area of the simple parcel theory; an even greater upper negative area, persistently nearly 18 km, was defined by the greatest height reached by the storm tops). On this occasion $-(\overline{Ri})$ was about 3 (this and some other features were included in Table 8.5).

On RHI displays intersecting the right rear flank of the severe storms echo-free vaults were seen to be a few kilometers across at heights of several kilometers, were occasionally traced to above 10 km, and were bounded on their rear sides by wall echoes extending down to the ground almost vertically. During the intensification of a storm the orientation of the wall echo, originally almost parallel to the direction of motion of the storm, appeared to change, so that whereas at first the echo-free vault appeared on a PPI display intersecting the storm at low levels as only a shallow notch in the right rear flank of the storm echo, eventually this notch became a deep indentation, behind which the echo curled into an appendage of the kind long recognized as a *hook* often accompanying storms which produce tornadoes (Fig.

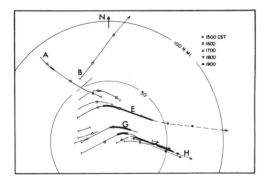

Fig. 8.48 The paths of several cumulonimbus near Oklahoma City on the afternoon of 26 May 1963, as determined by radar situated at the center of the range rings. The thickened lines indicate where the storms were severe and had hook echoes (191).

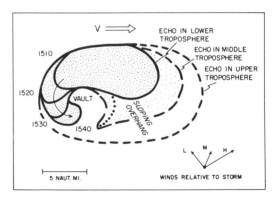

Fig. 8.50 A schematic diagram showing the formation of a hook echo at low levels behind the echo-free vault, during the intensification of the storm E of Fig. 8.48 (191). The directions of travel of the storm (**V**) and of the relative winds in the low (L), middle (M), and high (H) troposphere are indicated.

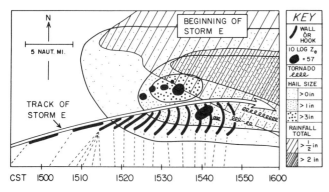

Fig. 8.51 The relation between features of the radar echo and of the weather experienced at the ground during the intensification of the storm E of Fig. 8.48 (191). The radar beam intersected the storm at heights between about 2 and 3 km; successive positions of the wall or the leading edge of the hook echo seen on the PPI display are indicated, and the dark areas are those over which the echo intensity at some time reached maximum values (in positions adjacent to the echo-free vault). It is thought that the largest hailstones were associated with the highest echo intensities, and that the tornadoes developed soon after their fall, beneath the echo-free vault ahead of the leading edge of the hook echo. Squalls of damaging severity also occurred in the path of the hook echo. The storms G and H which succeeded storm E became equally severe during periods of over an hour, in which the highest echo tops persistently reached over 18 km.

8.50 and 8.51). When such a hook echo has been examined at short range (and therefore with good resolution) it has been possible to detect evidence of cyclonic rotation at radii of several kilometers from a point near its extremity, which has been related to the existence of a *tornado cyclone* within which the much smaller tornado itself occurs (192).

On this occasion one of the storms (B of Fig. 8.48) traveled toward the northeast at about 15 m sec^{-1} and persisted for at least 2 hr, producing hailstones of diameter over 5 cm at several places along a path about 80 km long. This motion is in marked contrast to that of the nearer storms, and shows that even in the same general environment the velocity of a storm, at least across the direction of the middle tropospheric winds, has to be regarded as propagation at a speed different from that of any wind velocity. The possibility of such propagation is recognized in the tentative theories of three-dimensional cumulonimbus convection which are currently being developed (105). When the propagation is to the right, as seen in the storm of Pl. 8.17, the severe weather is experienced on the right flank of the storm (Section 8.5): the largest hailstones reach the ground in places some kilometers to the right of the swath of heaviest rain (often the stones are of almost uniform size and are neither accompanied nor followed by any rain), and the squall is strongest on this flank and spreads well beyond the region of precipitation, frequently leading to the formation of new storms at distances of 20 km or more. It seems that if tornadoes occur, they are usually some kilometers ahead and to the right of intense precipitation; they are in the path of the hook echo (Fig. 8.51), and evidently near the squall front, beneath the strong updraft which prevents the fall of any local precipitation and produces the echo-free vault directly below the storm dome. The towering cumulus which form on the right flank of the storm are over the front of the cold outflow, which represents a very intense intermediate-scale circulation, and grow through a lower troposphere in which the lapse rate may be much greater than when there has been previous widespread cumulus convection. In these particular circumstances the updraft speeds which are attained in the cumulus are likely to be substantially greater than normal, so that the first detectable precipitation in the clouds could be expected to appear at unusually high levels. On the occasion described the vertical extents of the first echoes detected in storm G were between about 8 and 9 km, implying that it was necessary for the cloud thickness to reach about 7 km before detectable precipitation developed. This is indeed substantially greater than that of about 5 km, which seems to be typical in ordinary circumstances well inland (Table 8.3). At the levels of the first echoes to be detected the temperature was about $-30°C$, and it is probable that nearly all the precipitation particles were frozen, although their growth may have been due primarily to coalescence among cloud droplets. In each of the three cells observed the rate of increase of echo intensity was consistent with the growth of ice particles by accretion in a droplet cloud of mass concentration a few grams per cubic meter; from the intensity and rate of descent of the echo in the first cell it was inferred (193) that the largest particles had diameters of a few millimeters and fall speeds of several meters per second.

It has recently been recognized that although the deviation of storm motion to the right is much the more common property of severe storms (in the northern hemisphere), deviations to the left are not rare, and on several occasions it has been observed that the radar echo of a storm has split into a diverging pair. Each of the resulting storms may become intense, but there is some evidence that the storm moving to the right is the more liable to be accompanied by tornadoes. In such a storm, as already described, the strong updraft is found on the right rear (usually southeast) side of the radar echo, the anvil cloud extends mainly forward of the updraft and perhaps somewhat to the left of the path of the storm, and the squall front near the ground extends outward to the right of the storm. In contrast, in the so-called left-moving storm the strong updraft is on the left forward

8   Cumulonimbus convection

(usually northwest) side of the radar echo, where the squall front is most sharply defined; the anvil cloud again extends mainly forward of the updraft but also at a large angle to the right of the path of the storm (194, 195).

Following the production of a pair of storms, the left-moving storm accelerates, but the speed of the right-moving storm decelerates until it is only about half that of the other.

8.9   Tornadoes and waterspouts

The funnel and other clouds associated with the tornado

The most obvious feature of the intense tornado is the trunklike funnel cloud, in which condensation occurs because of the pronounced fall of pressure associated with the great horizontal acceleration of air entering the circulation of the tornado. Estimates of the extreme surface wind speeds in tornadoes are almost all based on indirect evidence and fall in the range 50–100 m sec^{-1}. (As an example of a direct measurement, an anemometer record obtained during the passage of a tornado over Michigan in 1965 showed that the wind speed increased from about 13 to 67 m sec^{-1} in about 15 min and returned to 13 m sec^{-1} in between 3 and 4 min; 201.) If air near the ground is accelerated to a speed of 100 m sec^{-1}, then neglecting friction a pressure fall of about 50 mb may be anticipated, sufficient to produce condensation if the height of the adiabatic condensation level of the air is about 500 m. This height is considerably lower than the value of about 1200 m above the ground which is typically found in the potentially warm air ahead of the severe storms of the midwest (e.g., see Fig. 8.37, 8.46). The frequent lowering of funnel clouds to the ground and the previously mentioned presence of low scud or arch clouds in the vicinity of the tornado suggest that the zone of wind shift near a severe storm and the tornado which forms in it contain downdraft air derived from potentially warm air which has been moistened and cooled during passage through the region of heavy rain to one side of the strong updraft.

In recent years the spread of amateur cinematography has led to the collection of films of tornadoes from several locations during their passage through well-populated areas, and it has been possible from these to make fairly accurate measurements of the form and motion of the associated clouds and debris raised from the ground. In one such study (199) it appeared that each of several successive tornadoes developed below a persistent rotating cloud, the *wall cloud*, whose outer almost level base was between 400 and 500 m above the ground, about 1 km below the general level of the cloud base. At the ragged edges of the wall cloud there was obviously turbulent upward motion; from measurements of details identifiable on films the tangential speeds were found to be 10–15 m sec^{-1} and the upward speeds to be increasing approximately linearly with height, from about 10 m sec^{-1} near the cloud base to about 20 m sec^{-1} near its top. The vertical component of the (cyclonic) vorticity at the edge of the cloud which is implied by the tangential speeds is about 10^{-2} sec^{-1}.

During the development of a tornado the tip of a cone-shaped funnel cloud appeared projecting downward from the wall cloud and within about 30 sec lowered to the ground; its diameter at the level of the base of the wall cloud increased to a maximum of about 400 m during several minutes, after which it steadily shrank into a narrow cylindrical (*rope*) funnel, of diameter only about 30 m, as the tornado decayed. This trend in the shape of a funnel cloud appears to be typical; it has not yet been satisfactorily explained. The rope funnel may indicate an increase of relative humidity, accompanying a decrease of potential temperature, near the ground during the decay of the vortex, since otherwise the wind speeds implied by condensation so close to its axis seem implausibly great. In its early stages a funnel cloud is cone-shaped and its axis is almost vertical (Pl. 8.18); in the intense phase the shape of the cloud near the ground frequently is obscured by debris but it often seems more columnar, while in the dissipating phase the rope funnel may be sinuous and inclined at an angle of 45° or more to the vertical, but can still occasionally extend down to the ground, turning to become more nearly vertical close to the ground. (The more intense tornadoes have durations of between about 30 min and 1 hr; several may form in succession on the right flank of a tornado cyclone, curving cyclonically about its center and therefore crossing its path; 68.)

During the intense phase of a tornado it may be possible to follow on a film the paths of sufficient cloud details and pieces of ground debris to estimate the fields of tangential and vertical air motion at heights up to about 500 m and at horizontal distances up to a few hundred meters from the tornado axis, especially if the motion can be assumed to be steady throughout a period as long as 10–20 min. The most detailed study of this kind so far made is of a tornado at Dallas, Texas (202–4); the maximum vertical velocities inferred reached about 70 m sec^{-1} at about 50 m above the ground below the funnel cloud (whose tip was most of the time at a height of nearly 200 m), and only some tens of meters from the tornado axis. Above this level the zone of maximum upward speed was inclined outward from the axis, and at the surface of the funnel cloud there was little vertical motion. (When the tip of the funnel cloud was above the ground a region clear of dust and debris could sometimes be seen extending from the tip down to the ground, suggesting the presence there of a downdraft.) The maximum tangential speeds were in a zone inclined upward and slightly outward from the axis, reaching about 75 m sec^{-1} 70 m above the ground and 50 m from the axis. At a height of 300 m the angular momentum was almost constant between the position of the maximum tangential wind speed and the greatest distance of 400 m from the axis at which a speed was estimated. [Inferred three-dimensional trajectories relative to the tornado axis show that in this, as in other atmospheric circulations, some are essentially different from streamlines in a horizontal surface, and that there is a flow of air *through* the apparent disturbance. The curvature of some of the trajectories was *convex* toward the tornado axis even at distances from it as little as about 100 m. This feature and the observed vertical motions make doubtful the assumption of cyclostrophic wind flow often used to obtain from the tangential wind speeds the distribution of pressure and the pattern of air flow. However, the assumption (equivalent to using Bernouilli's equation for steady frictionless flow along trajectories) may give realistic results above a shallow boundary layer, and where the kinetic energy of vertical motion is small compared with that of the horizontal motion.]

The generation of the rotation in the tornado and other atmospheric vortices

Motion systems in which the air near the surface appears to rotate about a center occur on a great range of scales. In the large-scale vortices, the tropical storms and extratropical cyclones, the rotation has its source in the vertical component of vorticity in the atmosphere due to the earth's rotation, which is concentrated by the vertical stretching accompanying the horizontally converging flow. The diffusion of momentum by motions of smaller scales and the limited potential energy available during the phase of cyclogenesis restrict the maximum cyclonic vorticity attained over areas 100 km or more across to about 5×10^{-4} sec^{-1}, or a little more in the most intense tropical cyclones (whose energy is extracted continuously from the sea during their development). In the extratropical storm cyclonic vorticity is concentrated not only near its center, but also near the fronts, where over zones about 10 km wide it may reach 10^{-3} sec^{-1}. In these zones the large cyclonic vorticity may be associated with processes other than horizontal convergence; in particular, cold fronts often contain cumulonimbus, and what appears from observations at particular stations to be an intense large-scale front may actually be a local squall front.

In the squall front or wind-shift zone associated with intense cumulonimbus the vorticity may locally exceed 10^{-3} sec^{-1} in a zone a few kilometers across, and in and near the tornado the vorticity is of the order 1 sec^{-1} (corresponding to a velocity of 100 m sec^{-1} within 100 m of its axis). The vorticity of the tornado probably is derived from that present in the wind-shift zone, intensified by horizontal convergence beneath a strong updraft; it is remarkable that the convergence, associated with a vertical velocity which increases with height, is concentrated in a shallow and narrow layer near the axis of the tornado. The approximate conservation of angular momentum as air above the ground approaches the axis suggests that until this narrow zone is reached the flow is confluent with little horizontal convergence, and characterized by the horizontal circulation and the comparatively small vorticity which are observed in the region within a radius of about 1 km from the axis. The principal problem of tornado formation is to find the manner in which this large vorticity can be produced in the wind-shift zone of the cumulonimbus; it does not seem to be directly related to that present over larger neighboring regions. A plausible mechanism, not yet examined in recent theory (205), is the turning into the vertical of the horizontal vorticity corresponding to a large shear of wind in the vertical. It can be envisaged that this happens in a zone between intense updrafts and downdrafts.

Waterspouts

In general the whirlwinds with funnel clouds which form over the sea and are known as waterspouts are less intense and more frequent and widespread than tornadoes. However, the more vigorous have funnel clouds that descend a few hundred meters from the cloud base to near the sea surface, implying tangential wind speeds little different from those accompanying tornadoes, and occasionally when they have moved inland winds near the surface reaching about 50 m sec^{-1} have caused serious damage (207–9). The intense waterspouts are associated with cumulonimbus (and thunderstorms); they are commonly noted as occurring in a zone of wind shift, perhaps in a large-scale front (and in photographs they often appear below a low arch cloud near a region of heavy rain; 206; Pl. 8.19), or when there is a marked veer and increase of wind in the lowest 2–3 km (210). Usually intense waterspouts have cyclonic rotation (211).

Damaging whirlwinds sometimes occur inland when

there is cumulonimbus convection over the sea in strong winds. Thus minor tornadoes are more common over Great Britain in winter than in summer, and their annual frequency (212) is about two orders of magnitude greater than that of the intense summer tornadoes formed in frontal zones. Frequently they appear to have moved inland from the sea and occur at hours of the day when there is no ordinary convection over land and when there is an unusually large wind shear in the lowest kilometer. In these minor tornadoes, as in the intense waterspouts and summer tornadoes, the wind shear in the vertical may be an important source of vorticity.

On the other hand, most waterspouts and other whirlwinds that occur over land are much less violent, and their vorticity is probably derived from variations of wind in the horizontal which near and over land are associated with topographical features. Waterspouts usually have funnel clouds which are narrow, frequently do not reach the surface, and have a short duration. Moreover, they often form below cumulus which are not large, near showers of only moderate or slight intensity, or even well away from any precipitation (perhaps several at a time below a line of cloud), and their rotation may be in either sense (211). They seem to be more readily produced than tornadoes, especially in many coastal waters; for example, in the central and western Mediterranean (206, 210, 211), the Strait of Dover, and near the keys of Florida (209) they are not unfamiliar sights to local people, and are not particularly remarked upon when seen (partly because they rarely move inland).

An analysis of observations of waterspouts recorded in the meteorological logs of British merchant ships (206) suggests that they are seen over the open ocean with a frequency of up to about 3 in 1000 observations. However, it is probable that many occur in the lee of islands or capes; there the vorticity which is concentrated during the formation of waterspouts may be generated during the flow of airstreams past these obstacles; evidence of the shears and eddies in their wakes was noted in Chapter 7. Pronounced wakes behind hills and mountains are likely to be important for the production of waterspouts over some inland lakes (in Switzerland, e.g.) and for many minor tornadoes and funnel clouds observed overland in showery airstreams.

Waterspouts and whirlwinds which occur over or near low and level ground in peninsulas and beside estuaries may be generated by cumulus convection in shear zones associated with sea breezes. For example, a shear zone frequently forms off the southern tip of Florida between a general easterly flow and a westerly sea breeze across the east coast, and may be responsible for the waterspouts commonly seen near the keys (209). Again, in a general northwesterly airstream which flows across the narrow part of the Thames estuary there is frequently a southeasterly sea breeze across the coast of East Anglia in the outer reaches of the estuary. A shear zone with calm air and a locally high temperature near the surface occurs at an intermediate part of the coast, which is liable to be traversed by cumulus drifting from inland. The middle of the shear zone usually appears to be near the resort of Southend, where sunbathers on the beach and traffic on the coastal road are sometimes disturbed by brief but intense whirlwinds capable of overturning iron benches (213).

Whirlwinds of this intensity and scale are of the same kind as the dust devils mentioned in Chapter 7, which occur over deserts in light winds. Their vorticity is generated over comparatively featureless topography, but in some places where systematic observations have been made they have been found to be most frequent in the lee of small hills.

8.10 The evaporation of rain in cumulonimbus downdrafts

The departure from the saturated state in updrafts and downdrafts

It was shown in Section 5.4 that well above the condensation level in air ascending adiabatically at a speed U cm sec^{-1} the supersaturation ratio s'' is given approximately by the relation

$$s'' = f_7 U / N_g' \bar{r} \quad (5.40)$$

where f_7 is a function principally of temperature, and with $\theta_s \approx 20°C$ varies little from an average value of about 2×10^{-3} sec g^{-1}, and N_g' g^{-1} (typically of order 10^5 g^{-1} in cumulus) is the concentration of cloud droplets of average radius $\bar{r}$ cm. In typical conditions U is several meters per second and the average droplet radius $\bar{r}$ is about 10 μ, and s'' is of order 10^{-3}, or less than 1%. In these circumstances for most purposes the air can be considered to be just saturated, and its state in clouds can be compared directly with that given by the thermodynamic reference process.

However, in downdrafts which are sustained by the chilling of air due to the evaporation of rain, the vapor transfer proceeds under a considerable subsaturation.

If the air is to be given substantial vertical displacements, the chilling required demands the evaporation of concentrations of water about equivalent to those condensed in adiabatic ascent. Thus the same equation can be used to estimate the subsaturation involved by introducing appropriate values of N'_g and $\bar{r}$, and of a ventilation coefficient C_v. For $\bar{r} = 10^{-1}$ cm, $C_v \approx 10$ (Table 5.7), while in an intense rain (of about 400 mm hr^{-1}) $N'_g \approx 1$, whence in a downdraft even as small as 1 m sec^{-1},

$$s'' \approx f_7 U / N'_g \bar{r} C_v$$
$$\approx -2 \times 10^{-1}$$

that is, the relative humidity is only about 80%. The equation is valid only for small values of s'', but it is obvious that there are serious limitations to the rate of evaporation of large drops in downdrafts, and that when downdrafts do develop they are likely to be characterized by low relative humidities (consistent with the experience mentioned earlier that the relative humidity at screen level may fall to values below 70% during heavy thunderstorm rainfall). Further, it cannot be assumed that the changes of state of the descending air follow closely the simple pseudo-adiabatic reference process. The speed attained by a downdraft (and the proportion of the apparently available potential energy which is converted into kinetic energy) depends markedly not only on the particular atmospheric stratification present but also on the size distribution and concentration of the raindrops which enter it. Since, moreover, a substantial proportion of the condensed water in a cumulonimbus usually reaches the ground, it is unlikely that the mass flux of air in the cumulonimbus downdraft can, even in a well-organized storm, be a large fraction of that in the updraft. To consider these matters further the prevalent complication of the melting of frozen precipitation will be neglected, and attention will be concentrated upon the chilling of air in downdrafts containing rain.

The estimation of the vertical temperature distribution in downdrafts

It has usually been thought that the air in cumulonimbus downdrafts is derived mostly from levels below the cloud base, and although it was suggested that such air, after temporary descent, may play an important part in the behavior of intense storms, and in particular in the development of tornadoes (Section 8.9), it was previously emphasized that the lowest wet-bulb potential temperatures, representing potentially the most favorable source of air for downdrafts maintained by the evaporation of rain, are usually found in the middle troposphere near the top of the cumulus layer. Since the air there may be the predominant source of the cumulonimbus downdraft, we should consider the conditions that favor it sustaining intense downdrafts, which are known from aircraft observations (Section 8.3) to have speeds in the lower troposphere which are less than those of the most intense updrafts, but which sometimes exceed 15 m sec^{-1}.

A thorough examination of the evaporation in the downdrafts of cumulonimbus would involve the whole three-dimensional circulation of both air and particles, and even if some kind of steady state could be specified this would be extremely complicated. However, some insight into the conditions favorable for the occurrence of strong downdrafts can be obtained by calculating the vertical temperature distributions in horizontally uniform downdrafts into which rains of specified intensity and uniform raindrop size are introduced at the 500-mb level (214). Various other simplifications were introduced to ease calculations without losing any physical insight.

In the principal calculations the wet-bulb potential temperature was chosen to be 18°C, a value typical of the middle troposphere in the thundery weather of middle latitudes, and the air temperature at 500 mb to be about -12°C, corresponding to a state of saturation. Two additional calculations are illustrated in Fig. 8.52 and Table 8.7, for a strong downdraft of about 20 m sec^{-1} and for raindrops (at the 500-mb level) of initial radius 2 mm, and for a moderate downdraft of about 10 m sec^{-1} with drops of initial radius 0.5 mm. In these calculations a very low relative humidity at 500 mb was chosen, corresponding to a wet-bulb potential temperature of 15°C. It appeared, as might be anticipated, that the

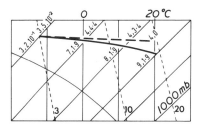

Fig. 8.52 Temperature and humidity in downdrafts produced in initially very dry air by the evaporation of rain; dashed line: moderate downdraft (speed about 10 m sec^{-1}), moderate rain of initial drop size 0.5 mm; solid line: strong downdraft (speed about 20 m sec^{-1}), intense rain of initial drop size 2 mm. The two temperature profiles are drawn on sections of a tephigram limited by the horizontal dry adiabatics corresponding to potential temperatures of 20 and 50°C. The isobars are drawn at intervals of 100 mb. The thin dashed lines are isopleths of saturation mixing ratio for 3, 10, and 20 g kg^{-1}. The lowermost curve is the saturated pseudo-adiabatic corresponding to the wet-bulb potential temperature of 15°C, which is preserved in the downdrafts. At each 100-mb level the pairs of figures entered beside the profiles give the relative humidity (in percent), followed by the drop radius [in units of 10^{-2} cm (upper curve) and in mm (lower curve)]. In both cases the downdraft of air is assumed to be very dry (R.H. = 3%) at a temperature of -12.2°C at the 500-mb level. (The conditions at lower levels are also specified in Table 8.7.)

Table 8.7 Temperature T and raindrop radius r in downdrafts (R.H. = 3% at 500 mb); p is pressure and T_s is wet-bulb temperature

	Initial drop radius 0.5 mm, moderate rain, moderate downdraft			Initial drop radius 2 mm, intense rain, strong downdraft	
p (mb)	T_s (°C)	T (°C)	r (mm)	T (°C)	r (mm)
500	−17.3	−12.2	0.5	−12.2	2
600	−7.8	1.3	0.44	0.8	1.98
700	−0.3	13	0.34	11.4	1.96
800	5.7	—	0	20.5	1.93

results are not particularly sensitive to the initial relative humidity at the 500-mb level.

The dependence of the downdraft magnitude on the rain characteristics and the atmospheric stratification

Although temperature distributions were calculated for the interiors of steady downdrafts, it is reasonable to suppose that they closely represent the general atmospheric stratification in which the specified rains (of initial intensity 5, 50, or 250 mm hr^{-1}) may produce drafts of the assumed magnitudes. This is because the buoyancy corresponding to only a small temperature difference between the downdraft and its surroundings is capable of providing an acceleration to a speed of 10–20 m sec^{-1}. If the temperature difference is ΔT°C, the vertical acceleration in a smooth steady motion according to the simple parcel theory is given by

$$dw/dt = g\Delta T/T$$

so that for an acceleration from zero velocity to 20 m sec^{-1} over a height interval of 5 km, ΔT is approximately 1°C. It must also be considered that an equivalent buoyancy is provided by a rain water mixing ratio of only about 4 g kg^{-1}, such as corresponds to a rainfall rate of 10 cm hr^{-1}, intermediate between the heavy and intense categories. Moreover, in any steady circulation a part of the kinetic energy of the downdraft is provided by the energy of horizontal motion with which the air enters the region of rain. Accordingly, the magnitude of the temperature difference between a steady downdraft and its general surroundings can be anticipated to be only about 1°C. Considering the difficulty of correctly measuring air temperature in rain from aircraft, this is not inconsistent with the values of 0–3°C in downdrafts of 2–8 m sec^{-1} found in the studies of thunderstorms over Florida (58).

It was found, in conformity with the earlier general discussion on the humidity in downdrafts, that only in an intense rain of small drops does the descent of air in the downdraft approximate to the saturated pseudo-adiabatic reference process, and then closely only if the downdraft is weak. In these cases the product $N_g'\bar{r}$ in Eq. 5.40 approaches 1 cm g^{-1}, but such a value is not likely to be attained, since most of the water in intense rains is in drops of radius considerably larger than 0.5 mm. In an intense rain of drops with radius about 2 mm the temperature profile even in a weak downdraft departs considerably from that in the reference process, and the air becomes rather dry (the relative humidity falling to about 60%). In a moderately intense (heavy) rain, regardless of the drop size, the temperature profile approximates to the dry adiabatic and the relative humidity becomes less than 20%. The curves in Fig. 8.52 show that the initial relative humidity does not have much influence on the calculated profiles. Although in heavy and intense rains the raindrop size had a noticeable influence on the calculated temperature profiles, the main conclusion drawn was that the principal factors in the production of strong downdrafts are the intensity of the rainfall and the steepness of the general lapse rate. Low values of the wet-bulb potential temperature and relative humidity aloft, especially in the lower middle troposphere, are significant (and will commonly be present in the stratifications favoring intense organized storms), but are not important. (If the general lapse rate is approximately equal to the dry-adiabatic lapse rate, then the microphysics of the evaporation process places little restriction on the downdraft magnitude, and even in a moderate rain of large drops strong downdrafts may be generated; on the other hand, the microphysics becomes unimportant and the descent approximates to the saturated adiabatic in a weak downdraft of speed less than about 2 m sec^{-1}.)

The calculations were not extended to levels near the ground, but the results have the obvious implication that it is air from the adiabatic layer beneath the level of the cloud bases which is most readily brought to the ground by the evaporation of rain. Downdrafts may reach the ground from the middle troposphere and produce strong squalls when the adiabatic layer is abnormally deep and the cloud bases are unusually high, as in the situations producing the andhis of northwest India. In other circumstances air from the middle troposphere cannot descend to the ground in a strong downdraft unless intense rain can enter it and the lapse rate above the adiabatic layer is abnormally large. These conditions are typical of occasions when intense cumulonimbus develop in sheared flows in which the air in the middle troposphere has not previously been modified by widespread cumulus convection, as exemplified by soundings in Fig. 8.46 (typical of those made near severe storms in the midwest of the United States). An abnormally large shear between the wind in the lower

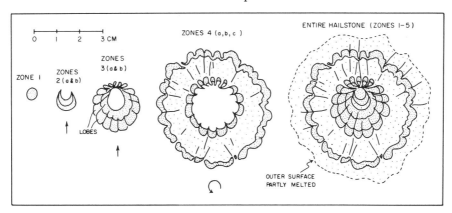

Fig. 8.53 Stages in the growth of the hailstone of which a thin section is shown in Pl. 8.22, inferred from the layered structure (219). The layers are grouped into several zones according to the inferred attitude of fall of the hailstone; arrows show the predominant direction of the air flow, the circular arrow indicating a random tumbling during the deposition of the last zones. Zones are subdivided according to whether there are distinct layers of milky rime (M, thin stippled) or transparent rime (T, dense stipple):

Zone and kind of rime		Attitude	Comments
1	M/T	Variable	Spherical embryo, diameter 5 mm
2a	M	One side persis-	Conical growth; accretion
2b	T	tently facing downward	on base only; individual large crystals elongated downward (Pl. 8.22)
3a	M	Same as	Accretion mainly on down-
3b	M	preceding	ward facing side; lobes begin to develop
4a	M	Variable	Accretion on all sides;
4b	T	(random	lobes grow radially out-
4c	M	tumbling)	ward; individual large
5	T		crystals elongated radially (Pl. 8.22)

middle troposphere and the winds at low levels (and the storm motion) may also represent available kinetic energy for a downdraft.

In contrast, during summer monsoons the continual deep cumulus convection leads to comparatively moist air and smaller lapse rates above the level of the cloud bases (as in Fig. 8.32), which are unfavorable conditions for the production of strong downdrafts in cumulonimbus. Then even on an occasion when there appears to be a large available potential energy for an updraft circulation, and sufficient shear between low and high levels to organize it, the impossibility of producing a strong downdraft circulation may significantly hinder the development of organized and intense cumulonimbus convection, and the value of $-(\overline{\text{Ri}})$ as ordinarily obtained may be an inadequate characteristic of the stratification when considering the storms which occur.

Similarly, in the thunderstorm situations investigated over Florida and Ohio (58) the average lapse rate in the lower middle troposphere was not much greater than the saturated adiabatic (Fig. 8.14 shows the combination of small shear, lapse rate, and wet-bulb depression typical of the summer season in Florida). The maximum measured rainfall rates over periods of 5 min (corresponding to a horizontal space dimension of about 2 km) reached values of 75–125 mm hr^{-1}, thus the steady downdraft speeds in these situations could not be expected to be more than a few meters per second. Those observed were mostly in the range 3–6 m sec^{-1}, occasionally reaching values between 6 and 10 m sec^{-1}. The speed of gusts at the ground in the downdraft outflows did not much exceed the rate of advance of the gust fronts, which on the average was only about 8 m sec^{-1}. [It was also observed, as might be anticipated from the previous discussion, that the relative humidity at screen level often fell rapidly at the ground during heavy rain; the magnitude of the fall, which depends on the previous value, is less significant than the minimum (or the associated value of θ_w), which is likely to be representative of conditions in the downdraft air as it approaches the ground. The minima recorded are not quoted, but would appear commonly to have been less than 70% and occasionally as little as 50%.]

Finally, the calculations mentioned have a bearing on reports that showers often evaporate completely before reaching the ground in deserts, or semi-arid continental regions well inland during the dry season, where the level of cumulus bases may be as much as 4 km above the ground (above the 600-mb level). If so, the rain must be composed of very small drops, for in these circumstances moderate or strong downdrafts develop and large raindrops can be expected to reach the ground with little change of size. Of course, the largest drops may compose only a small fraction of the rainfall, and reach the ground over such a small area in the immediate vicinity of the updraft that they are not usually noticed in sparsely inhabited regions, whereas shower clouds, and especially thunderstorms, can be observed from a distance, and the downdrafts may produce squalls which spread far from the rain area, as

8   Cumulonimbus convection

in the andhis of northwest India and the haboobs of the Sudan. Similarly, in the summer over the northern Rockies west of the Continental Divide squalls associated with cumulonimbus with high bases have speeds commonly reaching 20 m sec^{-1} and occasionally exceeding 25 m sec^{-1} (215). In this region lightning accompanied by little rain, or striking the ground to one side of the areas receiving showers, starts many forest fires whose establishment and initial spread may be helped by the fanning in the strong winds of the downdraft outflows. (Each year in the western United States lightning causes some thousands of forest fires—about two thirds of the total number—and the loss of about 100,000 acres of timber.)

8.11   Hail

Small hail

Hailstones grow among the supercooled droplets of cumulonimbus following the appearance of ice particles, which may be snow crystals, frozen droplets of various sizes, or fragments of ice produced by the shattering of freezing droplets or by the breaking away of pieces of the rime ice of other hailstones (Section 5.14). If these particles are very small, their initial growth is mainly by condensation, but as explained in Chapter 5 if the mass concentration of the cloud droplets is about 1 g m^{-3} or more, as is characteristic of cumulonimbus updrafts, and if the concentration of the ice particles is comparatively small, then when their diameters reach a few hundred microns their further growth is overwhelmingly due to the accretion of droplets. Under these conditions they grow into pellets of rime ice of low density (the ordinary rimes of Table 5.14), known as soft hail or graupel, in which the individual accreted droplets can be seen under the microscope, mostly with radii of about 10 μ. The pellets usually acquire the shape of a cone with an apex angle of about 70°, falling with the apex upward so that accretion proceeds on the rounded base. As a pellet grows and the accretion rate increases, so also does the density of the deposited ice, and its base may eventually become composed of one of the glazed rimes and appear glassy and translucent in contrast to the opaque white upper parts. Such hailstones commonly attain dimensions of several millimeters, and occasionally up to about 1 cm, and are the typical form of precipitation from cumulonimbus in cold weather, or in other circumstances when the temperature of the cloud base is about or below 0°C.

When the cloud base temperature is higher, and the accretion proceeds more rapidly in higher concentrations of droplets (particularly at air temperatures not much below 0°C), the soft hail pellets become completely covered with glazed rime before reaching this size. Evidently they begin to tumble, for they develop into roughly spherical stones with individually almost uniformly thick layers of glazed rimes of variable opacity (dependent upon the size and concentration of bubbles of air which come out of solution as the accreted liquid freezes). Such stones are distinguished as true hailstones. In ordinary cumulonimbus they often attain diameters of up to about 2 cm in the clouds, and not much less on reaching the ground, while in intense cumulonimbus a proportion may become large or giant hailstones (Table 8.8) several centimeters in diameter. When such hailstones are cut open or sliced through their centers the concentric layers of rime are displayed, and it is usually possible to find an embryo a few millimeters across. The embryo is often a conical soft hailstone; sometimes it has evidently partially melted before being refrozen (216), and sometimes the embryo consists wholly of clear ice, which could have been an ordinary raindrop or a completely melted soft hailstone.

The density of the ice accreted by a small hailstone can be estimated by regarding it as a sphere, adopting the approximate simple expressions (5.53) for the transfer of heat and vapor from the hailstone, and the empirical formula (5.81) which relates rime density to the radius and impact speed V_0 of the accreted droplets (practically

Table 8.8 An arbitrary scale for the use of voluntary observers in reporting the size of hailstones, employed in investigations of hailstorms in southern England. Similar scales have been used elsewhere

Kind of hail	Approximate (largest) diameter (cm)	Description
Ordinary, or small	<0.5	Grain
	0.5–1.0	Pea
	1.0–1.5	Mothball; small marble
	1.5–2.0	Cherry; marble
Large	2.0–2.5	Large marble
	2.5–3.0	Walnut
	3.0–4.0	Golf ball
	4.0–5.0	Small egg
Giant	5.0–6.2	Egg
	6.2–7.5	Tennis ball
	7.5–9.0	
	>9.0	

the fall speed V of the hailstone) and to the surface temperature of the hailstone. The density increases with increase of droplet radius and accretion rate (and therefore with droplet concentration and hailstone size), and so tends to be greater in cumulonimbus with high base temperatures. The *mean* density, the fall speed, and the rate of growth of a small hailstone therefore depend in a very complicated manner on the conditions encountered during its entire history; however, some interesting conclusions can be drawn (217) by estimating how the density of the accreted ice and the mean density change with hailstone size, according to the base temperature of the parent cumulonimbus. Considering first the growth in a cloud with a cold base, it is found that although hailstones with diameters as large as several millimeters may be a few degrees warmer than their environment, their equilibrium surface temperatures never reach 0°C while they are in the cloud; the rate of accretion is everywhere insufficient for this, the concentration of cloud water approaching zero as the air temperature rises toward 0°C. For the same reason the bulk of the growth in such a cloud occurs at air temperatures of $-10°C$ or less, and the mean density of a hailstone is likely to be low even if its diameter approaches 1 cm, and even if it originated as a frozen raindrop (at least if the growth continues by the accretion of cloud droplets). The most common hail clouds with such a low base temperature are the ordinary cumulonimbus of the cool airstreams in middle and high latitudes, and it is predominantly from these usually small clouds, containing updrafts whose maximum speeds reach only about 10 m sec^{-1}, that falls of soft hail are experienced. (Even inland the bases of these clouds are usually little more than 1 km above the surface, and the hail often reaches the ground with hardly any melting, since even near the ground the general wet-bulb temperature is only a little above 0°C, and is lower in the downdrafts which often accompany the hail. However, soft hail often falls with or is succeeded by some rain, which may be produced by the melting of smaller stones and the more or less rimed snowflakes grown in the weaker parts of the cloud updraft; in very cold weather the snow reaches the ground over a comparatively large area and may seem to be the only form of precipitation, but some soft hail usually falls more locally and briefly from the stronger parts of the updraft.) More intense cumulonimbus with bases which are much higher, and therefore have similarly low temperatures, occur in the interior of continents (e.g., near and in the lee of the northern Rockies); in their stronger updrafts larger hailstones of diameter 2 cm or more are grown. Such stones are likely to have an inner core (at least several millimeters in diameter) of opaque ice of low density, surrounded by layers of glazed rime which make the mean density close to that of pure ice.

Large and giant hailstones

Many large and giant hailstones are almost spherical, but as their size increases so also does the proportion having the flatter shapes of oblate spheroids (sometimes with a depression on one side, giving the shape of an apple) or ellipsoids (see, e.g., 88, 218). Of about 100 large hailstones collected from the intense storm over southeast England on 9 July 1959 discussed in Section 8.4 with largest dimensions between about 2 and 4 cm, about one fourth could be said to be approximately spherical and most of the remainder were oblate spheroids with the short axis between 60 and 80% of the others. Almost certainly many hailstones have become flatter during some melting in the air or on the ground before collection. Less common are much flatter lenticular or disclike shapes (sometimes likened to pieces of broken window glass or coins) and others which can only be described as irregular. These may arise from fragments broken away from more regular hailstones while they are still in the updraft of the hailstorm (Appendix 8.11).

The surfaces of large hailstones often have shallow rounded protuberances (*lobes*). On many giant hailstones the lobes are well developed and contain milkier ice than the intervening spaces, so that they are sometimes mistakenly supposed to be conglomerates of numerous smaller stones (Pl. 8.20). Such lobes may be extended into iciclelike fingers (Pl. 8.21), which occasionally are several centimeters long (218).

The size of the largest hailstones which fall during a storm has special interest, since it is some measure of the intensity of the storm updraft. Understandably, the size of hailstones is sometimes exaggerated by people whose property suffers damage, and may increase further with lapse of time. On the other hand, giant hailstones are sparse in number, and the worst damage is usually inflicted by the much more numerous smaller hailstones, while experience gained with voluntary observers, and during interrogations within a few days of a severe storm, has shown that most people can give accurate and consistent information about the character and duration of unusually large hail, naturally referring to various kinds of nuts, eggs, and fruit as measures of size. Fortunately people often weigh (or even try to preserve) specimens of large hail, so providing a more definite measure than one simply of circumference or dimensions. It appears that maximum weights of a few hundred grams to about 0.5 kg (equivalent spherical diameters of between 8 and 10 cm) are typical of the most severe hailstorms, but that hailstones weighing about 1 kg and perhaps even more occur very rarely (Appendix 8.12). Hailstones as large as 150–200 g occurred in the moderately severe storm of 5 September 1958 over southeast England (Table 8.5); such hailstones have fall speeds of 40–50 m sec^{-1} and are easily capable of breaking windows, cracking roof tiles, and severely denting the bodies of automobiles and caravans.

The internal structure of hailstones

According to the discussion in Section 5.13 rime ice can be considered to grow in two principal regimes, the dry and the wet, depending on whether the mean surface temperature is distinctly below or effectively at 0°C. There are three related principal forms of rime ice: (*a*) the ordinary opaque rimes containing air mainly enclosed from the atmosphere during the accretion; (*b*) glazed rimes, the ordinary glazed rimes, usually containing bubbles of air released from solution during the freezing of the accreted liquid; and (*c*) spongy rime, which during its growth contains also a proportion of unfrozen accreted liquid (Table 5.14).

Except perhaps in a small core, large and giant hailstones consist of glazed rimes. Occasionally there are reliable reports of large hailstones which disintegrate, or are even said to "splash" on striking the ground, suggesting that they contain a large proportion either of ordinary rime ice or of spongy ice (108; 191, p. 74; 236); generally, however, their glazed structure and hardness indicate that they have grown with a mean surface temperature close to 0°C and contain only a small proportion, if any, of spongy ice. Consistently, calorimetric methods of examining freshly fallen hailstones have hardly ever shown them to contain as much as 10% of unfrozen water. Of some tens of hailstones with maximum diameters of up to about 4 cm collected from each of two hailstorms in Oklahoma, those of diameter less than about 2 cm (the majority) were all hard throughout, but each of the larger contained a thin inner shell of diameter about 1 cm containing spongy rime, in which the liquid water amounted to between about 5 and 10% of the mass of the whole hailstone (237). In another series of measurements with hailstones collected from several hailstorms in Colorado, South Dakota, and Kenya (East Africa), the outer parts of the larger hailstones (diameter up to about 4 cm) were hard, and the proportion of internal liquid was a very small percentage (238). Small conical and mostly opaque hailstones from one storm contained a higher proportion (between 11 and 16%), probably produced by partial melting in the air or on the ground.

The frequent occurrence of lobes on large and giant hailstones is also evidence that their surfaces have not been so wet that there has been a significant degree of movement of accreted liquid away from the place of impact, for their development probably depends on the decrease in the efficiency of catch of cloud droplets with increase of collector size when its diameter exceeds about 1 cm (Fig. 2.7), so that small surface projections grow more rapidly than their surroundings. The implication of the lobes that the cloud water has frozen quickly where it was accreted is supported by the observation that during the growth of artificial hailstones (239) lobes were produced in both the dry and wet growth regimes, but were most prominent when the rate of accretion was near the critical value separating the regimes, and when the accreted droplets were small (median volume radius about 15 μ). The lobes were less prominent when the mean droplet size was made larger (up to about 100 μ) or the deposit of ice became more spongy, and the surface movement and seepage of liquid became greater; with the largest droplets the lobes were reduced to small and shallow surface irregularities.

Accordingly the typical hardness, constitution, and lobe structure of the large and giant hailstones which reach the ground are alone sufficient to indicate that generally they grow by the accretion of supercooled cloud droplets (rather than of unfrozen precipitation) in concentrations of up to several grams per cubic meter at which the accretion rate is near the critical value separating the dry and wet growth regimes, and which are provided by adiabatic or nearly adiabatic ascent of air into the upper parts of cumulonimbus with base temperatures well above 0°C. [It can, however, be envisaged that some of the hailstones in a cloud are composed partly or mainly of spongy ice with a high proportion of unfrozen water, which melts readily (perhaps completely) during fall to the ground, as considered later.]

Obviously more detailed and definite information about the conditions successively experienced by the hailstone during its growth might be obtained from an analysis of variations in properties of the ice inside it. These are most conveniently studied in a thin slice carefully cut through the growth center. Among the more obvious properties are the sizes, shapes, and orientation of the air bubbles and of the individual ice crystals. The location of the crystals in a thin slice can be established most simply by placing it between crossed polaroids and photographing it by transmitted white light, when changes of the alignment of crystal axes appear spectacularly as changes of color (see, e.g., 239, 240). In another technique one surface of a slice is allowed to evaporate through a fine network of tiny holes in a foil cover, leaving an array of *etch pits*. The shape of a pit indicates the local orientation of the (*c*-) axis of hexagonal symmetry of the ice, and the individual crystals are recognized by areas in which the pits have a uniform alignment (241). Other properties include the presence of various trace substances, which are found to be concentrated in particular shells. Probably the most informative properties will prove to be the concentrations (as a function of hailstone radius) of the isotopes deuterium and O^{18}, from which it is possible to infer the air temperatures at which the ice was accreted, and thereby the updraft speed and effective condensed water concentration encountered by the growing hailstone as a function of height in the cloud. In this way from the examination of a giant hailstone which fell in South Wales a height-radius trajectory similar to that marked *cdef* in Fig. 8.55 (derived during the following discussion) was deduced for the growth of the embryo of diameter 0.7 cm to a final diameter of about 6 cm (242; during successive ascents in the cloud the updraft speeds encountered were deduced to increase nearly uniformly from about 30 m sec^{-1} at 5 km to about 50 m sec^{-1} at 8 km).

In general the growth of hailstones occurs in a great variety of conditions, and correspondingly when they are examined a great variety of internal structure is found, further complicated by melting, freezing of trapped liquid, and modifications of crystal structure some time after accretion (especially on the ground before the hailstones were collected, and during subsequent treatment to preserve them). As yet there is no general agreement on internal structure of hailstones. In considering the features which are found by examining sections of hailstones, those whose significance is primarily meteorological may be distinguished from others which involve the aerodynamics of falling objects or the microphysics of phase transitions.

First, nearly all hailstones are composed of several distinct layers of alternately cloudy and clear ice (as shown in Pl. 8.22), the clear ice containing comparatively very few and large air bubbles. This structure has long been recognized and was supposed to show that the hailstones circulate several times between levels near and well above the 0°C level, where the clear and cloudy layers are produced by the accretion of rain and of snow, respectively (243). It now seems that the opacity of the cloudy rime ice indicates either the dry growth regime, or wet growth with a local surface temperature in the region of accretion which is appreciably below 0°C, usually requiring air temperatures below about −20°C (239, 242). Thus the microphysics of the freezing process complicate the meteorological interpretation of the layered structure, for while transitions between clear and cloudy ice may arise during wet growth from passages between the lower and upper parts of the region of supercooling, at air temperatures above about −20°C transitions in either sense may occur as a result of changes between the dry and the wet growth regimes produced by small variations in the cloud properties, not meteorologically significant.

Similarly, hailstones usually have several layers in which the individual ice crystals are numerous and small (dimension about 100 μ or less), or comparatively few and large (dimension a few millimeters or more, tending to be elongated in the direction of growth). These layers often correspond to the milky and clear layers, respectively (Pl. 8.22); however, although a change in crystal size almost always coincides with one in opacity, the converse is not true. The number of crystal layers is therefore fewer. It tends to increase with the size of the hailstone, usually being about three in large and about four in giant hailstones (but in both varying, even in one hail fall, between extremes of one and about eight).

The borders of layers are usually well defined, the transitions occurring over radial distances of less than 1 mm. The edges of the milky layers are more difficult to define on closer inspection, since then very thin layers containing bubbles become apparent, some of which may be incomplete. It seems that the transitions between the crystal layers are the more likely to indicate significant changes in the properties of the environment of the hailstones.

The boundaries of the internal layers evidently show the shape of the hailstone at several stages in its growth (Fig. 8.53). It is often apparent that lobes have grown radially outward from near the center, so that the surface protuberances imply the three-dimensional structure sketched in Fig. 8.54. The spaces between pronounced lobes in artificially grown hailstones were observed to be empty fissures (239), but in natural hailstones which have been examined they appear to have become at least partly filled with liquid which seeped in during wet growth, or (more probably) during periods of superficial melting (especially during fall from cloud, or on the ground before collection). Frequently this liquid later froze, with the formation of radial lines of large and often elongated air bubbles (Pl. 8.22).

Many of the layers in hailstones have an approximately uniform thickness of about 1 cm (as in Pl. 8.22). Since the cloud droplets are accreted predominantly on the side facing the relative air flow this implies that

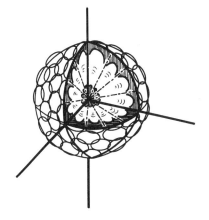

Fig. 8.54 A sketch of the three-dimensional arrangement of the lobes in a large hailstone (219). On its surface the lobes appear as convex protuberances, and in a thin section through its center their earlier form is revealed by the convolutions in the layer structure, with bulges which are convex outward.

during the growth of these layers the hailstone tumbled randomly so that every part of its surface spent about the same time facing downward, and raises uncertain aspects of the aerodynamics of hailstones. It seems plausible that an irregular spherical hailstone falling through turbulent air should experience random turning, but it is more difficult to explain the predominant distinctly oblate spheroidal shape of hailstones. When the same shape is indicated by internal layers it cannot be attributed to melting. It is likely that hailstones do not fall steadily in the vertical but spin and flutter irregularly while describing a helical path, with a short axis gyrating about the vertical like that of a coin in the last stages of its spin on a table. (Regular spiral motion has been observed at high but subcritical values of Re during the ascent of spherical balloons in still air and the rise or fall of solid spheres in liquids; 300.) Frequent overturning would seem to be an additional requirement for the production of oblate spheroids. Although such peculiar motion could appreciably affect the relative air speed, efficiency of catch, and transfer coefficients of hailstones, it is reasonable to regard the meteorological problem of the production of large and giant hailstones as not particularly dependent upon such complications, and sufficiently well examined in terms of the steady fall of spherical stones.

The growth of large and giant hailstones

Hail is grown inside cumulonimbus in the *zone of supercooling*, where the temperature is between 0 and some value approaching $-40°C$. In the conditions in which giant hailstones occur θ_s is about $20°C$, the thickness of this zone is about 5 km, and the concentration of condensed water in it may be close to the value of about 5 g m^{-3} produced by adiabatic ascent from a cloud base at a low level (Fig. 3.9). It is easily shown that if a hailstone is to reach giant size, with a fall speed of over 40 m sec^{-1} (Fig. 2.2), then the updraft speed in the zone of supercooling must somewhere exceed half this speed.

To make a study of the growth as simple as possible it is reasonable to assume that a hailstone is spherical and to take constant values for its density ρ_i (say 0.9 g cm^{-3}), its efficiency of catch E (say 0.8), and the concentration of cloud water m available for accretion (say 5 g m^{-3}) in the zone of supercooling, about the adiabatic value (m is therefore supposed to be unaffected by the growth of other hailstones). Considering that the drag coefficient may be almost constant and that the effects of variation in air density are small, it may be supposed that the fall speed V is given by

$$V^2 = k^2 R \quad (8.20)$$

where k^2 is a constant ($\approx 9 \times 10^6 \text{ cm sec}^{-2}$; cf. Eq. 2.29). Another simplifying assumption is that all the accreted water is retained, even though in very wet growth at high fall speeds some of the unfrozen liquid is shed into the wake of the hailstone, especially at air temperatures near $0°C$ (239). We consider an embryo hailstone which rises through the $0°C$ level with a radius R_0 and at some level below the top of the zone of supercooling attains a fall speed greater than the speed $U(z)$ of the updraft, so that it continues to grow during a subsequent descent and leaves the zone with a radius R_2. [Without being specific about the form of the updraft, it may be supposed that hailstones carried above the zone cease growth but are eventually carried aside in a horizontally diverging flow to fall outside the updraft, perhaps to reenter it at a lower level, and, as originally envisaged by Ferrel (243) in his explanation of the typical layered structure of the hailstone, to experience a sequence of further ascents and descents before finally falling to the ground.] It appears that to attain the largest values of R_2 and V_2 for a given maximum updraft speed and embryo size, the hailstone should just reach the top of the zone of supercooling in an updraft of constant speed U. In this case, neglecting the fall speed of the cloud particles, the accretion equation can be integrated to give

$$U \int_{R_0}^{R} \frac{dR}{V} - (R - R_0) = \frac{Em}{4\rho_i} \int_{z_0}^{z_1} dz \quad (8.21)$$

or, since $V = kR^{1/2}$,

$$\frac{2U}{k}(R^{1/2} - R_0^{1/2}) - (R - R_0) = \frac{Em}{4\rho_i} \int_{z_0}^{z_1} dz \quad (8.22)$$

If U_1 is the constant updraft speed for which the embryo is lifted just to the top of the zone of supercooling, of thickness Δ_z, and the fall speed $V_1 = U_1$ is that of the hailstone of radius $R = R_1$, then

$$\frac{2U_1}{k}\left(\frac{U_1}{k} - R_0^{1/2}\right) - \left(\frac{U_1^2}{k} - R_0\right) = b \quad (8.23)$$

where $b = Em\Delta z/4\rho_i$, which solved for U_1 leads to

$$\frac{2U_1}{k} = 2R_0^{1/2} + 2b^{1/2} \quad (8.24)$$

while for $R = R_2$ the integral on the right of Eq. 8.22 vanishes and the equation gives

$$\frac{2U_1}{k} = R_0^{1/2} + R_2^{1/2} \quad (8.25)$$

Thus from Eq. 8.24 and 8.25,

$$R_2^{1/2} = R_0^{1/2} + 2b^{1/2} \quad (8.26)$$

and

$$U_1 = k(R_1^{1/2} + b^{1/2}) \quad (8.27)$$
$$= V_1 + kb^{1/2} \quad (8.28)$$

From these results it is apparent that b is the increase

Table 8.9 Optimum growth of a spherical hailstone of initial radius R_0 (a very small value indicated by a zero), which is lifted by a constant updraft U_1 from the base to the top of a zone of supercooling of thickness Δz, before descending and attaining a final radius R_2 and fall speed V_2. Constant values are assumed for the density (0.9 g cm^{-3}) and efficiency of catch (0.8) of the hailstone, and for the concentration (5 g m^{-3}) of cloud water available for accretion; other conditions are specified in the text

	$\Delta z = 5$ km				$\Delta z = 3$ km			
R_0 (cm)	U_1 (m sec^{-1})	R_2 (cm)	V_2 (m sec^{-1})	V_2/U_1	U_1 (m sec^{-1})	R_2 (cm)	V_2 (m sec^{-1})	V_2/U_1
0	22.5	2.3	45	2	17.5	1.4	35	2
0.1	32	3.3	55	1.7	27	2.2	44	1.6
0.25	37.5	4.0	60	1.6	32.5	2.8	50	1.5
1	52.5	6.2	75	1.4	47.5	4.7	65	1.4

in radius of a particle which grows by accretion while falling from the top of the zone of supercooling when there is no updraft. With the values of the constants suggested above ($E = 0.8$, $m = 5$ g m^{-3}, $\rho_i = 0.9$ g cm^{-3}, $\Delta z = 5$ km) the value of b is about 0.5 cm. Table 8.9 lists some values of the growth from an embryo of radius R_0 in the updraft of constant speed U_1 which lifts the hailstone through the whole of the zone of supercooling, from which it eventually falls out with the radius R_2 and fall speed V_2 (the results are not particularly sensitive to the thickness of the zone of supercooling, as shown by the added values appropriate to a thickness of 3 km). Much of the increase in radius from R_0 to R_2 occurs in the upper part of the zone of supercooling; for example, in the uppermost 2 km of the zone 5 km deep the hailstone radius with the initial value 0.1 cm increases from 0.3 to 2.4 cm (and finally is 3.3 cm), while that with the initial value 1 cm increases from 1.6 to 5 cm (and finally is 6.2 cm).

When the embryo size is small the final fall speed $V_2 \approx 2U_1$, but as the embryo size increases the ratio V_2/U_1 decreases considerably (but is, of course, always greater than 1). By considering that the rate of increase of fall speed dV/dt is constant, it can readily be shown that the ratio V_2/U_1 obtained with the updraft constant is always greater than with an updraft which has a *maximum* speed of U_1 at a level L somewhere in the zone of supercooling (the ratio is 1 if L is at the base of the zone, and may be less than 1 if the hailstone begins to descend before reaching L). The condition $V_2 \approx 2U_1$ gives the largest possible hailstone which can be grown in an updraft of maximum speed U_1. It is not actually attainable because it arises when the embryo has a radius smaller than about 1 mm, in which circumstance the fall speed of the hailstone is overestimated by the formula used, partly because it is inappropriate at the smaller Reynolds numbers involved, but more seriously because the accreting particle then grows with a mean density much less than the assumed 0.9 g cm^{-3}. Moreover, when the hailstones reach giant size, even the values of 1.7–1.4 indicated for the ratio V_2/U_1 in Table 8.9 are probably not realizable, for the growth during descent through the lower part of the zone of supercooling is very wet, and it is likely that hailstones cannot retain all the cloud water which they collect there. For example, referring to the values quoted above for the growth of giant hailstones from embryos with initial radii R_0 of 0.1 and 1.0 cm, during their descent the radii at 3 km above the 0°C level, where the temperature is about -20°C, were, respectively, 2.4 and 5 cm (when the values of V_2/U_1 were 1.4 and 1.3, the final radii were 3.3 and 6.2 cm). However, according to Fig. 5.7 at the height where the temperature is -20°C the effective concentrations Em' of cloud water which could be frozen completely during accretion are, respectively, 1.5 and 1.0 g m^{-3}, if the hailstones had a smooth spherical shape, and about 2.0 g m^{-3} if they had a more natural, less regular shape. In the calculated growth the concentration Em was assumed to have the constant value 4.0 g m^{-3}; thus even at this level only half or less of the cloud water accreted could be frozen, and as the hailstones descended to lower levels this ratio steadily decreased (to zero at the 0°C level). The considerable calculated increases in size during this descent must therefore be regarded as unrealistic, and it appears that the maximum value of the ratio V_2/U_1 which could actually be attained is unlikely to exceed about 1.4. Accordingly, even without considering any melting that may have occurred below the 0°C level, the arrival at the ground of giant hailstones of diameter 5–10 cm implies that the updraft speeds in the parent zone of supercooling reached at least 30–50 m sec^{-1}. This is not surprising considering the stratifications characteristic of severe hailstorms and the great penetrations of their tops into the stratosphere; as already discussed these suggest that maximum updraft speeds of 70 m sec^{-1} or even more may be attained near the tropopause. Since the height of the -40°C level (the top of the zone of supercooling) is usually between 10 and 11 km, and since in the lower troposphere the buoyancy produced by adiabatic ascent is approximately proportional to height above cloud base, in the region of hail growth an updraft speed which increases nearly linearly with height (at a rate α between about 4 and 8 m sec^{-1} km^{-1}) is likely to be more typical of the severe hailstorm than one which is constant with height.

Before we begin to consider hailstone growth in such an updraft, however, it already appears that particles of the millimetric size required of the embryos of giant hailstones probably cannot be grown in air which ascends in the intense part of the updraft from the cloud base to the zone of supercooling, because of the brief time of this journey. Assuming that at the cloud base and in the lower part of the zone of supercooling, up to about 4 km higher, the updraft speeds are about 5 and 25 m sec^{-1}, respectively, this period is only about 300 sec. According to Eq. 8.15 this is hardly sufficient to grow a droplet of radius 1 mm even from a particle which already at the cloud base has a radius of 200 μ (assuming high values for the cloud water concentration and the efficiency of catch E). Possibly more rapid growth could occur in the forward part of the updraft where its speed (but probably also the cloud water concentration) is less. Again, reservations arise due to the extremely low concentration typical of giant hailstones (of order 10^{-3} m^{-3}; Fig. 8.3), and therefore of their embryos (implying an associated radar reflectivity factor Z_e of only about 10^{-1} mm^6 m^{-3} if they are droplets of radius 1 mm, and even less if they are small pellets of soft hail, but of course greater if the embryos are only a selected fraction of more abundant particles with a range of sizes). Nevertheless, the original conclusion on the difficulty of growing precipitation in very strong updrafts has some support from observations of the unusually great heights (and low temperatures) of the precipitation first detected by radar in the cumulus which grow to one side of the intense storm, and observed just above the tops of the echo-free vaults in the strongest storm updrafts. In the examples mentioned in the discussion of storms over Oklahoma in Section 8.8, these heights and temperatures were about 9 km and −30°C (Z_e about 3 mm^6 m^{-3}, implying that the largest particles were soft hail of radius about 1 mm and fall speed a few meters per second, in a concentration of order 1 m^{-3}), and 6 km, −5°C (Z_e < 40 mm^6 m^{-3}) to 10–12 km, −35 to −50°C (Z_e < 400 mm^6 m^{-3}). However, a more obvious source of embryos is the precipitation in the freshly growing flank clouds or in the forward part of the anvil cloud: much of this precipitation descends to reach the ground on one side of the storm as intense rain, but some enters the intense updraft in the middle troposphere and is swept aloft again, as discussed earlier in the chapter (and illustrated in Fig. 8.23, 8.26 and 8.27). The radar reflectivity factor Z_e in the lower part of the forward overhang commonly exceeds 10^4 mm^6 m^{-3}, and may even exceed 10^5 mm^6 m^{-3} (Fig. 8.23), so that it is inferred to contain ice particles of millimetric or even centimetric size, which all resume growth when they are lifted. However, it is unlikely that they were grown during a previous single ascent in the intense updraft, which if it continued to increase with height above the 0°C level provided a growth period for small cloud particles of even less than the 300 sec previously estimated, before the particles were carried above the −40°C level and subsequently into the anvil cloud. [Another observation which may be consistent with the difficulty of growing precipitation wholly within very strong updrafts was made on the occasion of the severe storms over Oklahoma referred to previously: the radar reflectivity factor Z_e in the anvils downwind of the storm tops decreased to 1 mm^6 m^{-3}, corresponding to the weakest detectable echo, at distances approaching 100 km in storms *I* and *J*, but at only about 20 km in the neighboring older and more intense storms *E* and *H* (191, pp. 188–90). The conclusion drawn was that the anvils of the latter storms contained comparatively fewer *small* precipitation particles (of fall speed < 1 m sec^{-1}). A value $Z_e = 1$ mm^6 m^{-3} can be provided by ice particles of low mean density and radius about 200 μ (fall speed about 0.5 m sec^{-1}), in a mass concentration of about 1 g m^{-3}. Particles of this size are probably about the largest that can be grown during a single ascent in a very intense updraft.]

It therefore seems that the source of the millimetric precipitation particles must ultimately be sought outside the most intense part of the updraft, probably on that flank of it where towering cumulus over the squall front successively renew the updraft, as indicated schematically in Fig. 8.27. The anvil cloud typically extends forward and somewhat toward the opposite side of the storm, consistent with other evidence that the newly generated precipitation may be carried not only forward of but also across the front of the storm, because either the actual or the relative wind it experiences has a component in this direction. Although the problem of the source of the hailstone embryos needs more study, it is reasonable to suppose that once recycling of precipitation has begun on the flank of the storm it continues, some of the particles moving progressively sideways, eventually to produce the overhang above the most intense updraft, and entering it as potential embryos with a range of sizes (it is also possible that once formed large or giant hailstones may fall through the updraft below the 0°C level, shedding potential embryos in the form of drips of melt water). By considering the growth of hailstone embryos in an updraft whose speed increases with height, it will be shown that a proportion of the recycling precipitation particles may reach giant size during only a few, or even only one passage through the intense part of updraft. (Some accounts of adventures in cumulonimbus, quoted in Appendix 8.13, suggest that balloonists and glider pilots have experienced the kind of recycling envisaged.)

It is assumed as before that the growth rate and fall speed of a precipitation particle, respectively, are given by the simplified relations

$$dR/dt = EmV/4\rho_i \quad (8.29)$$

and

$$V^2 = k^2 R \quad (8.20)$$

with constant values of ρ_i, k, and Em. The manner in which the particle reenters the updraft at a certain level and all variations in the horizontal are disregarded; the vertical speed U of the updraft in the zone of supercooling is assumed to increase steadily with the height z above some reference level:

$$U = \alpha z \qquad (8.30)$$

Then from the last three equations it follows that

$$\frac{d(U-V)}{dz} = \alpha - \frac{k^2 Em}{8\rho_i (U-V)} \qquad (8.31)$$

and therefore that if $(U - V)$ at the level where a precipitation particle reenters the updraft has the particular value $(U - V)^* = (k^2 Em/8\rho_i \alpha)$, it subsequently ascends steadily, the increase in its fall speed as it grows keeping pace with the speed of the surrounding updraft. With the values of the constants previously adopted, $(U - V)^*$ is about 12 m sec^{-1} for $\alpha = 4$ m sec^{-1} km^{-1} and about 6 m sec^{-1} for $\alpha = 8$ m sec^{-1} km^{-1}.

The reentering particles may be distinguished according to whether they have radii and fall speeds smaller or larger than certain values R^*, V^* which increase with the level of reentry (Fig. 8.55). The smaller accelerate as they rise; all arrive at the top of the zone of supercooling and cease growth (and can be regarded as possibly later again reentering the updraft). Those with the radius R^* just reach the top of the zone of supercooling before descending and eventually falling through its base with the largest size which can be attained for the particular level of reentry. Those initially larger begin their descent below the top of the zone and so attain a lesser final size. The largest possible hailstone is grown from the particle which reenters the updraft at the 0°C level with the appropriate radius R_0^*.

Figure 8.55 illustrates these kinds of growth, calculated after integrating Eq. 8.31 to give

$$x_1 - x_0 + a \log_e \left(\frac{x_1}{x_0}\right) = \alpha^2 (z_1 - z_0) \qquad (8.32)$$

where

$$U(z) - V(R) = \frac{x(z) + a}{\alpha}$$

$$a = \text{const} = \frac{k^2 Em}{8\rho_i}$$

and the subscripts 0 and 1 refer to values of the variables (x and z, and therefore also V and R) at the level where a particle enters the updraft and at some level it subsequently passes, respectively. The values used for the constants are those previously assumed ($k^2 = 9 \times 10^6$ cm sec^{-2}, $Em = 4$ g m^{-3}, and $\rho_i = 0.9$ g cm^{-3}, whence $a = 5$ cm sec^{-2}), and $\alpha = dU/dz$ has been chosen as 6 m sec^{-1} km^{-1}, with $U = 0$ at the ground; height and temperature are approximately consistent with adiabatic ascent and $\theta_s = 20°C$ (the base of the zone of super-

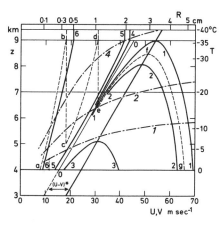

Fig. 8.55 A diagrammatic representation of the growth of hailstones under idealized conditions: the speed U of a steady updraft increases with height z at a constant rate α, here 6 m sec^{-1} km^{-1}, and the effective liquid water concentration Em in the zone of supercooling (here 4 g m^{-3}) is constant. Other simplifications introduced into the growth equations (8.29–8.32) are specified in the text.

The zone of supercooling in which the growth occurs is assumed to extend from 4 to 9 km; temperatures are shown on the right. The dot-dash lines labeled 1, 2, and 4 in italic figures show as a function of hailstone size (radius R or fall speed V) the levels at which liquid water in effective concentration Em' of 1, 2, and 4 g m^{-3} can be frozen as fast as it is accreted. These lines separate the dry- and the wet-growth regimes; they also imply the proportion of the collected cloud water which can be frozen, for example, the line 2 indicates that at a height of 7.3 km in the assumed growth conditions (with $Em = 4$ g m^{-3}) only half of the water collected by a hailstone of radius 4 cm can be frozen as it is accreted.

Various growth trajectories, numbered 0 to 6, result from the introduction of a hailstone embryo of a particular size (R_0, V_0 at the 0°C level) at various heights:

0: If $(U - V)$ has the particular value $(U - V)^*$, here 50/6 m sec^{-1}, it remains constant as the hailstone is lifted through the zone of supercooling.

1: If R_0 has a particular value such that $(U - V)_0$ is only about 1% less than $(U - V)^*$, the hailstone attains a fall speed equal to the speed of the updraft just before reaching the top of the zone of supercooling.

2: If $(U - V)_0$ is only a further 1% less than $(U - V)^*$, the final size at the 0°C level is considerably smaller.

3: If $(U - V)_0$ is 10% smaller than $(U - V)^*$, the growth is comparatively small, and so excessively wet that it is improbable.

4: If $(U - V)_0$ is greater than $(U - V)^*$, the hailstone is lifted above the top of the zone of supercooling and its growth ceases; for this growth trajectory $(U - V)_0$ is 2% more than $(U - V)^*$.

5: $(U - V)_0$ is 10% more than $(U - V)^*$.

6: Still smaller embryos are lifted rapidly through the zone of supercooling with little growth, and the growth is in the wet regime only near the 0°C level.

During the recycling of a hailstone through an updraft its growth occurs in successive stages; in this idealized model the growth trajectories are exemplified by the dashed line *abcdefg*, representing the growth of a hailstone which appears at the 0°C level as a small embryo and later reenters the updraft at successively higher levels after falling from the updraft outflow above the zone of supercooling.

cooling and of hailstone growth is at 4 km and its top is arbitrarily put at 9 km, where the temperature is near $-40°C$). For the chosen value of α, if $(U - V)^* = 50/6$ m sec^{-1}, particles rise steadily. A particle which enters the updraft at the 0°C level with a radius R_0^* such that $(U - V)$ is only about 1% smaller than $(U - V)^*$ attains a fall speed V equal to the speed of the updraft just before reaching the top of the zone of supercooling and eventually, on returning to the 0°C level, has the greatest size which can be reached during a single rise and fall in the assumed conditions (trajectory 1). Its fall speed V_2 is then 69.5 m sec^{-1}, while the maximum speed of the updraft U_1 in the zone of supercooling is 54 m sec^{-1}, so that the ratio (V_2/U_1) is nearly 1.3, not much less than can be realized in an updraft of constant speed (Table 8.9, $\Delta z = 5$ km). However, in that circumstance much of the later growth, and under the present conditions also the early growth, is excessively wet, and it is unlikely that all the cloud water collected could be retained. This can be seen in Fig. 8.55 by inspection of the isopleths (based on the mean values of the ventilation factor b in Fig. 5.3) giving approximately values of Em', the concentration of supercooled cloud water which can be frozen completely during its collection.

If $(U - V)$ initially is only a further 1% less than the equilibrium value $(U - V)^*$, that is, if the initial radius R_0 is only about 1% greater than the value R_0^* of nearly 3 mm required for optimum growth, then the particle does not rise so high, and the final size is considerably smaller, while the proportion of unfrozen water is still greater (trajectory 2 in Fig. 8.55). If $(U - V)$ initially is 10% smaller than $(U - V)^*$, that is, if R_0 is about 20% greater than R_0^*, then the final radius is only about 2 cm, compared with the optimum of over 5 cm, and the particle is at all times below the $-10°C$ level; the growth is so excessively wet that it is improbable that it could proceed at all as calculated, although it can be envisaged that shed drips of millimetric size might freeze and provide more suitable embryos for growth to large sizes.

The implication of these results is that if particles in a range of sizes enter the updraft, then those with a radius slightly larger than the value of R^* at the level of entry (some millimeters near the 0°C level, and 1 cm or more in the upper part of the zone of supercooling) may become large or giant hailstones, but the accreted ice contains a large proportion of unfrozen water, indeed, so much that the calculated growth may not be realizable.

Those particles which enter the updraft with a radius less than R^* cease growth when they reach the top of the zone of supercooling, with fall speeds less than the speed of the updraft there (54 m sec^{-1}). Such particles can descend only from some higher level where the updraft speed has diminished sufficiently; it can be supposed that it is primarily these particles which, after being moved aside in the outflow from the updraft, have the opportunity to reenter it at a lower level, in the manner indicated in Fig. 8.26 by the trajectory of a small particle. (It is reasonable to suppose that the smaller particles are carried farther horizontally in the outflow and therefore reenter the updraft at lower levels, although in a flow which is three dimensional and turbulent the particle trajectories must be more complicated, and particles may reenter at any level in a range of sizes; on the other hand many may not reenter at all, especially if they are small, being carried aloft beyond or to one side of the region of updraft at low levels.)

The larger of the particles under discussion attain radii of about 2 cm near the top of the zone of supercooling, and there nearly all the supercooled water which they collect can be frozen. However, if they grew during an uninterrupted ascent from a much lower level, the early stages of the growth would again have been excessively wet and probably not feasible. Thus not only on this account, but also if a layered structure of hard rimes is to be produced, such as is characteristic of large and giant hailstones which reach the ground, a cyclic growth has to be considered, in which the whole of the zone of supercooling may be traversed when the growing hailstone is small, but only its upper part may be traversed when the hailstone is large.

All the particles that enter the updraft with radii less than R^* reach the top of the zone of supercooling, and their growth is unlike those of larger size in that it is not as sensitive to the difference between $(U - V)^*$ and the initial value of $(U - V)$. This is shown in Fig. 8.55 by the trajectories numbered 4 and 5, representing the growth of particles which enter at the 0°C level with $(U - V)$ respectively 2 and 20% more than the equilibrium value $(U - V)^*$; these may be compared with trajectories 2 and 3, for $(U - V)$ respectively 2 and 10% less than $(U - V)^*$. However, particles which are comparatively small on entry into the updraft are lifted rapidly through the zone of supercooling with little growth, a shown by trajectory 6.

The dashed trajectory a-g represents the growth during a hypothetical recycling of an embryo (a small soft hailstone which might be at least partially melted) which enters the updraft at the 0°C level (height 4 km) with a fall speed of about 8 m sec^{-1}; it attains fall speeds of 18 and 31 m sec^{-1} after successive ascents before reaching a radius of nearly 3 cm at a height of 8.5 km during a third ascent and a final radius of nearly 5 cm at the 0°C level. The initial growth at millimetric radii, the renewed growth from radii of about 4–7 mm, and the final growth at radii beyond 4 cm are excessively wet (and the last probably not realizable). The hailstone might be expected to contain several layers of clear and cloudy ice, with transitions corresponding to the points where the trajectory in the diagram crosses either the line separating the dry and the wet growth regimes or the isotherm for an air temperature of about $-20°C$ (above which the ice deposited even in wet growth may contain small crystals and numerous small air bubbles). Clearly other hypothetical trajectories could be drawn to provide a different layer structure and a larger final size, and in particular to avoid periods of excessively wet growth, which seem not to have occurred in most large and giant hailstones collected from the ground. Whether such trajectories can actually occur must depend on the configuration of the updraft and the presence of a sufficiently large concentration of potential embryos, since the proportion that can closely follow the trajectory may be reduced during each reentry into the updraft. Such uncertainties, and the several simplifying assumptions made in the construction of the diagram, are unlikely to impair the following general inferences which can be drawn from it:

1. Of the hailstones which grow with not more than a small proportion of unfrozen water, the largest may have a fall speed about equal to the maximum speed of the updraft in the zone of supercooling.

2. In an updraft of the form typical of the severe storms of middle latitudes (in which the outflow from the updraft is predominantly above its inclined lower part), hailstones may make fresh ascents through the zone of supercooling after descending outside the cloud, or through the weaker forward parts of the updraft. Hailstone embryos of millimetric size probably enter the updraft near the $0°C$ level, but the later stages of the growth of large hard hailstones (when the radii exceed about 1 cm) must proceed mainly in the upper part of the zone of supercooling. Intermittently the growth may cease and the outermost layers may be chilled if the hailstones are lifted above the top of the zone.

3. During the ascents of a hard hailstone layers of spongy, clear, or transparent rime ice accreted in the lower or middle parts of the zone of supercooling are likely to be covered with layers of milky (opaque) rime ice deposited in its upper part.

4. The effective concentration Em of supercooled cloud droplets in the later stages of hailstone growth cannot exceed about 4 g m^{-3}. Since the mean radius of the cloud droplets is unlikely to exceed about 15 μ (see, e.g., Table 5.6), and the efficiency of catch E is therefore less than 0.8 for $R = 1$ cm and less than 0.6 for $R = 5$ cm (Fig. 2.7), this is not inconsistent with the presence of concentrations closely approaching the values of 4–7 g m^{-3} corresponding to adiabatic ascent of air from the cloud base (Fig. 3.9). Values less than the adiabatic may arise not only by mixing with other air, but by depletion due to the growth of hail, an effect whose estimation would require a detailed specification of the distribution of hail in the updraft.

5. A hailstone which descends through the updraft in the zone of supercooling, perhaps after recycling in the earlier stage of its growth, accretes a layer containing a high proportion of unfrozen water, especially in the lower part of the zone of supercooling, where it is unlikely that the layer could adhere to the hailstone. Drips of millimetric size from such hailstones could provide embryos which eventually grow into larger and more completely frozen hailstones. Some consolidation of layers of spongy ice may occur if hailstones are lifted above the top of the zone of supercooling (e.g., trajectory 4 of Fig. 8.55) or descend outside cloud through air with a wet-bulb temperature below $0°C$. It will appear (in the later discussion concerned mainly with the melting of hailstones during fall) that only thin layers can be consolidated in this way, but the freezing of an outer crust, which is not later completely melted at low levels, may allow the arrival at the ground of the rare large hailstones which contain so much unfrozen liquid that they are observed to disintegrate or splash on impact. In the lower part of the zone of supercooling there may be many hailstones of moderate size which contain a high proportion of unfrozen liquid, which melt and disintegrate before reaching the ground.

When the growth of hailstones is considered in a two-dimensional representation of an updraft it becomes apparent that particles entering the updraft at the same place but with different sizes are liable not only to attain different sizes and structures but also to take strikingly different paths through the storm. In particular, in the kind of updraft configuration drawn in Fig. 8.26 and represented as typical of the intense storms of middle latitudes, the recycling of particles seems unlikely to occur unless they remain on the forward side of the streamlines passing through the region of maximum updraft speed. On the other hand, small particles which after entering the front of the updraft cross this streamline and move toward the rear of the storm are likely to behave like those whose paths in Fig. 8.55 lie to the right of the equilibrium path 00. Probably they grow into hailstones which contain comparatively high proportions of unfrozen water and are liable to melt or disintegrate aloft; such hailstones or their residues reach the ground farther toward the rear of the storm. This kind of separation is indicated schematically in Fig. 8.26 and in more detail in a study (249) which attempts to trace the growth history of individual hailstones collected at the ground during the storm on which the diagram is based. The separation of particle paths occurs also normal to the direction of travel of the storm, because of the varying cross-component of the air velocity which the particles experience, leading to the skewed distribution of hail size and rainfall at the ground which is illustrated in Fig. 8.21 and 8.51. Large or giant hailstones may be the first or even the only precipitation experienced locally in some severe storms, as in the great hailfall at Potter (Appendix 8.12). Frequently observers of such falls have the impression that most of the hailstones have nearly the same size, and this is sometimes supported by photographs of hail lying on the ground immediately after the storm. The tendency of hailstones to become sorted according to size makes it difficult to interpret the size distribution spectra of hailstones collected from individual places after the passage of a storm (discussed in references 218 and 250) or to infer from them the spectra that exist aloft within storms, which would be useful in the study of the distribution of radar reflectivity within their radar echoes.

In investigations immediately after the severe hailstorms of 5 September 1958 and 9 July 1959 over southeast England, estimates were made (from a variety of evidence) of the average separation of the largest hailstones which fell at a number of places and of the duration of their fall, and hence also of their number and mass concentrations in the air near the ground. The number concentration N is displayed in Fig. 8.56 as a function of hailstone size; the individual values lie near a line which represents a distribution law ($\log N_D = 1.4 - D$ if N_D m^{-3} is regarded as the concentration of all hailstones of diameter D cm or more) of the same logarithmic form as that previously quoted for raindrops (Eq. 8.5), but with a very different slope. However, the observations do not provide a real size distribution, and will later be used only to estimate typical maximum concentrations in the cloud for small and for large hailstones, after considering the degree of melting likely to have occurred during their fall to the ground.

According to the previous discussion the production of large or giant hailstones in cumulonimbus depends upon two properties of the updraft:

1. A speed in the upper part of the zone of supercooling which reaches values about equal to the fall speeds of such hailstones (30–40 m sec^{-1} for large hailstones and 40–70 m sec^{-1} for giant hailstones).

2. A configuration which allows millimetric embryos, grown in a weaker part of the updraft, to fall into the intense part of the updraft in the zone of supercooling or not much below its base, so that they can be lifted through it again and continue their growth. Hailstones

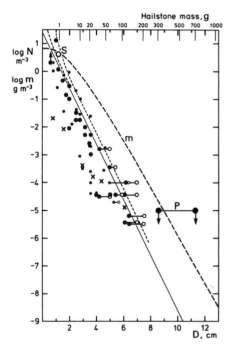

Fig. 8.56 The number concentration N in the air near the ground of the *largest* hailstones, of diameter approximately D, in several hailfalls (inferred from the estimated duration of the fall of the hailstones and their final average separation at the ground). Most values were obtained from places in southeast England affected by the severe hailstorms of 5 September 1958 (large circles) and 9 July 1959 (small circles). The hailstones whose largest diameters (open circles) reached several centimeters were distinctly aspherical, and the linked full circles show the estimated diameters of spheres of the same masses. The largest circles linked by line P represent values for the masses of hailstones which fell in a storm at Potter, Nebraska (Appendix 8.12); as indicated by the arrows, they are probably overestimates (because the reported average separation probably referred to somewhat smaller hailstones). The circle marked S represents a value for the extraordinary hailfall at Selden, Kansas (Appendix 8.14).

Individual values lie near the line $\log N = 1.4 - D$, which represents a size distribution law if N m^{-3} is regarded as the concentration of all hailstones of diameter D cm or more [the corresponding law for the mass concentration (m, g m^{-3}) is shown by the thick dashed line]. However, the law does not refer to any local population of hailstones simultaneously present in the air, and the diagram is used only to draw the inferences below, which are italicized. (Crosses mark points representing a kind of size spectrum averaged over the duration of a hailfall at a single place, inferred from a photograph of hailstones lying on a lawn after the storm of 5 September 1958. The circles can probably be regarded as the extreme right-hand points of a number of such "spectra," implying that concentrations of large or giant hailstones in a particular size range are greatest when they are the largest present.) Lines drawn to represent the observations are extrapolated to emphasize *the extremely small number and mass concentrations implied for giant hailstones*, especially for the largest reliably recorded.

In the warm weather in which severe hailstorms occur there is some melting of the hailstones before they reach the ground. Maximum size concentrations aloft, before melting, which correspond approximately to the solid straight line representing values near the ground, are shown by the thin dashed lines, for a small amount of melting (during fall in a downdraft) and a moderate amount of melting (during fall through clear air with no motion). They imply that *the mass concentrations aloft of hailstones of diameter about 1 cm reach values of order 1 g m^{-3}*.

which are hard, containing little unfrozen water, are more readily grown if the configuration of the updraft is such that they can be recycled through the updraft a few times at successively higher levels.

Evidently these conditions are always likely to be fulfilled in the typical intense storm of middle latitudes, in which the inflow to the updraft and its outflow aloft are both on the forward side of the storm (Fig. 8.26). Nevertheless, there is some evidence that in particular intense storms the fall of large and giant hail has ceased after an early phase. For example, the severe storm E of Fig. 8.48 (over Oklahoma) produced giant hail for only about 1 hr after its rapid intensification; it then remained intense for 2 hr, judging by the persistence of a radar echo with a hook and with a top at about 18 km, and the production of heavy rain, wind damage, and several tornadoes, but at the ground there were no hailstones with a diameter of more than about 2.5 cm. The largest hailstones (diameter about 8 cm) fell over a period of not more than about 15 min during the intensification. It appeared that this storm and its neighbors G and H each produced a swath of giant hail about 20 km long during the phase of intensification associated with the development of the hook echo and the veering in the direction of storm travel. Moreover, the heaviest rain began (east of the swath of largest hail) toward the end of the fall of the giant hail, and the damaging winds and tornadoes occurred after it.

This behavior was noted also in southeast England in the severe storm of 5 September 1958, and the storm of 9 July 1959 in the same region produced its largest hail during only about 30 min following its intensification, whereas lesser hail, equally heavy rainfall, and equally high radar echo tops were observed for a further 2 hr. It is not known whether this kind of behavior is more general, and therefore whether the occurrence of very long swaths of large or giant hail depends upon the continual growth of fresh storms to one side of their predecessors. The evidence is sufficient to suggest, however, that some small change in the concentrations of hailstone embryos, or in the configuration of the storm updraft, may make it impossible to produce hailstones of the optimum size corresponding to the intensity of the updraft.

Within the tropics the occurrence of small hail is more common than would appear from most climatological summaries, particularly over high ground, but falls of large hailstones are very rare and giant hailstones are practically unknown (101). This may be attributed partly to the prevalence of ordinary cumulonimbus, and therefore to a stratification in which the available potential energy is not large, and partly to the usual absence of a wind shear in which the cumulonimbus can be efficiently organized, as seen in Fig. 8.14. Probably, therefore, the cumulonimbus updrafts are impulsive, rather than persistent, and ordinarily attain speeds in the range 10–20 m sec^{-1}, insufficient for the growth of large hail. For the same reason frequently in middle latitudes only small hail and weak squalls are experienced in thunderstorms with heavy rain, even when the available potential energy may be large (as illustrated in the first column of Table 8.5.) On the other hand, when tropical cumulonimbus are efficiently organized into traveling squall storms and higher updraft speeds are likely to be attained, the configuration of the updraft differs from that typical of middle latitudes in that the outflow from the updraft lies to the rear of the storm, and over the downdraft rather than the foot of the updraft (Fig. 8.29). This configuration is unfavorable for the recycling of embryo hailstones and therefore for their growth to a large size. The most striking evidence of the importance of wind shear in organizing the convection, and in arranging a configuration favorable for the growth of large hail, is the cessation of severe hailstorms in northern India during the period of the summer monsoon, when the westerly jet streams in the high troposphere are displaced to regions north of the Himalayas.

The consolidation of hail with an outer layer of spongy rime

Consideration of the heat economy of a hailstone which is not growing by accretion is required in two kinds of circumstance: (a) when some of its internal liquid may be frozen while it is either in the cloud above the zone of supercooling or falling through clear air with a wet-bulb temperature below 0°C, and (b) when melting occurs during fall through cloud or air with a wet-bulb temperature above 0°C. The formulation is comparatively simple in the second circumstance, because the surface temperature of a melting hailstone can be assumed to be constant and 0°C. However, in (a) no such simplification is possible, and the heat transfer from the hailstone to the air involves both the internal transfer and the rapid change of air temperature and humidity which the hailstone experiences during its fall. Nevertheless, a simple estimate of the magnitude of the heat transfer shows that it is insufficient to freeze more than a small quantity of liquid inside a large hailstone, mainly because of the small time which is spent in falling several kilometers through clear air or in the cloud above the zone of supercooling.

From Eq. 5.48, 5.49, and 5.54, the rate of transfer $d(H + H_L)/dt$ of sensible and latent heat from the air to a falling hailstone is

$$d(H + H_L)/dt = 4\pi Rb(\mathrm{Re})^{1/2}(K\,\Delta T + DL_v\,\Delta\rho) \quad (8.33)$$

where ΔT and $\Delta\rho$ are given the appropriate signs. In considering the heat loss of a large hailstone, with an outer layer of spongy ice, which falls from the outflow of an updraft through still, clear air, it may be assumed that the surface temperature is initially 0°C and the air temperature about −40°C. The temperature difference soon decreases, and continues to diminish steadily, eventually becoming zero some distance above the 0°C level. The transfer of latent heat is probably small compared with that of sensible heat, and the total transfer during fall to the 0°C level is likely to be overestimated by neglecting $\Delta \rho$, assuming a very high conductivity inside the hailstone, and using an average value of 20°C for ΔT, so that

$$H + H_L \approx \frac{-80\pi R b (\text{Re})^{1/2} K \Delta z}{V} \quad (8.34)$$

where Δz, the distance fallen, is about 5 km. If the loss of heat is sufficient to freeze completely a surface layer of thickness ΔR in which initially ice and liquid water were present in approximately equal proportions, then

$$2\pi R^2 \rho_w L_f \Delta R \approx \frac{80\pi R b (\text{Re})^{1/2} K \Delta z}{V}$$

or

$$\Delta R \approx \frac{40 b (\text{Re})^{1/2} K \Delta z}{V \rho_w L_f R} \quad (8.35)$$

With $\rho_w = 1$ g cm^{-3}, $L_f = 80$ cal g^{-1}, and the approximate values $K = 5 \times 10^{-5}$ cal cm^{-1} sec^{-1} °C^{-1}, $\Delta z = 5$ km, then introducing

$$R = 1 \text{ cm}, \quad V = 25 \text{ m sec}^{-1}, \quad \text{Re} = 10^4, \quad b = 0.3$$

and

$$R = 5 \text{ cm}, \quad V = 60 \text{ m sec}^{-1}, \quad \text{Re} = 10^5, \quad b = 0.6$$

Eq. 8.35 gives for ΔR values of about 1.5 and 1 mm, respectively. Evidently, then, only a thin outer crust of a hailstone composed of spongy ice is likely to be consolidated in the few minutes required to fall through the upper troposphere. Since a hailstone might be lifted as much as 2 km above the zone of supercooling and therefore spend a similar period at air temperatures below −40°C, a much greater degree of consolidation might be produced in this way, but then the heat transfer during most of the free fall to the melting level would be reversed. Moreover, since the subsequent fall below the melting level occupies a comparable period, a considerable part of any consolidated layer is likely to melt before a hailstone reaches the ground. It appears therefore that if during the growth of a large hailstone a high proportion of the accreted water is unfrozen, then it can be expected still to be present inside the hailstone when it reaches the ground. The consolidation of a thin outer layer high in the cloud or outside it may be sufficient only to allow the hailstone to reach the ground with the size it attained in the cloud.

The melting of hailstones

Of the variety of conditions under which hailstones may melt during their fall through the lower troposphere, the following three deserve particular attention: the hailstones fall through clear air with no significant vertical motion; large hailstones fall through a cloud with an updraft; and hailstones fall in a downdraft.

Small internal transfers of heat are neglected, and the surface temperature of a melting hailstone is considered to be 0°C, since until the hailstone has been reduced to millimetric size the film of melt water is so thin (only about 10^{-1} mm, even on a stationary hailstone; 251) that its presence can be ignored (252). The rate of change of radius of a hailstone falling through unsaturated air is given by Eq. 8.33:

$$4\pi R^2 \rho_i L_f dR/dt = -4\pi R b (\text{Re})^{1/2} (K \Delta T + D L_v \Delta \rho) \quad (8.36)$$

where ΔT and $\Delta \rho$ are now the differences between the air temperature and 0°C, and between the vapor density in the air and the saturated vapor density at 0°C, generally also positive. Further,

$$dz/dt = U(z) - V \quad (8.37)$$

To a sufficient degree of accuracy suitable mean values can be used for the drag coefficient ($C_D = 0.6$) and density ($\rho_i = 0.92$ g cm^{-3}) of hailstones, and for the density and dynamic viscosity of the air (corresponding to a pressure of about 800 mb and a temperature of 10°C), so that

$$V^2 = 4 \times 10^6 R \quad \text{and} \quad (\text{Re})^{1/2} = 3.3(RV)^{1/2} = 150 R^{3/4}$$

For the first circumstance listed above, $U(z) = 0$; moreover, noting that for $(\text{Re}) < 10^4$ $b \approx \text{const} \approx 0.3$, and anticipating that for $(\text{Re}) > 10^4$ its value will hardly change during the melting, so that b can always be regarded as approximately a constant for a given initial radius of a hailstone, Eq. 8.36 leads to

$$R^{3/4} dR = \left(7.5 \times \frac{10^{-2} b}{\rho_i L_f}\right)(K \Delta T + D L_v \Delta \rho) dz$$

or

$$R_1^{7/4} - R_0^{7/4} = \frac{0.13 b}{\rho_i L_f} \int_{z_0}^{z_1} (K \Delta T + D L_v \Delta \rho) dz \quad (8.38)$$

where R_0 is the radius at the height z_0 where the wet-bulb temperature is 0°C and melting begins, and R_1 is the radius at the height z_1 of the ground. The integral on the right may be evaluated numerically for a given stratification; on occasions of severe hailstorms a wet-bulb temperature of 0°C is usually found about 3 km above the ground (corresponding to $\theta_w \approx 16$°C in the dry air above the moist layer), and the value of the term on the right of Eq. 8.38 is then about $-b$ cm$^{7/4}$. For

Table 8.10 The melting of a hailstone during fall through air without significant vertical motion, in a stratification typical of severe hailstorms; the radius R_1 at the ground after melting commenced at a height of about 3 km with the radius R_0, the mass m of melt water, and the corresponding number N of droplets of radius 1 mm

R_1 (cm)	R_0 (cm)	m (g)	N
(0)	(0.5)	0.4	100
0.5	0.8	1.1	270
1.0	1.2	2.3	550
2.0	2.2	7.5	1,800
3.0	3.2	22	5,000
4.0	4.2	40	9,000

example, according to representative soundings for some severe hailstorms over southeast England (Fig. 8.25) and Oklahoma it was $-0.9b$ on each occasion. Accepting this as a typical value Table 8.10 shows the size of hard hailstones aloft in relation to their size on reaching the ground. It appears that hailstones which aloft have a diameter less than about 1 cm are likely to melt completely before reaching the ground, but that large and giant hailstones lose only a surface layer about 2 mm thick. However, this would generate a large number of raindrops in the wake of big hailstones, and when, as occasionally happens, they arrive at the ground apparently without any rain it is possible that they have fallen through an updraft which carried away most of the shed melt water.

In this circumstance Eq. 8.36 and 8.37 have to be integrated numerically. It has been supposed that the hailstones fall through a cloud with $\theta_s = 24°C$ (0°C level at 5.3 km) and a base at 2.3 km (1.9 km above screen level, where θ_w is 24°C and the temperature is 34°C), and that the updraft U increases linearly with height above the ground at the rate $\alpha = 6$ m sec^{-1} km^{-1}, so that only large or giant hailstones with an initial fall speed of more than about 30 m sec^{-1} can fall through it. The degree of melting is increased slightly by assuming that captured cloud water is chilled to 0°C before it is shed. Then during the fall of a hard hailstone of initial radius 3 cm an outer coat of thickness 4 mm is melted, that is, about twice as much as during fall through clear air with no vertical motion. If the hailstone consists of spongy ice containing 50% unfrozen water, the decrease of radius is more than doubled, amounting to 1 cm. It can be concluded that only an outer coat a few millimeters thick has ever been melted away from giant hard hailstones which reach the ground. On the other hand, even large hailstones which have grown excessively wet, as shown by trajectory 3 in Fig. 8.55, cannot descend through a storm updraft without suffering much melting and probably also disintegration.

It remains to consider the third circumstance, in which high concentrations of hailstones leave the cloud and descend in a downdraft maintained by the chilling due to the evaporation and melting of the precipitation. The problem of inferring the temperature and relative humidity in the downdraft will be avoided by using the experience of forecasters in the United States; according to one study (253) the screen temperature in severe thunderstorms falls to about the value which on an aerological diagram lies on the saturated adiabatic through the point where the wet-bulb temperature on the appropriate sounding is 0°C. This is almost equivalent to the statement that the downdraft air is nearly saturated with θ_w equal to the value of about 16°C typically found on these occasions in the lower middle troposphere. This result is surprising in light of the previous conclusion that in only moderate downdrafts the evaporation of even intense rain cannot maintain a state of saturation. Moreover, a later study (254) found closer agreement between the screen-level wet-bulb temperature following a storm and a value of θ_w midway between that in the middle troposphere and that in the lowest kilometer away from the storm. However, for the purposes of a calculation conditions in a downdraft were chosen in which the air is saturated with $\theta_s = 16.5°C$ between the ground and a level 3 km higher where the temperature is 0°C. The shallowness and low temperature in this melting layer are about appropriate to the occasions of severe storms over southeast England and probably represent conditions more unfavorable for hailstone melting than are typical in the severe storms over the United States. Equations 8.36 and 8.37 were integrated numerically, assuming a downdraft of 10 m sec^{-1}, decreasing to zero in the lowest kilometer. The results show that a surface layer less than 1 mm thick melts away from a hard hailstone of initial radius 1 cm, as might be anticipated from Table 8.10. If the initial radius is 0.5 cm, the radius at the ground is about 0.4 cm, but over half of the mass of hailstone has melted. If a hailstone of this size consisted of spongy ice containing more than 50% unfrozen liquid, the melting would be practically complete before the ground was reached. This will prove interesting in the later discussion of the radar reflectivity and maximum rainfall intensity in severe hailstorms.

According to these calculations the chilling of air in downdrafts by the melting of hailstones, which was neglected in the discussion of downdraft production in Section 8.10, is usually small. Thus the melting of half of a mass concentration of several grams per cubic meter of ice in the form of hailstones of diameter 1 cm produces a decrease of the (potential) temperature of the air in a downdraft by about 1°C. This chilling is considerably less than that due to the evaporation of precipitation, providing a substantial fraction is evaporated before it reaches the ground. Even assuming that all the precipi-

8 Cumulonimbus convection

tation above the 0°C level is frozen and melts completely before reaching the ground, the contributions to the chilling by melting and by evaporation are equal only if the fraction of the precipitation aloft which is evaporated is as small as L_f/L_v, or about 1/8, whereas a typical *inferred* value in severe storms is nearly 1/2 (as shown, e.g., in Table 8.12). During the melting the surface temperature of the hail is close to 0°C, and the rate of evaporation previously *calculated* by assuming the precipitation particles to have the wet-bulb temperature of the air may therefore have been considerably overestimated over most of the layer below the 0°C level. The presence of a large proportion of frozen particles in precipitation appears to be unfavorable for the production of a strong downdraft.

The rarity of hail in low latitudes

Careful analysis of personal experience and unofficial reports shows that within the tropics small hail reaches the ground from thunderstorms more commonly than official statistics suggest, but large hail is exceptional (101). It was formerly thought that this might be due to a greater degree of melting of falling hail in warm climates, but the previous discussion shows that it must indicate that large hailstones are hardly ever grown in tropical cumulonimbus.

In the description of some severe storms over Oklahoma it was mentioned that there seems to be a tendency for large and giant hail to be produced in intense cumulonimbus there and elsewhere during an early phase, before a later phase characterized by more intense rainfall. Since in the later phase the storm tops have been equally high, updrafts of the speed required for the formation of large hail must be presumed still to have been present. The disappearance of the large hail, if indeed typical, must therefore be sought in some developing peculiarity of the configuration of the updraft, or in a progressive evolution of the size distribution of the precipitation particles which leads the depletion of cloud water by increased concentrations of small hail, leaving too little time available for the growth of large hail.

Such developments seem unlikely to provide an explanation for the rarity of large hail in the tropics; rather, they probably indicate the prevalence of unorganized cumulonimbus, which as in other latitudes have updrafts sufficiently intense and sustained to produce only small hail. Nevertheless, on the basis of the strength of their squalls at the ground and the long paths over which they are frequently observed to travel, it has been implied that in some tropical thunderstorms, in particular the West African disturbance lines, the cumulonimbus convection is well organized, prolonged, and intense. Large hail is conspicuously rare also in these storms. However, it may be important that the configuration of the flow through their updrafts, represented schematically in Fig. 8.29, is strikingly different from that in middle-latitude storms: the outflow of the updraft, instead of lying forward of the storm and at least partly over the updraft at lower levels, trails behind the storm and over the downdraft. Thus whereas in the middle-latitude storm the configuration allows the repeated cycling of some precipitation particles through the updraft, in the tropical storm the particles falling out of the updraft mostly enter the downdraft and do not have the same opportunity for successive periods of growth to a large size.

It can also be remarked that the squall storms of low latitudes occur in situations where the wind shear between the levels of inflow and outflow of the updraft is small, and less than that in the lower troposphere. It may then be that it is the downdraft, rather than the updraft, which is better organized than in ordinary cumulonimbus. The updraft in the upper troposphere may be impulsive, attaining maximum speeds and durations less than are required for the growth of large hail, even given a suitable updraft configuration.

8.12 The precipitation intensity and radar reflectivity in cumulonimbus

In the cumulonimbus motion system the important heat source is the condensation of water vapor (in the updraft) upon particles sufficiently small to be regarded as moving with the air. Although there is also (elsewhere) evaporation from such particles, the most important heat sink is probably that associated with evaporation from the more intense precipitation. The paths of the precipitation particles, which differ essentially from the air trajectories, and the magnitude and disposition of the associated heat sink relative to the heat source are important features of the organization of the cumulonimbus which are very difficult to observe directly. Of the indirect observations, those of precipitation at the ground, in particular of its local maximum rate, and of its areal extent, rate, and duration, are indicators of the intensity and scale of the cumulonimbus circulation, besides having a more immediate hydrological interest. Observations of the magnitude and distribution of radar reflectivity provide similar if less definite information and are at least potentially useful in forecasting weather and river behavior, since the information is available immediately for the whole of a large area, rather than from a scattered network of observing stations after a delay of hours or even days. Moreover, even the observations made with conventional radar allow inferences to be made about the nature and concentration of precipitation aloft, and, as already discussed, about the general form of the motion of the air and of the precipitation.

The amount, duration, and intensity of precipitation from cumulonimbus

From studies made in Florida and Ohio (255) it appears that ordinary cumulonimbus produce rain at the ground during a period of about 30 min (a similar period is implied by the duration of the radar echoes whose paths over southern England are shown in Fig. 8.18). The average area instantaneously occupied by rain at the ground was about 25 km^2, about an order of magnitude greater than that corresponding to the average dimension of about 1.5 km found by aircraft for the width of the updraft near the level of the cloud base. The average duration of rain at a point on the ground was also about 30 min (implying that the average speed of travel of the cumulonimbus was only several meters per second); the maximum rate of rainfall over a period of 5 min (corresponding to a space scale of about 1 km) reached 25 mm hr^{-1} frequently and about 100 mm hr^{-1} occasionally. Each of these rates is substantially less than the average upward flux of water vapor through the cloud base, equivalent (with a temperature of about 18°C and an observed updraft speed of 5 m sec^{-1}) to a rainfall rate of about 250 mm hr^{-1}.

The discrepancy is due mainly to the spread of the rainfall over an area larger than that covered by the updraft, but it is also due partly to the evaporation in the atmosphere of a fraction of the water condensed in the updraft. Attempts have been made to obtain values of this fraction by comparing an average rate of rainfall at the ground R with an estimate of the flux of water vapor into storm updrafts C. The former, made often from the records of special networks of rain gauges supplemented by radar observations, may well have been inaccurate, but not to the same degree as the latter (based upon comparatively sparse data from radiosonde soundings or measurements made near the cloud bases with aircraft), which are liable to be in error by a factor of two or even more. Accordingly, it is perhaps encouraging if the values inferred for the ratio R/C only infrequently exceed 1, as appears from those listed in Table 8.11. From an inspection of this table it seems that a typical value for the ratio in a well-organized storm with a pronounced downdraft is about 0.5. (Lesser values must be associated with many ordinary cumulo-

Table 8.11 Some estimated values of the rate R of precipitation at the ground from cumulonimbus, the flux C of water vapor into their updrafts, and of the ratio R/C. R and C in units of 10^9 g sec^{-1}

Kind of storm	R	C	R/C
Ordinary cumulonimbus, Florida and Ohio; no hail, or only small hail (255)[a]	0.1	0.5	0.2
Mature traveling (persistent) storms over Colorado (256)[b]			
With no reports of hail	3.3	3.4	1.0
	1.8	1.5	1.2
	0.4	0.7	0.6
With reports of small hail	1.5	2.7	0.6
	2.5	4.8	0.5
With reports of giant hail	1.3	2.3	0.6
	1.0	4.8	0.2
	2.0	5.8	0.3
Severe local storm over Oklahoma (257)[c]	4.7	8.8	0.5
Storm of 9 July 1959 over southern England, during intense phase with large hail[d]	8.5	12	0.7
Storm of 5 September 1958 over southeast England, during 2 hr after a phase with giant hail[e]	12	—	—

[a] The quoted values are averages over the duration of the storm (about 30 min). That for C might be reduced in other interpretations of the original data, in particular to 0.3 or less if as elsewhere it is intended to refer to the flux into the cloud bases.
[b] Dimensions of updraft near cloud bases determined by aircraft reconnaissance, that across the direction of travel L varying from about 10 to 25 km.
[c] Dimension L estimated from radar data as 20 km.
[d] Dimension L estimated from disposition of squall front at the ground to be between 20 and 30 km, and from apparent length of echo-free vault (15 km) to be equivalent of 20 km or more farther ahead of storm, considering component of wind from the right near the storm. Value adopted: 24 km.
[e] Estimated from records in the region of greatest total rainfall, where its unusual duration (of up to 2 hr in places) resulted from the arrival of a second storm, which had previously formed near the right flank of the first storm toward the end of the phase in which it produced giant hail.

nimbus, especially if they are small and have high bases, and might conceivably also arise if the speed of a cumulonimbus updraft, at least in its central parts, increased to such an extreme that hardly any aggregation of the cloud particles could occur before they were carried away in the anvil cloud. Considering also the previously mentioned observations that in some severe storms the greatest rainfalls followed an early phase in which giant hailstones occurred, there may be some significance in the tendency for the values of the ratio R/C entered in the middle of the table to decrease with the reported size of the largest hailstones. The aircraft observations quoted in Section 8.3 showed that in a number of anvil clouds explored, few of which are likely

8   Cumulonimbus convection

Table 8.12 Values of the fluxes of air and water substance associated with a severe storm over Oklahoma, estimated from special radiosonde and surface observations (257). The unit of flux is 10^9 g sec^{-1}

Flux of air in updraft	700
Flux of air in downdraft	400
Flux of vapor, updraft inflow	8.8
Flux of water, updraft outflow	1.2
(Precipitation rate aloft	7.6)
Flux of vapor, downdraft inflow	0.7
Flux of vapor, downdraft outflow	4.3
(Rate of evaporation into downdraft	3.6)
(Precipitation rate at ground	4.0)
Precipitation rate at ground, observed	4.7

to have been associated with intense cumulonimbus, a typical value for the average concentration of ice particles was a little more than 1 g m^{-3}, corresponding to a mixing ratio of over 2 g kg^{-1}. This is a considerable fraction of the mixing ratio of water vapor at the cloud bases in warm weather, which varies between about 10 and 15 g kg^{-1}.)

Approximate values of the various fluxes of air and water vapor for the storm over Oklahoma which is listed in Table 8.11 are probably representative of those in intense cumulonimbus. They are based on a simple model of the storm and radiosonde data, with assumed values of 20 km for the crosswind dimension of the drafts and 0.3 g m^{-3} for the concentration of ice particles in the anvil cloud. The estimated values of the fluxes and the actual average rate of rainfall (over a special network of observing stations) are given in Table 8.12.

The value of about 5×10^9 g sec^{-1} which appears in Table 8.11 as the magnitude of the areal rate of rainfall from a severe storm represents a rate of introduction of sensible heat into the upper troposphere which is very large, equivalent to the energy released by the explosion of a 1-megaton nuclear bomb every few minutes. The persistence of such storms over periods as long as several hours indicates that the scale of their circulations is much greater than would seem from the small areas which experience violent weather. After several hours, for example, the sensible heat released has the normal magnitude of the net radiative loss from the upper troposphere over an area of about (100 km)^2 of the same magnitude as the area covered by the outflow of the updraft.

The precipitation from a severe storm reaches the ground instantaneously over an area of as much as 10^3 km^2 or even more. In consequence of this great spread of the rainfall its average areal rate is unlikely to exceed 10 mm hr^{-1}. During the intense phase of the storm of 9 July 1959 over southern England precipitation reached the ground over a path about 100 km wide. Figure 8.24 shows the estimated typical fall of rain and the maximum hailstone size as a function of position in the path. The rainfall averaged over the whole path is only 4.7 mm; the maximum rainfall is about 20 mm, in a position where small hailstones fell, several kilometers to the left of places which experienced the largest hailstones, on the right flank of the storm. This maximum point rainfall is rather small for a severe storm, and is associated with the unusually high speed of travel of the storm. The total duration of the precipitation near this position was about 15 min, but most of it fell in only several minutes, implying that over this shorter period (corresponding to a distance of about 5 km in the direction of motion of the storm) the average rainfall rate exceeded 100 mm hr^{-1}. (The automatic rain gauges in England, as in most countries, rarely allow a rate to be evaluated over a period of less than 10 min; on this occasion the highest rates estimated from several records were 90–100 mm hr^{-1} over periods of 9–15 min, and one eye observation was made of 120 mm hr^{-1} over 10 min.)

In southern England rainfall rates between 100 and 150 mm hr^{-1} over a period of 10 min are recorded too frequently to be attributed only to severe persistent storms, the probability of occurrence at a point being once every few decades. The rate of 240 mm hr^{-1} listed in Table 8.5 is an example of the high values which may be attained below ordinary cumulonimbus (in this instance the tallest member of a large group). This particular value is not only twice as great as the maxima which occurred in the storm of 9 July 1959 but is probably about as great as any which occurred in the exceptionally severe storm over the same region on 5 September 1958. Among the most notable falls on that occasion were 65 mm in about 20 min, 50 mm in 30 min, and 130 mm in 2 hr. These are all classified as *very rare*, since their long duration has a probability of occurrence at a point less than once in 150 years. Thus it appears that in the persistent severe storm the maximum rainfall rate is not so much greater than in the ordinary cumulonimbus, but is maintained longer and leads to a higher total at individual places. This difference of character, indicative of the greater scale of the severe storm, is of course even more evident from the area over which the precipitation reaches the ground, and its total rate, which exceed the values associated with the ordinary cumulonimbus by up to about two orders of magnitude. Particularly striking among the persistent storms are those which move only very slowly or are practically stationary. In these storms the precipitation, although not especially intense, accumulates to phenomenal amounts, as described in Appendix 8.14.

In the United States the largest recorded falls over periods of 5 and 10 min have been at stations in regions east of the Rocky Mountains, and mostly amount to between 15 and 30 mm, and between 20 and 40 mm, respectively. On maps (258) showing the distribution of the largest recorded falls, particularly that for a period of 5 min, there is no evidence that the highest values are more prevalent in the regions of the midwest which are subject to the most violent storms; on the contrary, they appear to be more frequent near the Gulf coast, suggesting that they may be associated with ordinary cumulonimbus.

Because of the unsuitable design and resolution of conventional recording rain gauges very little is known about the values attained by the rate of rainfall over periods as short as 1 min, corresponding to a space scale of 1 km or less, which is not too small to be beneath the range of meteorological interest or beyond the resolution of a radar in favorable circumstances. However, it is evident to the unaided observer that the rainfall intensity in thunderstorms often fluctuates markedly and sometimes reaches maximum values over periods of about 1 min, accompanied by an impressive roar and a great reduction in the visual range, which occasionally even in England is reported to become as little as about 30 m. If the size distribution of the drops even in such intense rains is assumed to approximate to that given by Eq. 8.3 and 8.4, then these equations can be used to obtain a relation between the rainfall rate and the scattering coefficient σ of the drops at optical wavelengths, and hence also between the rainfall rate and the visual range. This relation appears to be consistent with the few available observations in which rainfall rate and visual range have been estimated simultaneously (considering the difficulty of obtaining representative measurements of either quantity), at least with rainfall rates up to about 80 mm hr^{-1}, for which the visual range is about 600 m (Appendix 8.15). The same relation indicates visual ranges of about 300 and 150 m for rainfall rates of 300 and 1000 mm hr^{-1}, respectively. These visual ranges are almost certainly too small, for the medium-volume diameters in the assumed drop size distributions are between 3 and 4 mm, close to the maximum stable value, and the distributions therefore probably considerably overestimate the concentrations of large drops. Moreover, in such intense rains the concentrations of the smaller droplets in the air near the ground may be significantly increased by splashing, and the viewing conditions are likely to be less favorable than assumed in the theory. Nevertheless, the reported reductions of visual range to less than 100 m evidently indicate that the rainfall intensity reaches values of several hundred millimeters per hour over periods as short as about 1 min.

The largest reported falls of rain over such periods were estimated from measurements made upon greatly enlarged sections of the traces of recording rain gauges. Elaborate means were taken to reduce errors, but the results cannot be regarded with much confidence; even so, the magnitudes are interesting. Of two recent reports, both from the United States, the first is from an occasion (the early hours of 10 July 1955) when slow-moving thunderstorms with radar echo tops over 15 km occurred over western Iowa, producing total falls of up to about 150 mm. At one observing station, where rain fell for over 3 hr at a rate of about 25 mm hr^{-1}, near the end of the fall the rate appeared to reach extreme values during two spells of about 12 sec, separated by nearly 1 min; over a period of 1.4 min including both spells the total fall was very nearly 1 in., corresponding to a rate of about 1000 mm hr^{-1} (261). The second report is of a storm, one of a number of intense thunderstorms over north central Maryland in the early afternoon of 4 July 1956, which produced a fall of 72 mm in 50 min, with intense darkness but only one notable lightning stroke. The fall apparently included 31 mm in 1 min, corresponding to a rate of nearly 1900 mm hr^{-1}, during which rain was reported as pouring from a roof "like the Niagara falls" (262).

The distribution within a storm of the concentration of precipitation, and of the rate at which it arrives at the ground, depends upon the rate at which the individual particles grow and aggregate, and upon the configuration of the air flow, which determines the trajectories of the particles. Figure 8.26, which contains some trajectories calculated from the pattern of the air flow inferred for a particular storm, shows how complicated they may be even in a flow assumed to be only two dimensional, and suggests that where they converge or intersect locally high concentrations of precipitation might arise even in a steady flow. Such effects have so far been considered quantitatively only in some simple kinematical models of two-dimensional circulations (263). Considering the particularly simple example of a cumulonimbus with an upright, axially symmetric updraft which has a maximum speed in the upper troposphere, it is evident that if the precipitation particles are carried above this level, their paths diverge horizontally, and that they may eventually reach the ground over a much larger area than that covered by the cloud base. If they arrive as raindrops, and therefore with fall speeds of several meters per second which have the same magnitude as the typical updraft speed at the cloud base, the mean areal rainfall rate must be less than the mean upward flux of water vapor through the cloud base (usually equivalent to about 250 mm hr^{-1}), even if there is no evaporation of precipitation. This appears to be characteristic of severe storms, in which the updrafts have high maximum speeds (more than about 30 m sec^{-1}), and in which large concentrations of particles of millimetric or smaller size, representing a considerable fraction of the water condensed in the cloud, are lifted into the updraft outflow. On the other hand, that fraction of the condensed water which is collected by precipitation particles whose growth proceeds near or below the level of the maximum updraft speed may reach the ground over a comparatively small area as intense precipitation whose rate is about equal to the vertical flux of water vapor at the cloud base. In the severe storm the more intense precipitation typically consists of heavy rain with small hail (see, e.g., Fig. 8.24), and probably develops as small wet or spongy hailstones (fall speeds up to about 30 m sec^{-1}), the bulk of whose growth occurs in the lower part of the zone of supercooling. During fall to the ground there may be much melting and disintegration of the hailstones, especially if their growth has been excessively wet, so that near the ground the mean fall speed is unlikely to be greater than about 10 m sec^{-1}. Consequently the concentration of the intense precipitation there is typically several grams per cubic meter (a concentration of 10 g m^{-3} corresponding to a precipitation rate of about 350 mm hr^{-1}).

The concentration which may be attained within the cloud, before the particles fall out of the updraft, is obviously partly dependent upon the efficiency with which the population of particles collects the condensed cloud water. However, it must be limited by the potential buoyancy (or kinetic energy) of the updraft in the zone of growth, since if the concentration were to become too great the weight of the precipitation would interfere with the updraft upon which its continued generation depends. Previously it was implied that in storms which grow on occasions of small wind shear, and have updrafts which are almost upright, this kind of interference does occur and is responsible for initiating downdrafts inside the clouds. In such storms the evaporation of precipitation, especially below the cloud base, subsequently contributes significantly if not predominantly to removal or reversal of the buoyancy of the air in the lower part of the updraft, and to the impulsive behavior of the cumulonimbus. In the persistent storm, in contrast, the configuration of the flow prevents the fall and evaporation of precipitation from affecting the supply of potentially warm air, and the direct effect of the precipitation upon the updraft can be considered to be restricted to the middle and upper levels, and to be due solely to the reduction in the buoyancy of the air by the weight of the precipitation.

Both the concentration of the precipitation aloft and its intensity at the ground can be expected to increase with the potential buoyancy, or speed, of the updraft at these levels, and therefore with the severity of the storm. In the most severe storms the temperature of the air in the updraft in the middle troposphere may exceed that in the environment by between about 7 and 10°C (see, e.g., Fig. 8.46). The corresponding extreme buoyancy, which is reduced a little by the mass concentration of small cloud particles, could be removed completely by the presence of precipitation in a concentration of about 25 g m^{-3}, so that actual concentrations approaching or even somewhat exceeding 10 g m^{-3} seem plausible.

When there is a steady state and the effects of horizontal gradients of vertical air motion and precipitation concentration are small, the changes of precipitation concentration in the vertical are governed by the equation

$$\frac{(1/m)\partial m}{\partial z} = \frac{w\,\partial \rho_a/\rho_a \partial z - \partial V/\partial z - S_m}{w + V} \quad (8.39)$$

Here m is the mass concentration of particles in a small size range whose fall speed can be regarded as V. (In the consideration of intense precipitation which follows, it will be assumed that it consists almost entirely of particles whose diameters vary from a few millimeters to 1–2 cm, with fall speeds varying only between about 10 and 20 m sec^{-1}, so that V can be regarded as an appropriate mean value; its sign is negative.) S_m is a source term representing their net generation during the processes of aggregation in the updraft, or evaporation after they have left the updraft. The first term in the numerator on the right of the equation arises simply from the compressibility of the atmosphere. [Changes with height of w (strictly of $\rho_a w$) are accompanied by a horizontal divergence of both the air and precipitation particles (assumed to have the same horizontal motion as the air), but they do not affect the mass concentration of the air or of the precipitation.]

During the growth of precipitation in an updraft the first term in the numerator on the right-hand side of the equation may be small compared with the second and third, but in some stages of the evolution of the precipitation all three terms are likely to have about the same magnitude, in particular when the precipitation has left the updraft and falls toward the ground (when the third term has a sign opposite to that of each of the others). The first term tends always to increase the concentration of precipitation as it approaches the ground, in the same proportion as the change of air density; over a given height interval the increase is therefore always about the same and not large, amounting to about 50% during fall to the ground from a height of 4 km. (This increase also tends to arise from the dependence of the fall speed V on the air density; e.g., if the particles remain the same size an increase of about half as much again results from the decrease in the fall speed, which is approximately inversely proportional to the square root of the air density.) The intense precipitation in cumulonimbus, which in the zone of supercooling consists mainly of wet and spongy small hailstones and drops of unfrozen water shed from them, leaves the updraft at a similar height, and during its fall to the ground its concentration tends to increase as melting and disintegration reduce its mean fall speed, and to decrease as evaporation occurs. The mean fall speed may be reduced from about

20 to about 10 m sec^{-1}; on the other hand, evaporation may remove up to about half of the mass flux of the precipitation (as indicated by the values listed in Table 8.12). It seems likely, therefore, that in the typical severe storm the concentration of the intense precipitation near the ground is greater than that with which it leaves the updraft, but by a factor of less than about two.

During the previous history of the precipitation, while the particles were within the updraft and growing, insofar as its concentration was affected by the source term and it had been present at higher levels in the cloud, the concentration could only have been less than that at the level where it eventually left the updraft. However, during at least one phase in the growth of the larger particles $(w + V)$ is positive but diminishing, and from a superficial examination of Eq. 8.39, or more usually of a simplified special form such as

$$m(w + V) = \text{const}$$

it has sometimes been concluded that their concentration might reach very high values as $(w + V)$ approaches zero. A more detailed study would probably show that although such an increase of concentration must occur if the particles are all of the same size, it is considerable only very near the level where $(w + V)$ becomes zero. Moreover, if the particles are introduced into the updraft at a variety of positions, and in a range of sizes, and if the updraft is not horizontally uniform, no one such level can be defined within a region several kilometers deep and across.

The concentration of the intense precipitation in the updraft of a persistent cumulonimbus cannot therefore be expected anywhere to be substantially greater than that with which it leaves the updraft or arrives at the ground. This conclusion could be examined with theoretical models in which a cumulonimbus circulation and the generation and evaporation of precipitation particles were represented in simple ways (it is supported, for example, by calculations discussed in reference 264). The validity of the calculated distributions and concentrations of precipitation, properties of fundamental interest in the study of cumulonimbus, could be tested by comparing the corresponding distributions and intensities of radar reflectivity with those typically observed, some of whose features are discussed in the following section. However, it already appears that the value of about 10 g m^{-3} estimated as the extreme likely to be attained by the concentration of precipitation in the updraft of the severe storm is consistent with the observation that the most intense precipitation at the ground (over a space dimension of a few kilometers in the direction of travel of the storm and therefore typical of a period of a few minutes at a place on the ground) has about the same concentration and a rate of a few hundred millimeters per hour.

Rain "gushes"

The intense prolonged rain from almost stationary cumulonimbus, examples of which are described in Appendix 8.14, is widely recognized as a recurrent phenomenon. Observers are also aware of very brief intensifications of thunderstorm rains, often called *rain gushes* (mainly in the United States) or *cloudbursts* (neither term has any official definition). It may help to promote their study if these phenomena are distinguished and tentatively defined as local rains of intensity more than about 50 mm hr^{-1}, whose duration, usually within longer periods of rainfall, is less than a few minutes (gushes), or about an hour or more (cloudbursts).

It has frequently been stated that a rain gush often quickly follows loud thunder from nearby lightning, as if it was somehow caused by the thunder. Perhaps the first to remark upon the association was the seventeenth-century physicist Robert Hooke, who wrote "that if it rained when it thundered, it immediately after the clap poured down much faster, much as if a gale of wind had suddenly shook a tree, all of whose leaves are full of drops of water" (287).

Meteorologists generally think that if a close relation exists it arises incidentally from the separation of electric charge during the development of precipitation. Especially in ordinary cumulonimbus, whose impulsive behavior leads to sudden and local developments, lightning discharges may well occur just before the arrival of the precipitation at the ground. In traveling storms, even with steadier behavior, lightning is similarly associated with the separation from the larger precipitation particles of the smaller cloud particles, which at least in middle latitudes are usually carried away *ahead* of the storm in the outflow of the updraft; accordingly it might be expected that lightning to the ground would commonly occur some distance in advance of the leading intense precipitation and therefore precede its arrival. This seems to have been confirmed by observation; for example, in studies made in New England with the help of voluntary observers, it was found that heavy rain usually began 3–4 min after the occurrence of the first lightning within 1 mi (288). The average speed of travel of the storms was about 30 mi hr^{-1}, indicating an average separation of 0.5–1 mi between the positions of the leading lightning discharges and the leading heavy rain, and of only about 1 min between the end of one of the first particularly loud peals of thunder and the arrival of the rain.

However, this frequent sequence of events on the approach of a storm is not an entirely satisfactory explanation of any association that may exist between thunder and intensified precipitation during a storm, and the possibility of a more direct physical relationship should perhaps not be dismissed, since some explanation also seems required for the occasionally observed extremes of rainfall rate, approaching or even exceeding 1000 mm hr^{-1}. The space scale corresponding to their duration is only about 0.5 km, so that they seem likely to be due to some transitory process effective on a scale about an order of magnitude smaller than that of the general storm updraft. Some small-scale or unsteady feature of the air motion could be responsible—for example, the rise through a region of precipitation of a fresh updraft with a circulation like that of a thermal. This could leave intercepted particles in a column in its wake, with a horizontal dimension of only a fraction of its greatest width, and therefore of the required magnitude (individual shafts of precipitation with a diameter of only a few hundred meters are sometimes seen beneath the base of ordinary cumulonimbus following the rise of separate cloud towers). However, the precipitation rate could be increased substantially in this way only by the growth of the precipitation particles in the updraft, which seems improbable considering that the mass concentrations implied by the extreme rates of rainfall reach values as high as 30 g m^{-3}, sufficient to remove all buoyancy of cloud air.

Observations with lightning detectors and a radar, scanning ordinary cumulonimbus overhead through a vertical plane slowly rotating in azimuth, have recently provided some evidence for intensification of radar echoes (the radar reflectivity factor 10 log Z_e of 20–30 sometimes increasing to about 50) within periods of 10–20 sec immediately following lightning discharges (289). The intense echoes were in columns a few kilometers tall and 0.5 km or less in diameter, and it seemed unlikely that they appeared by moving into the beam rather than by local sudden increases in the reflectivity of precipitation nearly overhead, plausibly near the paths of the discharges. About 2 min after the discharges gushes of rain, sometimes with small hail, fell at the radar site for about another 2 min, during which the precipitation rates, otherwise mostly less than 10 mm hr^{-1}, reached peak intensities, over a fraction of a minute, of 50–100 mm hr^{-1}. It was hypothesized that after lightning cloud droplets near the discharge paths become strongly charged by the capture of mobile ions, and subsequently coalesce with other droplets at a rate which is initially very much greater than in the early stages of the ordinary process of differential settling under gravity. However, it is not clear that raindrops could be grown with the rapidity suggested by the observations, or that the mass concentration of the raindrops should become substantially greater than that produced by the ordinary process, which has been regarded as rather efficient in converting condensed water from cloud particles into precipitation.

Another possibility is that the thunder, by changing the size spectrum of precipitation particles steadily sorted by fall at different speeds through a sheared flow, causes a temporary redistribution of the particle concentration, leading to an increase in a narrow region and a decrease in a neighboring volume. Within about 100 m of a lightning discharge, for example, small spongy hailstones in and below the zone of supercooling might be disintegrated into particles of millimetric rather than centimetric size by acoustically generated cavitation within their liquid enclosures (Appendix 8.11). For a short while after the discharge they would then arrive in a narrow swath at the ground as an addition to rain already falling there, while over a neighboring area there would be a temporary decrease in the precipitation rate. (It would be implied that decreases of the rate after thunder are as frequent as increases, and less common than no change at all, while a local decrease of radar reflectivity would be more noticeable than an increase.)

These speculative remarks only emphasize the difficulty of explaining the great intensity of rains which have been measured over periods as short as 1 min, and the paucity of observations of the magnitude and variability of precipitation rate on this scale, in ordinary as well as severe storms. However, since the space scale which they imply for the extreme mass concentrations of precipitation is only a few hundred meters in the horizontal, they need not be regarded as affecting the previous conclusion that in severe storms, whose behavior is steady on scales of tens of minutes and which are organized on scales of several kilometers or more, the typical mass concentration of precipitation does not exceed several grams per cubic meter.

The radar reflectivity of cumulonimbus

Apart from the general distribution of the radar reflectivity in cumulonimbus, which provides useful constraints in the inference of the pattern of the air flow in and around storms, there are three features of the echo which have particular interest. Of these, the first is the extreme height of the cloud tops; insofar as it is shown by the height of the echo tops and the atmospheric stratification is known, it is an indicator of the maximum speed attained by the updraft and the likely violence of the weather in the storm. Second, the radar reflectivity factor near the ground is a measure of the rainfall rate. Third, the maximum value of the radar reflectivity in the storm is also an indicator of its severity, and provides some information about small hail aloft and the possible presence of large or giant hail. The precautions which have to be taken to obtain usefully accurate measurements of the height of storm tops, mainly because of the width of radar beams, are outlined in Appendix 8.16. Some interesting aspects of the other two features will be discussed here.

The radar reflectivity near the ground and its relation to precipitation intensity

Studies have been made in many places to establish a relation between the rainfall rate R at the ground and the radar reflectivity factor Z near the ground, using either measurements of rainfall with rain gauges and observed values of Z or measurements of the size distribution of the drops and computed values of Z. As a result numerous values have been proposed for the constants a and b in a relation of the kind given in Eq. 8.6 and illustrated in Fig. 8.4; for small values of Z ($\approx 10^2$ mm^6 m^{-3}) the implied values of R are scattered over a range of about 5 db, while for high values of Z ($\approx 10^5$ mm^6 m^{-3}) the range increases to about 7 db (292). However, it is likely that much of the variability in the empirical relations arises from difficulties in the accurate radar measurement of Z and in obtaining representative measurements of R (especially if only a single rain gauge is used) or of the drop size distributions.

In the usual techniques of determining drop size distributions a sample of raindrops can be taken from a volume of not more than about 10 m^3, which may be insufficient to give a distribution representative of the larger drops (293) or of the rain in the larger (pulse) volume sampled by the radar, which even at short ranges is of order 10^6 m^3. When the rainfall has been inferred from radar observations over extended areas or times more consistent results have been obtained, the probable error decreasing to less than 2 db, then being limited by the accuracy of 2–3 db within which the radar measurements can be held by the most careful calibration and procedure (187), at least up to ranges of about 100 km. This may represent an improvement on the measurement of rainfall by other than an unusually dense network of ordinary rain gauges.

Evidently radar observation not only can indicate the rain intensity in a storm but can provide a useful measurement of the amount of rain as it accumulates. The reservations have to be made that the radar should have a narrow beam, which can discriminate the precipitation near the ground out to ranges of about 100 km, that it should be carefully maintained and operated with a standard performance, and that it should have one of the longer centimetric wavelengths, at which the attenuation of the beam in even the heavier rains does not raise any difficulties. The attenuation, which is approximately proportional to the rainfall intensity, is generally negligible at a wavelength of 10 cm, appreciable at a wavelength of 5 cm, and serious or intolerable at a wavelength of 3 cm; for example, the one-way attenuations of the beams during passage through rain of intensity 100 mm hr^{-1} are, respectively, about 0.1, 0.3, and 2 db km^{-1}. With radars using a wavelength of about 3 cm the effects of attenuation in distorting the echoes from severe storms are often obvious on RHI displays.

It must also be considered that relations between Z and R (e.g., Eq. 8.6) are based on observations which have rarely if ever included rainfall intensities of more than 100 mm hr^{-1} (corresponding to $Z \approx 3 \times 10^5$ mm^6 m^{-3}). They cannot reasonably be extrapolated to values of Z (or Z_e) exceeding about 10^6 mm^6 m^{-3}, since the precipitation is then likely to include hail as well as rain, with a consequent change in the form of the relation. It is a characteristic of severe hailstorms, as in that represented in Fig. 8.23, that such high values of the reflectivity factor are found several kilometers above the ground, and their significance is considered in the following section.

The radar reflectivity aloft and its relation to the presence of hail

The precipitation particles in a cumulonimbus containing a strong updraft are mostly lifted high into the zone of supercooling during the early stages of their growth, and must be expected to become at least partly frozen and to develop into hail. However, a considerable fraction of the cloud water accreted by small hailstones in the lower part of the zone of supercooling, and by large hailstones in its upper part, is likely to remain unfrozen, and low in the zone the growth of hailstones may be so excessively wet that they shed drips of liquid water. Hailstones can be expected to have dry surfaces only above the top of the zone of supercooling, and during descent through the weaker parts of the updraft (where there may be little cloud water to be accreted) or through clear air with a wet-bulb temperature below 0°C. Although locally the precipitation particles tend to become sorted according to size, generally they are present in a range of sizes and have more or less aspherical shapes, so that during fall they have a variety of orientations; moreover, they have a composition varying between completely solid and completely liquid, most of the hailstones having at least an outermost layer of mushy rime ice and a thin surface film of liquid. The scattering of radiation of centimetric wavelength by hailstones is strongly dependent upon their size, shape, orientation, and composition, and consequently only general inferences about the nature and concentration of precipitation can be made from measurements of radar reflectivity.

The radar reflectivity factor Z_e associated with solid spherical hailstones in a mass concentration of 1 g m^{-3} depends upon their size and the radar wavelength in the manner seen in Fig. 8.57. Theoretical calculations indicate that if the spheres have a sufficiently thick outer coat of liquid, they behave as liquid spheres of the same size, and can be said to be "electromagnetically wet,"

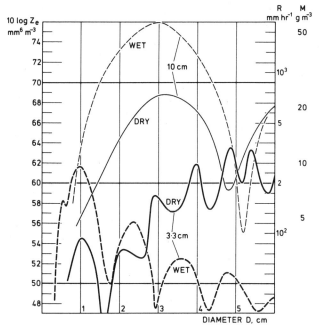

Fig. 8.57 The variation at two radar wavelengths of the reflectivity factor Z_e (mm^6 m^{-3}), expressed as $10 \log Z_e$, with the uniform diameter D (cm) of a mass concentration of 1 g m^{-3} of spherical hailstones, according to whether they are electromagnetically "dry" (behaving as pure ice) or "wet" (behaving as liquid water spheres of the same size; 187). The scales at the right show the mass concentration of rain M (g m^{-3}) and the rainfall rate at the ground R (mm hr^{-1}) corresponding to the same radar reflectivity ($Z = Z_e$), according to the relations (8.6) and (8.7).

Table 8.13 The approximate change (increases indicated as positive) in the backscatter cross section σ_i of a dry sphere of ice due to the addition of a surface film of liquid water of thickness 10^{-2} cm (summarized from reference 294)

Hailstone size (cm)	Change in backscatter cross section (db)		
	At 3.2 cm	At 4.7 cm	At 10 cm
Giant ($D > 5$)	− several	± a few	+ a few
Large ($2 < D < 5$)	− several	± a few	+ a few
Small ($1 < D < 2$)	− several	± a few	+ a few
Small ($0.2 < D < 1$)	+ several	+ several	+ several

Note: "A few" signifies mainly less than 4, and "several" mainly between about 4 and 8.

with the radar reflectivity factor also shown in Fig. 8.57. The required thickness of the coat is about 5×10^{-1}, 7×10^{-1}, and 1–2 cm for wavelengths of 3.2, 4.7, and 10 cm, respectively (294). In most circumstances a coat has an appreciable effect on the backscattering of an ice sphere when its thickness exceeds 10^{-4} cm. The thickness of the coat which develops during the melting of a suspended ice sphere is about 10^{-2} cm; excess liquid subsequently drips away. It seems reasonable to expect that the thickness would be about the same on a sphere melting during free fall. The calculations show that apart from some peculiarly large effects at particular combinations of values of the wavelength, the hailstone size, and the film thickness (which are not likely to have any importance when hailstones occur in a range of sizes), the effect of a film of thickness 10^{-2} cm on the backscattering cross section of a solid hailstone depends broadly on its size and the radar wavelength, as seen in Table 8.13. (The general large increase at the smallest sizes is related to the greater backscatter from liquid spheres of the same sizes, while the predominance of large decreases at other sizes with the shortest wavelength is related to the converse in that part of the Mie region where $\pi D/\lambda \gtrsim 2$, as can be seen in Fig. 8.57.)

At the shorter wavelengths the theoretical changes in the backscatter from a melting ice sphere have fairly satisfactory support from observations on individual spheres, chilled to temperatures well below 0°C and then suspended in the laboratory or lower troposphere, or allowed to fall from the upper troposphere through the melting level (187).

The presence of internal liquid, acquired during growth in the wet regime, may also substantially affect the backscatter from hailstones. Its effect has been studied in the laboratory using spheres of ice grown by accretion with shells of spongy ice in a wind tunnel, and snowballs drenched in water at 0°C and later allowed to freeze progressively during the measurements (295). The sphere diameters varied from about 1 to about 4 cm ($0.4 < \pi D/\lambda < 4$), the proportion L of the total mass which was liquid varied from about 0.05 to 0.55, and measurements were made with radar wavelengths λ of 3.2, 4.7, and 10 cm. It was found that the radar cross section σ was almost unaffected by the composition of a core of diameter up to about half of the total (and therefore containing only about an eighth of the total mass of a sphere, and hardly affecting the value of L). Nor was it obviously affected by the disposition of the layers of spongy ice, so that the value of L alone appeared to be the most important property of the artificial hailstone. Its significance depended mainly upon whether the value of $\pi D/\lambda$ fell within any of three ranges (in the first two of which σ_i is about 7 db below σ_l, the backscatter cross section of a liquid sphere of the same size):

1. $0.4 < \pi D/\lambda < 0.7$: σ rose above σ_i by about 3 to about 5 db as L increased from 0.1 to 0.3, thus remaining a few decibels below σ_l.

2. $0.7 < \pi D/\lambda < 1.6$: when L was less than 0.1, σ was within a few decibels of σ_i; in a narrow range of low values of L even as much as 10 db *below* σ_i, an effect not likely to be appreciable among hailstones with various sizes and compositions. More important, over the whole range σ was within a few decibels of (even sometimes slightly more than) σ_l when L was 0.2 or more.

3. $1.6 < \pi D/\lambda < 4$ (as $\pi D/\lambda$ increases through this range σ_i soon exceeds σ_l by several decibels, and eventually, at values near and greater than 4, exceeds σ_l by 10 db or more); in the higher part of this range σ was decreased by several decibels with a value of L as small as 0.05, and over the whole range $\sigma \approx \sigma_l$ if L was 0.1 or more.

The implication of these various results is that hailstones are likely to appear electromagnetically dry only if they are small or happen to have become large while remaining in the dry growth regime, and if they are above the 0°C level. Otherwise their backscatter cross sections are more likely to approximate to those of liquid spheres. In particular the largest radar reflectivity factor associated with a given mass concentration of hailstones can be expected when their size is such that $\pi D/\lambda \approx 1$, that is, when $D \approx 1$ cm at the shorter centimetric wavelengths and $D \approx 3$ cm at a wavelength of 10 cm. Now Fig. 8.56 shows that the mass concentrations aloft of large and giant hailstones (at least of those not too spongy to fail to reach the ground without melting and disintegration) are much less than those of the small hailstones, which in severe storms typically amount to a few grams per cubic meter, a large fraction of the total mass concentration of precipitation. Figure 8.57 therefore implies that small hailstones of diameter about 1 cm (which in the lower part of the zone of supercooling are well within the wet growth regime) are responsible for the extreme values of the radar reflectivity factor (10 log Z_e about 70) which have been observed in severe storms at all three of the conventional radar wavelengths. From Fig. 8.56 and 8.57 it is apparent that (large) hailstones of diameter about 3 cm, in concentrations of about 1 g m^{-3}, could produce an equally great value of Z_e at a wavelength of 10 cm. However, it seems that such large values of Z_e are observed no less frequently with 3-cm radars than with 10-cm radars, and in the particular example of a severe hailstorm observed simultaneously at all three wavelengths the maximum values of Z_e at the shorter wavelengths were within a few decibels of those at 10 cm, which could be expected only with hailstones of diameter about 1 cm.

A related implication of the two diagrams is that large and giant hailstones occur in concentrations too small to provide a distinctively intense radar echo, even if they are electromagnetically dry. In the hailstorm whose radar echo is represented in Fig. 8.23 the radar reflectivity factor in the echo near the 0°C level and within a few kilometers of the wall was observed as 10 log $Z_e \approx 54$ at wavelengths of 3.3 and 4.7 cm. It was known that in this region the precipitation consisted almost solely of hailstones with a diameter of about 4 cm, so that the intensity of the radar echo is consistent with their mass concentration of about 1 g m^{-3} inferred from indirect evidence (Fig. 8.56), but only if they were electromagnetically dry (Fig. 8.57). The radar reflectivity factor of hailstones of larger size could be expected to be less than this moderate value by nearly the ratio of the characteristic concentrations shown in Fig. 8.56. Nevertheless, at least in some middle-latitude regions the maximum echo intensity in a storm has been found to be correlated with the size of the largest hailstones reaching the ground. Thus from general experience in southern England (Appendix 8.3), and from more systematic studies in New England (296), it appears that small hail is likely at the ground when 10 log Z_e (at a wavelength of 10 cm) exceeds about 55, large hail when it exceeds about 60, and giant hail when it exceeds about 65 mm^6 m^{-3}. (A few studies of individual hailstorms and other experience in Oklahoma suggest that there these thresholds may be several decibels lower; 237.) Some indirect relation of this kind might be expected if the maximum radar reflectivity and the maximum hailstone size were both determined by the buoyancy developed in the updraft, the reflectivity being proportional to the mass concentration of small hail in the lower part of the zone of supercooling and the maximum size to the maximum updraft speed attained in its upper part. However, the increase of the reflectivity, by about 15 db, as the largest hailstones increase from small to giant size, is too large to be explained simply in this way, so that the relation only emphasizes some complexities in the laws governing the evolution of the size distribution of the precipitation particles, which are not yet understood.

It has usually been found (at least with the shorter radar wavelengths) that the more severe storms are characterized not only by a larger reflectivity factor at all heights (scaled according to the height of the storm tops), but by a maximum at a height of several kilometers which is not present in thunderstorms that produce only rain at the ground, as seen in Fig. 8.58. The maximum echo intensity is characteristically in the lower part of the zone of supercooling, but perhaps nearer its middle, at a temperature as low as −25°C, in the most severe storms. Between the 0°C level and the ground the echo intensity seems to decrease by at least several decibels, and although the actual decrease may be exaggerated a little by the difficulty of making an accurate measurement near the ground, even with an unobstructed horizon, a decrease of as much as 10 db can reasonably be explained by the partial melting and disintegration of small hail containing a high proportion of unfrozen water. It is interesting, however, that the maximum echo intensity in the more severe storms should occur well above the 0°C level. [From a more detailed analysis of the observations made in New England with a wavelength of 3 cm (Fig. 8.58) it was found that an increase of Z_e by more than 5 db between the 0°C level and a level 1.5 km higher was one of the most reliable indica-

8   Cumulonimbus convection

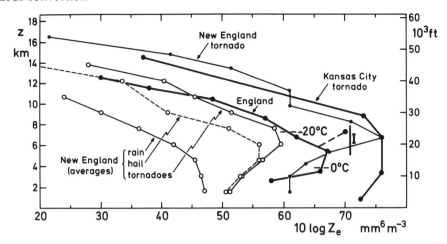

Fig. 8.58 Some examples of the distribution with height z of the greatest values of the radar reflectivity factor Z_e observed in cumulonimbus (not necessarily in true verticals). The lines joining open circles, on the left, show median values for storms observed in New England with a CPS-9 (3.2 cm) radar, and several hundred voluntary ground observers (297); the results are arranged according to whether within 10–15 min of the radar observation the precipitation was reported to be of rain only (182 examples), or to contain hailstones of diameter 1.25 cm or more (29 examples), or whether tornadoes occurred (11 examples). The line marked I shows the height range (temperatures -5 to $-25°C$) recorded for the most intense echoes ($10 \log Z_e \approx 70$), with a wavelength of 3.3 cm, in each of four severe hailstorms over the Po Valley of northern Italy.

An example of extremely high values observed with the CPS-9 radar in a tornadic storm appears on the right (298). The other two curves are also for individual storms: the severe hailstorm over southern England which is discussed in the text, observed with a wavelength of 4.7 cm while hailstones of diameter up to about 4 cm were falling, and a storm which produced hailstones of diameter 5 cm, observed from Kansas City, Missouri, with a wavelength of 10 cm (from an unpublished report by R.E. Hamilton). The extreme value of $10 \log Z_e = 70$ observed in the former storm is indicated by the point at a height of nearly 7 km.

The depletion of small cloud particles by the growth of hail

Considering that the mass flux of small hail leaving a storm (in a concentration of order 1 g m^{-3} and at a speed of about 20 m sec^{-1}) is a considerable fraction of the flux of water vapor into the storm (a concentration of about 10 g m^{-3} entering at about the same relative speed), it is obvious that the growth of the small hail must considerably reduce the concentration of small cloud particles. The concentration of liquid water which in general is available for the growth of hail can be only a fraction of the values corresponding to adiabatic ascent. However, the high concentrations of small ice particles found in cumulonimbus anvils (Section 8.3) suggest that in parts of the cloud there are similarly high concentrations of unfrozen droplets in the upper part of the zone of supercooling. Moreover, the largest hailstones probably are just those which by a favored path are separated from the smaller hailstones and are allowed to enter regions (near the echo-free vault) of high cloud droplet concentration. Thus the value of about 4 g m^{-3} assumed for the effective concentration of cloud water available for accretion in the previous discussion of the growth of large hailstones is probably reasonable, considering that in typical conditions the concentrations corresponding to adiabatic ascent are between about 5 and 6 g m^{-3} (Fig. 3.9). The neglect of depletion due to the growth of other hail need not then materially affect any of the conclusions drawn. This depletion would, of

tions of the severity of a storm (299).] It has been suggested that the decrease downward of the maximum echo intensity also above the 0°C level is due to the freezing of the outer parts of small hailstones leaving a tilted updraft at levels as high as several kilometers, and at the shorter wavelengths may be exaggerated by greater attenuation of the radar beams in the intervening precipitation at low levels (299). Alternatively, it may imply that in the updraft disintegration during excessively wet growth in the lower part of the zone of supercooling plays a part there in limiting the concentration of hailstones of diameter about 1 cm.

Above the level of the most intense echo the reflectivity factor could be expected to decrease by 10 db or more in the upper part of the zone of supercooling, where small hail is in the dry growth regime. Near and above the top of the zone of supercooling, where even the largest hailstones can be expected to be electromagnetically dry, the decrease of radar reflectivity with height is likely to be mainly above the level of the maximum updraft speed, where the larger hailstones progressively leave the updraft. However, a quantitative assessment of the variation of radar reflectivity with height, and a comparison with the observations, must await the construction of a detailed model of the updraft and of the growth and paths of the precipitation particles in at least two dimensions.

course, be included in a more complete discussion involving the formidable problem of the differing paths, in three dimensions, of the air and of the cloud particles.

(It is a matter of fundamental concern in a general theory of hail growth, and of the interference possible by the artificial provision of additional freezing nuclei.)

Appendix 8.1 Features of cumulus and cumulonimbus convection observed in special studies over Europe

The observations in Table 8.14 are selected from special studies in central Sweden (near Östersund, 63°N 15°E), in England (near Dunstable, 52°N 0°W), and in northern Italy (near Verona, 45°N 11°E). Cloud dimensions were determined by radar supplemented by single-theodolite observations and photography, or by double-theodolite ranging.

The table lists some typical summer afternoon values of the temperature T and the saturation potential temperature θ_s (in °C) at:

1. The level z_b of the cumulus bases (all heights in km).
2. The level z_s which the cumulus summits needed to surpass if shower formation was to occur, as signaled by their fibrillation (or by the detection of a 3-cm radar echo, implying in either case a radar reflectivity factor Z of at least 10 mm^6 m^{-3}), or, on some occasions of only very shallow clouds, simply by the visual observation of showers.
3. The level z_{max} of the tops of the tallest observed clouds (> Tr: z_{max} above the tropopause).

Also given are the wind direction DD (in tens of degrees) and speed vv (in m sec^{-1}) at some standard pressure levels, and the kind of airstream [P_m signifies polar maritime air which has been subject to convection over the ocean before entering the continent, where in the course of a few days it becomes modified into continental air (C), over the regions considered acquiring higher characteristic values of θ, θ_s, and θ_w, and of aerosol particle concentration].

Table 8.14

	z_b	$z_s - z_b$	z_s	z_{max}	\multicolumn{3}{c}{$DD\,vv$ at pressure (mb):}
					850	500	300

Observations made in central Sweden, 1954–55

9 Aug. 1954	1.5	3.1	4.6	9.0	16 10	17 13	17 20
T (°C)	6		−12	−41	P_m		
θ_s (°C)	14		14.2	17.5			
16 Aug.	1.3	1.5	2.8	2.8	34 10		
	6		−2	−2	Fresh P_m		
	12.8	13.2	13.2				
12 Sept.	1.5	1.1	2.6	2.6	27 17		
	3		−2	−2	Fresh P_m		
	11.5		12.4	12.4			
13 Sept.	1.2	4.0	5.2	8.4	18 2	21 5	13 10
	4		−22	−48	Stagnant P_m		
	12.2		10.8	12.5			
10 July 1955	2.4	4.4	6.8	11.1	26 2	23 7	25 12
	8		−19	−52	Old P_m (see Pl. 8.5)		
	18.0		18.0	19.5			

Observations made in England, 1957

11 June	1.5	1.7	3.2	4.6	32 7	32 10	
	1		−11	−20	P_m		
	8.2		7.8	8.8			
18 June	1.5	5.2	6.7	13.3	09 5	10 5	Light
	14		−23	Tr	C above shallow P_m		
	20.0		16.0				

Observations made in Italy, 1958

23 June	2.1	3.3	5.4	8.0	23 7	25 25	25 35
	6		−12	−33	P_m		
	16.0		17.0	17.2			
8 July	1.6	4.4	6.0	12.2	Light	08 5	08 7
	13		−15	>Tr			
	19.5		17.0				
9 July	1.6	4.6	6.2	8.4	09 2	09 10	09 25
	13		−18	−33			
	19.5		17.0	18.0			
21 July	1.5	4.9	6.4	8.8^a	21 5	27 17	27 35
	15		−18	−35	Old P_m		
	21.0		17.2	19.2			
				14.3^b			
				>Tr			

^{a} Afternoon over hills.
^{b} Late evening over foothills.

8 Cumulonimbus convection

Appendix 8.2   Observations of shower formation in and near the United States

The following paragraphs summarize some of the results of the more extensive and reliable of the observations of shower formation which have been made in the United States using airborne or ground-based radars. All the heights quoted are above sea level.

A   Studies with radar and double-theodolite ranging or cloud stereophotogrammetry

1   New Mexico, near Socorro (30; also discussed in 32)

Technique. Measurements were made of the range and height of prominent cloud tops, and films of the clouds and the radar display were examined to see if the clouds contained or subsequently produced radar echoes. "Many of the clouds continued to grow after the last measurement. This condition cannot be corrected for and is ignored."

Results. Of 399 clouds measured at ranges between 20 and 80 km, no echoes could be associated with about 55 clouds with tops measured in the height range between about 5.7 and 6.9 km; echoes were associated with 33 of 152 clouds with tops between 6.9 and 8.7 km and with nearly (but surprisingly not quite) all clouds with higher tops.

Comments. The technique, using a radar with a broad beam and not following individual clouds to their peak heights, may not have been well suited to determining the stage in their evolution at which the radar reflectivity factor of the cloud particles reached a definite value. The level of the cloud bases apparently varied by about 1 km from an average value of about 3.7 km, so that if the average minimum height of the cloud tops associated with shower formation was 8.7 km, consistent with an earlier study in the same place (31), the corresponding average minimum thickness was about 5 km.

2   Missouri (33)

Technique. Photographic records were made of the display of the radar used with a scanning cycle of period 3 min. Stereophotogrammetry was used to measure the development of visible clouds.

Results. From the description of the development of an individual group of clouds, and from more general experience of the heights of many echoes at the time of first detection, it appears that the usual minimum thickness of clouds which produced echoes was about 5.5 km.

B   Studies using ground-based radar alone

1   Ohio (26)

Technique. Photographic records of the display of the radar, used with a scanning cycle of period 3 min, were examined for newly detected echoes.

Results. At the time of first detection the radar echoes were usually about 1.5 km deep. Observations of the average temperatures at their tops and bases are presented for each of 12 days; the mean values were, respectively, 0.4 and 10.0°C. It can be inferred that the average heights of their tops varied between 2.9 and 5.4 km (between 4.2 and 5.0 km on 9 of the 12 days), with a mean value of about 4.4 km.

Comments. At the ranges of up to about 60 mi at which the radar was used the visible cloud tops are usually between several hundred meters and 1 km above the top of an intensifying radar echo. Assuming the latter value, and the average value of 1.5 km for the height of the cloud bases in this region, the minimum cloud thickness associated with the production of a radar echo appears usually to have been about 4 km.

2   Arizona (34)

Results. An analysis of the observations made on 7 days again shows that at the time of first detection depths were about 1.5 km; the average height of the tops appears to have varied with the condensation level of air in the lowest 2500 ft of the local afternoon radiosonde sounding, rising from about 17,000 to about 22,000 ft as the latter increased from about 8,000 to about 13,000 ft. Assuming that the latter was the level of the cloud bases, the mean heights and temperatures of the cloud base, and of the base and top of the first echo, were about 3.2 km, 11.5°C; 4.4 km, 4°C; and 6.4 km, −9.5°C, respectively.

Comments. The results suggest that the average minimum cloud thickness associated with the production of a radar echo was a little over 4 km (see also section C3 below).

3   Central Texas (about 50–250 mi from the Gulf coast) (35)

Technique. Photographic records were made of the radar display during a scanning cycle of 4–5 min and were examined for newly detected echoes at ranges between 25 and 100 mi.

Results. On 6 days the average values of the heights of the tops of echoes when first detected were 5.6, 3.5, 4.3, 5.2, 3.5, and 3.7 km, at which levels the clear air temperatures were $-5, 5, 1, -5, 7$, and $5°C$, respectively.

Comments. The scatter on any one day of the heights of the tops of the echoes when first detected was mostly about 2 km on either side of the average, perhaps partly because of the length of the scanning cycle and partly because the resolution in height was not better than 1 km at the greater ranges. A tentative examination of the large-scale weather situations suggests that the average values of about 3.6 km for the echo tops occurred when the low-level flow was from the Gulf of Mexico (and would correspond to a minimum cloud thickness for echo detection of about 3 km), while the average values of about 5 km occurred on days when the low-level winds were easterly or northeasterly and the air would previously have passed over the southern or central states (these values would imply a minimum cloud thickness for echo detection of between about 4 and 5 km).

C Studies from aircraft, using airborne radar

1 Over the Caribbean Sea near Puerto Rico (29, 32, 36)

Technique. The aircraft penetrated selected clouds, the height of whose tops were measured by an accompanying aircraft flying at their level (or making an estimate).

Results. Some of the results are displayed in Fig. 8.3.

2 Over the central United States

Results. The heights of the cloud bases were mostly between 5000 and 10,000 ft, but are not specified. The following table shows the numbers of clouds within which radar echoes were detected, as a function of the height of their tops:

Range of height of cloud tops		Number of clouds observed	Number of clouds producing radar echoes
thousands of ft	km		
12–14	3.7–4.3	15	2
14–18	4.3–5.5	83	16
18–22	5.5–6.7	87	22
22–26	6.7–7.9	20	7
above 26	above 7.9	4	4

Comments. The heights of the cloud bases are noted as "highly variable from one day to the next," probably implying that if there were a definite minimum cloud thickness associated with radar echo production it too may have varied substantially from day to day. It would not have been readily apparent even on individual days in this kind of investigation, in which the height of the tops of individual clouds was measured or estimated only in a brief period of encounter; however, the observations suggest that few clouds produced echoes unless their thickness exceeded about 4.5 km, supporting the high average value given for the same region in the observations described in section A2 above.

3 Over Arizona (37)

Results. The following tables show the number of clouds within which radar echoes were detected, as a function of the height of the tops and of the thickness of the clouds:

Range of height of cloud tops		Number of clouds observed	Number of clouds producing radar echoes
thousands of ft	km		
17–20	5.2–6.1	11	0
20–23	6.1–7.0	29	5
23–26	7.0–7.9	25	6
26–29	7.9–8.8	6	5
above 29	above 8.8	9	9
Range of cloud thickness			
5–8	1.5–2.4	15	0
8–11	2.4–3.3	18	2
11–14	3.3–4.3	20	2
14–17	4.3–5.2	19	13
over 17	above 5.2	8	8

Comments. The sudden increase in the proportion of clouds producing radar echoes when their thickness exceeds 14,000 ft (4.3 km) supports the value of the average minimum thickness inferred in section B2 above.

D Observations using a tracking radar and time-lapse photography

Type of radar and technique. An interesting series of observations was made from the island of Barbados in the Caribbean (38), in which a 3-cm radar was directed at offshore clouds which were observed visually to have prominent growing towers with tops between about 2.5 and 4.5 km. The most intense parts of the echo could be followed automatically, evidently moving in a regular manner. The clouds studied grew from fragments within a period of only 25–35 min.

Results. An echo when first detected was centered

8 Cumulonimbus convection

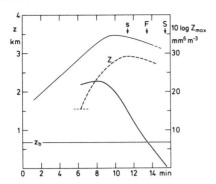

Fig. 8.59 A typical example of some features of shower formation observed over the ocean near Barbados, visually and with a 3-cm radar (38). The thin and the thick solid lines show as a function of time the height z of the visible top of a cloud tower (initially rising at 4 m sec^{-1}) and of the region of most intense radar echo in the rain which developed from it. The dashed lines show the radar reflectivity factor Z (mm^6 m^{-3}) of the most intense echo and the threshold at which an echo was first detectable. Also indicated are the height of the cloud base (z_b), and of the times at which visual observation showed the first signs of a shower aloft (s), of fibrillation in the tower tops (F), and of the shower reaching the sea (S). By 17 min the visible cloud tower had disappeared by evaporation.

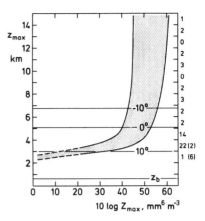

Fig. 8.60 The maximum (3-cm) radar reflectivity factor Z (mm^6 m^{-3}) in 62 tropical marine cumulus or cumulonimbus observed near Barbados (38), as a function of the maximum height z_{max} reached by the visible cloud tops. The individual values were scattered within the shaded area; the numbers on the right indicate the number of cloud tops observed in each height interval of 1 km, and those in parentheses the number in which the radar reflectivity did not reach a detectable value (Z 20 mm^6 m^{-3}). The average heights of some clear air temperatures and of the cloud bases (z_b) are indicated.

between 300 and 1200 m below the visible summits of the cloud tower (generally less than the tower diameter), and intensified at a rate between 3 and 10 db min^{-1}. The region of most intense echo soon began to descend, and the greatest intensity was usually recorded from near the cloud base. Figure 8.59 shows a typical example of the rise of the top of a cloud tower, the radar reflectivity factor Z of the most intense echo, and its change of height.

Figure 8.60 shows the maximum radar reflectivity factor attained in 62 clouds, as a function of the maximum height attained by their tops. All the values observed fell within the shaded area, which is extrapolated within the dashed borders to include cases in which no echoes could be detected.

Comments. Figure 8.60 shows that showers developed in all clouds whose thickness exceeded about 2.5 km, and also that the maximum value of the radar reflectivity attained increased by between 2 and 3 orders of magnitude as the thickness increased by 1 km. This maximum shows little tendency to increase as the level of the cloud tops rises beyond about 5 km, the height of the 0°C level.

Appendix 8.3   The digital representation of the distribution of radar echoes and their intensity

Figure 8.61 contains a digital representation of the distribution of the radar echoes from the thunderstorms over southeast England on the afternoon of 9 July 1959. By comparison a conventional PPI display contains an enormous amount of detail redundant or even misleading for most purposes. In the figure the maximum echo intensity observed with a 10-cm radar over any area of at least 4 km^2 is indicated within each 10-km square of the National Grid by a digit according to the code in Table 8.15, which includes approximate values of the corresponding rainfall intensity R (from Fig. 8.4) and the associated weather, according to a limited amount of experience in England. The digital display makes it immediately apparent that the severe weather is confined to the right flanks of two major storms, whereas in a PPI display of the radar at full receiver gain (which is often the only kind of radar display available to a weather forecaster) the more extensive and comparatively trivial rains elsewhere appear equally impressive. The situation is even less clear if only the conventional synoptic data are available, as indicated in Fig. 8.61 by the observations of surface wind, cloud form, and present and past weather, entered according to the international code. The profusion of thunderstorm symbols, representing past as well as present weather, leads to the impression of a chaotic outbreak of thunderstorms and does not show the organization and distribution of intensity revealed by the radar.

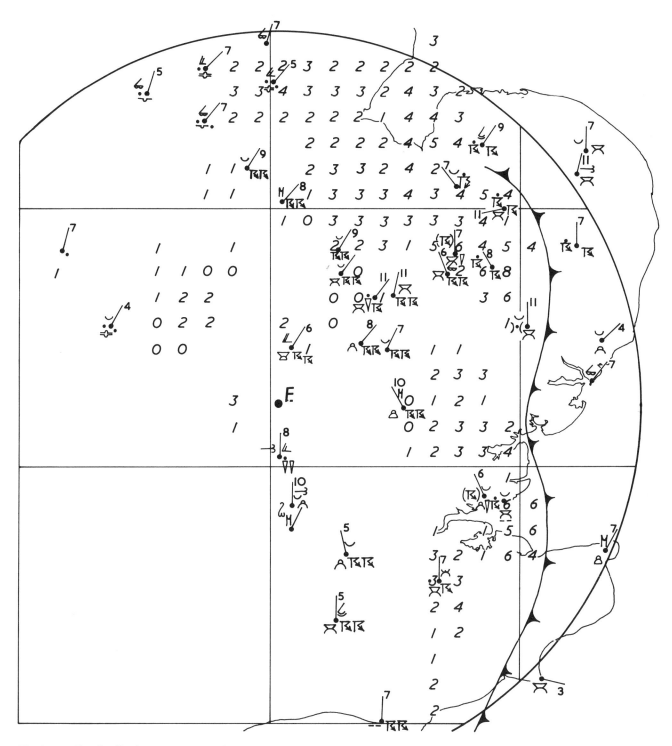

Fig. 8.61 The distribution at 1450 BST of radar echo intensity in the storms of Fig. 8.22, and the position of the extended squall front at the surface, shown by the barbed line.

The italic figures entered in the 10-km squares of the National Grid show the maximum 10-cm echo intensity present over any area of at least 4 km², according to the code of Table 8.15. Surface observations of wind (with speed in m sec⁻¹), cloud, and weather are entered according to the international code.

8 Cumulonimbus convection

Table 8.15 Code figures for 10-cm radar echo intensity (over areas of at least 4 km^2), after a small correction for incomplete filling of the radar beam, and the probable associated rain intensity and weather

Code figure	$10 \log Z_e$ (mm^6 m^{-3})	R (mm hr^{-1})	Weather
0	<25	1	Slight rain, drizzle
1	26–30	1–2	—
2	31–35	2–5	Moderate rain; slight shower
3	36–40	5–10	—
4	41–45	10–25	Heavy rain; moderate shower
5	46–50	25–50	Heavy shower with thunder
6	51–55	50–100	Thunderstorm
7	56–60	100–200	Thunderstorm with hail
8	61–65	200–400	Severe thunderstorm with large hail
9	>65	>400	Severe thunderstorm with giant hail

Another source of complication in the weather map is the great extent of the anvil clouds of severe storms, and the occurrence of nearly horizontal lightning discharges within them, which may reach as far as 100–200 km from the intense updrafts. Their paths can occasionally be observed on the displays of radars using long wavelengths (usually 10 or 23 cm), since the echoes from the free electrons in the discharge channels are temporarily detectable (usually for a fraction of a second) and may be more prominent than the echo from the precipitation in the anvil clouds (188). They can also sometimes be seen by eye (Pl. 8.23). Rain may reach the ground from an anvil cloud (especially if lower clouds are present), or in ordinary showers from low- or middle-level clouds, at distances just as great, so that a thunderstorm with rain may be correctly reported far from the updraft of a cumulonimbus which with radar can be recognized as an individual storm.

Appendix 8.4 Descriptions of arched squalls

The sudden onset of the squalls associated with intense cumulonimbus made them a serious hazard to sailing ships, and the descriptions of these storms by nineteenth-century mariners, who introduced the name "arched squalls," can hardly be bettered. Some excerpts are given below.

Piddington (98) stated that the most remarkable of the arched squalls in regularity, frequency, similarity, and severity are those of the Malacca Strait (between Sumatra and Malaya), followed by the "Nor'westers" of Bengal and the Pamperos of northeast Argentina and Uruguay. Of the first,

> They may be said (that is the heaviest of the wind invariably) to come from some point between NNW and WNW... a mass of black clouds collects and rapidly rises, forming a vast and magnificent arch, beneath which is always observed, even in the darkest night, a dull, gloomy, phosphoric light, like that transmitted through oiled paper by a candle, which at times becomes stronger, particularly on the approach of the arch to the zenith. Flashes of very pale sheet lightning are often observed crossing this space, which sometimes extends over 10 or 12 points of the horizon; the low grumbling of thunder, the falling of the rain, and even the distant roar of the wind, may, I think, be distinctly heard as the arch rises... the danger is from the wind, the first burst of which is always tremendous, and sufficient to dismast or upset the finest frigate, should she venture to meet it under any but storm-sail.

According to later comments on these and similar squalls in the East Indies the wind speed does not usually exceed about 20 m sec^{-1} (99).

Piddington also quotes a description of an arched squall experienced by the *Beagle* in the Rio de la Plata:

> Before mid-day the breeze was fresh from the NNW, but after noon it became moderate, and there was a gloominess and a close sultry feeling, which seemed to presage thunder and rain.... At about 3 o'clock the wind was light ... there was a heavy bank of clouds in the SW, and occasional lightning was visible even in day-light. There were gusts of heated wind. At 4 the breeze freshened up from NNW and obliged us to take in all light sails. Soon after 5 it became so dark towards the SW, and the lightning increased so much, that we shortened sail to the reefed top-sails and fore-sail; shortly before 6 the upper clouds in the SW quarter assumed a singular hard and rolled up or tufted appearance, like great bales of black cotton, and altered their forms so rapidly that I ordered sail to be shortened, and the top-sails to be furled, leaving set only a small new fore-sail. Gusts of hot wind came off the nearest land.... The wind changed quickly, and blew so heavily from the SW, that the fore-sail split to ribbons, and the ship was thrown almost on her beam ends!... The vessel was apparently capsizing, when top-masts and jib-boom went close to the caps, and she righted considerably. Two men were lost ... so loud was the sound of the tempest that I did not hear the masts break, though holding by the mizen rigging ... thunder, lightning, hail and rain came with it, but they were hardly noticed in the presence of such a formidable accompaniment! After 7 the clouds had almost all passed away; the wind settled into a SW gale, with a clear sky.

It seems that strictly it is the succeeding southwesterly wind which is called the *Pampero*, while the briefer squall accompanying the thunderstorms is known locally by the name *Turbonado*.

In a description of the second of two violent thunderstorms which interrupted the laying of a submarine cable in the middle of the Persian Gulf (100), it is said that after a warning that the storm had passed Bushire, 200 km to the northwest,

> at 3 p.m. black clouds were seen rising ... as the clouds approached they gathered into a peculiar form resembling the cap of a large mushroom, extending entirely across the heavens from one horizon to the other. The lower edge of the mushroom had a rounded and wrinkled margin, but was very sharply defined; the surface was composed of many similar strata, as if melted pitch had been poured out and allowed to solidify in numerous cakes, each rather smaller than the one below. Underneath all there was a dark chaos which soon enveloped the vessels, the wind still blowing aft. Suddenly there was a profound calm, and a few hundred yards ahead the squall was seen approaching. The sea, elsewhere covered with full-sized waves, became one dead level of creamy foam, the top of every wave being swept off into spray as soon as it rose. Small whirlwinds swept along the surface carrying up waterspouts towards the clouds, and for a few moments the darkness and the breathless calm contrasting with the threatening appearance of the approaching squall, and the lurid light of the sky still gleaming behind, formed a very impressive scene, which was heightened by the incessant roll of distant thunder. In another moment the squall struck, and the thermometer fell at once from 81° to 53°; torrents of rain swept the decks, accompanied with continuous flashes of lightning and peals of thunder.

Modern ships are less vulnerable to such squall storms, but more condensed and therefore usually less graphic descriptions by mariners continue to appear in specialized periodicals such as the *Marine Observer*. For example, there are recent accounts of squalls exceeding 25 m sec^{-1} near the south coast of Australia (134), in the Philippines (135), in the eastern Malacca Straits (136), in the Gulf of Oman (137), and in the Persian Gulf (138, 139). Near the Arabian shores in the month or two before the summer monsoon the squall storms are often accompanied by hail; the hailstones are sometimes very large, with maximum diameters estimated as 4 in. (10 cm; 138) and even 5 in. (about 12 cm; 139).

Appendix 8.5 The seasonal variation of hail damage to crops in the United States

Table 8.16 Monthly mean number of days of hail damage to wheat crops in several states, 1957–64 (158)

State	Apr.	May	June	July	Aug.	Sept.
Texas	8	24	25	5	0	0
New Mexico	0	5	10	3	0	0
Oklahoma	11	25	22	3	0	0
Kansas	8	26	29	14	0	0
Colorado	0	13	21	20	4	0
Nebraska	2	19	28	24	4	0
Wyoming	0	2	11	14	4	0
South Dakota	0	6	23	28	15	1
North Dakota	0	3	23	29	25	6
Montana	0	2	20	25	20	4

Appendix 8.6 The boundary layer wind jet

The large diurnal variation in the intensity of ordinary convection over land in fair weather produces a corresponding large variation in the magnitude of the coefficient of eddy viscosity which expresses the influence of friction due to flow over the ground. Accordingly there is a diurnal variation of wind, which has a maximum amplitude at a height of several hundred meters; in the afternoon the speed there is substantially less than the geostrophic (Chapter 9) and the wind is directed across the isobars toward low pressure, but following the cessation of the convection toward sunset the wind speeds well above the ground approach or even exceed the geostrophic value and tend to become directed toward high pressure. In steady large-scale flows the wind at these levels has a diurnal variation resembling a simple inertial oscillation, but the variation is more complicated when, as in general, the geostrophic wind and the magnitude of the eddy coefficient of viscosity both also vary with height and time (165–67, 169). It appears that observed diurnal wind variations are most pronounced over large regions of ground with a moderate slope near long mountain barriers (e.g., over the plains states east of the southern Rocky Mountains, and over the plateau between the Andes and the western coast of northern Chile and Peru). Over such regions there may be a large diurnal variation of (thermal) geostrophic wind within a convection layer of limited depth in a persistent flow along the slope. Such a variation seems necessary for a satisfactory explanation of the remarkable magnitude of the diurnal changes of wind speed over these regions (168). Combined with friction, the variation may reverse the sense of the intermediate-scale circulation which would normally be expected over the sloping region (but not over and near the high ground), leading to subsidence there during the day and upslope motion during the night, and thereby playing a significant part in an abnormal diurnal variation of thunderstorm formation.

After sunset the loss of heat from the sloping ground spreads upward through a layer of air whose thickness increases, probably rapidly early in the night; in this layer extreme wind shears develop which, in spite of the strong static stability accompanying the inversion of lapse rate, continually produce Richardson numbers below the threshold value of about 0.25 at which the flow becomes unstable (Chapter 9), and is intermittently subject to a vigorous turbulent stirring. The maximum wind speed is found at about the top of the layer, at a height which in the later part of the night is typically several hundred meters above the ground, and is often so pronounced that the shallow layer of strong wind has become known as a *boundary layer jet stream*.

The boundary layer jets have been recognized in all parts of the United States [and also over other continents, e.g., in the Sudan and in the northeasterlies at Fort Lamy (15°N, 20°E) in West Africa (165)], but they are most frequent and intense in the large-scale southerly flows which during the spring and summer enter the United States across the 30th parallel (where the period of the simple inertial oscillation is exactly one day), especially over northern Texas and the west of Oklahoma and Kansas. Here both by day and night the mean speed in a layer 1 km deep may be about 15 m sec^{-1}, and in regions far inland where storms remove the supply of moist air in a belt several hundred kilometers long on one day it is restored by the following day.

Appendix 8.7   Nocturnal thunderstorms

Cumulus and cumulonimbus are normally associated with convection from the earth's surface. When cumulonimbus occur over land at night, or over the sea in flows toward cooler waters, some other agency must be responsible for lifting and saturating air to form fresh cloud. One mechanism is the *lifting of air over the squall fronts* of cumulonimbus formed previously during ordinary convection, over land in the afternoon or over sea during flow toward warmer waters. The earlier widespread cumulus convection produces a stratification with considerable available potential energy, which may sustain efficiently organized cumulonimbus for some time. Figure 8.38 illustrates the persistence during the night of cumulonimbus convection which traveled across land far from the region in which it developed during the afternoon. Figure 7.10 is an example of a typical and common situation over the sea in which the cumulonimbus have a similar persistence. The cumulus convection in the rear of an intense trough weakens and eventually disappears after the flow has entered its forward side to return poleward to cooler waters, but cumulonimbus which have formed may persist for a considerable further distance. On the occasion illustrated in Fig. 7.10 the cumulonimbus convection in the flow which had reached about as far south as the 50th parallel continued where it became southwesterly and persisted until the air reached as far north as about the 65th parallel in the Norwegian Sea, along a path over which the sea surface temperature decreased by several degrees Celsius.

It is thought that over the open ocean the frequency of thunderstorms has a diurnal maximum during the night and a minimum in the middle of the day (171), but the observations usually quoted in support of this view are old and meager, and their interpretation has been doubted (172); moreover, they were made mostly in regions within several hundred kilometers of land, for example, in the "tropical Indian Ocean" from the Bay of Bengal to between Sri Lanka and Sumatra, and in the Atlantic between West Africa and Brazil, which may be reached by cumulonimbus systems formed over land during the daytime. On the other hand, it is known from experience in forecasting that cumulonimbus convection in cool flows over the sea is strongly suppressed by *intermediate-scale circulations* near land which becomes comparatively warm during the day and intensified at night when the land becomes comparatively cool. Such effects are particularly noticeable near coasts which are concave toward the sea (172) and in straits or enclosed seas, even when they are as large as the North Sea or the Baltic. At night, when chilled air drains down mountain slopes, these circulations produce lifting over low ground; especially in narrow valleys flanked by high ground, and in moist and slow-moving airstreams previously subject to convection the ascent may be sufficient to produce clouds which develop into large cumulus and cumulonimbus. In this way thunderstorms may occur during the night hours over low ground independently of any earlier cumulonimbus convection, although because of the difficulty of seeing in the dark it is often not possible to be sure of this or to follow the early stages of the cloud development without aids such as radar. However, such nocturnal thunderstorms have been described as frequent in some regions in low latitudes [e.g., in northern Bolivia (173), and near Lake Victoria in East Africa], and some have been observed in special summer studies in northern Italy. On some afternoons when cumulonimbus developed there over the Alps and the Appenines there were no clouds or only traces of small cumulus over the low ground of the Po Valley, but near sunset patches and belts of small cumulus formed in places over the plain and near the foothills, and locally developed into towering cumulus. On two successive evenings (20 and 21 July 1958; see Appendix 8.1) isolated cumulonimbus formed after dark, and for 1-2 hr became substantially taller and therefore probably more intense than any in the afternoon, with strong squalls, torrential rains, and hailstones 3 cm or more in diameter. On the first occasion, when the flow was easterly near the ground but westerly in the middle and upper troposphere, the cumulonimbus bases remained within an area only about 15 km across just north of the River Po, almost in the middle of the plain where the ground is practically level and featureless.

A third mechanism for the production of cumulonimbus at night is *large-scale ascent*, which is most pronounced and effective near fronts. It occurs predominantly on the eastern side of troughs, where the flow is poleward, and leads to condensation in layers of air which were subject to ordinary convection from the surface in regions up to several hundred kilometers away. If these regions are rather arid, then the layers may be deep with a lapse rate not much less than the dry adiabatic, so that when cloud eventually forms, usually in the lower middle troposphere, towering cumulus and even small cumulonimbus may develop quickly. (The cloud does not appear simultaneously everywhere over extensive areas because the water vapor is usually concentrated into bands, lying approximately in the wind direction, as a result of intermediate-scale circulations in the source region, and because the clouds first form in the crests of the more pronounced orographic waves, which are virtually always present.)

In some regions such cumulonimbus frequently produce slight or moderately intense thunderstorms. They are, for example, often located by sferic observations in the night or early morning over and near the Bay of Biscay north of Spain, and are regarded as an early warning of the development of more general thunderstorms in troughs entering western Europe. Such storms produce downdrafts which reach the surface, and may thus gain access to air of high wet-bulb potential temperature in a local layer of suppressed convection when the barrier has become weak, so that intense traveling cumulonimbus develop. Intense storms may also form directly when large-scale ascent both produces condensation in such a layer and completely removes the effectiveness of the suppressing barrier.

Appendix 8.8   Castellanus clouds

In air which ascends in large-scale flows after modification by ordinary convection over land, condensation may eventually occur in a layer in which the lapse rate is nearly the dry adiabatic. Cumuliform clouds then develop, representing a form of ordinary convection which does not extend down to the surface. It may occur at any time of the day or night, but sometimes has a more or less well-defined diurnal variation related to but out of phase with that imposed over the region where the airstream was modified (in particular, moistened) by day. The moisture in such airstreams is generally not uniformly distributed, tending to be concentrated in streaks of various widths which lie downwind of the topographical features responsible for intermediate-scale circulations.

The first clouds to form usually appear locally in oreigenic disturbances in the flow, well before the air generally has been brought close to saturation, but subsequently the large-scale ascent usually leads to bands of clouds which represent streak lines. In traveling large-scale motion systems the bands lie approximately in the direction of the relative flow at their level, often distinctly different from the direction of motion of the cloud details.

Within minutes of the formation of a cloud a convective overturning develops in the unstable stratification and the cloud becomes cumuliform, with towers of horizontal dimension comparable with the depth of the saturated layer. In flows from arid lands the clouds form in the middle troposphere and develop a distinctive castellated appearance, producing rows of towers which are much narrower than cumulus towers reaching the same level during convection from the surface (Pl. 8.24). Such clouds thus are distinguished from ordinary cumulus and described as *castellanus*. In their vertical development the cloud towers acquire upward velocities characteristic of small cumulus and frequently become detached from their bases before dissolving by mixing with the drier air above. (The mixing delays or even prevents the production of a general state of saturation and widespread cloud by the large-scale ascent; soundings made in regions from which castellanus are reported in the middle troposphere usually indicate a high relative humidity near their base level, but hardly ever a state of saturation.) Numerous small dissolving towers with ragged edges and tapering bases present a tufted appearance and are known as *floccus*. Sometimes the castellanus formed in the middle troposphere become larger and more persistent; they may glaciate (Pl. 8.25, 8.26) and produce trails of small hail or snow crystals, thunderstorms, and even occasionally showers which reach the ground.

In flows into frontal zones from moister regions over which the ordinary convection was confined to the lowest 1–2 km castellanus, regarded now more generally as cumulus not produced by local surface heating, form with bases in the lower troposphere, and may develop into clouds towering into the upper troposphere which in appearance are indistinguishable from large cumulus and cumulonimbus. In western Europe their arrival in the frontal zone is usually preceded by increasing amounts of high- and middle-level clouds, sometimes in several layers containing castellanus in places, which give the sky a chaotic appearance notorious as a precursor of intense thunderstorms.

Appendix 8.9 Cumulonimbus details; mamma

An often spectacular feature of the undersides of anvil clouds, especially near the region of updraft, is the presence of pendant bulges known as *mamma*. Irregular and small mamma are sometimes observed below the shelf cloud of a large cumulus, and can be attributed to local subsidence of both the shelf cloud and the clear air beneath, which produces instability in a shallow layer including the lower surface of the cloud. Larger and more widespread mamma appear in anvil clouds when ordinary convection results mainly from the chilling of the clear air below by the evaporation of copious but small particles of precipitation. In the circulations that arise it is the descending air which is visible, surrounded by clear air which penetrates upward into the anvil cloud without condensation. In the middle of the day the mamma appear dark and the spaces between them, reaching up into the anvil cloud, are notably more translucent (Pl. 8.27); in the evening it often happens that a receding anvil is illuminated by a setting sun, so that the mamma are brightly lit and the interstices are in shadow.

Mamma have surfaces which are less well-defined and smoother than those of cumulus towers or experimental thermals. This remarkable difference probably arises from the comparatively large size of the particles in the mamma, which have fall speeds comparable with the sinking speeds of the mamma (a few meters per second); as a result they are continuously falling from the (negatively) buoyant air into regions where the flow is not turbulent.

Mamma of the size, density, and detail shown in Pl. 8.14, near the dome of an intense cumulonimbus, are not so obviously associated with precipitation. It is possible that they represent a temporary descent below an equilibrium level of air streaming down from the dome, or a convection arising from the arrival in the anvil cloud of an inverted layer in which θ_w originally increased with height (this is not the normal circumstance, but it may happen in frontal zones).

Appendix 8.10 Some visual observations of tornadoes

There are few detailed observations of the winds, clouds, and weather in the immediate vicinity of tornadoes, mainly because of the observers' need to take refuge.

One of the most graphic accounts relates events in the severe storm near Oklahoma City on 26 May 1963, which is discussed in the text. Two meteorologists using radar near Tinker Air Force Base drove a few kilometers north during a power failure and observed the fall of very large hailstones south of an extensive region of heavy precipitation before returning and noticing near the base an alarming dark and turbulent circular region in the clouds nearly overhead (196). Delayed while seeking entry into the radar building they observed that there was no local precipitation and that the wind had increased to about 30 m sec^{-1} from the southwest. The circular cloud overhead was lower than the surrounding cloud and several hundred meters above the ground, with a smooth and unbroken outer boundary, within which fragmentary clouds seemed to have a turbulent motion superimposed on a steady clockwise rotation. The diameter of the whole cloud was about 100 m, and its angular velocity appeared to be constant except very near its center, corresponding to a speed of about 100 m sec^{-1} along its edge. It seemed from other accounts that subsequently a brief period of calm was followed by strong northwesterly winds, and that the cloud developed into a funnel cloud which lowered to the ground several kilometers farther east; severe wind damage was experienced there over a path about 7 km long. An article (197) immediately following the cited reference contains pictures of a similar circular cloud from which a tornado developed over Florida.

Another account of observations made during a car journey from Oklahoma City (198) emphasizes that the tornadoes seen occurred in the zone of wind shift and near the forward edge of a hook radar echo on the right rear flank of a severe storm. On approaching this zone from the southeast the southerly surface wind increased to 15–20 m sec^{-1} within about 3 km of a region of low cloud, from which rotating funnel clouds were seen to lower to the ground. Several kilometers behind the region of low cloud the winds were northwesterly, the air was cooler, and slight rain was falling. Very little lightning was seen near the tornadoes, although it was frequent in the regions of heavy rain to their north.

Many pictures of funnel clouds are contained in references 199 and 200; the latter also describes and illustrates the disastrous effects of tornadoes. A comprehensive illustrated account of tornadoes in Dallas, Texas, is contained in reference 203.

Appendix 8.11 Evidence of the breaking of large hailstones in the air

During the study of the layered structure of hailstones made apparent by cutting thin slices approximately through their growth centers it has occasionally been found that a sharply defined part of one of the internal layers has been missing, indicating the breaking away of this part at some stage in the growth of the hailstone (usually along a line in the slice more or less normal to rather than along a radius from the growth center). One example is given in reference 218 (Fig. 12g). In a study of 134 large and giant hailstones collected from storms in Oklahoma, evidence of such breaking was found in 15 of the stones, at times when their radii were variously between 2 mm and 2 cm (220); in at least one example the central embryo was irregular and may itself have been a fragment of some other stone. It is difficult to imagine that the hailstones sometimes reported by observers as like pieces of broken window glass or as shaped like coins could arise as anything but fragments of larger stones.

During the intensive studies of hailstorms in England in 1959, an observer at Spalding (Lincolnshire) reported a sparse fall of roughly spherical hailstones with diameters between about 3 and 5 cm, which with some noise each burst into several fragments in the air at heights up to about 50 m. The hail was not accompanied by any rain or thunder.

A number of explanations have been advanced for the breaking of hailstones: strain due to the freezing of internal liquid; collisions; and shock waves near lightning discharges, leading to cavitation in internal liquid (221, 222).

Appendix 8.12 Reports of giant hailstones of extreme size

In England hailstones weighing about 0.25 kg (equivalent spherical radius about 4 cm) have been recorded on several occasions during this century. In a storm over Suffolk in 1666 a carter is said to have had his head fractured by hail, "although he wore a thick felt hat," and it was claimed that one hailstone was 1 ft in circumference and that another weighed 1 lb 2 oz (about 0.5 kg, corresponding to an equivalent spherical circumference of about 1 ft); in spite of the compatibility of the two measures a Victorian compiler was recently supported as "justifiably incredulous" of this report (224). Inland over Europe, however, and in other regions where severe hailstorms are more frequent, such as the United States and northern India, a weight of about 0.5 kg is recognized as a normal upper limit, which apparently is occasionally considerably exceeded. For example, in the detailed early studies of hailstorms over Austria hailstones weighing 0.5 kg are mentioned as having occurred on several days in the years from 1897 to 1905 (e.g., 90), and on one such occasion a youth was killed by hail (89). The largest stones appear to have been reported in 1897, with a maximum diameter of 15 cm and weights of 0.8–1.1 kg (89), the latter corresponding to an equivalent spherical diameter of about 13 cm. Some hailstones in a storm over a village in the Spanish province of Valencia in 1935 were said to be larger than oranges; several weighed more than 1 kg, and the weight of one was 1.480 kg (225). During a severe hailstorm over Schleswig-Holstein in 1925 an ellipsoidal hailstone broke through the tiled roof of the house of a cigarette manufacturer (practiced in estimating small weights) and was found on the floor beneath with estimated greatest dimensions of $25 \times 14 \times 12$ cm and a weight of 4.5 lb (226). Assuming the shape of an exact ellipsoid with the quoted dimensions the corresponding weight is 2 kg, in close agreement with the estimate.

In the United States a hailstone which fell at Coffeyville (southeast Kansas) in 1970 had a weight of 766 g and a largest circumference of 44 cm; previously the largest reliably measured hailstones fell at Potter, Nebraska (227). There the storm began with "a peculiar hissing noise in the air" and the fall of hailstones the size of baseballs, which soon increased to the size of grapefruit. Most were "almost round" and up to 17 in. in circumference, weighing between 10 oz and 1.5 lb (between about 300 and 700 g, the latter weight corresponding to an equivalent spherical diameter of over 11 cm). The hailstones caused little damage apart from piercing a few roofs, because they were few and fell far apart; the average separation on the ground was about 4 m, corresponding to a concentration in the air of only about 10^{-5} m^{-3}, supposing the fall lasted less than 2 min. The hailstones were mostly smooth and composed of clear ice, but some evidently had pronounced lobes, being described as "jagged, having the appearance of one large stone with a number of little stones frozen on its outside."

Only one death from hail has been recorded in the United States (200), presumably because secure shelter has always been close at hand, but occasional accounts of numerous deaths of people, and particularly of large animals like sheep and cattle, in hailstorms over India and China give some credibility to reports of abnormally large hailstones in those regions. Thus whereas hospital treatment was needed for only about 200 of several thousand people taken unawares on beaches near Sydney, Australia, by a fall of hailstones as large as tennis balls on 1 January 1947 (228), in contrast 200 or more deaths have been attributed to individual hailstorms near Delhi, India, in 1888 (229) and in the province of Honan, China, in 1932 (230). In India, maximum hailstone weights of about 0.5 kg have been said to be exceptional, but the same author has listed reports of weights of 1–2 kg (231), and more recent accounts have quoted even larger values. For example, an official engineer in a district of the state of Hyderabad reported hailstorms on successive days in March 1939 which "lasted for 15 and 7 minutes respectively, followed by heavy wind and rain for about 1.5 hours. The largest stones weighed about 7.5 lb on the first day and 5 lb on the second ... roofs were blown away, tiles smashed to pieces ... sheep and cattle died by hundreds" (232). A weight of 7.5 lb is about 3.5 kg, and the corresponding equivalent spherical diameter is 19 cm. In another report of a hailstorm in India in 1939, which killed three people and many cattle, some hailstones were said to weigh as much as 3 lb (233). The largest stones mentioned anywhere in the meteorological literature were reported to have fallen at Yüwu in the Chinese province of Shansi, in the summer of 1902 (234). The occasion was recalled by an English missionary in these words: "The hail that fell that year was larger than my fist, which you may remember is not small, and two pieces that fell thrown over the scale weighed 7.5 catties (4.5 kg) each. 800 tiles on the roofs of house and Chapel were beaten almost to powder, bricks in the courtyard split into two halves and the exceptionally good harvest of grain ... was totally destroyed in less than 25 minutes. That was a sight never to be forgotten."

Although the largest sizes quoted may seem incredible (and imply fall speeds of as much as 90 m sec^{-1}), they should not be dismissed lightly, as for a time were reports that occasionally stones fall from the sky (the experts remarking that significantly such stones always fell at the feet of some simple countryman, not near a scientist). Even apparently fanciful stories, such as of hail the size of an elephant or even of a small hill, often have some rational interpretation [it is quite common for small hail to be swept by flooding rain water into accumulations which in ditches or barely perceptible depressions in open ground may lie in fused piles a meter or more deep (200), perhaps awaiting discovery hours after the storm]. A severe hailstorm, especially if accompanied by suddenly violent winds, can be a frightful experience whose details are deeply impressed on the mind. For example, in the locality worst affected by the moderately severe storm in the late afternoon of 5 September 1958 (Table 8.5), only about 30 km south of London, some of the people thought that the end of the world had come. A hospital suffered loss of electric power and pitch darkness relieved only by flashes of lightning, during which a sudden squall and giant hail smashed many windows on its weather side, scattering glass splinters on floors and beds, and allowing the following torrential rains to flood into the wards. At this time the forecaster on the television broadcast, without radar or other means of recognizing the severity of the storm, mentioned that "some thunderstorms" had come inland across the Hampshire coast, but would soon move away northeastward and die out, before turning to the main business of the forecast. In the afflicted area the power cut denied viewers such reassurance.

The following example of the vivid accounts of severe storms, full of significant detail, which can be given by acute laymen, is taken from passages in the autobiography of Benvenuto Cellini. He describes his return on horseback from Paris to Italy in the summer of 1545:

> We were one day distant from Lyons, and it was close upon the hour of twenty-two, when the heavens began to thunder with sharp, rattling claps, although the sky was quite clear at the time.* I was riding a cross-bow shot before my comrades. After the thunder the heavens made a noise so great and horrible that I thought the last day had come; so I reined in for a moment, while a shower of hail began to fall without a drop of water. At first the hail was somewhat larger than pellets from a pop-gun, and when they struck me, they hurt considerably. Little by little it increased in size, until the stones might be compared to balls from a cross-bow. My horse became restive with fright; so I wheeled round, and returned at a gallop to where I found my comrades taking refuge in a fir wood. The hail now grew to the size of big lemons. I began to sing a Miserere; and while I was devoutly uttering this psalm to God, there fell a stone so huge that it smashed the thick branch of the pine under which I had retired for safety. Another of the hailstones hit my horse upon the head, and almost stunned him; one struck me also, but not directly, else it would have killed me. In like manner, poor old Lionardo Tedaldi, who like me was kneeling on the ground, received so shrewd a blow that he fell grovelling upon all fours. When I saw that the fir bough offered no protection, and that I ought to act as well as to intone my Misereres, I began at once to wrap my mantle round my head. At the same time I cried to Lionardo who was shrieking for succour, "Jesus! Jesus!" that Jesus would help him if he helped himself. I had more trouble in looking after this man's safety than my own. The storm raged for some while, but at last it stopped; and we, who were pounded black and blue, scrambled as well as we could upon our horses. Pursuing the way to our lodging for the night, we showed our scratches and bruises to each other; but about a mile farther on we came upon a scene of devastation which surpassed what we had suffered, and defies description. All the trees were stripped of their leaves and shattered; the beasts in the fields lay dead; many of the herdsmen had also been killed; we observed large quantities of hailstones which could not have been grasped with two hands. Feeling then that we had come well out of a great peril, we acknowledged that our prayers to God and Misereres had helped us more than we could have helped ourselves. Returning thanks to God, therefore, we entered Lyons in the course of the next day, and tarried there eight days. At the end of this time, being refreshed in strength and spirits, we resumed our journey, and passed the mountains without mishap. (235)

*"*e l'aria era bianchissima*." The translator and Goethe prefer the version given to the alternative "and the air blazed with lightnings"; the reader will form his own opinion.

Appendix 8.13   The recycling of objects in cumulonimbus updrafts

The first clear exposition of the recycling of a hailstone between a high level in a storm cloud, in "the snow region, where it receives a coating of snow moistened by the small rain-drops carried up into the snow region before they freeze," and a lower level, in "the lower part of the snow region, where there is little snow, but mostly rain-drops not yet frozen, where it receives a coating of solid ice," was that of Ferrel (243). He envisaged an updraft symmetric about a vertical axis, in which the horizontally diverging flow aloft moved the hailstone out toward regions of weaker updraft and allowed it to descend, until it reached the horizontally converging air below and was carried inward to ascend again. "This process may be repeated a number of times, in each of which the hailstone, disregarding its gyratory motion all the while, describes a kind of orbit, not returning into itself, until it is carried out above so far from the centre, or the strength of the tornado becomes so much weakened, that it is no longer carried in toward, and up in, the central part, but falls to the earth a *hailstone with a snow-kernel and a number of alternating concentric coatings of solid ice and frozen wet snow.*"

This explanation of how a layered structure could arise even in a *steady* storm circulation seems in later times to have been modified to place the emphasis upon fluctuations of height in an *unsteady* updraft. For example, in the third (1940) edition of his book, Humphreys says: "After a time, each incipient hailstone gets into a weaker updraft, for this is always irregular and puffy, or else tumbles to the edge of the ascending column. In either case it then falls back into the region of liquid drops, where it gathers a layer of water, a portion of which, at once, is frozen by the low temperature of the kernel. But again it meets an upward gust, or falls back where the ascending draft is stronger, and again the cyclic journey from realm of rain to region of snow is begun; and each time—there may be several—the journey is completed, a new layer of ice and a fresh layer of snow are added" (244, pp. 359–60).

After seeing the relevance of the heat economy of a growing hailstone, first considered by Schumann in a paper published in 1938 (245), to the opacity of the accreted ice (and not liking the assumption of a fluctuating updraft, nor, unwisely, having read the much older literature), I tried to show that the typical alternation of the clear and cloudy layers of ice inside a hailstone represented fluctuations of the temperature of the accreting surface between 0°C and a lower value, which could arise during a single ascent and descent in a steady updraft (246, p. 56). However, it was necessary to appeal also to fluctuations in cloud water concentration to explain more than four layers, and subsequently I thought it probably necessary to return to the assumption of more than one up-and-down journey, to explain not only several or more layers, but also a large final size. The possibility of recycling at least once in a steady updraft, much as Ferrel imagined, was recognized only during the analysis of the remarkably steady hailstorm over southeast England in 1959, discussed in Section 8.4 (86).

In support of his idea Ferrel quoted accounts from balloonists in the United States and France who were drawn into cumulonimbus and apparently recycled several times before escaping. The first account is quoted from a letter written to the *New York Tribune* in 1857 by the balloonist Professor John Wise, describing a flight over Pennsylvania in June 1843. After ascending about 800 m from the ground the balloon moved toward the base of a "huge black cloud." In the letter the professor says that after entering it the updraft "would carry the balloon up to a point, where its force was expended by the outspreading of its vapour, whence the balloon would be thrown outward, fall down some distance, then be drawn into the vortex, again to be carried upwards to perform the same revolution, until I had gone through the cold furnace seven or eight times ... the last time of descent into this cloud brought the balloon through its base, where, instead of pellets of snow, there was encountered a drenching rain, with which I came into a clear field, and the storm passed on."

A lengthier account was published later (247), from which the following extracts are taken; they show that the balloon recycled through the zone of supercooling (whose base in this place and season is at an average height of about 3 km).

> The cloud appeared of a circular form as I entered it, considerably depressed in its lower surface, presenting a great concavity toward the earth, with its lower edges very ragged and falling downward with an agitated motion, and it was of a dark smoke color. Just before entering this cloud, I noticed, at some distance off, a storm-cloud from which there was apparently a heavy rain descending. The first sensations I experienced when entering this cloud were extremely unpleasant. A suffocating sensation immediately ensued its entrance, which was shortly followed by a sickness at the stomach, arising from the gyrating, swinging motion of my car, causing me to vomit several times in quick succession most violently; this vomiting, however, soon abated, and gave way to sensations that were truly calculated to neutralize more violent symptoms than a momentary squeamishness. The cold had now become intense, and everything around me of a fibrous nature became thickly covered with hoarfrost, my whiskers jutting out with it far beyond my face, and the cords running up from my car looking like glass rods, these being glazed with ice, and snow and hail was indiscriminately pelting all around me. The cloud at this point, which I presumed to be about the midst of it from the terrible ebullition going on, had not that black appearance I observed on entering it, but was of a light, milky color, and so dense just at this time that I could hardly see the balloon, which was sixteen feet above the car. From the intensity of the cold in this cloud I supposed that the gas would rapidly condense, and the balloon consequently descend and take me out of it. In this, however, I was doomed to disappointment, for I soon found myself whirling upward with a fearful rapidity, the balloon gyrating and the car describing a large circle in the cloud. A noise resembling the rushing of a thousand milldams, intermingled with a dismal moaning sound of wind, surrounded me in this terrible flight. Whether this noise was occasioned by the hail and snow which were so fearfully pelting the balloon I am unable to tell, as the moaning sound must evidently have had another source. I was in hope, when being hurled rapidly upward, that I should escape from the top of the cloud; but as in former expectations of an opposite release from this terrible place, disappointment was again my lot, and the congenial sunshine, invariably above, which had already been anticipated by its faint glimmer through the top of the cloud, soon vanished, with a violent downward surge of the balloon, as it appeared to me, of some hundred feet. The balloon subsided, only to be hurled upward again, when, having attained its maximum, it would again sink down with a swinging and fearful velocity, to be carried up again and let fall. This happened eight or ten times, all the time the storm raging with unabated fury, while the discharge of ballast would not let me out at the top of the cloud, nor the discharge of gas out of the bottom of it, though I had expended at least thirty pounds of the former in the first attempt, and not less than a thousand cubic feet of the latter, for the balloon had also become perforated with holes by the icicles that were formed where the melted snow ran on the cords at the point where they diverged from the balloon, and would by the surging and swinging motion pierce it through.
>
> I experienced all this time an almost irresistible inclination to sleep, notwithstanding a nauseating feeling in the stomach, causing me to vomit several times, and the terrible predicament I was placed in, until, after eating some snow and hail mixed, of which a considerable quantity had lodged on some canvas and paper lying in the bottom of the car, I felt somewhat easier in mind and in body (for it is no use to say that I cannot be agitated

and alarmed), and I grasped a firm hold of the sides of the car, determined to abide the result with as much composure as the nature of the case would admit; for I felt satisfied it could not last much longer, seeing that the balloon had become very much weakened by a great loss of gas. Once I saw the earth through a chasm in the cloud, but was hurled up once more after that, when, to my great joy, I fell clear out of it, after having been belched up and swallowed down repeatedly by this huge and terrific monster of the air for a space of twenty minutes, which seemed like an age. . . .

The density of this cloud did not appear alike all through it, as I could at times see the balloon very distinctly above me, also, occasionally, pieces of paper and whole newspapers, of which a considerable quantity were blown out of my car. I also noticed a violent convolutionary motion or action of the vapor of the cloud going on, and a promiscuous scattering of the hail and snow, as though it were projected from every point of the compass.

The following extract from the report of an experience by a sailplane pilot over Germany on 3 August 1938 (248) can be added to Ferrel's accounts:

I noticed a cumulus cloud tower overhead which was becoming darker and blacker at its base. With even spirals I climbed higher and higher. At the same time the lift was increasing. Soon it was 2 m sec^{-1} then 3 m sec^{-1}, and in no time I was about 700 m above the Wasserkuppe and close under the cloud. A quick glance to the left and right to check my 'chute connections and I was in the turbulent mass.

The ship climbed at 8 to 9 m sec^{-1} and while steadily circling I was thrown out of the top of the cloud tower. There before me I saw a gigantic, lofty cloud mass. With quick decision I noted my position and waded in. In a few seconds I was being pushed up with terrific force. It took me 70 seconds to go from 3 to 4 km.

Suddenly I felt a terrific jolt, another, and still more. There was another terrific jolt and a crash. I was tossed through the cockpit cowl with tremendous force. My head ached intensely. I heard the rending and tearing apart of plywood—followed by silence. Soon I collected my thoughts. I could tell the 'chute had opened by the sound the wind made against it. I couldn't see because of the darkness but I had the uncertain sensation that the 'chute had struck rising air currents and was ascending. Suddenly it became lighter and I saw over me the white canopy of my parachute. Then I fell into the center of a cyclone—and I knew this because of lack of condensation. I looked down as through a gigantic tube and saw a little piece of the earth which occasionally became cloud blanketed. After I had lost about 1 km of altitude, I was driven sideways into the cloud mass. To my astonishment, in a short time I found myself at the top of the cyclone cone again. This was repeated four times before I finally began to descend.

At last I came out of the clouds at about 750 m. I noticed small strips and splinters of my ship floating about me.

Appendix 8.14 Stationary local storms

When cumulonimbus form over land in light winds throughout the troposphere the precipitation chills the ground and often quickly leads to the cessation of the convection or to its displacement to some other favorable topographical feature. If the first clouds have formed early in the day over a prominent feature, they may be renewed after 1–2 hr, but in either case the cumulonimbus convection is obviously unsteady or intermittent. In the presence of a suitable wind shear it may become more organized, and the cumulonimbus may become more persistent, but then they usually travel away from their sources. Sometimes, however, they become slow-moving or practically stationary near a topographical feature, after forming over it or having approached it while already mature. Usually on such occasions, but not always, the wind in the low troposphere has a large component opposed to the winds in the upper troposphere, so that the cumulonimbus might be expected to be slow-moving, but frequently the storm becomes almost stationary near hills which are obviously not a heat source and whose importance seems to lie in their effectiveness as an obstacle. When the cumulonimbus become practically stationary, then even if the convection is not particularly intense, so that the rate of precipitation is only moderate for a thunderstorm rainfall, its continuation for several hours over the same area can have catastrophic effects on local communities, especially if they are beside rivers and on or at the foot of steep hill slopes. Such events have perhaps not been given the study they deserve, partly because of their rarity in a particular region and because it is generally only small communities which are so sited as to be at serious risk, but mainly because of the difficulty of obtaining sufficient relevant meteorological and hydrological information. Some recent examples from the United States, and from western Europe, are detailed in the following paragraphs, commencing with a description of some prolonged hailstorms whose small areal extents were particularly striking.

The Selden hailstorm

On 3 June 1959 hail began to fall in the late afternoon at Selden, a small town in northwestern Kansas (at 39.5°N, 100.6°W, almost halfway between the radiosonde stations at Dodge City, to the south, and North Platte, Nebraska, to the north). At first there were strong gusts and the hail broke many windows; the wind soon quietened, but the hail, with some rain, continued for 85 min (265). The hailstones were mostly of pea or marble size, "and many were soft." In and near the town they accumulated to a depth of 18 in. (45 cm), and a weighbridge recorded a weight of hail equivalent to a rainfall of about 30 cm, corresponding to an average precipitation rate of over 200 mm hr^{-1}. (Assuming a mean diameter of 1 cm and a mean fall speed of 14 m sec^{-1} the concentration of the hailstones in the air near the ground would have been about 4 g m^{-3}.) The great weight of hail upon flat or nearly flat roofs resulted in some damage to nearly every building. The area covered by the hail was elongated, about 14 km long and at most 10 km wide; aerial photographs show its boundary to be remarkably well-defined.

On this occasion a moist southerly airstream extended from western Oklahoma across Kansas into Nebraska, and scattered thunderstorms occurred over and east of the Rocky Mountains. The winds in the upper troposphere were westerly, and the evening sounding at Dodge City suggests that the afternoon cumulus convection in the lee of the mountains was suppressed. However, conditions were favorable for the development of local cumulominbus in the late afternoon over the west of Kansas and Nebraska. At screen level θ_w reached about 21°C, and the heights of the base and of the updraft outflow of the cumulonimbus were probably at about 3 and 10 km. At 10 km the upper westerlies were strong over Oklahoma and Kansas but weakened abruptly north of the 40th parallel. Near Selden their speed was about 30 m sec^{-1}; at Dodge City and at Goodland, about 100 km west of Selden, the shear between 300 m above the ground and 10 km was west-northwesterly, 32 and 30 m sec^{-1}, respectively, while at North Platte it was only 9 m sec^{-1}. Assuming a cloud base at about 700 mb with $\theta_w = 9°C$, consistent with the soundings and the screen temperatures, the (maximum) updraft speed at 10 km from the simple parcel theory was about 33 m sec^{-1} (sufficient to produce hail of diameter up to about 2 cm, as observed); accordingly, a mean Richardson number $-(\overline{Ri})$ for the updraft circulation was about 1, and the cumulonimbus near Selden could be expected to be efficiently organized. It is not clear why the storm should have been so slow-moving, nor is it evident why such a storm should have developed near Selden, where there is no obviously favorable topographical feature; however, the existence of such features in similar almost uniform terrain lying a few hundred kilometers southeast of Selden has already been illustrated indirectly in Pl. 8.18.

Prolonged hailstorms reported from Austria and southern England

A storm of the same kind as the Selden hailstorm was reported in Semmering, in a valley about 80 km south-west of Vienna, on 5 June 1947 (266). On this occasion a thunderstorm occurred during the eastward passage of a cold front zone. Rain and small hail began at 1340h; the rain continued for 7 hr, but the hail ceased at 1545, to be succeeded by other falls between 1600 and 1645 (in which hailstones reached the size of large cherries, weighing 4 g), and between 1750 and 1830. Afterward the scene was like a winter landscape, with hail 20 cm deep in places. The total precipitation collected in a gauge amounted to 320 mm.

A prolonged hailstorm occurred at Camelford, in Cornwall, southern England, on 8 June 1957 (267). About 140 mm of precipitation fell between 1230 and 1500 GMT, of which about 100 mm in the first hour included small hail (diameters up to about 1 cm) which was washed about in flood waters, and in places accumulated in congealed masses almost knee-deep. This and a later storm (most intense between 1700 and 1800 GMT) brought the rainfall total to about 200 mm over a small area of 200 km^2, mainly on the windward (southwestern) slope of Bodmin Moor, northeast of Camelford. Ordinary showers and thunderstorms were widespread in a southwesterly airstream, ahead of a slow-moving cyclone centered about 500 km to the west of Cornwall, in which the wind shear in the troposphere was very small.

A prolonged hailstorm in a more typical large-scale meteorological situation began at Tunbridge Wells (about 50 km southeast of London) about noon on 6 August 1956 as a thunderstorm with rain (268), and then produced very locally a continuous fall of small hailstones (of diameter up to 1.5 cm) for about 90 min. Hail choked drains; swept by flood water it accumulated feet deep in places and blocked roads, while 5 km away a cricket match continued without interruption in bright sunshine. Similar storms may have occurred during the same afternoon elsewhere in southeast England, judging by a record of a rainfall amount of 60 mm in 45 min at Throwley, on the north side of the North Downs, and by unofficial reports of severe flooding due to prolonged hail and heavy rain near Chelmsford (40 km northeast of London). It is probably significant that all these places have hills to the south and that over this part of England the wind at low levels was northerly or northeasterly (in the circulation of a weak cyclone centered over northern France), whereas above about 2 km the winds became southerly. The value of θ_w at the cloud bases was about 13°C, much below the values of about 20°C characteristic of severe storms at this season, but the available potential energy was sufficient to indicate a maximum updraft speed of about 25 m sec^{-1} (consistent with the size of the hailstones). The wind shear between levels near the ground and the level of the updraft outflows (about 7 km) was 20 m sec^{-1}, so that the mean Richardson number $-(\overline{Ri})$ for the updraft circulation was approximately 1.

Other prolonged hailstorms and rainstorms

In storms in which, as in the previous examples, hail falls continuously at one place for as long as an hour or more, the hailstones are all small and are accompanied by rain, or they fall during rains of greater duration. Slow-moving but persistent storms in which only rain reaches the ground are more frequent. It seems therefore that although the cumulonimbus on these occasions are well organized, the maximum speeds attained by their updrafts are only moderate (rarely as much as 30 m sec^{-1}), implying that the available potential energy is only moderately large, and probably, then, also that the wind shear between low and high levels in the troposphere is only moderately large. If, moreover, a favorable condition for a storm to be stationary is that the wind at low levels should be opposed to the winds at high levels, which outside the tropics are almost always westerly, it is understandable that a characteristic location for such storms is on the poleward side of weak cyclonic circulations near the ground, removed by a few hundred kilometers from frontal zones of pronounced wind shear and the local regions over which the air at low levels normally acquires its highest potential temperatures. Storms of this kind occasionally produce outstanding falls of rain in southern and eastern England, and more rarely in northwestern France, associated with cyclones whose centers usually travel along paths from near the Bay of Biscay through the English Channel and into the North Sea. Similar storms may occur near the Pyrenees and the Alps, especially on their southern and eastern sides, but heavy rains there apparently have not been given any comprehensive study from this point of view.

Although such rainstorms may have a duration of 12–24 hr or even more, they tend to have intense phases lasting several hours (the typical duration of intense traveling storms in Europe). The related cyclones usually move eastward or northeastward, and the nearly stationary cumulonimbus may then develop successively over favorable ground features in about the same position relative to the cyclone center. If so, it is only to be expected that when, as in the past, precipitation data consist almost entirely of rainfall totals from gauges read only once or twice a day, the most prominent characteristic of the rains is their disposition in a narrow belt parallel to the path of the cyclone center and a few hundred kilometers north of it. The belt may contain two or more small regions in which the rainfall greatly exceeds the general average. It frequently happens, however, that the cumulonimbus are not intense and produce no hail at the ground and sometimes little or no thunder and lightning; moreover, they often occur in the dark hours (supplied with potentially warm air produced during the previous day by sunshine over land to the southeast), and for this reason or because they are veiled by extensive cloud in the low and middle troposphere, they are not recognizable visually. The rains have therefore generally been regarded as essentially "cyclonic," and not as posing a problem in cumulonimbus dynamics. Analyses of the associated large-scale flow patterns have usually shown features such as horizontal convergence in the low troposphere near the rainstorms and horizontal divergence in the upper troposphere over the rainstorms, which are consistent with their existence but give no insight into their generation and maintenance.

Exceptional rainstorms in southern England

Nearly all of the several outstanding rainstorms which have occurred in southern England in the last few decades appear to have been of the kind under discussion. They have been responsible for falls of between about 10 and 20 cm of rain over areas of order 100 km^2, mainly during periods of several hours. All but one of the storms occurred with northeasterly surface winds, on the north side of depressions (or troughs of low pressure) centered in or near the English Channel.

Aerological data are available for the occasions of the two most recent storms. In both, the potentially warm air arrived locally from the southeast at about 2 km, above the cool northeasterly winds near the surface, and there are no soundings which at low levels are exactly representative of the storm areas. Near one (the Lynmouth storm of 1952) the wind shear between 2 km and the upper troposphere was 240° 12 m sec^{-1}, not inconsistent with a mean Richardson number for the updraft circulation of about 1, judging by the absence of reports of hail; there appears to have been an available potential energy (with $\theta_s \approx 16°C$ in the updraft), but its amount was too small to be estimated with any confidence. Near the other storm the corresponding wind shear was about 240° 15 m sec^{-1}, and the available potential energy (with $\theta_s \approx 19°C$ in the updraft) appears to have been sufficient to give a maximum updraft speed of 20 m sec^{-1}; again the absence of reports of hail suggests that the actual speeds are unlikely to have exceeded this value, and a mean Richardson number of about 1 is indicated. On both occasions the height of the storm tops was probably rather low, with the updraft outflows between about 8 and 9 km.

Exceptional rainstorms in northwest France

Sometimes the shallow depressions or troughs which form over the Bay of Biscay or France in thundery summer weather occupy positions farther to the south than previously described, leading to northeasterly winds across the northwest of France. Conditions may then become favorable for the development of prolonged rainstorms there, particularly near the northern slopes of the hills inland over Brittany, although apparently they occur more rarely than over southern England. (It was over these hills that sferic locations first indicated the intensification of a storm which was later detected by radars in mid-Channel and became the previously discussed severe hailstorm which traveled across England on 9 July 1959. On this occasion the axis of the trough at sea level extended from central Spain to the Low Countries, and the surface northeasterly winds were present over the whole of England as well as Brittany.) Nevertheless, the outstanding daily rainfall total for the north of France, of 25 cm near Saint-Malo on 17 September 1929 (269), occurred under just these circumstances, during a storm which appears in the following extracts from a report (270) by an observer at Dinard (a few kilometers to the west of Saint-Malo) to be the most prolonged thunderstorm mentioned in the meteorological literature:

> This most extraordinary thunderstorm started at about eight o'clock on Sunday evening, Sept. 15, 1929, and ended at about four o'clock on Tuesday afternoon, Sept. 17. The storm kept on all Sunday night without a break. The wind blew at gale force from the NE. The lightning was exceptionally vivid and lit up the whole of the coast. There was a continuous roar of thunder. The storm quietened down towards Monday morning. It started again with renewed violence at about ten o'clock on Monday morning, and torrential rain, accompanied by a stormy breeze, set in. Lightning struck the hotel conductor several times. The storm increased in violence towards night, and the power station at St. Brieux was struck, so that there was no electric light at St. Lunaire and candles had to be used in the hotel. Tuesday morning the weather was no better; vivid lightning and crashing thunder still continued. Women in the hotel were terrified and could not get any sleep. The rain fell in torrents. At about 4 o'clock in the afternoon the sky cleared and the thunder only rumbled in the distance and finally disappeared altogether.
>
> The damage done by the storm was considerable. So much rain had fallen that a bridge was swept away on the main Dinard to St. Lunaire road. A motor-car passing at the time was swept out to sea. All the streets in the village were torn up and shops and houses flooded.

A cold front moved south across the Channel early on 15 September and became slow-moving over northeast France but advanced over the Bay of Biscay and eventually entered southwest France on 17 September. A cool northeasterly airstream persisted across northwest France while a shallow depression developed over central France; a potentially warm southerly or southeasterly flow from the central and southern parts of the country ascended in the frontal zone and probably arrived over the hills inland of the northwest coast at heights above about 1500 m, with θ_w about 20°C, sufficient to provide a moderately large available potential energy for cumulonimbus convection (at screen level over southern and central France on 15 and 16 September the maximum temperatures exceeded 30°C and θ_w reached values between 20 and 24°C). Considering the general flow pattern and the substantial temperature gradient at the surface across France the winds in the upper troposphere over northwest France would almost certainly have been southwesterly and moderately strong.

During the period from 15 to 17 September scattered thunderstorms in the south of France were clearly separate from an isolated narrow zone extending from central Brittany across the Channel Islands to Cherbourg, from which there were reports of intermittent thundery rain, and in which the thunderstorm near Dinard appears to have been a principal isolated and stationary storm. On 16 and 17 September a total of over 10 cm of rain was reported from Jersey (about 100 km to the north), and on the sixteenth 14 cm from Cap Frébel (less than 20 km west of Dinard), of which 11 cm fell in 12 hr.

Exceptional rainstorms in the United States

A rainfall of over 20 cm in a day, typically over an area whose largest dimension is 20 km or more, is reported from some middle-latitude region in the northern hemisphere perhaps once in each summer; indeed, this is about the frequency in the United States alone, where such occurrences are usually discussed in official publications. Probably most if not all of these rains fall from nearly stationary cumulonimbus, although for the reasons already stated the evidence has not usually been conclusive, and the meteorological situation is usually complicated by the evident but not simple role of topographical features, and of the large-scale baroclinic developments usually required to provide the favorable wind distributions. It has frequently been remarked that the heaviest rainfall is often over low ground distinctly on the windward side of prominent high ground, so that the conventional attribution of the rain to the ascent of an airstream over an orographic obstacle is not adequate. The exceptional daily rainfalls occur mostly during the summer and fall in the Gulf states, where a large proportion is associated with tropical storms, not usually of hurricane intensity. In Texas, in particular, a topographic feature whose importance has been recognized is the Balcones escarpment of the plateau region, whose slope lies just west of several places which have experienced very heavy rains; for example, Taylor, Austin, Smithville, and Montell have experienced falls of about 50 cm in 24 hr, all of which occurred below and not on or above the slope (271).

The exceptional fall at Taylor was recorded during a period of three days in September 1921. A shallow depression (central pressure about 1008 mb) appeared to enter the extreme south of Texas and become stationary, maintaining a southeasterly flow of air at low levels from the Gulf of Mexico into southern and central Texas. At Taylor rain began at 0330h on 9 September; thunder was heard to the south at 1620, and after 1900 thunder and lightning were incessant all night, the thunder not ceasing until noon on the following day (272). Its total duration was therefore almost 20 hr. Nearly 60 cm of rain was recorded in 24 hr, of which 27 cm fell in three evening hours; it was estimated that locally as much as 75 cm may have fallen in 15 hr. The dimensions of the area enclosed by the isohyet for 20 cm were about 180×80 km (273), and mass of rainfall within it was about 450×10^{10} kg, so that clearly the total fall near Taylor was at least an order of magnitude greater than those which have been experienced in southern England. (In the Gulf states the supply of moist and potentially warm air—from over the sea—is not limited, as in Europe, by its day-to-day production over land in sunshine.) Disastrous floods caused 87 deaths near Taylor, and another 51 near San Antonio, where a similar storm produced falls of probably up to 50 cm.

Of the heavy rainfalls which have been reported from Florida, an outstanding example occurred on 21 January 1957, and was observed by radar from Miami (274). The storm formed a little south of Lake Okeechobee in southern Florida and moved very slowly westward, reaching a position just inland of the east coast by about 10h EST and then remaining practically stationary for the remainder of the day. Over and just south of the shore of the lake 15–18 cm of rain fell between 04 and 10h and later 40 cm fell between 11 and 16h several kilometers southwest of West Palm Beach, in a small region in which three gauges recorded a total for the day of 53 cm. Nearer the coast about 15 cm of rain and hailstones as large as golf balls fell between 20 and 22h, when the tops of the radar echo from the storm were well above 10 km. Aircraft pilots reported during the day that there was extensive cloud with tops a little above 3 km, but some clouds "extended to great elevations." It seems probable that most of the rain and the hail (of which there was apparently only the one report) fell intermittently during periods when the storm was particularly intense. Radar echoes from smaller showers which developed near and just east of the Atlantic coast moved westward, some merging with the echo of the main storm.

The sounding made at Miami at 1000 EST confirms that the cumulus layer extended up to about 3 km, but the middle troposphere was moist and cool, and according to the simple parcel theory low-level air ascending in cumulonimbus, with a mean θ_w of 19.6°C, could have reached a maximum speed of about 50 m sec^{-1} at 11 km. The winds in the low troposphere were east-southeasterly, 12 m sec^{-1} at 500 m and 20 m sec^{-1} at 2 km, but they were 10 m sec^{-1} or less above 3 km, veering upward west-northwesterly at 7 km; at higher levels the speeds were greater, reaching 17 m sec^{-1} at 9 km, and at 11 km the wind was northwest with a maximum speed of 43 m sec^{-1}. Thus the shear between low levels and the level of outflow from cumulonimbus is indicated as northwesterly, about 55 m sec^{-1}, and the mean Richardson number $-(\overline{Ri})$ for the updraft circulation was therefore very nearly 1. It is surprising that hail is mentioned only incidentally in the accounts of the storm; the main interest in the inhabited localities was in the phenomenal rainfall. Conditions were unfavorable for the production of a downdraft, but large hail would be expected from the calculated maximum updraft speed in the zone of supercooling. However, hailstones with diameters as great as 2 cm might have melted completely before reaching the ground if they were composed mainly of spongy ice.

Remarkably, rainfalls of 35–50 cm in 24 hr have twice again been experienced in the same locality, under similar conditions in which the winds were easterly in the low troposphere and westerly in the high troposphere (279).

In recent years intense persistent rainfalls have been reported from widely scattered parts of the United States: in Maine in 1959 (275), in Utah (276) and Oklahoma (277) in 1963, in Missouri in 1964 (276), in Florida in 1965 (279), near the border between southern New Mexico and Texas in 1966 (280), and in Montana in 1967 (281). On most of these occasions the presence of cumulonimbus has been recognized by the occurrence of thunderstorms (even in the winter season) or by radar observations, and the large-scale conditions have been of the kind previously described as favorable for prolonged local cumulonimbus convection. The potentially warm air has been in a southerly or southeasterly airstream beneath westerlies or northwesterlies in the upper troposphere, and has sometimes arrived in the storm area above a shallow intrusion of cold air from the east or northeast, north of the center of a small depression. However, detailed radar observations have sometimes shown the cumulonimbus behavior to be more complicated than has been suggested. For example, in a heavy fall of rain over a small area in Illinois, with amounts locally exceeding 25 cm in 16 hr, it appeared that several groups of cumulonimbus successively moved through the area from the northwest, with speeds diminishing from 6–20 m sec^{-1} during their approach to only about 5 m sec^{-1} during their passage (282). The winds were southwesterly near the ground and westerly aloft, increasing to about 20 m sec^{-1} in the high troposphere, and the reasons for the observed storm motions are obscure.

8 Cumulonimbus convection

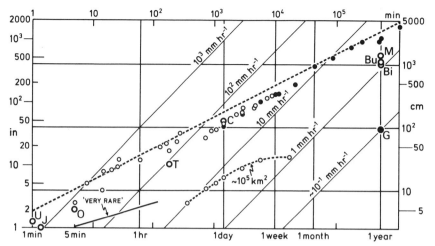

Fig. 8.62 Values of the greatest rainfalls recorded at individual stations, as a function of their duration, taken mostly (small circles) from reference 283. The straight dashed line represents the relation $R = 14.3 D^{1/2}$, where R (in.) is the amount of a fall of duration D (hr), which has been noticed to represent an upper limit to the values plotted (284), though not for any obvious physical reason. Another dashed line represents an upper limit for falls over an area of 5×10^4 mi^2 (about 10^5 km^2), from some observations made in the United States. The solid line on the lower left represents the lower limit of falls officially placed in the category of very rare in the British Isles.

The value marked G shows the mean annual global rainfall. The other marked values refer to rains in the following places: U, Unionville, Maryland (262); J, near Jefferson, Iowa (261); O, Demos, Ohio (5 September 1890; 286); T, Taylor, Texas (272); C, Cangamba, Angola (48.9 in., 7 Feb. 1929; 285); Bi, Bidundi, Cameroons (421 in. annual average over 5 yr; 286); Bu, Buenavista, Colombia (434 in., 1928; 286); M, Piui-Kikui, Maui (562 in., 1918; 286). The small full circles all refer to records from Cherrapunji, in Assam. All the records, except that from Maui, were of rains predominantly if not wholly from cumulonimbus.

World record rainfalls

The most intense recorded rainfalls, whatever their duration, are almost certainly due to cumulonimbus, those measured over periods of 1 hr or longer being from storms which are stationary and persistent, and those for periods of a day or more from nearly stationary storms which continually form over a favorable locality in a slow-moving or persistent large-scale motion system, such as a tropical storm or a monsoon depression (such rains have all been recorded in low latitudes, at places not far inland which are readily reached by moist airstreams from the oceans). Figure 8.62 shows some of the largest recorded falls as a function of their duration. Almost all of those for periods of more than a week are from Cherrapunji, at a height of 1300 m on the south slopes of the Khasi Hills of Assam, whose position is indicated in Fig. 8.31. It is thought that nearby places, for example, Manoyuram, a short distance to the south at 1100 m, may have even greater rains. These places are on the windward sides of the hills and receive their highest rainfall in the months of the summer monsoon, when the winds in the high troposphere are easterly. In this season, therefore, conditions may be favorable for almost stationary cumulonimbus over the hills nearly every day.

Appendix 8.15   Visual range in intense rains

The scattering coefficient τ_r of raindrops at optical wavelengths, which are very small compared with the drop sizes, is simply

$$\tau_r = 2\pi \int_0^\infty r^2 n(r)\, dr$$
$$= \frac{\pi}{2} \int_0^\infty D^2 n(D)\, dD \text{ cm}^{-1}$$

where $n(D)\,dD$ is the concentration of drops with diameters between D and $(D + dD)$.

If the size distribution of the raindrops is given by Eq. 8.4 and 8.5,

$$n(D) = n_0 e^{-\Lambda D}$$
$$n_0 = 0.08 \text{ cm}^{-4}$$
$$\Lambda = 41\, R^{-0.21} \text{ cm}^{-1}$$

where R is the rainfall rate (in mm hr^{-1}). Thus

$$\tau_r = \frac{0.08\pi}{2} \int_0^\infty D^2 e^{-\Lambda D} dD$$

$$= \left(\frac{0.08\pi}{2}\right)\left(\frac{2}{\Lambda^3}\right)$$

$$= 3.6 \times 10^{-6} R^{0.63} \text{ cm}^{-1}$$

Now from Eq. 4.19 the visual range V_R is

$$V_R = 3.9/\tau_r$$

using the conventionally accepted value for the least perceptible contrast between a distant dark body and the background of the sky near the horizon. From this relation values of the visual range have been evaluated for certain rates of rainfall, and are listed in Table 8.17, where they may be compared with some estimates based on recording rain gauge and eye observations.

Table 8.17 Some mean values of the visual range V_R in rains of intensity R, from eye and rain gauge observations in India (259) and Russia (260), and theoretical values corresponding to the drop size distributions given by Eq. 8.3 and 8.4, for which the median volume diameter is D_o

		V_R (km)		
			From observation	
R (mm hr^{-1})	D_o (mm)	From theory	(260)	(259)
1	0.9	11	12	—
10	1.4	2.5	2.5	~3
50	2.0	0.9	0.8	~0.9
80	2.3	0.7	0.6	~0.5
300	3.0	0.3	—	—
1,000	3.8	0.15	—	—

Appendix 8.16 Radar observations of cumulonimbus

The height of cumulonimbus tops

An important property of a cumulonimbus, and particularly of an intense storm, is the height which is reached by the cloud top. This height can be measured accurately (well within the desirable ± 0.5 km) by careful photogrammetry or by triangulation using coordinated observations with at least a pair of theodolites, and somewhat less accurately with one camera or theodolite and an approximate measure of range by radar observation. When the cumulonimbus is obscured visually the radar alone shows the height of the echo top on an RHI (range-height) display. In a rising cloud tower in which the precipitation is well developed, or in the dome of a cumulonimbus which protrudes into the lower stratosphere, the radar reflectivity a little below the level of the cloud top should be well above the lower limit of sensitivity of most radars (corresponding to a reflectivity factor Z_e of order $10^{-4} r^2$ mm^6 m^{-3}, where r is the range in kilometers), even at ranges of 200 km or more. If, for example, it is supposed that within a cloud top near the tropopause the mean undraft speed decreases with height at the rate of 20 m sec^{-1} km^{-1}, and that the precipitation consists of spherical ice particles in a size distribution defined by Eq. 8.18, up to a size at which the fall speed equals the updraft speed, then it can be calculated that the radar reflectivity factor Z_e is 10^4 mm^6 m^{-3} about 0.67 km below the cloud top, and 10^2 mm^6 m^{-3} only about 0.33 km below the cloud top. Since turbulent motions inside the cloud would tend to reduce these distances, the top of a vigorous cumulonimbus can be regarded as effectively at the level where the radar reflectivity factor becomes less than about 10^2 mm^6 m^{-3}, which might be expected to correspond closely to the level of the echo top observed on the radar display.

However, the distribution of echo intensity presented on the display is a distortion of the distribution of the radar reflectivity in the storm, and some precautions must be taken before echo tops can be regarded as giving an acceptable representation of cloud tops. First, it must be insured that the radar beam can be directed accurately in elevation; this requires checking that the antenna pedestal can be correctly leveled, and examining the emitted radiation at a rig a few decameters from the radar to confirm the position of the antenna when the axis of the beam is level. It may be necessary to add a cursor to the antenna so that it can be tilted accurately to direct the beam at certain angles during the setting up of the RHI display. Unless these measures are taken and tests are repeatedly made, the elevation angles may easily be in error by 0.5°, thereby introducing an error of about 1 km in the displayed height of an echo top at a range of 100 km. With these precautions, however, this kind of error can be removed (e.g., it was found that in the important range of 5-10°, the elevation angle of aircraft indicated by the MPS-4 radar used in the construction of Fig. 8.41 could be kept within $\pm 0.1°$ of the value simultaneously determined with a theodolite whose error did not exceed 0.03°).

A generally less important error may arise from the assumption usually made in setting up the display that the radar beam has the particular curvature associated with refraction in a standard atmosphere. The radar rays are regarded as straight lines over an earth's surface of radius of curvature $R' = fR$, where R is the actual radius of the earth and $f = 4/3$ (290; Eq. 2.67, Appendix 2.5). Occasionally the atmospheric stratification is sufficiently different from the implied standard for this assumption to cause considerable error. This is strikingly evident when inversions of lapse rate in the low troposphere, accompanied by a rapid decrease of water vapor density with height, deflect downward radiation emitted at small elevation angles, producing a radar "duct" and allowing the detection of targets well beyond the normal radar horizon. Such "super-refraction" is sometimes observed when scanning over regions occupied by downdraft outflows from thunderstorms. Excluding these obviously abnormal regions, under the conditions in which cumulonimbus occur over land in the afternoon and over the sea the errors in the inferred height of their echo tops arising from deviations of the atmospheric stratification from the standard are probably hardly ever more than 100 m at ranges up to 150 km, but they may reach 500 m at ranges of 300 km.

On the radar display more important and unavoidable distortions of the actual distribution of radar reflectivity in a storm arise because significant amounts of energy are radiated at considerable angles from the axis of the radar beam, and sometimes also because the beam is attenuated in the nearer precipitation. The beam is specified as having a "width" of twice the angle ϕ_0 from the axis at which the emitted radiation has half the intensity of that emitted along the axis. Most of the meteorological radars now in use have simple dish antennas in the form of a section of a paraboloid. With these the power decreases very rapidly as the angle ϕ from the axis increases beyond ϕ_0, but it has minor maxima in "side lobes," of which the first three are at $4\phi_0$, $6\phi_0$, and $8\phi_0$, where the emitted power is, respectively, about 24, 27, and 30 db below that on the axis. If radars are to have a useful resolution, ϕ_0 should not exceed about 1°, but clearly echo will be displayed as though on the axis of the beam if the radar reflectivity at an elevation as much as 4° lower is about 48 db or more above the minimum detectable. With radars of the normal sensitivity previously quoted this leads to gross distortion in the displayed reflectivity above regions when the radar reflectivity factor Z_e in the middle of cumulonimbus approaches its maximum observed values of about 10^7 mm^6 m^{-3}. Although usually it is only the first side lobe which causes this trouble, at short ranges or with more sensitive radars other side lobes may contribute, so that columns of echo extend up to the top of the display, to heights which are never reached by cumulonimbus. Fortunately the shape of the displayed echo often obviously indicates this kind of distortion, and in estimating echo heights it can be removed by appropriately reducing the sensitivity of the radar, without raising the minimum detectable reflectivity above the value of about 10^2 mm^6 m^{-3} which can be regarded as defining the top of a vigorous storm cloud. It then remains to correct the distortions introduced by the main lobe of the beam.

Calculations have been made of the distortions produced in the displayed echo using typical radar beam characteristics and models of actual distributions of radar reflectivity in cumulonimbus, and consideration has been given to the procedures by which the actual distributions might be recovered from those displayed (291). On the basis of the calculations it has been suggested that the displayed height of an echo top should be reduced by an amount which is a function of beam width and echo range, obtained from a nomogram or an empirical formula. However, at ranges beyond about 150 km the corrections exceed 3 km and are strongly dependent on the assumptions made about the actual radar reflectivity, concerning which there is hardly any reliable information on either the typical distribution or its variability. Accordingly, it appears from the formulas that there is little justification for anything but a common practice of reducing the indicated elevation by the half beam width ϕ_0, involving a height reduction of about 1 km at 50 km and 2.5 km at 150 km. Considering all sources of error, and presuming all the necessary precautions have been taken, it may be hoped that the corrected heights will be accurate to about ± 0.5 km at ranges up to 150 km, which at least at present must be regarded as the limit at which useful quantitative measurements can be made.

Details in the distribution of radar reflectivity; the echo-free vault

A true distribution of radar reflectivity is enlarged on the display by the distortion due to the beam width, and large gradients of reflectivity are smoothed, so that especially at the greater ranges significant detail in the pattern of reflectivity may be obscured. In particular the interesting maximum radar reflectivity within a volume of dimensions of a few kilometers in the heart of the storm may be reduced by a few decibels at ranges of only a few tens of kilometers, and by 10 db or more at a range of 150 km. Further, a vault in which precipitation is absent may be reduced in size to an apparently trivial feature or even be completely obscured unless the receiver gain is reduced. In the study of the intense hailstorm of 9 July 1959 over southeast England, in which the significance of the echo-free vault was appreciated for the first time, it was recognizable as a persistent feature because its greater horizontal dimension remained normal to the radar beams, while its narrow dimension of 2 km or less was easily within the resolution in range of the radars, which is determined by the pulse length and is only a few hundred meters. If the horizontal orientation of the length of the vault had been otherwise, the broad horizontal beams of the height-finding radars would have filled it with readily detectable echo. The distribution of the radar reflectivity factor in the storm, and in particular the region of its maximum value, was similarly made clearer because the reflectivity contours were arranged with their greater horizontal dimension approximately normal to radar beams, so that the distortion on the displays was not a serious interference.

References

1. Warner, J., and Twomey, S. 1967. The production of cloud nuclei by cane fires and the effect on cloud droplet concentration, *J. Atmos. Sci.*, **24**, 704–6.
2. Zaitsev, V.A. 1950. Liquid water content and distribution of drops in cumulus clouds, *Technical Translation TT-395 from Trudy Glavnoi Geofiz. Obs.*, **19**, 122–32 (Nat. Res. Council of Canada, Ottowa, 1953).
3. Battan, L.J., and Reitan, C.H. 1957. Droplet size measurements in convective clouds, in *Artificial Stimulation of Rain.* Pergamon Press, London, pp. 184–91.
4. Weickmann, H.K., and aufm Kampe, H.J. 1953. Physical properties of cumulus clouds, *J. Meteorol.*, **10**, 204–11.
5. Squires, P. 1958. The microstructure and colloidal stability of warm clouds, *Tellus*, **10**, 256–71.
6. Mordy, W.A. 1959. Computations of the growth by condensation of a population of cloud droplets, *Tellus*, **11**, 16–44.
7. Durbin, W.G. 1956. *Droplet Sampling in Cumulus Clouds.* M.R.P. 991, a paper of the Meteorological Research Committee, London. (A copy is available in the Meteorological Office Library.)
8. Durbin, W.G. 1959. Droplet sampling in cumulus clouds, *Tellus*, **11**, 202–15.
9. Murgatroyd, R.J., and Garrod, M.P. 1960. Observations of precipitation elements in cumulus clouds, *Q. J. R. Meteorol. Soc.*, **86**, 167–75.
10. Squires, P. 1958. The spatial variation of liquid water and droplet concentration in cumuli, *Tellus*, **10**, 372–80.
11. Weickmann, H.K., and aufm Kampe, H.J. 1953. Physical properties of cumulus clouds, *J. Meteorol.*, **10**, 204–11.
12. Braham, R.R., Battan, L.J., and Byers, H.R. 1957. *Artificial Nucleation of Cumulus Clouds.* Meteorological Monographs 2, no. 11, pp. 47–85, American Meteorological Society, Boston.
13. Browne, I.C., Day, G.J., and Ludlam, F.H. 1955. Observations of small shower clouds, *Meteorol. Mag.*, **84**, 72–76.
14. Brown, E.N., and Braham, R.R. 1959. Precipitation-particle measurements in trade wind cumuli, *J. Meteorol.*, **16**, 609–16.
15. Garcia-Prieto, P.R., Ludlam, F.H., and Saunders, P.M. 1960. The possibility of artificially increasing rainfall on Tenerife in the Canary Islands, *Weather*, **15**, 39–51.
16. Best, A.C. 1950. The size distribution of raindrops, *Q. J. R. Meteorol. Soc.*, **76**, 16–36.
17. Best, A.C. 1951. Drop-size distribution in cloud and fog, *Q. J. R. Meteorol. Soc.*, **77**, 418–26.
18. Khrgian, A. Kh, ed. 1961. *Cloud Physics.* Translated from the Russian by the Israel Program for Scientific Translations, and published by the Office of Technical Services, U.S. Department of Commerce, Washington, D.C.
19. Atlas, D., and Ludlam, F.H. 1961. Multi-wavelength radar reflectivity of hailstorms, *Q. J. R. Meteorol. Soc.*, **87**, 523–34.
20. Blair, T.A. 1928. Hailstones of great size at Potter, Nebr., *Mon. Weather Rev.*, **56**, 313.
21. Atlas, D. 1964. Advances in radar meteorology, *Adv. Geophys.*, **10**, 318–478.
22. Hardy, K.R. 1963. The development of raindrop-size distribution and implications related to the physics of precipitation, *J. Atmos. Sci.*, **20**, 299–312.
23. Fujiwara, M. 1965. Raindrop-size distributions from individual storms, *J. Atmos. Sci.*, **22**, 585–91.
24. Srivastava, R.C. 1967. On the role of coalescence between raindrops in shaping their size distribution, *J. Atmos. Sci.*, **24**, 287–92.
25. Plank, V.G., Atlas, D., and Paulsen, W.H. 1955. The nature and detectability of clouds and precipitation as determined by 1.25-cm radar, *J. Meteorol.*, **12**, 358–78.
26. Battan, L.J. 1953. Observations on the formation and spread of precipitation in convective clouds, *J. Meteorol.*, **10**, 311–24.
27. Braham, R.R. 1958. Cumulus cloud precipitation as revealed by radar—Arizona 1955, *J. Meteorol.*, **15**, 75–83.
28. Ludlam, F.H., and Saunders, P.M. 1956. Shower formation in large cumulus, *Tellus*, **8**, 424–42.
29. Byers, H.R., and Hall, R.K. 1955. A census of cumulus-cloud height versus precipitation in the vicinity of Puerto Rico during the winter and spring of 1953–1954, *J. Meteorol.*, **12**, 176–78.
30. Braham, R.R., Reynolds, S.E., and Harrell, J.H. 1951. Possibilities for cloud seeding as determined by a study of cloud height versus precipitation, *J. Meteorol.*, **8**, 416–18.
31. Workman, E.J., and Reynolds, S.E. 1949. Electrical activity as related to thunderstorm cell growth, *Bull. Am. Meteorol. Soc.*, **30**, 142–44.
32. Battan, L.J., and Braham, R.R. 1956. A study of convective precipitation based on cloud and radar observations, *J. Meteorol.*, **13**, 587–91.
33. Harrington, E.L. 1968. Initial precipitation echo in clouds that combine to form a thunderstorm, *Proc. 5th Conf. Severe Local Storms, St. Louis, Mo., 1967*, 193–97.
34. Battan, L.J. 1963. Relationship between cloud base and initial radar echo, *J. Appl. Meteorol.*, **2**, 333–36.

35. Clark, R.A. 1960. A study of convective precipitation revealed by radar observation, Texas, 1958–59, *J. Meteorol.*, **17**, 415–25.
36. Braham, R.R., Battan, L.J., and Byers, H.R. 1957. *Artificial Nucleation of Cumulus Clouds*. Meteorological Monographs, 2, no. 11, pp. 47–85, American Meteorological Society, Boston.
37. Morris, T.R. 1957. Precipitation in Arizona cumulus as a function of cloud size and temperature, *J. Meteorol.*, **14**, 281–83.
38. Saunders, P.M. 1965. Some characteristics of tropical marine showers, *J. Atmos. Sci.*, **22**, 167–75.
39. Scott, W.T. 1968. Analytic studies of cloud droplet coalescence I, *J. Atmos. Sci.*, **25**, 54–65.
40. Berry, E.X. 1967. Cloud droplet growth by collection, *J. Atmos. Sci.*, **24**, 688–701.
41. Berry, E.X., and Reinhardt, R.L. 1974. An analysis of cloud drop growth by collision, Parts I and II, *J. Atmos. Sci.*, **31**, 1814–31.
42. Berry, E.X., and Reinhardt, R.L. 1974. An analysis of cloud drop growth by collision, Parts III and IV, *J. Atmos. Sci.*, **31**, 2118–35.
43. Murgatroyd, R.J., and Garrod, M.P. 1960. Observations of precipitation elements in cumulus clouds, *Q. J. R. Meteorol. Soc.*, **86**, 167–75.
44. Day, G.J. 1956. *Further Observations of Large Cumuliform Clouds by the Meteorological Research Flight*. M.R.P. 980, a paper of the Meteorological Research Committee, London. (A copy is available in the Meteorological Office Library.)
45. Mossop, S.C. 1967. Comparisons between concentration of ice crystals in cloud and the concentration of ice nuclei, *J. Rech. Atmos.*, **3**, 119–24; Mossop, S.C., Ruskin, R.E., and Heffernan, K.J. 1968. Glaciation of a cumulus at approximately −4C, *J. Atmos. Sci.*, **25**, 889–99.
46. Koenig, L.R. 1963. The glaciating behavior of small cumulonimbus clouds, *J. Atmos. Sci.*, **20**, 29–47.
47. Braham, R.R. 1964. What is the role of ice in summer showers?, *J. Atmos. Sci.*, **21**, 640–45.
48. Koenig, L.R. 1966. Numerical test of the validity of the drop-freezing/splintering hypothesis of cloud glaciation, *J. Atmos. Sci.*, **23**, 726–40.
49. Thorkelsson, T. 1946. Cloud and shower, *Q. J. R. Meteorol. Soc.*, **72**, 332–34.
50. Craddock, J.M. 1949. The development of cumulus cloud, *Q. J. R. Meteorol. Soc.*, **75**, 147–53.
51. Howell, W.E. 1960. Cloud seeding in the American Tropics, in *Physics of Precipitation*. Geophysics Monograph 5, American Geophysical Union, Washington, D.C., pp. 412–23.
52. Jordan, C.L. 1958. Mean soundings for the West Indies area, *J. Meteorol.*, **15**, 91–97.
53. Saunders, P.M., and Ronne, F.C. 1962. A comparison between the height of cumulus clouds and the height of radar echoes received from them, *J. Appl. Meteorol.*, **1**, 296–302.
54. Saunders, P.M. 1962. The downdraught from a Florida thunderstorm, *Weather*, **17**, 390–400.
55. Jordan, C.L. 1961. On the maximum vertical extent of convective clouds over the central and southeastern United States, *Proc. 9th Weather Radar Conf., Kansas City, Mo.*, 96–101.
56. Plank, V.G. 1965. *The Cumulus and Meteorological Events of the Florida Peninsula during a Particular Summertime Period*. Environmental Research Paper 151, Air Force Cambridge Res. Labs., Bedford, Mass.
57. Saunders, P.M. 1962. Penetrative convection in stably stratified fluids, *Tellus*, **14**, 177–94.
58. Byers, H.R., and Braham, R.R. 1949. *The Thunderstorm*. U.S. Govt. Print. Off., Washington, D.C.
59. Staff members, N.S.S.P. 1963. Environmental and thunderstorm structures as shown by National Severe Storms Project observations in spring 1960 and 1961, *Mon. Weather Rev.*, **91**, 271–92.
60. Simpson, R.H. 1963. Comments on "Condensed water in the free atmosphere in air colder than −40°C," *J. Appl. Meteorol.*, **2**, 684–85.
61. Roys, G.P., and Kessler, E. 1966. *Measurements by Aircraft of Condensed Water in Great Plains Thunderstorms*. NSSP Report 19, U.S. Weather Bureau, Washington D.C., pp. 1–17.
62. Cole, A.E. 1960. Surface rates of precipitation, in *Handbook of Geophysics*. U.S.A.F. and Macmillan, New York, pp. 6-1–6-5.
63. McNaughten, I.I. 1959. *The Analysis of Measurements of Free Ice and Ice/Water Concentrations in the Atmosphere of the Equatorial Zone*. Tech. Note No. Mech. Eng. 283, R.A.E. Farnborough.
64. Jones, R.F. 1960. Size-distribution of ice crystals in cumulonimbus clouds, *Q. J. R. Meteorol. Soc.*, **86**, 187–94.
65. Jones, R.F. 1954. Five flights through a thunderstorm belt, *Q. J. R. Meteorol. Soc.*, **80**, 377–87.
66. Wichmann, H. 1951. Über das Vorkommen und Verhalten des Hagels in Gewitterwolken, *Ann. Meteor.*, **4**, 218–25.
67. Fujita, T. 1960. Mesometeorological study of pressure and wind fields beneath isolated radar echoes, *Proc. 8th Weather Radar Conf.*, 151–58.
68. Fujita, T. 1963. *Analytical Mesometeorology: A Review*. Meteorological Monographs 5, no. 27, American Meteorological Society, Boston, pp. 77–125.
69. Fujita, T. 1959. Precipitation and cold air production in mesoscale thunderstorm systems, *J. Meteorol.*, **16**, 454–66.
70. Middleton, G.V. 1966. Experiments on density and turbidity currents, *Can. J. Earth Sci.*, **3**, 523–46.
71. Simpson, J.E. 1969. A comparison between laboratory and atmospheric density currents, *Q. J. R. Meteorol. Soc.*, **95**, 758–65.
72. Benjamin, T.B. 1968. Gravity currents and related phenomena, *J. Fluid Mech.*, **31**, 209–48.
73. Farquharson, J.S. 1937. Haboobs and instability in the Sudan, *Q. J. R. Meteorol. Soc.*, **63**, 393–414; Freeman, M.H. 1952. *Duststorms of the Anglo-Egyptian Sudan*. Met. Rept. 11, Meteorological Office, London.
74. Hankin, E.H. 1913–21. On dust-raising winds and descending currents, *Mem. Indian Meteorol. Dept.*, **22**, pt. 6, 567–73.
75. Mitra, H., and Kulshreshtha, S.M. 1961. Radar observations of tropical dust-storms, *Proc. 9th Weather Radar Conf., Kansas City, Mo.*, pp. 56–65.
76. Ogura, Y., and Charney, J.G. 1960. A numerical model of thermal convection in the atmosphere, *Proc. Int. Symp. Numerical Weather Prediction, Tokyo*, 431–52.
77. Zipser, E.J. 1969. The role of organised unsaturated convective downdrafts in the structure and the rapid decay of an equatorial disturbance, *J. Appl. Meteorol.*, **8**, 799–814.
78. Sasaki, Y. 1960. Effects of condensation, evaporation and rainfall on the development of meso-scale disturbances: a numerical experiment, *Proc. Int. Symp. Numerical Weather Prediction, Tokyo*, 477–500.

79. Battan, L.J. 1953. Duration of convective radar cloud units, *Bull. Am. Meteorol. Soc.*, **34**, 227–28.
80. Auer, A.H., and Sand, W. 1966. Updraft measurements beneath the base of cumulus and cumulonimbus clouds, *J. Appl. Meteorol.*, **5**, 461–66.
81. Cunningham, R.M. 1959. Cumulus circulation, in *Recent Advances in Atmospheric Electricity*. Pergamon Press, London, pp. 361–67.
82. Davies-Jones, R.P., and Henderson, J.H. 1974. *Updraft Properties Deduced from Rawin-Soundings*. Tech. Memo. ERL NSSL-72, Nat. Severe Storms Lab., Norman, Okla.
83. Malkus, J.S. 1958. Tropical weather disturbances—why do so few become hurricanes?, *Weather*, **13**, 75–89.
84. Malkus, J.S. 1954. On the structure of some cumulonimbus clouds which penetrated the high tropical troposphere, *Tellus*, **6**, 351–66.
85. Saunders, P.M. 1962. The downdraught from a Florida thunderstorm, *Weather*, **17**, 390–400.
86. Browning, K.A., and Ludlam, F.H. 1962. Airflow in convective storms, *Q. J. R. Meteorol. Soc.*, **88**, 117–35.
87. Gibson, W.S. 1863. Hailstorms and their phenomena, in *Miscellanies*. Longham, Roberts and Green, London.
88. Weickmann, H. 1953. Observational data on the formation of precipitation in cumulonimbus clouds, in *Thunderstorm electricity*, H.R. Byers, ed. University of Chicago Press, Chicago, pp. 66–138.
89. Prohaska, K. 1905. Zugrichtung, Stärke und Geschwindigkeit der Hagelwetter, Dauer des Hagelfälles 1902 und im Mittel, *Meteorol. Z.*, **22**, 519–23. (This review refers to many of the older but interesting studies made on hailstorms in central Europe and elsewhere.)
90. Prohaska, K. 1907. Die Hagelfälle des 6 Juli 1905 in den Ostalpen, *Meteorol. Z.*, **24**, 193–200.
91. Marriott, W. 1892. Report on the thunderstorms of 1888 and 1889, *Q. J. R. Meteorol. Soc.*, **18**, 23–39.
92. Clark, J.E. 1920. The Surrey hailstorm of July 16, 1918, *Q. J. R. Meteorol. Soc.*, **46**, 271–88.
93. Douglas, C.K.M., and Moorhead, J.K. 1946. The relation between wind direction in the middle troposphere and the incidence of thundery conditions and rainfall in England in summer, *Q. J. R. Meteorol. Soc.*, **72**, 207–20.
94. Miller, R.C. 1959. Tornado-producing synoptic patterns, *Bull. Am. Meteorol. Soc.*, **40**, 465–72 (the quotation is from an unpublished manuscript by the same author).
95. Ludlam, F.H., and Macklin, W.C. 1960. The Horsham hailstorm of 5 September 1958, *Meteorol. Mag.* **89**, 245–51.
96. Ludlam, F.H. 1963. *Severe local storms: A Review*. Meteorological Monographs 5, no. 27, American Meteorological Society, Boston, pp. 1–30.
97. Roach, W.T. 1967. On the nature of the summit areas of severe storms in Oklahoma, *Q. J. R. Meteorol. Soc.*, **93**, 318–36.
98. Piddington, H. 1859. *The Sailor's Horn Book for the Law of Storms*, 3d ed. Fdk. Norgate, London.
99. Braak, C. 1931. Klimakunde von Hinterindien und Insulinde, in *Handbuch der Klimatologie*, 4, pt. R. Gebrüder Borntraeger, Berlin.
100. Clark, L. 1873. On the storms experienced by the submarine cable expedition in the Persian Gulf on Nov. 1st and 2nd, 1869, *Q. J. R. Meteorol. Soc.*, **1**, 117–19.
101. Frisby, E.M., and Sansom, H.W. 1967. Hail incidence in the tropics, *J. Appl. Meteorol.*, **6**, 339–54.
102. Hamilton, R.A., and Archbold, J.W. 1945. Meteorology of Nigeria and adjacent territory, *Q. J. R. Meteorol. Soc.*, **71**, 231–64.
103. Eldridge, R.H. 1957. A synoptic study of West African disturbance lines, *Q. J. R. Meteorol. Soc.*, **83**, 303–14.
104. Clackson, J.R. 1952. *Gusts in Nigeria*. Met. Note 1, British W. Afr. Met. Service.
105. Moncrieff, M.W., and Green, J.S.A. 1972. The propagation and transfer properties of steady convective overturning in shear, *Q. J. R. Meteorol. Soc.*, **98**, 336–52.
106. Schulze, B.R. 1965. Hail and thunderstorm frequency in South Africa, *Notos*, **14**, 67–71.
107. Barkhley, H. 1934. Thunderstorms in Australia, *Commonwealth of Australia Bull.*, **19**.
108. Carte, A.E. 1966. Features of Transvaal hailstorms, *Q. J. R. Meteorol. Soc.*, **92**, 290–96.
109. Knoch, K. 1930. Klimakunde von Südamerika, in *Handbuch der Klimatologie*, 2, pt. G. Gebrüder Borntraeger, Berlin.
110. Rönicke, G. 1965. Thunderstorm activity in the Andes of Northern Argentina, *J. Appl. Meteorol.*, **4**, 186–89.
111. Grandoso, H.N., and Iribarne, J.V. 1963. Evaluation of the first three years of a hail prevention experiment in Mendoza (Argentina), *ZAMP*, **14**, 549–53.
112. Sinha, K.L. 1952. Strong winds at Allahabad and their forewarnings, *Indian J. Meteorol. Geophys.*, **3**, 110–14.
113. Venkataraman, K.S., and Bhaskara Rao, N.S. 1966. Mesoscale study of summer thunderstorms in Delhi area, *Indian J. Meteorol. Geophys.*, **17**, 529–44.
114. Mitra, H., and Kulshreshtha, S.M. 1961. Radar observations of tropical duststorms, *Proc. 9th Weather Radar Conf.*, 56–69.
115. Kulshrestha, S.M., and Jain, P.S. 1967. Radar climatology of Delhi and neighbourhood: occurrence of severe weather, *Indian J. Meteorol. Geophys.*, **18**, 105–10.
116. Mull, S., and Kulshrestha, S.M. 1962. The severe hailstorm of 27 May 1959 near Sikar (Rajasthan): a synoptic and radar study, *Indian J. Meteorol. Geophys.*, **13** (spl. no.), 81–94.
117. Mitra, H., and Kulshreshtha, S.M. 1950. Severe squall, *Indian J. Meteorol. Geophys.*, **1**, 304.
118. Mitra, H., and Kulshreshtha, S.M. 1944. *Nor'westers of Bengal*. Tech. Note 10, India Met. Dept., Bombay.
119. Bhattacharyya, P., and De, A.C. 1966. Study of the heights of radar cloud tops in the Gangetic valley of West Bengal, *Indian J. Meteorol. Geophys.*, **17**, 591–600.
120. Subramanian, D.V., and Sehgal, U.N. 1967. Radar study of thunderstorm activity in north-east India during the pre-monsoon season, *Indian J. Meteorol. Geophys.*, **18**, 111–14.
121. Eliot, J. 1899. Hailstorms in India during the period 1883–97, with a discussion on their distribution, *Indian Meteorol. Mem.*, **6**, pt. 4, 237–315.
122. Deshpande, D.V. 1964. Heights of Cb clouds over India during the southwest monsoon season, *Indian J. Meteorol. Geophys.*, **15**, 47–54. Photogrammetric studies of cb up to 20 km over northeast India during a premonsoon period of little wind shear are reported in Cornford, S.G., and Spavins, C.S. 1973. Some measurements of cumulonimbus tops in the pre-monsoon season in north-east India, *Meteorol. Mag.*, **102**, 314–32.
123. Ghosh, B.P. 1967. A radar study on thunderstorms and convective clouds around New Delhi during southwest monsoon season, *Indian J. Meteorol. Geophys.*, **18**, 391–96.
124. Maheshwari, R.C., and Mathur, I.C. 1968. Radar reflectivity studies of Indian summer monsoon over N.W. India, *Proc. 13th Weather Radar Conf.*, 98–103.
125. Ramakrishnan, K.P. 1957. Squalls at Cochin, *Indian J. Meteorol. Geophys.*, **8**, 289–95.

126. Ramakrishnan, K.P., and Gopinath Rao, B.G. 1954. Some broad features of the occurrence of squalls in different parts of India, *Indian J. Meteorol. Geophys.*, **5**, 337–40.
127. Saxena, S.P., and Natarajan, R. 1966. A study of squalls at Ahmedabad airfield, *Indian J. Meteorol. Geophys.*, **66**, 71–76.
128. Sharma, K.K. 1966. Squalls at Nagpur, *Indian J. Meteorol. Geophys.*, **66**, 77–82.
129. Ramaswamy, C. 1956. On the sub-tropical jet stream and its role in the development of large-scale convection, *Tellus*, **8**, 26–60.
130. Venkateswara Rao, D., and Mukherjee, A.K. 1958. On forecasting hailstorms by the method of vectorial wind changes, *Indian J. Meteorol. Geophys.*, **9**, 313–22.
131. Frost, R. 1954. *Cumulus and cumulonimbus cloud over Malaya*. Met. Rept. 15, Meteorological Office, London.
132. Zobel, R.F., and Cornford, S.G. 1966. Cloud tops over Malaya during the south-west monsoon season, *Meteorol. Mag.*, **95**, 65–68.
133. Kerley, M.J. 1961. High-altitude observations between the United Kingdom and Nairobi, *Meteorol. Mag.*, **90**, 3–18.
134. *Marine Observer*. 1966. **36**, 171.
135. *Marine Observer*. 1966. **36**, 116.
136. *Marine Observer*. 1965. **35**, 56.
137. Rodewald, M. 1956. Ein Hagel-Orkan bei Kap Ras el Hadd und seine Ursache, *Der Seewart*, **17**, 122.
138. *Marine Observer*. 1963. **33**, 63.
139. *Marine Observer*. 1962. **32**, 112–13.
140. *Thunderstorm Rainfall*. 1947. Hydrometeorological Report 5, Hydromet. Section, U.S. Weather Bureau, Vicksburg, Miss.
141. Bowell, V.E.M., Brown, A.E., and Golde, R.H. 1966. *Thunderstorm Activity in Great Britain, 1955 to 1964*. Electrical Research Association, Leatherhead, Surrey.
142. Rowsell, E.H. 1956. Damaging hailstorms, *Meteorol. Mag.*, **85**, 344–46.
143. Dessens, H., et al. 1952 and subsequent years. *Annual Reports of the Association d'Etudes des moyens de lutte contre les fleaux atmosphériques*. Toulouse.
144. Lamb, H.H. 1957. *Tornadoes in England, 21 May 1950*. Geophys. Mem. 12, no. 99, Meteorological Office, London.
145. Dessens, J. 1965. Quelques tornades françaises recentes, *J. Rech. Atmos.*, **2**, 91–96.
146. Wegener, A. 1917. *Wind- und Wasserhosen in Europa*, Friedrich Vieweg und Sohn, Braunschweig.
147. Lacy, R.E. 1968. Tornadoes in Britain 1963–66, *Weather*, **23**, 116–24.
148. Flora, S.D. 1954. *Tornadoes of the United States*. University of Oklahoma Press, Norman.
149. Gentry, R.C. 1966. Nature and scope of hurricane damage, in *Hurricane Symposium*. Publication 1, American Society for Oceanography, Houston, Tex.
150. House, D.C. 1963. *Forecasting tornadoes and severe thunderstorms*. Meteorological Monographs 5, no. 27, American Meteorological Society, Boston, pp. 141–55.
151. *Tornado occurrences in the United States*. 1952. Tech. Paper 20, U.S. Weather Bureau, Washington, D.C.
152. Thom, H.C.S. 1963. Tornado probabilities, *Mon. Weather Rev.*, **91**, 730–36.
153. Gutnik, M. 1958. Climatology of the trade-wind inversion in the Caribbean, *Bull. Am. Meteorol. Soc.*, **39**, 410–20.
154. Storm Data, 1962. **4**, no. 5, U.S. Weather Bureau, Asheville, N.C.
155. Dodd, A.V. 1965. Dew point distribution in the contiguous United States, *Mon. Weather Rev.*, **93**, 113–22.
156. Lahey, J. F., Bryson, R.A., Wahl, E.W., Horn, L.H., and Henderson, V.D. 1958. *Atlas of 500 mb Wind Characteristics for the Northern Hemisphere*. University of Wisconsin Press, Madison.
157. Skaggs, R.H. 1967. On the association between tornadoes and 500-mb indicators of jet streams, *Mon. Weather Rev.*, **95**, 107–10.
158. Changnon, S.A., and Stout, G.E. 1967. Crop-hail intensities in central and northwest United States, *J. Appl. Meteorol.*, **6**, 542–48.
159. Frisby, E.M. 1963. Hailstorms of the Upper Great Plains of the United States, *J. Appl. Meteorol.*, **2**, 759–66.
160. Stout, G.E., Blackmer, R.H., and Wilk, K.E. 1960. Hail studies in Illinois relating to cloud physics, in *Physics of Precipitation*. Geophysics Monograph 5, American Geophysical Union, Washington, D.C., pp. 369–81; Changnon, S., Schickedanz, P., and Danford, H. 1967. Hail patterns in Illinois and South Dakota, *Proc. 5th Conf. Severe Local Storms*, 325–35.
161. Beckwith, W.B. 1957. Characteristics of Denver hailstorms, *Bull. Am. Met. Soc.*, **38**, 20–30; 1960. Analysis of hailstorms in the Denver network, 1949–1958, in *Physics of Precipitation*. Geophysics Monograph 5, American Geophysical Union, Washington, D.C., pp. 348–53.
162. Douglas, R.H., and Hitschfeld, W. 1959. Patterns of hailstorms in Alberta, *Q. J. R. Meteorol. Soc.*, **85**, 105–19; Summers, P.W., and Paul, A. 1967. Some climatological characteristics of hailfall in Central Albert, *Proc. 5th Conf. Severe Local Storms*, 315–24.
163. Bleeker, W., and Andre, M.J. 1951. On the diurnal variation of precipitation, particularly over central U.S.A., *Q. J. R. Meteorol. Soc.*, **77**, 260–71.
164. Means, L.L. 1954. A study of the mean southerly wind-maximum in low levels associated with a period of summer precipitation in the Middle West, *Bull. Am. Meteorol. Soc.*, **35**, 166–70.
165. Blackadar, A.K. 1957. Boundary layer wind maxima and their significance for the growth of nocturnal inversions, *Bull. Am. Meteorol. Soc.*, **38**, 283–90.
166. Buajitti, K., and Blackadar, A.K. 1957. Theoretical studies of diurnal wind structure variations in the planetary boundary layer, *Q. J. R. Meteorol. Soc.*, **83**, 486–500.
167. Hoecker, W.H. 1965. Comparative physical behaviour of southerly boundary-layer wind jets, *Mon. Weather Rev.*, **93**, 133–44.
168. Lettau, H.H. 1967. Small to large-scale features of planetary boundary layer structure over mountain slopes, *Proc. Symp. Mountain Meteorol., Fort Collins, Colo.*, 1–73.
169. Lhermitte, R.M. 1966. *Probing Air Motion by Doppler Analysis of Radar Clear Air Returns*. Nat. Severe Storms Lab., Rept. 26, E.S.S.A., Washington, D.C.
170. Newton, C.W., and Newton, H.R. 1959. Dynamical interactions between large convective clouds and environment with vertical shear, *J. Meteorol.*, **16**, 483–96.
171. Süring, R. 1951. *Lehrbuch der Meteorologie*, 5th ed. S. Hirzel, Leipzig, vol. 2, pp. 1040–41.
172. Neumann, J. 1951. Land breezes and nocturnal thunderstorms, *J. Meteorol.*, **8**, 60–67.
173. Brückner, W. 1951. Weather observations in Colombia, *Weather*, **6**, 54–58.
174. Bryson, R.A., and Lowry, W.P. 1955. Synoptic climatology of the Arizona summer precipitation singularity, *Bull. Am. Meteorol. Soc.*, **36**, 329–39.

175. Carlson, T.N., and Ludlam, F.H. 1967. Conditions for the occurrence of severe local storms, *Tellus*, **20**, 203–26.
176. (Staff members.) 1963. Environmental and thunderstorm structures as shown by the National Severe Storms Project observations in spring 1960 and 1961, *Mon. Weather Rev.*, **91**, 271–92.
177. Bonner, W.D. 1966. Case study of thunderstorm activity in relation to the low-level jet, *Mon. Weather Rev.*, **94**, 167–78.
178. De Jong, J.J.G. 1959. De verwoestende hagelbuien van 11 Juli '59 in Drente en Groningen, *Hemel en Dampkring*, **57**, 225–34.
179. Cunningham, R.M. 1960. Hailstorm structure viewed from 32,000 ft, in *Physics of Precipitation*. Geophysics Monograph 5, American Geophysical Union, Washington, D.C., pp. 325–32.
180. Roach, W.T. 1967. On the nature of the summit areas of severe storms in Oklahoma, *Q. J. R. Meteorol. Soc.*, **93**, 318–36.
181. Fujita, T., and Arnold, J. 1963. *Preliminary Result of Analysis of the Cumulonimbus Cloud of April 21, 1961*. Research Paper 16, Mesometeorology Project, Department of Geophysical Science, University of Chicago.
182. Plank, V.G. 1965. *The Cumulus and Meteorological Events of the Florida Peninsula during a Particular Summertime Period*. Environmental Research Paper 151, Air Force Cambridge Res. Labs., Bedford, Mass.
183. Clarke, R.H. 1962. Pressure oscillations and fallout downdrafts, *Q. J. R. Meteorol. Soc.*, **88**, 459–69.
184. Battan, L.J. 1959. *Radar Meteorology*. University of Chicago Press, Chicago.
185. Byers, H.R., and Braham, R.R. 1949. *The Thunderstorm*. U.S. Govt. Print. Off., Washington, D.C.
186. Wilson, J.W. 1966. *Movement and Predictability of Radar Echoes*. Tech. Memo. 28, Nat. Severe Storms Lab., Norman, Okla.
187. Atlas, D. 1964. Advances in radar meteorology, *Adv. Geophys.*, **10**, 318–478.
188. Atlas, D. 1963. *Radar Analysis of Severe Storms*. Meteorological Monographs 5, no. 27, American Meteorological Society, Boston, pp. 177–220.
191. Browning, K.A., and Fujita, T. 1965. *A Family Outbreak of Severe Local Storms—A Comprehensive Study of the Storms in Oklahoma on 26 May 1963*. Part 1, Special Report 32, Air Force Cambridge Res. Labs., Bedford, Mass.
192. Fujita, T. 1965. Formation and steering mechanisms of tornado cyclones and associated hook echoes, *Mon. Weather Rev.*, **93**, 67–78; see also further comments, 639–43.
193. Browning, K.A., and Atlas, D. 1965. Initiation of precipitation in vigorous convective clouds, *J. Atmos. Sci.*, **22**, 678–83.
194. Hammond, G.R. 1967. *Study of a Left Moving Thunderstorm of 23 April 1964*. Tech. Memo. 31, Nat. Severe Storms Lab., ESSA, Norman, Okla.
195. Fujita, T., and Grandoso, H. 1968. Split of a thunderstorm into anti-cyclonic and cyclonic storms and their motion as determined from numerical model experiments, *J. Atmos. Sci.*, **25**, 416–39.
196. Donaldson, R.J., and Lamkin, W.E. 1964. Visual observations beneath a developing tornado, *Mon. Weather Rev.*, **92**, 326–28.
197. Hoecker, W.H. 1964. Unusual tornado photographs, *Mon. Weather Rev.*, **92**, 328–29.
198. Ward, N.B. 1961. Radar and surface observations of the tornadoes of May 4, 1961, *Proc. 9th Weather Radar Conf., Kansas City, Mo.*, 175–80.
199. Fujita, T. 1960. *A Detailed Analysis of the Fargo Tornadoes of June 20, 1957*. Research Paper 42, U.S. Weather Bureau, Washington, D.C.
200. Flora, S.D. 1958. *Tornadoes of the United States*, rev. ed. University of Oklahoma Press, Norman.
201. Fujita, T. 1967. *Estimated Wind Speeds of the Palm Sunday Tornadoes*. Research Paper 53, Mesometeorology Project, Department of Geophysical Science, University of Chicago.
202. Hoecker, W.J. 1960. Wind speed and air flow patterns in the Dallas tornado of April 2, 1957, *Mon. Weather Rev.*, **88**, 167–80.
203. Hoecker, W.J. 1961. Three-dimensional pressure pattern of the Dallas tornado and some resultant implications, *Mon. Weather Rev.*, **89**, 533–42.
204. Hoecker, W.J., et al. 1960. *The Tornadoes at Dallas, Tex., April 2, 1957*. Research Paper 41, U.S. Weather Bureau, Washington, D.C.
205. Kuo, H.L. 1966. On the dynamics of convective atmospheric vortices, *J. Atmos. Sci.*, **23**, 25–42.
206. Gordon, A.H. 1951. Waterspouts, *Marine Observer*, **21**, 47–60.
207. Macky, W.A. 1953. The Easter tornadoes at Bermuda, *Weatherwise*, **6**, 74–75.
208. Price, S., and Sasaki, R.I. 1963. Some tornadoes, waterspouts, and other funnel clouds of Hawaii, *Mon. Weather Rev.*, **91**, 175–90.
209. Golden, J.H. 1968. Waterspouts at Lower Matecumbe Key, Florida, 2 September 1967, *Weather*, **23**, 102–14.
210. Kirk, T.H., and Dean, D.T.J. 1963. *Report on a Tornado at Malta*. Geophys. Mem. 14, no. 107, Meteorological Office, London.
211. Brooks, E.M. 1951. Tornadoes and related phenomena, in *Compendium of Meteorology*., American Meteorological Society, Boston, pp. 673–80.
212. Lacy, R.E. 1968. Tornadoes in Britain 1963–66, *Weather*, **23**, 116–24.
213. Lawrence, E.N. 1954. Whirlwind at Southend-on-Sea, August 10, 1953, *Meteorol. Mag.*, **83**, 4–9.
214. Kamburova, P., and Ludlam, F. H. 1966. Rainfall evaporation in thunderstorm downdraughts, *Q. J. R. Meteorol. Soc.*, **92**, 510–18.
215. Krumm, W.R. 1954. On the cause of downdrafts from dry thunderstorms over the plateau area of the United States, *Bull. Am. Meteorol. Soc.*, **35**, 122–25.
216. List, R. 1958. Kennzeichen atmosphärischer Eispartikeln, *Z. angew. Math. Phys.*, **9a**, 180–92, 217–34.
217. Browning, K.A., Ludlam, F.H., and Macklin, W.C. 1963. The density and structure of hailstones, *Q. J. R. Meteorol. Soc.*, **89**, 75–84.
218. Carte, A.E., and Kidder R.E. 1966. Transvaal hailstones, *Q. J. R. Meteorol. Soc.*, **92**, 382–91.
219. Browning, K.A. 1966. The lobe structure of giant hailstones, *Q. J. R. Meteorol. Soc.*, **92**, 1–14.
220. Browning, K.A. 1967. Hailstones breaking in mid-air, *Weather*, **22**, 331–34.
221. Vittori, O. 1960. Preliminary note on the effects of pressure waves upon hailstones, *Nubila*, **3**, 34–52.
222. Favreau, R.F., and Goyer, G.G. 1967. The effect of shock waves on a hailstone model, *J. Appl. Meteorol.*, **6**, 326–35.
223. Flora, S.D. 1956. *Hailstorms of the United States*. University of Oklahoma Press, Norman.

224. Schove, D.J. 1951. Hail in history, *Weather*, **6**, 17–21.
225. Rodes, L. 1938. Grêlons de dimension exceptionelle, *La Météorol.*, 3d ser., 320–21.
226. Heidke, P. 1925. Aussergewöhnlicher hagelfall zu Heidgraben, Schleswig-Holstein, *Meteorol. Z.*, **42**, 408.
227. Roos, D.v.d.S. 1972. A giant hailstone from Kansas in free fall, *J. Appl. Meteorol.*, **11**, 1008–11; Blair, T.A. 1928. Hailstones of great size at Potter, Nebr., *Mon. Weather Rev.*, **56**, 313.
228. Grelons monstrueux. 1947. *La Nature, Paris*, **75**, 43.
229. *Bull. Am. Meteorol. Soc.* 1930. **11**, 132.
230. *Bull. Am. Meteorol. Soc.* 1932. **13**, 158.
231. Eliot, J. 1899. Hailstorms in India during the period 1883–97, with a discussion on their distribution, *Mem. Indian Meteorol. Dept.*, **6**, pt. 4, 237–315.
232. Hariharan, P.S. 1950. Sizes of hailstones, *Indian J. Meteorol. Geophys.*, **1**, 73.
233. Sizes and shapes of hailstones in India. 1950. *Indian J. Meteorol. Geophys.*, **1**, 309–11.
234. Witt, H. 1914. Ausserordentlicher Hagelfall in China, *Meteorol. Z.*, **31**, 445–46.
235. Cellini, B. 1949. *The Life of Benvenuto Cellini*, trans. John Addington Symons. Phaidon, New York, pp. 325–26.
236. Knight, C.A., and Knight, N.C. 1968. The final freezing of spongy ice: hailstone collection techniques and interpretations of structures, *J. Appl. Meteorol.*, **7**, 875–81.
237. Browning, K.A., Hallet, J., Harrold, T.W., and Johnson, D. 1968. The collection and analysis of freshly fallen hailstones, *J. Appl. Meteorol.*, **7**, 603–12.
238. Gitlin, S.N., Goyer, G.G., and Henderson, T.J. 1968. The liquid water content of hailstones, *J. Atmos. Sci.*, **25**, 97–99.
239. Bailey, I.H., and Macklin, W.C. 1968. The surface configuration and internal structure of artificial hailstones, *Q. J. R. Meteorol. Soc.*, **94**, 1–11.
240. List, R. 1960. Growth and structure of graupel and hailstones, in *Physics of Precipitation*. Geophysics Monograph 5, American Geophysical Union, Washington, D.C., pp. 317–23; Ludlam, F.H. 1961. The hailstorm, *Weather*, **16**, 152–62.
241. Knight, C.A. 1966. Formation of crystallographic etch pits on ice, and its application to the study of hailstones, *J. Appl. Meteorol.*, **5**, 710–14.
242. Macklin, W.C., Merlivat, L., and Stevenson, C.M. 1970. The analysis of a hailstone, *Q. J. R. Meteorol. Soc.*, **96**, 472–86.
243. Ferrel, W. 1889. *A Popular Treatise on the Winds*. Wiley, New York, pp. 420–27.
244. Humphreys, W.J. 1940. *Physics of the Air*, 3d ed. McGraw-Hill, New York.
245. Schumann, T.E.W. 1938. The theory of hailstone formation, *Q. J. R. Meteorol. Soc.*, **64**, 3–17.
246. Ludlam, F.H. 1950. The composition of coagulation-elements in cumulonimbus, *Q. J. R. Meteorol. Soc.*, **76**, 52–58.
247. Blasius, W. 1875. *Storms: Their Nature, Classification and Laws*. Porter and Coates, Philadelphia.
248. Knopfle, H. 1948. Dismantling at 4 km, *Sailplane*, **16**, 9.
249. Browning, K. 1963. The growth of large hail within a steady updraught, *Q. J. R. Meteorol. Soc.*, **89**, 490–506.
250. Douglas, R.H. 1963. *Recent Hail Research: A Review*. Meteorological Monographs 5, no. 27, American Meteorological Society, Boston, pp. 157–67.
251. Atlas, D., Harper, W.G., Ludlam, F.H., and Macklin, W.C. 1960. Radar scatter by large hail, *Q. J. R. Meteorol. Soc.*, **86**, 468–82.
252. Macklin, W.C. 1963. Heat transfer from hailstones, *Q. J. R. Meteorol. Soc.*, **89**, 360–69.
253. Fawbush, E.J., and Miller, R.C. 1954. A basis for forecasting peak wind gusts in non-frontal thunderstorms, *Bull. Am. Meteorol. Soc.*, **35**, 14–19.
254. Foster, D.S. 1958. Thunderstorm gusts compared with computed downdraft speeds, *Mon. Weather Rev.*, **86**, 91–94.
255. Braham, R.R. 1952. The water and energy budgets of the thunderstorm and their relation to thunderstorm development, *J. Meteorol.*, **9**, 227–42.
256. Auer, A.H., and Marwitz, J.D. 1968. Estimates of air and moisture flux into hailstorms on the High Plains, *J. Appl. Meteorol.*, **7**, 196–98.
257. Newton, C.W. 1966. Circulations in large sheared cumulonimbus, *Tellus*, **18**, 699–712.
258. Jennings, A.H. 1963. *Maximum Recorded United States Point Rainfall*, rev. ed. Tech. Paper 2, U.S. Weather Bureau, Washington, D.C.
259. Rai Sircar, N.C., and Sikdar, D.N. 1963. On visibility at Bombay airport under different precipitation conditions, *Indian J. Meteorol. Geophys.*, **14**, 480–82.
260. Poliakova, E.A. 1960. Investigation of meteorological visibility during rain, *Glav. Geofiz. Obs., Trudy* (Leningrad), no. 100, 45–52.
261. Elford, C.R. 1956. A new one-minute rainfall record, *Mon. Weather Rev.*, **84**, 51–52.
262. Engelbrecht, H.H., and Brancato, G.N. 1959. World record one-minute rainfall at Unionville, Maryland, *Mon. Weather Rev.*, **87**, 303–6.
263. Kessler, E. 1963. Elementary theory of associations between atmospheric motions and distributions of water content, *Mon. Weather Rev.*, **91**, 13–27.
264. Srivastava, R.C., and Atlas, D. 1969. Growth, motion and concentration of precipitation particles in convective storms, *J. Atmos. Sci.*, **26**, 535–44.
265. Robb, A.D. 1959. Severe hail, Selden, Kansas, June 3, 1959, *Mon. Weather Rev.*, **87**, 301–3.
266. Gordon, A.H. 1948. Phenomenal rainfall at Semmering, Austria, *Meteorol. Mag.*, **77**, 246–50.
267. Bleasdale, A. 1957. Rainfall at Camelford, Cornwall, on June 8, 1957, *Meteorol. Mag.*, **86**, 339–43.
268. Booth, R.E. 1956. Severe hailstorm at Tunbridge Wells on August 6, 1956, *Meteorol. Mag.*, **85**, 297–99.
269. Sanson, J. 1949. *Climatologie appliquée*. Ed. Blondel la Rougery, Paris, vol. 1, p. 60.
270. Lazarus, E.H. 1930. Great thunderstorm of Sept. 15–17 near Dinard, France, *Q. J. R. Meteorol. Soc.*, **56**, 181–82.
271. Visher, S.S. 1941. Rainfalls of 10 inches, or more, during 24 hours in the United States, *Mon. Weather Rev.*, **69**, 353–56.
272. McAuliffe, J.P. 1921. Excessive rainfall and flood at Taylor, Tex., *Mon. Weather Rev.*, **49**, 496–97.
273. Bunnemeyer, B. 1921. The Texas floods of September, 1921, *Mon. Weather Rev.*, **49**, 491–94.
274. Hiser, H.W. 1958. Radar analysis of two severe storms in South Florida, *Bull. Am. Meteorol. Soc.*, **39**, 353–59; Sourbeer, R.H., and Gentry, R.C. 1961. Rainstorm in Southern Florida, 21 January 1957, *Mon. Weather Rev.*, **89**, 9–16.
275. Lautzenheizer, R.E., and Fay, R. 1966. Heavy rainfall at Island Falls, Maine, August 28, 1959, *Mon. Weather Rev.*, **94**, 711–14.

276. Richardson, E.A., Peck, E.L., and Green, S.D. 1964. Heavy precipitation storm in Northern Utah, January 29 to February 2, 1963, *Mon. Weather Rev.*, **92**, 317–25.
277. Staff, RFC, Tulsa, and DMO, Kansas City. 1964. Cloudburst at Tulsa, Oklahoma, July 27, 1963, *Mon. Weather Rev.*, **92**, 345–51.
278. Sayers-Duran, P., and Braham, R.R. 1965. An intense rainstorm at Fremont, Missouri, July 28–29, 1964, *Mon. Weather Rev.*, **93**, 387–91.
279. Carlson, T.N. 1967. Isentropic upslope motion and an instance of heavy rain over Southern Florida, *Mon. Weather Rev.*, **95**, 213–20.
280. Jetton, E.V., and Woods, C.E. 1967. Heavy rains in southeastern New Mexico and southwestern Texas, August 21–23, 1966, *Mon. Weather Rev.*, **95**, 221–26.
281. Dightman, R.A. 1968. Central Montana rainstorms and floods—June 6–15, 1967, *Mon. Weather Rev.*, **96**, 813–23.
282. Huff, F.A., and Changnon, S.A. 1964. A model 10-inch rainstorm, *J. Appl. Meteorol.*, **3**, 587–99.
283. Jennings, A.H. 1950. World's greatest observed point rainfalls, *Mon. Weather Rev.*, **78**, 4–5.
284. Fletcher, R.D. 1950. A relationship between maximum observed point and areal rainfall values, *Trans. Am. Geophys. Un.*, **31**, 344–48.
285. Monthly Weather Report. 1929. Feb. 1929, *Obs. Meteorol. Mag., Joao Capelo*.
286. Theaman, J.R. 1929. Excessive rainfall records, private compilation.
287. Gunther, R.T. 1930. *Early Science in Oxford*. Printed for the author, Oxford, vol. 6.
288. Shackford, C.R. 1960. Radar indications of a precipitation-lightning relationship in New England thunderstorms, *J. Meteorol.*, **17**, 15–19.
289. Moore, C.B., Vonnegut, B., Vrablik, E.A., and McCaig, D.A. 1964. Gushes of rain and hail after lightning, *J. Atmos. Sci.*, **21**, 646–65.
290. Battan, L.J. 1959. *Radar Meteorology*. University of Chicago Press, Chicago.
291. Donaldson, R.J. 1964. A demonstration of antenna beam errors in radar reflectivity patterns, *J. Appl. Meteorol.*, **3**, 611–23; 1965. Resolution of a radar antenna for distributed targets, *J. Appl. Meteorol.*, **4**, 727–40.
292. Stout, G.E., and Mueller, E.A. 1968. Survey of relationships between rainfall rate and radar reflectivity in the measurement of precipitation, *J. Appl. Meteorol.*, **7**, 465–74.
293. Joss, J., and Waldvogel, A. 1969. Raindrop size distribution and sampling size errors, *J. Atmos. Sci.*, **26**, 566–69.
294. Herman, B.M., and Battan, L.J. 1961. Calculations of Mie back-scattering from melting ice spheres, *J. Meteorol.*, **18**, 468–78.
295. Joss, J., and Aufdermaur, A.N. 1965. Experimental determination of the radar cross sections of artificial hailstones containing water, *J. Appl. Meteorol.*, **4**, 723–26.
296. Geotis, S.G. 1963. Some radar measurements of hailstorms, *J. Appl. Meteorol.*, **2**, 270–75.
297. Donaldson, R.J. 1961. Radar reflectivity profiles in thunderstorms, *J. Meteorol.*, **18**, 292–305.
298. Donaldson, R.J. 1959. Analysis of severe convective storms observed by radar—II, *J. Meteorol.*, **16**, 281–87.
299. Donaldson, R.J. 1965. Methods for identifying severe thunderstorms by radar: a guide and a bibliography, *Bull. Am. Meteorol. Soc.*, **46**, 174–93.
300. McCready, P.B. 1965. Comparison of some balloon techniques, *J. Appl. Meteorol.*, **4**, 504–8.

9 Large-scale slope convection

9.1 General aspects of the large-scale flow in the troposphere outside the tropics

It was seen in Chapter 1 that radiative equilibrium cannot be attained in the presence of water vapor in the atmosphere, and that the radiative imbalance in the troposphere implies a continual convection of energy from low into high latitudes, partly effected by ocean currents but mainly by airstreams in the troposphere. [Mean annual conditions were considered, although in midsummer the solar irradiation of the surface briefly reaches a maximum near the pole rather than in low latitudes, as it does throughout most of the year; however, the great heat capacity of the water near the surface of the oceans results only in a diminution of the latitudinal temperature gradient in the troposphere during the summer season, and not in a change in its sign (Fig. 9.1).] From the observed magnitude of the energy flux a mean meridional component of wind speed of about 10 m sec^{-1} in middle latitudes was inferred, supposing that the convection took the form of alternate cool airstreams from the pole, and warm airstreams toward the pole, and making a plausible assumption about the magnitude of their temperature difference.

A different form which the tropospheric convection could conceivably assume is one in which ascent occurs in all low latitudes and descent near the relatively cool pole, with poleward flow in the upper troposphere. However, on the rotating earth the tendency of the air in the horizontal branches of such a circulation to preserve its angular momentum about the earth's axis would result in the flows acquiring zonal components of velocity relative to the earth's surface, westerly aloft and easterly near the earth's surface (assuming the flows at these levels to begin with no relative motion), which in the absence of friction would become very large. Irrespective of the actual sign of the relative zonal motion, its westerly component would everywhere increase with height by an amount dependent on the magnitude of the friction near the surface and within the circulation.

Another effect of the earth's rotation (discussed below) is to impose a relation between the horizontal temperature gradient, the eddy viscosity in the vertical, and the wind shear in the vertical. This relation is satisfied by the (zonal) westerly shear of the *observed* mean tropospheric winds and the mean latitudinal temperature gradient, but for a realistic value of the eddy viscosity it appears that it cannot be consistent with the large shear implied by a symmetric circulation. Moreover, it can be shown by the perturbation method and numerical theoretical studies that the predominantly zonal flow of the symmetrical circulation which would be required to maintain thermal equilibrium and the less strongly sheared mean zonal flow which is observed are dynamically unstable. The flows can be anticipated to develop motion systems with horizontal scales of order 1000 km, in which airstreams make large meridional excursions, and whose structure can be shown to be of the kind required to effect the transports of heat and momentum which maintain the observed mean distribution of temperature and wind (Fig. 9.1). The large-scale convection outside the tropics therefore closely resembles the simple side-by-side alternation of cool and warm airstreams which was previously assumed, and is not an overturning in a single large cell. The transition between the two kinds of convective regime, and its dependence upon rotation rate and horizontal temperature gradient, can be strikingly illustrated by laboratory experiments with an annular vessel of fluid which is rotated about its vertical axis and whose inner wall is warmed (3). The atmospheric large-scale motion systems appear on conventional weather maps as wavelike distortions—troughs and ridges—of the height contours of isobaric surfaces, which have a mean zonal trend, often associated (especially at the lower levels) with the closed contours ordinarily regarded as defining cyclones and anticyclones (Fig. 9.2). They are not an

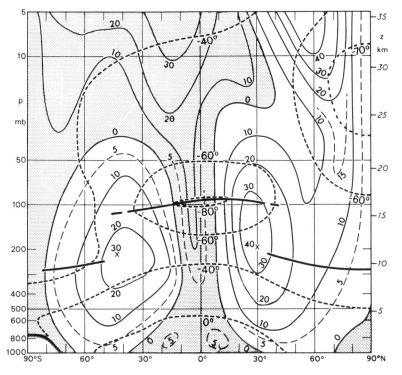

Fig. 9.1 Mean distributions with height (averaged over all longitudes) of wind speed (continuous lines labeled in m sec^{-1}, and lightly stippled where the direction is easterly) and of temperature (dashed lines labeled in °C), during midwinter in the northern hemisphere (2, 3). Heavy solid lines represent mean positions of tropopauses. The distributions are drawn against pressure p, and an approximate scale of height z is added on the right. The zonally averaged meridional temperature gradient in the northern hemisphere troposphere is nearly twice as great in midwinter as in midsummer. In January it is concentrated between about 20 and 80°N, the temperature difference between these latitudes amounting to about 30°C; in July the corresponding difference between 30°N and the pole is about 17°C.

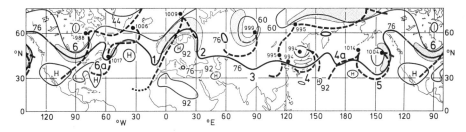

Fig. 9.2 Wavelike distortions (into troughs and ridges) typical of the generally westerly flow in the upper troposphere outside the tropics, and their relation to cyclones and fronts at the surface. The map shows the following features over much of the northern hemisphere at 1200 GMT, 9 July 1959:

(a) Contours of the height of the 500-mb surface, at intervals of 16 dkm (in dkm above 5 km); the contour for 576 dkm is shown by the bold line. These contours are approximately streamlines of the flow, directed with low height to the left. Regions of flow speed exceeding 20 m sec^{-1} are stippled. The dimensions and amplitudes of the troughs vary markedly; six major and two minor troughs are numbered (4a and 6a).

Trough 1 near the Greenwich meridian has already been mentioned in Chapter 8, in relation to an intense cumulonimbus over southeast England.

(b) Centers of cyclones at sea level are marked by dots, beside which are entered central pressures (in mb). The associated fronts are drawn as bold dashed lines.

On this summer occasion the cyclones and the flows in the upper troposphere are less intense than those typical of the winter season. The surface cyclones are mostly close to the regions of stronger winds aloft, consistent with their development in the more strongly sheared (more baroclinic) parts of the troposphere.

9   Large-scale slope convection

obvious form of convection until examined in detail or represented in other ways.

Clouds and rains occupy only small parts of the large-scale motion systems, and their theory can be developed without explicit reference to the processes of condensation and precipitation, the atmosphere being treated apparently as if it were dry. The effects of these processes are included, however, in the values adopted for some tropospheric properties, such as its depth and mean vertical distribution of temperature (stability). In the discussion of both the principal cloud systems and the incidental clouds of the large-scale slope convection, it is desirable to be aware of some features and concepts of the general theory. This chapter therefore begins with a brief review of the theory, and of some aspects of the behavior of large-scale motion systems.

9.2   Scale analysis and simple states of large-scale atmospheric motion

Geostrophic and thermal winds

In the equation expressing the conservation of momentum of a unit mass of air it is convenient to express the wind velocity, and consequently the acceleration, as that measured relative to an observer rotating with the earth. The relative acceleration $D\mathbf{v}/Dt$, following the motion, is then the resultant of the acceleration in an inertial frame, and two apparent accelerations which for meteorological purposes can be attributed solely to the rotation of the earth with angular velocity Ω ($= 7.3 \times 10^{-5}$ sec^{-1}) about its polar axis. One of these, the centrifugal acceleration, is independent of the relative motion and is included in measured values of the acceleration under gravity $\mathbf{g}$, which is directed normal to the earth's surface. The other, the Coriolis acceleration $-2\mathbf{v} \times \mathbf{\Omega}$, is normal to the velocity and parallel to the equatorial plane; it is important in motions whose horizontal scale is large or whose time scale is not small compared with a day. For practical purposes the transformed equation is usually resolved into components in a system of coordinates in which u, v, and w are the components of velocity along axes parallel (x eastward, y northward) and normal (z upward) to the earth's surface. (In the discussion of small-scale motion systems Cartesian coordinates are convenient, but in large-scale systems the curvature of the thin atmospheric shell which they occupy cannot be neglected, and there can be debate over the most appropriate choice of coordinates and the kind of map on which to represent the motion systems and their properties; 4). Considering that in the large-scale motion systems u and v are of order 10 m sec^{-1}, w is of order 1 cm sec^{-1}, and that the horizontal dimensions of the systems is of order 10^3 km, it follows that some terms containing products of the velocity components divided by the earth's radius can be omitted, and the following equations are then obtained:

$$\frac{Du}{Dt} - 2\Omega(v \sin \phi - w \cos \phi) = -\frac{1}{\rho}\frac{\partial p}{\partial x} + F_x \quad (9.1)$$

$$\frac{Dv}{Dt} + 2\Omega u \sin \phi = -\frac{1}{\rho}\frac{\partial p}{\partial y} + F_y \quad (9.2)$$

$$\frac{Dw}{Dt} - 2\Omega u \cos \phi = -\frac{1}{\rho}\frac{\partial p}{\partial z} + F_z - g \quad (9.3)$$

The terms which contain the latitude ϕ are components of the Coriolis acceleration, and the terms F_x, F_y, F_z represent frictional forces.

Two equilibrium states are evident from these *simplified* equations. The first is a state of rest in which there are no horizontal pressure gradients, implying that the pressure, density, and virtual temperature are functions only of z; the pressure is then the *hydrostatic*, with its vertical gradient given by

$$\partial p/\partial z = -g\rho$$

The *thickness* z_t between two isobaric surfaces in which the pressures are p_1 and p_2 ($p_2 > p_1$) is then proportional to a mean virtual temperature $\bar{T}_{vp}$ in the intervening layer:

$$z_t = \int_{p_1}^{p_2} \frac{RT_v}{g}\frac{dp}{p} = \bar{T}_{vp} R \log_e \frac{p_2}{p_1} \quad (9.4)$$

Second, if the Coriolis parameter $f = 2\Omega \sin \phi$ is regarded as constant, there can be an equilibrium state of horizontal, frictionless, and unaccelerated motion in which the Coriolis acceleration is equal and opposite to that produced by the pressure gradient. Such motion is called *geostrophic*, and a wind at any point can be resolved into geostrophic and ageostrophic components, of which the former, at a given latitude, can be regarded as a specification of the local horizontal pressure gradient. The geostrophic wind is directed along the isobars on a level surface, with low pressure to the left in the northern hemisphere. In geostrophic flow with speeds of the magnitude of the wind speeds encountered in the atmosphere the pressure is to a high degree of accuracy equal to the hydrostatic value: $Dw/Dt = F_z = 0$, and even with u as large as 100 m sec^{-1} the term $2\Omega u \cos\phi$ in Eq. 9.3 is only about 10^{-1} cm sec^{-2} or less, at least four orders of magnitude less than g. There is consequently a simple relation between the vector change of the geostrophic wind at two heights (the *thermal wind* $\mathbf{V}_t$) and a horizontal gradient of (virtual) temperature in the intervening layer. The mean decrease of temperature toward the poles implies that with increasing height the magnitude of the term $(1/\rho)\partial p/\partial y$, and therefore the westerly component of the geostrophic wind, becomes larger. Since the large-scale motion systems in the troposphere are associated primarily with the latitudinal temperature gradient, and because studies of the stability of sheared zonal flows form an important part of the theory of cyclones, it is interesting to establish quantitatively the relation mentioned.

It is convenient to use pressure as the height coordinate. The gradient of pressure in a horizontal surface corresponds to a small inclination to the horizontal of an isobaric surface; in geostrophic motion the components in this surface of the gravitational and the Coriolis accelerations are equal and opposite. In vector notation, the simplified equations (9.2, 9.3) for horizontal motion in an isobaric surface become

$$\mathbf{k} \times f\mathbf{V}_g + g\nabla_p h = 0 \qquad (9.5)$$

where $\mathbf{k}$ is a unit vector directed vertically upward, $\mathbf{V}_g$ is the geostrophic wind, h is the height of the isobaric surface, and $\nabla_p h$ is its slope. (If $V_g = 10$ m sec^{-1} the slope is of magnitude only 10^{-4} sec$^{-1} \times 10^3$ cm sec$^{-1}/10^3$ cm sec^{-2}, or 10^{-4}, corresponding to a rise of only 100 m in a distance of 1000 km.) Then

$$\mathbf{k} \times f\mathbf{V}_t = g\nabla_p z_t \qquad (9.6)$$

where z_t is given by Eq. 9.4. The thermal wind therefore blows along the isotherms of mean virtual temperature, with low temperature on the left in the northern hemisphere.

Properties characteristic of the large-scale tropospheric wind systems, some of which have already been inferred intuitively, which can be anticipated rather more rigorously (5), or which can be assumed from observation, are

> Horizontal dimension L of order 10^3 km;
> Vertical dimension H about 10 km;
> Horizontal wind speed V and velocity C of the wind relative to the systems, of order 10 m sec^{-1}; and
> Time scale L/C about 1 day.

By using these values in a scale analysis of the complete equations governing the motion (6) it can be shown that the pressure everywhere can be considered to be *hydrostatic* and that the motion is mostly *nearly geostrophic*. (On weather maps the height contours of isobaric surfaces or the isobars on level surfaces are therefore approximately streamlines, and the wind speed is approximately inversely proportional to their separation when they are drawn for equal intervals of h or of p. The relation also contains the sine of the latitude; the advantage of the isobaric representation is that the relation does not also involve the density of the air.)

The principal difficulties in the scale analysis are in assessing the importance of the terms representing the effects of friction. However, if they are estimated by using a large value of an observed coefficient of eddy diffusion (k_M of Eq. 7.5, of order 10^5 cm^2 sec^{-1}) in expressions of the kind $F_x = k_M \partial^2 u/\partial z^2$, it appears that in large-scale motion F_z is negligible. The only important effects of friction are associated with the horizontal components F_x and F_y within a shallow layer above the earth's surface (not deeper than about $H/10$), where the curvature of the vertical profile of wind speed is typically large. (Similar effects of dubious large-scale importance arise locally and intermittently in shallow layers at higher levels.) Within this boundary layer the frictional forces significantly disturb geostrophic balance, so that unless there is an abnormal thermal wind, the resultant winds have a component directed toward low pressure. (This disturbance has other effects: it tends to induce ascent in cyclones and descent in anticyclones, and in the atmosphere above the boundary layer produces a small mean drift toward high pressure, and therefore a continuous loss of kinetic energy there. In the presence of a purely latitudinal temperature gradient friction would be responsible for the meridional motion in a symmetric circulation.)

If F_z is disregarded, it appears from Eq. 9.3 that even the large vertical accelerations which sometimes occur in small-scale motion systems involve only small departures of the pressure from the hydrostatic value. For example, in cumulonimbus (Dw/Dt) is of order $(10$ m sec$^{-1})^2/2 \times 1/5$ km, or 1 cm sec^{-2}, which is three orders of magnitude less than g. In the large-scale motion systems this term is of order only 10^{-4} cm sec^{-2}, and is even less than the Coriolis term $2\Omega u \cos\phi$, of order 10^{-1} cm sec^{-2}. (In the reduction of measurements made during soundings the hydrostatic relation is always assumed, since even in extreme circumstances in either kind of motion system the errors which may thus be introduced are only comparable with those due to ordinary instrumental inaccuracy.) In the large-scale systems the scale analysis shows that the further (necessary) condition that $H \ll L$ allows the pressure everywhere to be considered equal to the hydrostatic value, even for calculating the horizontal accelerations. The vertical component of velocity in these motion systems can then be obtained indirectly by using the ("continuity") equation expressing the conservation of mass (and

9 Large-scale slope convection

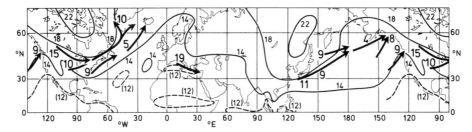

Fig. 9.3 Average midwinter distributions over most of the northern hemisphere of mean tropospheric stability $B = (1/\theta)\partial\theta/\partial z$, of principal cyclone paths, and of most frequent cyclogenesis. Isopleths of B refer to the layer between 850 mb and a level near the tropopause; they are labeled in units of 10^{-8} cm^{-1} (a value of 12 represents close approach to a saturated adiabatic lapse rate in low latitudes). *Principal* paths of cyclone centers at sea level (13) are shown as heavy lines with arrows. Shafts commence in regions of most frequent cyclogenesis and end in regions of maximum cyclone frequency. Large numbers beside the shafts indicate frequency of cyclogenesis; they are the total numbers of centers first appearing on daily maps during the month of January over a period of 20 yr, in areas defined by a length and width of 5° in a middle latitude.

changes of density determined by also using the equation of conservation of energy).

From inspection of Eq. 9.1 and 9.2 it is evident that outside the boundary layer, where F_x and F_y can be disregarded, the motion is mostly nearly geostrophic, for with $w \ll u, v$ these equations reduce to

$$\frac{D\mathbf{V}}{Dt} = \mathbf{k} \times f\mathbf{V} + g\nabla_p h \qquad (9.7)$$

The first term in this equation is of order $C^2/2L$, or nearly two orders of magnitude smaller than the second term, when the wind speed is about 10 m sec^{-1}. Thus $\mathbf{V} \approx \mathbf{V}_g$ (Eq. 9.5) over most of the large-scale systems; however, in those restricted regions of the upper troposphere where the wind speeds reach values of up to about 100 m sec^{-1} the ageostrophic component $\mathbf{V}'$ of the wind given by

$$\mathbf{k} \times f\mathbf{V}' = \frac{D\mathbf{V}}{Dt} \qquad (9.8)$$

may be large and easily observed.

The scale analysis draws attention to several non-dimensional numbers which are significant parameters of convection in a rotating fluid, and are therefore likely to enter discussion of the behavior of its motion systems. Among these are forms of appropriately defined Reynolds, Froude, Rayleigh, and Richardson numbers, which have already been mentioned in other contexts, and a Rossby number conventionally written as

$$\text{Ro} = \frac{C}{2Lf}$$

which may be regarded as a ratio between typical inertial accelerations ($DC/Dt \approx C^2/2L$) and Coriolis accelerations ($fV \approx fC$). The flow is nearly geostrophic if Ro $\ll 1$.

The parameter $B = (1/\theta)\,\partial\theta/\partial z$, which represents the stability with respect to disturbance in the vertical, appears in the definition of the Richardson number, and its mean value over the depth of the troposphere is a property often introduced into discussions empirically. Its distribution over the northern hemisphere in midwinter is illustrated in Fig. 9.3, which also shows the main regions of cyclogenesis and principal cyclone paths.

Vorticity; long waves

The vorticity, a vector property of fluid motion which can be regarded as double the mean velocity of rotation in the vicinity of a point, has an important role in the dynamics of fluids, analogous to that of angular momentum in the dynamics of rigid bodies. In particular, in the discussion of atmospheric motion it is often convenient to refer to the vertical component of the relative vorticity ζ_z, in which the subscript to the symbol will here be omitted:

$$\zeta = \frac{\partial v}{\partial x} - \frac{\partial u}{\partial y} \qquad (9.9)$$

The earth's rotation provides an additional component f in the vertical component of the absolute vorticity, $(\zeta + f)$. For large-scale motion which is horizontal and frictionless the equations of motion (9.1, 9.2) further simplify into

$$\frac{Du}{Dt} - fv = -\frac{1}{\rho}\frac{\partial p}{\partial x} \qquad (9.10)$$

and

$$\frac{Dv}{Dt} + fu = -\frac{1}{\rho}\frac{\partial p}{\partial y} \qquad (9.11)$$

Taking the partial derivative with respect to x in Eq. 9.11, and with respect to y in Eq. 9.10, then subtracting and rearranging, gives

$$\frac{D(\zeta + f)}{Dt} = -(\zeta + f)D - \left(\frac{\partial w}{\partial x}\frac{\partial v}{\partial z} - \frac{\partial w}{\partial y}\frac{\partial u}{\partial z}\right)$$
$$+ \frac{1}{\rho^2}\left(\frac{\partial \rho}{\partial y}\frac{\partial p}{\partial x} - \frac{\partial \rho}{\partial x}\frac{\partial p}{\partial y}\right) \quad (9.12)$$

where on the right $D = \partial u/\partial x + \partial v/\partial y$ is the *horizontal divergence*. Of the other terms on the right the second is often called the *tilting term*, since it expresses the turning into the vertical of horizontal vorticity associated with wind shear in the vertical. The last is called the *baroclinic term*; it is zero if the density is constant in isobaric surfaces, when the fluid is said to be *barotropic*.

From Eq. 9.12 it is evident that under the very restrictive conditions of horizontal, frictionless, barotropic flow with no horizontal divergence, there is conservation of absolute vorticity along the flow:

$$\frac{D(\zeta + f)}{Dt} = 0 \quad (9.13)$$

or

$$\frac{D\zeta}{Dt} = -\beta v \quad (9.14)$$

where $\beta = \partial f/\partial y = 2\Omega \cos \phi\, \partial \phi/\partial y$ varies slowly with y in middle latitudes (it decreases from 2.1×10^{-13} cm^{-1} sec^{-1} in latitude 20° to 1.0×10^{-13} cm^{-1} sec^{-1} in latitude 65°). With β regarded as constant a simple expression can be found for the eastward speed C of a train of sinusoidal transverse waves of wavelength L, superimposed on a steady westerly zonal flow whose speed u is constant (independent of x and y). The expression, which can be obtained simply from the kinematics of the waves (7), is

$$C = u - \beta(L/2\pi)^2 \quad (9.15)$$

There is no constraint to determine a wavelength. However, it appears that with u of magnitude 10 m sec^{-1} waves of length several thousand kilometers may be nearly stationary. At least in the northern hemisphere such nearly stationary long waves (cf. Fig. 6.2) seem to form continually where the general zonal current is disturbed by its passage over the large mountain ranges (the Himalayas and the Rocky Mountains), and from cold to comparatively warm surfaces near the east coasts of the major continents. Because it is observed that the most pronounced troughs, which lie approximately over the east coasts of North America and Asia, are present not only in the winter, but also in the summer season (though in weaker form) when the land and ocean reverse their roles as heat sources and sinks, it seems that the disturbance of the zonal current is primarily related to the mountain ranges. Weather forecasters pay much attention to the long waves, which are most prominent on maps of isobaric contours averaged over periods of several days, because the general character of the weather, including the occurrence and paths of the shorter waves accompanied by cyclones and the more intense weather phenomena, is closely related to position with respect to the long waves, and because usually at least a few days are required for their pattern to change substantially. However, the generation of the westerly flow outside the tropics, the latitudinal transports of heat and momentum and the convection which produces the weather systems depend on the baroclinic overturning considered in the following section.

9.3 The stability of large-scale sheared flows

Studies by the perturbation method

The perturbation method has been used extensively to study the stability of steady zonal flows to large-scale disturbances of small intensity (9). It is assumed that the flows are adiabatic, frictionless, and stably stratified, with the Richardson number large and the Rossby number small, so that the motion is nearly geostrophic and the shear of the wind in the vertical is the thermal wind, initially that corresponding to a steady decrease of temperature toward the pole. (The flow is considered to be unbounded in the zonal direction; the dimensions of the model disturbance, unlike those of atmospheric motion systems, are exactly defined.)

The equations which govern the motion soon after the imposition of a small disturbance are simplified and linearized by omitting terms containing squares and products of the deviations. The equations are still too difficult to solve in complete generality, but some simple kinds of solution can be found which are instructive. The interesting aspects of a solution are the rate of travel and amplification of a disturbance, and its structure and transfer properties, which can be compared with the properties of observed atmospheric systems.

The essential features of the relation between wavelength and amplification rate and of the structure of growing waves can be obtained with a very simple model in which the fluid is considered to be incompressible and to be confined between rigid upper and lower boundaries, with $\beta = 0$. In this model all the wave properties are symmetrical or antisymmetrical about the zonal axis in the middle of the fluid layer;

9   Large-scale slope convection

in particular, the waves have a phase speed equal to the undisturbed speed of flow in the middle of the layer. However, the results described here are those obtained (9) in a model in which a positive value of β is introduced, the fluid is considered to be compressible, and the rigid lid is replaced by a second, deep, layer of comparatively large static stability (crudely corresponding to a stratosphere).

Disturbances develop for nearly all initial wavelengths, traveling with the speed of the undisturbed flow at a height (the *steering level*, a term arising from forecasting practice) which is considerably below the middle level. The disturbances appear as a train of growing symmetrical troughs and ridges in the originally parallel zonal contours of the isobaric surfaces of the basic flow. If the basic flow is generally westerly, closed contours defining the centers of cyclones and anticyclones first appear at the lower boundary where the speed of the basic flow is least, and later spread to higher levels.

The amplification rate has a maximum for a definite wavelength (corresponding to a dominant wave), which depends mainly on the depth of the layer and less sensitively on the shear and the stability. The existence of a wavelength of maximum amplification rate leads to the concept that a flow subject to perturbations over a whole range of wavelengths will after some time be characterized by motion systems of the scale corresponding to the dominant wavelength. For typical magnitudes of the relevant parameters in middle latitudes ($f = 10^{-4} \sec^{-1}, \beta = 10^{-13} \text{cm}^{-1} \sec^{-1}, B = 10^{-7} \text{cm}^{-1}$, $H = 10$ km, and the difference of wind speed over the depth of the troposphere $\Delta U = 10$ m sec^{-1}, the dominant wavelength is about 5000 km. If the westerly flow is concentrated into a narrow range of latitudes, the curvature of the horizontal wind profile effectively increases β and reduces the dominant wavelength to as little as 2000 km, without materially altering the growth rate, which corresponds to a doubling of the amplitude in less than 3 days.

The structure of the dominant wave

The structure in the troposphere of the dominant growing wave, represented, for example, by the distribution of meridional and vertical velocity and the deviation of pressure and potential temperature from their undisturbed values, depends in some degree upon the constraints applied to the motion. For some simple conditions the structure is illustrated in Fig. 9.4, which shows the distribution of the perturbation pressure at the lower surface by isobars (which can be regarded as streamlines of the perturbation flow, at speeds inversely proportional to their spacing) and the form of isopleths

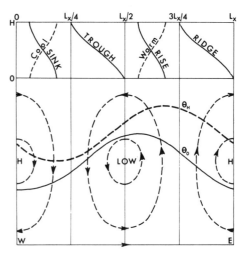

Fig. 9.4 The structure of the dominant amplifying baroclinic disturbance in a troposphere confined between rigid boundaries (with $\beta = 0$).

The lower part of the diagram shows isobars (dashed lines) of the perturbation pressure at the lower boundary; they may be regarded also as streamlines of the perturbation velocity, and arrows are added to indicate the sense of the circulations. Superimposed upon a zonal flow, the perturbed flow is represented by a series of sinusoidal streamlines or isobars, defining troughs and ridges with meridional axes. The perturbed form of two isopleths of potential temperature is also shown by two dashed curves labeled θ_0 (heavy line at the lower boundary) and θ_H (thin line at the upper boundary).

The upper part of the diagram shows in a zonal section the distribution with height of the phase of the perturbations of pressure, potential temperature, and vertical velocity; the sloping lines show the positions of the largest perturbations. The diagram covers one (zonal) wavelength L_x of the growing disturbance.

of potential temperature θ in the low and high troposphere (all parallel zonal lines in the unperturbed flow). The upper part of the diagram shows the variation with height of the positions at each level of the axes of the minima and maxima of pressure, potential temperature, and vertical velocity (which reaches its greatest value at the middle level). In a flow that was initially westerly at all levels the isobars or streamlines become sinusoidal with normally oriented trough and ridge lines at all levels; if the speed at the lower surface was initially zero, closed isobars appear there immediately, but otherwise they appear only at some later stage in the evolution of the disturbance, and as it grows they extend upward to successively higher levels. The disturbance moves eastward with the speed of the undisturbed flow near the middle of the layer (at the *steering level*); there the northward and upward components of velocity v and w, and the potential temperature θ, are all approximately in phase, implying a decrease of potential energy; by integration it is found that there is a net decrease over the whole system, which provides the kinetic energy of the intensifying disturbances. The axes of the troughs and ridges slope backward with height, by one quarter of a wavelength altogether, and the axes of the cool and warm tongues of air have a forward displacement which is half as great. At low levels the warmest air is a little in advance of the lowest pressure, while at high levels it is a little behind the highest pressure.

In the growing phase of the dominant short waves ($L \approx 2000$ km) the conversion of potential into kinetic energy is concentrated in the low troposphere, and their kinetic energy has a pronounced maximum near the surface (in the absence of friction). On the other hand, during the growth of the longer dominant waves (L several thousand kilometers), the conversion of potential into kinetic energy is more uniformly distributed with height throughout the troposphere, and the kinetic energy of the disturbances is increased, especially in the high troposphere, by transformation from the kinetic energy of the zonal flow. It is therefore sometimes stated that two distinct kinds of wave are to be anticipated in the atmosphere: short waves, which are readily discernible only near the surface, and long waves, which are hardly apparent in the low troposphere but which dominate the flow in the middle and high troposphere.

The evolution of baroclinic disturbances; frontal zones

In the perturbation method of the study of incipient cyclone waves it is reasonably assumed that the motion is adiabatic and frictionless, since the time scale of the growing disturbances, of not more than a few days, is small compared with the time scales of radiative and frictional adjustments, which, as mentioned in Section 6.4, are estimated to be about three weeks and several days, respectively. Further, the air is assumed to be unsaturated and to remain so, the stability parameter B referring to θ therefore being appropriate whether air moves up or down. Thus the vertical motion in the layer considered becomes divided into equal areas of upward and downward motion.

The continued evolution of baroclinic disturbances under various simple conditions has been studied by numerical methods. Common to these calculations are several features which could not be anticipated from the perturbation theory, principally the local concentration of regions of vertical motion into zones of strong horizontal temperature gradient, one to the north and east, and one to the south and west of each cyclone center. These *frontal zones* resemble those found in the atmosphere, and are zones of maximum thermal wind, which contribute to the production of a narrow, meandering belt of strong winds (a *jet stream*) in the upper troposphere.

In one study (10) a model was used representing flow in a zonal channel bounded laterally by smooth, rigid walls. The velocity of the undisturbed (westerly) current was assumed to be zero at the lower boundary, increasing in the middle of the channel to a maximum of 30 m sec^{-1} at the 300-mb level; at every level the speed decreased from a maximum at the central latitude, to become half of this value at each of the lateral boundaries, considered to be at latitudes 12 and 78°. Several calculations were made, which included both a lateral and vertical diffusion of properties and a frictional stress at the lower boundary proportional to the square of the wind speed near it. In all an initial perturbation was given the rather long wavelength of about 8000 km, for which the velocity and the growth rate could be anticipated from the perturbation theory to be small; however, in order to increase the growth rate arbitrary adjustments were made to the curvature of wind profiles in the horizontal. From the results of one calculated evolution (centers of low and high pressure reached their maximum intensity after 4.5 days, with surface pressures of 939 and 1053 mb) two diagrams have been selected to illustrate the form of the frontal zones which developed. Figure 9.5 shows the distribution of pressure at the

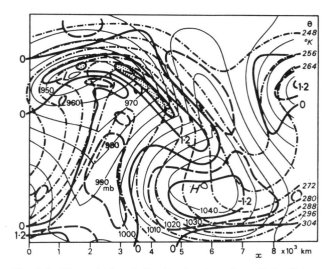

Fig. 9.5 The calculated distribution in a baroclinic disturbance (after 4 days of development) of *surface pressure*, shown by isobars at intervals of 10 mb (thin lines); of low-level *potential temperature* θ, shown by isopleths at intervals of 8°K (dot-dash lines with scale on the right); and of low-level *upward velocity* w, shown by isopleths labeled in cm sec^{-1} (bold lines, dashed for ± 0.6 cm sec^{-1}; light stippling shows where $w > 2.4$ cm sec^{-1}). The diagram shows a stage in the calculated evolution of a baroclinic disturbance in a zonal flow (in the direction x), using equations incorporating some lateral and vertical diffusion, and also friction due to flow over the lower (rigid) boundary (10). At this stage the low pressure in the cyclone center has reached nearly its minimum value, and two pronounced frontal zones have formed, a warm front ahead of the center and a cold front south of it. The drawing of sharp angles in the isobars crossing the frontal zones was justified by the smoothness of the neighboring pressure fields (defined by pressures at grid points 380 km apart).

9 Large-scale slope convection

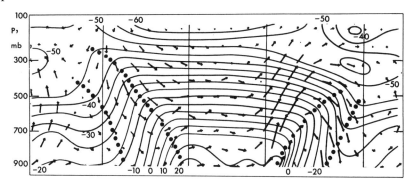

Fig. 9.6 The distribution of temperature (shown by isotherms drawn at intervals of 5°C) in a vertical section along the middle of the zonal channel in Fig. 9.5, illustrating the two frontal zones. Arrows show zonal air velocity, with the vertical component exaggerated by a large constant factor. The ordinate is linear in pressure, and therefore not a true scale of height. Approximate boundaries of the frontal zones (of largest horizontal temperature gradient) are indicated by lines of large dots.

surface and of potential temperature at the 900-mb level, and some features of the vertical motion at the 800-mb level, after 4 days of development. Frontal zones have appeared to the south of the cyclone center and ahead of it, where, with an associated pressure trough and maximum of upward velocity, the zone has become sharply defined, considering the scale of the whole disturbance. There is subsidence generally in the cool air, especially near the center of the anticyclone. Subsequently the calculated field of motion became much more complicated, with the formation of new cyclones and anticyclones. On the sixth day the principal frontal zones were oriented more nearly meridionally across the middle of the channel, along which a vertical section (Fig. 9.6) shows that they extended upward throughout most of the troposphere with an inclination of about 1 in 200 to the horizontal. Upward velocities occur in the lower troposphere ahead of the rear frontal zone and are more pronounced throughout the troposphere in the warm air near the leading frontal zone. Within the frontal zones, however, the vertical motion is generally downward.

It appears from other numerical studies that in the development of cyclones frontal zones are produced whose horizontal scale in the direction of the temperature gradient ("width") decreases ultimately to the separation of the grid points in the mesh over which a calculation is made.

The role of water in slope convection

Calculations of the evolution of baroclinic disturbances assuming adiabatic and frictionless flow, which imply the conservation of total energy, satisfactorily represent their intensification over a period of a few days. More elaborate models used in extended calculations include representations of energy transfers associated with friction, radiation, small-scale convection, and phase changes of water.

So far there has been little explicit consideration of the part played in the large-scale slope convection by the processes associated with presence of water vapor. In the preceding discussion of the development of baroclinic disturbances and of frontal zones, for example, phase changes of water have not been considered; the air has been assumed to be initially unsaturated and to remain so, its potential temperature appropriately being regarded as θ. Accordingly there was no necessity to distinguish between the thermodynamics of upward and downward motion. Nevertheless, condensation and precipitation have been implicit in the choice of a suitable value, after appeal to observation, for the parameter B which describes the vertical stability.

No satisfying analysis has been made of the mechanisms by which the observed stability is established. Intuitively it seems likely that in middle latitudes the dominating factors are (*a*) the slope convection involving only unsaturated air, which continually places potentially warm air over comparatively cool air, (*b*) the condensation (and precipitation) of water in the narrow frontal regions where the slope convection is intensified, and (*c*) cumulonimbus convection in the cool airstreams of the slope convection and in neighboring tropical regions. The radiative energy loss associated with the distribution of water vapor influences the *horizontal* temperature distribution, especially the meridional temperature gradient, which generates the slope convection, but it is likely to have little direct influence upon the *vertical* temperature gradient, since Fig. 1.3 suggests that at least in the mean the equivalent rate of cooling is uniformly distributed. Of the mechanisms listed, the importance of (*b*) and (*c*) is suggested by the large local contribution of the latent heat of precipitation to the heat source which is required to balance the mean annual radiative sink; this amounts to over half in middle latitudes (Table 9.4, Appendix 9.1). The cumulonimbus convection in middle latitudes is likely to be less important than frontal rain, since from the few studies which have been made it seems that the amount of shower rain is generally only a small fraction of the total (Appendix 9.2); moreover, the shower clouds (in contrast to the clouds of frontal zones) do not usually reach into the high troposphere (Appendix 9.3).

The flow relative to a baroclinic disturbance

By subtracting the velocity of translation of a baroclinic disturbance from the general field of horizontal motion the flow relative to the disturbance (the pattern of the relative streamlines) is obtained. Below a steering level the flow relative to a disturbance generally has a large component from its front toward its rear (from east to west); above the steering level the relative flow is in the opposite general direction. In the perturbation theory of cyclone development the relative streamlines at low and high levels have patterns similar to those of the isotherms, since temperature changes are due only to horizontal advection. On the other hand, at intermediate levels temperature changes are produced by both horizontal and vertical motion, the latter always acting in the opposing sense; the motion is inclined at about half the slope of the surfaces of constant potential temperature, and accordingly the local rate of temperature change is only about half that which would be produced by horizontal motion alone. Consistently, the regions of maximum upward and downward motion are ahead of and behind the region of minimum pressure, respectively, coinciding with the regions of strongest flow toward and away from the pole. The vertical

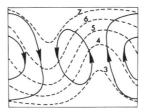

Fig. 9.7 Streamlines of the mid-tropospheric flow relative to a baroclinic disturbance in an early stage of its growth (solid lines), drawn on a surface of constant potential temperature whose height is shown by contours (dashed lines labeled in km). The vertical component of the motion is indicated by the flow across the contours.

component as well as the horizontal component of the motion can be illustrated by drawing the relative streamlines in a surface of constant potential temperature, and including contours of the height of the (isentropic) surface, as in Fig. 9.7. This refers to the isentropic surface which passes through the steering level where the disturbance has its maximum amplitude. The surface slopes upward into higher latitudes, and the motion across the contours clearly shows the sign of the vertical motion. The actual vertical velocities have only about half of the indicated magnitude, and the relative streamlines are not trajectories, as presumed in some analyses described elsewhere in this book, because the isentropic surfaces are themselves moving and the system is not in a steady state. As the disturbance evolves the isentropic surfaces become distorted (tending to move into the horizontal), and a closed cyclonic circulation develops on successively higher surfaces.

9.4 The observed properties of large-scale slope convection

It is not feasible to give an exhaustive description of the great complexity and variety of the observed behavior of the atmosphere. On any scale the aim must be to give an economical description of those features which are recurrent and by some criterion judged to be the most significant, expressed in terms of graphical and ultimately mathematical models.

On large scales the data are mainly the routine "synoptic" (synchronous) observations of the national weather services. These include radiosonde measurements of the distribution with height of wind, temperature, and relative humidity over sounding stations, generally made twice each day. At more numerous stations observations at more frequent intervals provide values of the same variables near the surface, together with measurements of pressure reduced to sea level, and less precise information about visual range, form and intensity of precipitation, and form, height, and amount of cloud. The sounding stations in the northern hemisphere have an average separation which varies from about 400 km over some densely populated regions, such as western Europe, to about 1000 km over large sparsely populated continental areas and to much more in low latitudes and over the oceans. The spatial density of the stations observing weather from the surface is everywhere about an order of magnitude greater. (Although in principle the acquired data could be supplemented greatly by observations from commercial aircraft, in practice almost no advantage is taken of this opportunity; better prospects lie in the systematic use of meteorological satellite observatories. These are now capable by day and night of making indirect soundings through cloud-free regions, and of observing the distribution of cloud and the temperature of cloud tops with a resolution of several kilometers or even better.) Thus over much of the hemisphere the observations satisfactorily define the structure of the tropospheric motion systems which have a horizontal scale of a few thousand kilometers, that is, the *major cyclone waves* and the *long waves*.

However, the resolution of the observations is adequate to define the structure of the smaller *wave cyclones* only over the continents of the northern hemisphere, and not over the oceans where they most frequently form and develop into the most intense cyclones. In the evolution of these essentially baroclinic disturbances an important part is played by the condensation and precipitation of water, processes which are of intrinsic interest in the study of clouds and rains. Now it is just in the definition of the distribution of water vapor and cloud that the routine soundings are least satisfactory. The radiosonde measurements of relative humidity in the upper troposphere are frequently inaccurate and sometimes misleading, while the soundings (especially in the abbreviated form in which they are usually communicated) at best give only indirect indications of the presence of clouds. Thus they can only supplement indirect weather observations (from the surface or satellites) in the analysis of the distribution of water substance in its vapor and condensed phases.

For these reasons there are still no satisfactorily comprehensive graphical (or theoretical) models of the typical genesis and evolution of wave cyclones (and smaller disturbances). The graphical cyclone model generally approved originated half a century ago, and has become known as the Norwegian model after the school of Norwegian meteorologists who devised it. Although originally renowned for their advanced theoretical concepts, they undertook for a while the task of producing daily weather forecasts; then, under the stimulus of the practical analysis of detailed weather maps, they rediscovered the frontal zones, known by much earlier observers to be characteristic of cyclones, and showed that they linked apparently separate cyclones (11). The frontal zones became regarded as part of a transition zone which is practically a surface of discontinuity—the *polar front*, a wavering but persistent boundary traceable around much of a hemisphere, separating cool and warm airstreams drawn, respectively, from polar and tropical sources, where *air masses* of almost uniform properties in the horizontal occupy very large areas.

Attempts were made to develop a theory of cyclogenesis as a manifestation of the dynamic instability of the polar front, but they were not successful. (Flows separated by such a surface can be shown to be unstable, but the disturbances which would develop do not convert potential into kinetic energy, and so are not of the kind sought.) Nevertheless, the Norwegian model was far superior to its predecessors in that it concentrated attention upon the concept of cyclone *evolution* and emphasized the interpretation of the principal kinds of cloud and weather in terms of the fundamental processes of large-scale ascent or descent and of heating or cooling of airstreams at the earth's surface. Moreover, although representing a drastic idealization of the great variability and complexity shown by weather maps, it possessed the virtues of simplicity and flexibility which allowed it to be applied in the day-to-day practice of analyzing weather maps and preparing weather forecasts in regions outside the tropics, so that eventually it became universally adopted, and still stands virtually unchallenged, although recognized as needing some modification or elaboration, principally in respect of tropospheric structure above the layer near the ground.

Skill in weather forecasting was until recently based upon subjective, or graphical, analysis of recent weather and extrapolation, modified in the light of long personal experience. Except in settled periods useful forecasting by such methods is limited to periods of about one day ahead. Now numerical techniques based on objective, or quantitative, analyses of the atmospheric structure as shown by radiosonde ascents, and the application of mathematical models, have made it possible to prepare useful forecasts of the general nature of the weather over much of the northern hemisphere for longer periods. It is widely believed that with improved data and technique these periods may eventually be extended to as much as a week or two. However, at least in western Europe it is still debatable whether the numerical methods have led to any significant improvement in the forecasting of detailed weather, including cloud and precipitation, for periods of as little as one day ahead. Serious errors are occasionally made in such forecasts, which may be due to poor analysis of the initial data arising from their sparsity over the neighboring ocean, from inadequate models of motion systems with horizontal scales less than 1000–2000 km (and of their associated cloud and precipitation distributions), or from progressive loss of experience in detecting the early stages of such systems as increasing reliance is placed on the numerical models. Whatever may be the reasons for the errors, they suggest that there is still need to consolidate and perhaps extend the range of the models of wave cyclones and other disturbances of scale smaller than that of the major cyclones and the long waves.

Long waves and cyclone families

At many places in middle latitudes, especially in the winter season, spells of weather with little or no rain tend to follow spells of several days of mainly westerly and often strong winds, with rains at intervals of 24–36 hr

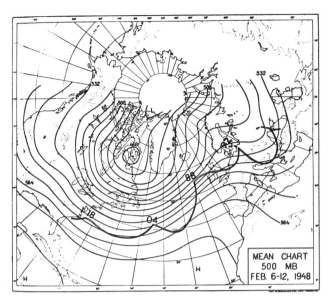

Fig. 9.8 A cyclone family on the eastern flank of a major trough (12). The trough is part of a long wave defined by mean contours of the 500-mb level during a winter week. The positions of the fronts at a time in midweek are shown, and the associated cyclone centers are marked by the last two figures of the central pressure (in mb). The distance between the cyclone centers in the west of the North Atlantic is about 2300 km.

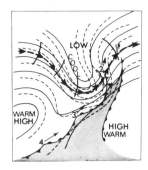

Fig. 9.9 A model of a long wave, shown at the 500-mb level (represented by height contours shown as thin lines), and a cyclone family (represented by the fronts at the surface, drawn in the conventional manner, and the sea-level cyclone centers, numbered from 1 to 4, with arrows indicating the velocity of the centers). The distribution of temperature at the 500-mb level is shown by isotherms (dashed lines), and the axis along which the winds at this level are strongest is marked by the heavy dashed line and chain of arrows. "The strong current is often characterised by weak disturbances which move very rapidly eastwards and is in some degree reflected in the wind and pressure distribution at the ground" (13, from which the model has been redrawn with slight modifications). These disturbances are shown by five short bowed trough lines which lie across the current. "A front is observed farther south at the ground which separates the cold air in the north and north-west and the warm air in the south (stippled). Along the surface front a series of polar front waves moves northeastwards."

accompanying the passage of about four successive cyclones. These cyclones are said to belong to a family; typically each forms near the southern end of the trough of a long wave, and as it intensifies travels northeast on the eastern flank of the trough (e.g., see Fig. 9.8). On occasion a surface chart shows as many as four cyclones simultaneously on this flank of a trough over an ocean, each in a different stage of development. The model seen in Fig. 9.9 suggests that there may be disturbances of correspondingly short wavelength but of small intensity in the upper troposphere also on the western flank of the trough of the long wave, there separated by large distances from the cyclones apparent near the surface in the early stages of their development. Cyclones or troughs of small amplitude can sometimes be observed also at the surface traveling southeast on the western flank of the major troughs, especially over the continents, but generally they remain weak. According to forecasting experience cyclogenesis on the eastern flank of the troughs may occur on the arrival of a weak disturbance from the other flank (14, p. 14), but the connection is difficult to trace and apparently no analyses of such events have been published. Families of cyclones are most regular and most frequently observed over the oceans and the lands near their eastern borders; over North America and Eurasia cyclones have a similar tendency to appear in series, but generally they appear irregularly and are not clearly related.

Behind the early members of a cyclone family there are limited excursions of cool air into lower latitudes, followed by its temporary return poleward ahead of the following member. Behind the last member, or behind one which becomes particularly intense, a broader cool airstream flows into the tropics. Such major excursions of cool air occur preferentially over particular parts of the northern hemisphere, especially into the Atlantic from the neighborhood of Greenland and North America, and into the Pacific along paths near the west coast of North America and the east coast of Asia. The cold front marking the leading edge of the excursion usually moves into the large subtropical regions of high pressure in the oceans, there weakening and eventually disappearing from analyses as the cool current becomes subject to large-scale subsidence (and horizontal divergence), and the cool air becomes shallow and modified by ordinary convection. The center of the anticyclone is usually transferred into the ridge of high pressure on the western flank of the current, which eventually becomes an easterly airstream on the southern side of the anticyclone. Subsequently a new family of cyclones and linking fronts develops well to the northwest of the anticyclone. [An example of a late stage in this sequence of events is shown over the eastern North Atlantic in Fig. 9.2. During the previous several days a series of cyclones had moved along paths from near Newfoundland to north of Norway, associated with the

9 Large-scale slope convection

major trough numbered 1. In the figure the cyclone with a central pressure of 1009 mb over northern Norway is the last of this family, and near the surface cool air has penetrated far south off the western coasts of Europe and North Africa, before turning westward south of the anticyclone which (at the surface as well as at the 500-mb level) is centered northwest of the Azores. A new family of cyclones and fronts has developed over North America and has extended toward Iceland, on the eastern flank of the next major trough, numbered 6.]

During the establishment of a broad cool airstream descending from the upper troposphere in high latitudes, a correspondingly large excursion of low-level tropical air has occurred in the opposite sense farther east, in association with the cyclone family. A large part of the mean latitudinal heat transfer is effected by these intermittent major exchanges of air between low and high latitudes.

The Norwegian cyclone model

Several features of the Norwegian cyclone model require comment. First, although the fronts are on detailed examination recognized as narrow transition zones, on the scale of the cyclone they are considered to be discontinuities. By convention, the front at the surface is drawn on the warm side of the zone of marked transition (this border is better defined than the other, the horizontal gradient of temperature being much smaller within the warm air mass than within the cool air mass).

Second, a front is regarded as a prerequisite for cyclone formation. Regions of precipitation are associated with the ascent of the warm air over the fronts, and the cyclone intensifies until the cold front appears to overtake the warm front and the air in the cyclone disappears from the surface (a process called *occlusion*; cf. Fig. 9.9). A strong perturbation aloft is indicated as appearing only when the development of the cyclone at the surface is well advanced.

Third, an *occluded front*, accompanied by a pronounced pressure trough, is drawn at the surface during the late stages of the intensification of the cyclone. It is present because in general there is a considerable contrast of temperature at low levels in the cool air masses ahead of and behind the *warm sector*, occupied by the warm air, due to the difference in their recent paths and the degree of modification by heat exchange with the underlying surface. Usually the air ahead of the warm front is returning poleward after having been warmed over the sea in a lower latitude, whereas the air behind the cold front is cooler, having more recently been drawn from a high latitude; the occluded front in the lower troposphere then retains the character of a cold front and is called a *cold occlusion*. However, in winter when the cool air behind the cold front enters a continent from the ocean, near the surface it may be considerably warmer than air which has previously been over the cold land. The occlusion may then be a *warm occlusion*. (Close to the ground there may sometimes even be a warming when the maritime air replaces the air in the warm sector.)

Fourth, a feature of the mature cyclone is the development of a trough of low pressure at sea level which extends from the center into the deep cold air in its rear. This trough is first apparent as a westward elongation of the isobars near the cyclone center; the trough usually rotates around the center until it is oriented toward the southwest. Marked troughs of this kind are usually accompanied by much cloud and rain, in contrast to the more scattered showers with clear intervals which in general characterize the cool airstream behind the cyclone, and some analysts draw attention to the phenomenon by prolonging the occlusion through the cyclone center into the trough as a *backbent occlusion*. However, no structure has been proposed for such a front, and the motion of the trough is too slow to make the association with any kind of front plausible. (The existence of backbent occlusions can be justified on the infrequent occasions when the center of an occluded cyclone is observed to be transferred quickly during fresh development to the tip of the remaining part of the warm sector.)

Mainly on the basis of observations made at the ground, but partly using soundings made into the middle troposphere by aircraft, an *indirect aerology* was developed which supplemented the cyclone model by classifying cloud and weather into several different basic types according to the sign of the large-scale vertical motion and of the heat exchange with the surface. In one such classification, for example, eight principal weather regions were thus defined in relation to the main fronts and air masses, and it was recommended that they be marked distinctively in analyses of surface weather maps:

1. Regions almost clear of low cloud, associated with widespread subsidence in *tropical air* masses.
2. Elongated zones of extensive low-level layer cloud or fog in streams of *tropical air* moving poleward ahead of cold fronts, and cooled from below.
3. Similar zones in which the low clouds are accompanied by precipitation, which is characteristically drizzle.
4. Regions of rain near *fronts*, which fall into polar air from deep layer clouds formed by the ascent of *tropical air*.
5. Regions in polar air at the surface, occupied aloft by dense clouds (altostratus) formed by the ascent of *tropical* air over *fronts*, with precipitation not reaching the surface.

6. Regions almost clear of cloud, associated with subsidence in *polar airstreams*.

7. Regions of showers in *polar airstreams* heated from below.

8. Regions of extensive low-level layer cloud or fog formed in *polar air*, cooled from below on returning poleward after having been warmed in a lower latitude.

In the recommended scheme of analysis new emphasis was placed on the recognition of regions 3 and 5. Drizzle, consisting of small, closely separated drops produced by the coalescence of cloud droplets, had come to be regarded as a form of precipitation essentially different from rain. Rain was attributed to the melting of snow or hail produced by the growth of ice particles in deep clouds consisting mainly of supercooled droplets. The distinction is not now considered to be definite, but drizzle may certainly be distinguished as characteristically reaching the ground only from comparatively shallow layer clouds with a low base, and therefore indicating the cooling of an airstream during flow into a higher latitude or otherwise over a progressively colder surface. However, special instructions were needed before some observers, particularly at reporting stations on hills and coasts, relinquished a habit of describing as drizzle the often small but widely separated drops which reach the ground on the fringes of areas of frontal rains, thereby easing the task of analysts.

The analyst provided with only sparse or inexpertly made surface observations, which often compose most or all of the data from large areas over oceans, may for two reasons especially value even only one report of a large amount or overcast of altostratus. First, in the numerical code in which the report is made, this cloud ($C_M = 1$) is defined as an extensive layer with little detail, through which the sun may be seen blurred, as though through ground glass (Pl. 6.2). This cloud of ice particles has such a distinctive appearance that it can hardly be mistaken for any other. It is quite unlike, for example, a dense layer of cloud in the middle troposphere which is composed mainly of droplets, and which has a well-defined lower surface showing some more or less irregular detail ($C_M = 7$). Unfortunately this kind of cloud may easily be confused by an inexpert observer with a layer of low cloud. Both kinds of deep cloud in the middle troposphere are physically distinct from the comparatively shallow layers in the lower or lower middle troposphere which originate as extensive shelf clouds (Section 7.9) during convection from the surface. Further, both have the important significance of indicating cloud formation by large-scale ascent during cyclogenesis. However, the former kind ($C_M = 1$) is usually the more extensive, and the other reason for paying special attention to a report of it is that this interpretation can be made with confidence and is often the first evidence of a new cyclogenesis.

Recently the Norwegian cyclone model, which was originally constructed from studies of observations over the North Atlantic and western Europe, has been supplemented by maps of the mean distributions over the North Atlantic of the flow aloft and other properties, such as the fluxes of sensible and latent heat from the ocean surface, which are listed in the legend to Fig. 9.10 (the mean distributions were obtained from analyses of about 50 cyclone evolutions, and expressed in coordinates using distance and orientation with respect to the center of a cyclone and its fronts at sea level). Figure 9.10 has been composed from those distributions specified as accompanying cyclones in the nascent wave and the occluded stages of their evolution. In the diagram the eight principal kinds of weather region listed in the classification given earlier can be located in various places, where they correspond to regions here indicated to be characterized by one kind of "predominant" weather.

It is noticeable that the borders of the regions of showers do not everywhere closely follow the isopleths of upward energy flux from the sea into the atmosphere: the maximum flux of both sensible and latent heat is located off the east coast of the United States, where large-scale subsidence behind the wave cyclone allows the development of only small cumulus. The mean convective flux of energy was found to be everywhere directed upward from the sea, although on individual occasions small downward fluxes of sensible heat (of up to 0.07 cal cm^{-2} min^{-1}) were inferred in narrow strong currents of tropical air moving into higher latitudes. (There was also everywhere a radiative flux of heat from the sea into the air, of magnitude between about 0.02 and 0.07 cal cm^{-2} min^{-1}.)

The mean upward convective flux of sensible heat from the ocean into the tropical air masses of low latitudes was everywhere less than about 0.1 cal cm^{-2} min^{-1}. This flux was evidently associated with the widespread occurrence of shallow cumulus, with tops at a pronounced inversion near the 800-mb level, and very dry air above. At about this level there was a maximum in the divergence of the upward radiative heat flux, equivalent to a rate of cooling of about 5°C day^{-1}. The energy loss by radiation (rather than by flow toward warmer waters) is generally responsible for the maintenance of the shallow layer of ordinary convection. South of about the 25th parallel, farther from the subtropical belt of high pressure, the layer of convection and the cumulus were somewhat deeper, so that slight showers were present in the region near the 20th parallel on about 80% of occasions.

The areas indicated to be occupied by intense showers probably extend farther to the west and northwest

9 Large-scale slope convection

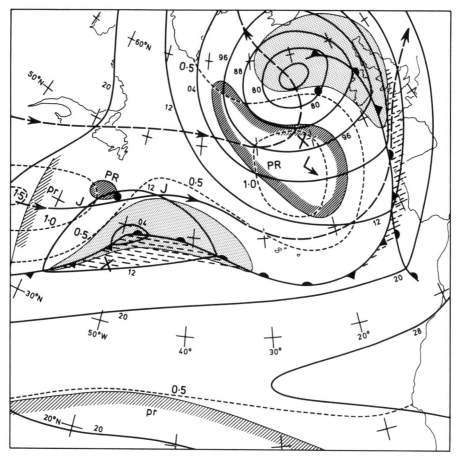

Fig. 9.10 Typical features of young ("wave") and mature ("occluded") cyclones over the North Atlantic (15). The following features appear:

1. Isobars at sea level, at intervals of 8 mb, labeled with the last two figures of the pressure (in mb).
2. The upward flux of energy from the sea surface in the form of latent and sensible heat, shown by isopleths labeled 0.5, 1.0, and 1.5 (units of cal cm^{-2} min^{-1}). South of the fronts, drawn in the conventional manner, the flux of sensible heat is everywhere small, and here also is almost everywhere upward, since even where the air moves toward cooler water there is usually cooling aloft, due either to large-scale upward motion or a net radiative heat loss a lowermost layer which contains some cloud.
3. Areas occupied by thick clouds at middle levels (shown by stippling), by low-level layer clouds or fogs (horizontal dashed lines), and by showers reported as intense (close hatching, marking only the border of the larger region, and labeled *PR*). In the larger region the lightning symbol (the sharply bent arrow) shows the position of maximum thunderstorm activity, probably the best indication of where the showers are most intense. In the rear of the wave cyclone, and in the far south, the borders of regions of slight showers are marked (light hatching, and labeled *pr*).
4. The heavy dashed lines in the upper part of the diagram show the shape of two contours of the 300-mb surface; the more southerly is approximately the axis along which the wind speeds are greatest, and is drawn as a solid line north of the wave cyclone and labeled *J* where there is a well-defined jet stream. The solid dots show the positions of local maxima of the vertical component of absolute cyclonic vorticity of the geostrophic wind near the cyclone centers (of about 2.3×10^{-4} sec^{-1} near the center of the mature cyclone and 1.5×10^{-4} sec^{-1} near the center of the nascent cyclone).
5. The crosses south of the cyclone centers mark the positions of the strongest reported surface winds.

than the regions of deep convection, since showers of hail and of heavy snow were included in the category "intense," but over cool waters they may fall from only moderately deep clouds. Probably the only region in which the convection reaches into the upper troposphere is that south of the center of the intense (occluded) cyclone. [It is interesting that in just this region is the (geographical) position of a pronounced maximum frequency of lightning (marked by a sharply bent arrow, the conventional symbol for lightning) found in a study of the distribution of the sferic sources in polar airstreams over the Northeast Atlantic (east of 40°W; 15). The observations, made by the special network of radio-location stations over Great Britain, were obtained over a limited winter period (a total of 107 daylight hours during three months of one winter). Nevertheless, the distributions found, with respect to geography and with respect to the positions of fronts, are probably significant and will be mentioned again.]

Figure 9.10 includes two contours of the height of the

300-mb surface, the more southerly of which is approximately the axis of the strongest geostrophic winds in this surface. The nascent wave cyclone and its upper tropospheric cloud system lie on the southern flank of a pronounced jet stream. The winds aloft over the occluded cyclone are not so strong and the jet axis is displaced well to the south of its center (because the thermal wind represents a large component of the geostrophic wind in the high troposphere the 300-mb contours tend to lie parallel to the fronts, especially where they are most intense). In a late stage of cyclone evolution, when the occluded front is weak and lies in the outermost part of the cyclonic circulation (but the central sea-level pressure is still nearly the minimum value attained), the cyclone is a great vortex with an almost upright axis, and there is little increase of wind with height over most of the region which it covers.

In the Norwegian model, as illustrated in Fig. 9.10, on the surface weather map the fronts are drawn through the center of lowest pressure in the cyclone. The observational evidence is not usually adequate, especially over the ocean, to substantiate this convention; where it is clear that the warm sector does not extend to the center it is often assumed that occlusion has begun and an occluded front is arbitrarily added to the analysis. Although the fronts cannot be regarded simply as partitions between air masses of different character, moving with the normal component of the wind, the extension of occlusions into cyclone centers implies that at a center adjacent elements of air can persistently be distinguished, and therefore virtually that there is a wind in the center with the velocity of the cyclone; according to experience, however, the surface wind in the central parts of cyclones is calm, or light and variable in direction, indicating that in a moving cyclone the air near the center is continuously replaced.

When cyclogenesis occurs in the presence of a well-defined front, the front is usually almost straight over a distance of 1000 km or more, and almost stationary, lying parallel to the winds at low levels. It has been suggested (16), and supported with a few detailed case studies, that the surface pressure minimum is likely to appear nearer the center of the baroclinic zone, at least a few hundred kilometers to the cold side of the surface position of the front, and that as the cyclonic circulation intensifies the front becomes distorted into the typical shape of the warm sector as the warm air is swept poleward ahead of the cyclone center.

Observed properties of cyclones

In discussions of the observed properties of baroclinic disturbances, particularly when referring to the surface weather charts which show the greatest amount of detail, attention is often concentrated on the centers of low pressure rather than the trough-ridge patterns which more properly represent the motion systems. This is partly because bad weather tends to become concentrated near a cyclone center, and partly because it is simple (although perhaps often not entirely satisfactory) to specify pattern and intensity by giving the position of a cyclone center and the sea-level pressure there. A major advantage of using the Norwegian model for analysis in weather forecasting practice is that the surface pattern of isobars and fronts nearly always (outside the tropics) adequately indicates the important features of the structure of the motion systems throughout the whole troposphere. However, that the cyclones must ultimately be regarded as inseparable parts of greater systems had been sensed over a century ago, as shown by the comment that "the less disturbed and ordinary condition of the atmosphere, though by far the more prevalent, had so slight a share of thought bestowed on it, that there was no clue, apparently, to the common alternations or changes of wind and weather; and even the ablest philosophers almost derided the idea of foretelling them, except on occasions of great storms" (11).

The wavelength of baroclinic disturbances

It is not possible to devise an entirely satisfactory procedure for identifying and measuring the horizontal dimensions of individual motion systems from the complicated pattern of streamlines on a weather map, because of the difficulty of defining their boundaries. On the other hand, the exact mathematical analysis of the spectral form of the components of the pattern provides results difficult to interpret physically.

In Section 6.4 (and Fig. 6.2) it was shown that the kinetic energy of the troposphere is associated predominantly with the long waves of horizontal dimension several thousand kilometers. It is apparent, however, from an inspection of daily weather maps that these large disturbances are associated with the intensification of members of a family of cyclone waves which individually have a much shorter wavelength. The various expressions of the Norwegian cyclone model are not very explicit about wavelength, but in accompanying descriptions it has been remarked, for example, that waves on the polar front "which develop into real cyclones usually have a wavelength of 1500 to 2500 km" (4, p. 741), and a similar value is implied by the model of Fig. 9.9. On charts of surface weather it is possible to estimate a wavelength by measuring the distance between neighboring centers of low pressure, or the distance between the positions on either side of a center where the associated front is stationary (usually the two methods applied to one situation lead to practically the same result). These estimates are likely to be signifi-

Table 9.1 The distribution of estimated wavelengths L of cyclones appearing over or near the Atlantic and Pacific oceans during the winter months January–March 1958, estimated from inspection of published surface weather maps. Cyclones likely to have developed in cloudy frontal zones are described as *wave cyclones*

L ($\times 10^3$ km)	Number of cyclones	Number of wave cyclones
1.1–1.5	1	7
1.6–2.1	8	38
2.2–2.6	8	31
2.7–3.2	12	19
3.3–3.7	10	4
3.8–4.3	6	2
4.4–4.8	4	1
4.9	4	0
Total	53	102

cant if they are made excluding occasions when during the later stages in the evolution of one cyclone it approaches (and eventually appears to absorb) another. The results of a set of such estimates are given in Table 9.1, arranged according to whether or not the new cyclones appeared likely to have developed in the cloudy frontal zone of a preceding cyclone (if so, they are described as wave cyclones). A judgment of the distinction from ordinary weather charts must depend strongly on the models favored by the chart analysts, but it is noticeably more difficult to make over the western Pacific than over the Atlantic. However, the table suggests that the wave cyclones are the more numerous and usually have shorter wavelengths, nearly all in the range 1500–3000 km. It appeared that the wave cyclones did not always intensify while remaining distinguishable in the analysis.

In a more systematic study (17) data from the northern hemisphere during a short winter period were examined for a relation between the wavelength L of a wave of small amplitude (excluding the long waves) in the contours of the 500-mb surface, and a measure s of the tropospheric stability B and the value of the Coriolis parameter f appropriate to the region in which the wave was observed. These variables appear in the expression for the dominant wavelength according to the perturbation theory in the combination $B^{1/2}/f$.

The stability s was defined as the difference ΔT (in °C) between the temperature at 500 mb and the temperature at that level in air lifted dry adiabatically from 850 mb, averaged over an area about 10° of latitude wide along one 500-mb contour defining a wave. The waves were selected to have small amplitudes, so that an appropriate value for f could readily be chosen. The wavelengths observed were mostly between about 1500 km (a lower limit) and 6000 km, with a maximum frequency between 1500 and 3000 km consistent with the data previously given. The analysis showed that variations in s and f were both significant in the expected sense, but it was found that $s^{1/2}/f$ varied little for wavelengths less than about 4000 km, suggesting that for these short wavelengths other factors are important.

The shorter wavelengths of the wave cyclones are usually attributed to one of three effects. First, in the presence of clouds the appropriate value of the stability may be significantly reduced. This explanation seems implausible, since the fraction of cloud in the volume of the frontal zones in which the wave cyclones develop is small, especially in the first stages of cyclogenesis. The cloud present is usually in a few layers, with large intervening spaces, and even when viewed from above by a satellite the band of extensive cloud cover associated with a trailing cold front usually appears to be 200–300 km broad and only occasionally as much as 700 km broad, which is only a fraction of the width of the strongly baroclinic zones (in those zones in which cyclogenesis occurs the concentration of the isobaric temperature gradient varies between about 15°C over a width of 1000 km and 35°C over a width of 2000 km).

Moreover, the wave cyclones of small wavelength may develop in shallow baroclinic zones which occupy only about the lower half of the troposphere. It seems, however, that systems of *high* clouds are typically the first to appear, and there is some evidence that the amplitudes of wave cyclones increase upward in the troposphere, rather than decrease.

Finally, the short cyclone waves may be associated with strongly baroclinic (frontal) zones in which the curvature of the lateral profile of wind speed is important; the half-width of the strong currents accompanying frontal zones varies from a minimum of about 300 to several hundred kilometers (7); typically, the curvature $-\partial^2 u/\partial y^2$ of the wind profile across the currents may be about 3β at 700 mb and 10β or even more in the upper troposphere. Such values reduce the dominant wavelength L_x of the perturbation theory into the range typical of wave cyclones. However, in the theory as so far developed (9) the kinetic energy of such waves is concentrated in the low troposphere, a conclusion for which there is little if any supporting evidence (see, e.g., 12).

The initial perturbation

In North America cyclogenesis is almost always associated with the arrival of a trough in the upper troposphere (traveling from the west) over a trailing cold front (7). The frontal zone is initially well marked only in the low troposphere, but it lies in a broad baroclinic zone; during the cyclogenesis it intensifies, steepens, and extends upward throughout most of the troposphere.

Usually, therefore, the cyclogenesis can be related to a preexisting disturbance of considerable amplitude. Over the North Atlantic the evidence of such preexisting disturbances is not definite, perhaps mainly because of the poor resolution of the observations, but if they occur they are certainly comparatively weak. Nevertheless, it is probable that disturbances of detectable intensity in the varying pattern of the large-scale flow are to be found preceding practically all cyclogeneses. The comparatively great frequency of cyclogenesis in particular geographical regions, for example, in the lee of extensive mountain ranges and near the east coasts of the major continents, can be attributed partly to the direct influence of the topography on the large-scale flow and partly to the related irregularity produced in the distribution of potentially warm air (especially for saturated ascent) at low levels. In the southern hemisphere, where the intrusion of topographic features into the zone of prevalent cyclogenesis (near the 45th parallel) is less marked, the frequency of cyclogenesis is more evenly distributed. However, it still attains noticeable maxima near the east coasts of South America and Australia (45).

The general circulation

Descriptions of the general circulation of the atmosphere refer to a variety of characteristic features of the fields of the motion and variables of state, from typical patterns of their instantaneous distributions on all scales to various kinds of time- and space-averaged statistics, such as the simple mean annual and zonal distributions of wind and temperature seen in Fig. 9.1. The statistics often obscure the presence of the characteristic motion systems because of the great variability with which they appear in time and space, and accordingly the physical interpretation of the statistics is not usually obvious. However, apart from the practical value of some of the statistics as convenient climatological summaries, the statistics as well as the characteristic properties of individual motion systems must be essentially interrelated aspects of a satisfying theory of atmospheric behavior. At present the most promising method of constructing such a theory appears to lie in the tentative formulation of governing equations explicitly expressing the physical conservation laws, their numerical solution under appropriate boundary conditions, and comparison of results with analyses of observations. However, in its generality this is a task of such enormous magnitude that there is great scope for modeling based upon apparently very drastic simplifications, partly to obtain more insight into the fundamentals of such a complicated physical system, possibly with the prospect of more economical use of numerical labor, and partly as a guide to the reformulation of equations which lead to unacceptable discrepancies between prediction and observation.

Recently impressive progress has been made with the method of numerical computation, yielding evolutions which in many respects satisfactorily represent the evolution of individual large-scale motion systems outside the tropics, as well as statistical distributions of motion and temperature in the troposphere and lower stratosphere and of rainfall at the surface (19, 20). However, the apparent success of the method in some measure rests on the incorporation of strong constraints obtained from observation, as already mentioned; for example, even in a model in which evaporation, condensation, and precipitation are considered (19), the radiative energy transfers are those inferred from the observed mean distributions of cloudiness and of the absorbing gases, in particular of water vapor, upon which the transfers essentially depend.

Similarly, in an example of the other kind of approach (5) the principal features of the seasonal distributions of wind and temperature in the troposphere are derived by regarding the cyclone waves and long waves as the principal agents of the transfer of momentum and heat, whose efficiency can be represented by transfer coefficients of the magnitude indicated by the perturbation theory. However, the results depend on an assumed appropriate value of the stability B, which is obtained from observation and not from any theory.

Without any further discussion of the relation between the properties of individual motion systems, which are the concern of this book, and the average states of the atmosphere which are established in their presence, some concepts of mean circulations and representations of climatological distributions are now introduced to supplement and to provide a background for the discussions of the various kinds of motion system which have been defined. Generally the occurrence of an individual motion system of a particular scale and the intensity which it attains appear to be under the dominant control of the systems of larger scale within which it arises, but of course *collectively* some small-scale systems (including molecular motions) have equal importance in modifying the properties of the large-scale systems.

The collaboration of small- and large-scale convection

The circulation of air through the major motion systems which occupy the whole depth of the troposphere, and which may represent either slope convection or cumulonimbus convection, can be projected schematically upon a vertical section, as in Fig. 9.11. The pattern of vertical motion is very asymmetric, with the upward velocities large compared with the downward velocities, and concentrated into small parts of the systems where

9 Large-scale slope convection

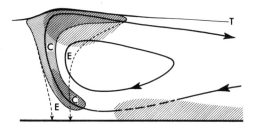

Fig. 9.11 A schematic representation of the flow in deep tropospheric convection (large-scale slope convection or cumulonimbus convection), projected into the vertical. The hatched area at the surface represents the comparatively shallow layer of small-scale convection, within which the air in the descending branch of the deep convection has its potential temperature raised, mainly by the addition of water vapor evaporated at the earth's surface. Air from this layer enters a cumulonimbus or frontal zone and ascends into the upper troposphere, with condensation C of water vapor. Precipitation falls from the cloud (dashed lines) and there is some evaporation E outside it. Small-scale convection may develop in the cloud system (cross-hatching). The descending branch of the circulation is left open on the right; it proceeds through several other motion systems with descent at a slow rate determined by a radiative loss of potential temperature. The tropopause is indicated by the line marked T.

condensation of water vapor occurs, producing belts of frontal cloud or thunderclouds. In and near these regions the flow is most rapid, and the changes of state following the motion may be nearly those in the appropriate dry or (pseudo-) saturated adiabatic process.

In the diagram air is shown rising from the low into the high troposphere through cloud. It is clear from the evidence in Chapter 8 that this passage is made by much of the air which enters the cumulonimbus, although a considerable proportion may leave the ascending branch of the circulation at intermediate levels. Probably a much higher proportion does so in the frontal clouds of slope convection, especially of that air which enters the cloud system with a rather low wet-bulb potential temperature (over cold continents or in high latitudes over the oceans, where θ_w at low levels may be less than θ_s near the tropopause). By overlooking this complication, it is clear that while ascent to high levels can be rapid (less than an hour in cumulonimbus, and about a day in frontal clouds), the descent required to complete the circulation must be comparatively very slow. Because of the efficient precipitation of water vapor condensed during large upward displacements, air leaving the ascending branch of a deep circulation soon becomes cloud free, and the rate of descent in the same latitude is limited to a small value by the rate of radiative heat loss. This, which in the mean is practically equal to the rate of absorption of solar radiation by the troposphere and the earth's surface, is equivalent to a tropospheric cooling rate of about $1°C\ day^{-1}$ (Section 1.2 and Fig. 1.3). Since the difference between the actual temperature of the low-level air and its temperature after the condensation and precipitation of its content of water vapor (which determines its potential temperature after leaving the ascending branch of the circulation) is about 20°C in middle latitudes and up to twice as much in low latitudes, a rather long descent period of about 20 days or more is implied.

This period might be considerably reduced if air ascends from a middle or high latitude and returns to low levels in a lower latitude. Nevertheless, it is clear that in slope convection the descent cannot be completed in the same motion system in which the ascent occurred: even the largest motion systems, with a horizontal extent of several thousand kilometers, are traversed by air in the upper and lower troposphere within a few days, since the relative speed of the winds there is typically 15 m sec^{-1} or more. In Fig. 9.11, therefore, the arms of the descending part of the circulation have been left open, to indicate that the descent continues in other systems, and that there is no definite boundary on that side.

Air approaching the surface enters the layer of small-scale convection, where its potential energy is restored during the transfer of sensible and latent heat from the surface. This layer has a variable but generally small vertical extent; it occupies only a small fraction of the depth of the troposphere and the air in it can be prepared for a fresh ascent within one major system.

Figure 9.11 is reminiscent of Fig. 7.15, which represented the circulations of small-scale convection and their relation to the comparatively shallow superadiabatic layer. Together they indicate the increasing importance of motion systems of smaller scale (and ultimately of molecular scale) upon closer approach to the earth's surface.

The mean meridional circulation

Because the meridional components of the winds averaged over all longitudes and a season do not much exceed 1 m sec^{-1} in low latitudes and are less than 1 m sec^{-1} elsewhere, and because of some deficiencies in the data, the details of the mean meridional circulation in the atmosphere are still uncertain. However, the principal features appear to be well established, as illustrated for the winter season in the northern hemisphere in Fig. 9.12.

The mean circulation in low latitudes, occupying more than half of the volume of the troposphere, consists of a cell in the direct sense (with ascent of the warmer air, concentrated near the equator, and conversion of potential into kinetic energy). It is associated with deep convection in cumulonimbus, or systems of cumulonimbus, which produce a maximum of rainfall near the equator. The descending branch of the circulation in

The observed properties of large-scale slope convection 9.4

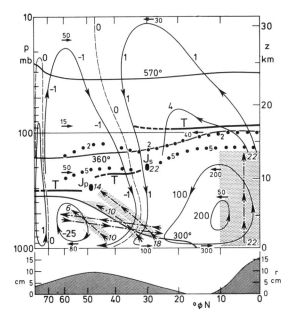

Fig. 9.12 Some features of the mean meridional motion in the troposphere and lower stratosphere during the winter of the northern hemisphere. The horizontal scale is proportional to the sine of the latitude, to emphasize the proportion of the atmosphere contained within latitude belts. The ordinate is pressure p; an approximate scale of height z is marked on the right. Streamlines of the mean meridional mass circulation are labeled in units of 10^{12} g sec^{-1}; dashed lines marked zero separate three principal cells. Typical values of mean horizontal speed are indicated by short arrows labeled in cm sec^{-1}. Mean positions of the tropopause T are shown by bold lines, and mean positions of the axes of the subtropical and polar front jet streams are shown by oval dots labeled J_s and J_p, respectively. Isopleths of potential temperature are entered for 300, 360, and 570°K.

In the troposphere ascent from low to high levels occurs in cloud systems, predominantly in cumulonimbus in low latitudes and in frontal clouds of slope convection elsewhere. These kinds of ascent are indicated schematically by arrows in stippled regions (italic numbers give typical values of θ_s). In middle latitudes the less steep paths of dry ascent and descent in slope convection are indicated by lines of arrows crossing the stippling. Typical values of monthly rainfall r (at places not far inland) are shown at the foot of the diagram. The lines of dots are isopleths of mean mixing ratio of water vapor, labeled 2 and 5 in units of 10^{-6}. Observational evidence accumulates that in the stratosphere the mixing ratio is nearly uniform, varying between about 2×10^{-6} and 3×10^{-6} (56).

the cell is over the subtropical arid lands and the oceanic regions of mainly very shallow ordinary convection and persistent large anticyclones. Middle- and high-latitude zones are occupied by a cell which is indirect when the mean winds are computed in these geographical coordinates, although slope convection is prevalent. This peculiarity arises because in the fully developed asymmetric motion systems of slope convection subsidence (of cool air) is predominant in a lower latitude than the most intense ascent (of warm air). Averaged with respect to distance from the axes of the jet streams associated with the motion systems, the mean circulation is direct; the same difference in the sense of the circulation, according to the coordinate system, can be demonstrated in steady experimental convective circulations established in a rotating cylindrical annulus of liquid (21).

The two major tropospheric circulation cells, together with a third (direct) cell confined to high latitudes, extend upward into the lower stratosphere. There both the major circulation cells are indirect, consistent with the view that the kinetic energy of the large-scale motion systems is generated and transferred upward by the motion systems of the troposphere, especially the long waves. The vertical motions are comparatively small, and in contrast to circumstances in the extratropical troposphere ascent relative to the almost horizontal isentropic surfaces occurs during motion toward low latitudes, and descent during motions toward the poles.

A feature of the tropospheric field of motion which is indicated in Fig. 9.12 but whose existence has so far been mentioned only incidentally in the discussion of cumulonimbus (Section 8.6) is the subtropical jet stream, whose axis is at about the 200-mb level and in midwinter meanders around the hemisphere about a mean position near 30°N (Fig. 9.13). Whereas the jet streams associated with frontal zones in the slope convection wander over a large range of latitudes, the subtropical jet stream is comparatively very steady in position, and it attains greater speeds (often over 75 m sec^{-1} and sometimes over 100 m sec^{-1}). Consequently it dominates the pattern of the mean zonal wind speed (Fig. 9.1) not only in the winter but also in the summer, when it is weaker, farther north, and less easily traceable around the hemisphere. The related region of large horizontal temperature gradient below the axis of this jet stream is present only in the upper troposphere, whereas below a well developed polar front jet stream it occupies the whole troposphere. The subtropical jet stream lies near the northern limit of the direct cell of low latitudes, and it is usually attributed to a tendency for zonal rings of air in the high troposphere, moving poleward in the mean circulation from near the equator, to preserve angular momentum about the earth's axis (2). However, although locally the speed in the jet axis reaches the value of about 120 m sec^{-1} corresponding to conservation of angular momentum along a path in the high troposphere which originates in the zone between the equator and 5°N, the mean speed along the axis is only about half as great.

During the northern hemisphere summer the middle-latitude indirect cell retains approximately its winter position and intensity. The direct cell of low latitudes becomes very weak and shifts northward by between 20 and 30° latitude, or even disappears, while the intense direct cell of the other hemisphere extends across the

9 Large-scale slope convection

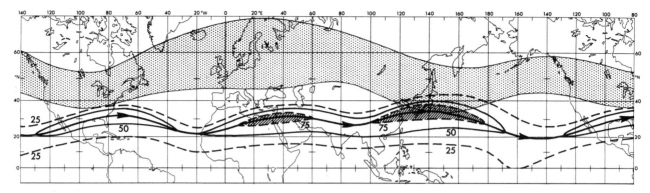

Fig. 9.13 The geographical positions of the subtropical and polar front jet streams in the northern hemisphere during midwinter. The heavy line with arrows shows the mean position of the subtropical jet stream during a winter season (23). Isotachs of mean wind speed at the 200-mb level, with respect to the position of the jet axis are drawn at intervals of 25 m sec^{-1}; in the hatched areas it exceeded 75 m sec^{-1}, reaching maxima of about 80 m sec^{-1} over northern Arabia and 95 m sec^{-1} over Japan. The subtropical jet stream can readily be traced around the hemisphere and is rather steady in position and speed; in contrast the polar front jet streams are associated with individual moving trough-ridge systems, and they occur over a greater range of latitudes. The area within which they are most frequent is stippled.

equator. In each of the extreme seasons the mean meridional circulation is strongly affected by the intense monsoons over Asia, which occupies only about one fifth of the area of the hemisphere. (In midsummer, for example, a widespread mean subsidence in latitudes near 25°N is probably eliminated from the zonal average by the ascending branch of a circulation associated with the cumulonimbus convection which is prevalent over northern India.) Thus doubts arise about the advantage of averaging over all longitudes and the possibility of providing comprehensible explanations of the resulting patterns. Much effort has been expended over a long period in attempts to explain the principal features of annual or seasonal mean zonal and meridional distributions of wind and temperature, with a lack of any completely satisfying degree of success. This failure may be an inevitable consequence of such distributions being essentially only a statistic of the very complicated motion systems which arise over a rotating earth with irregular surface properties. It is significant that the most encouraging progress has been in the production of similar statistics from recent experiments in the numerical calculation of atmospheric motion over prolonged periods, using simplified governing equations but a computation grid of resolution sufficient to give a good representation of individual large-scale motion systems. This kind of study seems to be the only one capable of quantitative development and able to provide reasonable answers to such problems as the prediction of the spread of material injected into some part of the atmosphere, or the results of natural or artificial changes in external conditions. On the other hand, for many problems inspection of the answers may provide no more satisfying physical insight than inspection of the observations, so that there is still justification for seeking analytical solutions to some aspects of atmospheric behavior for which it may be possible to construct sufficiently simple models. Transports of momentum, energy, and matter in the atmosphere, their relation to the observed mean distribution of motion and temperature, and issues of the kind mentioned concerning the significance of the mean distributions and the feasibility of a comprehensive and illuminating theory have recently been given extensive and perceptive survey (22).

Some climatic data

From previous discussion it will be evident that the general distributions of the potential temperatures θ and θ_w in the convective boundary layer are properties important in the consideration of the distribution of deep convection, in both its principal forms: cumulonimbus convection and large-scale slope convection. Although these potential temperatures typically decrease with height in the boundary layer, especially rapidly near the surface, and although it is an average value in a layer at least about 1 km deep which is significant, nevertheless, these averages are generally only about 1 or 2°C less than representative values at screen level, which are obtained in observations over the sea throughout the day and night, and over land near the time of maximum screen-level temperature during afternoons in the warm seasons. Consequently the distributions of these screen-level averages, available from many more places than the sounding stations, are also significant. As they are not included explicitly in conventional summaries of data, examples of mean summer distributions over the northern hemisphere are given in Appendix 9.6.

9.5 The large-scale cloud and precipitation systems of slope convection

The distribution of cloud over the whole globe is now surveyed daily in remarkable detail from meteorological satellites. The routine observations are mostly made in daylight, but with infrared sensors observations can be made during both day and night, and by the measurement of effective emission temperature they can provide useful estimates of cloud temperature and height (e.g., 24).

The daylight pictures, such as Pl. 9.1, with their resolution of a few kilometers, show cloud patterns of great complexity. However, an outstanding characteristic is the tendency of clouds to appear in *clusters*, of diameter up to about 1000 km, and in *bands* or *belts* and smaller *streaks* lying approximately along the direction of the flow at cloud levels on a great range of scales (lengths up to several thousand kilometers). The clouds of ordinary convection are clustered over land in the intermediate-scale circulations associated with topographical irregularities. Over the sea they occur in clusters and cells of diameter up to about 50 km (Section 7.6), and in clusters or chains of up to several hundred kilometers in length (25) associated with the downdraft fronts of shower clouds (Section 8.3), over which spreading anvil clouds often greatly increase the area of apparently unbroken cover. In low latitudes large clusters are grouped into very long bands, often several hundred kilometers wide, which mark the positions of the intertropical convergence zone. In middle and high latitudes, on the other hand, the most obvious features are rather narrower and more continuous bands, with a mainly streaky structure, which lie in the frontal zones of the long waves. These extend eastward and poleward, usually from latitudes between about 20 and 30° to latitudes of about 60° or more.

The typical streakiness on the larger scales is due mainly to the strong deformation of the flow in the baroclinic zones, where the prevalent layer clouds contain marked shears of wind in both the vertical and lateral directions. Figure 9.14 shows an example of the pronounced tendency, even in motion confined to an isobaric surface, for a marked sheet of fluid to become drawn out by the large-scale flow into long filaments. Since the greatest upward displacements tend to lead toward maxima of horizontal velocity this tendency can be anticipated to be even more pronounced in the three-dimensional flows which produce cloud systems. In considering the form of cloud patterns this tendency has to be related to the irregularities in the distribution of water vapor which are introduced into the layer of ordinary convection, and eventually into the middle and upper troposphere, by topographical features of all scales. Whereas introduced temperature gradients are generally smoothed dynamically, there is no such strong control over the distribution of water vapor. Moreover, topographical (mainly orographic) features impose disturbances upon the large-scale flow above the boundary layer, and thereby help to insure that at least over land condensation never begins smoothly on the large scale.

Over land the first clouds to form during large-scale ascent appear where orographic disturbances locally lift air sufficiently in the moister parts of the flow. This may happen in the middle troposphere where the flow is drawn from the convective boundary layer, in and above which θ_w generally decreases with height; ordinary convection then usually develops following condensation, and the clouds may become groups of castellanus (Appendix 8.8) which extend in long streaks downwind, before the towers evaporate during mixing with less moist parts of the flow. Thus the large-scale saturation of the general flow may be delayed. In the high troposphere very long plumes of ice cloud may extend away from individual cumulonimbus, and downwind of orographic disturbances in which the air is locally brought near or beyond saturation with respect to liquid water. In these and other ways cloud systems associated with large-scale ascent acquire streaky internal patterns and irregular, fringed edges—except perhaps along that side which lies along the flow within a frontal zone and marks the axis of a pronounced confluence, across which there is often a very large gradient of relative humidity.

The deformation in unsteady flows is accompanied by a tendency for the average separation of a group of neighboring particles to increase with time. The rate at which their dispersion proceeds depends on the scale s of the particle separation and the scales of the motion systems present: at any stage those which are most effective have about the same scale s. However, even in an atmospheric flow containing only the large-scale motion systems it is evident that particles initially

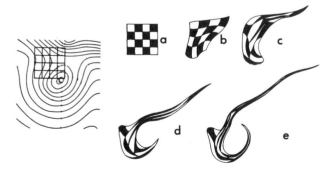

Fig. 9.14 An example of deformation in an atmospheric flow, computed with a barotropic model of motion in the 500-mb surface (26). The diagrams show the initial streamline pattern, with squares of side 300 km superimposed, and the deformation of the squares at successive 6-hr intervals.

9 Large-scale slope convection

only narrowly separated must eventually be widely dispersed, and correspondingly that the time during which the dispersion becomes large must impose a limit on the period over which the concept of a trajectory can keep its immediate physical significance. The rate of dispersion, suitably defined, probably varies by orders of magnitude according to the separation of the particles, the local intensity of the shear and deformation of the flow, and the local form of the spectral distribution of energy with scale of motion. Quantitative observations of dispersion have been made only over small times and distances, mostly less than a few hours and 100 km, but recently up to 24 hr and about 1000 km (27). However, from these observations and from estimates which have been made (28), especially by numerical computations with models of the large-scale flow in middle latitudes (29), it appears that in the free atmosphere trajectories can justifiably be considered over periods of several days during which air traverses an individual large-scale motion system, but probably not over the longer periods required for sink from the high troposphere to low levels in the descending branches of the circulations of several such systems. Over these longer periods trajectories originating in a small region can be imagined to become interwoven into a tangled skein eventually covering a whole hemisphere, gradually losing their identity by diffusion accompanying unsteady small-scale motions.

The relative flow and bands of frontal cloud in long waves

Generally the velocity of an individual motion system and the flow within it can be regarded as nearly steady during a period τ which is short compared with the recognizable life T of the system, but at least as long as the time t taken for air to traverse the system, especially in those more rapid parts of the flow which pass through an associated region of condensation. Consequently some insight into the nature of a system often can be obtained very economically, without the great labor of analyzing its whole evolution, by examining the motion relative to the system, assuming that the relative streamlines represent trajectories (they are not trajectories with respect to the ground, but these can readily be derived if they are required). This kind of analysis was used in Section 8.4 to obtain a graphical model of the flow in intense cumulonimbus (in which t, τ, and T are, respectively, about 15 min, 1 hr, and a few hours). Insofar as the assumption of steady flow is justified, the successive states observed along the trajectories are related by the physical conservation laws, since these states are experienced by particular samples of air moving through the system.

Observations of wind are rarely sufficiently abundant for it to be possible to infer with confidence the form of the trajectories from such information alone, and it is helpful to refer to some constraints upon the motion. Apart from those which are simply kinematic, the most useful are based on the assumption that the motion is adiabatic, with no loss of energy by radiation and no diffusion of properties by motion systems too small to be identified. Along the trajectories there is then conservation of potential temperature (defined by the appropriate dry or saturated adiabatic reference processes), so that the motion is in isentropic surfaces whose form can be defined by contours of pressure (or of height). Similarly, there is conservation of the mixing ratio of water vapor, most conveniently expressed by the pressure at which saturation occurs in adiabatic expansion.

The technique can be applied to illustrate the structure of large-scale motion systems, in which the principal departures from adiabatic conditions in the free atmosphere are due to slow radiative loss of energy and to small-scale motions in the cloud systems. In the large-scale systems the periods t, τ, and T are, respectively, about 1–2 days, a few days, and several days. Since the radiative loss of energy in 1–2 days reduces θ by only about 1°C, and most of the volume of an entire system is in the cloudless part of the free atmosphere, the derived trajectories are sufficiently representative to be instructive. Where they are drawn through the convective boundary layer the assumption of adiabatic motion cannot be employed; there the trajectories have to be regarded not only as approximate but also as significant only in some statistical sense. Nevertheless, it is generally obvious that along them the potential temperature and, particularly, the mixing ratio of water vapor are not conserved; however, the observed increases are a further interesting source of information and do not detract from the value of this kind of analysis. (Some examples of such analyses have been used already in Section 8.7 to illustrate the large-scale flows favorable for the development of intense cumulonimbus convection.)

The analyses can be refined by considering other constraints on the motion, exploiting principally the conservation of the potential vorticity and of energy which is characteristic of adiabatic and frictionless motion (30, 31). The potential vorticity Z (32) can be expressed to a good degree of approximation as

$$Z = -\frac{(f + \zeta_\theta)\partial\theta}{\partial p}$$

(Since under the assumed conditions a fractional change in the vertical component of the absolute vorticity is associated with a horizontal divergence, which produces a corresponding fractional change in the thickness Δp of the layer between two isentropic surfaces whose potential temperature difference $\Delta\theta$ remains constant, it follows that $DZ/Dt = 0$.) In the troposphere the magnitude of Z varies between nearly zero in very small regions on the right of the axis of jet streams and other intense flows, and values of mainly $10-20 \times 10^{-9}$ °C cm sec g^{-1} [a unit smaller by a factor of 10^5 than the unit of °C $(100 \text{ mb})^{-1}$ sec^{-1} which may be more convenient for mental assessment from observational data]. Locally it increases to between 20 and 40×10^{-9} °C cm sec g^{-1} in frontal zones and on the left flanks of jet streams. Similarly, large values are characteristic of the lowest kilometer of the stratosphere, increasing upward to 300 or more units several kilometers higher, where the radiatively dominated stratification is very stable. Within the troposphere and lowest kilometer of the stratosphere the magnitude of the potential vorticity generally does not vary spatially by a factor of more than about four. The variations in time along adiabatic trajectories of unsaturated air, inferred from the observed fields of wind and temperature, have been found occasionally to be equally large over a period of 24 hr. Such extreme variation has to be attributed to errors in the observations, the analysis, or the assumptions of adiabatic and frictionless motion; they are mostly observed in regions of large gradients of wind speed and potential temperature, where all three kinds of error may be significant. However, such occasional large errors do not spoil the general value of using the conservation of potential vorticity as a constraint in analysis (and upon study they may prove valuable indicators of nonadiabatic and turbulent processes).

It is interesting to compare analyses of the same large-scale motion system using, on the one hand, the conventional representation of the height contours of an isobaric surface (and therefore showing the geostrophic winds), and, on the other hand, the representation of the (actual) relative flow in an isentropic surface. Figure 9.15, for example, shows the two kinds of representation of the structure in the middle troposphere of the major trough which is numbered 1 in Fig. 9.2 (some aspects of which have already been discussed in Section 8.7). In the chart for the 700-mb level, Fig. 9.15b, descent in the rear of the trough and ascent in its southwestern quadrant is implied by the component of the geostrophic wind across the contours. This vertical motion is indicated more strikingly (and in the original analysis, which shows also the speed of the relative wind, *quantitatively*) by the flow across the isobars which define the configuration of the isentropic surface. This surface is low in the (cold) trough, but generally slopes upward from the lower troposphere in the south into the upper troposphere in the north. Moreover, whereas the isobaric chart gives the misleading impression that over most of its length the air flows across the axis of the trough from west to east, the isentropic chart

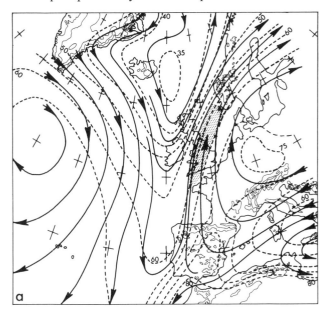

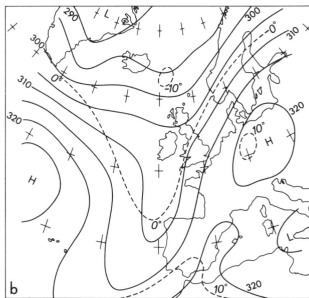

Fig. 9.15 Representations of the structure of the large-scale trough numbered 1 in Fig. 9.2, by charts of (*a*) streamlines of the relative flow in the isentropic surface $\theta = 37$°C, and (*b*) contours (in dkm) of the 700-mb isobaric surface.

shows that on the eastern flank of the trough there is a confluence of two distinct airstreams. One of these enters the southeastern quadrant of the trough and ascends as it flows northward and accelerates, and the other enters the northwestern quadrant and generally descends as it flows southward. Part of this latter stream, however, moves around the trough and flows northward, without much vertical motion, beside the other stream. The confluence line lies in the frontal zone, marked by a crowding of the isobars, which indicates a steepening of the isentropic surface. A frontal cloud system forms

9 Large-scale slope convection

over Great Britain in the airstream which has been drawn from low latitudes, in that part of the flow which has been moistened by ordinary convection over Spain. Except over Spain and the extreme north of Africa the air in this isentropic surface had not entered the layer of ordinary convection and was generally rather dry (saturation level about 500 mb or above), remaining unsaturated even in the region of ascent. In the cloud system the air rose above the isentropic surface considered, as its potential temperature increased during the condensation of water vapor. After widespread condensation has begun it is more difficult to construct trajectories, since in the early stages of the formation of a cloud system small-scale convection commonly develops; moreover, in the later stages precipitation often falls into unsaturated layers and at least partly evaporates. In these circumstances neither θ in the unsaturated air nor θ_w in the cloudy layers is conserved, and the motion can be inferred only indirectly and with less confidence.

From a number of similar analyses of the relative flow in isentropic surfaces Fig. 9.16 has been composed to represent the typical form of the large-scale motion in a section through an intense major trough-ridge system, in which the airstream producing the associated cloud belt (in the selected surface of high potential temperature) is drawn from the convective boundary layer in a low latitude, and rises into the upper troposphere in middle latitudes. (The flow of air in this manner, from latitudes as low as 20°N on the southern flank of an oceanic subtropical anticyclone, into the southern end of a band of frontal cloud, is strikingly illustrated in analyses of cloud patterns and motions in pictures from geosynchronous satellites; 33.) On isentropic surfaces of higher potential temperature, which lie at higher levels, the flow may be entirely above the convective boundary layer and remain generally unsaturated (but be penetrated from below by the cloud system). On the other hand, isentropic surfaces of successively lower potential temperature enter the convective boundary layer in successively higher latitudes where the representative values of θ_w are less, and the moist airstreams ascend to lower levels in the belt of frontal cloud. Over the ocean it might be expected that in middle latitudes this cloud would extend upward continuously from near the surface into the upper troposphere. When the trough approaches or lies overland, however, the flow which enters its southeastern quadrant before ascending into the cloud system has usually passed over diverse topography, so that in the convective boundary layer the poleward decrease of θ and θ_w is not regular. Instead, the flow may be divided into several streams in which θ, θ_w, and the depth of the layer of ordinary convection tend to acquire discrete values characteristic of the terrain over which

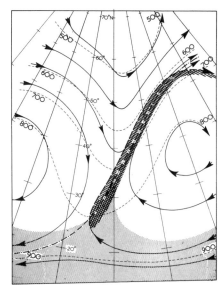

Fig. 9.16 A schematic representation of relative flow in a major trough of large-scale slope convection over an ocean, in a surface of constant potential temperature θ (about 30°C). The height of the surface is shown by dashed lines (in mb). The cold frontal zone, in which the narrowing of the space between two isobars shows a steepening of the isentropic surface, contains a (dot-dash) line of confluence between two principal airstreams. The stippled zone in the south shows where trajectories (of the *mean* flow) lie within the layer of small-scale convection, in which strictly θ increases along the flow (but where the main modification of the air is a moistening accompanied by a large increase of θ_w). The air is nearly everywhere unsaturated, but the hatched area marks a band of clouds which rises above the isentropic surface. The clouds form in low latitudes, where they are liable to develop into more or less deep clouds of ordinary convection (in the region shown by cross-hatching), but appear subsequently as middle-level layer clouds, and eventually as ice clouds in the high troposphere of middle latitudes (where they lie near and to the right of the axis of the jet stream). These clouds evaporate after the flow at their level has turned to become northwesterly, and the airstream begins to subside. The flow east of the line of confluence, projected upon a zonal vertical plane, composes the circulation illustrated in Fig. 9.11.

the air has been modified. On the occasion of the trough shown in Fig. 9.15, for example, it was possible to distinguish three layers of cloud over Great Britain, which formed in air that had been modified over the Sahara, Spain, and France, respectively (Section 8.7). In these circumstances a *multilayer* cloud system is observed, which may appear to become continuous where the intervening spaces become filled with precipitation in the form of snow.

In Fig. 9.16 it is assumed that the major trough has only one associated cyclone, centered in a high latitude, and that the structure of the trough and the shape of its cloud system have not been distorted by the formation of later members of a cyclone family in the frontal zone. Even when this occurs, it is evident that on the large scale the frontal zone extends from low into high latitudes and is essentially a cold front moving slowly eastward. The associated bands of streaky cloud mark the positions of the several long waves on mosaics of satellite pictures covering a whole hemisphere. During the early stages of cyclone formation in frontal zones the warm front can be regarded as only part of a minor wavelike distortion in a major cold front zone, but generally as a cyclone matures a warm front with the characteristic shape illustrated in the Norwegian cyclone models becomes a prominent feature of the cloud pattern for a while, as seen in the cloud system approaching the west coast of the United States in Pl. 9.1 and in the systems over the Pacific in Pl. 9.3.

Ascent to the upper troposphere in frontal zones

Figure 9.17 shows the typical distribution of temperature in a meridional vertical section through a well-defined frontal zone, and the corresponding distribution of zonal geostrophic wind.

The region occupied by layer clouds is displaced distinctly away from the frontal zone, within the warmer air. The frontal zone often contains air of remarkably low relative humidity, corresponding to a depression of the dew point below the air temperature of as much as 30°C. The correspondingly low values of the mixing ratio of water vapor which are observed, even at low levels, indicate that the air has recently descended from the high troposphere, generally in the northwesterly flow in the rear of the major trough. The zone of transition between saturated air in the cloudy region and the air of very low relative humidity frequently is very narrow.

The routine observations of relative humidity in the high troposphere are not reliable, and its distribution has not been included in Fig. 9.17. However, it is evident that when high cloud is present near the axis of the jet stream over the polar front the associated value there of θ_s (about 16°C) is consistent, as implied in Fig. 9.16, with (nearly adiabatic) ascent into that region of air from the convective boundary layer in low latitudes (near 20°N, where at this season in the Caribbean Sea θ_w reaches about 20°C near the surface. Over the ocean the comparative sparsity of reliable sounding data, particularly of wind and relative humidity in the upper troposphere, makes it difficult to demonstrate convincingly that air does follow paths of this kind. It is interesting, however, to consider the horizontal acceleration which may accompany such ascent.

It is supposed, first, that the large-scale motion system is stationary, so that the relative speed v_r of the ascending air is the observed speed v relative to the earth. Assuming that the motion system is in a steady state (so that local derivatives, in particular $\partial p/\partial t$, are zero), and that the motion of the rising air is adiabatic and frictionless, the conservation of energy in a unit mass of air traveling along a trajectory is expressed (31) by the equation

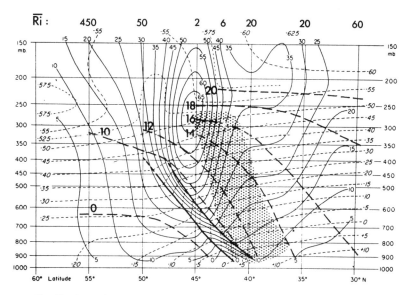

Fig. 9.17 The typical mean distribution of temperature (thin dashed lines), of potential temperature θ_s (heavy dashed lines), all labeled in °C, and of zonal geostrophic (westerly) wind speed (continuous lines, in m sec^{-1}), in a meridional section through a well-defined frontal zone (boundaries marked by bold lines, dashed near the 400-mb level where they become indistinct). (Constructed from data near 80°W, and drawn using distance from the front as horizontal coordinate at each pressure level up to 400 mb, and from the front at the 400-mb level for higher regions; 34). The stippled zone shows the region found from other studies to be occupied by layer clouds. Some values of the mean Richardson number $\overline{Ri}$ for a column between the 900- and 300-mb levels are entered along the top of the diagram.

9 Large-scale slope convection

$$v^2/2 + gz + c_v T + p/\rho = \text{const} + L(r_0 - r_s) \quad (9.16)$$

where z = (geopotential) height above some reference level
c_v = specific heat of air at constant volume
p = air pressure
ρ = air density
L = latent heat of condensation
r_0 = initial mixing ratio of water vapor
r_s = mixing ratio of saturated water vapor at temperature T

The individual terms can be identified with the kinetic energy, the potential energy, the internal energy, the "elastic energy," and, finally, the energy available from the latent heat of condensation of water vapor, respectively. (It is assumed that during condensation the water vapor is just saturated at the air temperature T.) This equation can be simplified to

$$v^2/2 + gz + c_p T + Lr_s = \text{const} \quad (9.17)$$

where c_p is the specific heat of air at constant pressure.

The identical changes of state (defined by values of T, p, and r_s) occur during upward adiabatic displacement of air in a horizontally uniform hydrostatic atmosphere with the same vertical distributions of pressure and temperature, corresponding to the chosen adiabatic (or pseudo-adiabatic) reference process. There the variables of state are by definition associated not with the height z but with a height z' above the same reference level, according to an equation

$$gz' + c_p T + Lr_s = \text{const} \quad (9.18)$$

which can be regarded as an integrated form of Eq. 3.25.

Thus along a trajectory, by subtraction of Eq. 9.18 from Eq. 9.17,

$$v^2/2 + g(z - z') = \text{const}$$

and in particular

$$v_1^2 - v_0^2 = 2g(\Delta z' - \Delta z) \quad (9.19)$$

where v_0 is an initial speed, v_1 a speed attained after an ascent Δz in the slope convection, and $\Delta z'$ the change of height which produces the same change of state in an atmosphere with the lapse rate of the reference process. This result has been called the *extended parcel theory*, since it represents an extension of the simple parcel theory (Section 7.2). Incidentally, it avoids the dubious concept of an undisturbed "environment": in the practically hydrostatic pressure distribution of the large-scale motion system the only relevant property of the atmosphere is its state in the column below the parcel, wherever the parcel may be.

In saturated air which ascends from the 900- to the 300-mb level with $\theta_s = 16°C$, $\Delta z'$ is about 8400 m. In latitude 55°N over the North Atlantic the normal height of the 300-mb surface is about 8800 m, and the normal thickness of the layer between the 900- and 300-mb levels is about 7900 m. Thus according to Eq. 9.19 air from the boundary layer in latitude 20° which reached the 300-mb level in latitude 55°, by frictionless adiabatic ascent in a stationary motion system, should arrive with a speed corresponding to a value of $(\Delta z' - \Delta z)$ of about 500 m, that is, with a speed approaching 100 m sec^{-1}. This is considerably greater than the maximum wind speeds usually observed in the polar front jet streams of middle latitudes.

However, the calculated speed has to be reduced if the motion system is moving steadily eastward and the trajectory reaches a latitude higher than that in which the ascent began (35). The simplest and perhaps the most reasonable circumstance to consider is that in which the motion system moves zonally eastward with a steady angular velocity ω_r relative to the earth. Then it can be shown that Eq. 9.17 becomes

$$v_r^2/2 + g(z - h_r) + c_p T + Lr_s = \text{const} \quad (9.20)$$

where h_r, a function of latitude ϕ only, is the contour height of an isobaric surface with a slope required to provide everywhere a geostrophic wind with the eastward speed $c(\phi) = \omega_r R \cos \phi$ of the motion system, where R is the radius of the earth. [If h_r is subtracted from z_p, the height of an isobaric surface, then the resultant distribution of $(z_p - h_r)$ can be regarded as defining the geostrophic wind in the isobaric surface *relative to the motion system*.]

Given the value of ω_r, and therefore of $c(\phi)$, then h_r can be obtained by the integration of Eq. 9.5:

$$gh_r = R^2 \Omega \omega_r \cos^2 \phi + \text{const} \quad (9.21)$$

If the eastward speed of the motion system is expressed as L degrees of longitude per day—about the speed in meters per second in latitude 40°—and if h_r is in meters, then

$$h_r = 61.2L \cos^2 \phi + \text{const} \quad (9.22)$$

In the moving system, therefore, Eq. 9.19 becomes modified into

$$\Delta(v_r^2) = 2g(\Delta z' - \Delta z + \Delta h_r) \quad (9.23)$$

If the motion, as usual, is eastward, then h_r decreases with latitude, and if the trajectory terminates in a higher latitude than it began, the term Δh_r is negative, and the acceleration along the trajectory is accordingly reduced. The value of L typically varies between about 2 for long waves and an extreme of about 10 for intense major trough-ridge systems of comparatively short wavelength.

Considering again the former presumed saturated ascent from the 900-mb level in latitude 20° to the 300-mb

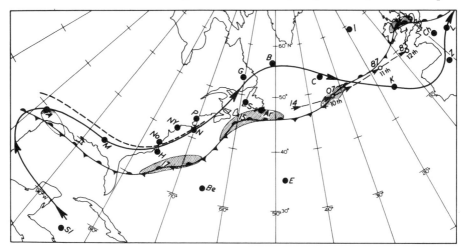

Fig. 9.18 A representative inferred trajectory of air from the convective boundary layer in trade winds entering the Gulf of Mexico, which ascended in a frontal zone to produce middle and high clouds over the Gulf states before entering a jet stream, traversing the Atlantic, and reaching western Europe (9–12 July 1961). The position of the edge of the high cloud system over the United States at 1200 GMT, 10 July, is shown by the dashed line, and the surface position of the front at that time is shown in the conventional manner. Also marked are areas of rain (stippled), and the last two figures of the central sea-level pressure associated with several shallow wave cyclones. The path of one cyclone which intensified in mid-Atlantic (during the establishment there of the northwesterly jet stream) is shown by arrows through open circles marking the position of the center at sea level at 1200 GMT on 10, 11, and 12 July. The solid dots with identifying letters mark the positions of sounding stations referred to in the text and other figures.

level in latitude 55°, the numerical values of Δh_r corresponding to $L = 2$ and $L = 10$ deg day^{-1} are, respectively, about 65 and 320 m. Thus the former requires only a small reduction in the value of 500 m obtained for $\Delta(v_r^2)$ from Eq. 9.19, leading to a final speed of about 90 m sec^{-1}, while the latter requires a substantial reduction, leading to a final speed of a little less than 60 m sec^{-1}. Thus the range of the observed speeds of polar front jet streams is consistent with the presumed ascent and the range of eastward velocities of the major trough-ridge systems. More definite evidence that this kind of ascent and the corresponding accelerations actually occur can be found from the study of individual systems. Illustrations will be given from two analyses of flow over the North Atlantic, the first in a midsummer situation and the second on a winter occasion when the jet stream attained unusually great speeds near the British Isles.

First, during 8–11 July 1961 the axis of a pronounced trough in the middle troposphere lay approximately along the eastern seaboard of the United States, moving eastward at a speed of about 2° of longitude per day. On this and the previous two days an almost stationary front with shallow wave cyclones extended across the Atlantic from the British Isles into the eastern flank of this trough, and moved slowly south over the coasts of the Gulf states. The airstream of the trade winds flowed into the Gulf of Mexico and toward the front, producing extensive cloud over the Gulf states, with a sharply defined northern edge to a system of middle and high clouds (Fig. 9.18). Over the Gulf the wet-bulb potential temperature in the convective boundary layer was 24°C according to ship observations and about 20°C at the 850-mb level (Fig. 9.19); over southern Texas soundings show that the moist air extended up to about the 800-mb level with θ_w between 21 and 23°C. However, as this air moved northward and into the region of large-scale ascent near the front the stratification in the troposphere became favorable for deep

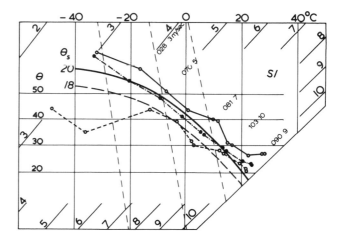

Fig. 9.19 Sounding at Swan Island in the Caribbean (*SI* in Fig. 9.18), 0000 GMT, 9 July 1961, showing the character of the trade wind airstream. The dot-dash line joining the solid circles shows the distribution of θ_w; it decreases from 23°C near the surface to 20°C at the top of the moist layer, near the 850-mb level.

9 Large-scale slope convection

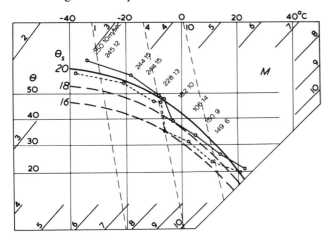

Fig. 9.20 Sounding at Montgomery (*M* on Fig. 9.18), 1200 GMT, 11 July 1961. Intermittent slight rain was reported at the ground.

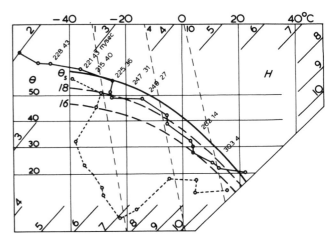

Fig. 9.21 Sounding at Cape Hatteras (*H* on Fig. 9.18), 1200 GMT, 10 July 1961. There were no low clouds beneath an overcast of thick cloud in the upper troposphere.

ordinary convection, and the clouds produced over the Gulf states contained large cumulus and cumulonimbus, with some thunderstorms as well as slight rains from layer clouds. Already over these states the potentially warm and moist air appeared on most of the soundings to be above the 500- or 400-mb level, with θ_w between 20 and 21°C (Fig. 9.20). Where this air flowed into the Atlantic over the east coast of the United States it was clearly present above the 400-mb level with θ_w about 20°C (Fig. 9.21). It was therefore assumed that air moved along the trajectory drawn in Fig. 9.18, passing over the Gulf states in about latitude 32°N, where it was saturated at the 500-mb level (about 5830 m) with $\theta_s = 20°C$ and had a speed of about 10 m sec^{-1}. (Because the stratification above the 500-mb level was nearly the saturated adiabatic and the wind speed varied little with height, the precise choice of a level of origin over the Gulf states does not materially affect the subsequent calculations.)

Late on 9 July the trough off the east coast and the associated flow of warm air toward the northeast began to intensify; late on the following day strong high-level northwesterlies and a wave cyclone developed rapidly in mid-Atlantic. The generally westerly winds aloft were replaced by a strong northwesterly jet stream, in which the maximum speed was over 70 m sec^{-1}, and the cyclone deepened and moved to the mouth of the English Channel. By 12 July a new pronounced trough and cyclone had thus become established near longitude 10°W. During the progress eastward of this trough on 12 July the wind at the 300-mb level over central France backed to southwest and increased from about 30 to 60 m sec^{-1}. Near the axis of the jet stream which thus extended across the Atlantic into Europe θ_s was consistently observed to be about 20°C, and the trajectory drawn in Fig. 9.18 is based on soundings which showed this air to be distinctly warmer than that occupying most of the troposphere beneath, as shown, for example, in Fig. 9.22.

At a number of soundings made near the trajectory the wind speed v observed near the 300- or 250-mb level, wherever θ_s was 20°C, was compared with the speed v_c calculated using Eq. 9.23. [Since in an atmosphere with $\theta_s = 20°C$ the thicknesses of the layers from 500 to 300 and from 500 to 250 mb are, respectively, 3750 and 4980 m, the term $(\Delta z' - \Delta z)$ was found simply by subtracting the reported 300-mb height from $(5830 + 3750)$ m, or the reported 250-mb height from $(5830 + 4980)$ m.] In such a comparison two particular kinds of error have to be considered, those due to incorrect reports of the height of a pressure level and those arising from incorrect observation (or assessment, by interpolation between reporting levels) of wind veloc-

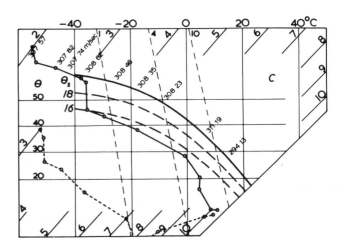

Fig. 9.22 Sounding at weather ship *C* (*C* on Fig. 9.18), 1200 GMT, 11 July 1961.

The large-scale cloud and precipitation systems of slope convection 9.5

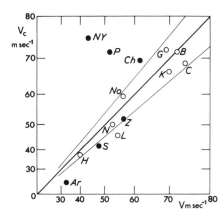

Fig. 9.23 A comparison of the wind speeds observed v and calculated from the extended parcel theory v_c, in the jet stream of 10–12 July 1961. The letters identify the stations (marked on Fig. 9.18) from which soundings were used in the calculations. The comparison is made at the level where θ_s was observed to be 20°C, except at Argentia (Ar), Stephenville (S), and Zaragoza (Z), all on the right flank of the jet stream, where θ_s was between 20.7 and 21°C. The soundings at Norfolk (No), Portland (P), and New York (NY) were judged to have been made, respectively, about 50, 100, and 150 km northwest of the edge of the high cloud which probably marked the axis of the jet stream (Fig. 9.18), and the calculated speed is increasingly an overestimate. Soundings known to have been made away from the axis of the jet stream are marked by solid circles. The separation of the thin diagonals corresponds to the effect of an error, probably extreme, in the observations of geopotential height and wind speed (reaching ±50 m at high speeds and in the high troposphere).

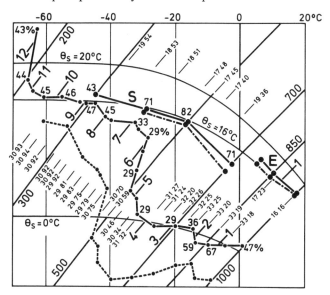

Fig. 9.24 A sounding from weather ship I, 12 February 1962, showing an unusually strong jet stream in the high troposphere. In the air of the jet stream, whose maximum speed is over 90 m sec^{-1}, θ_s is about 16°C. Air with θ_w about 16°C was present in the upper troposphere in a sounding from Stephenville (S on Fig. 9.18), and later near the surface in a sounding from weather ship E at 35°N 48°W (Fig. 9.18). On the included parts of the latter two soundings distributions of θ_w are shown by dot-dash lines. Against the soundings are entered winds (in tens of degrees and meters per second), heights (in km), and relative humidities with respect to liquid water (in %, not corrected for lag).

ity. From general experience the errors of the former kind are thought mostly to lie between ±30 m in the high troposphere, and those of the latter kind to be nearly as much at the higher wind speeds (when expressed in the same units of potential energy). Thus the combined errors probably lie within ±50 m, a range indicated in Fig. 9.23, which is likely to be less than that due to the uncertain validity of the assumptions made in obtaining a representative trajectory and in deriving Eq. 9.23. In spite of these uncertainties, this diagram shows that there is a close correspondence between calculated and observed wind speeds at positions close to the trajectory, whereas the calculated speed is a large overestimate at positions a little farther poleward.

The second illustration of the application of the extended parcel theory, 12–17 February 1962, is from a time when unusually intense northwesterly jet streams were observed over and near the British Isles. Throughout this period the geostrophic wind at the 300-mb level, averaged over a distance of a few hundred kilometers across the direction of the strongest flow, was persistently greater than 75 m sec^{-1}, and at individual sounding stations the maximum wind speed observed intermittently reached values between 90 and 100 m sec^{-1}.

Figure 9.24 is an example of a sounding made at the beginning of the period, which shows extreme speed.

Between 10 and 15 February the axis of a major intense trough moved slowly away from the eastern seaboard of the United States, at an average speed of about 3° of longitude per day. An associated frontal zone lay from northwest Europe across the Atlantic into the Caribbean Sea, and wave cyclones which formed off the coast of the United States moved to near Iceland before becoming very intense while traveling into northern Europe; a second pronounced trough was maintained lying across Europe and into the Mediterranean. A typical trajectory, corresponding to that drawn in Fig. 9.18, commenced in the convective boundary layer well east of Florida, turned northward and rose in the frontal zone near Bermuda (Be in Fig. 9.18), and then in the high troposphere passed over the east coast of Newfoundland and near the south Cape of Greenland before entering western Europe from the northwest. At the beginning of the trajectory near the 20th parallel the sea temperature was a little above 20°C and θ_w at ships' screen level was 1 or 2° below 20°C, but a representative value of θ_w in the lowest kilometer farther north appears to have been between 16 and 18°C, as shown by the sounding included in Fig. 9.24, made at weather ship E in 35°N during the close approach of the (cold) frontal zone. On numerous soundings the potentially warm and moist air could be found in the upper troposphere over Newfoundland and in the high troposphere over weather ship B and near the British Isles, nearly saturated with θ_s between about 15 and 17°C (Fig. 9.24; on the sounding shown from ship I the reported relative humidity increases rapidly above the 350-mb level and reaches a maximum of 47%; corrected for lag of the sensor, the values near the 300-mb level are at least several percent greater, and therefore represent approximate saturation with respect to ice).

If the ascent of the moist air is considered to be adiabatic (with $\theta_s = 16°C$) from the 900-mb level (height 1 km; relative air speed 20 m sec^{-1}) in latitude 35°N, to the 300-mb level in latitude 59°N, then $\Delta z' = 8340$ m, whereas $\Delta z = 7860$ m. Also, with $L = 3°$ of longitude per day, $\Delta h_r = -75$ m. Accordingly $(\Delta z - \Delta z' + \Delta h_r) = 405$ m, and from Eq. 9.23 the calculated wind speed at the 300-mb level at weather ship I is 91 m sec^{-1}, almost exactly equal to the reported value of 92 m sec^{-1}. Such close agreement must be regarded as partly fortuitous, but is not uncommonly found, especially when the observed wind is very strong and the particular choice of values of L and of the initial relative speed are not particularly important.

Deep ordinary convection in frontal zones

In Fig 9.17, representing the structure of the polar front, the mean stratification for some distance south of the front is unfavorable for deep ordinary convection: θ_s increases upward from the surface. However, this stable structure is related to the season and meridian for which the cross section was constructed, and perhaps also to the abnormally high latitude in which the front is shown. Stratifications favorable for deep ordinary convection occur in summer over land ahead of cold fronts (and near warm fronts) and throughout the year ahead of cold fronts over and near the open ocean, especially in the lower latitudes. Thus in the winter period discussed in the last paragraphs, for example, thunderstorms occurred during the passage of the cold front over Florida, in latitudes near 25°N. There θ_w was about 19°C in the lowest kilometer of the potentially warm air entering the frontal zone, where θ_s at higher levels was less than 19°C up to the 300-mb level (with a minimum value of about 17°C at about the 500-mb level). In latitudes 32 and 35°N, however, the stratification near the frontal zone, as shown by later soundings from Bermuda and the weather ship E, was more nearly neutral, θ_s throughout the middle troposphere remaining about 17°C but θ_w in the lowest kilometer decreasing northward to nearly the same value. Climato-

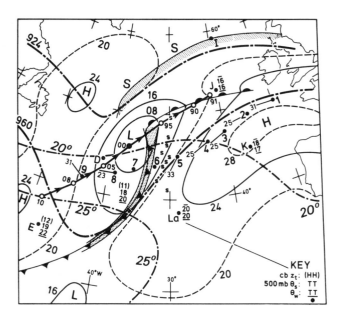

Fig. 9.25 The synoptic situation at 1200 GMT, 5 September 1970, on the occasion of the clouds in Plate 9.2, associated with a cyclone and its cold front. The path of the surface center of the cyclone is shown by arrowed lines between circles marking positions at intervals of 12 hr, labeled with the last two figures of the central pressure (in mb); the position at the map time is shown by a solid circle marked L. Sea-level isobars, similarly labeled, are drawn as solid thin lines for intervals of 8 mb, with two dashed lines for an intermediate pressure (1020 mb). Fronts at the surface are entered in the conventional manner. Two contours of the 300-mb surface are shown by thick dot-dash lines, labeled 960 and 924 dkm. Two isotherms of sea-surface temperature are drawn as dot-dash lines, labeled 20 and 25°C. Sources of sferics found by the British location system are marked by the letters s (at this extreme distance the ranges are clearly in error, but their association with the band of cumulonimbus at the cold front is evident).

Successive positions of an aircraft crossing the area in the high troposphere are shown, relative to the cold fronts, by small solid circles numbered from 1 to 10, with arrows and numbers giving the direction, and speed in m sec^{-1}, of the observed wind. Plates 9.2a and b were taken from position 6.

Stippled zones mark the positions of two prominent cloud features, the principal cumulonimbus band of the cold front and the edge SS of the system of high cloud north of the warm front, which can be seen in the satellite picture of Pl. 9.2d.

Finally, data are given from analyses of soundings made at some marked stations [the weather ships J, K, D, E, and Lagens (La)] on the close approach of the cold front, according to the key at the bottom of the diagram [<u>TT</u> = the mean value of θ_w (°C) in the lowest km; TT = the value of θ_s (°C) at 500 mb; (HH) = the height (km of the top of the corresponding positive area, likely to be the height z_t of the tops of cumulonimbus anvils associated with the cold front—the dash entered at J, K, and La indicates that there was no positive area and deep cumulonimbus convection was unlikely)]. The position of the weather ship I in latitude 59°N is also marked.

logical data for Bermuda show that in the winter season the most intense rainfalls occur during thunderstorms and showery rains associated with frontal zones; thunderstorms are experienced on one or two days each month, during about one third of all cold front passages. Such storms are said to be prevalent near the fronts of developing wave cyclones.

Generally the stratification favorable for deep ordinary convection arises in frontal zones because during large-scale ascent there the middle troposphere, where the air is comparatively dry with a lower value of θ_w, cools more than the air in the convective boundary layer. Examples of this destabilization were considered in the discussion of cumulonimbus formation over land (in Chapter 8); the process is more frequent over the oceans, and especially over their warmer western waters, in the vicinity of cold fronts during the first stages of cyclogenesis, when the large-scale ascent and its concentration into the frontal zones is most intense. (When the cyclones mature and the fronts near their centers move into the cooler surface waters of the eastern parts of the oceans, deep ordinary convection disappears from the frontal zones.)

From aircraft flying at high levels over the oceans the cold fronts are often seen as long chains of cumulonimbus (Pl. 9.2), especially in the western parts of the ocean and in latitudes lower than about the 50th parallel, where the height of the anvil tops may exceed 10 km. Farther east and poleward cumulonimbus are less frequently associated with cold fronts, and when they occur their tops are lower and their intensity less. Nevertheless, the fronts are notorious for the suddenness of onset and strength of the squalls which accompany their passage at the surface, often with a short period of heavy rain and perhaps hail, characteristics which all suggest the presence of cumulonimbus.

Consistently, the observations of sferics locating systems (Appendix 9.4) confirm the frequent occurrence of thunderstorms over and near the oceans in frontal zones and developing wave cyclones. For example, observations of sferic sources over the Northeast Atlantic during a single winter soon after the establishment of the British system were sufficient to show that in wave cyclones the maximum frequency of sferics is near the apex of the wave on the surface position of the front, and that more sferics occur near the warm front than the cold front (Fig. 9.25). On the other hand, in the more mature cyclone they were found to occur most frequently near the point of occlusion and along the cold front (Fig. 9.26). Figure 9.27 suggests that in this part of the Atlantic the frequency of sferics associated with cold fronts is greatest near latitude 45°N and decreases rapidly toward both higher and lower latitudes.

Generally similar results have since been obtained

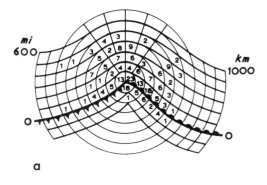

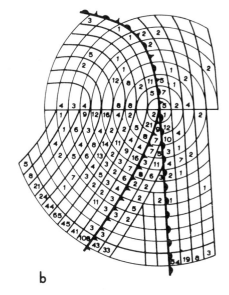

Fig. 9.26 The distribution of sferic sources in (*a*) wave cyclones and (*b*) partly occluded cyclones over the Northeast Atlantic during a winter season (37). The numbers represent the number of sources located during 107 hr of observation, in a coordinate system using distance from the fronts and from the center of the cyclone or the apex of the warm sector (at sea level). In (*b*) the numbers in the bottom row show the number of sources within each zone parallel to the fronts. In each diagram the zones are 100 mi wide.

by sferics location systems covering areas including the southeastern United States, the eastern Caribbean Sea and Bermuda (38), Australia and neighboring seas (39), and eastern Asia and the neighboring Pacific (40). In studies with all these systems it was found that sferics over the sea in winter are associated especially with regions of cyclogenesis and frontogenesis. Over the western Pacific the areas of most frequent sferic location are just south and southwest of Japan, in latitudes between about 27 and 33°N, coincident with the regions of most frequent cyclogenesis shown in Fig. 9.3.

The prevalence of deep ordinary convection in the cold front zones of developing cyclones, especially in the low latitudes whence the supply of the potentially warmest air in slope convection is drawn, together with

9 Large-scale slope convection

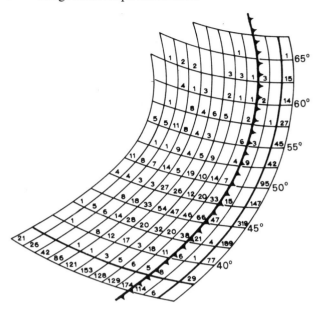

Fig. 9.27 The distribution of sferics sources observed near major cold fronts over the Northeast Atlantic during a winter season (37). The numbers represent $10P$, where P is the percentage of all sources, in coordinates of degrees of latitude and distance (drawn at intervals of 100 mi) from the front. The sums of the numbers within the columns and rows are entered in the bottom row and the column on the right, respectively.

the difficulty of locating this air in the middle rather than the upper troposphere of higher latitudes (only partly attributable to the sparsity of soundings over the oceans), suggests that much of the air which rises from the low to the high troposphere (and into its jet stream) ascends in cumulonimbus before reaching middle latitudes, and does not follow a path of the small inclination typical of most trajectories in slope convection. Providing that the ascent in the cumulonimbus follows the chosen moist reference process, this does not affect the earlier discussion of the energetics of the air motion. It has already been remarked in Chapter 8 that even in ordinary isolated cumulonimbus there is an approximate equality of θ_w in the air entering the cloud base and in the air leaving it, in the outflow marked by the upper parts of the anvil cloud. In the cumulonimbus chain of the frontal zone this approximate equality is even more probable, because within a broad cloud belt some of the ascent is likely to be through surroundings saturated by previous convection. Further, although the ascent may not be strictly adiabatic, it may happen, as discussed in Chapter 8, that because of neglect of the release of the latent heat of fusion and of precipitation, the changes of state may still be well represented by the chosen pseudo-adiabatic reference process. Now in the derivation of Eq. 9.23 there is no reference to the form of the path taken by the air considered, and it may be anticipated still to be valid. Evidently, for example, if the air rises vertically in cumulonimbus through a neutral stratification no work need be done during the ascent, and a subsequent acceleration is determined by the change of geopotential energy and latitude along the path taken in the high troposphere.

It has been noted that an intensification of the jet stream downwind of an outbreak of severe storms is occasionally observed over the United States, and a similar intensification may occur over the ocean during cyclogenesis accompanied by thunderstorms, as on the occasion illustrated in Fig. 9.25, but such developments have not been analyzed in relation to the cumulonimbus convection. The local transfer to the high troposphere of some potentially warm air in cumulonimbus implies a decrease in the potential energy available for the slope convection, but one which is probably too small to have a considerable effect upon its intensity. In both middle and low latitudes paths of the same kind, through cumulonimbus, are implied for the travel of moist air from near the surface into the middle of the jet streams of the high troposphere. In middle latitudes, however, the journey occupies only a day or even less, whereas in low latitudes at least several days are required for air from the cumulonimbus of the intertropical convergence zone to reach the subtropical jet stream.

Cloud and precipitation systems in cyclones

Layer cloud systems

Satellite pictures, such as Pl. 9.1–9.3, confirm that the general horizontal distribution of clouds in cyclones is well represented by the Norwegian models (summarized, e.g., in Fig. 9.10), which were developed from the analysis of surface observations. The patterns and extent of the cloud systems apparent in the satellite pictures are related to the stage of evolution of the cyclone, and graphical models have been constructed to show their mutual development (42). Most cyclogeneses occur in a baroclinic region which contains an extensive band of cloud associated with the trailing cold front of a preceding and now mature cyclone. Otherwise, however, it seems that when organized clouds first appear during cyclogenesis they are often cirrus in the high troposphere (43). Consistently, during the genesis of a wave cyclone the first visible sign of the development of the disturbance, before any widespread precipitation is observed at the surface, is a characteristic broadening of the band of frontal cloud. This broadening is the result of the formation, or extension and thickening, of clouds in the upper troposphere, in a system which bulges poleward, often with a well-defined convex edge marking a ridge in the flow aloft (Pl. 9.3, 9.4). Corresponding decreases of the effective radiative emission temperature in the atmospheric "window" have been observed from satellites, over areas centered several hundred kilometers to the northeast of the positions at the surface where cyclogenesis first became apparent from conventional observations only about a day later (24). Such evidence supports the view expressed in Section 9.4 that the intensity of young wave cyclones is greatest in the high troposphere.

It is more difficult to synthesize into a satisfactory model the numerous observations, made mostly with aircraft and with radar, of the *vertical* distribution of cloud and precipitation in cyclones. The early Norwegian models have long been recognized as deficient in several respects. In particular, they represent the layer clouds as extending upward continuously from a frontal *surface* into the high troposphere, whereas it is more usual to find a few separate layers of cloud in a region which in the middle and upper troposphere is distinctly above a frontal *zone*, and generally on the lower-latitude side of the axis of a jet stream (45), as indicated in Fig. 9.17. The air in the frontal zone often has a remarkably low relative humidity, especially in the middle troposphere, but in regions of widespread precipitation there may be thick layer clouds in and below frontal zones in the lower troposphere (46).

Sharply defined and gently curved poleward edges to extensive layers of high cloud have been found to lie along the axes of middle-latitude jet streams (47), supporting the form of the trajectories previously inferred for cloudy air ascending from low latitudes, which have been prolonged into these jet streams. When the solar elevation is not too high, and when the angle of view is also favorable, these cloud edges may be strikingly apparent in satellite pictures, the clouds showing bright against a shadow which they throw upon a lower overcast. On other occasions an edge is detectable only by a change in the brightness and texture of the clouds. Thus an edge is barely discernible in the original pictures of Pl. 9.3 and is obvious in Pl. 9.4. The presence of a broad shadow thrown by an edge implies that there is a distinct and deep cloudless layer between the upper and lower cloud systems, probably more marked in the young than in the mature cyclone.

Nevertheless, considering also that the clouds which are most readily formed, and which are likely to be the most abundant sources of condensed and eventually precipitated water, are those produced in airstreams ascending from the convective boundary layer rather than from the comparatively dry free atmosphere, it seems reasonable to regard the cyclone tentatively as containing two principal systems of layer clouds. Their form, and some of the most significant of the paths of air relative to the cyclone, are illustrated schematically in Fig. 9.28, which represents an intensifying warm-sector cyclone.

One of the principal cloud systems is associated with air drawn from the warm sector, which ascends near the cold front: much of this airstream reaches into the upper troposphere and produces a system of high clouds, which advances forward of the cyclone center and evaporates while subsiding in the northwesterly winds on the east flank of the ridge aloft. Most of the precipitation from this cloud system reaches the surface in a narrow belt near the surface position of the cold front. The other principal cloud system is produced by ascent near the warm front, mainly on its poleward side, of an airstream which enters the cyclone from the east, rising into the middle troposphere near the apex of the warm sector and then subsiding as it flows around and through the cyclone center, and away from its rear at low levels. Both of the airstreams which produce these principal cloud systems consist of originally potentially cold air which has recently descended from the free atmosphere and has been modified by convection from the surface. The first, which ascends mainly near the cold front, has usually spent several days in a low latitude, probably passing around the south side of an anticyclone before returning poleward, and in the terminology of the Norwegian classification it would be characterized as "tropical" air. The second, however, has only recently penetrated into a middle latitude, and by comparison has been only partially modified; it would be described as "polar" or "returning polar" air.

Away from the center of the cyclone the moist layers in each airstream are typically 1–2 km thick. The stratification within the layers tends to have two levels of maximum relative humidity or minimum dew point depression, and therefore two levels at which the upward displacement of air required to produce saturation is a minimum: one near the base and one near the top of the layer previously occupied by cumulus clouds (Chapter 7). Consequently, during the large-scale ascent, on approach to the cyclone center and its frontal zones, clouds first form as thin layers in the low and middle troposphere, and thicken progressively. Generally the horizontal distribution of the clouds is irregular; various layers of cloud, separated vertically and horizontally, appear when the airstreams have been modified by convection over land areas with variable surface characteristics. Over and near land the top of the cumulus layer in the warmer months is often as high as 3–4 km, and the first clouds to form in the system *B* (in Fig. 9.28) of the comparatively cool air then often appear in the middle troposphere (as shallow droplet clouds), some time after the arrival and thickening of the ice clouds, of the system *A*, in the upper troposphere (see Pl. 6.2). These droplet clouds, in contrast to the ice clouds, represent systems in the process of formation, and not of decay. However, they appear in patches or bands where topographical disturbances imposed on the large-scale ascent locally produce saturation, and individually

9 Large-scale slope convection

may at first occupy only small areas and be evanescent, especially when subject to small-scale convection (because the accompanying stratification was originally favorable, or, when the clouds above are high or tenuous, is soon made so by radiative energy transfers). Even in airstreams which have been modified over the sea, with only large-scale variation of surface temperature, the first clouds to form usually appear in narrow bands or patches, implying the existence or recent presence of other kinds of intermediate-scale circulations. However, on closer approach to the frontal zones the moist airstreams deepen, and as the large-scale ascent continues the clouds become widespread and continuous in layers a few kilometers thick, and precipitation develops. The apparent distribution of cloud, already complicated by the development of ordinary convection in the cloud layers and its extension upward into previously clear air, becomes further confused by condensation in air earlier moistened by the evaporation of precipitation, and is obscured where there is precipitation in the form of snow. Over the regions of widespread and most intense precipitation near the cyclone center almost the whole depth of the troposphere may appear to become occupied by cloud.

Cumulus systems; the dry tongue

In the systems of cumuliform clouds which cover most of both the cool and warm airstreams in a hemisphere the thickness of the clouds away from the frontal zones varies as represented in the Norwegian models. It is least in high latitudes and in the anticyclones on the tropical sides of the warm sectors, and greatest in the cool airstreams which have moved into lower latitudes behind the centers of intense cyclones, usually reaching a maximum where these streams begin to return poleward. The clouds show the clustering or arrangement into narrow bands discussed in Chapter 7, but a more prominent and larger-scale feature of their distribution, not previously mentioned, is their absence or stunted development in a long zone, usually 200–300 km broad, which appears behind the cloud band of a major cold front. This comparatively clear zone extends poleward east of the associated cyclone center, tending to become narrower and to spiral inward as it approaches the center. Examples of these zones can be seen in Pl. 9.1, 9.2, and 9.3. They are associated with tongues of air which in the low troposphere is abnormally dry, having descended rapidly from the high troposphere or the low stratosphere through the frontal zone on the western and southern flanks of the major trough. The abnormal dryness has the effect explained in Chapter 7 of raising the level of the cumulus bases in the layer of ordinary convection and of lowering the level of their tops, perhaps thereby leading eventually to their complete

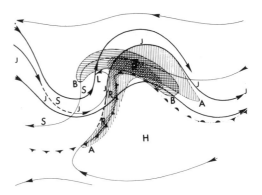

Fig. 9.28 A model of the principal large-scale cloud system and some significant paths of air in a maturing wave cyclone. The arrowed lines are trajectories relative to the cyclone center, whose position at the surface is marked by the letter L. Where the relative flow is in the low troposphere, mainly from east to west, the lines are dashed; where it is in the high troposphere, and generally from west to east, the lines are heavy. The position of the axis of the jet stream is marked by the letters J. The regions of most rapid ascent and descent are marked by the letters R and S, respectively. Two principal cloud systems are indicated: that formed by the ascent of the potentially warmest air is marked by the letters A and vertical hatching, and that formed by the ascent of comparatively cool air is marked by the letters B and horizontal hatching. The region of continuous precipitation at the surface is shown by diagonal hatching. The positions of the fronts at the surface are marked in the conventional manner.

disappearance. A relative trajectory in this narrow and shallow zone of rapid and prolonged descent, the *dry tongue*, is entered in the model of Fig. 9.28; it commences at a high level on the poleward side of the jet stream in the northwest and reaches the low troposphere south of the center of the cyclone before turning poleward on its eastern side. This remarkable flow appears as the counterpart of the similarly restricted flow through a cloud system from the low troposphere to levels in the vicinity of the tropopause on the eastern flank of the major trough. The vertical velocities in the two flows have about the same magnitude if those in the moist flow are averaged over the same appropriate horizontal width, attaining maxima in the middle troposphere of 10 cm sec^{-1} over regions about 200 km across. The flows probably develop only during the intense phase of slope convection, generally recognized by the associated pronounced cyclogenesis.

A dry tongue of rapid descent over much of the depth of the troposphere, in contrast to the narrow zone of deep saturated ascent, is a phenomenon less readily anticipated. At least some of its features, however, have been reproduced in a recent theoretical study of frontogenesis produced by a simple deformation field independent of height (51). During the frontogenesis an initially negligible thermal wind increases, and accelerations occur parallel to a developing frontal zone. These imply ageostrophic motions, and circulation, in a plane normal to the frontal zone. Typically only about one day is required to produce a frontal zone of width only a few hundred kilometers. The front near the surface, which forms on the warm side of the axis of dilatation, soon becomes concentrated into a virtual discontinuity. In the development of the frontal zone near the tropopause, on the cool side of the axis, the tendency to produce a discontinuity arises only if there is a maximum or minimum of potential temperature on the tropopause. In general during the frontogenesis in the models the tropopause becomes deformed; ascent occurs on the warm side of the axis of a developing jet stream, raising the tropopause there a little, and descent occurs on the cool side, leading to a lowering of the tropopause and a marked downward protrusion of a tongue of air from the low stratosphere. The Richardson number tends to become small (<1) in a zone in and below the lowest part of the tongue. The same tendency in natural situations is associated with the prevalence of small-scale motions in this region, arising from a dynamic instability of strongly sheared flows, which will be discussed later. (Evidently they eventually limit the narrowness of frontal zones.)

Such theoretical studies, and analyses of observations including radiosonde soundings (52), which in spite of the defects of humidity sensors are well able to detect very low relative humidities when they occur in the low and middle troposphere, show that the development of dry tongues is a general aspect of cyclogenesis and frontogenesis, and that they are not associated only with cold fronts, as might be supposed from a casual inspection of cloud distributions in satellite pictures. Pronounced tongues appear in the middle troposphere in warm frontal zones, ahead of and below the associated cover of middle and high cloud.

The form of the dry tongue has been clarified by analyses of the structure of the flow in the rear of large-scale troughs containing fronts (48, 49). Because the air is practically everywhere unsaturated and the interesting part of the motion is completed within 1–2 days, the analyses have been made by assuming adiabatic motion and studying the flow in thin laminae between isentropic surfaces (of constant θ), using the conservation of mass, energy, potential vorticity, and the mixing ratio of water vapor as constraints on the shape of the trajectories inferred from wind observations. Detailed examination of original radiosonde records shows that the atmosphere is composed of a number of well-defined layers, within each of which the lapse rate of temperature is almost constant (incidentally justifying the conventional procedure of representing a sounding by plotting reported significant points on an aerological diagram and joining them by straight segments). When the records from a network of soundings are analyzed it becomes clear that many of these layers are not merely local features but can be traced over large areas for periods of at least a few days (50). On vertical cross sections showing the distribution of potential temperature this structure is evident in the pattern of the isopleths, which are crowded together in the more stably stratified layers and especially in frontal zones and in the stratosphere. Individual isopleths extend into the frontal zones from either side and pass from the troposphere into the comparatively very stably stratified lower stratosphere, so that a front and a tropopause appear more obviously as features produced within a field of motion, and not as partitions impervious to the flow. In analysis on their scale it has become customary to define their positions by surfaces across which the potential vorticity is virtually discontinuous, bounding regions of small lapse rate of tempera-

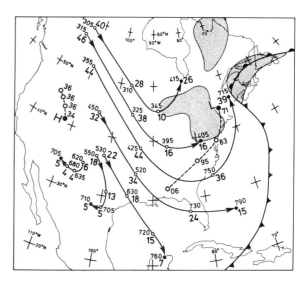

Fig. 9.29 Trajectories of air in a major trough over North America, followed during the intensification of a cyclone near the east coast (49), illustrating the origin of the dry tongue. Along the paths of air (solid lines) and of the centers of the major cyclone L and the major anticyclone H (heavy dashed lines) circles mark positions at intervals of 12 hr; final positions are shown by solid circles. Also, at this time the fronts at the surface are shown in the conventional manner, and three major areas of precipitation are shown by stippling.

The circles on the paths of the surface pressure centers are labeled with the last two figures of the central pressure (in mb). Those on the trajectories are labeled with the speed (in m sec^{-1}, italic numbers) and the pressure (in mb); all the trajectories lie in the isentropic surface $\theta = 305°\mathrm{K}$. The two northernmost trajectories originated in the stratosphere; the adjacent one, which on this occasion approximately defines the axis of the dry tongue and along which there was the greatest and most rapid descent, originated at a level a little below the tropopause. In the rear of the cyclone there is evidently general subsidence and deceleration of the flow throughout the troposphere, accompanied by a striking diffluence of the trajectories. The maximum sinking speed is in the middle troposphere where the air originally near the axis of the northwesterly jet stream approaches the axis of the trough. Only in some places where the flow returns poleward does the subsidence cease, to be followed by slight upward motion and acceleration.

ture. According to this convention the boundaries of frontal zones in (at least) the upper troposphere may be drawn to be continuous with tropopauses, and where these zones contain air of low relative humidity it evidently could have descended nearly adiabatically from the high troposphere or low stratosphere. The analyses of the motion in major troughs in which such frontal zones develop show that such descent does occur. Examples of trajectories over North America during the intensification of a cyclone on the eastern flank of a trough, as seen in Fig. 9.29, illustrate how the most pronounced descent is described by a trajectory which approaches the cold front near the surface before turning poleward toward the cyclone center; the path of this trajectory, when drawn relative to the center and the front, clearly corresponds to the axis of the dry tongue so typically seen in satellite pictures as a dark, comparatively cloud-free zone in the cloud systems of an intensifying cyclone.

The interchange of air between stratosphere and troposphere; stratospheric humidity

It is apparent from a mean vertical section that continual rapid interchanges of air between the stratosphere and the lower troposphere occurring during slope convection can involve only a shallow layer at the base of the stratosphere in middle and high latitudes. Whereas it is evident that potentially warm moist air in low latitudes might during ascent with constant θ_s reach heights near the mean level of the tropopause in a latitude as low as about 30°, with an equivalent potential temperature θ there not exceeding about 330°K, the conditions for nearly adiabatic and unsaturated descent from the stratosphere into the lower troposphere are more restrictive. Adiabatic descent from the mean position of the tropopause to the surface is not possible at all, and only air with a potential temperature θ less than about 320°K, in about the lowest kilometer of the stratosphere, could enter the lower troposphere. Analyses of individual motion systems confirm that it is generally air of potential temperature between about 295 and 305°K which composes the dry tongues. It descends within about 36 hr from just above the low tropopauses found on the poleward side of the middle-latitude jet streams to levels between about 1 and 4 km above the surface; there it is subsequently drawn into and modified by the small- and large-scale circulations of the lower troposphere.

It is clear also from the typical distributions of potential vorticity Z that interchanges of stratospheric and tropospheric air by motions which are adiabatic and frictionless (in which Z is conserved), or nearly so, must be confined to narrow regions on the poleward side of the axes of jet streams. Only there (and in the frontal zones below the jet streams), where there are large cyclonic horizontal wind shears, does Z in the troposphere attain the large values (of about 40×10^{-9} °C cm sec g^{-1}) which are characteristic of the lower stratosphere. Although the rapid interchanges (those effected within 1–2 days) are thus limited to narrow regions, the associated mass exchanges are significant. In a dry tongue the mass flux of stratospheric air entering the lower troposphere is about 10^{17} g day^{-1} (48). Such a flux maintained continuously by each of five cyclones simultaneously present in a hemisphere is equivalent to the removal in about 3 months of the entire mass of the lowest kilometer (the lowest 50 mb, representing 50 g over each square centimeter) of the stratosphere over a hemisphere north of the 30th parallel, that is, an area of about a quarter of the surface area of the globe (5×10^{18} cm^2).

The mechanisms of compensating transformation of tropospheric into stratospheric air have not been established. The flows into the stratosphere may be either slow, with radiative processes playing an essential part in the modification of the air, or rapid and under a primarily dynamical control. Those of the latter kind, like the flows descending from the stratosphere, can occur only in the restricted regions on the poleward side of jet streams, and from a very few analyses seem to originate in the high troposphere on the eastern flank of troughs.

The presence of slow transfers and their mechanisms are more difficult to establish. Their occurrence is manifest in the annual variation of the height and pressure of the tropopause. This variation is not apparent in most statistical treatments of temperature distribution, partly because of the arbitrary nature of the criteria for defining and reporting the height of the tropopause, but mainly because only the readily available data from standard pressure levels are used (as in the construction of Fig. 9.1). However, an examination of the annual variations of the vertical distribution of temperature at some individual sounding stations indicates that in middle and high latitudes the mean height of the tropopause has an annual variation of about 1 km (or 50 mb), reaching twice as much in the interior of the continents, with a minimum in the winter and a sharp rise after the spring to a maximum in the late summer (53, 54). Evidently during the period of this rise the rate of decrease of the mass of the stratosphere is comparable with the mean flux out of the stratosphere in the dry tongues of the slope convection. During the remainder of the year there must be a compensating flux in the reverse sense.

In low latitudes the comparatively high tropical tropopause has about the same mean annual height variation of 1 km, corresponding to a smaller pressure variation of about 20 mb (see also, e.g., 56). Remarkably, at least in the northern hemisphere, the height is a *minimum* in the late summer, which seems to conflict with the concept that the tropopause represents the effective upper limit to deep convection, which is most intense in the summer.

The transfers of air between the stratosphere and troposphere play a crucial part in the general distribution of trace substances, of which the most important are water vapor, whose source is at the surface, and ozone, whose source is in the stratosphere, for the distributions of these substances contribute to the determination of the position and sharpness of the tropopause separating the two regimes. Other important trace substances are those producing aerosol particles, isotopes which are potentially valuable in the tracing of air motions, and, at least in the recent past, the radioactive debris of nuclear explosions in the stratosphere. The necessity to understand the diffusion of this debris, and the pattern of its deposition on the ground, provided a strong stimulus to the study of circulations and exchanges of mass between troposphere and stratosphere (54). In the context of the present discussions one aspect of particular interest is the mean distribution of water vapor in the stratosphere, which has not been established beyond doubt by observation (52), but which by an increasing consensus of soundings using different methods appears to be one in which the mixing ratio is almost uniform with height (up to 30 km) and latitude, varying between about 2×10^{-6} and 3×10^{-6} (56). Observations made at Washington, D.C. ($39°N$) and Trinidad ($11°N$) provide some evidence of a seasonal variation (diminishing upward to become indiscernible at 20 km), with the smallest values occurring in the winter and spring, and also of similar departures of mean monthly values from the seasonal trend.

Both kinds of variation seem to be related to changes in the temperature of the tropopause at the low latitude, which appears to be the frost point of the air in the stratosphere, a plausible inference from the kind of mean circulation illustrated in Fig. 9.12 if air passes through the tropopause with the removal of vapor condensed in the troposphere. However, it is difficult to visualize the mechanisms by which the passage of the air and the removal of its condensate are effected.

According to Fig. 9.12, the winter mean meridional circulation in low latitudes of the northern hemisphere transports air from the troposphere into the stratosphere at the rate of about 4×10^{17} g day^{-1}, and there is a corresponding mean sink into the troposphere in latitudes near the 30th parallel, where the tropopause is ill defined. (During the summer season the fluxes through the tropopause in low latitudes probably have the same magnitude but are part of a cell in which the mean circulation is reversed; although the mean flow remains upward between the equator and $30°N$, in the lower stratosphere it is toward the south and the pronounced subsidence is between the 10th and 30th parallels of the southern hemisphere.) The equivalent mean vertical velocities at the height of the tropopause are very small—not more than a few tens of meters per day.

A persistent ascent near the equator would at first seem to imply a prevalent state of saturation accompanied by an overcast of high cloud. Over large areas, however, the sky appears to be clear of high cloud, suggesting that the mean upward velocities may be the resultant of intermittent ascent associated with cumulonimbus or cumulonimbus systems, with intervening periods of little vertical motion or subsidence. High clouds observed in very low latitudes are probably all produced by cumulonimbus convection; even extensive thin layers well away from regions of cumulonimbus formation are likely to be the residues of anvil clouds. Consistently with the form of the mean circulation, they are more frequently observed near the 10th parallels than in broad zones, usually clear of high cloud, which extend from about the 15th parallels poleward (57), to the nearer flanks of the subtropical jet streams (where there is another maximum in the frequency, related to cloud formation in circulations of the high troposphere associated with the jet streams). However, the high clouds of low latitudes are generally in layers lying at least a few kilometers below the tropopause (58); those which have been observed above aircraft flying at heights near 15 km have been reported as very tenuous (e.g., 59). Moreover, it seems that the outflows from cumulonimbus are generally also a few kilometers below the tropopause; only very locally over particularly favorable regions (principally northern India and neighboring parts of Southeast Asia) are the outflows at levels as high as the tropopause. It seems unlikely that these outflows could determine the general level of the tropopause in low latitudes, and more probable that the tropopause is produced within more widespread regions of slow ascent (as in a numerical model discussed in detail in reference 60). In these regions there is an upward transport of water vapor, but there may be no associated cloud formation and indeed evaporation of high clouds composed of cumulonimbus residues, because at these levels clouds are subject to a long-wave radiative warming. It is significant that even in apparently cloud-free conditions the air near the tropopause has been found to be practically saturated with respect to ice (in special soundings with balloon-borne frost point hygrometers; see Fig. 9.30), and that the presence there of clouds or of aerosol too tenuous to be visible had been inferred from soundings with balloon-borne radiometers (61); these have shown rates of radiative energy loss which when compared with calculated values are nearly equal over most of the troposphere, but significantly smaller between about the 350- and the 100-mb levels.

9 Large-scale slope convection

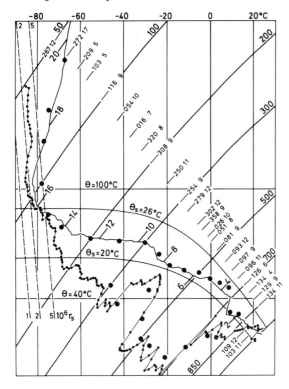

Fig. 9.30 A sounding of temperature and of dew or frost point made in a low latitude (from Trinidad, at 11°N), with balloon-borne instruments including a frost point hygrometer (56). Interesting features are the sudden decrease of humidity above the trade wind inversion, marking the top of the layer of ordinary convection, the presence of a shallow damp layer in the upper troposphere, and the state of approximate saturation with respect to ice near the tropopause. The solid dots are temperatures and dew points reported from the local routine radiosonde sounding made some hours later on the same day (the winds entered are also from this sounding). There is a generally close correspondence between the two soundings. Throughout the day no high cloud was observed above some cumulus. The moist layer in the middle troposphere had probably arrived from a part of the nearby mainland where θ_w near the ground had been raised to about 25°C, allowing deep cumulus convection.

Within the stratosphere diffusion in the horizontal is effected readily by the unsteady large-scale motions driven by those generated in the troposphere, but it is comparatively slow in the vertical because of the great stability of the stratification (as illustrated in the model of reference 60, which, however, has insufficient resolution in the horizontal to include exchanges between the low stratosphere and the troposphere in middle and high latitudes). That the mixing ratio of water vapor throughout most of the stratosphere seems to be determined by the temperature of the tropopause in low latitudes suggests that the transfers between troposphere and stratosphere in middle latitudes, where the temperature of the tropopause is comparatively high, are not a significant source of stratospheric water vapor. This might be because part of the transfers involves air recently drawn from the stratosphere, or because transfers from the moist (cloudy) part of the troposphere are into only a shallow layer at the base of the stratosphere which is again exchanged before a significant upward diffusion of water vapor occurs. It appears reasonable to regard the lowest part, about 1 km deep, of the stratosphere in middle latitudes as a transition layer in which there are continual significant exchanges with tropospheric air which may be considerably moister, but which is not a source of water vapor for diffusion into higher parts of the stratosphere. The view of this lowest part of the stratosphere as a transition layer is supported by several kinds of flight experience; for example, in it there is a marked variability of mixing ratio (between about 2 and 5×10^{-6}) from day to day, in contrast to a small variability at higher levels (57, 62). Second, there is commonly no abrupt change of frost point or of its gradient during horizontal or vertical passage across the tropopause; on the other hand, there is sometimes a sudden decrease in the lapse rate of frost point (to the small value characteristic of most of the stratosphere) about 1 km above the tropopause (63). Third, there is evidence of moist intrusions into the lower stratosphere. Thus distinct increases in the frost point of the lower stratosphere (corresponding to increases of mixing ratio from about 2×10^{-6} to 5×10^{-6}) have been observed in cloudless zones several hundred kilometers wide during flight across the positions of cold fronts at the surface (64). Moreover, although the tops of layers of high cloud are usually 1 km or more below the tropopause, they have sometimes (in 8 of a series of 78 observations) been found to extend several hundred meters into the stratosphere (65). (On rare occasions clouds over England have been reported a few kilometers above the tropopause, but it is not certain that they were composed of water rather than of dust; some of the observations were made in the late summer of 1953, during the period when the layers of volcanic dust described in Appendix 4.1 were present in the stratosphere.)

Precipitation systems

The extent and intensity of the precipitation in a cyclone appear to be highly variable but probably depend mainly on the stage of evolution and intensity of the cyclone and, on the other hand, on its geographical position, which controls the general temperature (and therefore vapor content) of the air that ascends from the low troposphere, and the effects of topographical features of a great range of scales. Excluding the effects of the last and of other small-scale motion systems (including showers produced by ordinary convection in airstreams warmed at the surface), and referring for example to the model of Fig. 9.28 or to the first studies of rainfall and the relative flow in cyclones (74), it seems reasonable to regard the cyclone as containing two principal precipitating cloud systems, one producing a narrow belt of precipitation near the cold front and the other a comparatively broad area of less intense precipitation near and ahead of the warm front. Over regions of low temperature at the surface (wet-bulb temperature less than 0°C, and therefore dry-bulb temperature less than about 3°C) the precipitation usually consists entirely of snow; in warmer regions the precipitation at the surface is of rain or drizzle, perhaps with some small hail near cold fronts in middle latitudes. Drizzle alone is characteristic of clouds of limited thickness (between about 500 and 1500 m) with very low bases (not more than a few hundred meters above the surface; otherwise small drops which fall from the clouds are likely to evaporate; see Section 5.10). Drizzle represents a stage in the development of precipitation by coalescence of cloud droplets, and it often accompanies rain near warm fronts where the base of thicker clouds may become very low. Drizzle alone typically occurs in airstreams which are cooled during travel from the oceans poleward or across cold land, in the principal weather region 3 of the Norwegian cyclone model (Section 9.4), and is associated with fogs in light winds and very low shallow clouds in fresh or strong winds. In these airstreams the rate of cooling of shallow layers is several degrees per day. This rate of cooling is equivalent to that produced by ascent at a speed of only about 1 cm sec^{-1}; consequently, the intensity of drizzle is small compared with that of the rains which are produced in cyclones by the large-scale ascent of deeper layers at rates typically of several centimeters per second, and the drizzle region does not represent an important precipitation system.

Cloud and precipitation near cold fronts. In the early stages of cyclogenesis and near the centers of maturing cyclones over warm surfaces the clouds associated with cold fronts occupy much of the depth of the troposphere and usually contain cumulonimbus convection (of only moderate or weak intensity over the oceans). In the general large-scale upward motion the cumulus are closely packed, and shelf clouds, including the anvil clouds, tend to be persistent and to thicken. Precipitation forms in cumulus in the manner discussed in Chapter 8, and the intermittent development of cumulonimbus over squall fronts produces a long belt of continuous cloud and precipitation in which the positions of the more active clouds may be visible from their protruding tops or detectable by radar as regions of more intense precipitation. The rate of the intense precipitation is related to the temperature and the flux of the air entering the cumulonimbus bases; the flux depends at least partly on the potential buoyancy of the air. From a large-scale point of view the general precipitation rate is related to the motion of the moist layer of the air in the warm sector relative to the front, an ageostrophic motion to which friction near the surface may make a large contribution.

When a cold front enters a region in which large-scale subsidence predominates the frontal zone becomes confined to the low troposphere, and the associated dense clouds become restricted to a narrow band at low levels. Such bands are observed even near the centers of cyclones which are mature or decaying, particularly when they have reached cool regions. They are also seen when the general pattern of the flow is such that the wind ahead of the fronts is westerly or has a component toward lower latitudes, and when cold fronts enter low latitudes far from cyclone centers; in these circumstances the frontal cloud is often a narrow belt of cumulus which may produce showers but in whose tops the temperature is barely if at all below 0°C, so that there may be extensive shelf clouds but no glaciation. The front then often appears on the PPI display of a conventional radar as a long chain of small shower echoes which are mostly distinctly separated. The front illustrated in Pl. 9.5 was moving southeast across southern England and was associated with a weakening cyclone centered east of Scotland. It was marked by a long belt of cumulus orientated toward 245°, practically parallel to the strong winds ahead of it at the level of the cloud bases (velocity 245° 25 m sec^{-1} at 1.0 km). The clouds were mostly below an inversion at 2.5 km, above which the relative humidity decreased to about 10%. The potential buoyancy of the warm air was very small; nevertheless, the clouds locally towered to 3.6 km (temperature about -3°C) with the formation of showers of moderate intensity. In the warm air the relative wind toward the front was mainly in the layer several hundred meters deep in which friction during flow over the ground produced a backing of 20–30°, and it seems likely that it was the energy of this relative flow (at several meters per second), rather than buoyancy acquired in the cumulus layer, which was responsible for the vertical development of the clouds.

As a cold front weakens the associated band of low cloud becomes more narrow and shallow, and eventually disappears, probably rapidly once it is no longer sufficiently well developed to produce showers and a squall front of cool air. Remarkably, in middle latitudes there can be a pronounced deterioration of the weather during the passage of such a front even when the cloud tops reach heights of only about 3 km. The rain may be at the rate of 20–30 mm hr^{-1} for a few minutes, perhaps with small hail and thunder, and preceded by a large and rapid change of the direction of the wind near the ground, with squalls of up to 15–20 m sec^{-1} (75). On

9 Large-scale slope convection

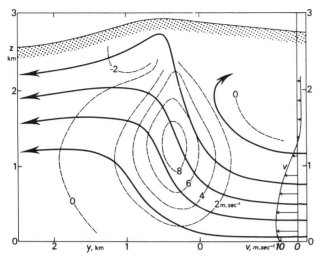

Fig. 9.31a The relative flow near the leading edge at the surface in a vertical section transverse to a cold front over England (76). The solid lines are streamlines; the dashed lines are isopleths of vertical velocity (in m sec^{-1}). The distribution of the horizontal speed v relative to the front in the warm air (2 km ahead of the leading edge of the frontal zone at the surface) is shown on the right, and the stippled zone at the top of the diagram represents the top of the radar echo.

some occasions there is little or no potential buoyancy in the warm air ahead of the front, and yet from above the clouds can be seen to have towering cumuliform tops, and from the occurrence of hail they evidently contain updrafts of at least several meters per second. In such circumstances the production of a cool downdraft by the evaporation of precipitation from the band of shower clouds, or from shelf clouds in an outflow with a component toward their rear, may play an essential part in the energetics of the circulation transverse to the front (a possibility envisaged in Section 8.5).

A detailed study has been made of the air flow and the development of precipitation in one front of this kind, using special radiosonde ascents and other observations from a radar station in the west of England (including data obtained with a Doppler radar, from which horizontal and vertical components of the wind in the precipitation were inferred; 76). The front moved southeast over England during a winter afternoon, preceded by a narrow belt of intense rain and some small hail and thunder, and a sudden large veer and decrease of the wind near the ground. A satellite picture (Pl. 9.6) of the front as it approached the British Isles on the previous morning shows several features typical of the structure of the front observed during its passage over the radar station. Most prominent is the long, narrow bright band of cloud over the leading edge of the frontal zone at the surface, where the inferred relative flow was as shown in Fig. 9.31a. The relative flow toward the cloud band was confined practically to the lowest 0.5 km

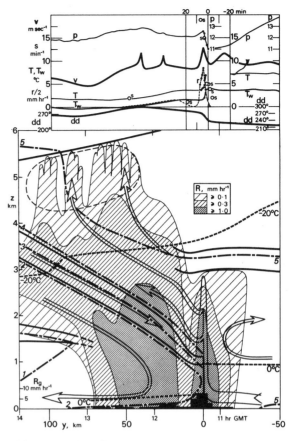

Fig. 9.31b Relative flow and some other features over a broader region of slight precipitation, inferred mainly from Doppler radar observations of the precipitation (76). The arrows are again streamlines; the heavy lines are boundaries of the frontal and other stably stratified zones. The heavy dot-dash lines are isopleths of θ_w (in °C, italic numbers). The shaded zones indicate regions in which the precipitation rates aloft exceeded 0.1, 0.3, and 1 mm hr^{-1}. Isotherms for 0 and -20°C are included. The average precipitation rate R_g at the ground is shown at the base of the diagram, and a space scale has been added consistent with the observed speed of the front (12 m sec^{-1}). The panel at the top shows surface weather during the passage of the front at Ascot, 120 km southeast of the radar station.

Key: dd = the direction of the wind; v = the speed of the wind (direction and speed are arbitrarily smoothed, but the latter shows correctly four maxima); T, T_w = the dry- and wet-bulb temperatures; p = the sea-level pressure; and r = the rainfall rate (averaged over 1 min). In addition, the circles marked s show the number of lightning discharges within about 30 km during intervals of 1 min. (The dots indicate some values of p, T, T_w, and r at successive intervals of 1 min.)

in which friction produced a backing of the wind; analyses of several sequences of observations during a period of nearly 30 min showed, consistently with the virtual absence of any cellular structure in the radar echo associated with the cloud band, that the flow at least throughout this period was almost steady and two dimensional. The diagram shows that the vertical component of the air velocity in the cloud band reached a maximum of about 8 m sec^{-1} over a region less than 1 km wide, and that there was probably no significant acceleration of the warm air under buoyancy, since the mean kinetic energy of the inflow hardly changed, or decreased a little, in the flow through the cloud and in the outflow which was predominantly into the rear of the front. At least locally and intermittently a part of this outflow must have been in a shelf projecting a little ahead of the updraft, and the flow could not have been strictly two dimensional, for occasionally a vault could be detected in the radar echo, a region 1–2 km wide including the core of the updraft where the echo intensity was about 15 db lower than on either side (there the equivalent reflectivity factor Z_e reached values between 10^4 and 10^5 mm^6 m^{-3}). As discussed in Sections 8.4 and 8.8, a vault is formed where precipitation from a shelf cloud projecting forward falls into the updraft but mostly fails to enter its core before it is again carried aloft. In this manner precipitation particles may be recycled through the updraft and those which freeze are able to reach a larger size than would otherwise be possible in a limited depth of cloud. On this occasion the 0°C level was at about 0.9 km in the warm air, barely above the cloud base, but the cloud tops reached only 3 km (temperature about −15°C), and a recycling appears necessary for the production of the small, hard hailstones observed (diameters up to nearly 1 cm and fall speeds probably up to 12–15 m sec^{-1}). The intensity of the rain at the front suggests that it was produced by the melting of small hail; also, the high density of the larger hailstones which reached the ground (indicated by the noise of their impact on roofs and windows) suggests that their embryos were frozen large drops grown by coalescence, and that the cloud droplets subsequently accreted were rather large (even though the cloud was shallow), as might be anticipated in a cloud formed in maritime air. However, the presence also of snow aloft was indicated by an intensification near the 0°C level of the radar echo in the regions of less intense precipitation, a phenomenon discussed in Section 9.6.

Another interesting feature of this front was that the outflow from the narrow zone of vigorous ascent, in a layer about 1 km deep, continued to ascend (at about 20 m sec^{-1}) over the sloping frontal zone rearward of the front. Snow fell from the cloud in this layer: precipitation reached the ground (as slight sleet or rain) in a zone extending 40–80 km behind the band of intense precipitation. As the air in the cloud ascended the flow became more nearly parallel to the front, and eventually a well-defined rear edge to the cloud was produced, about 200 km behind the band, as can be seen in Pl. 9.6. The front was not far from the center of the associated maturing cyclone (in the northern North Sea). It appears to have been intermediate in structure between those near intensifying cyclone centers, which are characterized by deep layers of cloud occupying most of the troposphere (with a rear edge of high cloud several hundred kilometers from the front at the surface), and those weaker fronts characterized by a shallow band of cumuliform cloud with little or no shelf cloud (of the kind previously discussed and illustrated in Pl. 9.5). There was a general subsidence of the cool air, concentrated in the cold front zone (where the downward speeds reached 20 cm sec^{-1}), and consequently in the inferred transverse circulation (Fig. 9.31b) there was a relative flow (at speeds of several meters per second) toward the surface position of the cold front in the upper part of the cool air, and away from it a layer extending upward about 0.5 km from the surface.

Cloud and precipitation near warm fronts. According to the model of Fig. 9.28, near the warm front of a cyclone the two principal cloud systems overlap. In the cool airstream which approaches from the east the air has previously been modified by ordinary convection, and moistened, to a depth of a few kilometers. As the flow enters the cyclone and perhaps turns poleward the convection ceases, but eventually the large-scale ascent restores condensation and an extensive layer cloud develops which may eventually reach a thickness of a few kilometers. This cloud is usually subject to a small-scale convective stirring; it occupies a layer in which the previous convection leaves a nearly neutral stratification, and in which the vertical stretching (and radiative exchanges, if the air above is clear or contains only high cloud) tends to increase the lapse rate of temperature. Flight in such clouds is therefore moderately bumpy, and their upper surfaces are irregularly domed or waved, so that they are usually described as stratocumulus. Their properties in respect of the vertical distribution of liquid water concentration and of the size distribution of cloud droplets probably are not significantly different from those of cumulus (in an airstream of similar temperature and content of aerosol particles), which have been studied much more extensively (see Section 8.1). This expectation seems to be supported by a comparatively small amount of observational data.

Thus, for example, in an analysis of samples of droplets taken from volumes of a few tens of cubic centimeters in cumulus, stratocumulus, and fogs, it was found that the size distributions were all well represented by Eq. 8.1:

$$1 - F = \exp\left[-(r/a)^c\right]$$

in which F is the fraction of the concentration m of liquid water held in droplets of radius less than r, c is a constant with a mean value of 3.3, and a is a scaling radius; the value of a was related to m but did not vary significantly with the kind of cloud sampled (77).

Most of the observations of droplet size distributions and liquid water concentrations were made (often at unspecified levels) in stratocumulus clouds of thickness between a few hundred meters and about 1 km (78–82). They show that the total droplet concentration N varies between about 100 cm^{-3} in maritime air and several hundred cubic centimeters inland, and that the median volume radius r_0 and the liquid water concentration m increase with height above the cloud base to a level near the cloud tops; values quoted for r_0 vary from a few microns to 10–13 μ, and those for m are usually in the range 0.2–0.6 g m^{-3}. The ratio m/m_γ, where m_γ is the concentration corresponding to adiabatic ascent from the cloud base, is usually between 1/4 and 1/2 (80, 81). The size distributions of the cloud droplets are generally broad, and even in the limited volumes sampled the maximum droplet radius frequently exceeds 20 μ. [Similar size distributions have been obtained in models which assume that under a turbulent diffusion a constant total concentration of droplets is maintained, of which individuals spend a variable time within the cloud growing by condensation (83, 84). Reasonable constant values must also be assumed for a size distribution of condensation nuclei, a supersaturation, and an eddy diffusivity; models with a more satisfactory physical basis, and which could consider an evolution of the size distribution toward the production of precipitation, must await improved understanding of the motion within the clouds.]

From the preceding observations it might be anticipated that in maritime airstreams precipitation is likely to form in stratocumulus, just as in cumulus (Section 8.1), if the thickness of the clouds exceeds 1 km, and precipitation is virtually certain to form if the thickness reaches 2 km. It is well known that in low latitudes rain commonly develops in clouds which are only about 2 km deep and whose temperature is wholly above 0°C. However, most of the reported observations (e.g., 85) refer to showers from cumulus, and even when the clouds have been described as stratocumulus they have almost certainly been extensive shelf clouds with cumulus present. [An exception appears to be a report (86) of a special flight made over the Mariana Islands of the West Pacific to explore an extensive layer of cloud producing intermittent moderate showers; the height of the base was 300 m, and that of a "perfectly level" top, under a clear sky, was 1500 m (temperature 17°C); conditions inside the layer were only "very slightly" bumpy.]

In middle latitudes stratocumulus alone is much more frequent, but it is difficult to obtain evidence of the formation of rain in wholly liquid clouds, since if their thickness exceeds about 2 km, the temperature of their tops is likely to be below 0°C, or there may be other clouds above, so that ice particles may form in or fall into the layers. Rare instances have been reported of locally prolonged drizzle or rain below layer clouds which from aircraft observations and soundings have been found to have tops with heights below 2 km and temperatures above 0°C (e.g. 87; reference 88 reports that over southern England a research aircraft found only liquid particles in a layer cloud about 2.5 km thick—at temperatures above $-6°C$—from which rain reached the ground at rates of less than 1 mm hr^{-1}). From analyses of routine soundings made in Northern Ireland (89) and in the west of England (90) it appears that in Northern Ireland continuous slight rain (intensity < 0.5 mm hr^{-1}) at the ground occurs occasionally below cloud layers with temperatures wholly above 0°C when the cloud thickness exceeds 1.2 km; the probability of precipitation is nearly 50% (nearly 80% if drizzle is included) when the thickness is between 1.5 and 2.0 km. Over England continuous rain occurred only below clouds more than 2.2 km thick; over Northern Ireland continuous moderate rain (intensity between 0.5 and 4 mm hr^{-1}) was observed only when the cloud thickness exceeded 2.5 km.

The rate of widespread continuous rainfall from layer clouds is much less than the intensity attained in the showers of ordinary convection, since it is limited to the rate of condensation produced by large-scale ascent at speeds of order only centimeters per second, or by advective and radiative cooling. An estimate of the thickness of a layer cloud produced by large-scale ascent, which is required to give precipitation at a particular rate, may be made by assuming that air in the cloud is saturated in a vertical column having given profiles of temperature and ascent speed, and that the rate of precipitation is the integral throughout the column of the rate of condensation in adiabatic ascent (this rate is a function of pressure and temperature; some values are tabulated in Appendix 3.2). If the speed of rise of the air changes linearly from zero at the surface and at a height of 10 km to a maximum of 10 cm sec^{-1} at 5 km, a profile representative of deep large-scale ascent near a front, and if θ_s is constant with height, then the rate of precipitation as a function of the thickness of a layer cloud (or the height of its top, as it is assumed to have a low base) is a function also of θ_s (or of surface temperature), approximately as seen in Fig. 9.32. [With $\theta_s \approx$ 20°C the rate of precipitation corresponding to a mean upward speed of 1 cm sec^{-1} in a cloud 1200 m thick, with a base near the surface, is about 0.1 mm hr^{-1}. An increase of this rate to a few millimeters per hour requires an increase in the mean upward speed by a corresponding factor, that is, to a value of a small fraction of 1 m sec^{-1}. It is interesting that such a comparatively large value could be provided, over a zone several

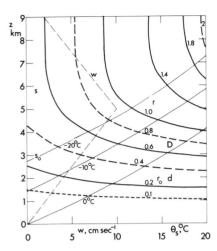

Fig. 9.32 In a column of saturated air, with a base at a pressure of 1000 mb, θ_s is constant and the local rate of ascent $w(z)$ is specified by the thin dashed line. The diagram shows as a function of θ_s the thickness z of the column in which the total rate of condensation is equivalent to a rate of precipitation R, by the isopleths (heavy lines, some of which are dashed) of R (R in mm hr^{-1} of liquid water). (The thickness z is approximately also the level of the top of a layer of low cloud with a base within 500 m of the surface.) Three isotherms (for 0, -10, and $-20°C$) are entered, and typical rates of widespread precipitation associated with its various forms are indicated by the placing of the letters s (moderate snow) and s_0 (slight snow); r (moderate rain) and r_0 (slight rain); and D (heavy drizzle) and d (moderate drizzle).

kilometers wide, across the length of a roll vortex in the boundary layer, of the kind described in Section 7.6. Since such vortices move slowly across the direction of wind they could be responsible for intermittent moderate "showers" from tropical maritime layer clouds with level tops, as described in a report (86) previously mentioned.]

From Fig. 9.32 it appears that near fronts large-scale conditions are suitable for the development of widespread precipitation of slight intensity in layer clouds about 2 km thick, and precipitation of moderate intensity in clouds of thickness between about 3 km in warm climates and several kilometers in cool climates. Moreover, the previous discussion indicates that in such clouds the microphysical conditions are favorable for the production of rain by coalescence of cloud droplets, at least over and near the oceans.

The diagram also shows, however, that in cool and temperate climates frontal cloud layers sufficiently thick to produce moderate precipitation at the ground have upper parts in which the temperature is considerably below 0°C. Significant concentrations of ice particles are therefore likely to form in the cloud tops. Near frontal zones, especially close to the centers of intensifying cyclones, these concentrations may be increased and maintained by the fall of ice particles from the upper troposphere (the particles may form locally or arrive after traveling several hundred kilometers in the out-flows from frontal cumulonimbus). Then account must be taken of the presence of the ice phase in considering the development of precipitation, although in maritime airstreams it is not likely to have any important effect on the distribution or intensity of the precipitation which reaches the surface. On the other hand, in airstreams which have moved over a continent and become rich in aerosol particles the ice phase may play an essential part in the generation of widespread precipitation, for then the thickness of a layer cloud which is required for the production of even slight precipitation by the coalescence of cloud particles probably increases to several kilometers.

The importance of the ice crystals in the generation of precipitation lies in their small concentrations, which given sufficient growth periods allow them to reach the size of precipitation particles by condensation alone (Section 5.12). In the later stages of their evolution accretion with other crystals to produce snowflakes or, more significantly, with cloud droplets (before or after melting) may substantially increase their eventual size. A typical total concentration of raindrops in slight or moderate rainfall is of order 10^{-4} cm^{-3} (as may be confirmed by reference to observation—e.g., as summarized in Fig. 8.3—or by calculating the concentration of small raindrops, of fall speed several meters per second, required to provide a precipitation rate of about 1 mm hr^{-1}). The minimum concentrations of ice crystal embryos required for the efficient generation of rain are therefore probably one or two orders of magnitude greater, considering that in a steady state the embryos are supplied in an airstream of upward speed of about 10 cm sec^{-1}, while the raindrops leave the cloud with fall speeds of several meters per second. Such concentrations of embryos can be anticipated to arise by the action of freezing nuclei in droplet clouds, provided the temperature in their tops is at least as low as -15 to $-20°C$ (Fig. 5.5). It is, of course, possible that concentrations produced by freezing nuclei may be increased by some multiplication or fragmentation process (Section 5.14), but inside layer clouds the most plausible such process is the breaking of the frail arms of dendritic crystals, which grow only in the temperature range between about -12 and $-16°C$ and at the high supersaturations with respect to ice which represent approximate saturation with respect to liquid water (Fig. 5.6). It may therefore be significant that the probability of rain or snow reaching the ground from layer clouds in cool climates (and of detecting ice particles during flight through clouds) increases rapidly as the temperature in the cloud tops decreases below about $-12°C$, but in the statistics of the observations usually presented it is uncertain whether some other variable, for example, the cloud thickness, is not more directly involved. It seems possible, however, that in continental airstreams the large-scale conditions sometimes lead to the formation of layer clouds a few kilometers thick and are favorable for the production of widespread light precipitation, which still fails to develop because the coalescence process is ineffective and because the temperatures in the cloud tops are not sufficiently low to provide adequate concentrations of frozen droplets. Such clouds are ideal for experiments in the stimulation of precipitation by the introduction of artificial ice nuclei from aircraft, and the best evidence of their occurrence is in the experience gained in a few research projects designed to explore the feasibility of cloud modification by this method. According to observations made for this purpose in central and western North America (over the states of Ohio and Washington) they occur infrequently; layer clouds whose upper parts at least contained supercooled droplets and which were more than 1 km thick were encountered only on the fringes of and within precipitation systems. Such clouds from which snow was not already falling could readily be made to glaciate over small treated areas and to produce temporary falls of slight snow (91, 92).

Similar clouds more than 2 km thick were rarely found, and those layer clouds which reached levels where the air temperature was as low as about $-15°C$ almost invariably had tops which were naturally glaciated or contained readily detectable concentrations of ice particles. An analysis of long series of routine aircraft soundings over Germany (93) indicates that these properties are more generally typical. Again, flight experience in layer clouds in the lower troposphere (e.g., over the northern United States east of the Rockies; 94) has shown that airframe icing, indicating the presence of considerable concentrations of supercooled cloud droplets, is encountered mainly at air temperatures between -5 and $-11°C$, and rarely at temperatures as low as $-20°C$ (and then probably only in convection clouds, which are sometimes embedded in layer clouds; in the strong updrafts of cumulus and cumulonimbus icing occurs at temperatures as low as $-40°C$). It thus appears that in thick layer clouds whose tops are at temperatures as low as about $-15°C$ the development of the ice phase normally leads to the production of precipitation, and moreover to the disappearance of droplets from the upper parts of the clouds because the ice particles become sufficiently numerous to lower the vapor pressure below that representing saturation with respect to liquid water.

It can readily be shown that in a layer of high supersaturation with respect to ice, extending 1–2 km above the $0°C$ level, the development of the ice phase by condensation alone can lead to precipitation at the equivalent rainfall rate of about 1 mm hr^{-1} typically associated with large-scale upward motion. For example, in Fig. 9.32 it is apparent that a corresponding rate of condensation is provided by an upward speed of magnitude 10 cm sec^{-1}; moreover, from Table 5.12 it appears that crystals settling through such a layer may attain millimetric sizes, masses of magnitude 10^{-4} g, and fall speeds of nearly 1 m sec^{-1}. A concentration of about 10^{-3} cm^{-3} provides the required precipitation rate. (The masses of individual crystals, however, are considerably less than those of small raindrops, suggesting that if rain is eventually produced in this way there must be aggregation into snowflakes or accretion of droplets.)

Further, it has been shown that with snowflake size-distribution spectra of the kind observed at the ground, steady snowfall rates of magnitude 1 mm hr^{-1} imply that a state of saturation with respect to liquid water cannot be maintained at temperatures more than several degrees below $0°C$ (126). [The size distributions of snow particles, obtained as averages from several falls and based almost entirely on observations of crystal aggregates of melted diameters larger than 100–200 μ, are reasonably well represented by the exponential relation (8.3)

$$n(D) = N_0 e^{-\Lambda D}$$

previously given for rain, where D is the diameter of the spherical drop of the same mass (96). However, both N_0 and Λ are functions of the precipitation rate R, again expressed in mm hr^{-1} of liquid water:

$$N_0 = 38 \times 10^{-3} R^{-0.87} \text{ cm}^{-4}$$

$$\Lambda = 25.5 R^{-0.48} \text{ cm}^{-1}$$

These expressions probably lead to underestimates of the concentrations of the small single crystals in snow, but for a typical large-scale precipitation rate of 1 mm hr^{-1} they indicate a total particle concentration N_0/Λ of about 10^{-3} cm^{-3}, and imply an even greater concentration of crystal embryos, as estimated above.] This result seems consistent with observation, and it emphasizes the problem of the origin of the crystal embryos in those layer clouds from which there are prolonged falls of snow, but which do not extend upward to levels where the temperature is at least as low as $-15°C$. In these clouds some process of crystal multiplication seems necessary, effective at relative humidities less than those representing saturation with respect to liquid water, since ordinary freezing nuclei are unlikely to provide an adequate source of crystal embryos.

During the production of snow in clouds which extend upward into the high troposphere there are probably always adequate concentrations of slowly settling embryos, and the relative humidity can be anticipated to become adjusted to some value dependent upon the local large-scale vertical velocity and size distribution of ice particles, usually increasing downward until there is saturation with respect to liquid water near the $0°C$ level. According to the model of Fig. 9.28, in the upper clouds near the warm front (even well ahead of the cyclone center) these particles may have originated in the freezing of droplets in cumulonimbus associated with the cold front, rather than in clouds formed in the high troposphere by the large-scale upward motion. This large-scale motion may play a significant part in delaying the descent and evaporation of the ice particles, and allowing them to be carried large distances forward of the cold front (a descent of several kilometers at a speed of a fraction of 1 m sec^{-1} results in a travel of about 1000 km during several hours in a wind of mean speed 50 m sec^{-1}). The cloud system of the warm front then appears as a large-scale anvil cloud, and the stunted forms of the ice particles found near its fringes should be interpreted as indicating the partial evaporation of particles which may earlier have possessed more sharply defined crystal faces and complex forms, rather than in the usual way as indicating very slow growth (97). Consistently, the mass concentration of condensed water and rate of precipitation in the extensive ice clouds associated with warm fronts are usually found to be very small, for example, less than 10^{-1} or even 10^{-2} g m^{-3} and mm hr^{-1}, respectively (e.g., 98).

Long-range meteorological research aircraft, equipped to observe the concentration and form of cloud particles and precipitation, have been in service for about two decades; however, inability to measure the property of basic importance, large-scale vertical air velocity, may have been responsible for the little use made of such aircraft to study the structure of large-scale precipitating cloud systems. Recently the introduction of techniques for the accurate measurement of horizontal winds, from whose distribution vertical velocity may be estimated, and the concerted exploitation of Doppler radar and other ground-based observing devices have stimulated fresh systematic studies (99). An example of an analysis of conventional data combined with aircraft soundings and reconnaissance is seen in Fig. 9.33, where a vertical cross section through the cloud systems is associated with a cyclone which had previously intensified over the eastern coastal regions of the United States, and at the time of the analysis was over New England with a central sea-level pressure of 985 mb (100). Such analyses have shown a variety of structures difficult to assemble into any general model, probably mainly because it has not been possible to recognize their relation to the part of the cyclone explored, the stage in its evolution, and the effects of the local large-scale geography. However, the analysis partly represented in Fig. 9.33 and Table 9.2 contains the following features which are probably typical of those precipitating clouds which near warm fronts occupy most of the depth of the troposphere:

First, the extensive ice clouds in the high troposphere are tenuous; there is very little condensation upon their particles even in the central parts of the precipitation

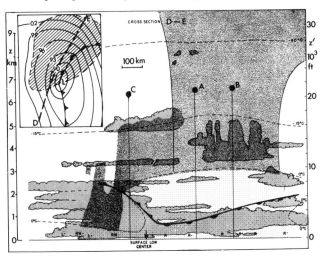

Fig. 9.33 A vertical cross section through the cloud systems of a mature cyclone near the coast of New England (adapted from reference 100). The pattern of sea-level isobars (labeled with the last two figures of the pressure in mb) and surface fronts, the extent of the area of precipitation at the surface (hatched) and the horizontal extent of the cross section (the heavy dashed line DE, nearly 1500 km long) are shown in the small inset in the upper left. In the main part of the diagram isotherms labeled 0, -15, and $-40°C$ are drawn as dashed lines. Droplet clouds are shaded with coarse stippling; fine irregular stippling represents ice crystal clouds or snow (in the darkest areas cloud droplets and snow coexist), and open stippling represents rain. Scales of height z (in km) and z' (in thousands of ft) are marked at the sides. Calculations were made of the growth of precipitation particles during descent along the paths marked A, B, and C, constrained to be made consistent with observations of particle size, form, and concentration. Those made for the path A are summarized in Table 9.2.

system, because the large-scale upward air velocity is small and the temperature is low (the turning into the vertical of the isopleths of Fig. 9.32 as they approach the top of the diagram has the same implication). In this study it was particularly remarked that the whole of the system of ice clouds in the upper troposphere, which had a sharply defined rear edge with the appearance of cumulonimbus anvils in places well to the southeast of the flight path, could have originated in cumulonimbus accompanied by thunderstorms, which had occurred near the cold front and the cyclone center during the period of intensification of the cyclone (until several hours before the flight reconnaissance).

Second, the precipitating cloud system contained regions of cumuliform droplet clouds in the middle troposphere, beneath which the rain at the surface became moderate to heavy and contained comparatively large drops, of diameter up to 2.5 mm. (The presence of regions of more intense precipitation near warm fronts will be discussed later.) Elsewhere the rain at the surface was slight and more uniform, with the largest drops less than 1 mm in diameter; their history is shown

9 Large-scale slope convection

Table 9.2 The conditions observed near the path marked A in Fig. 9.33, in the precipitation system of a cyclone, and the observed form and estimated growth of a typical precipitation particle descending along this path (adapted from reference 100)

z	T	$\theta_s(°C)$	U	t	D	$10^6 M$	N_{rel}
6.1	−21	15	84	0	0.2	1	11

In uniform cirrostratus; sun dimly visible with faint halo. About equal concentrations of platelike and columnar crystals of average large dimension 0.2 mm. Only slight supersaturation with respect to ice.

5.0	−13	15	88	50	0.3	2	10
4.3	−9	15	100	90	0.6	9	7

Thin droplet clouds; slight airframe icing and riming of the ice crystals, developing into spatial dendritic forms.

2.7	0	14	100	130	3.2	105	5

Thin melting zone, appearing as the base of altostratus. Shallow convective droplet clouds just below.

2.3	2	14	100	—	0.6	120	1

Slight evaporation of raindrops above lowest cloud layer.

1.5	5	12	100	—	0.6	100	1
0.3	3	4	100	145	0.8	300	1

Dense low layer clouds, containing up to 0.8 g m^{-3} of condensed water, between 0.3 and 1.5 km.

Key:
- z = height (km)
- T = temperature (°C)
- U = relative humidity with respect to liquid water (%)
- t = particle descent time (min)
- D = diameter (mm)
- M = mass (g)
- N_{rel} = relative concentration

by the estimates, partly based on direct observation, which are listed in Table 9.2. Most of their growth occurred in the layer of low cloud; the total time of descent from the upper troposphere to the surface was over 2 hr, and so their path could not have been vertical, as indicated schematically by the line marked A in Fig. 9.33.

Third, there are interesting optical and radar phenomena associated with a shallow layer in which the slowly settling ice and snow particles melt into comparatively rapidly falling raindrops. In the conditions illustrated in Table 9.2 it was inferred that there was no considerable aggregation of ice crystals into snowflakes, and that during the melting of each large crystal its fall speed increased by a factor of about 5. The concentration of the precipitation particles consequently decreased by a corresponding factor. The change was mainly responsible for the observation that the visual range under the melting zone was about an order of magnitude greater than within the ice cloud, which therefore appeared from beneath as a layer of altostratus with a base near the 0°C level.

It has been found by observation that in a horizontal path the optical attenuation coefficient τ (Section 2.8, Eq. 2.44) due to scattering from snow crystals only, which we may call τ_s, is given by

$$\tau_s = 2.5R$$

where τ_s is in km^{-1} and the precipitation intensity R is expressed in the familiar units of mm hr^{-1}, the rate of accumulation of liquid water (101). (The rate of accumulation of fresh unrimed snow is an order of magnitude greater, its mean density varying with crystal form but usually lying in the range 0.05–0.1 g cm^{-3}; 102.) This relation was derived for values of R between about 0.2 and 4 mm hr^{-1}. Then from Eq. 4.19 the visual range V_R in the horizontal is

$$V_R = 3.9/\tau_s$$
$$= 1.5/R \text{ km}$$

Accordingly, in snow of intensity 1 mm hr^{-1} falling through "clean" air the horizontal visual range is about 1.5 km, compared with over 10 km in rainfall of the same intensity (Table 8.17). Considering that snow crystals and snowflakes are likely to present a greater geometrical cross section normal to a path directed vertically, or obliquely upward, the change of visual range along a line of view entering snow aloft, and hence the appearance of the snow as a diffuse cloud, is likely to be more marked than this comparison suggests.

The definition of the base of snow clouds in the upper troposphere depends on the depth of the layer in which their particles evaporate. From Table 5.12 it can be anticipated that for particles of diameter up to about 1 mm this depth is a few hundred meters, since the evaporation commonly proceeds under vapor pressure differences somewhat greater than those assumed in the calculations represented. The evaporation chills air at the base of the snow clouds and increases the lapse rate in a shallow layer below, as illustrated in the sounding of Fig. 2.12b, often producing shallow convection manifest by mamma.

The apparent definition of the base of snow clouds which produce rain depends on the depth of the layer in which their particles melt. The melting begins when the particles enter air with a wet-bulb temperature above 0°C, and its subsequent progress depends in a complicated way upon the stratification of temperature and humidity in the layer beneath, and upon the changes in the form, fall speed, and ventilation coefficient of the melting particles. If the air is unsaturated, the dew point in the upper part of the layer may at first be below 0°C, so that there is some evaporation from the melting particles, but while the wet-bulb temperature of the air is above 0°C the air is chilled and the melting proceeds. At lower levels, where the dew point is above 0°C, the melting is assisted by condensation upon the particles. In the layer of melting there is a tendency to produce a saturated isothermal state with a temperature of 0°C, which extends downward as the precipitation continues. The depth of the layer of melting depends on the rate of heat transfer to the precipitation particles rather than the capacity of the air in the layer to supply the heat required to melt all the precipitation which passes through it; before they have melted completely the particles reach levels where the air temperature is considerably above 0°C. To estimate the depth of the layer of melting it is usually supposed that the air is already saturated with a given constant lapse rate, and the particles are assumed to remain spherical with constant mass and with reasonable values of fall speed and ventilation coefficient during their progressive melting, so that the heat transfer to the particles can be calculated from Eq. 5.48 and 5.49. In this way it is found that with ordinary lapse rates and melted particle diameters of 1–3 mm the depth of the melting layer is between a few and several hundred meters (e.g., 103). The result is consistent with the previously mentioned flight observation that a layer of snow crystals which melted into small raindrops appeared as an altostratus cloud with a base at about the 0°C level. [Such an apparent cloud base at the 0°C level is a common occurrence; when the lowest layer of droplet clouds usually present in a rain area does not form a complete overcast a second cloud layer, of altostratus or nimbostratus (coded $C_M = 2$), with a base at a height of a few kilometers is usually reported from the surface. The phenomenon is occasionally even more striking in an anvil cloud beside a weak or decaying shower, especially when it is illuminated by sunshine (104); sometimes a rainbow reaching up to the bright part of the anvil is the only clear evidence of the shower beneath it.] Similar calculations (105) show that the layer of melting can be anticipated to lower at a rate of magnitude 100 m hr^{-1} through still air when the precipitation intensity is 1 mm hr^{-1}; such descent is occasionally observed but is more often obscured by a large-scale upward motion or advection of warmer air. Over land prolonged intense precipitation through air of small average lapse rate occasionally causes the level of the 0°C wet-bulb temperature to descend by several hundred meters and the 0°C level to descend as much as 2 km and to reach the surface, so that a period of rain may be accompanied by a decrease in the screen-level temperature of several degrees and be succeeded by a fall of snow (106).

Finally, a striking phenomenon associated with the layer of melting is the locally enhanced radar echo from the precipitation, leading to the well-known narrow horizontal "bright band" observed on RHI displays (107). There are measurements of the distribution with height of both the radar reflectivity factor Z_e and (with Doppler radar) of characteristics of the distribution of fall speeds of the precipitation particles, but very few observations (with the required accuracy and resolution) of temperature and humidity, which would allow the confident relation of the changes in the physical state of the particles to the changes in their radar reflectivity. However, when snowflakes begin to melt and become covered with a film or beads of liquid water, without much change in occupied volume or fall speed, it can be anticipated that Z_e *increases* by a factor of about 5 (108, p. 338), and that this change is likely to occur within a layer extending little more than 100 m below the level where the wet-bulb temperature first reaches 0°C. After complete melting during a further fall of a few hundred meters, the decrease of concentration consequent upon the increase of fall speed implies a *decrease* of Z_e by about the same or a somewhat smaller factor (see Section 2.6). However, according to observation (97, p. 465) the maximum reflectivity factor Z_e in the bright band is generally about a factor of 10 greater than that (almost constant with height) of the rain in the layer extending 1 km below the bright band, and still greater than that of the snow a few hundred meters above the 0°C level. Thus it is usually inferred that there is considerable aggregation of snow crystals and flakes in the latter zone, before melting begins, and in the upper part of the bright band, and some fragmentation of the melted aggregates in the lower part of the bright band (108, p. 410). Estimates of the magnitude of these effects, however, are subject to large errors because of uncertainty about the scattering cross sections of aspherical and partly melted snowflakes.

Cloud and precipitation in numerical prediction models

It was mentioned in Section 9.3 that baroclinic numerical prediction models successfully reproduce many of the observed features of cyclone evolution. When the formulation of the models includes representations of the condensation and precipitation of water, and when their horizontal resolution is adequate (a separation between grid points of only a few hundred kilometers), the calculated evolutions include the generation of families of cyclones, with narrow bands of rain and associated frontal zones whose structures are very similar to those observed. Striking examples are illustrated in reference 109.

An interesting feature of these models is the necessity to include criteria for the onset of ordinary convection, and methods of representing its short-period effects. These methods have to be adapted to the particular kind of modeling chosen, and the resulting effects may not reproduce closely those of the natural process (nor is it possible in most published discussions to recognize in what respects or degree they differ). The necessity arises because in the calculated large-scale flows, especially if they are subject to moistening and warming from below,

9 Large-scale slope convection

regions develop in which the lapse rate exceeds that in the adopted reference process for adiabatic ascent. Since generally the stratification in the atmosphere is stable and the circulations are in the thermodynamically direct sense, involving the ascent of the potentially warmer air and the descent of the potentially cooler air, the stability tends to reduce horizontal temperature differences and to brake the development of the large-scale circulations. Further, small-scale convection rapidly produces a stratification which is neutral or stable for large-scale vertical motion, and subsequently efficiently controls the *depth* of the affected layer. Unless some such limitation is imposed upon a model atmosphere, usually represented on only a few levels, parts of the large-scale motion systems in which the stratification becomes unstable may intensify at unrealistically large rates. (It is convenient to make the adjustments gradually, to a degree depending on the closeness of approach to a state of saturation once the calculated relative humidity has exceeded some critical value less than 100%; in this way, for example, small-scale convection may be introduced when the lapse rate is about halfway between the dry and wet adiabatics, before large-scale saturation develops, and more in accord with the lapse rate previously described as characteristic of cumulus convection; 110.) Moreover, since in most models the most rapid amplification in statically unstable regions is associated with the shortest horizontal wavelengths, if such regions are allowed to occur in the initially specified or calculated states of large-scale flows, then motion systems whose horizontal scale is comparable with the separation of the grid points are liable to appear and eventually to dominate the field of flow (110), since the period required to double their amplitude is typically only a few hours. Such unrealistic systems can be attributed to defects in the physical formulation of the model or to mathematical difficulties introduced by the finite-difference methods of computation. They are suppressed by introducing a convenient "convective adjustment" to prevent the generation of regions of unstable stratification.

The generation of unstable stratifications in the unsaturated free atmosphere by differential advection. In an atmosphere in sheared but adiabatic flow with a lapse rate $-\partial T/\partial z = \gamma$, the rate of change of temperature with height following the motion involves $\Gamma = \Gamma_d$ or Γ_s, the appropriate (dry or saturated) adiabatic lapse rate:

$$\frac{DT}{Dt} = -w\Gamma$$

Differentiating with respect to z, we can obtain

$$\frac{D\gamma}{Dt} = \frac{\partial u}{\partial z}\frac{\partial T}{\partial x} + \frac{\partial v}{\partial z}\frac{\partial T}{\partial y} + (\Gamma - \gamma)\frac{\partial w}{\partial z}$$

Since $\Gamma_d = $ const while to a good approximation $D\Gamma_s/Dt = w\partial\Gamma_s/\partial z$, this relation may be written

$$\frac{D}{Dt}(\gamma - \Gamma) = \frac{\partial u}{\partial z}\frac{\partial T}{\partial x} + \frac{\partial v}{\partial z}\frac{\partial T}{\partial y} - (\gamma - \Gamma)\frac{\partial w}{\partial z}$$

The last term on the right has the obvious significance that if the vertical velocity increases upward, that is, if there is "stretching" in the vertical, then the lapse rate tends to increase. In unsaturated air it may eventually exceed the saturated adiabatic lapse rate, but it cannot ever attain the value of the dry adiabatic lapse rate.

The second pair of terms on the right of the equation are said to represent the effect of differential advection in a sheared flow containing horizontal temperature gradients; their significance is less obvious. However, it is evident that these terms are small unless the shear in the flow is large, and since if the motion is then also horizontal and geostrophic

$$\frac{\partial u}{\partial z} \approx -\frac{g}{fT}\frac{\partial T}{\partial y} \quad \text{and} \quad \frac{\partial v}{\partial z} \approx \frac{g}{fT}\frac{\partial T}{\partial x}$$

the sum of the terms remains small. [In geostrophic and adiabatic motion, strictly it is not γ but $\partial\theta/\partial p$ which is conserved (112). This is almost obvious from a consideration of the displacement of the isopleths of θ, or of T, at the base and top of an isobaric slab of infinitesimal thickness: the displacement of those at the top is that of those at the base, with an added component which in geostrophic flow is parallel to the isotherms—along the thermal wind—and therefore leaves $\partial\theta/\partial p$ or $\partial T/\partial p$ unchanged.]

It can be shown theoretically, and by analyses of real flows, that significantly rapid increases of the lapse rate of unsaturated air (leading within periods of less than a day to values approaching the dry adiabatic lapse rate) are liable to occur in the upper troposphere on the warm flanks of frontal zones and of jet streams (e.g., 113). Their production is associated with thermally indirect circulations (tending to displace potentially cool air over potentially warm air) in regions where there is a considerable horizontal temperature gradient in the same plane. (In air passing through such regions the rate of change of the lapse rate in unsaturated air may be equivalent to conversion from an isothermal state to a dry adiabatic lapse rate in a day.) This conclusion seems to be supported by observations of lapse rates approaching the dry adiabatic value in layers about 1 km deep on the warm side of jet streams associated with frontal zones. The layers have tops at about the level of the strongest winds, and extend over regions a few hundred kilometers wide from the positions of the axis of jet streams toward their warm sides. Such layers are evident on cross sections showing mean conditions with respect to the axis of a jet stream (in which the mean lapse rate over a layer about 1.5 km deep may reach $8°C\,km^{-1}$; 114, 115) and appear in analyses of individual occasions, in which the lapse rate is frequently found to reach 8°C km^{-1}, and occasionally to exceed 9.0°C km^{-1}, at levels in the upper troposphere where the saturated adiabatic lapse rate is about 9°C km^{-1} (e.g., 116, 117). The production of such large lapse rates by differential advection in unsaturated air may play a part in the generation and maintenance of some forms of cirrus clouds, as discussed later.

The generation of unstable stratifications in the free atmosphere following condensation. If θ_w in a layer of air decreases upward, and the whole layer is saturated by large-scale adiabatic ascent, the stratification becomes unstable even if originally the lapse rate was less than the saturated adiabatic. Since θ_w normally decreases, upward large-scale ascent, particularly if it is accompanied by vertical stretching (as is usual in the lower troposphere), is liable to lead to ordinary convection above any level at which it produces condensation. Earlier in this chapter and in Chapter 8 deep ordinary convection was seen to be commonly produced in this way near cold fronts, leading to the development of intense cumulonimbus in geographically favorable regions; generally differential advection can be said to play a part by leading dry air over moist air. Ordinary convection on a smaller scale thus becomes a typical feature of large-scale cloud systems during the early stages of their development (as indicated schematically in Fig. 9.11 and 9.16). Consequently the precipitation at the surface from these systems is often not steady on their own large scales, and when the clouds are observed from the air (Fig. 9.33) or their radar echoes are examined the presence of the ordinary convection is easily recognized (Fig. 9.31b; 118, 119, e.g.).

9.6 Small-scale clouds and cloud systems in large-scale flows

Small-scale features in the precipitation systems of slope convection

From studies made with radar and the records of autographic rain gauges (118, 120) it is evident that only in a broad leading belt of the precipitation system associated with a cyclone is the precipitation fairly uniformly distributed over large horizontal areas. Nearer the warm front the precipitation usually develops in the middle troposphere as snow, mostly in cells which apparently are associated with ordinary convection, having dimensions in the horizontal and vertical which are typically about 1 km (121; Pl. 9.7). Each cell produces a long trail of snow as its particles descend slowly through the pronounced wind shear; the form of the trail becomes distorted and its width increases as it lowers, since the particles follow different paths according to their size and fall speed. Eventually the snow, or the rain into which it melts, reaches the ground in swaths which may be a few tens of kilometers wide and several tens of kilometers long, and which may overlap, so that precipitation at places on the ground may be continuous although variable in intensity. The horizontal velocity of a trail has been shown to be practically the same at all heights, and within about 2 m sec^{-1} of that of the wind at a level near the base of the cells, above which there is little horizontal displacement or increase of horizontal area of the radar echoes (121). This level has consequently become known as the *generating level*, since this observation is consistent with the view that the particles are grown in a cloud within which the speed of the relative motion (including that of the updraft) is small compared with the wind speed (only about 1 m sec^{-1}), and that the particles pour out of the cloud.

Sometimes the cells are not organized into any obvious pattern, but usually they are grouped into parallel rows, or lines of clusters (with one long axis), typically at least 100 km long. At the generating level the distance between neighboring cells is usually a few kilometers, and that between neighboring rows or clusters is more variable, even on the same occasion, and much greater: between about 10 and 50 km. These distances can be anticipated to be proportional to the depth of the cells, that is, to the depth of the layer of convection, but it seems that a relation has not been sought. Over an area containing a number of rows or irregularly scattered cells the fraction occupied by the cells at the generating level is usually less than one tenth, so that typical rates of precipitation at the ground are consistent with upward velocities of about 1 m sec^{-1} in the generating clouds.

Frequently there is no evidence of appreciable growth of the precipitation particles below the cells, and from a comparison of the form of the trails with the observed winds, or from direct measurements with Doppler radars, it can be inferred that the fall speeds of the particles are usually about 1 m sec^{-1}, and occasionally reach as much as 2 m sec^{-1} (121–25). Such fall speeds are appreciably greater than those of single snow crystals (see Section 2.6), and sometimes they have been interpreted as implying aggregation into snowflakes. However, the air temperature is often too low (below −10°C) for this kind of aggregation to be plausible, and it is more probable that the larger of the fall speeds are due to the riming of crystals in the parent cumuliform clouds, since if the magnitude of the updraft speeds is 1 m sec^{-1} the clouds are likely to contain supercooled droplets. The more vigorous of the clouds may even produce some soft hail as well as snow.

A striking property of the cells is their persistence: frequently individual cells can be followed by radar for 1–2 hr with no sign of decay, and consistently the vertical extent (often a few kilometers) of the associated trails shows that the snow is generated continuously over periods at least as long. Thus the cells appear as a weak form of the persistent nearly steady cumulonimbus rather than as a kind of ordinary cumulus convection, and the nature of the responsible circulations may be suspected to depend in some way on the presence of the precipitation and associated downdrafts near the clouds, considering also that there is usually (121) a shear of the wind in the vertical which may play an organizing role. Even the weak shears observed (distinctly away from the warm front zones) imply a velocity difference of a few meters per second over a depth of about 1 km through the generating level, which exceeds the typical updraft speed.

Although the cells are often presumed to arise in unstable stratifications produced by the arrival of unsaturated and potentially cool air over saturated layers (i.e., by differential advection combined with large-scale ascent; 118–20, 123, 124, 129), in Canadian studies it was found (according to analyses based on routine radiosonde ascents made about 200 km away from the observing radar) that the lapse rate γ at the generating level was usually between 5 and 8°C km^{-1} and nearly always less than the saturated pseudo-adiabatic Γ_s with respect to liquid water (122). On only 3 of 14 occasions was $\gamma > \Gamma_s$; the ratio $(\Gamma_s - \gamma)/\Gamma_s$ varied between 0.5 and -0.1, about an average value of 0.15. This result, if representative, does not necessarily imply that the updrafts in the cells cannot be buoyant, for it is possible that there is a release of latent heat additional to that considered in the reference process with respect to liquid water, if the vapor in the air entering an updraft is initially supersaturated with respect to ice and if during the growth of ice particles it is brought close to a state of saturation with respect to ice (this possibility will be considered again in the subsequent discussion of cirrus clouds). It can be shown that in this way a buoyant ascent of several hundred meters can occur when the general lapse rate is between 6 and 8°C km^{-1}, provided the initial temperature is about -15°C and the initial vapor pressure is near that representing saturation with respect to liquid water (122). Nevertheless, the probability that the buoyancy is small supports the view that the kinetic energy of the updrafts may be derived at least partly from that of the mean horizontal flow.

The initial orientation of the trail of precipitation from a cell can be expected to be in the direction of the wind shear near the cell base, and the cell probably propagates in this direction at a small speed (about 1 m sec^{-1}) not readily detectable. A relation might also be anticipated between the orientation of cell rows and the direction of this shear, but none was apparent in a study of their occurrence on five occasions inland over Canada, and the angle between the orientation and the direction of motion of the individual cells varied over the large range between 15 and 70° (121). However, in a few recent studies made in the United States (129) and in England it has been found that rows of cells or of cell clusters are usually aligned approximately parallel to the surface position of the warm front, and therefore also approximately parallel to the thermal wind and the wind shear in a deep layer, as shown schematically in Fig. 9.34 (120). From these studies it also appears that the cell clusters may be organized on an even larger scale, into bands several hundred kilometers long and between about 50 and 100 km broad and apart. Such organization may be related to some unidentified dynamical process, or to inhomogeneity in the distribution of water vapor within the moist airstream entering a cyclone at low levels, which as already remarked is a general cause of streakiness in the cloud patterns of flows which have been modified by convection over land. The bands form near the surface position of the warm front and travel approximately with the relative flow in the middle troposphere, so that they move away from the front; eventually the small-scale motion in the constituent cells weakens and far ahead of the front at the surface only a slight modulation of the large-scale speed of ascent may be detectable, which usually has an insignificant effect on the precipitation intensity.

Figure 9.34 includes bands of clustered cells also in

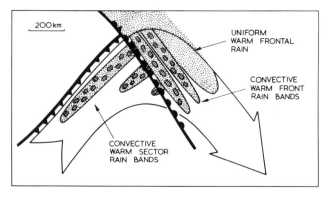

Fig. 9.34 A schematic representation of the principal precipitation-producing relative flow in a partly occluded cyclone (heavy arrow), and of the pattern of the precipitation (120). Areas of moderate and of more intense precipitation are indicated by stippling and hatching, respectively, the intense precipitation being associated with elongated clusters of precipitation cells.

the warm sector, aligned approximately parallel to the cold front, and therefore at a large angle to the bands above the warm front. Such bands may be traceable to distances a few hundred kilometers ahead of the surface position of the warm front and may overlap the bands above the front (119). It is probable that the warm sector bands usually represent convection from near the surface and, in contrast to the bands above the warm front, over land their position and the intensity of the associated precipitation may be strongly affected by topography. They may appear where precipitating clouds form continually over a topographical feature and travel away for a long distance downwind, producing bands of precipitation which are stationary for some hours, or which move only slowly. Along the tracts of country covered by such bands the precipitation which falls during the passage of a cyclone may greatly exceed that on neighboring areas; in mild weather the mean rainfall rates in the bands may reach several millimeters per hour and the total falls may exceed 50 mm. Such nearly stationary rainbands have been observed not only over Great Britain but also over Japan (126, 127). A few such rainbands parallel to a cold front and in the zone extending up to 500 km ahead of it are said to be common in wave cyclones over the ocean near Japan (128); the rainfall beneath them is sometimes more intense than that beneath the band directly associated with the cold front and the arrival of the cool airstream.

Although the orientation of bands and rows of cells over the ocean can be expected to be in the direction of the wind relative to the system at the level of the cells, over land the formation of cells, even in the middle troposphere, over particular orographic features could lead to the rows and bands lying along *streak lines*. The pattern of these lines, and their local orientations, could in a moving and developing cyclone differ considerably from those of other characteristic flow lines: the streamlines of the actual and of the relative flow, and trajectories.

Cirrus clouds

The texture of ice clouds

Separated ice clouds, with individual horizontal dimensions less than about 10 km, as distinct from the extensive ice clouds associated with large-scale ascent or prolonged cumulonimbus convection, are produced in the upper troposphere by small-scale motion systems of comparable horizontal dimensions. The associated vertical displacements are practically always less than 1 km, and typically are only a few hundred meters; if they are much less, then they are not likely to lead to condensation and cloud formation. Since ice nuclei generally are effective only when a state of saturation with respect to liquid water is closely approached, and because at the low temperatures of the upper troposphere this involves an ascent of at least a few hundred meters beyond the level of saturation with respect to ice (Appendix 2.2), the formation of the ice clouds commonly occurs when in the undisturbed air there is a widespread state of near saturation, or even of slight supersaturation, with respect to ice. (The large-scale convection tends to produce this state over large areas in the high troposphere.) Thus the evaporation of ice clouds is characteristically much slower than that of droplet clouds, and the regions of descent and turbulence in the motion systems are not so clearly indicated by sharply defined edges and details in the clouds. (Occasionally the prevalence of supersaturation with respect to ice is strikingly shown by the persistence of aircraft condensation trails, which may even thicken as they spread horizontally in otherwise clear air, mainly as they are distorted by wind shear but also as they are diffused by small-scale motions.)

A droplet cloud marks the presence or very recent existence of a cloud-forming motion system, because unless the cloud is renewed it usually evaporates completely within minutes; exceptions are the shelf clouds of cumulus convection, which may persist in a shallow, nearly saturated layer hours after the parent cumulus have disappeared. In contrast, ice clouds may persist for comparatively long periods after their formation, especially if they were produced by a large upward displacement and consequently the concentration of condensed water is large compared with the vapor density representing the subsaturation with respect to ice in their surroundings. The extreme examples are the dense anvil clouds which may survive their parent cumulonimbus by a day or more, and in strong winds of the upper troposphere may drift a few thousand kilometers from their place of formation (over such long periods radiative exchanges may play a significant part in their evaporation). Probably most of the ice clouds of very low latitudes are the products of cumulonimbus.

The distortion and diffusion accompanying the long duration of ice clouds tends to obscure the kind of air motion which was responsible for their formation. Only in freshly formed clouds is there ever a regular pattern which provides a definite indication. Even so, there has been little study of these patterns, and there is no general agreement on their interpretation or explanation of the variable texture of the clouds.

Three principal kinds of texture can be recognized:

1. Virtually no detail, so that the cloud is amorphous with very diffuse edges difficult or impossible to define. Such cloud may be so tenuous that it is not perceptible from the ground, although its presence may be suspected from measurements of variations in long-wave radiative fluxes; it may be seen in thin layers when looking toward the horizon from nearly their own levels, but it cannot then readily be distinguished from haze or dust.

2. Compact and not much different, although usually less dense, than the texture of droplet clouds. The ice cloud is usually shallow, and details, particularly dapples or waves, surrounded by clear air, may at first be well defined, although their edges usually soon become irregular, diffuse, and finely fibrous (Pl. 9.8).

3. Coarsely fibrous. The fibers may be gathered into long bundles which are often dense and obviously inclined out of the horizontal, tapering inward and downward from denser and broader parts near the tops (Pl. 9.8).

The first kind of texture can be ascribed to aging before complete evaporation. The second seems to imply that the particle size is not much different from that characteristic of droplet clouds. In the third the particles are evidently sufficiently large to be regarded as composing precipitation, and the bundles of fibers are called *fallstreaks*. Similar fallstreaks often appear beneath supercooled droplet clouds (Pl. 8.25), but in the high troposphere they usually have no obvious parent cloud. (As the temperature of shallow droplet clouds increases toward 0°C the decreasing concentrations of active ice nuclei, and the decreasing depth of the layer below the clouds which may be supersaturated with respect to ice, lead to any precipitation of crystals becoming increasingly slight in intensity, shallow in depth, and tenuous in appearance. Often when the temperature of the droplet clouds enters the range between about -20 and $-10°C$ the crystal trails can be detected only when seen against clear sky near the sun, where the illumination is particularly favorable and the droplet clouds are strikingly more brilliant, mainly because of their efficient forward scattering. Sometimes the presence of crystals is revealed or confirmed by optical phenomena, especially the sun pillar.)

Occasionally the ice nuclei appear to become activated in only a small part of an extensive droplet cloud, or they may be introduced artificially—deliberately or during the passage of aircraft (130)—and the crystal growth leads to the evaporation of neighboring droplets and the production of a hole in the cloud above the fallstreaks.

Small-scale convection in cloud layers, which may be stimulated by the release of additional latent heat during crystal growth, often leads to the diffusion of crystals, and perhaps products of their fragmentation, into regions much broader than those in which they first appeared. Holes may therefore enlarge to as much as several kilometers across before the crystal concentrations decrease to some small value at which they are no longer significant.

The shallowness or absence of any parent cloud shows that the ice particles of typical fallstreaks attain their large size by condensation, and therefore that the crystal concentrations must be much smaller than those of cloud droplets. In the high troposphere, at temperatures of about $-40°C$ or below, it appears from the table in Appendix 2.2 that a likely upper limit to the concentration of water vapor condensed in a shallow cloud is about 50×10^{-9} g cm^{-3}, corresponding to a reduction from a state of saturation with respect to liquid water to one with respect to ice. In many small-scale motion systems the updraft speeds may reach or exceed 1 m sec^{-1}. If such speeds are maintained during a sufficiently large upward displacement saturation with respect to liquid water is likely to be reached or closely approached (Section 5.12), and nearly all the aerosol particles, whose concentration in the upper troposphere is about 10^2 cm^{-3}, can be anticipated to act as condensation nuclei (Section 5.3) and subsequently, at the low temperatures considered, to produce ice particles in about the same concentration. The average mass of the individual ice particle is then not likely to be much greater than 10^{-9} g, corresponding to an average dimension of little more than 10 μ and a fall speed of only a few centimeters per second (Table 5.10). Such ice particles are little larger than ordinary cloud droplets, and the texture of the newly formed cloud which they compose must be similar to that of droplet clouds. However, the ability of the particles to diffuse without evaporation into the descending parts of small-scale motions, which may arise during radiative destabilization or otherwise, ensures that edges and details soon become diffuse, as noted in 2 above and illustrated in Pl. 9.8.

In contrast, the particles of fallstreaks have sizes which are much larger and concentrations which are correspondingly much smaller. Those of the high cirrus, which almost always have no distinct parent cloud, are individual crystals or tufts of crystals sprouting from a common center, which from direct sampling have been found to consist of prisms up to about 0.5 mm long (Section 5.12; Pl. 5.1). They grow by condensation, and not any kind of aggregation. Their fall speeds can be estimated to reach about 0.5 m sec^{-1} (Table 5.10), and values of several tens of centimeters per second can be inferred from observations of the downward extension of the trails (131) or of their inclination in layers of known wind shear. Since the mass of the large particles may be more than 10^{-6} g their concentrations (Table 5.13) are likely to be as much as three orders of magnitude smaller than those typical of cirrus with a compact texture. One experienced observer, using for special reconnaissance a slow aircraft with an open cockpit (and a ceiling of over 9 km), reports the fallstreak particles to be large and well separated, giving "the impression of crystal rain" glittering in the sunshine (132, p. 49), probably implying a concentration of less than 10^{-2} cm^{-3}. (Concentrations of this or one order greater have been measured on a few occasions; 132.) The particles may grow substantially while descending several hundred meters through air supersaturated with respect to ice; their fall speeds may increase by an order of magnitude (Table 5.13). If their downward flux is maintained approximately steadily, their concentration is likely to decrease by the same factor, so that the estimate from the observation is not inconsistent with that first made. Both estimates imply that during the formation and maintenance of cirrus fallstreaks only a very small fraction of the aerosol particles act as crystal nuclei. The most probable explanation seems to be that ascent in the updraft of the small-scale motion system causing the cloud formation ceases a little before saturation with respect to liquid water is attained. A very small proportion of the aerosol particles may produce ice crystals by heterogeneous nucleation when saturation with respect to the liquid is closely approached, so that there is a large dilution of the water-soluble components of particles with solid components able to act as freezing nuclei (at low temperatures, and at the low relative humidities with respect to liquid water, which represent small supersaturations with respect to ice, the watery parts may be saturated solutions that depress the freezing temperature by as much as 10–20°C). If so, fallstreak formation is unlikely to occur at temperatures considerably below $-40°C$, since following such a dilution a large proportion of the aerosol particles could then be anticipated to produce crystals by homogeneous nucleation; moreover, the vapor density, and accordingly the concentration of vapor which may be condensed and the potential ice particle size, become appreciably smaller at such temperatures. The few available observations of the height and temperature at the tops of cirrus fallstreaks appear to support this conclusion. (Many and probably more reliable measurements can now readily be made from the ground with laser rangefinders.)

The form of cirrus fallstreaks

If ice particles of constant fall speed settle from a source region, moving with practically the speed of the wind at its level, and descend through a layer of constant shear of wind with height, then the trail produced has the shape of part of a parabola: its local inclination to the vertical increases with distance below the source, or with the time taken for a particle to make the descent. This is approximately the typical shape of the cirrus fallstreak, which nearly always lags behind its head, and is usually described as having the form of a comma. [It is cirrus *uncinus* in the International Classification, or, in ordinary language, the "hooked" cirrus or the "mare's tail." It is also the "curl cloud" which the daring spirit Ariel was ready to ride on his master Prospero's errands (133); evidently it was recognized as having a good shape for mounting, and had long been known for its distance of travel and its speed—implying an appreciation of its height, for its angular speed is not extraordinary.] With the evaporation of the particles near the base of a trail it becomes more nearly horizontal, while near the top there is sometimes a tendency for the trail to curl forward, which may be associated with a rapid increase in the fall speeds of growing particles or with a local increase in the magnitude of the wind shear in the vertical.

The true inclination of a fallstreak and its orientation in the horizontal are often difficult to determine visually because of the effects of perspective. In particular the orientation, which usefully shows the direction of the wind shear, can often be determined reliably only when it or the fallstreak's extrapolated length passes through the zenith.

The density and size of cirrus fallstreaks are highly variable. Their diameter generally is only a few hundred meters and their depth less than 1 km, but occasionally they may be comparatively dense and deep, and a number may merge below into broad streaks which extend horizontally over tens of kilometers. Usually the nearly horizontal parts, the "tails," are considerably longer than the nearly upright parts, the "heads." Frequently their inclinations and orientations change abruptly a few hundred meters below their tops, within a layer perhaps less than 100 deep (Pl. 9.9). Sometimes deep fallstreaks have more than one such twist, and a zigzag shape. When there are numbers of fallstreaks in the sky they usually have similar twists at about the same levels, demonstrating that the responsible wind shears are characteristic of widespread, nearly horizontal layers.

The persistence of cirrus fallstreaks

The initial formation and the development of cirrus fallstreaks are seldom observed, except perhaps near hills (Pl. 9.10). Usually when fallstreaks arrive in view they seem to be in slow decay, becoming more diffuse and more nearly horizontal. It appears that while their development may occur within about 20 min, their duration and decay occupy a considerably longer period, during which they may travel a few hundred kilometers from the place in which they formed. The development of a trail several hundred meters long demands that the supply of the crystals, and therefore presumably an updraft, should be maintained in a traveling source region, not more than about 1 km across, for a period of at least 20 min and often substantially more. A kind of motion system with appropriate dimensions and duration has not been specified, and a process of regeneration which is dependent on the presence of the trail itself after it has been formed in some more local or evanescent motion system seems likely. The existence of such a process was previously remarked as indicated by the extraordinary persistence of the convection associated with the precipitating cells of the middle troposphere, which are observed within the large-scale cloud systems of slope convection. Similar precipitating clouds are recognized visually as a form of castellanus, and have been illustrated in Pl. 8.25 and 8.26. The cumuliform heads of such clouds can often be seen to be renewed on the upshear (and usually upwind or rear) side of the cloud, above the trail of precipitation. Updrafts in the clear air beside inclined fallstreaks have been exploited by a sailplane pilot, who reported that above their lower ends "the air ascended in the neighbourhood of the fallstreaks, and reached upward speeds of up to 2 m sec^{-1} for a period almost up to an hour. A new cloud formed at the head of the updraught beside the old, sinking one, which after some time gave further ice precipitation which produced a third cumulus, and so on" (134, p. 264).

The form and dynamics of the kind of motion system which contains such updrafts and leads to the regeneration of precipitating cloud has not been investigated. In the middle troposphere the renewed parent cloud contains supercooled droplets and is always cumuliform, while the general lapse rate sometimes exceeds the saturated adiabatic with respect to liquid water, so that buoyant convection above the saturation level is plausible. As previously remarked, it may also occur over limited depths in stratifications which are apparently stable, if the development of the ice phase leads to the release of significant amounts of latent heat not considered in the reference process. Nevertheless, it was suggested that in the presence of the wind shear and a downdraft accompanying the precipitation, some kind of motion system may be established whose kinetic energy is drawn at least partly from that of the mean flow.

In the high troposphere persistent fallstreaks are often seen with no trace of a distinct mother cloud. On some occasions, however, fresh clouds can be observed to form intermittently above fallstreak trails; the trails beneath may become distorted in a way which suggests that the responsible updrafts extend down into the trails and may even originate within the trails (Pl. 9.10). It has therefore to be considered whether the fall of crystals into air which is supersaturated with respect to ice could by the release of latent heat produce a warming sufficient to lead to an ascent and the formation of fresh cloud.

The simplest conditions to consider are those in which the temperature of the air is raised by an amount ΔT during the isobaric reduction of the vapor mixing ratio by an amount Δr corresponding to the difference between the states of saturation with respect to liquid water at the temperature T and to ice at the temperature $(T + \Delta T)$. Then

$$c_p \Delta T = L_s \Delta r$$

where c_p is the specific heat of air at constant pressure

Table 9.3 The variation with temperature T of ΔT, the temperature rise during an isobaric condensation which reduces a state of saturation with respect to liquid water to one with respect to ice (at the temperature $T + \Delta T$). The air pressure is assumed to be that appropriate to $\theta_s = 18°C$, except at $-70°C$, where it is 160 mb ($\theta_s = 22°C$)

		Δz for $\gamma =$			
$T(°C)$	$\Delta T(°C)$	7	8	9	$\Delta z'$
-10	0.51	180	280	640	150
-20	0.65	230	360	810	270
-30	0.54	190	300	670	370
-40	0.33	120	210	410	440
-50	0.16	60	90	200	490
-60	0.06	20	30	60	510
-70	0.02	10	10	20	505

Key:
Δz = dry adiabatic ascent required to remove temperature excess ΔT in the atmosphere (m)
$\Delta z'$ = dry adiabatic ascent required to restore a state of saturation with respect to liquid water (m)
γ = lapse rate (°C)

($\approx 10^3$ J kg^{-1} °K^{-1}) and L_s is the latent heat of sublimation ($\approx 2.8 \times 10^6$ J kg^{-1}). If $\Delta r'$ is the difference between the saturation mixing ratios over liquid water and that r_s over ice at the same temperature T (extracted from tables), then Δr is less by an amount which according to the Clausius-Clapeyron relation (2.3) is

$$\Delta r' - \Delta r \approx \Delta T r_s L_s / R_v T^2$$

where $R_v = 461$ J kg^{-1} °K^{-1}. Thus

$$\Delta T = \frac{L_s \Delta r'}{c_p + r_s L_s^2 / R_v T^2}$$

The value of ΔT is only a fraction of 1°C, varying principally with the temperature, as shown by Table 9.3 for conditions typical of warm airstreams aloft. However, it appears possible that the implied warming of the air in the upper part of fallstreak trails could lead to volumes with diameters of a few hundred meters rising out of the trails, leaving the crystals behind, and acquiring ascent speeds of about 1 m sec^{-1} before reaching new equilibrium levels a few hundred meters above their original heights. (If the volumes behave like thermals, Eq. 7.19 implies that a temperature excess of about 0.33°C would be sufficient to produce a limiting speed of ascent of 1 m sec^{-1}.) The distance Δz to the new equilibrium level can be estimated by assuming the rise to occur adiabatically without condensation; it varies according to the general lapse rate γ as shown in the table, which includes the adiabatic ascent $\Delta z'$ required to restore a state of saturation with respect to liquid water (from Table 2.14, Appendix 2.2). The lapse rate γ typically observed in the regions occupied by cirrus is about 8°C km^{-1}. Considering that some penetration of the equilibrium level is likely, the calculated values of $\Delta z'$ seem to imply that saturation with respect to liquid water may readily be attained at temperatures of about −20°C and may be closely approached at temperatures down to near −40°C, but not at temperatures considerably lower.

These conclusions have been obtained under very simple assumptions. The maximum degree of warming, corresponding to a change from a state of saturation with respect to liquid water to one with respect to ice, has been assumed without considering whether the crystal concentration in the fallstreaks is sufficient to effect this change. Moreover, at least while the warmed air begins to rise out of the crystal trails the process is neither isobaric nor adiabatic. However, a more elaborate treatment hardly seems justified, while the conclusions seem generally to be consistent with observed features.

First, in the middle troposphere, at temperatures between about −10 and −30°C and especially when the general lapse rate is large, the envisaged process seems readily capable of producing cumuliform droplet clouds and of regenerating ice particle precipitation through the action of freezing nuclei sparse compared with the condensation nuclei. The process could be effective even if the general vapor density below the droplet clouds corresponded not to a state of saturation with respect to the liquid, but only to a high supersaturation with respect to ice.

Second, at lower temperatures, down to about −40°C, the attainment of saturation with respect to liquid water in the rising air becomes increasingly improbable. At these temperatures, characteristic of the lower cirrus levels, a high supersaturation with respect to ice over large areas may seem improbable. Nevertheless, the observed growth of cirrus fallstreaks and their extension downward over depths commonly exceeding 1 km is evidence that large supersaturations can occur: they must exist in at least the upper part of such layers. (Shallow layers at levels where cirrus occurred during the day have been observed during special nighttime hygrometer soundings to be nearly saturated with respect to liquid water; 135.) Moreover, traces of shallow cloud with a compact texture—*shred clouds* (Pl. 9.11)—sometimes form near the tops of fallstreaks, usually above their inclined parts. Their heights above the fallstreaks are difficult to define accurately, but they usually appear to be between about 250 and 450 m (134) and therefore are closely comparable with the appropriate values of $\Delta z'$ given in Table 9.3. Iridescence has been seen in shred clouds, sometimes at temperatures certainly above −40°C and occasionally at temperatures inferred to be below −40°C (see legend to Pl. 9.10). Such clouds are therefore probably at first composed mainly of droplets,

and the observations support the view that shred clouds in general form during the attainment of saturation with respect to liquid water.

When the general lapse rate at the cloud levels is unusually large the fallstreak bundles may be dense, and accompanied not by shred clouds, only, but more conspicuously by cumuliform growths which rise out of their upper parts, or which may have a fairly well-defined base clearly separate from their tops. Both the shred clouds and the cumuliform growths, if formed at saturation with respect to liquid water and at temperatures near or somewhat below −40°C, could be anticipated to be poor sources of fresh fallstreaks. This expectation is supported by prolonged observations of shred clouds, which usually become diffuse after their formation and sometimes evaporate within minutes. Also, detached cumuliform growths, beneath which the fallstreak trails frequently are distorted in a manner which suggests the rise of air out of them (Pl. 9.21), are usually sources of only tenuous fallstreaks.

Nevertheless, a marked cumuliform head, occasionally even with pileus formation, and a well-developed fallstreak trail are both parts of the typical form of the persistent dense cirrus cloud, sketched in Fig. 9.35. (It closely resembles the snow-producing cell or the precipitating castellanus cloud of the middle troposphere, but it does not have an upper part of distinctive texture which obviously is composed mainly of supercooled droplets.) However, as indicated in the sketch, the fallstreak regeneration may occur in air on the fringes of the updraft which produces the cumuliform top, where the upward displacement of air is less and does not quite lead to the attainment of saturation with respect to liquid water.

The flow indicated in the sketch assumes the presence of a wind shear, so that the fallstreak is inclined. Because the mass concentration of the precipitation in the fallstreak is probably too small to interfere seriously with generation of buoyancy for ascent, as may happen when precipitation develops in cumulonimbus, the shear may not play an essential part in the organization of a persistent circulation and a long-lasting cloud. On the other hand, in the kind of flow sketched it is evidently possible in a moderate shear for a large part of the kinetic energy of the updraught to be derived from the mean kinetic energy of the airstream.

Third, Table 9.3 implies that in ordinary fallstreak cirrus, which have no cumuliform head, the regeneration of the trails may proceed at temperatures near −40°C by a circulation similar to that drawn in Fig. 9.36, but without the attainment anywhere of saturation with respect to the liquid, and therefore in a hardly detectable manner, without the formation of any secondary shred or cumuliform cloud. If the temperature is much below −40°C, or if there is no general large supersaturation

9 Large-scale slope convection

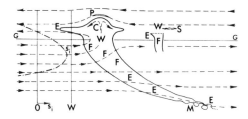

Fig. 9.35 The typical form of the cirrus fallstreak cloud, drawn in a vertical section along its direction of travel, and a tentative representation of the relative streamlines in this plane (air motion in the third dimension is implied). The horizontal scale is compressed: the depth of the cloud is about 1 km, but its horizontal extent may be several kilometers. All parts of the cloud travel with the velocity of the wind at the level GG, the so-called generating level.

The main diagram refers to a dense cloud with a cumuliform top C and anvil-like extensions, which forms when the general lapse rate is unusually large. The ice particles in this compact part of the cloud form in regions W, where saturation with respect to liquid water is attained, are numerous and small, and follow the air motion. Rarely, pileus P form over the cumuliform head, in which updraft speeds may exceed 1 m sec^{-1}. When formed impulsively the base of the cumuliform cloud may be clearly separate from and above the main fallstreak trail, and in less vigorous circulations only a shred cloud S may appear (Pl. 9.11), as drawn on the right. Frequently saturation with respect to liquid water is not attained, and only the trail of sparse large particles is present; the principal regions of their formation are labeled F. On the left of the diagram is drawn the distribution with height z of the supersaturation s_i (with respect to ice) in air approaching the cloud, and two values corresponding to zero and to saturation W with respect to liquid water are indicated.

In the base of the trail the cooling associated with evaporation E often leads to the formation of mamma M. The mamma are downward extensions of the lower end of the trail (Pl. 9.12). Generally they have smooth surfaces, unlike the shred clouds and cumuliform heads over the upper part of the trail, probably because particles continuously fall into the smooth flow just in advance of the modified air (Appendix 8.9).

with respect to ice, regeneration by the process envisaged is clearly impossible. This conclusion is not inconsistent with a few special observations partly presented in reference 134, according to which the temperature at the tops of cirrus fallstreaks is usually near $-40°C$ and rarely if ever more than several degrees lower. [This result seems to be supported by some photogrammetric observations of cirrus uncinus made in Japan (136), where the average height was nearly 1 km greater than

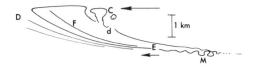

Fig. 9.36 The typical form of a well-developed cirrus fallstreak cloud. The direction (and shear) of the wind is indicated by the arrows, with lengths proportional to the speed, and the scale by the line indicating a layer of depth 1 km in which the air generally is supersaturated with respect to ice.

over northwestern Europe.] The apparent support for the conclusion may be fortuitous, for on the right flanks of both the polar front and subtropical jet streams, probably the regions of most prevalent cirrus formation, the temperature lies typically in the range between about -40 and $-50°C$. It seems, however, that most of the reports from low latitudes of cirrus at heights above 12 km, and therefore at temperatures below about $-50°C$, refer not to fallstreak or other well-defined forms but to thin amorphous layers (e.g., 137, p. 78; 138).

In summary it may be said that a warming accompanying the growth of crystals entering air which has a high supersaturation with respect to ice is a process likely to play a significant part in the production and maintenance of the fallstreak forms of separate cirrus clouds. The origin of the clouds must be sought in small-scale disturbances—typically with a horizontal dimension of less than about 1 km and a duration of about 10 min—which locally and intermittently lead to a state of near saturation with respect to liquid water, and thereby initiate the growth of sparse crystals in a source region traveling with about the velocity of the wind. These disturbances probably arise from the instability of strongly sheared flows, as will be mentioned in the subsequent discussion of billow clouds and their relation to topography.

The orientation of cirrus fallstreaks

Assuming that the horizontal component of the motion of ice particles falling in a trail is that of the wind at their level, and that the wind is uniform in the horizontal near the trail, the local orientation of the trail (in the horizontal) is parallel to the direction of the wind shear in the vertical. Generally a fallstreak cloud is associated with a motion system of similar scale, which implies the presence of significant local disturbances in the wind field. These are sometimes suggested by differences in the form of the individual narrow fallstreaks on the sides of dense clouds, but interpretation in terms of air motion is hampered by the difficulty of distinguishing such fallstreaks, and of separating the effects of perspective and of condensation and evaporation (e.g., a tapering downward and inward of an initially broad trail can be due simply to a more rapid evaporation of its outer parts). However, unless the wind shear is small it is usually readily possible to determine representative fallstreak orientations, and to confirm by detailed conventional wind soundings with balloons that they correspond to the wind shears at the appropriate heights. Typically the orientation of cirrus fallstreaks is nearly uniform in each of a few layers up to several hundred meters deep, changing abruptly in comparatively very shallow transition layers, and thus illustrating the previously noted tendency of the free atmosphere to be

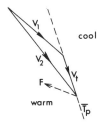

Fig. 9.37 A vector diagram illustrating the veer (a clockwise change of direction) of geostrophic wind with height associated with the advection of warmer air (in the northern hemisphere). The vectors V_1 and V_2 represent the winds at the boundaries of a layer in which the thermal wind V_t is parallel to the pecked isotherm $\bar{T}_p$ of mean temperature (with respect to pressure) in the layer (Eq. 9.6). During the advection of warmer air the direction of V_t is as shown, and the level 2 is above the level 1; during the advection of cooler air V_t is in the opposite direction and the geostrophic wind accordingly backs with height. Particles falling through the layer from a source moving with the wind at its top trail in the direction opposite to that of the shear in the layer. During the advection of warm air this would be in the direction opposite to that of V_t if the winds were geostrophic. However, a typical orientation of the upper part of a cirrus fallstreak, like that shown by the arrow F, indicates that the winds back with height.

stratified into distinct layers (within which properties such as wind shear and lapse rate are nearly constant).

The upper parts of fallstreaks near the edges of the cirrus systems of slope convection have a characteristic orientation in which the clouds trail behind their tops, showing the wind to be increasing upward in the direction of motion (consistent with the tops being at or below the level of maximum wind speed). More remarkable, the fallstreaks are directed inward toward the system, thus indicating (in the northern hemisphere) a wind which backs with height near the forward, advancing, edge of a system, and one which veers with height near its rear edge. Since the forward edge is in a region of advection of warmer air, corresponding to locally rising temperature, while near the rear edge there is advection of cooler air, the changes of wind direction with height are in the opposite sense to those implied by the thermal winds, and show that near the cloud tops the general wind is markedly ageostrophic (Fig. 9.37).

The characteristic appearance and orientation of cirrus fallstreaks ahead of a warm front, associated with warm air advection and the approach of the eastern flank of a system of widespread high clouds, are illustrated in Pl. 6.1 and 9.9, where belts of detached cirrus, containing developing fallstreaks, moved across the sky with a well-defined leading edge lying from 350 to 170°. This orientation was veered by about 30° from the direction of the winds at the level of the cloud tops, and the edge moved eastward through the zenith at 10–14 m sec^{-1}, approximately the speed in this direction of the cloud tops (12–14 m sec^{-1}) and of the large-scale motion system (about 12 m sec^{-1}). Thus, as well as could be determined by observations made over a small area and time, the edge of the cloud system lay along streamlines of the large-scale relative flow near the level of the cloud tops.

The initial orientation of the fallstreaks was toward 280°, that is, into the system, turning abruptly toward 340–350° at a level between 300 and 400 m below the height of the fallstreak tops (Pl. 9.9). These orientations were consistent with measurements made of winds and cloud heights by theodolite triangulation.

The same phenomenon is seen in Pl. 9.13, where a warm front approached from the west and the associated warm ridge in the high troposphere moved eastward at about 12 m sec^{-1}. Special observations were made of high clouds in the northerlies on its eastern flank (again from Östersund in central Sweden). The cirrus system advanced with a well-defined edge oriented toward 028°; it consisted of rows of detached clouds with tops at heights measured to be about 9.4 km, and with velocities of about 008° 30 m sec^{-1}. A backing of the wind with height was found between 8.5 and 9.5 km on successive wind soundings, with a shear vector parallel to the main orientation of the fallstreaks (toward 355°). The component of the cloud velocities contributing to the advance of the cloud system was about 11 m sec^{-1} and therefore again hardly different from that of the large-scale motion system. [However, it was evident that the advance of the cloud system was effected over the mountainous region to the west partly by the continual formation of fresh clouds beyond its leading edge (Pl. 9.13a); there the edge advanced at an average speed of about 30 m sec^{-1}. On reaching low ground its advance slowed; it moved 55 km in the following hour and 95 km in the next two, an average speed of about 13 m sec^{-1}.]

The large-scale significance of cirrus ahead of warm fronts

Sometimes systems of high cloud advance with a very ill-defined and diffuse edge, so that an observer on the ground is at first aware of only a barely perceptible thickening of an amorphous overcast (it is on such occasions that the finest halo phenomena occur). Since at high levels the relative flow is generally from west to east, this appearance might be expected to accompany the gradual evaporation on a system's eastern side of cloud which had become diffuse after forming in its central or western parts. However, it is recognized as more typical for the leading clouds of the system associated with a warm front, traveling at high speeds with a large component of motion from the pole, to consist of fallstreak forms of cirrus such as those described in the last paragraphs. Thus in the Norwegian model of the warm front a distinction is made between a leading zone of cirrus uncinus and the cirrostratus which is part of the "real frontal" cloud; it is said that the individual elements of cirrus clouds are constantly dissolving, but they are replenished from the horizon so that the cloud amount continues to increase. This view that the leading clouds mark the edge of a zone of evaporation of a large-scale cloud system is consistent with the previous discussion, in which the flow (toward lower latitudes) on the east flank of the warm ridge in the upper troposphere was represented as characterized by large-scale subsidence (e.g., in the model of Fig. 9.28 and by the inferred and calculated trajectories of Fig. 9.29). It is not consistent, however, with the observation that fallstreak cirrus *form* on this flank. Moreover, although it might be anticipated that air within a system of high clouds would be approximately saturated with respect to ice and would become subsaturated in the flow approaching and then passing beyond the leading edge of the cloud system, observations that fallstreaks near this edge develop and extend downward, by as much as 1–2 km, imply that in a layer of air at least several hundred meters deep there is a state of supersaturation with respect to ice. Similarly, the observed persistence of the fallstreaks for periods of up to a few hours, and during travel over distances of up to a few hundred kilometers, implies a process of regeneration which seems to require a general state of supersaturation near the level of the cloud tops over the whole extent of the zone occupied by the fallstreaks, typically several hundred kilometers along and a few hundred kilometers across the direction of the flow.

The formation of fallstreak clouds can be associated with motion systems of their own small scale, but it appears that their development and persistence depend on some disturbance of much larger scale. Near the rear edge of a system of high cloud due to slope convection this could be the flow in the trough-ridge system itself, since it is directed poleward, accompanied by general ascent, and could readily lead to the production of widespread supersaturation with respect to ice before the general formation of cloud. The presence of fallstreak cirrus ahead of the warm front, however, suggests the existence of disturbances of somewhat smaller scale, at least on the east flank of the warm ridge, which there restore upward motion near the level of the jet stream and probably on the warm side of its axis.

It may be relevant that when the flow at high levels in a middle-latitude trough-ridge system is represented on an isobaric chart it often appears that the jet stream has pronounced maxima of speed on the flanks of the troughs and ridges, and speed minima near their axes. (The geostrophic wind has a more pronounced minimum on the axis of the ridge, made obvious by a marked separation of the contours as the axis is approached. The ratio V_g/V between the speeds of the geostrophic and the actual winds on the axis of a jet stream may be as large as 2 on the axis of the ridge and as small as 1/2 on the axis of the trough; 2, p. 201.) The regions of speed maxima are known as *jet streaks*. They have been attributed to inertial oscillations about the geostrophic wind, superimposed upon the general flow (the oscillations considered involve the centrifugal and Coriolis accelerations V^2/r and fV, where r is the radius of curvature of trajectories in a perturbed constant zonal geostrophic flow of speed $\bar{u}$. The period of such oscillations is $2\pi/f$, and only about half that found by assuming that trajectories pass through successive observed jet streaks; the amplitude is proportional to the magnitude of the initial ageostrophic perturbation; 140, p. 114). Jet streaks tend to move eastward through the large-scale pattern, but at a speed much smaller than the wind speed (2, p. 206). There are therefore large accelerations on entering and leaving a jet streak along trajectories assumed to lie in an isobaric surface. Associated ageostrophic components of the wind can be anticipated, largest on the axis of a jet stream, decreasing on either side and above and below, and directed toward low pressure in the accelerating flow entering a jet streak and toward high pressure in the decelerating flow leaving it.

These inferred components, across the general flow, have been regarded as implying circulations normal to a jet stream, and in particular ascent below the level of maximum wind on the warm side of the entrance to a jet streak (and descent on the warm side of the exit). Such upward motion might be responsible for a renewal of cloud and the development of fallstreak cirrus in a narrow zone on the poleward side of the eastern flank of a system of high clouds. However, the inference of the circulations and of the zone of upward motion is based on an oversimplified representation of the flow in jet streams. The motion is unsteady, three-dimensional, and more nearly isentropic than isobaric; when it is examined in isentropic surfaces the intensity of the jet streaks may be less marked (the maximum wind speeds tending to occur at lower pressures and higher levels in the ridges than in the troughs). Moreover, the axis of the jet stream may be less clearly defined, the sharp peak of the profile of wind speed on an isobaric surface being replaced by one in which the maximum wind speed varies little over a zone extending a few hundred kilometers across the flow (cf, e.g., 139, p. 112). Indications of accelerations are therefore less apparent, and a jet stream axis which obviously sharply separates two flows of distinctly different potential vorticity, in only one of which (the warmer) cloud formation is likely, is less distinct. Rather, it appears that the clarification of the distributions of acceleration and of vertical motion, and their relation to the distribution of cloud, requires further study and perhaps more detailed data than are provided by the conventional observations.

Nevertheless, there is observational evidence for the existence of ageostrophic winds of the kind inferred from the ordinary isobaric analyses. For example, in a study of routine wind soundings made over the British Isles within about 300 km of the axes of jet streams (identified by concentrations of approximately straight contours of the 300-mb surface) it was found that the mean speeds of the components of the observed winds normal to the axes were about 5 m sec^{-1} at the level of maximum wind speed, decreasing to zero at a level about 150 mb (3 km) lower (140). The components had the sign expected: they were directed toward low pressure in the entrances to jet streaks, and toward high pressure in the exits.

Further, the backing of the wind with height near the level of the tops of cirrus fallstreaks, which has been exemplified in the case studies detailed earlier and described as characteristic of cirrus appearing ahead of warm fronts in the northern hemisphere, is evidence of ageostrophic winds in a layer below the level of maximum wind speed which is a common feature of soundings made into northwesterly jet streams. Thus in an analysis of such soundings over Great Britain in 1949 and 1950, during a period when the wind directions were reported to the nearest whole degree, a backing with height was found in nearly half of all soundings (in 32 of 81). In the layer extending 150 mb (about 3 km) below the level of maximum wind speed, on the average the wind veered with height by 5°, and by 1.5° even in those soundings in which there was backing in a shallower layer. In the layer extending downward 50 mb (about 1 km) below the level of maximum speed the average value of the change in direction was 3.4° when there was a veer upward and 2.7° when there was a backing with height (representing an added component normal to the general flow of about 3 m sec^{-1}). The latter value is naturally considerably less than those of up to about 10°, in layers 1–2 km deep, which have been found in observations made near the edges of cirrus systems, partly because it is based on winds determined as averages over layers which were about 1 km deep and centered on standard pressure levels 50 mb apart. The results confirm, however, that a backing with height of several degrees in a layer about 1 km deep below the level of maximum wind speed is a common occurrence in northwesterly jet streams. That this feature, and similar evidently ageostrophic winds in jet streams in general, are not familiar phenomena is probably due mainly to the convention in forecasting practice of averaging winds over deep layers and reporting their direction in whole tens of degrees. Detailed analyses of series of special soundings made at short intervals (1–3 hr), and from groups of stations separated by only about 100 km, have more conclusively shown that the large-scale trough-ridge systems contain marked and coherent ageostrophic structure with scales of a few hours, a few kilometers in the vertical, and a few hundred kilometers in the horizontal (e.g., 141: intense wind shears were found in layers several hundred meters thick—up to about 15 m sec^{-1} below jet streams, compared with thermal winds of about 1 m sec^{-1}).

It can readily be inferred that horizontal accelerations near the level of the jet stream axes typically imply ageostrophic winds of greater magnitude than the thermal wind, and can therefore produce a change of wind direction with height in the reverse sense to that corresponding to the thermal wind. The latter usually amounts to about 5 m sec^{-1} km^{-1}, but it may be less on the warm side of the jet stream near the level of its axis, where the average horizontal temperature gradient is small. The associated components of the thermal wind across the general direction of the flow are therefore less than ± 5 m sec^{-1} in a layer extending for 2 km below the level of the maximum wind speed, considering that the angle between isotherms and contours is generally less than 30°. On the other hand, a typical rate of change of velocity at the level of maximum wind speed is from 50 to 70 m sec^{-1} in several hundred kilometers, or 10^{-1} cm sec^{-2}, which from Eq. 9.8 corresponds to an ageostrophic wind of about 10 m sec^{-1}, directed across the flow. If the acceleration at a level 2 km lower is comparatively small, the contribution to the shear across the flow evidently may exceed that of the thermal wind, and on the warm side of the entrance to jet streaks may therefore in the layer considered produce a change of wind direction of the opposite sign and of magnitude several degrees.

Wave clouds

The presence of water in the atmosphere and the occurrence of its phase changes have an indirect influence on the properties of all kinds of motion systems by controlling the stratification of the airstreams in which they occur. In some motion systems, for example, those of cumulonimbus convection, the water plays a direct and essential role; in others, however, the presence of water has practically no immediate significance, and if clouds appear they can be regarded only as serving to mark a part of the flow. Among such clouds are very small cumulus, and the wave and billow clouds discussed in the following paragraphs.

Some frequently observed cloud patterns are obviously associated with disturbances introduced into airstreams during their flow over irregular land, because they remain practically stationary. Irregularities may occur over level surfaces in the distribution of surface roughness or of potential temperature, but only the latter ever have much importance, and then predominantly in the distribution of the intensity of ordinary convection (as

9 Large-scale slope convection

already described). The disturbances now to be considered arise from the obstacles presented by surface relief. These disturbances are very common, even over terrain which is not rugged, and may accompany ordinary convection, but usually the vertical displacements of air which they produce are small—only a few decameters. Even so, among the shallow cloud systems of middle latitudes it is unusual not to be able to detect their presence from details in the cloud patterns, while occasionally they are so pronounced as to dominate the cloud pattern or even be entirely responsible for the formation of clouds. Such clouds usually extend between several and a few tens of kilometers in the direction of the wind, and have edges and arched upper surfaces which are smooth. From their appearance they have become well known as lenticular or *wave clouds*, and from their persistence of position have been inferred to be *orographic*, or *oreigenic*, which is a preferable word (142). Sometimes a succession of clouds is observed in the lee of a particular obstacle, occasionally at more than one level, showing that there are regular vertical oscillations in the airstream. The theory of oreigenic disturbances has the principal aim of defining the factors that determine their amplitude and wavelength.

The theory of oreigenic disturbances

During the flow of a stably stratified airstream over irregular surfaces, obstacles behave as sources of gravity waves which propagate upward and outward. Similar waves may be excited in other ways—for example, by the growth of cumulus or of disturbances due to shearing instability—but probably they nearly always have comparatively trivial amplitudes away from their sources; however, as mentioned in Chapter 7, they may produce clouds (pileus). In the waves vertical displacements produce restorative buoyancy forces. The phase speed of the waves is typically about 10 m sec^{-1}, and almost independent of the horizontal wavelength in the range between a few and 100 km. Depending upon the details of the stratification the wave energy may be transmitted upward to levels where it is absorbed by transfer to that of the mean flow, or dissipated by transfer to (turbulent) motions of smaller scale (and eventually by viscosity); or it may be partially or wholly reflected down from some level, so that in two-dimensional flow the waves may by an optical analogy be said to be "trapped" within a layer or "duct," and extend indefinitely downstream of an obstacle.

Many contributions to the theory of oreigenic disturbances have been made, based on simplifications involving all or most of the following assumptions:

1. The effect of the earth's rotation is negligible, limiting the scale in the wind direction (the x axis) to a few hundred kilometers.

2. The flow is steady, frictionless, adiabatic, and unsaturated.

3. The flow is two dimensional, perpendicular to a symmetrical barrier in the form of a ridge (usually of height

$$h = \frac{h_m b^2}{b^2 + x^2} \qquad (9.24)$$

and therefore of peak height h_m and half-width b), which provides a spectrum of perturbations from its Fourier components.

4. The height of the barrier and the amplitude of the disturbances are small (the perturbation method is used and the governing equations simplified, principally by the omission of nonlinear terms).

5. As a lower boundary condition, the flow at the ground is along the profile chosen for the ridge.

6. As an upper boundary condition, the kinetic energy of the perturbation vanishes at a great height.

The restrictions introduced by these and some other assumptions have been given much discussion (see, e.g., 143). Generally, once the simplified governing equations and boundary conditions have been established, a search has been made for stationary solutions specifying the disturbed flow. Difficulties arise, some purely mathematical and others which raise doubts whether physically realistic interpretations can be made; in particular, it appears that there may not always be a unique or a stationary solution. However, it has been shown that if the evolution of the flow from a state of rest or other simple initial conditions is considered, then it eventually approximates to the stationary solution when one exists (143). Further, if the state of the undisturbed flow is changed abruptly, the new solution for the disturbed flow soon becomes dominant near the obstacle (after about an hour, a period determined by the phase speed of the waves). It also seems that disturbances which progress upwind may modify the vertical distribution of the properties of the airstream, but, in accordance with virtually all observations, vertical oscillations can occur only downstream of the obstacle. (Very rarely, a few waves, apparently similar to lee waves, of small amplitude and wavelength and accompanied by very shallow clouds, have been observed extending a small distance—over the sea—upstream of an obstacle; 145.) Usually it is assumed that some process of energy dissipation is effective, and that the state of the undisturbed airstream can be specified by the vertical profiles of the wind speed U and the potential temperature θ well upstream of the obstacle. The speed U_0 at the surface is important, and is difficult to define in natural conditions—the speed measured at a height of a few hundred meters often is regarded as appropriate.

Even using the simplified governing equations the

solution is sensitive to the form of the upstream profiles and to the shape of the obstacle. Considering the variety of natural profiles and their unsteadiness, the complexity of hilly terrain expressed in the shape and extent of its particular features, and the pecularities of the flow close to the ground, it is evident that no concise theory can be applicable to all natural circumstances. On the other hand, many of the principal features of the observed flows are reproduced in solutions of the simplified and linearized governing equations, with models in which the barrier has a simple shape, and the airstream has a simple kind of stratification in one deep layer or in each of a few superimposed layers which may include a stratosphere (146–51). (In the more detailed models the solutions may not be obtainable analytically but are readily calculated by numerical methods.) Remarkably, also, the simplified equations lead in some circumstances to solutions which for two-dimensional flow are probably valid even for disturbances of large amplitude, in which the streamlines are strongly tilted out of the horizontal, or may even be upright or contain closed circulations.

The simplified governing equation for the displacement Δz of a streamline from its original level contains the terms

$$\partial^2 (\Delta z)/\partial z^2 = (k^2 - l^2)\Delta z$$

where $2\pi/k$ is the horizontal wavelength of the periodic disturbances assumed possible,

$$l^2 = \frac{gB}{U^2} - \frac{U^{-1}\partial^2 U}{\partial z^2}$$

where

$$B = \frac{\theta^{-1}\partial\theta}{\partial z}$$

and $U(z)$, $\theta(z)$ are the undisturbed velocity and potential temperature. These terms are always important, and are often the only ones of importance in more general equations. Accordingly, the property of an airstream which has come to be regarded as of prime importance in respect of the occurrence of lee waves is the distribution with height of the parameter l^2.

Hill waves and lee waves

From solutions for the disturbed flow it appears that it can be regarded as composed of two parts:

1. A *hill disturbance* or *hill wave*, which is always present, mainly in a wedge-shaped region extending upward and outward from the barrier.
2. A train of lee waves, which may occur behind the barrier, and can extend indefinitely downstream over level ground (sometimes—in frictionless two-dimensional flow—with no change of amplitude, their energy remaining confined within a tropospheric layer).

The *wavelength* of the lee waves depends only on the properties of the airstream; there may be more than one wavelength present simultaneously. Their amplitudes depend on the properties of both the airstream and the barrier. Usually lee waves of wavelength about 10 km are found in the lower troposphere; wavelengths between about 15 and 30 km are more conspicuous in the upper troposphere and lower stratosphere. It has been shown (149) that the wavelength of lee waves observed in the lower troposphere is usually close to that given by assuming that the profile of $l(z)$ up to a level of maximum wind speed in the upper troposphere can be approximated by the simple exponential profile

$$l = l_0 \exp(-cz) \quad (9.25)$$

where l_0 and c are suitably chosen; when lee waves are evident the values of l_0 and c usually lie in the ranges between 0.7 and 1.3 km^{-1} and 0.1 and 0.3 km^{-1}, respectively. The wavelengths of lee waves can then be obtained by reference to Fig. 9.38, which includes a second wavelength possible when $l_0/c > 5.5$; three or more wavelengths may be present if $l_0/c > 8.6$, but in the troposphere it is usually only one of the shorter and most often the shortest wavelength which is dominant.

In the troposphere wavelengths increase with the mean wind speed (as is apparent from Fig. 9.38, since a general increase of wind speed reduces l_0 and an increase of shear increases c). It has been shown (148) that the dominant wavelength λ is given approximately by combining the mean speed $\bar{U}$ in the lowest several kilometers with the frequency of vertical oscillation of a displaced parcel corresponding to the mean stability $[\lambda = 2\pi\bar{U}(gB)^{-1/2}]$.

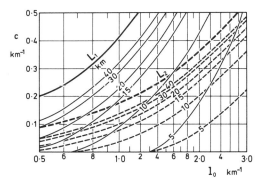

Fig. 9.38 The wavelength of tropospheric lee waves according to a model of a two-dimensional flow in which the undisturbed vertical profile of l has the form $l = l_0 \exp(-cz)$, given as a function of the parameters l_0 and c. Two (or more) wavelengths may be present simultaneously if $l_0/c > 5.5$, but usually the shorter, of wavelength L_1 (indicated by the solid lines), will be dominant. If $l_0/c < 2.5$, no lee waves exist in this model. The diagram often gives approximately correct values of wavelength when natural lee waves occur; then the observed profile of l throughout the troposphere can usually be represented reasonably well by the assumed form of profile (149).

9 Large-scale slope convection

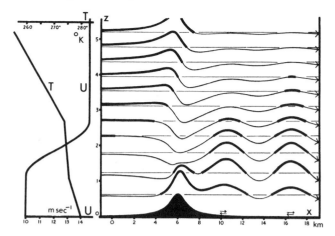

Fig. 9.39 An example of a hill wave and lee waves in the lower troposphere in two-dimensional flow near a ridge, calculated from a model (143) in which the parameter l^2 is constant in each of two layers (0.33 km^{-2} above, and 2.12 km^{-2} below about 1.5 km, corresponding to the undisturbed profiles of temperature T and wind speed U shown on the left). The streamlines are thickened where they lie above their original level. Near the second lee wave the disturbance due to the hill wave has practically disappeared. The lee waves have a maximum amplitude at a height of about 1.5 km in the layer of greatest stability and die away exponentially upward. Typically, the distance between the crest of the first lee wave and the axis of the ridge is less than one wavelength. The amplitude of the hill wave is increasing with height; its energy must eventually be absorbed in some way not apparent from this simple model.

In conditions favorable for lee waves the hill wave is enhanced, and in both the vertical displacement of streamlines from their original level may exceed the height of the ridge. Figure 9.39 illustrates an airstream with lee waves in the lower troposphere, the solution of a model in which the airstream consists effectively of only two layers in each of which l^2 is constant.

According to this model the condition for the occurrence of lee waves which persist indefinitely downstream with an amplitude that has a maximum in the lower troposphere (characteristics suggested by many observations) is that l^2 has a value of l_1^2 in a layer near the ground which must be greater (by at least $\pi^2/4d^2$, where d is the depth of the lower layer) than its value l_2^2 in an upper layer (in the most favorable circumstances this condition is only just fulfilled). The wavelength of the lee waves is then given by a value of k which is between l_1 and l_2. (In the solution illustrated in Fig. 9.39 the airstream consists of three layers; in the uppermost $l_2^2 = 0.33$ km^{-2}, corresponding to $U = 15$ m sec^{-1} and the lapse rate $\gamma = 8°C$ km^{-1}, while in the lower two $l_1^2 = 2.12$ km^{-2}. In the lowest layer $U = 10$ m sec^{-1} and $\gamma = 4°C$ km^{-1}; continuous profiles of U and temperature T are obtained by a suitable choice of the term $U^{-1}\partial^2 U/\partial z^2$ in the intervening layer.)

A sufficient upward decrease of l^2 is a common property of the extratropical troposphere because of the typical increase of wind speed with height; a decrease of stability with height may contribute. Moreover, the numerical evaluation of models allowing the specification of the properties of an airstream at several or more levels up to well within the stratosphere has shown (148, 152) that lee waves may occur, and extend to great heights, also in other kinds of profile of l^2.

According to the simple two-layer model the amplitude of the lee waves decreases exponentially upward from a layer of greatest stability in which it has a maximum. In the atmosphere a layer of large stability in the lower troposphere usually occurs above a lowermost layer modified by cumulus convection, often with a base marked by an inversion of lapse rate and a sudden upward decrease of relative humidity to very low values (as illustrated in Chapter 7). Since the relative humidity typically then has a maximum immediately below the inversion, implying a minimum vertical displacement required to produce condensation, wave clouds can be anticipated to be prevalent at about the levels where the lee waves have a maximum amplitude. The typical smoothness of their surfaces, especially the upper (which is likely to define streamlines), is a striking indication of the stability of the air. (Figure 9.40 shows that the inversion often noticed on radiosonde soundings on occasions of lee wave occurrence may sometimes be exaggerated or even produced by the sloping and not vertical paths followed by the balloons.)

The more general models suggest that some lee waves maintain a large amplitude up to and into the lower stratosphere, although radiating energy upward and therefore weakening downstream, and that the interference of waves of different length may produce disturbances of comparatively small scale in the lower stratosphere, as suggested by the rumpling of stratospheric haze layers (Pl. 4.5).

When there is no loss of energy from a layer which contains lee waves, and they do not die away downstream (as in Fig. 9.39), their phase lines are vertical. Nearly upright phase lines appear in the solutions of more complicated models, partly because in most illustrations the horizontal scale is greatly compressed; closer inspection shows that locally they may tilt either downstream or upstream, but generally in the latter sense, consistent with an upward radiation of energy from the barrier and a downward flux of momentum toward it, so that it exerts a drag on the airstream. Another feature common to all models is that the flow down the lee slope of the ridge is beneath a trough and is more intense than the flow up the windward slope; indeed, in the lee of the ridge there is no shelter from a strong wind.

Since theory suggests that lee waves can occur in a large variety of airstreams, it is important to consider circumstances which commonly restrict their amplitude, and those which occasionally, especially in particular localities, lead to large amplitudes, extraordinary clouds, and peculiar and even violent surface winds.

Factors affecting the amplitude of lee waves

In the theories based on the perturbation method, and in particular the two-dimensional, two-layer model, the amplitude $\Delta z(z)$ of lee waves at the height z depends on a complicated function, containing several factors which are not all independent but which considered individually indicate some principal circumstances which control the development of the lee waves (143, 153). They include expressions representing the shape and size of the barrier, terms $\rho_0/\rho(z)^{1/2}$ and $U_0/U(z)$ (where ρ_0 and U_0 are the density and the velocity of the undisturbed flow at the surface), and an expression which is a function of the stratification of the airstream, which can be related to the vertical profile of the parameter l^2.

A barrier of the conventionally assumed symmetrical shape (Eq. 9.24) affects the amplitude proportionally to the expression $h_m b \exp(-kb)$. The linear dependence on h_m results from the method of analysis, and it is more interesting to note that for barriers all of the same maximum height the remainder of the expression has a maximum value when $b = k^{-1}$. On the other hand, for barriers all of the same shape but of variable size, a pronounced maximum occurs when $b = 2k^{-1}$. Evidently the shape of barriers is important; the largest natural obstacles may not be the most effective, and over irregular terrain the effective features are likely to vary from occasion to occasion, depending on both the direction of the airstream and the details of its stratification. (Moreover, since the disturbances produced by several obstacles are additive, waves over irregular terrain may locally be either suppressed or enhanced, and where pronounced they may have no obvious relation to any prominent neighboring ground feature.)

For a typical tropospheric lee-wave length of 10 km the half-width of the most effective ridges is only a few kilometers, which is narrow for natural ridges 1 km or more high, unless perhaps they have a steep escarpment on one side. The most striking phenomena are observed in the lee of such escarpments. Numerical calculations based on linearized theory and assuming two-dimensional flow (154) have shown that ridges may more effectively excite lee waves over a wider range of wavelengths if they are asymmetrical, and that when the lee slope is the steeper the first lee trough is closer to it, perhaps favoring a faster flow down it. It may also be significant that in a numerical calculation of a time-dependent flow over a very wide ridge of rectangular section, using equations not linearized, it was found that the disturbance developing near the lee edge was much more intense than that over the windward edge (155, 156). The reasons for this behavior are not clear; in the linear theory the two flow patterns should be similar, with the sign of the vertical velocities reversed.

There are some other features of the shape of natural barriers and the kinds of flow near the ground which restrict the development of the disturbances. Foremost is a limited extent of a barrier across the airstream.

Studies of the disturbances produced by isolated obstacles confirm the intuitive anticipation that their amplitudes are considerably less than those due to extensive ridges of the same height lying across an airstream; moreover, the amplitude of lee waves always decreases downstream as their energy spreads sideways. One study (146) suggests that the lee waves should have an arched horizontal form, convex toward the obstacle. [Horseshoe-shaped regions of updraft are said to have been observed in the lee of isolated mountains (143), and a crescent-shaped wave cloud is one of the many characteristic forms which appear in the lee of the conical Mount Fujiyama (160). However, lee-wave clouds of this particular shape seem to be very rare.] In another study a two-layer model was considered, and the disturbances produced by an isolated hill of circular or oval horizontal section were found by superimposing solutions for oblique flows across two-dimensional obstacles (150). The flows thus derived have no qualitative differences from those produced by a long ridge of the same height lying across the airstream, but the lee waves have a comparatively small amplitude, especially well away from the hill top. The amplitude of the hill wave decreases more rapidly with height (proportionally to $z^{-1/2}$), and that of the lee waves decreases downstream (proportionally to x^{-1}), in comparison with the values in an airstream in which there would be no change of amplitude of the lee waves behind a two-dimensional ridge. Further, the lee waves are confined to a broadening region extending downwind of an isolated hill (within a wedge of semi-angle typically between several degrees and $30°$).

Another complication associated with the shape of the ground, frequent in rugged natural terrain, is that because of the presence of sharp-edged features, particularly where a lee slope descends steeply in an escarpment, the lowest streamlines in the undisturbed flow can leave the surface. They may enclose a region of stagnant air, or a lee eddy in which the direction of the wind at the ground is locally reversed; the effect is to smooth the shape of the mountain to the changed form of the lowest streamline, and thus usually to increase the wavelength and reduce the amplitude of disturbances at higher levels, although probably not drastically. The separation of the flow from the ground is the more likely the less stable is the lowermost layer. Its occurrence and form may be affected by disturbances generated upstream, and by the presence of anabatic or

9 Large-scale slope convection

katabatic winds (produced by radiative heating or cooling of the hill slopes); it may occur intermittently with the shedding of eddies into the airstream downwind of the obstacle (Section 7.6; Pl. 7.20).

Of the factors that affect wave amplitudes, other than the form of the flow near the ground, the term $U_0/U(z)$ indicates, as might be expected, that the development of lee waves is favored by a strong wind at (or near) the surface. On the other hand, the increase with height of the parameter l^2 which is required in the two-layer model for the production of an extensive train of lee waves is most readily insured by an upward increase of the wind speed. It appears on examination that the most favorable circumstance for the development of such trains is that the variation of l^2 with height should be associated mainly with changes of stability, and that the required condition of a sufficient decrease should only just be satisfied.

By considering models with three or more layers in the lowermost of which the stability is small ($l \approx 0$), it can be shown that lee waves may occur in the presence of such a layer (Fig. 9.40) but particularly if its depth is comparable with or exceeds the height of the barrier, its effect is to increase the wavelength of lee waves and, more significantly, to decrease their amplitudes: when the layer is deep, hill waves are likely to become weak and lee waves to disappear. The prevalence of ordinary convection over land is probably a major reason why pronounced hill- and lee-wave phenomena are not observed much more frequently, at least during the daytime and where the relief is not pronounced. Nevertheless, although the effects of lee waves of small amplitude on the distribution of cumulus are difficult to observe from the ground, their presence can sometimes be detected, especially early and late in the day, and many examples of long trains of (daytime) lee waves observed from satellites appear to be associated with systems of moderately deep cumulus, rather than with clouds produced by the waves themselves, either at the top of a layer moistened by previous convection or in the middle troposphere.

In the model with an exponential profile of l in the troposphere, represented by the expression in Eq. 9.25, the intensity of lee waves in the troposphere, expressed by the maximum vertical velocity $|W_1|$ or $|W_2|$ in the waves of the first (L_1) or second (L_2) possible wavelengths (Fig. 9.38), is given approximately by

$$|W_1| = \left(2.5 + \frac{0.7}{cL_1}\right) a \left(\frac{\rho_0}{\rho_1}\right)^{1/2}$$

at
$$z_1 = c^{-1} \log_e \frac{l_0}{l_0 - 2.2c}$$

$$|W_2| = 3.2 a \left(\frac{\rho_0}{\rho_2}\right)^{1/2}$$

at
$$z_2 = c^{-1} \log_e \frac{l_0}{l_0 - 5.5c}$$

where $a = h_m c U_0$, and ρ_1, ρ_2 are air densities at the heights z_1, z_2 (146). According to these formulas, which illustrate the complication of the relation between wave intensity and hill shape even for a simple kind of stratification, lee waves of the shortest wavelength L_1 are dominant in the lower troposphere, and those of longer wavelength are dominant in the upper troposphere. For the most usual circumstance in which lee waves are evident (a substantial decrease of l^2 with height, due mainly to a large wind shear), both formulas for the vertical velocities can be further simplified, within the limits with which any reasonable comparison can be made with observed values of vertical velocity, to

$$|W| \approx 3 h_m c U_0$$

Thus even with ridges only a few hundred meters high and surface winds of several meters per second, for a common value of c (about 0.2 km^{-1}) vertical velocities of a few meters per second at a height of a few kilometers are implied, readily exploitable by sailplanes (whose sinking speed in still air is about 1 m sec^{-1}).

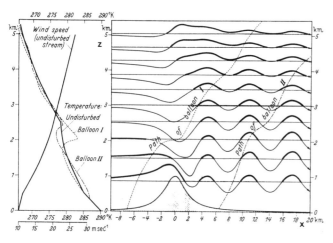

Fig. 9.40 Another example of lee waves, calculated (147) for a flow in which the undisturbed wind and temperature profiles (shown by solid lines on the left) are arranged so that there is a lowermost layer, extending up to 0.5 km, with an adiabatic lapse rate ($l^2 = 0$). Above 2.5 km $l^2 = 0.75$ km^{-2}; in the intermediate layer l^2 is a function of z, chosen to avoid discontinuities in relevant variables at the interfaces. The general features of the solution are similar to those of the two-layer model; note, however, that at 5 km the disturbed streamline remains above its original level for about 20 km in the lee of the ridge, so that a wave cloud of similarly great extent would be possible. On the other hand, an extensive low layer cloud upstream of the ridge would become thickest near the crest and descend some distance down the lee slope before probably becoming broken into a train of lee-wave clouds. Two hypothetical paths of sounding balloons are shown to illustrate how their records may represent a stable layer as an inversion.

Strong lee slope winds and rotors

In many calculated flows in which the hill disturbances attain large amplitudes there is descent in the lower troposphere over a ridge which becomes even more pronounced over its lee slopes (156; a tendency for an intensified flow over the surface in the lee of ridge crests is apparent in Figs. 9.40 and 9.41). The combination of strong winds down a lee slope and a first lee wave of very large amplitude resembles the "hydraulic jump" formed by the flow of a fluid with a free surface over an obstacle in a channel (e.g., of a river over a weir) and has been reproduced in model experiments with stratified liquids (143).

One interesting feature of such experiments is that in two-dimensional flow the stream may be brought to rest in a shallow layer upstream of the obstacle. In a model atmosphere in which l is constant in a lowermost layer, the layer will similarly be "blocked" if $h_m > \pi/l$ (157). In a convective boundary layer l may be too small for there to be any significant effect. However, in more stably stratified airstreams l may exceed 1 km^{-1}, and especially behind ridges which are high the flow at the surface l may have descended from a height of a few kilometers. It arrives with a comparatively high potential temperature, which is characteristic of *foehn winds* behind mountain barriers (the warmth is conventionally attributed to a release of latent heat accompanying precipitation in a surface flow crossing the mountains).

When conditions are favorable for strong lee slope winds and intense lee waves *rotors* may develop. These are disturbances whose amplitude becomes so great, and in which streamlines become tilted at such a large angle to the horizontal, that air at one level becomes displaced to a position above air originally at a higher level. Typical patterns of the associated streamlines are drawn in Fig. 9.41. In that shown in Fig. 9.41b a part of the fluid has become separated from the mainstream. Rotors may develop near the ground and locally reverse the direction of the surface wind, especially in the lee of steep slopes when the flow in a strong current descends without separation and the amplitude of lee waves increases rapidly with height.

When lee waves occur the stratification, at least above a shallow layer near the ground, is alway originally stable, and probably markedly so where the wave amplitude becomes large. The stratification therefore becomes unstable inside a rotor, and in some rotors may be more so than in any other natural circumstance, even if the air remains unsaturated: the lapse rate may greatly exceed the dry adiabatic, and intense small-scale convection must arise. Since the region occupied by this turbulent motion occupies only a small part of the whole oreigenic disturbance it may not disrupt the flow, but it is likely to introduce some degree of unsteadiness if the flow is not already unsteady. (Although it has generally been supposed that rotors can be anticipated whenever lee waves attain abnormally large amplitudes, and when the hill waves are also likely to be intense, it has now become evident that more study is required of the hill waves themselves and of airstreams in which periodic or steady disturbances cannot occur; see, e.g., 159.)

Observations of waves by radar and other techniques

Striking oreigenic disturbances which occur near major mountains, especially in the winter season, have been carefully observed since they were recognized as a powerful aid to soaring flight and a serious danger to commercial aviation (143, p. 20; 156). In particular there have been many well-illustrated accounts of wave clouds and associated air motions (see, e.g., 160–69, and the bibliography in reference 143), mainly from locations in central and western Europe and the western United States. Recently there has been renewed study of these phenomena (170), partly because of concern that disturbances might extend into the stratosphere (where they are only rarely manifest as nacreous clouds) sufficiently often to affect the operation of a new generation of (supersonic) transport aircraft.

In almost all studies the form of the air flow has been inferred from observations made along lines. Various sensors have been mounted on tracked carriers, including radiosondes and *constant-volume balloons* (which react only sluggishly against air motions causing displacement from an equilibrium level, and therefore virtually follow air trajectories; 171), *sailplanes* (sensitive to vertical

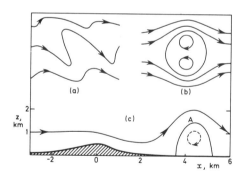

Fig. 9.41 Typical calculated forms of streamlines in rotors: (a) in an open rotor; (b) in a closed rotor (either of these kinds of flow may occur well above the ground in stationary lee waves of large amplitude); (c) in a calculated two-dimensional flow with lee waves (of wavelength 6 km) downwind of an asymmetrical ridge 1 km high (157). Only the flow at the lower levels (the disturbances decrease upward) and up to the first lee wave is shown. The lee wave has a large amplitude; the flow separates from the ground and ascends steeply toward the crest of the wave, and beneath it the direction of flow near the ground is reversed. In a cloud formed around the region A the flow could be very turbulent in its lower, statically unstable part, but smooth in its stable upper part.

9   Large-scale slope convection

velocities of less than 1 m sec^{-1}), and *powered aircraft* (from which vertical velocities are less easily measured, and which are more usefully employed in measuring potential temperature over nearly horizontal paths inaccessible to sailplanes; 170). These methods of exploration may be supplemented by observations of cloud location and behavior, but they have the disadvantage that the construction of the three-dimensional flow pattern has to be based on data only very sparsely distributed, so that it is convenient and virtually necessary to assume steady flow. This assumption frequently may be justified over periods of a few hours during which the changes in the large-scale properties of an airstream are small, but it precludes detailed study of some disturbed flows which may be complicated, very sensitive to the upstream conditions, or even essentially unsteady.

A potentially more powerful method is the use of a *radar* to observe well-distributed artificial or natural targets (174). The possibility of managing without artificial targets arises from the characteristic large-scale stratification of the atmosphere, in which large vertical gradients of wind speed and potential temperature tend to occur in shallow transition layers. Especially in the distorted flow of oreigenic disturbances these shallow transition layers locally become liable to a shearing instability (discussed later in the description of billow clouds) which produces turbulent motion, accompanied by irregularities of refractive index detectable by a sufficiently sensitive radar at ranges of up to about 50 km (see Section 6.4). The presence of hill and lee waves is then manifest on the RHI displays by the form of these layers of echo, as illustrated in Pl. 9.14. This picture was obtained with a radar in the lee of the Welsh mountains, on a day when two major trains (over 100 km long) of lee-wave troughs were clearly shown in a morning satellite picture by gaps in an almost complete cover of low layer cloud. Many of the extensive trains of lee waves occasionally visible in satellite pictures seem to be detectable because of the pattern which they impose on widespread shallow cumulus or low layer cloud. To a ground observer the pattern would not necessarily be obvious, nor would the clouds appear lenticular. (Examples are given in references 172 and 173.)

The large antenna of a suitable radar cannot be maneuvered rapidly, but a complete revolution providing RHI displays at azimuth intervals of 10° can be made in about 45 min to explore the form of each detectable layer within a large volume, and the steadiness of the forms on particular bearings can be examined over much shorter intervals. Evidently by this method a more thorough survey can be made of lee-wave properties, especially if it is supplemented by conventional techniques (174–76).

The forms of wave clouds

Several kinds of wave cloud can be distinguished; the number present and their degree of development vary greatly from occasion to occasion, but they can be represented schematically as in Fig. 9.42, and are described in the following paragraphs as simple isolated *wave* or lenticular *clouds*, the *foehn wall clouds*, the train of *lee-wave clouds*, the *roll* and *puff* clouds of the rotor, the *great hill-wave* cloud, and *nacreous-wave clouds*.

The most favorable large-scale conditions for the

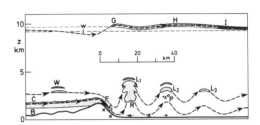

Fig. 9.42 A schematic representation of various oreigenic phenomena, not always all present simultaneously. The dashed arrowed lines are streamlines of the flow in a vertical section, approximately across the crest of a hill range in the lee of whose escarpment the low ground is practically level. Several distinctive kinds of cloud are stippled and labeled as follows:

- W = Isolated and irregularly distributed *wave clouds*, sometimes in thin layers stacked almost vertically, formed where the interference of hill or lee waves locally increases the vertical displacement of air.
- C = Clouds, often extensive at the top of a lowermost layer of air now or previously modified by ordinary convection. Part of this layer, B, may be *blocked*, not crossing the high ground. The cloud may thicken during ascent toward the ridge crest, and near it may precipitate; especially over high ground in cold weather it is liable to produce snow. It only partly descends the lee slopes of the ridge, hanging over the crest as wall clouds.
- F = The *wall cloud* characteristic of foehn weather, perhaps with fringes S of snow (sometimes partly blown up from the ground by strong winds), which evaporates less readily than the droplet cloud.
- L, R, P = A regular train of smooth *lee-wave clouds* L of diminishing size. Near the crest there may be rotors in which turbulent *roll clouds* R or evanescent *puff clouds* P form; especially in the first lee wave of largest amplitude they may extend up into the smooth lenticular clouds. The surface winds may reach hurricane force down the lee slope of the escarpment and be strong *between* the lee-wave crests (in the positions marked x), but fall light and variable, or even reverse direction, *below* the crests.
- G, H, I = A cloud may form in the upper troposphere in the lee of the ridge crest, and extend unbroken for a long distance downwind as a *great hill wave cloud* G. Sometimes the presence of lee waves is shown by undulations H. Cloud usually forms at the saturation level w with respect to liquid water; if the air does not soon descend below the saturation level i with respect to ice (several hundred meters lower) a very long trail I of ice cloud develops.

observation of extensive wave clouds are those in which the hill- and lee-wave amplitudes are large and the airstream is moist, but the troposphere is not mostly occupied by deep clouds in which the presence of waves is unlikely to produce visible effects. Generally these conditions are found near slow-moving cold fronts in winter airstreams which enter land masses from the west or with a poleward component, with winds strong at low levels but increasing upward without much change of direction, and in which ordinary convection, perhaps previously deep, has become feeble or suppressed.

Isolated wave clouds. Isolated hill-wave clouds are common on the crests (*cap clouds*) or slopes of individual large mountains, above the peaks, or immediately in their lee (with usually no more than a trace of a second lee-wave cloud). The clouds in the lee may sometimes have the crescent shape previously mentioned as observed near Mt. Fujiyama, but much more commonly they and the clouds above the peaks appear as more or less regular discs, whose true horizontal outlines are difficult to judge when they are seen at low elevations. The cloud texture may be practically without detail (Pl. 9.15), indicating smooth flow, or may have fine structure, perhaps locally with billows, indicating some small-scale motion. Radiative transfers often lead to internal convection and a dappled structure developing toward the lee edges of wave clouds, when they are sufficiently broad for the passage of air through them to occupy a period of several minutes or longer, and there is little cloud above or below their level. After the dappled pattern has been established it persists in the distribution of relative humidity in clear air, and it may reappear at the windward edge of other clouds.

In flows with a large poleward component ahead of cold fronts, and therefore normally with large-scale upward motion which increases with the approach of the fronts, the first clouds to appear in the middle troposphere (on the eastern fringes of the main frontal cloud systems) are often shallow isolated wave clouds, even over topography with apparently only minor features (hills and ridges rising only about 100 m above the general ground level). Ridges of limited length are observed to be more effective when their crest lines are concave toward the wind. The wave clouds are likely to appear at levels between about 3 and 5 km, in air which in the last 1–2 days has risen in slope convection from the top of a layer of shallow cumulus convection over land in latitudes lower by at least several degrees. Nearer the fronts, and when condensation begins well below the top of modified air in which there was a deep adiabatic layer, castellanus clouds are more likely to grow when cloud forms in oreigenic disturbances (Appendix 8.8; Pl. 8.24). In the warm seasons systems of both kinds of clouds may appear with a noticeable periodicity, partly due to a local enhancement of oreigenic disturbances early and late in the day (and their suppression by the local ordinary convection in the middle of the day), and partly due to the diurnal variation of the modification (in particular, the moistening) of the air by previous ordinary convection in lower latitudes. These influences need not be in phase, but it often happens that cloud systems locally well developed in early and late hours disappear completely in the middle of the day, behavior puzzling to forecasters alert for the development of thundery weather.

Foehn wall cloud. An extensive layer of low cloud may thicken during a slow ascent toward a high ridge; it may eventually precipitate, particularly near the crest and when the airstream is maritime (or the weather is cold so that the cloud is supercooled and liable to produce snow). The cloud can then be expected to evaporate in the lee of the crest, leaving a broad cloud-free region, but this also often happens independently of the occurrence of precipitation when the lee slope is steep and favorable for the excitement of intense lee waves; the flow at low levels upstream of the ridge may be partially blocked and fail to cross the ridge, especially when it extends far across the airstream. In strong winds the cloud then characteristically covers the ridge crest and seen from the valley floor seems to pour rapidly a short distance down the lee slope, often like a waterfall (160, Fig. 25–27), before abruptly evaporating. This crest cloud accompanies other manifestations of pronounced foehn weather on the low ground, so that it has become known as the *foehn wall* or *wall cloud*.

Trains of lee wave clouds. These appear in favorable conditions in the lee of long ridges (Pl. 9.16). The upper surface of individual wave clouds is always arched upward and nearly always sharply defined; the form of the lower surface is best seen from an aircraft flying at about the same height. Usually it is level (indicating a uniform condensation level in a moist layer), or is also arched upward (indicating a rapid decrease of relative humidity below a moist layer). The level base is characteristic of wave clouds formed in the upper part of the adiabatic layer produced during ordinary convection, and the arched base is characteristic of shallow wave clouds formed in air from within a cumulus layer.

The position and details of the form of lee-wave clouds are liable to shift and change a little, but often they are remarkably steady. Sometimes, however, a periodic adjustment of the flow is indicated by a slow movement of a wave cloud downwind, over a distance of 1–2 km in several minutes, followed by a sudden apparent "leap" upwind (within several seconds), due to reformation in the original position. This behavior may be repeated over and over again for an hour or more. Similar periodic changes in lee flows are suggested by the experience of sailplane pilots and consistent variations in the direction, strength, and gustiness of the wind at places on the ground (152; 177, p. 346; 178).

Occasions when lee waves produce regular long trains of clouds at only one or two levels in the middle or upper troposphere (Pl. 9.17) seem to be rare compared with those in which they impose extensive patterns on the distribution of low-level clouds, which are not readily observed or photographed from the ground. Near fronts, however, when the whole troposphere may be moist, trains of wave clouds are sometimes observed which from the air can be seen to consist of several cloud layers, clearly separated (at least at their edges), and extending in stacks up to cirrus levels. As many as 25 well-defined layers have been observed between 2 and 10 km (168), evidently indicating a corresponding variability with height of the relative humidity rather than of the lee-wave amplitude. Flight in such clouds has been found to be smooth with gradual transition between updrafts and downdrafts, whose maximum speeds are typically several meters per second. Consistent with simple theory, it has been observed (168) that when these extensive lee-wave clouds occur the wind has nearly the same direction at all levels, is strong (speed greater than 10 m sec^{-1}) at low levels, and is usually increasing (never decreasing) with height; the stability is small above (corresponding approximately to the saturated adiabatic lapse rate) and great below a height of 2–4 km (except sometimes in a shallow lowermost layer modified by ordinary convection). A strong inversion is frequently found at about the level of the crest of a ridge responsible for pronounced lee waves; however, it is probable that an inversion has no special significance. It may simply be a normal product of previous large-scale subsidence and ordinary convection, representing only the equivalent of a somewhat less stable stratification distributed over a deeper layer.

Roll and puff clouds. Bar or *roll clouds* form in the upper parts of rotors in lee waves with large amplitudes, appearing in almost continuous long lines in the lee of a ridge and parallel to it. The bases are often nearly level and the tops are cumuliform (Pl. 9.18; see also the illustrations in references 143, 161, 164, 165, and 169). Details on the sides of a cloud may have a striking rolling motion, but the direction of the flow at the cloud base is rarely if ever reversed, so that perhaps contrary to a first impression the cloud does not rotate as a whole. Often there is only one roll cloud, but occasionally there are two or three, and rarely as many as five, from whose spacing the wavelength of the lee waves may be determined. The bases of roll clouds are usually at about or distinctly above the level of the crest of the ridge; sometimes a roll cloud, especially the first of a series, which itself has much detail in ceaseless irregular motion, extends upward into wave clouds which in contrast have very smooth surfaces. On other occasions there is a distinct gap between a roll cloud and a wave cloud above it, or a roll cloud may appear alone or below a more extensive wave cloud in the upper troposphere.

A roll cloud may not be continuous; it may appear only intermittently as short-lived arched and clawlike trails (Pl. 9.19), or small *puff* clouds, in agitated motion. Whether or not there is cloud the air motion in the rotor is very turbulent and is dangerous to aircraft which enter it, as graphically described in the reports of sailplane pilots (Appendix 9.5).

Great hill-wave clouds. The parts of the troposphere most liable to contain clouds, or to have a high relative humidity, are a lowermost layer containing ordinary convection, or layers in the lower and middle troposphere (especially in airstreams returning toward higher latitudes) which have recently been moistened by ordinary convection, and an uppermost layer in the high troposphere containing an airstream which has recently ascended in slope convection. Consequently, the levels most likely to be occupied by clouds in which oreigenic disturbances impose their patterns, or where intense disturbances may produce clouds, are in the lower and middle troposphere and the high troposphere. In particular, clouds in and near the fringes of the high cloud systems produced by slope convection, where observation from the ground is often helped by a shallowness and small amount or absence of low cloud, can commonly be seen to have forms affected or dominated by oreigenic disturbances. Since in the disturbances which produce wave clouds the vertical velocity is typically at least 1 m sec^{-1} the clouds form at saturation with respect to liquid water (Section 5.12); the upwind edges are sharply defined and near them the cloud may consist predominantly of droplets and be iridescent. However, at temperatures of about $-30°C$ the cloud becomes transformed within minutes into an ice cloud of comparatively sparse particles; at temperatures below $-40°C$ the transformation is much more rapid, practically all of the droplets freezing by homogeneous nucleation as they grow, and the concentration of solute from the condensation nucleus becomes small. High wave clouds therefore develop a characteristic asymmetry, since the ice particles evaporate only when the air has descended several hundred meters below the level at which cloud formed. Moreover, a wave cloud formed at temperatures in the range between about -30 and $-40°C$ often has a well-defined internal lee edge where droplets evaporate, as air descends rapidly through the level of saturation with respect to liquid water, before the transformation into an ice cloud is complete. Beyond this edge extends a more tenuous and diffuse "tail" of ice particles, which may reach a great distance if the air generally in its undisturbed state is nearly saturated with respect to ice (Pl. 6.4; 162, Fig. 16a; 168, Photo 1).

Lee waves may be manifest as undulations in such tails (Pl. 6.4), which sometimes link successive droplet clouds (168, Photos 2, 3). Frequently, however, such undulations are not apparent in extensive high wave clouds, including, apparently, the Moazagotl in the lee of the Sudeten mountains. This cloud has been described as occasionally at least 1 km and sometimes as much as 4 km thick, and as extending as far as 150–200 km from its windward edge, and more than 300 km across the wind direction (161, 163). It seems likely, in spite of the great width, that at least for a few hundred kilometers air in the high clouds described does not descend below its original level (a possibility suggested also by some calculated flows; 152). Oreigenic clouds which extend continuously so far from the responsible obstacle may be called *great hill-wave clouds*. It is in waves of large amplitude near and below the windward edges of great hill-wave clouds produced by major escarpments that sailplane flights to record heights have been made [e.g., to 11.4 km near the Riesengebirge in 1940 (163), and to between 13 and 14 km over the Owens Valley in the lee of the Sierra Nevada; such heights are near the limit safely attainable in aircraft or clothing not pressurized, and heights exceeding this limit have sometimes led to fatal accidents].

Even ridges of only moderate height and extent may produce high wave clouds. They are most easily recognized when prolonged observations from the ground on occasions of little low cloud show that isolated high clouds, or the windward and often apparently denser edges of bands of high clouds, form continually in particular parts of the sky (188–90).

From several years of visual observations made at Dunstable it appeared that when the winds over England are uniformly westerly or northwesterly throughout the troposphere and increasing with height (but not only when the winds at low or high levels are unusually strong), high clouds (developing into cirrus) may form over or near particular ranges of hills, such as the Cotswolds, whose peaks are not more than 250 m above the level of the surrounding ground. These oreigenic clouds were usually in the region extending 100–200 km poleward of extensive systems of high clouds whose edges were near the axes of jet streams, but they occurred also in deep cool or warm flows well away from jet streams. They were much more frequently observed in the winter than in other seasons, and although observations were not made at night, it seemed that they were much more likely to appear in the early and late hours of the day than during the night, or during the middle hours of the day when convection was in progress.

Evidently, consistent with theory, the amplitude of hill waves and the likelihood of wave cloud formation decreases as the depth of a layer of ordinary convection increases or, more probably, as the depth of the adiabatic layer in which (in contrast to the cumulus layer) the mean stability is small increases. The unfavorable circumstance at night is likely to be a decrease of wind speed at low levels. The intensification of hill waves during the late hours of the day, at least in the low troposphere, is well known to sailplane pilots (204).

On most occasions of hill-wave cloud formation in the high troposphere the air generally can be anticipated to have been nearly saturated with respect to ice, as often indicated by radiosonde observations of the relative humidity (which, however, are inaccurate at the low temperatures concerned) and occasionally as confirmed by frost points measured by research aircraft. The upward displacement of air in the hill waves required to produce the clouds was therefore probably at least 400 m (Appendix 2.2), and thus considerably greater than the height of the peaks of the responsible hills. This seems all the more remarkable considering that typically the width of the high hill-wave clouds was at most about 20 km, so that although the responsible hill ranges extend much greater distances across the wind it seems unlikely that they were behaving as effectively as obstacles of the same peak heights in a two-dimensional flow.

Occasionally the high hill-wave clouds form almost continuously for periods of up to several hours, and in the lee of the region of formation there is little or no evaporation, but only a more or less rapid transformation into a cloud composed wholly of ice. The clouds produced may then extend for hundreds of kilometers as streamers whose length is very much greater than their width, called *hill-wave plumes*. It appears either that there is a general state of supersaturation with respect to ice or that after passing through the hill wave at least a proportion of the air does not descend to its original level.

The occurrence of extensive regions of supersaturation with respect to ice was inferred in previous discussion from the duration of individual fallstreak forms of cirrus, and although their existence has rarely been confirmed by frost point observations it may be presumed that they are equally likely to be characteristic of occasions when hill-wave plumes develop. However, it is rare for fallstreak forms and plumes to occur simultaneously; rather, the plumes contain no (or only very feebly developed) fallstreaks, which if present are typically aligned along the length of the plumes. The plumes generally have a compact texture with ribs which are aligned across the length of the plumes and whose spacing typically is between a few and several kilometers, probably increasing with the thickness of the cloud and the wind speed at its level (188, Pl. vi; 189, Pl. 37; and Pl. 9.20 and 9.21). It therefore seems possible that the remarkable persistence of the clouds is associated with the development of some kind of small-scale instability (probably that described later as responsible for billow clouds) in the hill-wave cloud during or soon after its formation, which leads to an irreversible overturning in which some cloudy air remains above its original level.

Nacreous-wave clouds. In midwinter in high latitudes the beautiful nacreous-wave clouds occur on rare occasions in the lower stratosphere; their optical properties were described in Section 2.8 in the discussion of iridescence. During the winter the stratosphere in these latitudes cools markedly and the tropopause becomes ill-defined (Fig. 9.1); the temperature between about 20 and 30 km may fall to $-80°C$ or below, and in strong westerly flows which extend upward into the stratosphere the temperature may evidently in some locations be subject to a further lowering in pronounced hill and lee waves, and reduced to the abnormally low values of about $-90°C$ which are required for condensation to occur in the presence of the typical stratospheric mixing ratio of 2×10^{-6} (Fig. 9.12 and legend). These wave clouds have been observed in Scotland, Alaska, and (by the writer) in Iceland, and, at least during periods of careful watch, with a frequency of about once every year in Scandinavia and in Antarctica (194). The large-scale conditions under which these remarkable clouds occur, and the physical nature of their particles and of the noniridescent veils which trail far downwind of the hill waves in a manner suggestive of the change of phase associated with the cirrus of the troposphere plume clouds, have not yet been subject to satisfying analysis.

Billows

Kelvin-Helmholtz instability

When the shear in a layer of fluid exceeds some threshold value, depending on the degree of static stability imposed by the stratification of density, the flow becomes unstable and liable to the development of motion systems which draw their kinetic energy from that of the mean flow. By the action of pressure forces they redistribute, tending to make the momentum of the mean flow more nearly uniform. They develop in the form of growing waves of well-defined horizontal scale, but if their amplitude has become large, motions on a range of smaller scales appear, and the resemblance to ocean waves which break has prompted the use of the word *billows* to describe them.

Such theoretical studies of this so-called Kelvin-Helmholtz instability have employed various simplifying assumptions. For example, at first the shear was assumed to be concentrated at a horizontal interface separating incompressible fluids of constant density and velocity, and the motion has always been considered to be two dimensional. Results obtained for an incompressible fluid in which the initial velocity and density vary linearly across a transition layer (195) are likely to be at least qualitatively valid for the more complicated vertical profiles of velocity and density which arise in natural fluids, and in particular the atmosphere. They indicate two necessary conditions for the instability. First, there should be a point of inflection in the velocity profiles (implying a maximum in the horizontal component of vorticity); this condition also arose in the discussion of the generation of roll vortices in a boundary layer (Section 7.6). It is always likely to be fulfilled in the free atmosphere, where intense shears are encountered in layers between comparatively deep regions in which the shear is small. Second, the Richardson number in the transition layer should be less than some critical value.

The relevance of the Richardson number can be illustrated by referring to a layer of depth h in which wind speed and potential temperature increase linearly upward by amounts ΔU and $\Delta \theta$, respectively. It is assumed that the disturbances eventually invert the profile of potential temperature in the layer (consequences of the static instability thus arising are not considered) and make the wind speed within it uniform. In such a transformation the work done against gravity during adiabatic interchange of unit masses originally at the top and the base of the layer is $2(g\Delta\theta/2\theta)\Delta z$, and the decrease of the kinetic energy of the mean flow is

$$\frac{1}{2}\left[U^2 + (U + \Delta U)^2 - 2\left(\frac{U + U + \Delta U}{2}\right)^2\right]$$
$$= \frac{(\Delta U)^2}{4} = \frac{(\partial U/\partial z)^2 (\Delta z)^2}{4}$$

If this is to exceed the work done, the necessary condition is

$$\frac{(g/\theta)\Delta\theta}{\Delta z} < \frac{(\partial U/\partial z)^2}{4} \quad \text{or} \quad \text{Ri} < \frac{1}{4}$$

On the other hand, if the final distribution of potential temperature as well as that of the wind speed is considered to be uniform, the criterion becomes $\text{Ri} < \frac{1}{2}$.

In most analytical studies the early growth of the disturbances is considered and the flow is regarded as smooth; then only the momentum can be redistributed and the critical value of the Richardson number is indicated to be $\frac{1}{4}$.

From such studies it can be anticipated that in a transition layer of thickness h the wavelength λ of the disturbance with the greatest growth rate is approximately $2\pi h$, and that the time scale t of its exponential growth is given approximately by $t^{-1} = \alpha(1 - 4\overline{\text{Ri}})^{1/2}\Delta U/h$, where $\overline{Ri}$ is the uniform value of the Richardson number in the transition layer, and $\alpha \approx 0.3$.

An early study (196), in which development beyond the initial stages was calculated, suggested that the amplifying waves become asymmetrical and eventually roll the surface of separation or transition layer into separate vortices, between which the original shear is much reduced.

Laboratory experiments (197) and observations of billows in shallow layers of the ocean thermocline (198) confirm many features of these theoretical results. In particular, the waves are observed to grow into rolls, in which the vorticity is concentrated, while the original layer of shear between is stretched and narrowed. It is also observed that secondary rolls of smaller scale sometimes appear, and that soon after the principal rolls have attained a large amplitude there is a sudden generation of "turbulent" motions on a range of smaller scales, perhaps mainly as a result of static instability arising within the rolls. These small-scale motions spread to occupy a layer which in the thermocline has been observed eventually to become about four times as deep as the original layer of strong shear (corresponding to a mean Richardson number of about 1) before they decay (following the transfer of their kinetic energy to the smallest scales, where it is dissipated by viscosity) and the episode of disturbed flow ends.

In the atmosphere the occurrence and some of the properties of Kelvin-Helmholtz instability, again generally consistent with theory, have been inferred from observations of clouds (199) and from observations with aircraft and with radars of sufficient sensitivity to detect echoes from the more intense fluctuations of refractive index associated with motions of centimetric scale, produced in the later stages of billow evolution (200, 201).

Billow clouds of the high troposphere

Billow clouds can be divided into several kinds according to the circumstances in which they arise, or their characteristic appearance. Those that occupy a large part of the depth of an unstable layer, and which at some stage in their evolution show the development of the rolls, occur only on the upper surfaces of clouds formed by some other process; they appear as short-lived detail among widespread layers of ice clouds, in hill-wave clouds (Pl. 9.22), and in the tops of moderately large cumulus or in small cumulus (whose base is just below the top of a layer of convection) when there is a fresh or strong wind in the free atmosphere above.

Generally the instability arises in unsaturated air, and if clouds form they appear only in the upper parts of the billows. Among the most spectacular of billow clouds are those characteristic of the high troposphere, asso-

ciated with the most intense and deep shear layers and the jet streams produced by frontogenesis during slope convection. Near the jet streams the Richardson number, which in the free atmosphere is generally much greater than 1, is commonly reduced to about 1 even when averaged over deep layers (Fig. 9.17). However, in the layers of small Richardson number the relative humidity is characteristically low, and vertical displacements approaching or even exceeding 1 km may be required to produce condensation (Fig. 9.43). Thus the billow development which is common in these regions only rarely produces cloud, and more typically is responsible for bumpy flight conditions in clear air, or "clear-air turbulence." Billows in the high troposphere of middle latitudes are probably more prevalent than is generally recognized. In one series of radar observations made in the west of England billows with amplitudes exceeding 100 m were detected at ranges of a few tens of kilometers on 11 out of 30 days, during 5% of the total observing time of nearly 200 hr (201). Apart from its practical significance in aviation, Kelvin-Helmholtz instability in the high troposphere is probably mainly responsible for the virtually incessant scintillation of stars seen from the ground (Appendix 2.6) and may be an appreciable or even significant internal source of kinetic energy dissipation (202).

Clouds form in the billows very rarely, and in their upper parts probably only when the shear is unusually

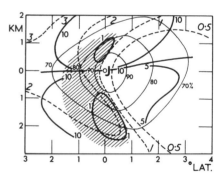

Fig. 9.43 The average distribution of the Richardson number (heavy lines) and some other properties of the high troposphere in northwesterly jet streams over western Europe, drawn with respect to the position J of the axis of the jet stream (199). The wind speed is shown by thin lines which are isopleths of the percentage of the speed on the axis. The dashed lines, labeled with values in km, show the upward displacement necessary to produce saturation with respect to liquid water; on the right within the isopleth for 0.5 km there is approximate saturation with respect to ice and ice clouds are often present. The hatched region is that in which aircraft are especially liable to encounter clear-air turbulence, and it may be presumed to be that in which Kelvin-Helmholtz instability commonly produces billows in layers whose thickness is normally a few and rarely several hundred meters. The relative humidity in this region is so low that clouds only very rarely form in the billows, and then only in their upper parts.

intense and the unstable layer is deep. In these circumstances the upward air speeds during cloud formation can be expected to have the magnitude of 1 m sec^{-1} and the clouds to form at saturation with respect to liquid water before becoming transformed into ice clouds (the temperature is typically near or below −40°C).

Consistently, the clouds most often appear as a group of several arched strips of compact and smooth texture (but occasionally there are as many as 20 or even more; 199, Fig. 16). The individual clouds are aligned across the shear vector, which is usually but not always at right angles to the direction of motion (Pl. 9.23). They extend sideways for a distance of about one wavelength, which is usually a few and sometimes as much as several kilometers, consistent with a thickness of the unstable layer which is usually a few hundred meters and sometimes as much as 1 km. As the billows amplify individual billow clouds form both ahead of and behind the first members of a group, indicating a characteristic feature of motion in which sinusoidal components grow at different rates from a spatially concentrated disturbance of small amplitude. The clouds move with the velocity of the billows, which is the mean velocity of the unstable layer. The relative velocity at the cloud level is therefore in the direction of the shear, and usually forward (Fig. 9.44); as the clouds become ice clouds, and small-scale motions develop, their leading edges become frayed and tendrils extend across the intervening spaces (Pl. 9.23). Typically the time scale of the growth of intense billows which is indicated by observations is about 10 min, of the same magnitude as that obtained by using values of $\overline{\text{Ri}} = 0.2$ and $\Delta U/h = 30$ m sec^{-1} km^{-1} in the theoretical formula previously quoted. The entire duration of the disturbed flow is probably rarely as long as 1 hr.

Sometimes ragged clouds appear also below the fraying arches, often as shreds but occasionally as more cumuliform fragments. They are probably due to condensation in air which has risen from the lower part of the billows in a convective overturning late in their evolution. These ragged *shred* or *puff clouds* are occasionally the only clouds to appear (Pl. 9.24), often singly rather than in the groups which indicate the kind of motion system responsible for their origin. They usually diffuse and evaporate within minutes of their formation, whereas the irregular residues of groups of billow clouds are more persistent, sometimes lasting for more than an hour. In time-lapse films the puff clouds show striking internal commotion, but the turbulent flow which also accompanies the later stages of the evolution of the ordinary billow clouds has often decayed and disappeared before their residues evaporate. The diffused, irregular residues may travel long distances, and on arrival within an observer's view offer no indications of their manner of formation.

Billow clouds sometimes form repeatedly near prominent topographical features, but on occasions when the winds are strong and the billows develop in air of low relative humidity the first billow cloud may appear at distances of up to about 100 km from the region in which the billows originated, a distance so great as to obscure the common circumstance that this region is in a hill wave. The Richardson number deviates from its general value during inclined and accelerated flow through a hill wave; in particular, it can be shown that following steady two-dimensional flow through a hill wave the Richardson number decreases where the flow is decelerated (assuming that the shear is directed forward; deceleration implies that the wave amplitude increases with height) and inclined upward (203). During a large-scale reduction in the value of a Richardson number, the threshold value for the generation of billows is therefore likely to be passed first in the crests of pronounced hill waves. Consistently, since billows are probably the predominant source of bumpy flight conditions in the upper troposphere, there is a recognized but not altogether clear or reliable relation between clear-air turbulence and topographical features. The small-scale wavy detail which is sometimes visible from the ground on the surfaces of wave clouds, especially the lower surfaces when they are illuminated by the sun at a very low elevation, are also probably generated by a Kelvin-Helmholtz instability which in a wave of limited horizontal extent is suppressed before it reaches an advanced stage of development. Occasionally, when the lighting is favorable, separate shallow billow clouds can be seen beneath hill-wave clouds.

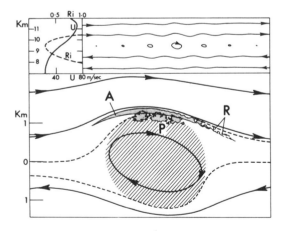

Fig. 9.44 Some features probably representative of Kelvin-Helmholtz instability in the high troposphere (199). Above: The form of the relative streamlines in a group of amplifying billows, and (left) typical profiles of wind speed U and Richardson number Ri. Below: The probable form of the relative streamlines in a mature billow, and the location of an arched billow cloud A with a ragged fringe R of ice particles and of puff clouds P forming (perhaps alone) in air overturning in the hatched region. The turbulent motion generated by the overturning may be carried into the region between the dashed lines, and even spread beyond them, before eventually dying away.

Billow clouds of the lower troposphere

There are two other meteorological regimes, both in the lower troposphere, in which Kelvin-Helmholtz instability may play a significant role in the formation of clouds, or in modifying their form.

The first of these is the region near the top of a layer of ordinary convection when the general wind is at least moderately strong. As previously remarked, the mean wind in a layer of ordinary convection characteristically has little shear, and if its top is well defined, the conditions for Kelvin-Helmholtz instability may continually arise in the shallow layer of transition into the free atmosphere: clear-air turbulence is commonly detected by radar at the top of cloud-free layers of convection. Further, among the first clouds to appear as the layer of convection overland deepens during invasions of maritime air, some frequently form in ragged groups of three or four individuals which can be envisaged to compose a wave train with a wavelength of about 1 km or less. Small and frequently ragged rudimentary cumulus are called *fractocumulus*, but those of the kind mentioned have details with a rolling motion typical of billow clouds, and might be distinguished as *cumulus billows*.

Second, Kelvin-Helmholtz instability may play a role in organizing the pattern of some of the shallow *altocumulus* droplet clouds of the lower middle troposphere. The processes responsible for the formation and persistence of these clouds do not appear to have been examined in any detail. Almost certainly they usually arise when air from the top of a layer which has been subject to moderately deep cumulus convection is lifted in slope convection. Shallow altocumulus are characteristic of the eastern fringes of cloud systems ahead of cold fronts, first appearing in patches localized by hill waves (especially early and late in the day). Near the fronts the layers are thick, other clouds may be present both above and below, and internal convection may be vigorous and is associated with differential advection. In thinner layers with no cloud above, radiative exchanges produce a net cooling and a slow convection. If the wind shear at the cloud level is small, a cellular (dappled) pattern appears, but otherwise rolls transverse to the shear vector develop, with clear spaces between. In the presence of moderately intense shears the rolls seem all to have the same orientation, but when the shear is weaker its direction may vary considerably during flow over hills, so that the orientation may be different in different parts of the sky, or may change within one group of billows as it travels. The original formation of the roll pattern may be governed by Kelvin-Helmholtz instability, but the clouds commonly persist for periods as long as an hour in a mature and apparently steadily maintained condition.

Appendix 9.1 The mean annual latitudinal distribution of components of the energy balance of the atmosphere

The mean annual local rate of loss of energy R_a of the atmosphere by radiation is

$$R_a = I_a - S_a$$

where I_a is the net long-wave radiative loss and S_a is the absorption of the solar short-wave radiation. This loss is balanced in the equation expressing the heat economy

$$R_a = Lr + H + A$$

by the local release of latent heat associated with the annual precipitation r, the upward flux of sensible heat H from the earth's surface, and the horizontal convergence A (representing an energy gain) of the advective heat flux (which is usually obtained as a residual following the estimation of the other terms). Table 9.4 presents the assumed latitudinal variation of the annual mean of the various terms in the northern hemisphere (compiled from reference 66). The table includes estimates of the annual evaporation E from the earth's surface. The values of the fluxes upward from the surface are given in degrees Celsius per day of equivalent

Table 9.4 Estimated annual mean rates in the northern hemisphere of evaporation E from the earth's surface, of rainfall r, of the provision to the atmosphere of latent heat Lr and sensible heat H, and of the rate of cooling of the atmosphere equivalent to the mean radiative energy loss R_a, all as functions of latitude

Latitude zone (°N)	E (cm)	r (cm)	$E - r$ (cm)	Warming (°C day^{-1}) equivalent to:		Cooling (°C day^{-1}) equivalent to:	Residual
				Lr	H	R_a	A
80–90	4	12	−8	0.1	−0.1	0.9	0.9
70–80	15	19	−4	0.1	0	0.9	0.8
60–70	33	41	−8	0.3	0.1	0.9	0.5
50–60	47	79	−32	0.5	0.2	0.8	0.1
40–50	64	91	−37	0.6	0.2	0.7	−0.1
30–40	100	87	13	0.6	0.3	0.7	−0.1
20–30	125	79	46	0.5	0.3	0.8	0
10–20	139	115	24	0.8	0.2	0.8	−0.1
0–10	123	193	70	1.3	0.1	0.8	−0.6
0–90	94	101	—	0.7	0.2	0.9	—
Globe	100	100	—	0.7	0.1	—	—

9 Large-scale slope convection

warming, assuming that the whole depth of the atmosphere is involved, although the energy represented is introduced directly into the troposphere only.

Precipitation amounts over the oceans (which occupy nearly two thirds of the surface area of the northern hemisphere) are estimates based on measurements over islands or neighboring lands, and other indirect information; they are liable to considerable error.

Appendix 9.2 The relative proportions and typical rates of middle-latitude precipitation from frontal zones and from showers

Studies of the amounts of rainfall which at particular inland places are associated with frontal zones or with showers are made laborious by the necessity to relate each rainfall over a lengthy period to the accompanying large-scale weather pattern; moreover, because of the strong effects of local topography upon the intensity and distribution of both kinds of rainfall the results are not necessarily representative of the rainfall over larger areas, especially belts between particular latitudes, which include much ocean surface. (From indirect experience it seems that in middle latitudes the total precipitation over the open ocean is considerably less than that over islands or the western parts of the continents; e.g., see 67.) The most recent and careful study of this kind (68) was made by inspection of the autographic records from seven stations in northern England and of the analyses of the surface weather maps published in the British Daily Weather Report. According to these analyses, the individual rains during a period of five years were associated with nine kinds of weather listed in Table 9.5, of which the first three may be regarded as frontal rains and the second three as shower rains in the cool airstreams of the large-scale slope convection. Of the seven stations the first two in the table are on the northwest coast, the third is about 50 km inland, and the following three are on higher ground on the western side of the hills in central northern England; the last is in northeast England and usually in the lee of the hills. The table shows that during the five years the rains in the several categories vary from station to station in the sense which might be anticipated, but not by much when expressed as percentages of the total at each station.

Considering that the rainfall associated with the so-called polar lows probably more often falls in unrecognized frontal zones than from cumulonimbus, and that the rainfall from warm airstreams is probably mostly associated with ascent over topographical features and is not representative of oceanic regions, the implication of the table is that over the whole region the shower rainfall represents only about one third of that which falls in frontal zones.

Among earlier studies, one made of the rainfall during one year at four stations in the same part of England led to similar results (69). On the other hand, a more detailed study using autographic records obtained at low islands off the northwest coast of Great Britain during a period of three years concluded that at these places (probably more representative of the open ocean) nearly 90% of the total rainfall was associated with fronts and less than 10% with showers (70). However, it seems that some rains were associated with fronts not marked on the ordinary synoptic charts but inferred to have been present from an examination of autographic records, especially in airstreams south of intense cyclones with centers well north of Great Britain. Many of these inferred fronts are likely to have been squall fronts associated with traveling cumulonimbus, whose rains would have been attributed to cold fronts but more properly to showers. This possibility is supported by the proportion of the total rainfall allocated to cold fronts (38%), which is high in comparison with that given by other studies, including the one whose results are summarized in Table 9.5.

Other interesting results of the earlier study include the conclusion that in southeast England, farthest from the ocean, frontal rains amounted to little more than half (57%) of the total, evidently partly because these rains are intensified locally by orographic effects in the more hilly regions, whereas the shower rains are considerably increased in the more sheltered southeast by solar heating of the ground (mainly in summer).

It was also found that in the outer islands there was little variation of the mean hourly rate of fall of prolonged (even if intermittent) rains from warm fronts about an average value of 0.7 mm hr^{-1}. On the other hand, at an inland place on high ground (Eskdalemuir Observatory) the predominant rate of rainfall was about 2 mm hr^{-1}, and the mean of the extreme rates was 7 mm hr^{-1}. In contrast, although the mean of the average rates of rainfall in several places was everywhere between 1 and 2 mm hr^{-1} for all kinds of fronts, cold fronts (and occlusions) were observed to be characterized by great variations in the rate of rainfall during their passage, with extreme rates generally between 15 and 20 mm hr^{-1} in winter, but sometimes reaching 100 mm hr^{-1} or more.

A steady rainfall without evaporation can be related to a mean rate of ascent in a cloud (71). Most of the

Table 9.5 The average annual rainfall (cm) at seven stations in northern England during the period 1956–60, divided into nine categories according to the weather situation, and the percentage P of the total rainfall in each category (68)

Station	Warm front	Cold front	Occlusion	Cool airstream			Warm airstream	Polar low	Thunderstorm
				Polar maritime	Polar continental	Arctic maritime			
Speke	20	13	12	20	—	1	11	9	2
Ringway	18	12	11	18	—	1	11	8	2
Keele	18	13	12	16	1	1	10	9	3
Nelson	23	16	16	26	—	—	18	10	2
Greenfold	28	19	20	32	1	1	20	11	3
Rotherham	17	7	13	10	1	1	6	9	2
Average P	22	15	14				13	10	3
		51			23				

rainwater is provided by condensation in the lower troposphere, so that a very approximate relation can be obtained by equating the rainfall rate to a large fraction of the upward flux of water vapor. Over a middle-latitude ocean of surface temperature about 10°C the concentration of water vapor at the base of a cloud in the low troposphere is several grams per cubic meter. Then if w (cm sec^{-1}) is the magnitude of the vertical velocity in the low troposphere and R (mm hr^{-1}) is the rainfall rate,

$$w \times 5 \times 10^{-6} \approx R/3.6 \times 10^4$$

or

$$w \approx 5R$$

The mean rate of rainfall of about 0.7 mm hr^{-1} in warm front zones is therefore consistent with a typical large-scale upward velocity of a few centimeters per second in the low troposphere. Inland this rainfall rate is considerably increased locally by ascent during the passage of airstreams over hilly ground. The characteristically irregular and often large rates of rainfall associated with cold front zones imply that there the vertical velocities locally exceed 1 m sec^{-1}, and therefore that convection clouds are probably present, producing rates of rainfall typical of showers (Chapter 8).

Appendix 9.3   The frequency distribution of the heights of shower-cloud tops over the North Atlantic

Figure 9.45 is a histogram showing the number N of occasions on which the tallest cumulus or cumulonimbus observed over the Northeast Atlantic were reported to have tops lying within pressure intervals of 50 mb, between the 850- and 400-mb levels. The observations, totaling nearly 300, were made during a period of about two years by reconnaissance aircraft of the British Meteorological Office, in the course of a series of flights on which special attention was given to the heights of the bases (Fig. 7.9) and tops of cumuliform clouds. The clouds were reported as cumuliform or as cumulonimbus with glaciated tops or anvils, but showers were observed on virtually all occasions when the cloud tops reached above the 800-mb level.

The observations were made during soundings between a level near the sea surface and the 500-mb level, by either ascent or descent. The heights of cloud tops above the 500-mb level thus are estimates; however, the great number of occasions on which the tops were estimated to be only a little above this level is not likely to have been due simply to a persistent tendency to

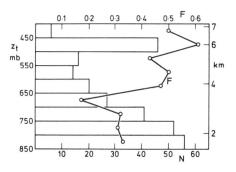

Fig. 9.45 The number N of occasions in a series of flight observations over the Northeast Atlantic when the level z_t of the tallest clouds of ordinary convection lay within the pressure intervals marked. An approximate scale of height (in km) is added on the right. Also shown is the fraction F of occasions when the surface wind had a southerly component.

underestimate the heights, since they were consistently supported by observations during parts of the flights made at the 500-mb level over distances of a few hundred kilometers, and tops were never estimated to reach the 400-mb level (i.e., little more than 1 km above the flight level). Moreover, on most occasions when the tops were estimated to be above the 500-mb level, θ_s at 500 mb was within $\pm 1°C$ of its value at the level of the cloud base, suggesting that little further cloud growth was likely. Consequently it is thought that the diagram correctly indicates that the 450-mb level is rarely reached by the shower clouds of the cool airstreams in slope convection over the ocean.

On the overwhelming majority of occasions the surface wind was from a westerly point. The diagram shows that when the clouds were small the surface wind usually had a northerly component. Remarkably, however, on more than half the occasions when the cloud tops exceeded about 4 km the surface wind had a *southerly* component. This feature again illustrates the effect of pronounced large-scale subsidence (associated mainly with the airstreams moving into lower latitudes) in suppressing the depth of the layer of ordinary convection.

According to some observations made by aircraft over the west of the North Atlantic the height of the (tallest) tops of the clouds of ordinary convection in cool airstreams is in general rather lower there than over the east of the ocean: in one series of observations over the sea between about the 40th and 50th parallels the average height was just over 3 km, and the extreme only 4.8 km (72), while in another series of nearly 700 observations, made over the route between Washington, D.C., and Bermuda (32°N), the average height was between 2 and 3 km, with an extreme of about 6 km (73). Remarkably, it was found in a study of the latter series that the height of the tops was better (and usefully) correlated with the mean relative vorticity of the observed 700-mb level winds over the route than with any other single meteorological variable. When the relative vorticity was anticyclonic the tops were almost always below 2.5 km, and the highest tops (about 6 km) were associated with the largest cyclonic vorticity (about $5 \times 10^{-5} \text{ sec}^{-1}$). This result may be regarded as another expression of the control by the large-scale vertical motion.

Appendix 9.4 The location of thunderstorms by observations of atmospherics

The radio-wave energy radiated by a lightning stroke is most intense at frequencies near 10^4 Hz. Receivers tuned to this frequency can determine the apparent direction of arrival of the energy, observed as an *atmospheric* or *sferic*, and therefore the bearing of the source, at great distances. A network of a few receivers, used to determine the bearing of sferics received virtually simultaneously, comprises a thunderstorm location system whose performance depends in large measure on the techniques employed (36). The British system, which has been in routine use since the 1940s with only minor refinements, has proved that the observations obtained, although subject to some limitations, are a valuable and cheap aid in large-scale analysis. With this and some other mostly experimental systems it has been shown almost conclusively that sferics are received only from places where a ground observer would report a thunderstorm, and probably almost entirely from lightning discharges to the ground, because the energy radiated by these discharges is about an order of magnitude greater than that from discharges within a thundercloud. Moreover, although in the normal mode of operation not all thunderstorms may be detected, a sferics location system provides a more comprehensive survey of storm distribution over a large area than the networks of conventional observing stations at the surface. For example, it has been found that the British system detects about ten times as many thunderstorms during daylight hours over the whole of Europe as the 2000 routine reporting stations. After the position of a source has been specified by this system to the nearest 0.5° of latitude and longitude, it is estimated that half of the positions are in error by less than 75 km; the system has about this acceptable degree of accuracy (for large-scale analysis) at ranges up to nearly 2000 km.

Appendix 9.5 Experience of hazardous mountain wave phenomena

The oreigenic phenomena which can be damagingly severe when well developed are the strong surface winds on lee slopes and beneath the troughs of lee waves, the intense smooth downdrafts which may quickly cause a perilous loss of flying height over mountainous terrain, and the extremely bumpy flight conditions on traverses of rotors (which may be encountered suddenly at seemingly safe distances from mountain ridges), or turbulent layers within mountain and lee waves.

Strong and often gusty surface winds are a well-known feature of foehn weather in the lee of long ridges (158), even of only moderate height, for example, the northern Pennines in England, on whose southwest-facing escarpment the ground descends 600 m to the valley floor in a horizontal distance of about 3.5 km (the separation of lee roll clouds over the low ground is typically about 7 km). Occasionally in winter northeasterly airstreams there may be a steady gale of about 20 m sec^{-1} down the smooth parts of this slope, whose speed is rather greater than that experienced on the summit of the ridge (about once in several years the slope wind attains hurricane force—about 40 m sec^{-1}). In surprising contrast, the wind falls light and variable near the foot of the slope, reverses under the bar, and beyond another calm region becomes a gusty northeasterly, but only about half as strong as on the lee slope (164). The phenomenon occurs only when the direction of the wind is within about 30° of a line normal to the main crest line of the ridge, and when its general speed near the ground on the windward side of the ridge is at least several meters per second. It is always accompanied by a wall cloud lying on or slightly above the ridge crest, which is known as the *helm cloud*. In the calm zone below the first bar cloud a characteristic roar can be heard up to 1–2 km away from the region of violent wind. This roar of the nearby strong wind is mentioned in many other accounts of foehn winds (including those like the *bora* of the eastern Adriatic which lash the seas in narrow zones in the lee of steep cliffs; these winds seem especially prevalent when the high ground is covered with ice or snow, where the nature of the flow is probably affected by a strong contrast of temperature between land and sea). The slopes worst affected by the *helm wind* are inhabited only by a few hill farmers, who suffer a scourge, hardly understood by lowlanders:

> The desiccating and shrivelling effect of the strong surface current in spring time is to be feared ... those who wish to ascend Crossfell in March find themselves bending low with ears covered, wearing every available wind proof, as they breathlessly toil up the steep slope against the remorseless thrust of a force 9 torrent of air. Behind them, the farmer's wife shivers in the brilliant sun and finds her heavy blankets stretched horizontally from the wire on which they are hung to dry, or flapping and tearing at their edges when the wind in its retreat becomes gusty. The farmer looks gloomily at the roofs of his Dutch barns, at the dusty ploughland, the cowering sheep and shrivelled pasture and the snowy slopes above; for when "t'helm's on" nothing will grow and if snow is on the way he must toil upward to see that every gate leading to the fell must be left open for the sheep to find their way down. Neglect of this in the past has led to terrible losses as soon as the snow has begun to drift against the walls barring the way to lower ground. (164)

Similar foehn winds occur in every hilly or mountainous part of the world outside very low latitudes, and are described by a great variety of local names (180). Among the most intense, frequent, and widely known are the *chinooks* associated with ranges in the Rocky Mountains of North America. During the winter, the onset of a chinook in a westerly airstream, which replaces stagnant cold air in the lee of these mountains, is often accompanied by extraordinary rapid rises of screen-level temperature (up to 20°C in a few minutes) and decreases of relative humidity (to 15% or less); a wavering narrow turbulent transition zone between the cold and warm air, made visible by a mixing fog, has been observed (181). At the feet of the lee slopes of the Front Range in Colorado extreme mean speeds of 25 m sec^{-1}, with frequent gusts to 40–50 m sec^{-1}, occur about once a year (mainly in the late or early hours of midwinter days; 182). Such winds destroy walls, roofs, and windows; often (perhaps by felling powerline supports and overturning vehicles or parked aircraft) they cause and fan serious fires. Violent foehn winds which produce the same hazards, especially rapidly spreading brush and timber fires, occur in many other regions [e.g., the Santa Ana and similar winds in California (183) and slope winds over forested regions near the southeast coast of France].

A particular rare combination of properties in an airstream sometimes causes extraordinary surface winds in the lee of a ridge at places unaccustomed to such visitations, where the damage caused is the greater because of the consequent lack of precautions in the siting and strengthening of constructions. Two examples are the destruction of banana plantations near the north coast of Tenerife in 1937 and the damage in two narrow zones parallel to and in the lee of the Pennines of northern England early on 16 February 1962. On the latter occasion most of the damage was to trees in open country, but it seems that the city of Sheffield, about 40 km from the ridge crest, lay below the trough of a second lee wave of large amplitude (187). In a gale not otherwise exceptional for the region, gusts of about 45 m sec^{-1} occurred in the city; of about 160,000 dwellings, 100,000 suffered some damage and 34,000 suffered damage classified as moderate.

According to the experience of pilots, especially sailplane pilots, the flow in mountain waves is mostly remarkably smooth, and while soaring or maintaining the position of a sailplane it is often possible not to touch the controls for as long as several minutes. However, at all levels up to great heights and even in the (lower) stratosphere turbulent flow is encountered, the width of the transition zone sometimes being remarkably narrow, perhaps only a few tens of meters. Much of the turbulence is of only slight or moderate intensity, containing disturbances of comparatively small scale and intensity, but occasionally it becomes impossible to control the aircraft, and especially perhaps under additional stresses imposed by the pilot's responses, the aircraft may break up.

The following accounts by B.L. Edgar are examples of the extreme conditions experienced by sailplane pilots:

> The turbulence became more severe as we progressed under the roll cloud. The magnitude of the vertical change of the position between the two aircraft was very great. The rate at which these changes occurred was amazing. Too, the distances between the two aircraft would change with no apparent change in speed or attitude. The towplane would suddenly loom larger and we would be rapidly overtaking it with the sailplane.
>
> We released under the roll cloud when the tow plane dropped, pitched forward and disappeared from sight. The release was made before the tow rope could become tight and jerk the nose of the sailplane. The time was 14:48 and altitude 11,000 feet. The maximum accelerometer readings on tow were found under the roll cloud with $a = +5$ and $-2G$.
>
> The towplane came along beside us. To have the towplane flying beside us off to our right, really gave us a picture as to what was going on. We would suddenly rise and the tow plane would drop out of sight. Soon the towplane would come back up and we would be on our way down. It reminded me of two elevators operating side by side.

On another occasion:

> The flight path went into the very top of the little cloud puff. It seemed to swell up before the nose in the last moment. I looked at the needle and ball. Suddenly and instantaneously the needle went off center. I followed with correction but it swung violently the other way. The shearing action was terrific. I was forced sideways in my seat, first to the left, then to the right. At the same time when this shearing force shoved me to the right, a fantastic positive G load shoved me down into the seat. This positive load continued. Just as I was blacking out, it felt like a violent roll to the left with a loud explosion followed instantaneously with a violent negative G load.
>
> I was unable to see after blacking out from the positive G load. However, I was conscious and I felt my head hit the canopy with the negative load. There was a lot of noise and I felt like I was taking quite a beating at this time. I was too stunned to make any attempt to bail out.
>
> Just as suddenly as all of this violence started, it became quiet except for the sound of the wind whistling by. I felt I was falling free of all wreckage except for something holding both feet.

Similar encounters occasionally are responsible for disasters to powered aircraft or may be suspected to have been their cause. One such encounter, for example, almost certainly led to the breakup in the air and subsequent crash of a passenger jet aircraft (a Boeing 707) approaching Mt. Fujiyama from the lee side in fair weather in 1966 (184). However, the hazards of flight in the strong downdrafts near some lee slopes, and in rotors farther away, have become well known, and notorious locations are now generally carefully avoided. Consequently, of 20 catastrophic accidents to civil transport aircraft in the period 1950–66 and in which extremely bumpy flight conditions played a significant part, all but the one just mentioned were probably (and 17 were certainly) associated with intense cumulonimbus (185). Nevertheless, the frequency of very bumpy flight experienced by large commercial aircraft in clear air, nearly all associated with turbulent flow in mountain waves in the upper troposphere, is not negligible: brief encounters described as "severe," sufficient to cause momentary loss of control and the vertical accelerations of up to 2 g, are (often unexpectedly) experienced about once in several hundred hours of flight time (186).

A striking example of the kind of damage which can be inflicted is seen in Pl. 9.25, a picture of a large B-52 aircraft returning from research flights in which searches were made for turbulent flow on 10 January 1964. During the return to base when at about the level and not far from the crests of the Sangre de Cristo range in southern Colorado, changes of wind velocity inferred to have reached about 30 m sec^{-1} over distances of only a few hundred meters tore away most of the tailfin (about 12 m tall). The crew members were fortunate to be able to land the aircraft safely. (This range, whose crests reach about 5 km, is notorious for the strong downdrafts on its lee side in westerly winds.)

Appendix 9.6 Climatological data

Figure 9.46 shows distributions over the northern hemisphere of mean values of θ and θ_w for an early afternoon in July 1959. The data used in the analyses were for 1200 local time, or for the time of maximum screen temperature, supplemented in places by long-period averages. Isopleths are drawn mostly at intervals of 4°C, and those for θ over land only.

At this time of day the values are mainly between 1 and 2°C higher than those representative of the adiabatic layer during ordinary convection. However, over land in hot calm weather the differences may be considerably larger, partly as a result of the inaccuracy of screen thermometers which are not specially ventilated.

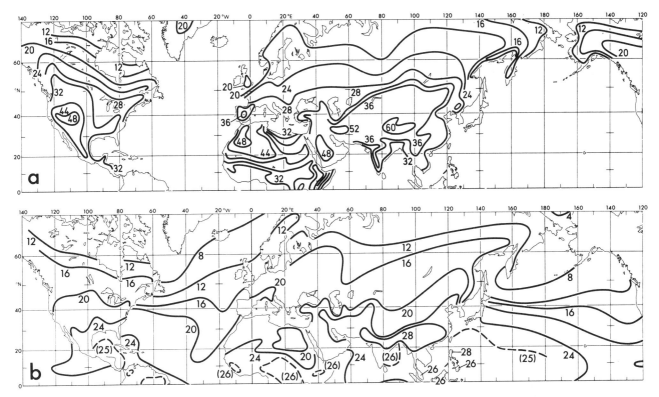

Fig. 9.46 Distributions of mean values of (a) θ and (b) θ_w at screen level during an early afternoon in July 1959.

References

1. Mintz, Y. 1954. The observed zonal circulation of the atmosphere, *Bull. Am. Meteorol. Soc.*, **35**, 208–14; Batten, E.S. 1961. Wind systems in the mesosphere and lower ionosphere, *J. Meteorol.*, **18**, 283–91; Goldie, N., Moore, J.G., and Austin, E.E. 1958. *Upper Air Temperature over the World*. Geophys. Mem. *13*, no. 101, Meteorological Office, London.
2. Palmén, E., and Newton, C.W. 1969. *Atmospheric Circulation Systems: Their Structure and Physical Interpretation*. International Geophysics Series, Vol. 13, Academic Press, New York.
3. Fultz, D. 1961. Developments in controlled experiments on larger scale geophysical problems, *Adv. Geophys.*, **7**, 1–103.
4. Godske, C.L., Bergeron, T., Bjerknes, J., and Bundgaard, R.C. 1957. *Dynamic Meteorology and Weather Forecasting*. American Meteorological Society, Boston (see p. 245).
5. Green, J.S.A. 1970. Transfer properties of the large-scale eddies and the general circulation of the atmosphere, *Q. J. R. Meteorol. Soc.*, **96**, 157–85.
6. See, e.g., Burger, A.P. 1958. Scale considerations of planetary motions of the atmosphere, *Tellus*, **10**, 195–205; Phillips, N. 1963. Geostrophic motion, *Rev. Geophys.*, **1**, 123–76.
7. Petterssen, S. 1956. *Weather Analysis and Forecasting*, 2d ed. Vol. 1, *Motion and Motion Systems*. McGraw-Hill, New York (see ch. 8–10).
8. Klein, W.H. 1957. *Principal Tracks and Mean Frequencies of Cyclones and Anticyclones in the Northern Hemisphere*. Research Paper 40, U.S. Weather Bureau, Washington, D.C.
9. Green J.S.A. 1960. A problem in baroclinic stability, *Q. J. R. Meteorol. Soc.*, **86**, 237–51.
10. Edelmann, W. 1963. *On the Behaviour of Disturbances in a Baroclinic Channel; Research in Objective Weather Forecasting by Staff Members*. Summary Report 2, Contract AF 61(052)-373, Res. Div. Deutsches Wetterdienstes, Offenbach.
11. Ludlam, F.H. 1970. *The Cyclone Problem: A History of Models of the Cyclonic Storm*. Inaugural Lectures, 1966–67 and 1967–68, Imperial College of Science and Technology, London, pp. 19–49.
12. Berggren, R., Bolin, B., and Rossby, C.-G. 1949. An aerological study of zonal motion, its perturbations and break-down, *Tellus*, **1**, no. 2, 14–37.
13. Rossby, C.-G. 1959. Current problems in meteorology, in *The Atmosphere and the Sea in Motion*. Rockefeller Institute Press, New York, pp. 9–50.
14. George, J.J. 1960. *Weather Forecasting for Aeronautics*. Academic Press, New York.
15. Petterssen, S., Bradbury, D.L., and Pedersen, K. 1962. The Norwegian cyclone models in relation to heat and cold sources, *Geofys. Publ.* (Oslo), **24**, 243–80.
16. Postma, K.R. 1948. The formation and the development

of occluding cyclones; a study of surface-weather maps, *Med. en Verh.*, K.N.M.I., ser. B., **1**, no. 10, 1–57 (de Bildt, Netherlands).
17. Breistein, P.M., and Parry, H.D. 1953. *Some Dynamical Aspects of Static Stability*. Tech. Rept. 2, Contract AF 19(604)-390, Department of Meteorology, University of Chicago.
18. Taljaard, J.J. 1967. Development, distribution and movement of cyclones and anticyclones in the southern hemisphere during the IGY, *J. Appl. Meteorol.*, **6**, 973–87.
19. Manabe, S., and Bryan, K. 1969. Climate and the ocean circulation, *Mon. Weather Rev.*, **97**, 739–827.
20. Manabe, S., Smagorinsky, J., Holloway, J.L., and Stone, H.M. 1970. Simulated climatology of a general circulation model with a hydrological cycle: III. Effects of increased horizontal computational resolution, *Mon. Weather Rev.*, **98**, 175–212.
21. Riehl, H., and Fultz, D. 1957. Jet streams and long waves in a steady rotating-dishpan experiment: structure of the circulation, *Q. J. R. Meteorol. Soc.*, **83**, 215–31.
22. Corby, G.A., ed. 1971. *The Global Circulation of the Atmosphere*. Royal Meteorological Society, London.
23. Krishnamurti, T.N. 1961. The subtropical jet stream of winter, *J. Meteorol.*, **18**, 172–91.
24. Shenk, W.E. 1970. Meteorological satellite infrared views of cloud growth associated with the development of secondary cyclones, *Mon. Weather Rev.*, **98**, 861–68.
25. Zipser, E.J. 1969. The role of organised unsaturated convective downdrafts in the structure and rapid decay of an equatorial disturbance, *J. Appl. Meteorol.*, **8**, 799–814.
26. Welander, P. 1955. Studies on the general development of motion in a two-dimensional, ideal fluid, *Tellus*, **7**, 141–56.
27. Angell, J.K., Allen, P.W., and Jessup, E.A. 1971. Mesoscale relative diffusion estimates from tetroon flights, *J. Appl. Meteorol.*, **10**, 43–46.
28. Robinson, G.D. 1967. Some current projects for global meteorological observation and experiment, *Q. J. R. Meteorol. Soc.*, **93**, 409–18.
29. Djurić, D. 1966. The role of deformation in the large-scale dispersion of clusters, *Q. J. R. Meteorol. Soc.*, **92**, 231–38.
30. Danielson, E.F. 1961. Trajectories: isobaric, isentropic and actual, *J. Meteorol.*, **18**, 479–86.
31. Green, J.S.A., Ludlam, F.H., and McIlveen, J.F.R. 1966. Isentropic relative-flow analysis and the parcel theory, *Q. J. R. Meteorol. Soc.*, **92**, 210–19.
32. Eliassen, A., and Kleinschmidt, E. 1957. Dynamic meteorology, in *Encyclopedia of Physics*, **48**, Geophysics II, Springer-Verlag, Berlin, pp. 1–154.
33. Leese, J.A., Novak, C.S., and Clark, B.B. 1971. An automated technique for obtaining cloud motion from geosynchronous satellite data using cross correlation, *J. Appl. Meteorol.*, **10**, 118–32.
34. Palmén, E., and Newton, C.W. 1948. A study of the mean wind and temperature distribution in the vicinity of the polar front in winter, *J. Meteorol.*, **5**, 220–26.
35. Betts, A.K., and McIlveen, J.F.R. 1969. The energy formula in a moving reference frame, *Q. J. R. Meteorol. Soc.*, **95**, 639–42.
36. *Atmospherics Techniques*. 1955. Tech. Note 12, World Meteorological Organisation, Geneva.
37. Berson, F.A., and Petterssen, S. 1943. *Atmospherics in relation to fronts and air masses*. Meteorological Research Paper 130, Meteorological Office, London. (A copy is available in the Meteorological Office Library.)
38. *Final Report on the AWS Sferics Evaluation Project (1951)*. 1954. Air Weather Service Tech. Rep. 105–102, Air Weather Service, Washington, D.C., p. 152.
39. Phillips, E.F. 1958. An analysis of atmospherics for June, July and August, 1957, *Australian Meteorol. Mag.*, no. 22, pp. 4–9.
40. Kimpara, A. 1955. Atmospherics in the Far East, *Proc. Res. Inst. Atmospherics* (Nagoya University), **3**, 1–28.
41. Parmenter, F.C. 1970. Picture of the Month—Pacific cyclones, viewed by ATS 1, *Mon. Weather Rev.*, **98**, 83–84.
42. Conover, J.H., Lanterman, W.S., and Schaefer, V.J. 1969. Major cloud systems, in *World Survey of Climatology*, vol. 4. Elsevier, New York, pp. 205–45; Widger, W.K. 1964. A synthesis of interpretations of extratropical vortex patterns as seen by TIROS, *Mon. Weather Rev.*, **92**, 263–82.
43. Conover, J.H., and Wollaston, S.H. 1949. Cloud systems of a winter cyclone, *J. Meteorol.*, **6**, 249–60.
44. Parmenter, F.C. 1968. Picture of the month, *Mon. Weather Rev.*, **96**, 54–55.
45. Endlich, R.M., and McLean, G.S. 1957. The structure of the jet stream core, *J. Meteorol.*, **14**, 543–52; McLean, G.S. 1957. Cloud distribution in the vicinity of jet streams, *Bull. Am. Meteorol. Soc.*, **38**, 579–83.
46. Sawyer, J.S. 1958. Temperature, humidity and cloud near fronts in the middle and upper troposphere, *Q. J. R. Meteorol. Soc.*, **84**, 375–88.
47. Whitney, L.F., Timchalk, A., and Gray, T.I. 1966. On locating jet streams from TIROS photographs, *Mon. Weather Rev.*, **94**, 127–38.
48. Staley, D.O. 1960. Evaluation of potential-vorticity changes near the tropopause and the related vertical motions, vertical advection of vorticity, and transfer of radioactive debris from stratosphere to troposphere, *J. Meteorol.*, **17**, 591–620.
49. Danielson, E.F. 1966. *Research in Four-Dimensional Diagnosis of Cyclonic Storm Cloud Systems*. Sci. Rept. 1, Contract AF 19(628)-4762, Pennsylvania State University, University Park.
50. Danielson, E.F. 1959. The laminar structure of the atmosphere and its relation to the concept of a tropopause, *Arch. Meteorol., Geophys. Bioklim.*, A, **11**, 293–332.
51. Hoskins, B.J. 1971. Atmospheric frontogenesis models: some solutions, *Q. J. R. Meteorol. Soc.*, **97**, 139–53; 1972. Non-Boussinesq effects and further development in a model of upper tropospheric frontogenesis, *Q. J. R. Meteorol. Soc.*, **98**, 532–41.
52. Vuorela, L.A. 1953. On the air flow associated with the invasion of upper tropical air over Northwestern Europe, *Geophysica*, **4**, 105–30.
53. Goldie, N., Moore, J.G., and Austin, E.E. 1958. *Upper Air Temperature over the World*. Geophys. Mem. 13, no. 101, Meteorological Office, London (see pp. 12–16).
54. See, e.g., *Meteorological Aspects of Atmospheric Radioactivity*. 1965. Tech. Note 68, World Meteorological Organisation, Geneva.
55. Murgatroyd, R.J. 1971. The structure and dynamics of the stratosphere, in reference 53, pp. 159–95.
56. Mastenbrook, H.J. 1968. Water vapour distribution in the stratosphere and high troposphere, *J. Atmos. Sci.*, **25**, 299–311.
57. Murgatroyd, R.J. 1965. Ozone and water vapour in the upper troposphere and lower stratosphere, in *Meteoro-*

logical Aspects of Atmospheric Radioactivity. Tech. Note 68, World Meteorological Organisation, Geneva, pp. 68–94.
58. Graves, M.E. 1968. Aircraft reports of cirriform clouds on certain high latitude routes and California to Honolulu, *Mon. Weather Rev.*, **96**, 809–12.
59. Kerley, M.J. 1961. High-altitude observations between the United Kingdom and Nairobi, *Meteorol. Mag.*, **90**, 3–18.
60. Manabe, S., and Hunt, B.G. 1968. Experiments with a stratospheric general circulation model: I. Radiative and dynamic aspects, *Mon. Weather Rev.*, **96**, 477–502; II. Large-scale diffusion of tracers in the stratosphere, *Mon. Weather Rev.*, **96**, 503–39.
61. Cox, S.K. 1969. Observational evidence of anomalous infrared cooling in a clear tropical atmosphere, *J. Atmos. Sci.*, **26**, 1347–49.
62. Helliwell, N.C., Mackenzie, J.K., and Kerley, M.J. 1957. Some further observations from aircraft of frost point and temperature up to 50,000 ft, *Q. J. R. Meteorol. Soc.*, **83**, 257–62.
63. Murgatroyd, R.J., Goldsmith, P., and Hollings, W.E.H. 1955. Some recent measurements of humidity from aircraft up to heights of about 50,000 ft over southern England, *Q. J. R. Meteorol. Soc.*, **81**, 533–37.
64. Helliwell, N.C. 1960. *Airborne Measurements of the Latitudinal Variation of Frost-Point, Temperature and Wind.* Sci. Paper 1, Meteorological Office, London.
65. Murgatroyd, R.J., and Goldsmith, P. 1956. *High Cloud over Southern England.* Prof. Notes, 7, no. 119, Meteorological Office, London.
66. Sellers, W.D. 1965. *Physical Climatology.* University of Chicago Press, Chicago.
67. Tucker, G.B. 1961. Precipitation over the North Atlantic Ocean, *Q. J. R. Meteorol. Soc.*, **87**, 147–58.
68. Shaw, E.M. 1962. An analysis of the origins of precipitation in Northern England, 1956–60, *Q. J. R. Meteorol. Soc.*, **88**, 539–47.
69. Poulter, R.M. 1936. Configuration, air mass and rainfall, *Q. J. R. Meteorol. Soc.*, **62**, 49–76.
70. Goldie, A.H.R. 1936. *Rainfall at Fronts of Depressions.* Geophys. Mem. 7, no. 69, Meteorological Office, London.
71. Bannon, J.K. 1948. The estimation of large-scale vertical currents from the rate of rainfall, *Q. J. R. Meteorol. Soc.*, **74**, 57–66.
72. Godske, C.L., Bergeron, T., Bjerknes, J., and Bundgaard, R.C. 1957. *Dynamic Meteorology and Weather Forecasting.* American Meteorological Society, Boston (see pp. 522–25).
73. Crutcher, H.L., Hunter, J.C., and Sanders, R.A. 1950. Forecasting the heights of cumulus cloud-tops on the Washington-Bermuda Airways route, *Bull. Am. Meteorol. Soc.*, **31**, 1–7.
74. Lempfert, R.G.K., and Shaw, W.N. 1906. *The Life History of Surface Air Currents. A Study of the Surface Trajectories of Moving Air.* M.O. 174, Meteorological Office, London (Reprinted in 1955. *Selected Meteorological Papers of Sir Napier Shaw, F.R.S.* Macdonald & Co., London, 15–131.); Shaw, Sir Napier. 1911. *Forecasting Weather.* Constable & Co., London (3d ed., 1940). (This leads to a cyclone model discussed in reference 33.)
75. Kessler, E., and Wexler, R. 1960. Observations of a cold front, 1 October 1958, *Bull. Am. Meteorol. Soc.*, **41**, 253–57.
76. Browning, K.A., and Harrold, T.W. 1970. Air motion and precipitation growth at a cold front, *Q. J. R. Meteorol. Soc.*, **96**, 369–89.
77. Best, A.C. 1951. Drop-size distribution in cloud and fog, *Q. J. R. Meteorol. Soc.*, **77**, 418–26.
78. Neiburger, M. 1949. Reflection, absorption, and transmission of insolation by stratus cloud, *J. Meteorol.*, **6**, 98–104.
79. Frith, R. 1951. The size of cloud particles in stratocumulus cloud, *Q. J. R. Meteorol. Soc.*, **77**, 441–44.
80. Kline, D.B., and Walker, J.A. 1951. *Meteorological Analysis of Icing Conditions in Low-Altitude Stratiform Clouds.* Tech. Note 2306, no. 7, National Advisory Commission on Aeronautics, Washington, D.C.
81. Lewis, W. 1951. Meteorological aspects of aircraft icing, in *Compendium of Meteorology.* American Meteorological Society, Boston, pp. 1197–1203.
82. Khrgian, A. Kh., ed. 1961. *Cloud Physics.* Translated from the Russian by the Israel Program for Scientific Translations, and published by the Office of Technical Services, U.S. Department of Commerce, Washington, D.C.
83. Best, A.C. 1951. The size of cloud droplets in layer-type cloud, *Q. J. R. Meteorol. Soc.*, **77**, 241–48; 1952. Effect of turbulence and condensation on drop-size distribution in cloud, *Q. J. R. Meteorol. Soc.*, **78**, 28–36.
84. Mason, B.J. 1960. The evolution of droplet spectra in stratus cloud, *J. Meteorol.*, **17**, 459–62.
85. Mordy, W.A., and Eber, L.E. 1954. Observations of rainfall from warm clouds, *Q. J. R. Meteorol. Soc.*, **80**, 48–57.
86. Kotsch, W.J. 1947. An example of colloidal instability of clouds in tropical latitudes, *Bull. Am. Meteorol. Soc.*, **28**, 87–89.
87. Kinshott, E.J. 1950. Unusual fall of rain at Upavon on November 4, 1949, *Meteorol. Mag.*, **79**, 205–6.
88. Cornford, S.G. 1966. A note on some measurements from aircraft of precipitation within frontal clouds, *Q. J. R. Meteorol. Soc.*, **92**, 105–13.
89. Mason, B.J., and Howorth, B.P. 1952. Some characteristics of stratiform clouds over Northern Ireland in relation to their precipitation, *Q. J. R. Meteorol. Soc.*, **78**, 226–30.
90. Stewart, J.B. 1964. Precipitation from layer cloud, *Q. J. R. Meteorol. Soc.*, **90**, 287–97.
91. Coons, R.D., Gentry, R.C., and Gunn, R. 1948. *First Partial Report on the Artificial Production of Precipitation. Stratiform Clouds—Ohio, 1948.* Research Paper 30, U.S. Weather Bureau, Washington, D.C.
92. Hall, F. 1957. *The Weather Bureau ACN Project.* Meteorological Monographs 2, no. 11, American Meteorological Society, Boston, pp. 24–46.
93. Peppler, W. 1940. Unterkühlte Wasserwolken und Eiswolken, *Forsch. Erfahr. Reichsamt Wetterdienst*, **B**, no. 1.
94. Kline, D.B., and Walker, J.A. 1951. *Meteorological Analysis of Icing Conditions Encountered in Low-Altitude Stratiform Clouds.* Tech. Note 2306, no. 7, National Advisory Commission on Aeronautics.
95. Wexler, R., and Atlas, D. 1958. Moisture supply and growth of stratiform precipitation, *J. Meteorol.*, **15**, 531–38.
96. Gunn, K.L.S., and Marshall, J.S. 1958. The distribution with size of aggregate snowflakes, *J. Meteorol.*, **15**, 452–61.
97. Mason, B.J. 1971. *The Physics of Clouds*, 2d ed. Clarendon Press, Oxford, p. 245.
98. Cornford, S.G. 1966. A note on some measurements from

aircraft of precipitation within frontal clouds, *Q. J. R. Meteorol. Soc.*, **92**, 105–13.
99. Mason, B.J. 1969. Some outstanding problems in cloud physics—the interaction of microphysical and dynamical processes, *Q. J. R. Meteorol. Soc.*, **95**, 449–85.
100. Cunningham, R.M. 1952. *The Distribution and Growth of Hydrometeors around a Deep Cyclone.* Tech. Rept. 18, Contract DA-36-039-sc-124, Department of Meteorology, M.I.T., Cambridge. Some aspects of this work are discussed in Weickmann, H.K. 1957. Physics of precipitation, *Meteorol. Res. Rev.*, **3**, no. 19, 226–55. Similar studies are described in Cunningham, R.M. 1951. Some observations of natural precipitation processes, *Bull. Am. Meteorol. Soc.*, **32**, 334–43.
101. Warner, C., and Gunn, K.L.S. 1969. Measurement of snowfall by optical attenuation, *J. Appl. Meteorol.*, **8**, 110–21.
102. Power, B.A., Summers, P.W., and d'Avignon, J. 1964. Snow crystal forms and riming effects as related to snowfall density and general storm conditions, *J. Atmos. Sci.*, **21**, 300–305.
103. Wexler, R. 1955. *The Melting Layer.* Meteorological Radar Studies 3, Contract AF 19(604)-50, Blue Hill Met. Obsy., Harvard University, Cambridge, Mass.
104. Vonnegut, B., and Moore, C.B. 1960. Visual analogue of radar bright band phenomenon, *Weather*, **15**, 277–79.
105. Wexler, R. 1957. *Advection and the Melting Layer.* Meteorological Radar Studies 4, Contract AF 19(604)-50, Blue Hill Met. Obsy., Harvard University, Cambridge, Mass.
106. Wexler, R., Reed, R.J., and Honig, J. 1954. Atmospheric cooling by melting snow, *Bull. Am. Meteorol. Soc.*, **35**, 48–51; Lumb, F.E. 1961. The problem of forecasting the downward penetration of snow, *Meteorol. Mag.*, **90**, 310–19.
107. Battan, L.J. 1959. *Radar Meteorology.* University of Chicago Press, Chicago.
108. Atlas, D. 1964. Advances in radar meteorology, *Adv. Geophys.*, **10**, 317–478.
109. Gilchrist, A. 1971. An example of synoptic development in a general circulation model, *Q. J. R. Meteorol. Soc.*, **97**, 340–47.
110. Gadd, A.J., and Keers, J.F. 1970. Surface exchanges of sensible and latent heat in a 10-level model atmosphere, *Q. J. R. Meteorol. Soc.*, **96**, 297–308.
111. Benwell, G.R.R., and Timpson, M.S. 1968. Further work with the Bushby-Timpson 10-level model, *Q. J. R. Meteorol. Soc.*, **94**, 12–24.
112. Panofsky, H. 1958. *Introduction to Dynamic Meteorology.* Pennsylvania State University, University Park (see pp. 106, 107).
113. Todsen, M. 1964. *A Computation of the Vertical Circulation in a Frontal Zone from the Quasi-Geostrophic Equations.* Tech. Note 4, Institute of Theoretical Meteorology, University of Oslo, Norway.
114. Endlich, R.M., and McLean, G.S. 1965. Jet-stream structure over the Central United States determined from aircraft observations, *J. Appl. Meteorol.*, **4**, 83–90.
115. Briggs, J., and Roach, W.T. 1963. Aircraft observations near jet streams, *Q. J. R. Meteorol. Soc.*, **89**, 225–47.
116. Ludlam, F.H. 1967. Characteristics of billow clouds and their relation to clear-air turbulence, *Q. J. R. Meteorol. Soc.*, **93**, 419–35.
117. Defant, F. 1959. On hydrodynamic instability caused by an approach of subtropical and polar-front jet stream in northern latitudes before the onset of strong cyclogenesis, in *The Atmosphere and the Sea in Motion.* Rockefeller Institute Press, New York (see p. 312).
118. Wexler, R., and Atlas, D. 1959. Precipitation generating cells, *J. Meteorol.*, **16**, 327–32.
119. Browning, K.A., and Harrold, T.W. 1969. Air motion and precipitation growth in a wave depression, *Q. J. R. Meteorol. Soc.*, **95**, 288–309.
120. Browning, K.A. 1971. Radar measurements of air motion near fronts, Part Two: Some categories of frontal air motion, *Weather*, **26**, 320–40.
121. Langleben, M.P. 1956. The plan pattern of snow echoes at the generating level, *J. Meteorol.*, **13**, 554–60.
122. Douglas, R.H., Gunn, K.L.S., and Marshall, J.S. 1957. Pattern in the vertical of snow generation, *J. Meteorol.*, **14**, 95–114.
123. Wexler, R. 1955. Radar analysis of precipitation streamers observed 25 February 1954, *J. Meteorol.*, **12**, 391–93.
124. Wexler, R., and Atlas, D. 1959. Precipitation generating cells, *J. Meteorol.*, **16**, 327–32.
125. Atlas, D. 1964. Advances in radar meteorology, *Adv. Geophys.*, **10**, 317–488.
126. Yanasigawa, Z. 1961. An analysis of stationary rainbands as observed by radar, *Papers Meteorol. Geophys., Meteorol. Res. Inst., Tokyo*, **12**, 294–309.
127. Miyazawa, S. 1968. A mesoclimatological study on heavy snowfall—a synoptic study on the mesoscale disturbances, *Papers Meteorol. Geophys., Meteorol. Res. Inst., Tokyo*, **19**, 487–550.
128. Nozumi, Y., and Arakawa, H. 1968. Prefrontal rain bands located in the warm sector of subtropical cyclones over the ocean, *J. Geophys. Res.*, **73**, 487–92.
129. Kreitzberg, C.W., and Brown, H.A. 1970. Mesoscale weather systems within an occulusion, *J. Appl. Meteorol.*, **9**, 417–32.
130. Ludlam, F.H. 1956. Fall-streak holes, *Weather*, **11**, 89–90 (illustrations follow the article, and references to several others are quoted).
131. Ludlam, F.H. 1948. The forms of ice-clouds, *Q. J. R. Meteorol. Soc.*, **74**, 39–56.
132. Weickmann, H. 1948. *The Ice Phase in the Atmosphere.* Library Translation 273, Royal Aircraft Establishment, Farnborough, England. (Much of the material in this monograph is summarized in Weickmann, H.K. 1945. Formen und Bildung atmosphärischer Eiskristalle, *Beitr. Phys. freien Atmos.*, **28**, 12–52. The difficulties of measuring cirrus crystal concentrations and a few results using airborne apparatus are discussed in McTaggart-Cowan, J.D., Lala, G.G., and Vonnegut, B. 1970. The design, construction and use of an ice crystal counter for ice crystal cloud studies by aircarft, *J. Appl. Meteorol.*, **9**, 294–99. Recent observations of the shape, size, and concentration of particles in cirrus fallstreaks, and estimates of their fall speeds, have been given in the following references; the magnitudes are generally consistent with those assumed in the text: Heymsfield, A.J. 1972. Ice crystal terminal velocities, *J. Atmos. Sci.*, **29**, 1348–57; Heymsfield, A.J., and Krollenberg, R.G. 1972. Properties of cirrus generating cells, *J. Atmos. Sci.* **29**, 1358–66.)
133. Shakespeare, W. 1623. *The Tempest,* Act I, Scene ii, l. 192.
134. Ludlam, F.H. 1956. The forms of ice clouds: II, *Q. J. R. Meteorol. Soc.*, **82**, 257–65.
135. Harimaya, T. 1968. On the shape of cirrus uncinus clouds: a numerical computation—studies of cirrus clouds: Part III, *J. Meteorol. Soc. Japan*, **46**, 272–78.

136. Yagi, T. 1969. On the relation between the shape of cirrus clouds and the static stability of the cloud level—studies of cirrus clouds: Part IV, *J. Meteorol. Soc. Japan*, **47**, 59–64.
137. ———. 1959. Forecasting cirrus, *Meteorol. Mag.*, **88**, 74–83.
138. Kerley, M.J. 1961. High-altitude observations between the United Kingdom and Nairobi, *Meteorol. Mag.*, **90**, 3–18.
139. Reiter, E. 1969. Tropospheric circulation and jet streams, in *Climate of the Free Atmosphere*, D. F. Rex, ed. *World Survey of Climatology*, vol. 4, Elsevier, New York, pp. 85–203.
140. Murray, R., and Daniels, S.M. 1953. Transverse flow at entrance and exit to jet streams, *Q. J. R. Meteorol. Soc.*, **79**, 236–41.
141. Kreitzberg, C.W. 1968. The mesoscale wind field in an occlusion, *J. Appl. Meteorol.*, **7**, 53–67.
142. Bergeron, T. 1968. Cloud physics research and the future fresh-water supply of the world, *Proc. Int. Conf. Cloud Phys.*, 26–30 Aug. 1968, Toronto, Canada, 744–57.
143. Alaka, M.A., ed. 1960. *The Airflow over Mountains*. Tech. Note 34, World Meteorological Organisation, Geneva, p. 135.
144. Bretherton, F.P. 1969. Momentum transport by gravity waves, *Q. J. R. Meteorol. Soc.*, **95**, 213–43.
145. Edinger, J.G. 1966. Wave clouds in the maritime layer upwind of Pt. Sal, California, *J. Appl. Meteorol.*, **5**, 804–9.
146. Wurtele, M.G. 1957. The three-dimensional lee wave, *Beitr. Phys. freien Atmos.*, **29**, 242–52.
147. Scorer, R.S. 1958. *Natural Aerodynamics*. Pergamon Press, London, pp. 229–51.
148. Sawyer, J.S. 1960. Numerical calculation of the displacements of a stratified airstream crossing a ridge of small height, *Q. J. R. Meteorol. Soc.*, **86**, 326–45.
149. Foldvik, A. 1962. Two-dimensional mountain waves—a method for the rapid computation of lee wavelengths and vertical velocities, *Q. J. R. Meteorol. Soc.*, **88**, 271–85.
150. Scorer, R.S. 1967. Causes and consequences of standing waves, *Proc. Symp. Mountain Meteorol.* 26 June 1967, Department of Atmospheric Science, Colorado State University, Fort Collins.
151. Pao, Y.-H. 1969. Inviscid flows of stably stratified fluids over barriers, *Q. J. R. Meteorol. Soc.*, **95**, 104–19.
152. Danielson, E.F., and Bleck, R. 1970. Tropospheric and stratospheric ducting of stationary mountain lee waves, *J. Atmos. Sci.*, **27**, 758–72.
153. Corby, G.A., and Wallington, C.E. 1956. Airflow over mountains: the lee-wave amplitude, *Q. J. R. Meteorol. Soc.*, **82**, 266–74.
154. Wallington, C.E. 1958. A numerical study of the topographical factor in lee-wave amplitude, *Q. J. R. Meteorol. Soc.*, **84**, 428–33.
155. Foldvik, A., and Wurtele, M.G. 1967. The computation of the transient gravity wave, *Geophys. J. R. Astrol. Soc.*, **13**, 167–85.
156. Wurtele, M.G. 1970. Meteorological conditions surrounding the Paradise Airline crash of 1 March 1964, *J. Appl. Meteorol.*, **9**, 787–95.
157. Scorer, R.S., and Klieforth, H. 1959. Theory of mountain waves of large amplitude, *Q. J. R. Meteorol. Soc.*, **85**, 131–43.
158. Defant, F. 1951. Local winds, in *Compendium of Meteorology*. American Meteorological Society, Boston, pp. 655–72.
159. Vergeiner, I. 1971. An operational linear lee wave model for arbitrary basic flow and two-dimensional topography, *Q. J. R. Meteorol. Soc.*, **97**, 30–60.
160. Abe, M. 1941. Mountain clouds, their forms and connected air current, Part II, *Bull. Central Meteorol. Obsy. Japan*, **7**, no. 3, 93–145.
161. Kuettner, J. 1939. Moazagotl und Föhnwelle, *Beitr. Phys. freien Atmos.*, **25**, 79–114. ["Moazagotl" is a name originally given in the local dialect to an extensive high cloud which in southerly airstreams forms above the Riesengebirge—part of the Sudeten range of mountains (with peaks to 1.5 km) which lies almost E-W at 51°N, along the border between East Germany, Poland, and Czechoslovakia. The name is probably derived from that of Gottlieb Matz, a local weather prophet who used the appearance of the cloud to forecast bad weather in the following 24 hours.]
162. Kuettner, J. 1939. Zur Entstehung der Föhnwelle, *Beitr. Phys. freien Atmos.*, **25**, 251–99.
163. Krug-Pielsticker, U. 1942. Beobachtungen der hohen Föhnwelle an den Ostalpen, *Beitr. Phys. freien Atmos.* **27**, 79–114.
164. Manley, G. 1945. The helm wind of Crossfell, 1937–1939, *Q. J. R. Meteorol. Soc.*, **71**, 197–215.
165. Holmboe, J., and Klieforth, H. 1957. *Investigations of Mountain Lee Waves and the Air Flow over the Sierra Nevada*. Final Report, Contract AF 19(604)-728, Department of Meteorology, University of California, Los Angeles.
166. Colson, DeVer. 1952. Results of double-theodolite observations at Bishop, California, in connection with the "Bishop-wave" phenomena, *Bull. Am. Meteorol. Soc.*, **33**, 107–16.
167. Colson, DeVer. 1954. Meteorological problems in forecasting mountain waves, *Bull. Am. Meteorol. Soc.*, **35**, 363–71.
168. Larsson, L. 1954. Observations of lee wave clouds in the Jämtland Mountains, Sweden, *Tellus*, **6**, 124–38.
169. Gerbier, N., and Berenger, M. 1961. Experimental studies of lee waves in the French Alps, *Q. J. R. Meteorol. Soc.*, **87**, 13–23.
170. Vergeiner, I., and Lilly, D.K. 1970. The dynamic structure of lee wave flow as obtained from balloon and airplane observations, *Mon. Weather Rev.*, **98**, 220–32.
171. Hanna, S.R., and Hoecker, W.H. 1971. The response of constant-density balloons to sinusoidal variations of vertical wind speeds, *J. Appl. Meteorol.*, **10**, 601–4.
172. Döös, B.R. 1962. A theoretical analysis of lee wave clouds observed by Tiros I, *Tellus*, **14**, 301–9; Corby, G.A. 1966. Essa II sees mountain waves over Britain, *Weather*, **21**, 440–42; Cohen, A., and Doron, E. 1967. Mountain lee waves in the Middle East: theoretical calculations compared with satellite pictures, *J. Appl. Meteorol.*, **6**, 669–73.
173. Brandli, H.W., and Webb, J.A. 1970. Essa 8 APT shows lee waves near Aleutian islands, *Mon. Weather Rev.*, **98**, 406–7.
174. Starr, J.R., and Browning, K.A. 1972. Observations of large-amplitude lee waves by high-power radar, *Q. J. R. Meteorol. Soc.*, **98**, 73–85.
175. Corby, G.A. 1957. A preliminary study of atmospheric waves using radiosonde data, *Q. J. R. Meteorol. Soc.*, **83**, 49–60.
176. Reid, S.J. 1970. The use of radiosonde data to examine lee waves and other small-scale motions in the atmo-

sphere, Ph. D. thesis, University of London.
177. Scorer, R.S. 1955. Theory of airflow over mountains: IV—Separation of flow from the surface, *Q. J. R. Meteorol. Soc.*, **81**, 340–50.
178. Wallington, C.E. 1961. *Meteorology for Glider Pilots.* John Murray, London (2d ed., 1966; see p. 230).
179. Jefferson, G.J. 1961. Orographic clouds over the Matterhorn, *Weather*, **16**, 402–4.
180. Brinkmann, W.A.R. 1971. What is a foehn?, *Weather*, **26**, 230–39; Suzuki, S., and Yabuki, K. 1956. The air-flow crossing over the mountain range, *Geophys. Mag.*, Tokyo, **27**, 273–91.
181. Walker, E.R. 1966. A well-marked warm sector, *Weather*, **21**, 410–13 (see pp. 412–13).
182. Julian, L.T., and Julian, P.R. 1969. Boulder's winds, *Weatherwise*, **22**, 109–12, 126.
183. Sergius, L.A. 1952. Forecasting the weather—the Santa Ana, *Weatherwise*, **5**, 66–68; Sergius, L.A., Ellis, G.R., and Ogden, R.M. 1962. The Santa Ana winds of southern California, *Weatherwise*, **15**, 102–5, 121; Koutnik, W. 1968. Newhall winds of the San Fernando Valley, *Weatherwise*, **21**, 186–89, 202.
184. ———. 1967. *Report on the Accident to Boeing 707 G-APFE at the Foot of Mt. Fuji, Japan, on 5 March 1966.* Board of Trade translation of the Report of the Japanese Commission of Investigation, H.M.S.O., London.
185. Burham, J. 1970. Atmospheric gusts—a review of the results of some recent research of the Royal Aircraft Establishment, *Mon. Weather Rev.*, **98**, 723–34.
186. Roach, W.T. 1972. *Clear Air Turbulence.* Merrow, Watford, England.
187. Aanensen, C.J., ed. 1965. *Gales in Yorkshire in February 1962.* Geophys. Mem. 14, no. 108, Meteorological Office, London.
188. Ludlam, F.H. 1952. Orographic cirrus clouds, *Q. J. R. Meteorol. Soc.*, **78**, 554–62.
189. Ludlam, F.H. 1952. Hill-wave cirrus, *Weather*, **7**, 300–306.
190. Austin, A.R.I. 1952. Wave clouds over southern England, *Weather*, **7**, 50–53.
191. Ludlam, F.H. 1955. Hill-wave clouds, *Discovery*, March 1955, pp. 114–19.
192. Jefferson, G.J. 1961. Orographic clouds over the Matterhorn, *Weather*, **16**, 402–4.
193. Roper, R.D. 1952. Evening waves, *Q. J. R. Meteorol. Soc.*, **78**, 415–19.
194. George, D.J. 1971. Mother-of-pearl clouds in Antarctica, *Weather*, **26**, 7–12. For a general review of knowledge on stratospheric clouds see Hesstvedt, E. 1969. The physics of nacreous and noctilucent clouds, in *Stratospheric Circulation*, W.L. Webb, ed. Academic Press, New York, pp. 209–18.
195. Miles, J.W., and Howard, L.N. 1964. Note on a heterogeneous shear flow, *J. Fluid Mech.*, **20**, 331–36.
196. Rosenhead, L. 1932. The formation of vortices from a surface of discontinuity, *Proc. Roy. Soc.*, A, **134**, 170–92.
197. Thorpe, S.A. 1968. A method of producing shear in a stratified fluid, *J. Fluid Mech.*, **32**, 693–704.
198. Woods, J.D. 1968. Wave-induced shear instability in the summer thermocline, *J. Fluid Mech.*, **32**, 791–800.
199. Ludlam, F.H. 1967. Characteristics of billow clouds and their relation to clear air turbulence, *Q. J. R. Meteorol. Soc.*, **93**, 419–35.
200. Browning, K.A., Bryant, G.W., Starr, J.R., and Axford, D.N. 1973. Air motion within Kelvin-Helmholtz billows determined from simultaneous Doppler radar and aircraft measurements, *Q. J. R. Meteorol. Soc.*, **99**, 608–18.
201. Browning, K.A. 1971. Structure of the atmosphere in the vicinity of large-amplitude Kelvin-Helmholtz billows, *Q. J. R. Meteorol. Soc.*, **97**, 283–99.
202. Trout, D., and Panofsky, H.A. 1969. Energy dissipation near the tropopause, *Tellus*, **21**, 355–58.
203. Scorer, R.S. 1969. Billow mechanics, *Radio Sci.*, **4**, 1299–1308.
204. Ludlam, F.H., and Scorer, R.S. 1957. *Cloud Study.* John Murray, London (see e.g., Pl. 44).
205. Colson, DeVer. 1954. Wave-cloud formation at Denver, *Weatherwise*, **7**, 34–35.

Conclusion

The later chapters of this book presented descriptions of some of the great variety of atmospheric motion systems in which distinctive clouds appear. In some of these, for example, lee waves, the clouds have virtually no effect on the motion, and merely manifest its presence and nature. In others, including the circulations in frontal zones, the clouds and precipitation substantially modify the local motion, probably without significantly affecting the motion on the large (cyclonic) scale which is responsible for them. In yet others the thermodynamics of transformations associated with condensation, precipitation, and evaporation play a fundamental part in the organization of the motion, and even in the intensity and form of the general circulation of the atmosphere. Outstanding among these are the cumulonimbus of deep convection, which are the principal working parts of the atmospheric engine in low latitudes. (The electrical phenomena of these thunderclouds are spectacular but energetically insignificant, and so have not been discussed.) Hardly less important are the widespread small cumulus, which affect the heat economy through their albedo. That the albedo does not have the high value which would accompany a complete cover of cloud is a consequence of the readiness with which precipitation can develop in deep clouds, behavior which is intimately linked with their microphysical properties.

For these reasons, and because the principal aims in the application of cloud physics are predicting or artificially controlling the time, place, and kind of precipitation, it is the physics of the rain clouds which attracts most interest. Inevitably, processes of a whole range of scales are of concern. [For example, the explanation even of the familiar cirrus fallstreak seems to demand the complex interaction of special *nucleation* properties, leading to (*a*) a widespread state of supersaturation over ice in a favorable *large-scale* situation, and (*b*) condensation on sparse crystal nuclei when near saturation (but not saturation) with respect to the liquid is attained in a *small-scale* motion system, the production of (*c*) a motion system dependent on the condensation, precipitation, and evaporation of crystals in surroundings of suitable lapse rate and wind shear and (*d*) a motion system of intermediate scale allowing the persistence of the fallstreak mechanism over distances of hundreds of kilometers.]

Remarkably, however, the microphysical behavior of cloud particles has come to be regarded as composing a *static* or *physical* meteorology as distinct from a *dynamics* which is a more general physics. With some exceptions the processes involved in the statics are well identified and understood, and subject to comparatively simple laws (this cannot yet be said, e.g., of the reliable assessment of the radiative properties of water vapor of high density, or of phenomena such as the extensive development of the ice phase in some clouds of barely supercooled droplets, and twilight colors and the brightness of clouds, which are dependent on the multiple scattering of light). These laws enter the *dynamics*, whose difficulty and comparatively poorly advanced state is due to the complexity which arises when three space dimensions and an evolution in time are involved, when several processes are at work simultaneously, and when it is not possible to separate clearly motion systems of different scales in space and time.

The difficulty even of describing three-dimensional fluid motions which evolve in time was discussed in Chapter 6. Further, when condensation of vapor and the evaporation of precipitation are important sources and sinks of heat (as in fronts and in cumulonimbus), the position and intensity of the latter depend on the paths and hence the sizes of the precipitation particles. Since there is always a significant spread of sizes, the

Conclusion

dynamics can be regarded as that of an engine of extraordinary complexity, whose working depends on the relative motion of an intermingling set of fluids.

Presuming the formulation of appropriate governing equations (composing a *model*) and boundary conditions, it is possible by numerical methods and the use of electronic machines to calculate the evolution of even such complicated phenomena. Modern meteorology is dominated by such procedures, employed mainly in efforts to improve weather forecasting for periods between about one day and one week. Much of this work is based on tentative adaptation of models, rather than systematic experimentation in which results are analyzed to examine the efforts of altering one by one the various parameters or representations which have been employed. However, ultimately the success of a model has to be judged by the comparison of two patterns, one calculated and one observed. Hardly enough attention has been given to the problem of deciding which features of a pattern are significant, and giving them quantitative expression. Nevertheless, superficially the comparison in respect of the large-scale motion systems appears to be straightforward, since at least outside the tropics the routine synoptic observations are about adequate to define initial and subsequent states, or the climatic averages which should be reproduced in an extended calculation based on a model.

There are no such comprehensive observations to define the typical evolutions of the motion systems of smaller scale, and in particular of those in which cloud and precipitation play important parts. For such systems the states observed in a small number of special investigations have to be combined with a general experience, often based on eye observations alone, in order to assess their typical structure and evolution. Progress in cloud dynamics will for some time depend heavily upon the latter kind of limited and subjective experience, which also is a main source of inspiration for the construction of simple but penetrative models. A principal aim in the preparation of this book was the assembly of a body of such experience.

Index

Subject Index

Absorption and emission, *see* Radiation
Adiabatic process, 57-59, *see also* Lapse rate
Aerological diagram, 59-62
Aerosol
  composition, tropospheric, 68-69
  gases and vapors, 65-66
  mass concentration, 66, 70
  number concentration
    Aitken nuclei, 66
    change through aggregation, 18
    continental versus maritime sources, 88
  removal from atmosphere, 67-68
  settling speed, 16-17, 64
  size and size distribution, 16, 18, 21-22, 67-68
  solution droplet, 71-72
  sources, global production, 64-66
  stratospheric, 70-71
  see also Aggregation, Dust, Nuclei, Optical phenomena, Scattering
Aggregation, aerosol, 16, 67
  Brownian motion, 17-18, 68
  growth rate, 18, 22
  size distribution, effect on, 67-68
  see also Coalescence
Air properties
  gas laws, 12
  physical data, 41
  see also Refraction, Water vapor
Aircraft observation (clouds)
  cumulus, 161-65
    humidity in, 162, 164
    liquid water concentration, 163
  droplets, 185-86
  ice particles, 203
  photography, 179
  thunderstorms, 208-10
  visual range, 187
  see also Clear-air turbulence, Sailplane soaring

Airflow properties, large-scale
  deformation and dispersion, 331-32
  divergence
    related to vorticity change, 314-15
    sea breeze, 153
  vorticity
    cumulus tops, correlation with, 386
    potential, conservation of, 332-33
    see also Kinetic energy, Trajectories, Vertical motion, Vertical shear
Altocumulus, 383
Altostratus, 323
Anvil (cumulonimbus), 206-8
  extent and character, 246-47
  height
    related to sounding, 206
    tropics, 231
  mamma, 290
  orientation, and storm movement, 249-50
  precipitation particles, 219
Attenuation, electromagnetic radiation, 9-10, 73-74, *see also* Radar measurement, Scattering.

Baroclinic instability, 315-17, 326
Billows, 381-83, *see also* Clear-air turbulence
Birds and insects, circulation indicators, 144-45, 148-49, 153
Boundary layer (low-level) jet, *see* Jet stream
Brownian motion, *see* Aggregation, Diffusion
Buoyancy, *see* Convective cloud theory, Cumulus.

Castellanus, 289-90
Cirrus
  height, in tropics, 347
  jet stream association, 343, 366-68
  oreigenic, lee wave, 379
  persistence and texture, 361-63
  warm front, 367-69

  see also Anvil, Fallstreak
Clear-air turbulence
  jet stream vicinity, 381
  mountain waves and rotors, 375, 387-88
  oreigenic billows, 382
  radar detection, 48, 122, 381
  Richardson number criterion, 380-81
  scintillation of starlight, 47-48
  see also Kinetic energy, Shearing instability, Weather hazards
Cloud and precipitation systems
  airstreams in cyclone, related to, 342-44, 350-51
  cold frontal, 349-51
  dry tongue clear zone, 344-46
  jet stream boundary, 343
  Norwegian model, 322-25
  slope convection, major trough, 333-35
  small-scale cells and bands, 359-61
  streaks, deformation, 331
  warm frontal, 351-57
Cloud forms and characteristics, 123-27, *see also* Altocumulus, Altostratus, Anvil, Castellanus, Cirrus, Convection patterns, Cumulonimbus, Cumulus, Mamma, Nacreous cloud, Noctilucent cloud, Optical phenomena, Oreigenic wave, Stratocumulus
Coalescence
  collision and capture efficiency, 21-24
  droplet growth
    aerosol abundance, control by, 88, 201-2
    single droplet, 196-98
    stochastic theory, 198-202
  precipitation growth, layer cloud, 354-56
  raindrop growth time, 25
  see also Aggregation, Droplet size, Hailstone growth, Shower formation
Condensation
  droplet growth, 82-83
    adiabatic expansion, 85, 89

397

Index

coalescence compared, 90-91
rate, and vertical motion, 63
trail, formation and persistence, 52-53
see also Ice particle, Liquid water concentration, Saturation
Convection patterns
  cellular, 8, 151
    open (downdraft outflow), 213-14
  laboratory experiments, 7, 149-50
  rows, 148-50
    embedded in frontal cloud, 359-61
    wind profile, related to, 149
  topographic associations, 151-54
    island effects, 152-54
    sea breeze, 152-53
    uniform terrain, 158-59
  see also Shower, Thunderstorm
Convective cloud theory
  free and forced convection, 129
  heat transfer, vertical, 170-71
  numerical modeling, 175-76
  parcel, 132-33
    cloud tops, compared with, 133, 223
  perturbation, 169-70
  thermals and plumes, 154-56
    actual cloud, similarity to, 157-61
  see also Freezing, Mixing
Conductive capacity of media, 130
Corona, 33-35
Cumulonimbus
  cells, ordinary
    draft speeds, 209-10
    duration, 214
    evolution stages, 208
  distinctions from cumulus, 118, 210
  downdraft, 219-20
    hail melt in, 269-70
    humidity and temperature, 252-54
    outflow, suppression of convection, 213-14
    rain evaporation in, 92, 252-56
    speed of, 209-10, 255
  glaciation growth surge, 204-5
  lightning-rainfall association, 275-76
  movement (of storms), 247-50
  organized and severe, 182
    airflow features, 218-20, 224
    cell persistence, 216-17
    condensation-evaporation cycle, 216-224
    observed storms (summary), 222
    radar features, 218
    tropical disturbance line, 227
    vertical shear influence, 222-25, 243-44
  precipitation
    cell stage, relation to, 210
    characteristics at ground, 271
    inside clouds, 208-9
  squall front, 210-17
    arched squalls, 215, 286-87
    cell development over, 213-15, 249
    pressure and temperature changes, 210-13
    wind gusts, 225-26, 256
    top heights, 222-23, 229-31

towers, rise rate, 207
  successive impulses, 214-15
turbulence, 209-10
  kinetic energy spectra, 121
updraft speed, 209, 219, 222-23
  below cloud base, 215
water and ice concentration, 209
see also Anvil, Cumulonimbus (synoptic conditions), Hailfall, Hailstone growth, Rainfall, Thunderstorm occurrence
Cumulonimbus (synoptic conditions), 239
  airflow over heated ground, 242, 245
  dry-line convergence, 235, 239-44
  environment stratification, 205-6
  inversion, role in explosive instability, 233-35, 239-46
  nocturnal thunderstorm factors, 236-38, 288-89
  vertical shear, organizing role, 222-25, 243-44
  see also Thunderstorm
Cumulus
  base height, 138-41
    diurnal variation, 140-41, 159, 193-94
    land and sea, 139-41
  buoyancy, 133, 166
  liquid water concentration, 162-65
  modification of environment, 177
  shelf cloud generation, 141, 172-73
  sky cover, diurnal variation, 159
  temperature in, 162-64
  towers, 159-61
    growth, cloud-modified environment, 167-69
  vertical velocity
    cloud base, 145, 149
    in cloud, 164, 166
  see also Convection patterns, Convective cloud theory, Droplet size, Shower formation, Vertical motion
Cyclogenesis
  cloud development during, 341-43
  geographical regions, favored, 314
  upper waves, relation to, 321, 326-27
Cyclone, 118
  Norwegian model, 322-25
  numerical simulation, 317-18
  relation to jet stream, 324-25
  tracks, principal, 314
  see also Cloud and precipitation systems, Fronts, Shower, Wave

Deformation, *see* Airflow properties
Diffusion
  Brownian motion, 17-18, 68
  coefficient, water vapor, 42
  in numerical modeling of thermals, 175-76
Divergence, *see* Airflow properties
Downdraft, *see* Cumulonimbus
Drag coefficient, *see* Reynolds number
Drizzle, 349
Droplet
  depletion by hail growth, 280
  evaporation, distance of fall, 95

freezing temperature, nucleated, 96
number concentration, 16
  continental versus maritime air, 88
  near cloud base, 7
sea spray, 64
see also Aerosol, Coalescence, Condensation, Refraction, Scattering, Shower formation
Droplet size
  aerosol solution, 71
  cloud base, near, 88
  cumulus cloud
    continental versus maritime air, 183, 187-88, 201
    mixing effects, cloud edge, 183, 187
    spectra, 185, 188, 199
    spectra, evolution of, 182-85, 198-202
  lee wave cloud, 34
  optical inference, 34, 38
  spectrum change after saturation, 85
  stratocumulus cloud, 351-52
  vertical displacement, dependence on, 34, 199
Dry line, *see* Fronts
Dust, 64-65
  devil, 143
  freezing nuclei, 65, 97
  interplanetary, aerosol source, 69-70, 98
  long-distance transport, 65, 98
  storms, cumulonimbus, 212, 228
  stratospheric (volcanic), 77-79

Energy
  conservation (extended parcel theory), 335-36
  thermodynamic diagram, 62
  see also Kinetic energy
Energy transfer
  conductive capacity of media, 130
  cumulonimbus outflow modification, 213-14
  meridional (slope convection), 5-6, 8
  near surface, 134-36
  surface to atmosphere, 130-32
    extreme, in cold outbreaks, 138
    oceanic cyclones, shower association, 323-24
    sensible and latent heat, ratio, 136-37
    zonal mean, 383
  tradewind trajectory, 137-38
  vertical, by convection
    beneath cumulus, 142
    cumulonimbus, 231, 272
    cumulus, 129-31, 170-71
    ordinary, 7-8
  see also Radiation
Entrainment, *see* Mixing
Evaporation
  ice particle, 108-9
  rain, in downdraft, 92, 252-56
  water drop, 94-95
Extinction, *see* Scattering

Fall speed
  aerosol, 64
  Doppler measurement, 33

Index

hailstone, 19-21, 93
ice particle, 21, 100, 104
microscopic particles (Stokes' law), 16-17
  see also Reynolds number
Fallstreak, 104
  forms, wind shear effect, 363, 366-67
  generating cells, 360, 364-66
  particle growth, 362-63
Fibrillation, see Shower formation
Freezing
  buoyancy enhancement by, 174-75, 204-5
  instability, snow generating cells, 360, 364-66
  saturated adiabatic process, 58-59
  see also Ice particle, Nuclei, Shower formation
Friction, see Planetary boundary layer
Fronts (frontal zones)
  circulation, cold front, 350-51
  cyclone model, Norwegian, 322-25
  dry line, 235, 240-44
  humidity, upper levels, 344-45
  numerical simulation, 317-18
  thunderstorm distribution (oceanic), 340-42
  wind distribution, upper levels, 335
  see also Cloud and precipitation systems

General circulation
  air mass exchange and modification, 321-22
  airflow through convection systems, 327-28
  mean atmospheric structure, 310-11, 328-30
  mean energy balance, components of, 383
  numerical modeling of, 327
  see also Energy transfer, Monsoon, Radiation, Slope convection
Glaciation, see Freezing
Glory, 35

Hailfall
  relation to cumulonimbus, 217, 219, 249
  swath dimensions, 221
  see also Hailstorm occurrence
Hailstone characteristics
  density and roughness, 19-21
  fragmentation in air, 291
  internal structure, 258-60, 293-94
  shapes, large hail, 257, 260
  size
    extremes observed, 291-93
    parcel theory compared, 223
    spectra and number concentration, 266
  see also Fall speed, Reflectivity
Hailstone growth
  accretion
    cloud droplet depletion by, 280
    collision efficiency, 24
    supercooled water, 19
    embryo source, 262
    heat economy, 105-6, 267-68

ventilation coefficient, 92-93
melt during fall, 268-70
recycling and size sorting, 219, 262-66
small hail, 256-57
soft and spongy hail, 107-8
storm characteristics for
  cloud base temperature, 257
  cloud water, supercooling zone, 260, 265
  updraft form and speed, 261, 265-66
theory and calculations, 260-64
tropics (unsuitability for), 267, 270
Hailstorm occurrence
  excessive accumulations, 295-97
  historical accounts, 221
  in Europe, 231-32
  in India, 229
  in North America, 236, 287
  in southern hemisphere, 227-28
Halo, 39-41
Haze, 67, 85, see also Sea breeze
Heat transfer, see Energy transfer
Humidity
  effects on atmospheric properties
    aerosol size, 71
    particle persistence (contrails), 53
    scattering, 29
    visibility, 74
  frontal zone, upper levels, 344-45
  in cumulonimbus downdraft, 253-55
  jet stream vicinity, 381
  mixing ratio and specific humidity, 13
  stratification, lee waves, 378
  tropopause vicinity, tropics, 347-48
  see also Hygrometer, Refraction, Saturation, Temperature, Vapor pressure
Hygrometer
  humidity measurement, 14, 43-44, 162, 164
  psychrometric theory, 14-15

Ice fog, 53
Ice particle
  concentration, 96-99, 209
    multiplication (fragmentation), 109-11, 203-4
    variation with temperature, 97
  evaporation, 108-9
  forms, and cloud temperature, 99-100
  growth
    accretion, 105-8
    condensation, 99-104
    on sublimation nuclei, 101-3
    rate, 108
  nucleation, 95-98
  persistence, 53, 361, 363
  role in precipitation, 202-4, 353-55
  size distribution (snow), 354
  size, optical estimation, 35
  see also Fall speed, Hailstone characteristics, Nuclei, Rime, Snow
Instability, see Baroclinic instability, Convective cloud theory, Shearing instability
Inversion (temperature)

modification
  along tradewind trajectory, 137-38
  by convection and radiation, 172
  radio and radar ducting, 47, 302
  instability suppression and release, 233-35, 239-46
Iridescence, 35-37
Isentropic relative flow, see Trajectories

Jet stream, 335
  ageostrophic transverse flow, 368-69
  layer destabilization by, 358-59
  ascent of low-level air into, 337-40
  cirrus associated, 343, 366-68
  cyclone, relation to, 324-25
  humidity distribution, 381
  low-level, diurnal variation, 236-37, 287-88
  potential vorticity near, 332-33
  turbulence (clear air), 381
  see also Wave

Kelvin-Helmholtz instability, see Shearing instability
Kinetic energy
  cascade, motion scales, 113
  cumulonimbus, 121
    related to vertical shear, 224-25
  dissipation, turbulent, 48, 122, 381
  generation, jet stream, 337-40
  spectra, various motion systems, 120-23

Lapse rate, 2
  dew point, 57
  differential advection, increase by, 358-59
  dry adiabatic, 57
  radiation effect upon, 2-3
  saturated adiabatic, 58-59
  snow-melt isothermal layer, 356-57
  see also Inversion, Stratification of atmosphere
Latent heat (evaporation, fusion, sublimation), 42
Lee wave, see Oreigenic wave
Lightning
  radio location (sferics), 246, 386
  rainfall association, 275-76
Liquid water concentration
  adiabatic expansion, 58
  observed values compared, 165, 185-87
  cumulonimbus, 209
  cumulus, 162-65
  environment mixing effect, 167-68
  see also Droplet, Ice particle, Mixing

Mamma (cumulonimbus), 290
Mirage, 26
Mixing
  cloud water diminution by, 165, 167-68
  entrainment, convective cloud, 173-75
  thermals and plumes, 156-57
  isobaric
    cloudy and clear air, 54-56
    cumulus boundary, 163-65, 171-72

Index

fog production, 52-53, 91
Monsoon
    influence on thunderstorms, 228-31, 238
    mean meridional circulation, 329-30
Motion systems, 113-14
    classification (types), 116-19
    observation, problems of, 115-16
    scales, size and time, 113-14, 119
    see also Cloud forms, Kinetic energy

Nacreous (mother-of-pearl) cloud, 35-37, 380
Noctilucent cloud
    aerosol particles, 70-71
    polarization of light, 27-28
Nuclei, 68
    activation
        condensation 87-88
        heterogeneous and homogeneous, 95-98
        temperature dependence, 97-98
    artificial, 97, 98
    concentration, 66, 98-99
        continental versus maritime, 201-2
    freezing
        dust, 65, 97
        precipitation, role in, 353-54
    salt, large droplet growth, 184, 188
    see also Droplet, Ice particle

Optical phenomena
    aerosol effects, 34, 36
    color of sky, sun, distant objects, 28
    visual contrast of objects, 73
    see also Corona, Glory, Halo, Iridescence, Mirage, Rainbow, Scintillation, Twilight
Ordinary convection, 6-8, 118
Oreigenic wave, 370-71
    barrier shape effects, 373-74
    cloud
        droplet size, 34
        forms, 376-80
        persistence downwind, 37, 378-80
    observation methods, 375-76
    rotors, 375
    stratospheric, 78-79, 372, 380
    wavelength and amplitude, 371-74
    see also Clear-air turbulence, Weather hazards

Parcel theory, see Convective cloud theory, Energy
Photography
    hailstone internal structure, 258
    mensuration of cumulus clouds, 179
    tornado studies, 250-51
Planetary boundary layer, 6
    frictional influence in, 13, 313, 349-50
    kinetic energy, and dissipation, 120-22
    organized circulations in, 144-45, 148-54
    shearing instability, 383
    vertical fluxes in, 130-31
Potential vorticity, 332-33
Precipitation, 92
    evaporation in downdraft, 92, 252-56

generating cells and rainbands, 359-61
numerical prediction models, 357-58
particle growth, frontal clouds, 355-56
showers versus frontal zones, 384-85
topographic influence, 361
see also Cumulonimbus, Drizzle, Fallstreak, Hailfall, Rainfall, Shower, Snow, Stratocumulus
Psychrometer (theory), see Hygrometer

Radar (measurement characteristics), 30-32
    attenuation, 9-10, 31
    backscatter, ice and water, 31-32
    calibration, 31
    cloud-top errors, 46, 301-2
    discrimination of detail, 302-3
    Doppler radar, 32-33
    rain-rate estimation, 277
    refraction effects, 302
        ducting (trapping), 26, 47
        elevation and range correction, 46-47
    see also Reflectivity
Radar-observed phenomena
    birds and insects in circulations, 148
    clear-air turbulence, 48, 122, 381
    convective circulations (mantle echoes), 147-48
    lee waves, 376
    severe cumulonimbus features, 218, 248-50
    showers, first echo and top rise, 192-94, 283-84
Radiation
    atmospheric cooling rate, 4, 383
    cloud absorption, 9-10
    cloud layer, modification of inversion, 172-73
    isolation, 5
        reduction by volcanic dust, 79
    ozone effect, 3-4, 9
    spectrum, electromagnetic, 9
    terrestrial, 5-6
        radiative-convective equilibrium, 1-3
    water vapor, 2-4, 9-10
    see also Attenuation, Scattering
Rainbow, 37-39
Raindrop, see Coalescence, Evaporation
Rainfall
    characteristics, cumulonimbus, 271
    excessive rates and accumulations, 209, 272-76, 295-300
    rate, contributing factors
        cloud physical, 273-76
        storm movement, 225
        stratocumulus thickness, 352-53
        upward vapor flux, 271-72, 385
    see also Reflectivity
Ray path (radio, optical), 45-47
Reflectivity (radar), 31-32
    bright band, snow melt, 357
    cloud top height relation, 284
    coalescence, increase during, 202
    drop-size relationship, 32-33, 188-89
    hail size and wetness, 277-79
    hailstorm identification (profile), 279-80
    lightning discharge, intensification, 276

mapping, digital, 284-86
rainfall rate relationships, 189-90, 277
sensitivity, 192
Refraction, 25-26
    ducting, radio wave, 26, 47
    see also Radar, Ray path
Refractive index, 25
    humidity measurement, 162, 164
    ice, 36
    scattering effect, 26
    water droplets, 27, 29
Reynolds number, 16
    coalescence dependence on, 22-24
    drag coefficient relations, 19-21, 45
    fall speed, particle, 16
Richardson number, 224-25, 243-44, see also Clear-air turbulence
Rime, 105-7

Sailplane (soaring)
    cumulonimbus updraft observations, 210
    cumulus rows, 148
    lee waves, 374, 378, 379, 387-88
    sea-breeze front, 152-53
Saturation, 12-13
    dew and frost points, 13, 42-43
    lifting required for, 57
    supersaturation, 82
        aerosol solution droplets, 72
        in rising air, 86-89
Scattering, 26
    aerosol, 9-10, 29-30, 72-75
    extinction, 29-30
    molecular, 9
    polarization, 27-28
    water droplet, functions for, 27, 33
    see also Refraction
Scintillation (of starlight), 47-48
Sea-air transfer, see Energy transfer
Sea breeze, 152-53, 212
Settling speed, particle, 16-19, see also Aerosol, Fall speed
Shearing instability (Kelvin-Helmholtz waves), 380-83, see also Clear-air turbulence
Shower
    cloud top heights, Atlantic, 385-86
    distribution in oceanic cyclones, 323-24
    evaporation aloft, 255
    rainfall versus frontal, 384-85
Shower formation
    droplet spectra evolution, 182-90
    fibrillation, 190-94
    ice phase role in, 202-4
    minimum cloud depth, 281-84
        continental versus maritime, 190, 194-96
Slope convection, 6, 118, 327-28, see also Cloud and precipitation systems, Energy transfer, Trajectories
Snow, 354-57, see also Fallstreak
Squall front and arch, see Cumulonimbus
Squall line, 237-38
Stability, see Convective cloud theory, Stratification of atmosphere
Steam fog, 142-43

Stratification of atmosphere
  cumulonimbus, 205-6
  cumulus and tower layers, 145-47
  modification by cumulus, 177
  subcloud layers, 141-44
  tropospheric mean stability, 314
  see also Lapse rate, Vertical motion
Stratocumulus (cloud and precipitation), 351-55
  cumulogenitus, 141, 172-73
Stratosphere
  aerosol, 70-71
  clouds in, 27-28, 35-37, 380
  exchange with troposphere, 346-47
  radiative equilibrium, 3-4
  water vapor in, 329, 347-48
  see also Dust, Oreigenic wave

Temperature (definitions and properties)
  dew point, lapse with height, 57
  potential, 57
  saturation potential, 59
  virtual, 13
  wet-bulb, 14-16, 56, 60, 61
  wet-bulb potential, 60
  see also Inversion, Lapse rate
Tephigram, 61-62
Thermal wind, see Vertical shear
Thermodynamic properties and physical data
  air, 41
  conductive capacity of media, 130
  thermodynamic diagram, 59-62
  water, 42-43
  see also Temperature
Thunderstorm (geographical occurrence and factors)
  nocturnal, 236-38, 288-89
  oceanic cyclones, 324, 340-42
  regional factors
    monsoon, 228-31, 238
    orography (intense storms), 232-37
    vertical shear, 226, 230-31, 235
  see also Cumulonimbus
Tornado
  funnel and wind characteristics, 250-51
  jet stream related, 235-36
  radar reflectivity indicator, 280
  regional occurrence, 231-33
  visual features, 290-91
Trajectories, large-scale
  ascent into jet stream, 337-40
  behind upper-level trough, 344-46
  cyclone, principal airstreams, 343-44
  isentropic relative flow
    overrunning maritime air, 240-41
    potential vorticity conservation, 332-33
    slope convection, long wave, 319, 333-34
Troposphere
  exchange with stratosphere, 346-47
  radiation-convection equilibrium, 3-4
Turbulence, see Clear-air turbulence, Cumulonimbus, Weather hazards
Twilight, 75-79

Updraft, see Cumulonimbus, Cumulus

Vapor pressure, 12-13, 42-43
  aerosol solution droplets, 71
  see also Refraction
Vertical flux of properties, see Energy transfer
Vertical motion, large-scale
  control upon convection, 140, 177-79, 386
  general atmospheric circulation, 329
  horizontal acceleration associated, 335-40
  numerical simulation, disturbance, 317-18
  waves, related to, 316, 319, 333-34, 344-46
  see also Trajectories
Vertical shear
  convection rows, 149, 170
  cumulonimbus organization, role in, 222-25, 243-44
    prolonged heavy rainstorms, 296-99
    tropics (hail rarity), 267, 270
    intense cumulonimbus (climatology), 226, 230-31, 235-36
  thermal wind, 313
    departures from, jet stream, 358-59, 369
  see also Fallstreak, Shearing instability
Visual range, 73-75
  intense rainstorms, 273, 300-301
  snow, 356-57
Volcanic eruption, atmospheric effects, 64-65, 77-79
Vorticity, see Airflow properties

Warm front, see Cirrus, Cloud and precipitation systems, Precipitation
Water vapor, see Humidity, Radiation, Vapor pressure
Waterspout, 251-52
Wave, classes of, 117-18, see also Billows, Oreigenic wave
Wave, large-scale
  airflow relative to, 319, 344-46
  length, dominant
    observed, 325-26
    theoretical, 315-17
  relation to surface disturbances, 316-17, 324
    cyclogenesis, 321, 326-27
    cyclone family, 320-22
  speed, related to wind, 315
  subtropical jet stream, 329-30
  topography relations, 315
  see also Vertical motion
Weather distribution, see Cloud and precipitation systems, Shower, Thunderstorm
Weather hazards
  flood, 299
  hail, 287, 292-93, 299
  turbulence (aircraft), 209-10, 378, 387-88
  violent surface wind

  lee waves, 375-76, 386-87
  squall arch (mariner accounts), 286-87
  see also Rainfall, Tornado
Weather modification
  artificial nuclei, 97, 98
  cumulus growth, artificial glaciation, 174-75
  layer cloud seeding, 354
Wind
  ageostrophic, 314, 367-69
  geostrophic, 312-13
  mean zonal, 311
  see also Cumulonimbus, Jet stream, Vertical shear, Wave, Weather hazards

Name Index

Note: Numbers in boldface refer to plate numbers.

Aanensen, C. J., 387
Abe, M., 373, 375, 377
Ackerman, B., 165
Alaka, M. A., 370, 372, 373, 375, 378
Albrecht, F., 144
Alkezweeny, A. J., 110
Allen, P. W., 332
Andre, M. J., 236-38
Angell, J. K., 148, 332
Arakawa, H., 361
Archbold, J. W., 226
Arnold, J., 247
Arons, A. B., 84, 85
Asai, T., 170
Atlas, D., 30-32, 121, 122, 147, 149, 188, 189, 192, 202, 249, 268, 275, 277, 278, 286, 357, 359, 360; **10**
Auer, A. H., 215
Aufdermaur, A. N., 278
aufm Kampe, H. J., 53, 97, 183, 186, 187
Austin, A. R. I., 34, 35, 379
Austin, E. E., 346
Axford, D. N., 381

Bagnold, R. A., 65
Bailey, I. H., 20, 21, 24, 203, 258, 259, 260
Ball, F. K., 135, 137, 177
Bannon, J. K., 384
Barkhley, H., 227
Bartlett, B. M., 109
Battan, L. J., 30, 165, 183, 188, 192, 214, 247, 282, 283, 302, 357
Bean, B. R., 47
Beard, K. V., 19, 45
Beckwith, W. B., 236
Benson, C. S., 36

401

Index

Berenger, M., 375, 378
Bergeron, T., 123, 312, 325, 370, 386
Berggren, R., 321, 326
Berry, E. X., 119-201
Berson, F. A., 341, 342
Best, A. C., 188, 189, 352
Betts, A. K., 177, 336
Bhattacharyya, P., 229
Bibilashvili, N. Sh., 166
Bigg, E. K., 70, 97
Bignell, K. J., 10
Bjerknes, J., 123, 312, 325, 386
Blackadar, A. K., 237, 288
Blackmer, R. H., 236
Blair, T. A., 188, 292
Blanchard, D. C., 67
Blanchard, M. B., 70
Blasius, W., 294
Bleasdale, A., 296
Bleck, R., 372, 378, 379
Bleeker, W., 236-38
Bolin, B., 321, 326
Bonner, W. D., 244
Booth, R. E., 296
Bouruard, A. D., 97
Bowell, V. E. M., 232
Braak, C., 227, 286
Bradbury, D. L., 324, 327
Braham, R. R., 181, 188, 192, 203, 208, 214, 254, 255, 271, 283
Brancato, G. N., 273
Brandli, H. W., 376
Breistein, P. M., 326
Brier, G. W., 174
Briggs, H., 358
Brinkmann, W. A. R., 387
Brooks, C. R., 34, 35
Brooks, E. M., 251, 252
Brown, A. E., 232
Brown, E. N., 188
Brown, H. A., 360
Brown, R. A., 134, 135
Browne, I. C., 188
Browning, K. A., 20, 108, 216, 217, 219, 220, 248, 249, 255, 257-59, 262, 265, 279, 291, 294, 350, 359-61, 376, 381; **39, 40, 48, 53**
Brückner, W., 289
Bryan, K., 327
Bryant, G. W., 99, 100, 381
Bryant, H. C., 35
Bryson, R. A., 238
Buajitti, K., 237
Budyko, M. I., 79
Bullrich, K., 29, 67
Bundgaard, R. C., 123, 312, 325, 386
Bunker, A. F., 139
Burger, A. P., 313
Burham, J., 388
Burns, A., 121
Burrows, D. A., 109
Byers, H. R., 181, 185, 188, 208, 214, 254, 255, 283

Campen, C. M., 162-64
Carlson, T. N., 160, 239, 243-45, 299

Carte, A. E., 225, 227, 257, 258, 266, 291
Cellini, B., 293
Chamberlain, A. C., 68
Champion, R. J. B., 26
Chandrasekar, S., 7
Changnon, S. A., 237, 281, 287
Chichester, F., 154
Chien, C. W., 83
Chleck, D., 14
Chopra, K. P., 154
Church, J. F., 53
Chuvaev, A. P., 165
Clackson, J. R., 227
Clark, B. B., 334
Clark, J. E., 221
Clark, L., 287
Clarke, R. H., **43**
Clayton, H. H., 124
Cohen, A., 376
Cole, A. E., 209
Colson, DeVer, 375; **55**
Condron, T. P., 75
Conover, J. H., 148, 342
Coons, R. D., 354
Corby, G. A., 330, 373, 376
Cornford, S. G., 231, 352
Cottis, R. E., 109
Cox, A. J., 35
Cox, L. C., 17, 18
Cox, S. K., 10, 347, 348
Craig, R. A., 4
Crutcher, H. L., 179, 386
Cudrey, R. A., 67
Cunningham, R. M., 162-64, 215, 246, 355, 356; **33**

Daniels, S. M., 368
Danielsen, E. F., 345, 372, 378, 379
d'Arignon, J., 356
Day, G. J., 165, 188
De, A. C., 229
Dean, D. T. J., 251, 252
Defant, F., 359, 387
Defoe, Daniel, 123
DeJong, J. J. G., 245
Delafield, H. J., 17, 18
Deshpande, D. V., 229, 231
Dessens, J., 232
Dickson, C. R., 148
Dietze, G., 40
Dightman, R. A., 281
Djurić, D., 332
Dodd, A. V., 235
Donaldson, R. J., 280, 291, 302
Döös, B. R., 376
Doron, E., 376
Douglas, C. K. M., 221
Douglas, R. H., 236, 266, 359, 360
Downie, C. S., 52, 53
Draginis, M., 165
Dufour, L., 56
Durbin, W. G., 185
Durst, C. S., 144, 152
Dutton, E. J., 47
Dye, J. E., 110
Dyer, A. J., 131, 136

Ebdon, R. A., 79
Eber, L. E., 352
Eddy, J. A., 70, 75, 77
Edelmann, W., 317
Edinger, J. G., 370
Eldridge, R. H., 226
Eldridge, R. J., 74
Elford, C. R., 273, 300
Eliassen, A., 332
Eliot, J., 229, 292
Ellis, G. R., 387
Endlich, R. M., 343, 358
Eng Young, R. G., 20
Engelbrecht, H. H., 273
Englemann, R. A., 23, 24
Eriksson, E., 64, 66
Estoque, M. A., 152; **14**
Exner, F. M., 26, 39

Faller, A. J., 149, 150
Farlow, N. H., 70
Farquharson, J. S., 212
Favreau, R. F., 291
Fawbush, E. J., 269
Fay, R., 299
Fechtig, H., 70
Fenn, R. W., 68
Ferrel, W., 259, 260, 293
Ferry, G. V., 70
Fessenkov, V. G., 69
Feuerstein, M., 70
Findeisen, W., 97, 98
Fletcher, R. D., 300
Flora, S. D., 232, 291, 292
Foldvik, A., 371, 373
Foster, D. S., 269
Fowler, C. W., 26
Fraser, A. B., 172, 177
Freeman, M. H., 212
Friedlander, S. K., 68
Friedman, M., 44
Friend, J. P., 70
Frisby, E. M., 226, 236, 267, 270
Frith, R., 352
Fuchs, N. A., 17, 18, 23
Fujita, T., 210, 215, 247-50, 258, 262, 291
Fullum, E. F., 70
Fultz, D., 311, 329

Gadd, A. J., 176, 358
Garcia-Prieto, P. R., 153, 188, 189
Garrod, M. P., 184, 186, 188
Gates, D. M., 10
Gentry R. C., 232, 299, 354
George, D. J., 380
George, J. J., 321
Geotis, S. G., 279
Gerbier, N., 375, 378
Ghosh, B. P., 229
Giblett, M. A., 144, 152
Gibson, W. S., 221
Gilchrest, A., 357
Gitlin, S. M., 258
Glass, M., 160
Glover, K. M., 122
Godske, C. L., 123, 312, 325, 386

402

Golde, R. H., 232
Golden, J. H., 251, 252
Goldie, A. H. R., 384
Goldie, N., 346
Goldsmith, P., 17, 18, 348
Goody, R. M., 51, 70
Gopinath Rao, B. G., 230
Gordon, A. H., 251, 252
Goyer, G. G., 258, 291
Grandoso, H. N., 228, 250
Grant, D. R., 145
Graves, M. E., 347, 348
Gray, T. I., 343
Green, H. L., 65
Green, J. S. A., 224, 249, 313, 327, 335
Green, S. D., 299
Gruner, P., 76
Gunn, K., 23
Gunn, K. L. S., 354, 356, 359, 360
Gunn, R., 19, 45, 92, 354
Gunther, R. T., 275
Gutnik, M., 233

Hall, F., 354
Hall, R. K., 195, 283
Hallett, J., 36, 37, 110, 258, 279
Haman, K., 177
Hamilton, R. A., 226
Hammond, G. R., 250
Hankin, E. H., 144, 212
Hanna, S. R., 375
Hardy, K. R., 121, 122, 148, 180, 189; **10**
Hariharan, P. S., 292
Harimaya, T., 365
Harper, W. G., 268
Harrold, T. W., 258, 279, 350, 359-61; **48**
Harvey, R. L., 21
Haurwitz, B., 139
Hawk, N. E., 21
Heidke, P., 292
Helliwell, N. C., 43, 348
Hemenway, C. L., 70
Henderson, T. J., 258
Herlofson, N., 139, 179
Herman, B. M., 278
Hesstvedt, E., 35, 380
Heymsfield, A. J., 362
Hidy, G. M., 68
Hiser, H. W., 299
Hitschfeld, W., 23, 236
Hobbs, P. V., 24, 109, 110
Hoecker, W. H., 250, 291, 375
Hoffer, T. E., 96, 99
Hofstede, G. F., 176
Hollings, W. E. H., 348
Holloway, J. L., 117, 327
Holmboe, J., 375, 378
Honig, J., 357
Hoskins, B. J., 344
House, D. C., 232
Howard, L. N., 380
Howell, H. B., 28, 33
Howell, W. E., 126
Howorth, B. P., 352
Hubert, L. F., 151, 154
Huff, F. A., 281

Humphreys, W. J., 26, 39, 293, 294
Hunt, B. G., 347, 348
Hunter, J. C., 179, 386
Hwang, H. J., 121

Iribarne, J. V., 228
Ives, R. L., 143

Jacobs, L., 78
Jaenicke, R., 67
Jain, P. S., 229
James, D. G., 144
James, R. W., 139, 179
Jayaweera, K. O. L. F., 21, 100
Jeans, J., 15; **2**
Jefferson, G. J., **54**
Jennings, A. H., 273, 300
Jessup, E. A., 332
Jetton, E. V., 280
Johnson, D., 258, 279
Jones, R. F., 209
Jordan, C. L., 206
Joss, J., 277, 278
Julian, L. T., 387
Julian, P. R., 387
Junge, C. E., 65-68, 70

Kalma, J. D., 176
Kamburova, P., 253
Kao, S.-K., 121
Karve, C. S., 93
Kassander, A. R., 179
Kaylor, R. E., 150
Keers, J. F., 176, 358
Keith, C. H., 84, 85
Kerker, M., 34
Kerley, M. J., 43, 231, 347, 348, 366
Kessler, E., 209, 273, 349
Khrgian, A. Kh., 189, 352
Kidder, R. E., 257, 266, 291
Kimpara, A., 341
Kinshott, E. J., 352
Kinzer, G. D., 19, 45, 92
Kirk, T. H., 251, 252
Kleinschmidt, E., 332
Kliefoth, H., 375, 378
Kline, D. B., 352, 354
Knight, C. A., 258
Knight, N. C., 258
Knighting, E., 139, 179
Knoch, K., 227
Knollenberg, R. G., 362
Knopfle, H., 295
Koenig, L. R., 203
Kondrat'yev, K., 176
Koschmeider, E. L., 7
Koto, B., 65
Kotsch, W. J., 352, 353
Koutnik, W., 387
Kramers, H., 93
Kreitzberg, C. W., 360, 369
Krishna Rao, P., 5
Krishnamurti, T. N., 330
Krug-Pielsticker, U., 375, 379
Krumm, W. R., 256
Kryukova, G. T., 165

Kuettner, J., 149, 375, 378, 379
Kulshreshtha, S. M., 212, 228, 229

Lacy, R. E., 232, 252
Lala, G. G., 362
Lamb, H. H., 79, 232
Lamkin, W. E., 291
Lane, W. R., 65
Langleben, M. P., 359, 360
Lanterman, W. S., 342
Lapcheva, V. F., 166
Larsson, L., 375, 378, 379
Latham, J., 110
Lautzenheizer, R. E., 229
Lawrence, E. N., 252
Lazarus, E. H., 298
Leese, J. A., 334
Lempfert, R. G. K., 349
LeMone, M. A., 148
Lettau, H. H., 288
Lewis, R. E. J., 36, 37
Lewis, W., 352
Lhermitte, R. M., 288
Liljequist, G. H., 38, 39
Lilly, D. K., 150, 177, 375, 376
List, R., 20, 256, 258
List, R. J., 12, 13, 41, 42, 56
Lovell, A. C. B., 69
Low, R. D. H., 71
Lowan, A. N., 27
Lowman, P. D., **16, 26**
Lowry, W. P., 238
Ludlam, F. H., 20, 34, 35, 37, 43, 108, 127, 128, 139, 153, 160, 161, 180, 188, 189, 193, 194, 216, 217, 219, 220, 222, 223, 235, 239, 243-45, 253, 257, 258, 268, 294, 320, 325, 335, 359, 362, 364-66, 379-82; **60**
Lumb, F. E., 176, 357
Lumley, J. L., 122
Lyons, W. A., 2

Mackenzie, J. K., 43, 348
Macklin, W. C., 19-21, 24, 93, 101, 105-8, 110, 203, 257-60, 268
Macky, W. A., 251
Magono, C., 19, 99; **7**
Maheshwari, R. C., 230
Malkus, J. S., 135-39, 148, 149, 153, 160, 164, 166, 179
Manabe, S., 3, 4, 117, 327, 347, 348
Manley, G., 375, 378, 387
Marriott, W., 221
Marshall, J. S., 354, 359, 360
Marshall, W. R., 92
Marvin, C. F., 61
Mason, B. J., 19, 21-24, 65, 95, 96, 99, 100, 107, 110, 352, 355, 357
Mastenbrook, H. J., 43, 329, 346-48
Mathur, I. C., 230
Matthews, B. J., 19
Matvejev, L. T., 175
McAuliff, J. P., 299, 300
McCaig, D. A., 276
McCready, P. B., 260
McDonald, J. E., 10, 100

Index

McIlveen, J. F. R., 43, 335, 336
McLean, G. S., 343, 358
McNaughten, I. I., 209
McTaggart-Cowan, J. D., 362
Means, L. L., 236
Merlivat, L., 259
Middleton, G. V., 212
Middleton, W. E. K., 73, 75
Miles, J. W., 380
Miller, R. C., 221, 269
Minnaert, M., 26, 39
Mitra, H., 212, 228, 229
Miyazawa, S., 361
Möller, F., 4
Mollo-Christensen, E. L., 149
Moncrieff, M. W., 224, 249
Monteith, J. L., 131, 176
Moore, C. B., 276, 357
Moore, J. G., 346
Moorhead, J. K., 221
Mordy, W., 153, 184, 185, 352
Mossop, S. C., 65, 96, 98, 99, 109, 110, 203
Mueller, E. A., 277
Mukherjee, A. K., 231
Murgatroyd, R. J., 184, 186, 188, 347, 348
Murray, F. W., 13
Murray, R., 368

Naito, K., 121, 147
Nakaya, U., 109
Natarajan, R., 230
Neiburger, M., 24, 83, 352
Neuberger, H., 26, 39
Neumann, J., 152, 288, 289
Newkirk, G., 70, 75, 77
Newton, H. R., 238
Newton, C. W., 178, 238, 272, 311, 329, 335, 368
Nitta, T., 121
Novak, C. S., 334
Nozumi, Y., 361

O'Connell, D. J. K., 75
Ogden, R. M., 387
Ogura, Y., 175
Oliver, M. B., 53
Oliver, V. J., 53
Ono, A., 70
Oort, A. H., 121, 122
Orville, H. D., 160, 176

Pack, D. H., 148
Packer, D. M., 74
Palmén, E., 178, 311, 329, 335, 368
Panofsky, H. A., 122, 358, 381
Pao, Y.-H., 371
Parmenter, F. C., **46**
Parry, H. D., 326
Paul, A., 236
Paulsen, W. H., 192, 202
Pease, S. R., **2**
Peck, E. L., 299
Pedersen, K., 324, 327
Peppler, W., 354
Pernter, J. M., 26, 39

Petterssen, S., 139, 179, 315, 324, 326, 327, 341, 342
Phillips, E. F., 341
Phillips, N. A., 175, 313
Pidding, H., 286
Plank, V. G., 148, 149, 158, 159, 162-64, 192, 202, 206, 247; **35**
Poliakova, E. A., 301
Postma, K. R., 325
Poulter, R. M., 384
Power, B. A., 356
Price, S., 179, 251
Priestley, C. H. B., 8, 130, 131, 134, 142
Prohaska, K., 221, 232, 292
Prospero, J. M., 65
Protheroe, W. M., 48
Pruppacher, H. R., 19, 45
Pskovski, Y. P., 77

Querfeld, C. W., 38

Raghavan, K., 153
Rai Sircar, 301
Ramakrishnan, K. P., 230
Ramaswamy, C., 231
Ranz, W. E., 92
Rao, N. S., 228
Ray, D., 8, 151
Reed, R. J., 357
Reid, S. J., 376
Reinhardt, R. L., 200, 201, 199
Reitan, C. H., 165, 183, 188
Reiter, E. R., 121, 368
Reynolds, S. E., 100, 101, 282
Rhyne, R. H., 121
Richardson, E. A., 299
Riehl, H., 148, 179, 329
Roach, W. T., 223, 246, 358, 388
Robb, A. D., 295
Robertson, C. E., 109
Robinson, G. D., 332
Rodes, L., 292
Rodewald, M., 287
Rönicke, G., 227
Ronne, F. C., 206
Roos, D. v. d. S., 292
Rosenhead, L., 381
Rosinski, J., 69
Rossby, C.-G., 314, 321, 326
Rowsell, E. H., 232
Roys, G. P., 209
Ryan, B. F., 107

Sand, W., 215
Sanders, R. A., 179, 386
Sansom, H .W., 226, 267, 270
Sanson, J., 298
Sasaki, R. I., 251
Saunders, P. M., 52, 53, 58, 59, 139, 153, 159-61, 166, 179, 188, 189, 193, 194, 206, 207, 211, 283, 284, 303
Sawyer, J. S., 343, 371, 372
Saxena, S. P., 230
Sceicz, G., 131
Schaefer, V. J., 100, 342
Schove, D. J., 292

Schulz, G., 97, 98
Schulze, B. R., 227
Schumann, T. E. W., 294
Scott, W. T., 198, 199
Scorer, R. S., 34, 35, 117, 123, 126, 127, 144, 156, 371, 373-75, 378, 379, 382; **12**
Sehgal, U. N., 229
Sekera, Z., 73
Sellers, W. E., 131, 383
Sergius, L. A., 387
Severynse, G. T., 67
Shackford, C. R., 275
Shafrir, U., 24
Shah, G. M., 79
Sharma, K. K., 230
Shaw, C. C., 10
Shaw, E. M., 384, 385
Shaw, Sir Napier, 154, 349
Shenk, W. E., 331, 343
Sikdar, D. N., 301
Silverman, B. A., 52, 53
Simpson, G. C., 35
Simpson, J. E., 153, 212; **24**
Simpson, J. S., 173, 174
Simpson, R. H., 174, 209
Sims, L. L., 179
Sinclair, P. C., 143
Sinha, K. L., 228
Sinton, W. M., 2
Skaggs, R. H., 236
Skrivanek, R. A., 70
Smagorinsky, J., 117, 327
Smith, W. R., 66
Smith-Johannsen, R. I., 24
Snow, R. H., 69
Sobermann, R. K., 70
Soulage, G., 98
Sourbeer, R. H., 299, 308
Spavins, C. S., 229, 231
Squires, P., 165, 183, 186
Srivastava, R. C., 275
Staley, D. O., 172, 345, 346
Stanhill, G., 176
Starr, J. R., 23, 107, 376, 381; **53**
Steiner, R., 121
Stevenson, C. M., 65, 259
Stewart, J. B., 352
Stommel, H., 139
Stone, H. M., 327
Störmer, C., 36, 37; **2**
Stout, G. E., 236, 277, 287
Strickler, R. F., 3, 4
Subramanian, D. V., 229
Sulakvelidze, G. K., 166
Summers, P. W., 236, 356
Süring, R., 288
Sutton, O. G., 130
Suzuki, S., 387
Symons, G. J., 78
Szeicz, G., 176

Tatro, P. R., 149
Taylor, A., 121
Taylor, J. H., 53
Telford, J. W., 11, 144, 161, 164-66
Tennekes, H., 23

404

Theaman, J. R., 300
Thom, H. C. S., 233
Thompson, W. J., 70
Thorpe, S. A., 381; **58**
Timchalk, A., 343
Todsen, M., 358
Townsend, A. A., 48
Trout, D., 381
Tucker, G. B., 384
Turner, J. S., 155, 157, 173, 176
Twomey, S., 65, 67, 88, 183
Tyler, J. N., 26

van de Heurel, 99, 100
van de Hulst, H. C., 26, 28, 29, 34, 35, 37, 38, 77
van der Hoven, I., 121
Venkataraman, K. S., 228
Venkateswara Rao, D., 231
Vergeiner, I., 375, 376
Visher, S. S., 298
Vittori, O., 291
Volz, F. E., 39, 70, 79
Vonnegut, B., 67, 276, 357, 362
Vrablick, E. A., 276
Vuorela, L. A., 345, 347

Waldram, J. M., 74
Waldvogel, A., 277
Walker, E. R., 387
Walker, J. A., 352, 354
Wallington, C. E., 373, 378
Walton, W., 23
Ward, N. B., 291
Warner, C., 356
Warner, J., 143, 144, 161, 164-66, 183
Webb, E. K., 134
Webb, J. A., 376
Wegener, A., 39, 232
Weickmann, H., 20, 39, 97, 99, 183, 186, 187, 221, 257, 355, 356, 362
Welander, P., 331
Wendell, L. L., 121
Went, F. W., 66
Wetherald, R. T., 3
Wexler, H., 65, 78, 79, 123
Wexler, R., 349, 357, 359, 360
Whipple, F. J. W., 69
Whipple, F. L., 70
Whitney, L. F., 343
Wichmann, H., 210, 215
Widger, W. K., 342
Wiggert, V., 173

Wilk, K. E., 236
Wilkins, E. M., 122
Williams, G. C., 37
Williams, N. R., 143
Willmarth, W. W., 21
Wilson, J. W., 247
Winston, J. S., 5
Witt, G., 70
Witt, H., 292
Wolf, M., 79
Wollaston, S. H., 342
Woo Lee, C., 99; **7**
Woodcock, A. H., 23, 149
Woods, C. E., 280
Woods, J. D., 22-24, 381
Woodward, B., 156
Workman, E. J., 282
Wurtele, M. G., 371, 373-75
Wyatt, S. T., 70

Yabuki, K., 387
Yagi, T., 366
Yanasigawa, Z., 354, 361

Zaitsev, V. A., 183
Zipser, E. J., 214
Zobel, R. F., 137, 231

PLATES

Plate 2.1 Radar echoes over Oklahoma, as shown on displays at the ESSA National Severe Storms Laboratory. *Above*: The RHI display of an MPS-4 radar, the range markers are at intervals of 10 nautical miles and the height markers at intervals of 20,000 ft (about 6 km). In the storm on the right the values of $10 \log Z_e$ corresponding to the successive echo boundaries are 23, 33, 43, and 53 (Z_e in $mm^6 \, m^{-3}$). On part of the PPI display of a WSR-57 radar the range markers are at intervals of 20 nautical miles. The innermost white core of the storm is where $10 \log Z_e$ exceeds 50; the other echo boundaries, including those to two gray areas of which the outer is hardly discernible, represent successive reductions of 10 db in Z_e.

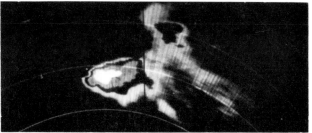

2.1

Plate 2.2 A digital display of average echo intensity ($10 \log Z_e$) corresponding to the echoes on the superimposed PPI display (both extending to a maximum range of 100 nautical miles).

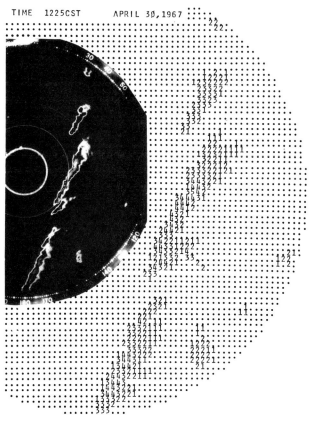

2.2

2.3

Plate 3.1 Traces of cumulus (near the top of the picture) and extensive steam fog over Lake Michigan, about 35 km southeast of Milwaukee on the west shore, 1215 CST, 30 January 1971 (13). The view is toward the north-northeast, across the direction of the westerly winds, from a height of 1200 m. The steam fog is in long rows of nearly upright plumes to a height mainly of 10 m, but locally to as much as 100 m, or even 500 m (near the level of the cloud bases). The implications of this structure are discussed briefly in Section 7.5. (Photograph by R.W. Pease.)

Plate 2.3 Nacreous clouds seen toward the west from Oslo, after sunset, at about 1800 MET on 19 February 1932. These clouds were part of a large group lying between about 40 and 100 km west of Oslo, showing fine colors. Numerous photogrammetric measurements of the heights of cloud details gave results between 23 and 25 km, with a mean value of 24.1 km (56). (Photograph by Per Störmer.)

2.4

Plate 3.2 A view looking east from the Milwaukee shoreline at cumulus and steam fog over Lake Michigan, in fresh winds away from the observer, 1505 CST, 31 January 1971 (13); the air temperature at screen level overland was $-21\,°C$. A few kilometers offshore some steam devils (with diameters up to about 20 m) rose nearly vertically into the cumulus, aligned in rows at a height of several hundred meters. Individual steam devils lasted only a few minutes; they were observed to be rotating slowly (at up to several revolutions per minute), mostly in a cyclonic sense.

Plate 2.4 A brilliant undersun, seen in the central Alps for a few minutes in the early afternoon (solar elevation about 15°), after the sudden clearance of cloud accompanied by snow showers. During this period stellar plate crystals of diameter mostly greater than 1 mm were falling. Remarkably, while the undersun was most brilliant the end of the lower shaft of light was tinged blue, and the upper extension, hardly visible in the picture, had a pale but distinct red color, suggesting that refraction played a part in the phenomenon (66).

3.1

3.2

3.3

Plate 3.3 Condensation trails produced by U.S.A.F. World War II bomber (B-28) aircraft, each with four piston engines.

4.1

4.2

Plate 4.1 View from about 1600 m above sea level on the north flank of the Canary Island of Tenerife, looking west-northwest (almost directly upsun) toward the island of La Palma (130 km distant). The top of the cloudless moist trade wind layer, at a height of about 1350 m, is seen as a bright, level haze top; in the very clean air above it the visual range is at least 200 km.

Plate 4.2 The shadow of the isolated Appenine mountain Monte Cimone, seen looking east near sunset from a height of about 2100 m on the peak itself. The shadow is made visible by the backscatter of sunlight from particles in the haze layer over the plains.

4.3

Plate 4.3 Stratospheric haze photographed from its height of about 15 km, above cirrus between 10.5 and 11.5 km (which, although thicker, appears as the narrow band CC), cumulus (with tops to 3.5 km), and stratocumulus. The picture was taken over Worcestershire at about 1500 GMT, 30 July 1953, by F/Lts. Douglas and Smith, R.A.F.

4.4

4.5

Plate 4.4 Stratospheric haze waves (upper part of the picture) a few minutes after sunset at Dunstable, England, 17 August 1953. The haze was still colorless, not becoming pink until 15 min later, although the old condensation trail in the middle of the picture and other cirrus had become pink 10 to 15 min earlier. The group of pronounced haze undulations in the center of the picture (where the wavelength is about 2 km) has a well-defined upwind edge, whose position on this occasion was 20–40 km nearer than the Cotswold hills, which lie in the same direction.

Plate 4.5 Unusually dense stratospheric haze, after sunset at Ascot, England, on 15 November 1964. The dark streak almost parallel to the horizon low in the picture is cirrus beneath the haze layer. The pattern of corrugations in the haze layer was almost motionless.

Plate 5.1 Typical examples of particles collected from detached ice clouds in the high troposphere (23):

(*a–c*) From an isolated cloud extending between about 8500 ($-37°$C) and 9000 m ($-41°$C). The small crystals in (*a*) were collected from near the cloud tops, and the large crystals in (*b*) and (*c*) from near the cloud bases. The large particles are long hexagonal prisms with hollows tapering inward from their ends. Sometimes the crystals appear as straight pairs ("twins"), but many consist of fragile clusters ("tufts") of several crystals radiating from a common center, liable to disintegrate during collection. Flight was slightly bumpy in and below the cloud and an updraft above its middle was estimated to have a speed of about 2 m sec^{-1}. The only optical phenomenon observed was a brilliant halo of 22° radius.

(*d*) Another example of a crystal tuft from a cirrus cloud, collected at 9200 m ($-40°$C).

The scale of the pictures is shown by the line drawn in the upper part of (*d*): its length is 300 μ.

Crystals collected from amorphous layers of ice cloud were generally smaller, single, and with no or only small air enclosures.

Unlike the particles of other kinds of cloud, which form at the cloud base and attain their largest sizes near the cloud tops, the crystals of detached ice clouds form near the level of the tops, grow by condensation while falling through air which is supersaturated with respect to ice but generally cloud-free, and are largest at some lower level.

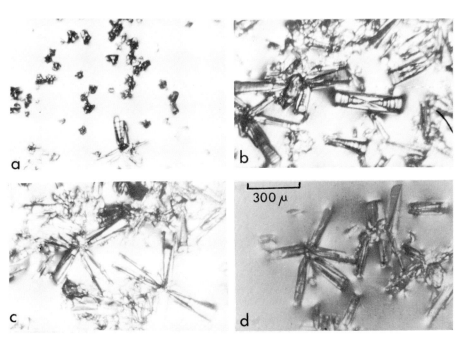

Plate 6.1 Fallstreak forms of cirrus; view westward from near Östersund, central Sweden, at 1200 MET, 6 July 1955 (see also Pl. 9.9; these clouds are discussed in Section 9.6).

The fallstreaks formed over mountains at the edge of a cirrus system approaching from the west, individually moving from 325° at 20 m sec^{-2}. Their tops were at about 8.6 km and they trailed toward 280° for about 4 km before turning toward 340° at about 8.3 km. (Fallstreaks with this orientation can be seen in the distance, beyond the developing examples in the upper center of the picture, and the evaporating residues in the upper right. In the middle distance there are also growing patches of shallow fibrous cloud which are not producing fallstreaks.)

The particles in the upper parts of the fallstreaks were inferred to have a mean fall speed of about 15 cm sec^{-1}, and the fallstreaks to have a development period of about 1 hr. Since cirrus fallstreaks move at least several tens of kilometers during their development, it is difficult to follow their evolution (see, however, Pl. 9.10).

6.1

6.2

Plate 6.2 Altostratus: an extensive, thick, and diffuse layer of ice cloud in the upper troposphere, sufficiently dense to obscure the sun (the dark details in the upper part of the picture are lower droplet clouds). Thick altostratus is a cloud characteristic of large-scale slope convection.

6.3

Plate 6.3 Longitudinal rolls of low cloud over the English Channel, on the morning of 23 April 1962 (view southward from the Cornish coast). The low-level winds were easterly, 4–7 m sec^{-1}; the clouds were formed in air cooled over the sea after leaving the Continent, and the stratification was distinctly stable in the lowest few hundred meters, and slightly stable above. The level of the cloud bases is about 300 m above the sea, and the clouds have a maximum thickness of nearly 300 m. The most prominent roll cloud is near the coast; another can be seen faintly about twice as far away; a third which was present does not appear in the picture. The separation of the rolls is about ten times the height of their tops, and they appeared to be aligned in a direction distinctly backed from the general wind direction (cf. Section 7.6).

6.4

Plate 6.4 Wave clouds in the high troposphere near Denver, Colorado, in the early afternoon of 4 March 1962, looking southwest toward the eastern slopes of the Rocky Mountain range. The wave clouds are at a height of about 9.5 km, where the temperature is about $-55°$C and the wind 305° 35 m sec^{-1}. Near this level the undisturbed air has a relative humidity probably somewhat exceeding that representing saturation with respect to ice. However, clouds form only where air is lifted rapidly in hill—or lee—waves to become saturated with respect to liquid water; at the low temperatures prevailing the cloud particles are all frozen and do not evaporate when they are carried leeward, so that long plumes of ice cloud trail downwind from the first-formed clouds. The undulating form of the nearest plume clearly shows the presence of lee waves with a wavelength of about 12 km and an amplitude of about 250 m, implying maximum vertical velocities in the waves of about 3 m sec^{-1}. It is probable that the clouds in this picture form in pronounced first lee waves (rather than hill waves), and that the group of cumuliform clouds beneath the most prominent is in a rotor.

Plate 7.1 Features of ordinary convection evident in radar echoes from associated irregularities in the refractive index of the air, seen on the displays of radars near the coast of Virginia (27).

(*a–f*) At the top are two series of pictures of the RHI displays of radars using wavelengths of 3.2, 10.7, and 71.5 cm, directed on the same azimuth of 340° on 8 June 1966: first, at 1246 EST and at full receiver gain (*a–c*), and second, 3 min later with the gain reduced by 12 db (*d–f*). Range markers are at intervals of 5 naut mi (about 9 km) and the height marker is at 20,000 ft (about 6 km). In all displays a large cumulus appears, whose tallest tower reaches a height of about 4 km. The echo from the cloud is received mainly from near its boundaries, except on reduced gain at 3.2 cm, where it is generally from very small precipitation particles inside the cloud (Z of order 10^{-1} mm^6 m^{-3}). The comparative intensity at the longer wavelengths of the echoes from the cloud boundaries demonstrates that they are not due to scattering from particles [considering also the differing beam widths, the power received from the refractive index fluctuations is about 15 db *greater*, whereas that received from the particles is about 5 db (at 10.7 cm) and 40 db (at 71.5 cm) *less* than at 3.2 cm].

(*g*) The RHI display of the 10.7-cm radar, directed toward the north during the afternoon of 13 May 1966, with a test signal at a range of 11 naut mi.

(*h*) A contemporary PPI display at an elevation of 2°. The prominent echoes (apart from "ground clutter" mainly within 10 naut mi) are received from refractive index fluctuations near the upper boundaries of convective circulations whose tops reach a height of about 1.5 km. The narrow radar beam (of width 0.5°, corresponding to a transverse dimension of about 200 m at a range of 15 naut mi) intersects the lower parts of the domelike boundaries, so that on the PPI display the individual circulations are outlined by roughly circular or oval rings of echo. The rings are about 2 km across and tend to be arranged touching in rows lying approximately along the direction of the (westerly) wind, separated by distances of about 3 km across the wind direction.

(*i–k*) PPI displays of the 10.7-cm radar at an elevation of 1.5°, on the afternoon of 9 June 1966, with successive reductions of 12, 18, and 30 db from full receiver gain. Range markers are at intervals of 10 naut mi, and the outlines of the local shores are shown in (*j*).

On the first display (*i*) there is extensive striated echo associated with convection mainly inland (west) of the radar. On the second display (*j*) a narrow band of intense echo which marks a sea-breeze front can be seen more clearly, separating rows of echoes which are aligned approximately along the wind direction. West of the front rows about 2 km apart, composed of small cells about 1 km across, are in a southwesterly airstream in which convection reaches a height of nearly 1 km. East of the front rows are in a southerly sea breeze in which the convection reaches a height of about 0.5 km. At the greatly reduced gain of the third display (*k*) echoes from fluctuations of refractive index disappear, and the front is marked by clustered "dot" echoes from individual targets whose radar cross section is of the order 10 cm^2 and which are therefore identified with soaring birds. Similar echoes are apparent near the ground on either side of the 11-mi marker in the RHI display (*g*).

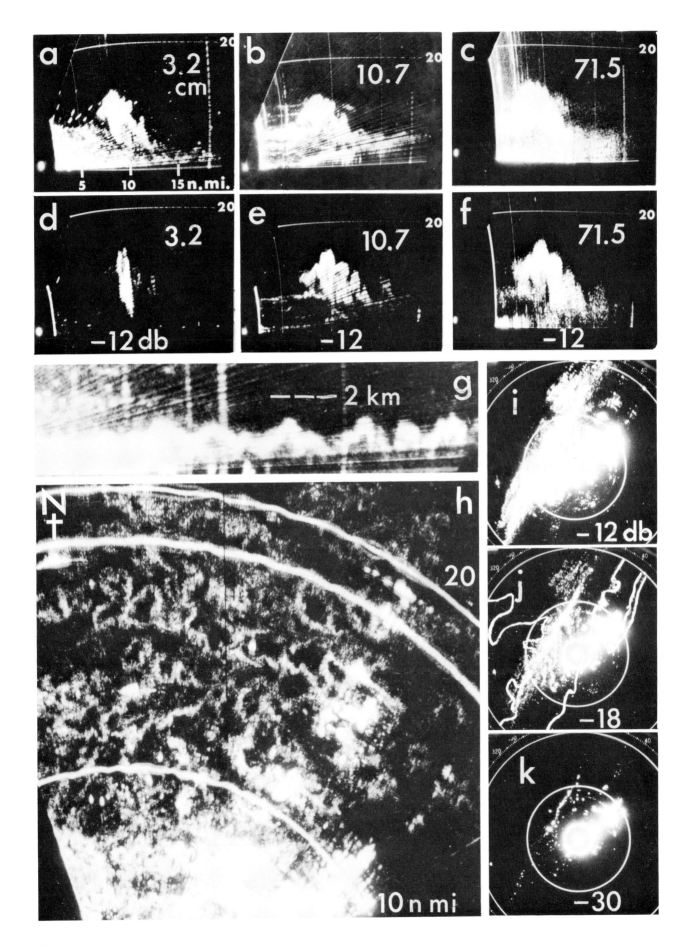

7.1

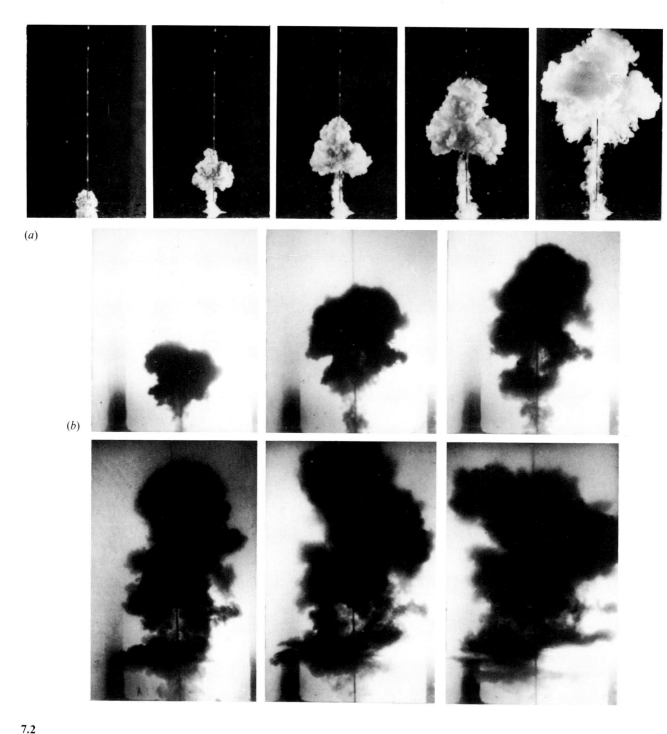

Plate 7.2 Stages in the growth of experimental thermals sinking in a tank of water (54); the pictures are inverted to make the resemblance to rising cumulus towers more apparent. In the upper sequence (a) the dense fluid was marked with a white precipitate and sank through water of uniform density, corresponding to a neutral stratification. In the lower sequence (b) the dense fluid was a dark solution, released into a tank of water in which the density was uniform only in the top quarter, and increased downward in the remainder, corresponding to a stratification which was at first neutral, and then stable. The width of the thermal increased at first, but then hardly changed, the thermal leaving behind a trail of "eroded" material. The inertia of the leading part of the thermal allowed it to penetrate to a level where its buoyancy was reversed; subsequently it retreated and spread sideways (last picture).

(a)

7.3

(b)

Plate 7.3 Stages in the development of the "mushroom clouds" of nuclear bombs, exploded above the sea (in tests over the Pacific). The first cloud from such explosions forms when the heated air has cooled sufficiently and closely resembles an experimental thermal (a). In this example moist air of the adiabatic layer below the small cumulus was subsequently drawn into the circulation of the thermal and produced the stem of the mushroom cloud (b). Its form shows that the flow was smooth at low levels but became turbulent higher up the stem. There is a marked contrast also between the crenellated outline of the turbulent main cloud and the smooth texture of the *pileus* ("cap cloud") which formed over it in the upper troposphere. The pileus is produced by the smooth lifting and cooling of air lying above the main cloud, and it often has a laminated structure indicating that the vertical profile of relative humidity is not regular. Usually it retains a smooth texture, showing that the lapse rate is less than the saturated adiabatic, while the head of the main cloud advances and eventually appears to penetrate and then absorb it. Similar clouds often form over rising towers of cumulus and cumulonimbus.

Plate 7.4 Cumulus over Florida (looking southward from a height of 10 km; 44). The cumulus are uniformly scattered with no definite pattern. There are large cumulonimbus with anvil clouds near the horizon.

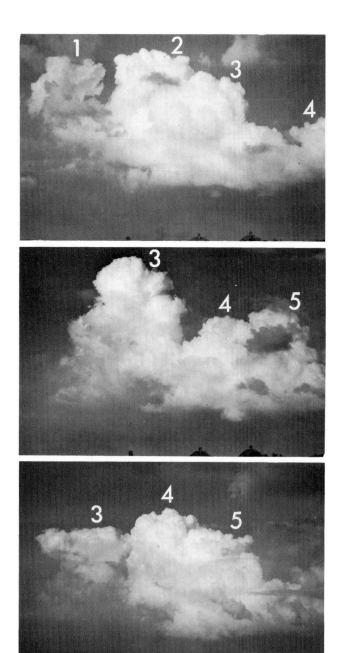

Plate 7.6 Large cumulus, base 1.3 km above sea level (0.9 km above the ground), with towers to over 4 km, and shelf clouds with almost level tops at 3.3 km (seen looking northeast from 3.7 km). Some decaying frozen ("glaciated") towers of shower clouds can be seen in the middle distance. (Photograph by L. Larsson.)

7.5

Plate 7.5 Photographs at intervals of 8 min showing the rise of a cumulus tower, numbered 3, to a peak height of 5 km, where its diameter $2r$ was about 2 km (second picture). The maximum rate of rise of the tower top was about 6 m sec^{-1}. The tower subsequently subsided, its top becoming flattened at a height of about 4 km (third picture), and soon afterward evaporated completely. Previous and succeeding towers are numbered 1, 2, 4, and 5.

7.4

7.6

Plate 7.7 Rows of cumulus in the trade winds over the Caribbean Sea. The view is toward the south and the easterly trade wind blows across the picture from left to right. The tops of the tallest clouds reach about 2 km above the sea. (Photograph by Dr. J. Simpson.)

Plate 7.8 Cumulus in rows over Florida. (Photograph from a manned satellite; 96, p. 186.) The view is northward and includes the whole of the Florida peninsula. Over the peninsula the low-level winds were easterly, 5–8 m sec^{-1}; in the south the wind was easterly throughout the troposphere, with a maximum speed of about 15 m sec^{-1} in the layer from 8 to 12 km. In the convection over the sea the cumulus base was at about 600 m, and the top of the cumulus layer was at about 5 km. The convection intensified over the land during the day, and in the picture there is a clear ring around Florida except near the south and southwest coasts, downwind of the neighboring Bahama Islands. Other interesting features are the clear sky over Lake Okeechobee (in the center of the picture) and the enhanced cumulus development in a surrounding arc (probably associated with a lake-breeze front which is absent downwind of the lake), and the three cumulonimbus near Key West, in the lower left of the picture. The easternmost has only recently grown and its bulging top is beginning to spread.

[In this and succeeding legends to photographs taken from manned satellites reference is frequently made to a page in the listed reference 96 or to a plate in reference 97, in which the photograph can be found printed on a larger scale and in color; such pictures show more detail and are often easier to interpret than the half-tone pictures printed in this book.]

7.7

7.8

7.9

Plate 7.9 Cumulus in rows over the sea. (Photograph from a manned satellite; 96, p. 116.) The view is southeastward over the open Atlantic east of the Lesser Antilles. The convection in the easterly trade wind is vigorous; the height of the cumulus bases is about 600 m and the top of the cumulus layer is at about 4.5 km (about the height of the 0°C level); some shower clouds have produced small anvils (upper right and top left). There is a tendency for spaces clear of cumulus to form near the larger clouds.

In the lower half of the picture the cumulus are more clearly organized into lines approximately along the east-southeasterly direction of the low-level winds, but the arrangement is irregular. The more pronounced lines containing the larger clouds do not have quite the same orientation as the more closely packed rows of small clouds, and they fork or bow to enclose "cells" a few tens of kilometers across.

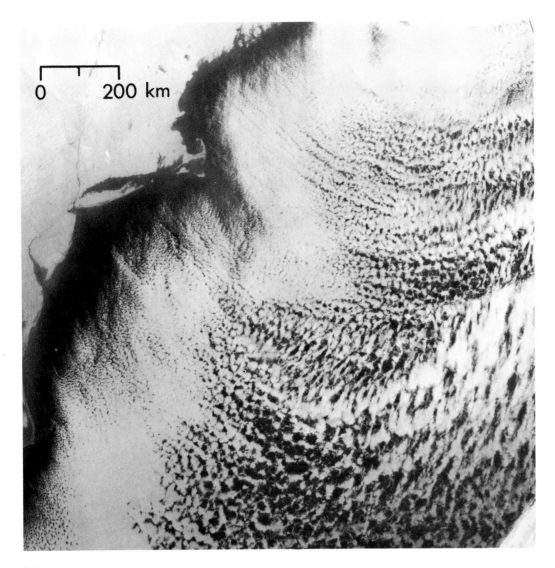

7.10

Plate 7.10 Cumulus and cumulonimbus formed over the Atlantic in a cold airstream flowing off the eastern United States. (Part of a nighttime infrared image provided by the satellite NOAA 2, with a resolution of about 1 km, on 18 December 1972.) The view is everywhere almost vertically downward; north is toward the top of the picture. A wealth of remarkable detail can be seen in the clouds, which appear lighter in tone the lower their temperature. The bright cloud in the extreme southeast corner is the edge of the cirrus system associated with the receding cold front.

Cumulus are in rows, along the wind, which form about 100 km from the coast. The rows have patterns evidently related to topography at the coast and just inland, including transverse groupings which may indicate lee waves. The separation of the rows is initially only a few kilometers. Several hundred kilometers from the coast the scale changes suddenly, probably when the clouds have become deep enough to produce showers; clear spaces appear, some enclosing almost regular "open cells," with diameters of about 30 km. Toward the right of the picture the shower clouds are probably several kilometers deep, with anvils up to about 100 km long aligned approximately parallel to the front and the mean tropospheric thermal wind.

Plate 7.11 Cumulus and shelf clouds off the east coast of Brazil, from a height of 7.5 km. The clouds are in the east-southeasterly trade winds of the South Atlantic, well south of the intertropical convergence zone (in about latitude 5°N). The height of the cloud bases is about 700 m, and their tops reach 2500–3000 m (temperature about 10°C). The larger clouds are producing showers, and extensive shelf clouds at about 1800 m, which spread out on the northwest sides of the parent cumulus clusters. The thinner and older shelf clouds have a smaller albedo than the cumulus. (Photography by R. Cunningham.)

Plate 7.12 Cumulus and shelf clouds over the sea. The view is southward from a height of 5 km, over the sea about 200 km east of Recife, Brazil, in latitude about 7°S. The cumulus are in the easterly trade winds of the South Atlantic; the height of the cloud bases is about 650 m, and the larger clouds, which are producing showers, have tops at about 2500 m, where the temperature is about 13°C. The extensive shelf clouds are up to about 200 m thick; their tops are at 1600 m and are humped by a small-scale internal convection due to radiative exchanges. (Photograph from meteorological reconnaissance aircraft used in studies of trade wind clouds by R. Cunningham.)

7.11

7.12

7.13

Plate 7.13 Cumulus and extensive shelf clouds. The view is steeply downward over the ocean southeast of Madagascar. The brighter cumulus tops can be seen in places protruding above the shelf clouds, especially in the upper left and upper right of the picture, where the greater density and sharply defined edges of the shelf clouds indicate their recent formation. The older and thinner shelf clouds have a finely dappled pattern with a scale of a few hundred meters or even less, corresponding to a slow internal convection produced by radiative exchanges. Adjoining shelf clouds are typically separated by narrow channels of clear air locally distorted into eddies, as in some previous plates. (Photograph from a manned satellite; 96, p. 246.)

7.14

Plate 7.14 A panoramic view of a belt of large cumulus and cumulonimbus lying over a sea-breeze front inland of the south coast of England, 1400 GMT, 18 June 1975. The picture was taken from 1400 m above the coast at the position marked A on Fig. 7.21.

Plate 7.16 A view toward the south coast of England from above Petersfield (corresponding to the position P on Fig. 7.19), 1724 GMT, 11 August 1963. To the right of the tailplane of the aircraft are the ragged fringe clouds of the sea-breeze front, and above and beyond is the main belt of cumulus formed by the ascent of air from landward of the front.

7.16

7.15

Plate 7.15 A view westward along the south coast of England, 1422 GMT, 18 June 1957 (see also Pl. 7.14). The picture was taken from 1500 m above the coastline, at the position marked B on Fig. 7.22. Large cumulus and cumulonimbus lie in a belt over the sea-breeze front, between 15 and 20 km inland. The inland air has passed across London before reaching the front, and produces a wall of haze where it ascends into the clouds; the air to seaward of the front is comparatively very clear.

7.17

Plate 7.17 Cumulus near a sea-breeze front about 40 km inland of the south coast of England. The bases of the cumulus inland of the front, seen at the top of the picture, were at a height of about 1700 m, about 900 m above the bases of the cumulus in the center of the picture, formed by the ascent of air on the seaward side of the front. (Photograph by P.M. Minton from a sailplane at a height of 1500 m, looking toward the coast from just inland of the front; 45.)

7.18

Plate 7.18 Effects of islands on the distribution of trade wind cumulus. The view is westward toward the island of Hawaii; less distinctly seen in the right center of the picture are the islands of Maui, Kahoolawe, Lanai, Molokai, and Oahu. The island of Hawaii is surrounded by a marked almost complete clear ring, which on the windward (northeastern) side is bounded by a line of enhanced cumulus development about 30 km from the coast. Other cumulus are massed over the island slopes, and a pronounced line of cumulus just inland of the northwest coast probably marks the position of a sea-breeze front. The smaller islands have lesser effects upon the cumulus distribution. (Photograph from a manned satellite, 2248 GMT, 24 August 1965; 96, p. 200.)

7.19

Plate 7.19 Cumulus lines in the lee of islands in the Caribbean Sea. The view is southward, and includes Acklins Island—bottom right, Great Inagua—lower left, Cuba—upper right, and the westernmost peninsula of Haiti—upper left. In the easterlies approaching the islands the cumulus bases were at about 600 m ($\theta_s = 25°C$), and the top of the cumulus layer over the sea was at a height of about 4.5 km. The cumulus convection was intensified over the islands during the day, and at the time of the photograph, just before local noon, there is a well-marked clear ring around the coast of Cuba. The cumulus tend to be arranged in rows approximately along the wind direction at low levels. In the lowest kilometer the wind speed was 5–8 m sec^{-1}, and over Cuba in the middle troposphere it was somewhat veered and stronger, so that there the larger clouds appear to lean forward and to the right. Long and nearly straight lines of cumulus are conspicuous downwind of Little and Great Inagua and Haiti (top center). (Photograph from a manned satellite, 1640 GMT, 25 August 1965.)

Plate 8.1 The fibrillation and glaciation of cloud towers during thundercloud development near Verona, Italy. Time in minutes after noon is indicated near picture edges, and azimuths at intervals of 10°, including 030°, are marked at the base of some pictures. Figure 8.13 contains further information.

The clouds were over hills north of the place of observation. A little before noon, when towers reached nearly 6 km, the first radar echo was detected (Fig. 8.13); the first picture shows tops of the clouds in the group A reaching 6000 m at a range of 30 km; they produced weak radar echoes. (The apparently taller cloud B in the center of the picture is nearer (range 24 km); its top reached only 5800 m and it did not produce a radar echo.)

Soon after the formation of showers the towers of the clouds in group A rose much higher; the tower C reached 7800 m (second picture), and became fibrillated (third picture). Succeeding towers E, F, and G reached 8100 m and their fibrillated residues were denser and more persistent. Subsequently tower H rose at about 7 m sec^{-1} to much greater heights, 8600 m at 1222, 10,500 m at 1226, and 11,000 m at 1231 (and it glaciated). Similar towers rose to merge and produce an expanding anvil cloud whose level top was at about 10,500 m. Another group K of nearer towers (range 24 km) shows a similar rapid transformation; tops reached 7600 m at 1231, 9000 m at 1236, and 10,300 m at 1238 before spreading and merging with the original anvil cloud (last picture).

7.20

Plate 7.20 Extensive cumulus and shelf clouds showing an eddy centered about 100 km in the lee (to the south) of the mountainous island of Tenerife in the Canary Islands, which can be seen faintly in the top left corner of the picture. (Photograph from a manned satellite, December 1965; 97, Pl. 65.)

8.1

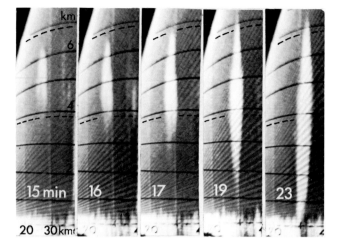

8.2

8.4

Plate 8.2 Development of radar echo from a growing cumulus, as seen on an RHI display. Ranges of 20 and 30 km are indicated by white vertical lines, and heights at intervals of 1 km by superimposed dark lines. The heights of the levels where the air temperature was 0 and $-20°C$ are shown by dashed lines. The time of each picture in minutes after 2100h is shown in white near its lower left corner.

In its early stages of development the echo appears forked, with the shape of an inverted V, and its top rises at about 4 m sec^{-1}; its base descends to the ground at an average speed of nearly 10 m sec^{-1}.

Plate 8.4 Typical example of a shred cloud of droplets (top left) above glaciated residues of a small shower cloud. On this occasion the shred is a remnant of a shelf cloud; others can be seen on the right and in the center of the picture.

8.3

Plate 8.3 Typical glaciated residues of a small shower cloud formed inland in a stream of maritime air over central Sweden. The height of the cloud bases was 1.6 km (temperature 2°C); the greatest height reached by the tops of the cloud which produced a shower and glaciated was 4.1 km (temperature $-11°C$). The cumulus on the left was a large cloud 30 min before and after the time of the picture, but its tops just failed to reach the height of 3.3 km (temperature $-7°C$) which on this occasion was associated with shower formation and subsequent glaciation.

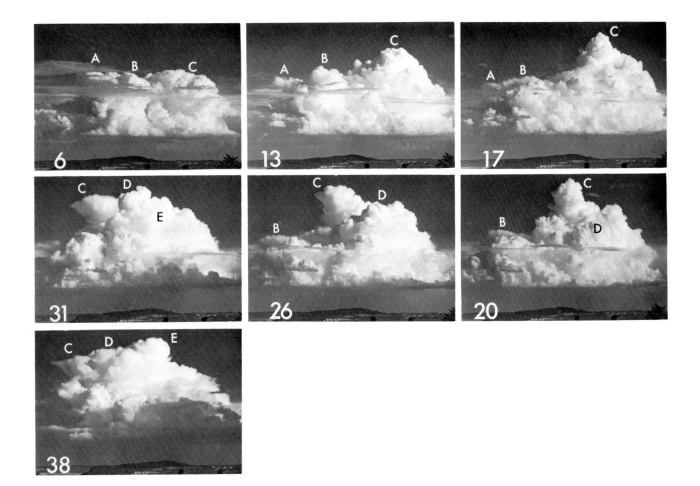

Plate 8.5 A typical example of the growth surge which rapidly leads to the development of a large anvil cloud with its top near the tropopause, following shower formation and glaciation in cumulus towers in the middle troposphere (seen south of Östersund, central Sweden; 10 July 1955). At 1712 MET a tower of a group of congestus clouds was observed to reach a height of 5.1 km; succeeding towers in the same group reached greater heights, and at about 1720 one reached the fibrillation level at about 6.8 km (point O on Fig. 7.27). At 1727 a tower reached 7.5 km after ascending past the fibrillation level at 4 m sec^{-1} (A on Fig. 7.27) and left glaciated residues; this tower and its residues are marked A in the sequence of pictures (time in minutes after 1727 is entered in the lower left of each). A subsequent tower, B, also glaciated, and a third, C, reached a peak height of about 10 km; these towers rose through the upper troposphere at a speed of between 4 and 5 m sec^{-1} (B and C on Fig. 7.27). They were succeeded by other towers (including D and E), which attained greater rising speeds (of about 7 m sec^{-1}, as shown by D and E on Fig. 7.27), and peak heights of 11 km; the residues of these towers persisted and spread to form an anvil cloud which in the last picture is seen extending to the left of and away from the observer. A shower became distinctly visible below the cloud bases at 1747 (fourth picture).

8.6

Plate 8.6 Consistent with the frequent growth surge of cumulus clouds in which showers form, during cumulonimbus convection there are usually two distinct layers in which shelf clouds appear: one in the middle troposphere, at the top of the cumulus layer, and another near the tropopause, where the cumulonimbus updrafts spread out (there the shelf clouds appear thicker, partly because precipitation falls from them; they are also more persistent and extensive, because evaporation is much slower at their lower temperatures). The towers of the cumulus and the cumulonimbus protrude above the level of their shelf clouds.

This photograph was taken at almost the same time as Pl. 7.6, toward hills beyond a cool lake. Cumulus bases were at 1.3 km ($\theta_s = 11.5°C$); the top of the cumulus layer was marked by extensive thin shelf clouds at 3.3 km ($T = -11°C$; $\theta_s = 9°C$), which persisted while drifting away from their parent cumulus (there was no convection over the cool lake). Cumulus which reached 5.2 km ($T = -23°C$; $\theta_s = 11°C$) produced showers and developed into cumulonimbus whose towers reached 8.4 km and whose anvil (shelf) clouds spread out with tops at 7.7 km ($\theta_s = 11.5°C$). Winds were light at all heights, and the stratification was typical of widespread cumulonimbus convection. The principal characteristics of the convection and of the stratification can be inferred from this picture alone.

8.7

Plate 8.7 The dust cloud of an approaching haboob, which has buttresses and projecting lobes similar to those of experimental density currents. (Photograph by Flight International).

8.8

Plate 8.8 A typical arch cloud over a squall front approaching an airfield in Florida. (Photograph by James H. Mayer, Patrick Air Force Base.)

8.9

Plate 8.9 Arch and scud clouds ahead of an approaching hailstorm. The parallel arches surround the region of heavy rain and small hail, perhaps giving a misleading impression of rotation about a vertical axis; they indicate a laminated structure in the humidity of the air entering the cumulonimbus over the cold outflow. The strong horizontal divergence near the ground in the cold outflow causes the visible edge of the precipitation to extend forward at low levels.

8.10

Plate 8.10 Near its forward edge an arch cloud has a smooth surface and looks like a wave cloud, but small-scale convection develops and becomes especially apparent in the underside of the arch cloud when observed, as here, from a position between the squall front and the precipitation. The lumpy structure and obvious slope upward at a large angle to the horizontal give the cloud a very distinctive appearance. (Photograph by M.C. Gillman.)

8.11

Plate 8.11 View looking south from just west of the dry line (position 2 of Fig. 8.47), 1435 CST, 26 May 1962. There is intense cumulus and cumulonimbus development just east of the dry line; although from this angle the clouds appear to compose a continuous chain, they probably consist of separated clusters. Cumulonimbus at B (see also Fig. 8.47) are beginning to develop anvils.

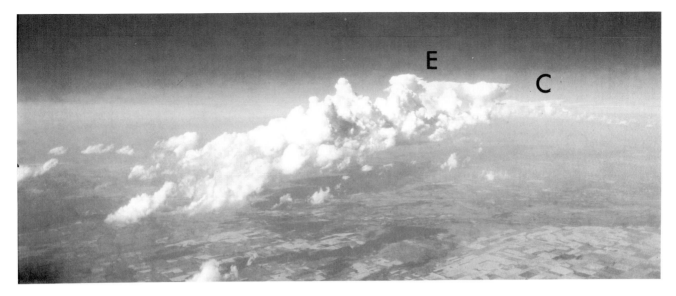

8.12

Plate 8.12 View looking northeast from 7500 m (above position 4, Fig. 8.47), 26 May 1962. The clouds in the group E are beginning to produce anvils and partially obscure the more distant cumulonimbus of group C.

8.13

Plate 8.13 A view looking east-northeast from about 10 km of the rear of a storm over Colorado (179) (on the right) and (in the top center) a similar cumulonimbus about 70 km away. Beneath the rearward-protruding shelves of the nearer cloud are deep fallstreaks of precipitation, consisting mostly of small hailstones (white patches on the ground in the lower center of the picture are accumulations of hail in the wake of the storm).

Each of the cumulonimbus had towering cumulus on its right-hand (southern) flank, with a well-defined level base about 2800 m above the ground.

8.14

8.15

Plate 8.14 A view from a height of about 20 km of anvil cloud spreading east-northeast (at a height of about 13 km) from an intense cumulonimbus over Oklahoma. The turbulent dome of the cumulonimbus is on the extreme right of the picture; also on the right large mamma can be seen pendant from the outflow into the anvil cloud. (Photograph by U.S. Air Force.)

Plate 8.15 A closer view of the dome of the cumulonimbus of Pl. 8.14, reaching up in the foreground to within 2–3 km of the aircraft. (Photograph by U.S. Air Force.)

Plate 8.16 An example of a cumulonimbus with an extensive anvil (182). The view is southward from about 11 km over Florida. The principal cumulonimbus towers are over the southern tip of the peninsula, about 90 km away, and reach a height of 16.5 km; the visible part of the anvil cloud is about 70 km long.

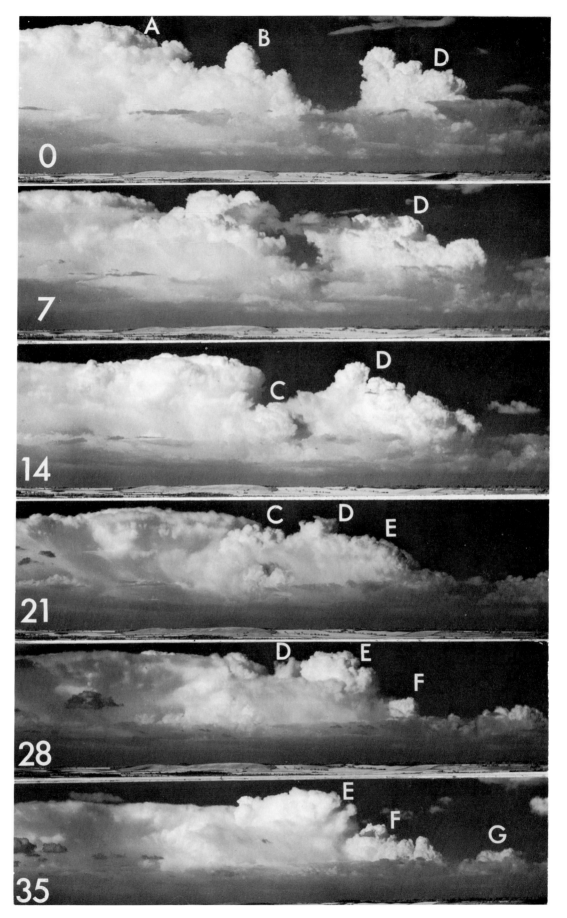

8.17

Plate 8.17 A receding organized but not very intense cumulonimbus, showing one stage in the progression of the main updraft to the right of the wind direction by the successive development of cumulus in the belt over its downdraft front on the right flank of the storm (cf. Fig. 8.27). The view is toward the northeast over England, looking almost in the general direction of the winds in the troposphere (about 210°). The pictures were taken at intervals of 7 min after 1650 GMT; the range of the main cumulonimbus towers increases from 55 to 70 km. Although the individual cloud towers move across the field of view to the left, the position of the principal cumulonimbus towers shifts to the right, with a mean storm velocity of 240° 7 m sec^{-1}.

Some of the principal groups of towers are identified by letters. The cumulus towers D eventually fibrillate (fourth and fifth pictures) before a growth surge leads to their becoming the principal cumulonimbus towers.

8.18

Plate 8.18 A broad cone-shaped funnel cloud in a tornado over northern France (late afternoon, 4 May 1961, near Evreux). (Photograph from *Paris-Normandie*, Evreux.)

8.19

Plate 8.19 A waterspout observed about 8 km south-southwest of a ship in the eastern Mediterranean in winter. Squally showers were occurring in the rear of a cyclone centered a few hundred kilometers farther north.

The funnel cloud is beneath an arch cloud (base height about 570 m) beyond which the sky is clear, and is nearly 2 km away from the edge of a region of intense precipitation from cumulonimbus.

Most waterspouts occur with only slightly unstable stratifications, often below cumulus not large enough to precipitate. Some like that shown here, however, seem to develop in about the position relative to cumulonimbus which is characteristic of tornadoes. In winter, when cool airstreams over continents reach warm waters, especially near mature cyclone centers, the available potential energy may become rather large. In the present example θ_s was only 11°C at cloud base, but was little more than 8°C in the middle troposphere, and the available potential energy was estimated to be sufficient to give a maximum updraft speed (according to the simple parcel theory) of about 35 m sec^{-1} (and a maximum cloud top height of about 9 km). (Photograph by Capt. A.L. LoRe of the *Exilona*.)

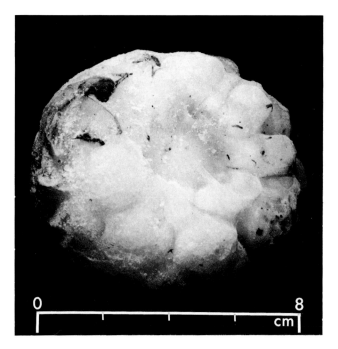

8.20

Plate 8.20 The hailstones shown in this and the two following plates were collected from a storm (G of Fig. 8.48) near Oklahoma City. The hailstone shown here has pronounced lobes in which the ice is milkier than in the spaces between, and so has the appearance sometimes mistakenly supposed to indicate that it is a conglomerate of smaller hailstones (219). The dark indentation of diameter about 2 cm near the middle of the hailstone surface interrupted the layer structure, and was therefore probably produced by melting after the fall of the stone from the cloud in which it grew. Such indentations on one side are a common feature of large hailstones collected from the ground.

8.21

Plate 8.21 A giant hailstone with pronounced and rather jagged protuberances. The internal structure indicated that similar knobs probably covered the surface before they were removed from one side during melting (219).

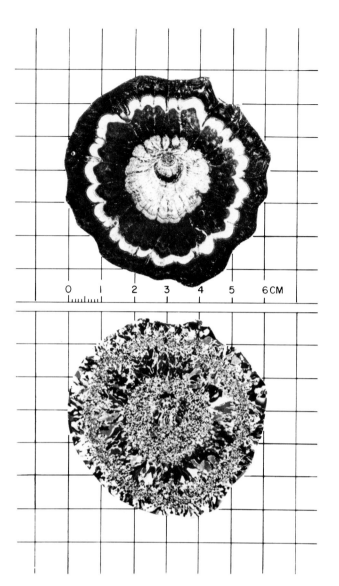

8.22

Plate 8.22 Views of a thin section of a giant hailstone showing (above) the layers of clear ice (black) and milky ice (white) by reflected light, and (below) the crystal fabric revealed by transmitted polarized light (219). The photographic technique in making the upper picture has almost certainly greatly exaggerated the contrast between the clear and milky layers and the sharpness of their boundaries.

Plate 8.23 Photograph taken at 2015 GMT, 5 September 1958, looking northeast from London. Long lightning discharges extend toward the zenith in the anvil cloud of a receding severe storm. The longer of similar discharges, with the most extensive branching, were observed to have unusual durations of up to some seconds, and sometimes to cover nearly the whole sky; many could be seen in the distance to have one or more channels down to the ground. (Photograph by T.H.A. Summers.)

Plate 8.24 Castellanus and floccus clouds over southeast England (looking southeast from Dunstable), 100 GMT, 29 June 1957. The clouds had a base height of just over 3 km, and were in bands lying approximately in the direction of the wind (220°) at that level. In the layer up to 4 km the potential temperatures θ and θ_w were, respectively, about 35°C and between 16 and 18°C, values typical of air which arrives over England after modification in the layer of ordinary convection over Spain. Some of the nearer clouds have towered to about 4 km but are beginning to dissolve and already have only fragmentary bases. A little farther away it can be seen that the towers sprout from a common base of much greater horizontal dimension, a characteristic feature of these clouds, and in the distance a still broader cloud band has produced larger cumuliform clouds which are more persistent and retain well-defined bases. Later in the day castellanus produced showers and thunderstorms, and on the following day more intense thunderstorms with large hail occurred during the passage of a cold front zone.

8.23

8.24

8.25

8.26

Plate 8.25 Glaciated castellanus clouds, seen looking west-southwest along the north side of the island of Tenerife in mid-afternoon, 21 March 1958. The mountain on the left is Teide, which rises to 3700 m above sea level. On the right extensive trade wind cumulus have tops at 1200 m. The view is toward the sun, and the parts of the middle-level clouds composed predominantly of droplets (which scatter light forward more effectively than the ice particles) are brilliant compared with the glaciated parts. The castellanus form at a height of about 5.5 km (where $T \approx -10°C$ and $\theta \approx 42°C$), in air drawn from the adiabatic layer over the Sahara.

Plate 8.26 Castellanus clouds very similar to those shown in Pl. 8.25, also formed in air from the adiabatic layer over the Sahara, but on its arrival over southern England (on the eastern flank of the system of high clouds in a trough advancing from the west). The clouds formed at 6 km ($T \approx -15°C$, $\theta \approx 43°C$, $\theta \approx 18°C$); some towered to over 8 km and glaciated, with trails of precipitation falling into very dry air in the middle troposphere and evaporating. Typically, the small cumulus were in a cool easterly stream below a strong inversion. (Nearer the frontal zone in such a trough castellanus may occur at several levels, in airstreams from the adiabatic layers over the Sahara, Spain, and France, giving the sky a chaotic appearance which notoriously heralds severe thunderstorms.)

8.27

Plate 8.27 Mamma in the lower middle troposphere, probably associated with well-developed castellanus, in a frontal zone with thunderstorms over southeast Australia (183).

9.1

Plate 9.1 A mosaic of pictures covering a large part of the northern hemisphere, taken by a satellite on 10 October 1969 (everywhere at about local noon). Several characteristic kinds of cloud pattern are shown:

1. The *long cloud bands* of middle latitudes which are associated mainly with the cold front zones of cyclones. One cyclone is centered near the south of Greenland; its cold front reaches south to southern Newfoundland, and there becomes a warm front extending into another cyclone centered near the southern end of Hudson Bay. Cool air streaming away from the west coast of the United States turns toward the northwest over the ocean, where a more intense cyclone is centered near 50°N 155°W. Its cold front and major cloud band extend to the top left corner of the picture (in latitude 20°N). The slight alteration of the otherwise smooth curve of this band suggests that a wave cyclone may be developing near 40°N.

2. West of the last-mentioned cold front is the typical *speckled pattern* of the cells and clusters of the clouds of ordinary convection, in the cool airstream behind and south of the center of the main cyclone.

3. Also prominent is a northward intrusion of a zone almost free of these clouds, between the front and the center of the cyclone. A similar nearly cloudless zone can be seen behind the cold front of the cyclone near Greenland.

4. Clouds of ordinary convection occur also over land, in clusters associated with intermediate-scale topographical features, especially near the coasts and over the high ground of Central America.

5. Typical *spiral-vortex* patterns of the clouds of ordinary convection, and of the cover of high clouds, which are associated with tropical cyclones, can be found more or less well-developed in three places in low latitudes: near 30°N 50°W on the right edge of the picture, near 30°N 71°W, and near 20°N 109°W.

6. A long nearly zonal band of deep clouds of ordinary convection, typical of the intertropical convergence zone, lies over the ocean in the lower left, with a northern margin near the 10th parallel.

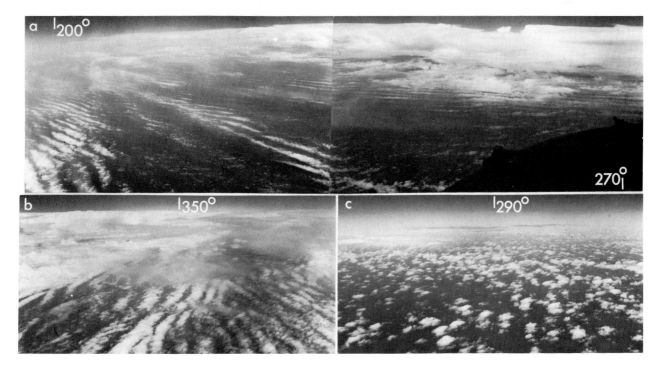

9.2

Plate 9.2 Clouds in the cyclone of Fig. 9.25.

(*a, b*) Views toward the main band of cumulonimbus, from 10 km above position 6; anvil tops were at 12 km. In the foreground are shallow cumulus in the south-southwesterly flow at low levels, where the geostrophic wind was 210° 25 m sec^{-1}. The orientation of the bands of cumulus is backed by 10–20° from the geostrophic wind; individual rows of small clouds lie across the bands, more nearly in the direction of the geostrophic wind. The separation of the bands is several times greater than the depth of the moist layer beneath. Although the clouds are cumuliform the air at ships' deck level was about 1°C *warmer* than the sea surface, and the visual range was only 5–10 km (farther northeast, near the 50th parallel, where the surface cooling of the airstream was greater, the visual range was less than 2 km). However, the stratification above a shallow surface layer was nearly neutral after previous convection, and the bands probably mark longitudinal vortices in the boundary layer, of the kind discussed in section 7.6. With progressive cooling as the airstream advanced northward the cover of low clouds became complete. In the ascent on the large scale near the cold front, or over the squall fronts of the associated cumulonimbus, the cloud developed into towering cumulus.

(*c*) This picture was taken at position 7, 250 km west of the main cumulonimbus band, on the edge of the zone of towering cumulus. Apart from dark patches of residual shelf cloud the clouds are confined to the low troposphere and are scattered cumulus of ordinary convection in a cool airstream warmed by the sea.

(*d*) A satellite picture: positions 6 and 7, from which the aircraft pictures were taken, are marked on either side of the cumulonimbus in the cold front zone; the surface center of the cyclone is marked *L*. Between the positions *SS* the edge of a system of high cloud throws a shadow upon broken low clouds; the edge passes directly over weather ship *I* west of Scotland. At midday the axis of a jet stream lay over this ship at 10 km, where θ_s was 19°C and the wind was 260° 94 m sec^{-1}, consistent with the air having ascended near the cold front south of latitude 45°N (Eq. 9.23).

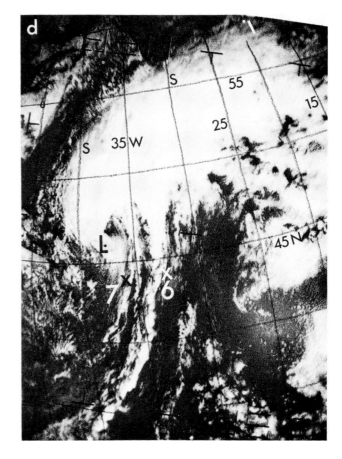

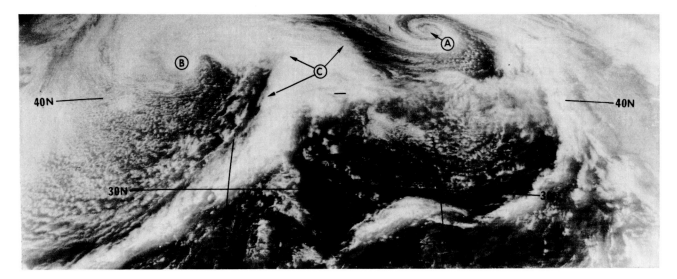

9.3

Plate 9.3 A picture from the geosynchronous satellite ATS 1, showing cloud patterns typically associated with large cyclones, here centered at A and B over the North Pacific (41). Near the places marked C the prominent band of clouds near the cold front of cyclone B extends in an arch (convex toward the north) across lower clouds, which reach westward toward the cyclone center. Other features, which generally conform to those in the Norwegian cyclone model, include the speckled patterns of the clouds of ordinary convection, often grouped into clusters and lines. Nearly cloud-free zones, immediately west of the bands of dense clouds associated with the cold fronts, extend northward into spirals at least partly surrounding the cyclone centers.

Plate 9.4 Cloud systems in an intensifying wave cyclone (44). The cyclone has a surface center (pressure 999 mb) at L. In the following 24 hr it deepened and moved to the position x near 30°W. The associated band of frontal cloud extends across the picture, and between A and B consists mainly of high clouds, which have a pattern of transverse narrow striations about 15 km apart (most prominent near C). These high clouds pass across and above a broad band of low cloud with a brighter and more irregular texture, which stretches from south of B to the north and then to the southwest of the cyclone center, where it becomes broken into curved narrow bands of cumulus. Northeast of the center the system of high clouds, which has a sharply defined edge, throws a distinct narrow shadow upon the low clouds. The edge of the high cloud seems to begin forming as cirrus near the position A, but traces of high cloud can be discerned still farther west, where they are probably part of the outflow from a weakening tropical storm (centered near the position T).

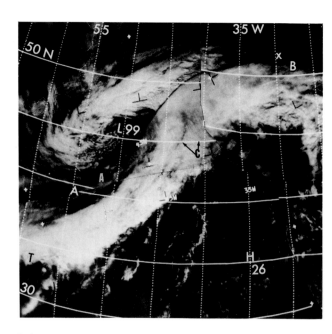

9.4

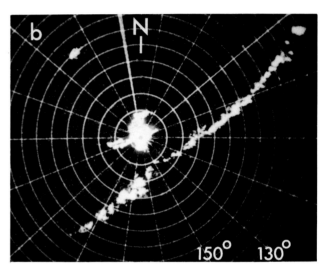

9.5

Plate 9.5 Clouds associated with a weak cold front over England, seen (*a*) visually and (*b*) on a radar PPI display. On the radar display the range markers are at intervals of 5 mi; the front appears as a line, about 200 km long, of echoes from small showers, which are mostly distinctly separated. The line moved southeast with the front, which at the surface was accompanied by a sudden fall of 4°C in the screen temperature and a veer and decrease of wind. The other picture was taken about 90 min later, and shows the cumuliform tops of the shower clouds, visible from the radar after the recession of extensive shelf cloud.

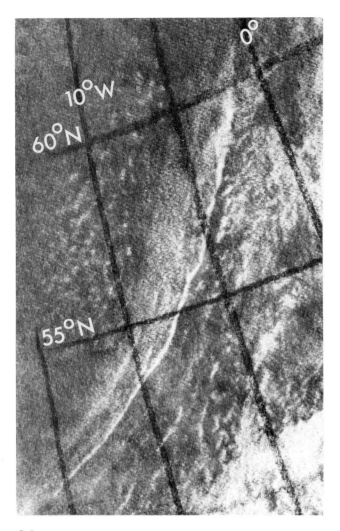

9.6

Plate 9.7 Precipitation cells in the middle troposphere, as seen on a time-height record from a vertically pointing 8-mm radar, operated by the British Meteorological Office. (The white lines are height markers at intervals of 6000 ft and time markers at intervals of 5 min.) The cells, with a generating level just above 5 km, produced inclined fallstreaks with mammilated lower ends, which merged into a complete cover of cloud (appearing from the ground as a layer of thin altostratus, through which the position of the sun was just visible). The wind at all levels shown was from about 240°, with the speed v indicated. The temperature T is marked at two levels. The horizontal scale is consistent with the wind speed at the generating level. The form of the trails of ice particles suggests that the wind shear between 4 and 7 km was uniform (it was about 4 m sec^{-1} km^{-1}), and that the mean fall speed of the particles was about 0.5 m sec^{-1}. The nearly upright parts of the cell heads and of the mamma imply greater speeds—of a few meters per second—for the associated vertical air motions.

Plate 9.6 Some features of a cold front cloud system seen from a satellite, whose structure was studied during its passage across England on the following day (76). There are extensive shallow low clouds in the strong southwesterly flow ahead of the long narrow band of bright cloud (10–20 km wide), which marks the positions of the leading edge of the cold front near the surface. The contrast between the leading band of cumuliform cloud and the broad zone of (mostly higher) frontal layer cloud following it, which are probably features of many cold fronts, is unusually enhanced by the very low elevation (about 10°) and favorable azimuth (southeast) of the sun. The granular appearance of the layer cloud suggests that there is small-scale convection (as also indicated in Fig. 9.31b).

Plate 9.8 High clouds showing different kinds of characteristic texture. In the upper part of the picture is a shallow maturing cloud with a *compact* texture, which near its middle has acquired a finely dappled pattern (probably after radiative destabilization). The cloud is evidently composed mainly of small ice particles, whose diffusion has begun to blur the dappled pattern. Near the edges of the cloud there are numerous short, fine fallstreaks. Near the bottom of the picture is a smaller and younger cloud also with a compact texture, in which there are still clear spaces between the rows of dapples. The visual appearance of the cloud indicates only that the particles are small and not their phase; sometimes, however, similar clouds seen near the sun are iridescent, suggesting that they are at first composed mainly of droplets and have formed following the attainment locally of saturation with respect to liquid water. On the left is a cloud producing long tenuous *fallstreaks*, which consist of comparatively sparse and large ice crystals.

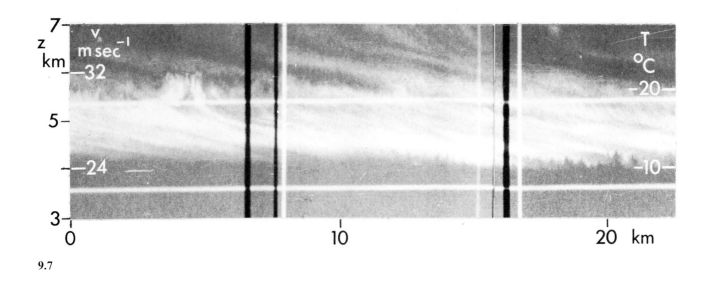

9.7

9.8

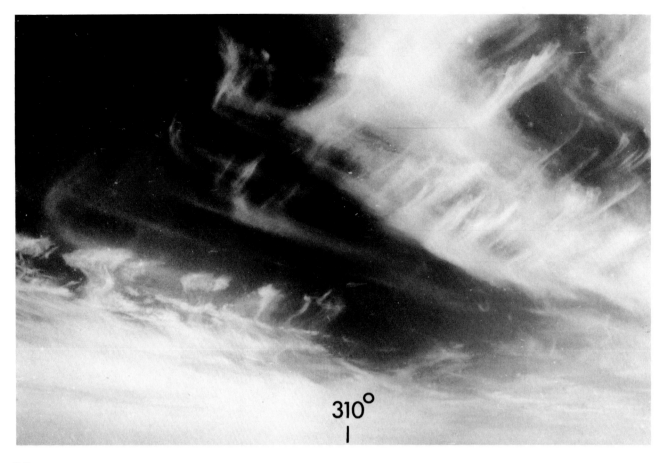

Plate 9.9 Cirrus fallstreaks seen looking northwest from near Östersund, central Sweden, 1315 MET, 6 July 1955 (Pl. 6.1 is an earlier picture of cirrus in the same system). The upper parts of the fallstreaks were at about 8.6 km ($-36°C$), and trail almost horizontally for about 4 km toward the azimuth 280° before turning sharply toward 340° at 8.3 km (as discussed in the section on the orientation of cirrus fallstreaks).

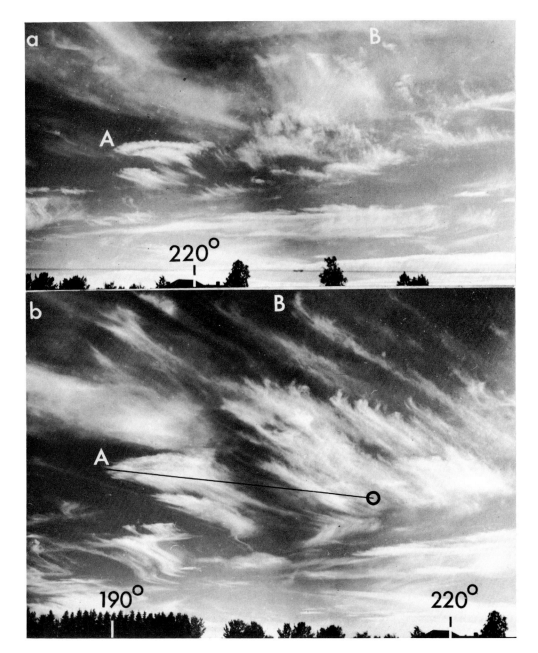

9.10

Plate 9.10 Successive pictures of developing cirrus fallstreaks: (*a*) at 1630 and (*b*) at 1648 MET, 20 July 1954, looking southwest from near Östersund, central Sweden. The tops of the nearer fallstreaks were at 8.5 km (330 mb, −40°C). Prominent among the clouds are the two tops *A* and *B*; the displacement of the former between the two pictures is shown by the dark line in *b*.

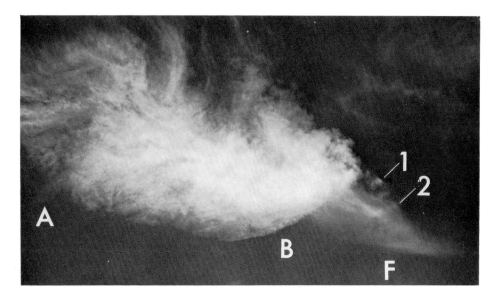

9.11

Plate 9.11 One of a system of cirrus clouds ahead of a weak warm front, with tops at 10.6 km ($-47°$C). The shallow compact part A was continually renewed at its rear edge B by the spreading of cumuliform growths which formed above the trail F of fallstreaks. Within a period of about 5 min the upper part of the trail would become distorted in a manner suggesting the local ascent of air out of it, and the compact secondary cloud would form about 400 m above the trail. Two or three such clouds might be present at one time, in various stages of evolution. One is developing at the position 1 and another is about to form at 2. (Faint iridescence was seen in these clouds within about $1.5°$ of the sun. There were no color phenomena in large and aged patches of cloud such as that labeled AB.)

9.12

Plate 9.12 A fallstreak trail with well developed mamma near its lower end (cf. Fig. 9.36).

9.13

Plate 9.13 Successive views of an approaching cirrus system. It advanced with a well-defined edge oriented toward 028°: in (*a*) the range of the nearest details was about 70 km. Patches of cloud which are denser and probably lower have recently formed ahead of the edge of the main system. This is composed of fallstreak cirrus which typically point into the system, toward about 300°, before turning northward, as more obviously seen in (*b*), taken 2 hr later.

Plate 9.14 The RHI display of a radar directed along the azimuth of 260° from Defford, east of mountains in Wales (174). [The 11-cm radar is high powered, with a large antenna, and so is very sensitive: at a range of 10 km the minimum detectable value of η (Eq. 2.52) for targets filling the pulse volume is about 10^{-17} cm^{-1}. It is therefore able to detect slightly turbulent layers at ranges of some tens of kilometers, as discussed in Section 6.4.] The intense echoes extending to a range of 16 km are "ground clutter" detected in side lobes. Three distinct layers of echo in clear air have an undulating form which shows the presence of lee waves.

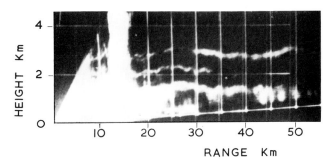

9.14

9.15

Plate 9.15 A wave cloud with a smooth texture and regular outlines in the lee of the Matterhorn (which reaches nearly 4.5 km). Later the flow over and near the Matterhorn and neighboring mountains became unsteady, and the associated clouds more extensive and irregular, as shown in Pl. 9.18 and in more detail in reference 192. (Photograph by G. J. Jefferson.)

Plate 9.16 Wave clouds at about 3 km, seen looking south from near Reykjavik, Iceland. Hills at the western end of the range southeast of Reykjavik can be seen on the horizon. A belt of cumulus with bases at about 500 m lies in the lee of the range; their growth has probably been enhanced in the first lee wave.

9.16

Plate 9.17 A train of lee-wave clouds in the lower middle troposphere, well above the layer occupied by cumulus convection. (Photograph from the Clarke Collection, Royal Meteorological Society.)

9.17

9.18

Plate 9.18 Wave phenomena over the Owens Valley and the Sierra Nevada. The phenomena in this region are famed for their intensity. The snow-capped Sierra Nevada (on the right) form an unbroken range about 600 km long, and here reach an average height of nearly 4 km; the slope upward from the coast is gradual, but the escarpment above the Owens Valley (elevation about 1.2 km) is one of the greatest in the world.

This photograph was taken from a twin piston-engine aircraft (Lockheed P-38) on an occasion when the aircraft soared between 4 and 9 km for more than 1 hr in the lee wave, with the engines dead and the propellers feathered; indicated rates of ascent exceeded 15 m sec^{-1} and the maximum updraft speed was estimated to be about 35 m sec^{-1} (166). A wall cloud reaches the mountain crests; the roll cloud on the left is about 4 km above the valley floor. Strong westerly surface winds down the lee slopes and along the valley floor raise plumes of dust, which is carried steeply upward into the roll cloud. Isolated shallow wave clouds formed above the roll cloud, at about 12 km. (Photograph by Robert Symons.)

9.19

Plate 9.19 This photograph was taken within a few minutes of Pl. 9.15. The lower ragged clouds shown in this picture heralded an increase, to almost an overcast, of cloud at about 3.6 km. Simultaneously the large smooth wave cloud above the Matterhorn developed much irregular detail. The low clouds moved up and over the mountain, occasionally just covering its peak. Others soon formed in positions considerably nearer, and all are likely to have been part of a flow in the lee of the ridges several kilometers northwest of the observer, which suddenly developed rotors or otherwise became very disturbed and unsteady. (Photograph by G.J. Jefferson.)

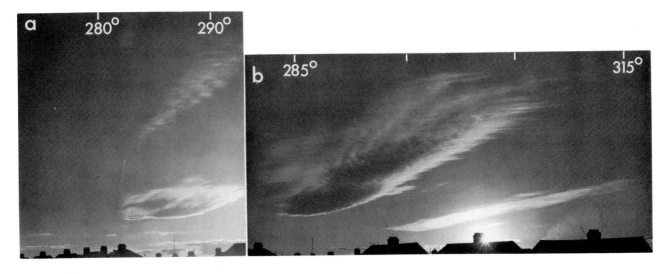

9.20

Plate 9.20 Stages in the growth of hill-wave plumes, seen looking west from Dunstable, eastern England.

(a) 1922 GMT. At 1900 GMT a hill-wave cloud with a sharply defined windward edge began to form at 8 km (temperature $-26°C$) over a ridge (which rises to 200 m at a range of 100 km). The rear edge traveled eastward with the speed of the wind at its level, the cloud steadily extending into a hill-wave plume. Traces of cloud formed at the same level over nearer small hills (upper part of picture).

(b) 1948 GMT. A similar hill-wave plume lies farther north. At first brilliantly iridescent, its colors were now faint, and there was fibrous and probably frozen detail in the nearer ends of both clouds. Typically, closely spaced transverse ribs were a prominent feature.

(c) 2005 GMT. The clouds are lit from beneath by the setting sun, enhancing the ribbed detail. The plumes soon attained maximum lengths of over 100 km, but by 2100 GMT their formation ceased.

Plate 9.21 Successive photographs illustrating the formation of a hill-wave plume over the Canary Islands, among cirrus of the subtropical jet stream. The views are west-northwest from 2350 m above sea level on the island of Tenerife, and show also the island of La Palma, with cumulus banked on its far side (the central ridge of the island, at a distance of about 130 km, rises to about 2 km). Throughout the day cloud formed intermittently a few tens of kilometers in the lee of La Palma, leading to the formation of a long and almost unbroken plume of cirrus extending east-northeast (at a height of about 12 km and temperature of $-60°C$). The upper photograph was taken at 1820 GMT and the lower after an interval of 6 min. The clouds numbered 1, 2, and 3 formed in succession.

9.21

(a)

(b)

9.22

Plate 9.22 (a) Three stages in the growth of Kelvin-Helmholtz billows in a laboratory experiment (197); an increasing shear between a layer of dyed brine and a layer of clear water was produced by suddenly tilting a long closed horizontal tube containing the layers (here represented as if it were horizontal).

(b) Billows observed near Denver, Colorado, looking west toward the Rocky Mountains (205). The billows lasted only a few minutes in a cloud which continued to move rapidly from right to left beneath a northwesterly jet stream. The billow clouds are likely to have been in a layer of intense shear (about 20 m sec^{-1} km^{-1}) above a height of 8 km. (Photograph by Paul E. Branstine.)

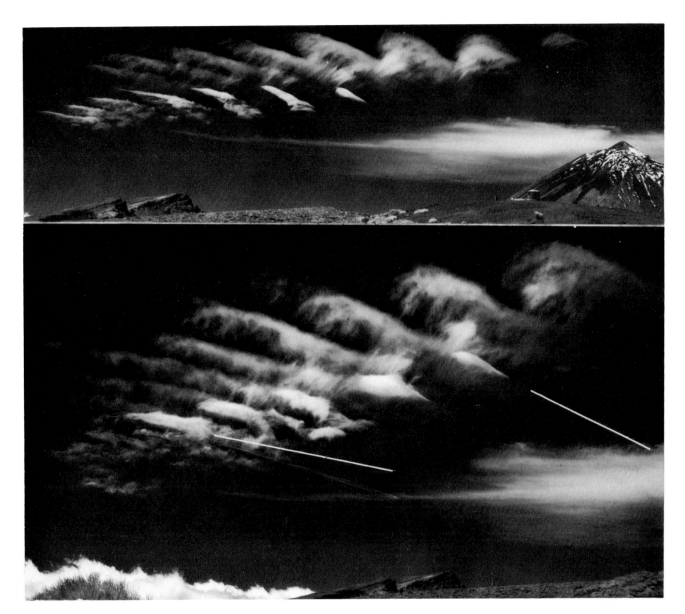

9.23

Plate 9.23 Billow clouds in the lee of the mountain Teide (upper right) in the Canary Islands, seen looking west-southwest and southwest (below, from the observatory Izana; height 2350 m). The white lines in the lower picture show the apparent direction of travel of the clouds (from 260°) in the interval of 10 min between the pictures. The individual billow clouds were elongated toward 250°; on this occasion, as also indicated by the orientation of the cloud tendrils, the shear vector evidently lay nearly across rather than along the direction of motion.

 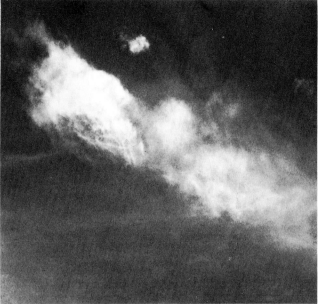

9.24

Plate 9.24 A group of high-level puff clouds soon after their formation in the lee of small hills over England (left). On the right is a close-up view of a well-developed puff cloud, seen at a high elevation. The clouds were at 9.6 km (temperature $-44°C$), nearly below the axis of a jet stream (199).

9.25

Plate 9.25 This picture shows the damage to the tall tailfin of a military aircraft, suffered during flight in clear air near mountain crests. (Official U.S. Air Force photograph.)